Physics of Nanostructured Solid State Devices

Supriyo Bandyopadhyay

Physics of Nanostructured Solid State Devices

Springer

Supriyo Bandyopadhyay
Department of Electrical and Computer Engineering
Virginia Commonwealth University
601 West Main Street
Richmond, VA, USA

ISBN 978-1-4899-9629-9 ISBN 978-1-4614-1141-3 (eBook)
DOI 10.1007/978-1-4614-1141-3
Springer New York Dordrecht Heidelberg London

Softcover reprint of the hardcover 1st edition 2012

Printed on acid-free paper

Springer is part of Springer Science+Business Media (www.springer.com)

Dedicated to the memory of my father

Preface

This textbook is intended to introduce a first year graduate student of electrical engineering and/or applied physics to concepts that are critical to understanding the behavior of charge carriers (electrons and holes) in modern nanostructured solid state devices. The student is assumed to have undergraduate background in solid state physics, solid state devices, and quantum mechanics. Many of the topics discussed here are specific to ultrasmall (nanostructured) devices, but some are more general and apply to any solid. This material is the result of the author's teaching a graduate level introductory course on electron theory of solids in three US universities over a period spanning nearly 25 years. It has been his experience that once students are able to grasp the concepts presented here and become comfortable with them, they are able to handle more difficult and specialized topics quite easily.

This book is organized into nine chapters. The first chapter reviews the steady-state "drift–diffusion model" of charge transport in solids that electrical engineers typically learn in their first undergraduate solid state device course. Physics undergraduates are less exposed to this topic, but should be able to grasp the concept easily. This chapter introduces the basic drift–diffusion model, starting with two important assumptions about the nature of charge conduction in solids, and emphasizes the notion that this model is valid only as long as nonlocal transport effects are absent. It ends with an introduction to the so-called "equations of state" (also known as drift–diffusion equations) that are used to compute the carrier concentration and current density in a solid state device self-consistently. Only steady-state transport is considered.

Chapter 2 discusses a more sophisticated charge transport model based on the Boltzmann Transport Equation (BTE). It derives this equation from conservation principles, and then uses it to deduce the generalized moment equation (or the hydrodynamic balance equation) which governs charge transport in the presence of both local and nonlocal effects. The steady-state drift–diffusion equations of Chap. 1 are shown to be special cases of the last equation. Two methods of solving the BTE—the relaxation time approximation and the Monte Carlo (MC) simulation method—are discussed and some analytical results are obtained from the relaxation time approximation. This chapter also discusses linear response transport or ohmic conduction and finds an expression for the linear response conductivity. It then distinguishes chemical potential from electrostatic potential inside a device and discusses a few important thermodynamic concepts pertinent to charge transport. Overall, the purpose of this chapter is to provide a sound basis for understanding both linear and nonlinear (or hot-carrier) transport in solids.

Chapter 3 reviews basic concepts in quantum mechanics, operators and their applications, energy quantization in quantum-confined systems (i.e., nanostructures) such as quantum wells, wires and dots, and ends with a description of time-independent and time-dependent perturbation theory. The purpose is to present the essential tools needed to understand and appreciate the quantum foundations of nanostructured solid state devices. As an example of applying quantum mechanics to a solid state device, time-independent perturbation theory is employed to elucidate the operation of an electro-optic modulator based on the quantum-confined Stark effect. Performance metrics of this device are calculated using perturbation theory, thereby exemplifying how quantum mechanics plays a critical role in device operation.

Chapter 4 presents methods of calculating the bandstructure (energy versus wavevector relation) of a crystalline solid based on time-independent perturbation theory. Since bandstructure plays a vital role in the operation of many devices, particularly optical devices, it is included in this book. More importantly, it is one more application of time-independent perturbation theory to an actual problem. We discuss four different bandstructure calculation methods—the nearly-free electron method, the orthogonalized plane wave (OPW) expansion method, the tight-binding approximation (TBA) and the $\vec{k} \cdot \vec{p}$ theory—three of which invoke time-independent perturbation theory. Band structure results are then applied to calculate the density of states of electrons and holes as a function of particle energy in bulk (three-dimensional systems) and nanostructured systems such as quantum wells (quasi two-dimensional) and wires (quasi one-dimensional). Based on the density of states, analytical expression are derived for equilibrium carrier concentrations in three-, two- and one-dimensional structures. This chapter ends with a derivation of the phonon and photon density of states in a crystal.

Chapter 5 illustrates the application of time-dependent perturbation theory to transport physics. It uses this theory to derive Fermi's Golden Rule which provides a useful prescription to calculate the rate with which an electron scatters as it travels through a solid and interacts with various entities such as impurities and phonons. These rates appear in the BTE and are first mentioned in Chap. 2, but derived here. In order to make the student comfortable in applying Fermi's Golden Rule to various problems, the scattering rate of an electron interacting with nonpolar acoustic phonons is derived in three-, two-, and one-dimensional systems as a function of the electron's kinetic energy. Such exercises are extremely instructive and reveal important physics associated with carrier interactions with the environment in both bulk- and quantum-confined systems. Once students master the technique of calculating the scattering rate due to any one type of interaction, they should be able to calculate the rate associated with any other type of interaction since the basic principle is the same. A few advanced topics such as phonon confinement and phonon bottleneck effects are also discussed.

Chapter 6 discusses electron–photon interactions and their impact on the optical properties of solids, while addressing the general concepts of absorption, spontaneous emission, and stimulated emission of light. Fermi's Golden Rule is utilized to calculate absorption and luminescence intensities as a function of photon energy in three-, two and one-dimensional systems. The basic physics behind the operation of some solid state optical devices, such as light emitting diodes (LED) and lasers, is also discussed. This chapter also discusses excitons since they impact optical properties of solids, and could produce nonlinear optical effects. Finally, some special topics of current interest in nanophotonics, such as polariton lasers, photonic crystals, and negative refraction are discussed. This is the only chapter that deals with "optical properties" of nanostructured solids; the rest are mainly focused on "transport properties."

Chapter 7 discusses the behavior of an electron in a magnetic field. It first introduces the Dirac equation and the Pauli equation to account for an electron's spin explicitly and then focuses on the "spinless" electron in order to discuss effects unrelated to the spin. The important concepts of magnetic vector potential and "gauge" are introduced, and the Schrödinger equation is solved in two- and one-dimensional systems to find the wavefunctions and energy eigenstates. Solution of the Schrödinger equation in a two-dimensional system leads to the idea of Landau level quantization, as well as its observable effects such as Shubnikov–deHaas conductance oscillations. In the context of one-dimensional systems, the concept of edge states is introduced along with hybrid magneto-electric states. A device application of the physics, namely the operation of a magneto-optical device based on quantum-confined Lorentz effect (QCLE) (a magnetic analog of the quantum-confined Stark effect) is discussed. This chapter concludes with a few basic remarks regarding the integer and fractional quantum Hall effect (FQHE).

Chapter 8 introduces some popular quantum transport formalisms such as the scattering matrix formalism, the Landau–Vlasov equation (which can be viewed as a quantum-mechanical equivalent of the collisionless BTE), the nonequilibrium Green's function approach, the Wigner distribution function, the Tsu–Esaki formalism, and the Landauer–Büttiker approach for linear response transport. Applications of these formalisms are presented in Chap. 9.

In Chap. 9, some of the quantum transport formalisms developed in Chap. 8 are applied to actual quantum devices. The Tsu–Esaki formula is used to calculate the tunneling current in resonant tunneling devices and numerous mesoscopic devices and phenomena are treated with the Landauer–Büttiker formalism. This chapter is intended to show how quantum mechanics impacts the operation of nanostructured devices.

The author is grateful to his numerous graduate students who had this course and often made valuable contributions to developing the course material. There are too many of them to name here. He is also indebted to an anonymous reviewer who made helpful comments while reviewing the draft.

As always, in spite of the author's best efforts at proof reading, it is possible that some typographical errors have eluded detection. The author will be immensely grateful if they are brought to his notice by e-mailing him at sbandy@vcu.edu.

Welcome to the world of electron physics in nanostructured devices!

Table of Constants

Table 1 Table of universal constants

Free electron mass (m_0)	9.1×10^{-31} Kg.
Dielectric constant of free space (ϵ_0)	8.854×10^{-12} Farads/meter
Electronic charge (e)	1.61×10^{-19} Coulombs
Reduced Planck constant ($\hbar$)	1.05×10^{-34} Joules-sec
Bohr radius of ground state in H atom (a_0)	0.529 Å= 5.29×10^{-11} meters
Bohr magneton (μ_B)	9.27×10^{-24} Joules/Tesla

Contents

Chapter 1
Charge and Current in Solids: The Classical Drift–Diffusion Model

1.1 Introduction

All solid state devices, including those that have nanoscale dimensions, can be broadly classified into two main categories: (1) *transport devices* where transport of electric charge (resulting in current flow) takes place and is responsible for the device's operation, and (2) *optical devices* where transport of charge may or may not take place, but such transport, if present, is not primarily responsible for device action. Instead, what governs device behavior is transition of charge from one energy state to another resulting in the emission or absorption of light or photons. An example of the first class of devices is the ubiquitous complementary metal oxide semiconductor (CMOS) transistor, while an example of the second class of devices is a semiconductor light emitting diode. Both classes play important roles in our lives.

This book is primarily concerned with the electron theory of solids that undergirds the operation of the first class of devices, namely transport devices, although we will also have occasion to delve into the second class of devices in one chapter. Needless to say, the operation of transport devices is governed by the physics of charge transport in solids. Therefore, to understand their operation at the fundamental level, we have to understand how charge transport takes place within a solid, typically a semiconductor. Over the years, increasingly sophisticated models of charge transport have been developed. This was necessitated by the decreasing size of devices; smaller devices need more sophisticated models. As a result, a hierarchy of charge transport models has emerged, which is broadly as follows:

- *Drift–diffusion model*: It is the easiest to understand and historically the most widely used. Its shortcoming is that it cannot capture so-called *nonlocal* transport effects that frequently occur in submicron (nanoscale) devices, heterostructured devices, and devices with strong doping modulation. Within such devices, the electric field that drives charge transport may change very rapidly in space, resulting in "nonlocality" whereby the behavior of a charge carrier at any location in the device not only depends on its local surroundings but also on events

S. Bandyopadhyay, *Physics of Nanostructured Solid State Devices*,
DOI 10.1007/978-1-4614-1141-3_1,

taking place at remote sites. In other words, the behaviors at different locations within the device are interconnected. Such nonlocal behavior is beyond the drift–diffusion model.

- *Boltzmann transport model*: This is more sophisticated than the drift–diffusion model since it can handle *nonlocal* transport effects. However, it cannot capture quantum mechanical effects, such as those arising from the interference of electron waves. These effects sometimes manifest themselves in ultrasmall devices at low temperatures and is at the heart of "quantum devices" such as resonant tunneling diodes, whose operations are based on quantum-mechanical principles. As long as such effects are absent, the Boltzmann Transport model is applicable and therefore it is widely used to model charge transport in nanostructured devices at room temperature.
- *Quantum transport models*: This is the most sophisticated and the most powerful of all models since it can handle both classical and quantum-mechanical transport, but unfortunately it is also the most difficult. Unlike the previous two models which are purely classical and treat charge carriers as classical particles, quantum transport models account for the quantum-mechanical wave nature of charge carriers. The wave nature is manifested whenever a device's critical dimensions are smaller than the distance an electron can travel before losing quantum-mechanical phase coherence. Today, many nanostructured devices call for a full quantum-mechanical treatment. They exhibit effects such as current modulation arising from interference of electron waves, which cannot be modeled with any classical prescription.

The drift–diffusion model, which is the subject of this chapter, is predicated on two basic assumptions about the nature of charge carriers in a solid. They are:

1. *Assumption 1*: Charge carriers (electrons and holes) are classical particles, like billiard balls, that travel through a solid obeying the classical Newton's laws of motion.[1]
2. *Assumption 2*: While traveling in a solid, charge carriers are subjected to forces due to external electric and/or magnetic fields, as well as instantaneous forces due to scattering. The latter tend to restore equilibrium of the charge carriers with their local surroundings.[2] The electric and magnetic fields are assumed to vary slowly enough in time and space that a carrier scatters many times before these fields change significantly. As a result, a carrier always remains at local equilibrium with its surroundings and its behavior is determined *solely* by local conditions.

If both these assumptions hold, then the drift–diffusion model is adequate to describe charge transport in a solid. However, if there are rapid doping and/or

---

[1]Relativistic effects that are beyond Newton's laws are typically unimportant in most solid state devices since charge carriers travel in them with speeds much less than the speed of light in vacuum.

[2]This is *local* equilibrium which is different from *global* equilibrium. Under global equilibrium conditions, no current flows through the solid.

compositional variations within a device (e.g., in a heterostructure), then the internal electric fields can change very quickly in space. In that case, assumption 2 may be violated. This assumption is also violated if the device is so small (smaller than the mean free path of carriers) that a carrier can traverse the entire device without suffering too many collisions with other carriers, impurities or the vibrating atomic lattice (phonons). This type of transport is called nearly collisionless (or "quasi-ballistic") transport. Since there is not enough scattering to restore local equilibrium for the carrier, assumption 2 can be violated. When this happens, nonlocal effects are manifested in the behavior of charge carriers and these effects cannot be captured within the drift–diffusion model. We will see an example of this in the next chapter.

Finally, at very low temperatures and in very small devices, even assumption 1 may be violated. When the device dimension is smaller than the inelastic mean free path of carriers at the operating temperature, a carrier can traverse the entire device without suffering a single inelastic scattering event.[3] Inelastic scattering events destroy the quantum-mechanical wave nature of carriers, so that when they are absent, the wave nature is preserved and carriers can no longer be treated as classical particles obeying Newton's laws. Instead, they must be treated as waves propagating through the device according to the laws of quantum mechanics (Schrödinger equation). Since the inelastic mean free path increases rapidly with decreasing temperature in most solids, this scenario is quite likely to occur in submicron devices at temperatures of a few Kelvin.

If carriers behave like waves, they can interfere like waves,[4] resulting in a host of "quantum interference effects" that have collectively come to be known as *phase coherent effects* or *mesoscopic effects*. The term "mesoscopic" was coined in the early 1980s to describe a regime between the truly microscopic (single atoms or molecules) and macroscopic. Devices of mesoscopic size will have dimensions comparable to the so-called "phase breaking length" of carriers, which is the distance a carrier can travel before phase randomizing collisions randomize an electron wave's phase and restore the electron's classical particle nature. Only inelastic collisions randomize the phase; elastic collisions do not. Therefore, the phase-breaking length is basically the inelastic mean free path. The inelastic mean free path depends on many parameters, such as the ambient temperature, material purity, electron concentration, spin orbit interaction in the material, etc., but under favorable conditions (i.e., at low enough temperatures), it can be longer than 1 μm in high-purity semiconductors. Hence, mesoscopic effects are not that rare! In the mesoscopic or phase coherent regime of charge transport, neither drift diffusion nor

---

[3]The inelastic mean free path is usually longer than the elastic mean free path at low temperatures. A carrier can traverse a device without suffering inelastic collisions, but may suffer elastic collisions. Such a carrier is still capable of manifesting quantum-mechanical interference effects since its phase has not been randomized.

[4]An electron can interfere only with itself; two different Fermi particles do not interfere with each other. However, an electron interfering with itself can result in observable effects that cannot be captured by the drift–diffusion model or the Boltzmann transport model, and will require full application of quantum transport models.

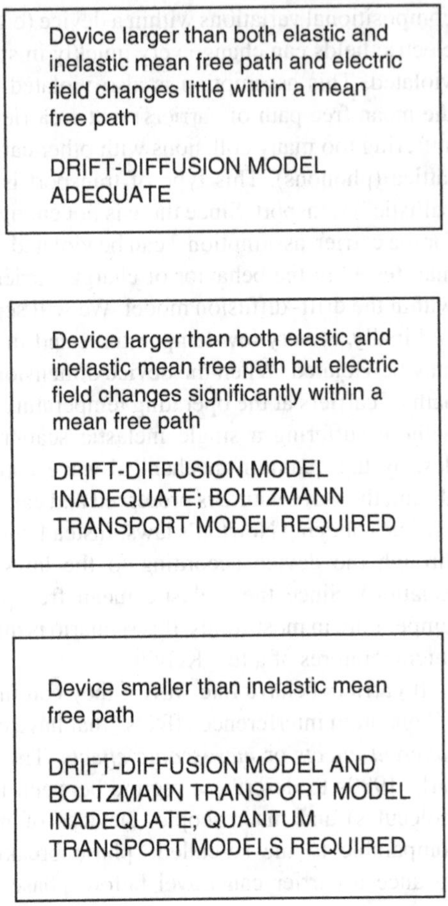
Device larger than both elastic and inelastic mean free path and electric field changes little within a mean free path
DRIFT-DIFFUSION MODEL ADEQUATE
Device larger than both elastic and inelastic mean free path but electric field changes significantly within a mean free path
DRIFT-DIFFUSION MODEL INADEQUATE; BOLTZMANN TRANSPORT MODEL REQUIRED
Device smaller than inelastic mean free path
DRIFT-DIFFUSION MODEL AND BOLTZMANN TRANSPORT MODEL INADEQUATE; QUANTUM TRANSPORT MODELS REQUIRED

Fig. 1.1 Validity regimes for three different transport models

Boltzmann transport models have any validity since they are both based on classical physics and cannot describe quantum-mechanical effects. In this regime, the only reliable recourse will be the quantum transport models.

The validity regimes for different transport models are indicated in Fig. 1.1.

In the remainder of this chapter, we will discuss the drift–diffusion model. Graduate students or senior undergraduate students with some background in solid state devices, physics and/or materials, are probably already familiar with the essential elements. We revisit them below.

1.2 Drift–Diffusion Model

In the *drift–diffusion* model, electric current in a solid is viewed as being the result of two effects: (1) electrons (and/or holes) *drifting* under an applied electric field (caused by an applied voltage across the device), and (2) the same particles *diffusing* from a region of higher concentration to a region of lower concentration as a result of interparticle collisions. These two phenomena—drift and diffusion—are solely responsible for current flow. Hence the name "drift–diffusion model."

We just mentioned that drift is caused by an applied electric field, but this immediately raises a question. The term "drift" usually implies that the charge carriers are moving or drifting with a *constant* velocity. But why should the velocity be constant in an electric field? After all, the electric field results in a force on the electron equal in magnitude to the product of the electron's charge and the strength of the electric field. This force should *accelerate* the electron, causing its velocity to increase continuously with time. So, how does the electron reach a constant "drift velocity?" It happens because there are other forces that act on the electron in a solid. The electron typically collides with other electrons, impurities, and even the vibrating atoms in the solid (phonons), which slows it down. In the end, these scattering forces act like a frictional force whose magnitude is roughly proportional to the electron's velocity. The frictional force always opposes the force due to the accelerating electric field. When these two forces exactly balance, the net acceleration vanishes and the electron reaches a steady-state velocity and begins to "drift." This picture is sometimes referred to as the Drude model in solid state physics.

Diffusion, on the other hand, is a more complicated issue. It has its origin in interparticle collisions which has the net effect of driving charge carriers from a region of higher concentration to one of lower concentration, resulting in current flow. However, it is more complicated because diffusion is not a single particle concept. Hence, it cannot be derived from simple Newton's laws. *Many* particles are required to make up a "concentration" and collisions between them drive diffusion. Hence, diffusion is inherently a multiparticle concept. We will examine the implication of this later in this chapter.

1.3 The Drude Model and Ohm's Law

We mentioned at the outset that the drift–diffusion model is predicated on two assumptions, namely (1) charge carriers are classical particles that obey Newton's laws, and (2) they scatter many times in a solid before the electric field driving them changes significantly. Under assumption (2), scattering enforces local equilibrium. Hence, the frictional force due to scattering depends on *local conditions alone*, such as the local carrier velocity (Fig. 1.2).

Consider now a single electron that obeys the above two assumptions. Newton's second law (assumption 1) mandates that the electron's velocity $\vec{v}_n(\vec{r},t)$ at position $\vec{r}$ and at the instant of time t obeys the equation

$$m_n^* \frac{d\vec{v}_n(\vec{r},t)}{dt} = -e[\vec{\mathcal{E}}(\vec{r},t) - \vec{v}_n(\vec{r},t) \times \vec{B}(\vec{r},t)] + F_{scat}(\vec{r},t), \qquad (1.1)$$

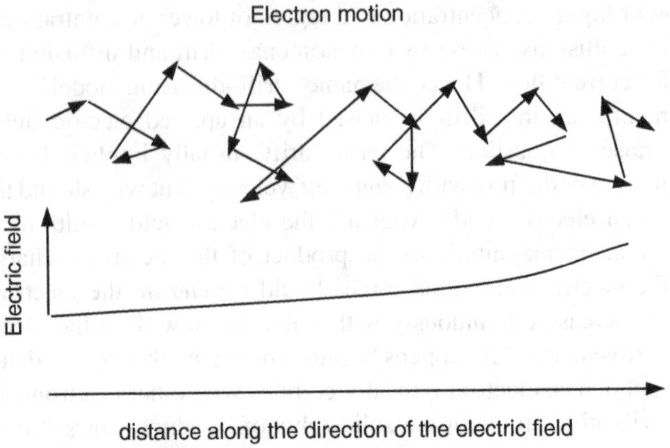

Fig. 1.2 Pictorial depiction of the conditions under which assumptions 1 and 2 that undergird the drift–diffusion model are valid. Scattering causes the electron motion to be a random walk. Note that the carrier scatters many times before the electric field changes appreciably

where $\vec{\mathcal{E}}(\vec{r},t)$ is the electric field acting on the electron at position $\vec{r}$ at the instant t, and $\vec{B}(\vec{r},t)$ is the magnetic flux density at that position at that instant of time. The electron's charge is $-e$ and $F_{\text{scat}}(\vec{r},t)$ is the local scattering force that the electron experiences at location $\vec{r}$ at the instant t. Furthermore, $m^*(\vec{r},t)$ is the local effective mass of the electron, which, we assume to be time dependent for the sake of generality.[5] In a crystal, (1.1) is not exactly the correct equation for Newton's law. In the correct equation, we have to replace $m^*\vec{v}$ with the crystal momentum $\hbar\vec{k}$, where $\vec{k}$ is the electron's wavevector. At this point however, we will not worry about this subtlety and assume that the two quantities are the same.

Since scattering forces act like frictional forces that damp an electron's velocity, and since frictional forces are usually proportional to a moving body's velocity but act in the direction opposite to the velocity, we can write (Drude's assumption)

$$F_{\text{scat}}(\vec{r},t) = -\frac{m_n^*\vec{v}_n(\vec{r},t)}{\tau_n(\vec{r},t)}, \tag{1.2}$$

[5]The concept of "effective mass" will be discussed in more detail in Chap. 4 when we discuss bandstructures of solids. For the time being, suffice it to say that in a crystalline solid, an electron experiences a periodic electrostatic force because of the periodically placed background ions. This force needs to be accounted for in Newton's law. Its presence will make matters immensely complicated, but fortunately we can get by without accounting for it as long as we replace the real mass of the electron with an effective mass. The effective mass depends on the energy-wavevector relation of the electron in the crystal and is generally different from the real mass. The use of the effective mass also allows us to use Newton's classical law to describe the motion of an electron in a solid, which, in truth, should be described using quantum-mechanical laws and not Newton's laws.

where $\tau_n(\vec{r},t)$ is the local (instantaneous) time constant associated with relaxation of electron velocity due to scattering. Note that we implicitly invoked assumption (2) when we wrote down the last equation because we assumed that the local frictional force depends only on the local effective mass, local velocity, and local time constant. Drude's assumption is valid only under weak electric fields when the acceleration of the electron is small, because only then can the frictional force instantaneously adjust to the local velocity.

Let us now restrict ourselves only to *steady-state* conditions when all variables become time independent. In that case, the derivative with respect to time in (1.1) will vanish[6] and we will immediately get (assuming that there is no magnetic field)

$$\vec{v}_n(\vec{r})\Big|_{\text{steady-state}} = -\frac{e\tau_n(\vec{r})}{m_n^*}\vec{\mathcal{E}}(\vec{r}) = -\mu_n(\vec{r})\vec{\mathcal{E}}(\vec{r}), \tag{1.3}$$

where $\mu_n(\vec{r})\ (= e\tau_n(\vec{r})/m_n^*)$ is the local electron *drift mobility*.[7] Since we restricted ourselves to steady-state conditions, we dropped the variable t in (1.3). The negative sign in the last equation is a result of the negative charge of the electron. Because of this negative sign, the electron velocity is always directed *opposite* to the electric field.

Equation (1.3) is the well-known equation of the *Drude model* of charge conduction and is the defining equation for drift mobility at low electric field strengths when Drude's assumption holds. At high electric fields, Drude's assumption will break down. Effectively, that will make $\tau_n(\vec{r})$ depend on $\vec{\mathcal{E}}(\vec{r})$, which will make the relationship between $\vec{v}_n(\vec{r})\Big|_{\text{steady-state}}$ and $\vec{\mathcal{E}}(\vec{r})$ *nonlinear*. When that happens, the drift mobility will no longer be constant, but will depend on the local electric field $\vec{\mathcal{E}}(\vec{r})$. As long as that is all it depends on, our "local" model will still hold and the drift–diffusion model will remain valid. However, sometimes the mobility may depend on other factors that are associated with what is going on at remote locations. If and when that happens—i.e., nonlocal effects take hold—the entire drift–diffusion model collapses, as we have stated before. In this chapter, we will not worry about such situations.

The local electron current density in a solid $\vec{J}_n(\vec{r})$ is, by definition, related to the electron velocity as

$$\vec{J}_n(\vec{r}) = -n(\vec{r})e\vec{v}_n(\vec{r}), \tag{1.4}$$

where $n(\vec{r})$ is the local concentration of electrons. The negative sign on the right-hand-side merely reflects the fact that the direction of conventional current (defined as the direction of flow of positive charges) is opposite to the direction of the electron velocity.

---

[6]This happens when the electron is no longer accelerating or decelerating, which means that the external and frictional forces exactly balance.

[7]The perceptive reader would notice that we have defined the drift mobility while ignoring any magnetic field. Therefore, the drift mobility is not a good transport parameter if a magnetic field is present.

We deduce from the last two equations that the steady-state current density will be

$$\vec{J}_n(\vec{r})\Big|_{\text{steady-state}} = n(\vec{r})e\mu_n(\vec{r})\vec{\mathcal{E}}(\vec{r}) = \sigma_n(\vec{r})\vec{\mathcal{E}}(\vec{r}), \tag{1.5}$$

where

$$\sigma_n(\vec{r}) = n(\vec{r})e\mu_n(\vec{r}). \tag{1.6}$$

Here, we have attached the subscript "n" to the mobility μ and the velocity v just to remind us that it is the mobility and velocity of electrons, as opposed to the mobility and velocity of the other type of charge carriers in a semiconductor (holes). The quantity $\sigma_n(\vec{r})$ is the local conductivity of electrons. At low electric field strengths, both $\mu_n(\vec{r})$ and $\sigma_n(\vec{r})$ will be independent of the electric field strength $\vec{\mathcal{E}}(\vec{r})$. In that case, (1.5) will be nothing but *Ohm's law*. Therefore, the Drude model directly gives us Ohm's law.

Special Topic 1: Hall resistance and longitudinal resistance

Consider the resistor (a rectangular sample) shown in Fig. 1.3. A magnetic field is directed along the z-direction and an electric field is applied along the x-direction that causes a current to flow in that direction. Because the magnetic field exerts a Lorentz force on the electron in the y-direction, the latter's trajectory will bend as shown (so that its velocity develops a y-component) and consequently electrons will pile up at one edge of the resistor. This charge "pile up" will cause an electric field along the y-direction, which, in turn can cause a current to flow in the y-direction if we electrically connect (or shunt) the two edges of the resistor.

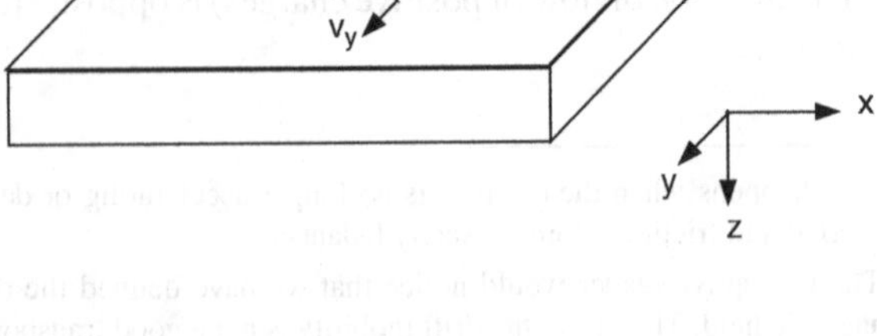

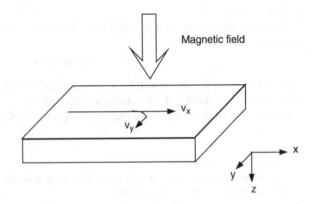

Fig. 1.3 A resistor where a magnetic field is applied along the z-direction and an electron is injected with a velocity along the x-direction

The steady-state version of (1.1) in this case will be

$$e[\vec{\mathcal{E}}(\vec{r}) - \vec{v}_n(\vec{r}) \times \vec{B}(\vec{r})] = -\frac{m_n^* \vec{v}_n(\vec{r})}{\tau_n(\vec{r})}. \tag{1.7}$$

This vector equation can be written as three scalar equations for the three components of the velocity:

$$\begin{aligned} v_x(\vec{r}) &= -\frac{e\tau_n(\vec{r})}{m_n^*}[E_x(\vec{r}) - v_y B(\vec{r})] = -\mu_n(\vec{r})[E_x(\vec{r}) - v_y B(\vec{r})] \\ v_y(\vec{r}) &= -\frac{e\tau_n(\vec{r})}{m_n^*}[E_y(\vec{r}) + v_x B(\vec{r})] = -\mu_n(\vec{r})[E_y(\vec{r}) + v_x B(\vec{r})] \\ v_z(\vec{r}) &= -\frac{e\tau_n(\vec{r})}{m_n^*}E_z(\vec{r}) = -\mu_n(\vec{r})E_z(\vec{r}), \end{aligned} \tag{1.8}$$

where

$$\begin{aligned} \vec{\mathcal{E}}(\vec{r}) &= E_x(\vec{r})\hat{x} + E_y(\vec{r})\hat{y} + E_z(\vec{r})\hat{z} \\ \vec{v}_n(\vec{r}) &= v_x(\vec{r})\hat{x} + v_y(\vec{r})\hat{y} + v_z(\vec{r})\hat{z} \\ \vec{B}(\vec{r}) &= B(\vec{r})\hat{z}. \end{aligned} \tag{1.9}$$

The first two relations in (1.8) can be combined and written in a matrix form as

$$\begin{bmatrix} -\frac{1}{\mu_n(\vec{r})} & B(\vec{r}) \\ -B(\vec{r}) & -\frac{1}{\mu_n(\vec{r})} \end{bmatrix} \begin{bmatrix} v_x(\vec{r}) \\ v_y(\vec{r}) \end{bmatrix} = \begin{bmatrix} E_x(\vec{r}) \\ E_y(\vec{r}) \end{bmatrix}. \tag{1.10}$$

Using (1.4), the last equation can be written as

$$\begin{bmatrix} \frac{1}{\mu_n(\vec{r})} & -B(\vec{r}) \\ B(\vec{r}) & \frac{1}{\mu_n(\vec{r})} \end{bmatrix} \begin{bmatrix} J_x(\vec{r})/en(\vec{r}) \\ J_y(\vec{r})/en(\vec{r}) \end{bmatrix} = \begin{bmatrix} E_x(\vec{r}) \\ E_y(\vec{r}) \end{bmatrix}. \tag{1.11}$$

Finally, using (1.6), we can write

$$\frac{1}{\sigma_n(\vec{r})} \begin{bmatrix} 1 & -\mu_n(\vec{r})B(\vec{r}) \\ \mu_n(\vec{r})B(\vec{r}) & 1 \end{bmatrix} \begin{bmatrix} J_x(\vec{r}) \\ J_y(\vec{r}) \end{bmatrix} = \begin{bmatrix} E_x(\vec{r}) \\ E_y(\vec{r}) \end{bmatrix}. \tag{1.12}$$

The resistivity tensor $[\rho(\vec{r})]$ is defined as

$$\begin{bmatrix} \rho_{xx}(\vec{r}) & \rho_{xy}(\vec{r}) \\ \rho_{yx}(\vec{r}) & \rho_{yy}(\vec{r}) \end{bmatrix} \begin{bmatrix} J_x(\vec{r}) \\ J_y(\vec{r}) \end{bmatrix} = \begin{bmatrix} E_x(\vec{r}) \\ E_y(\vec{r}) \end{bmatrix}, \tag{1.13}$$

which yields

$$\rho_{xx}(\vec{r}) = \rho_{yy}(\vec{r}) = \frac{1}{\sigma_n(\vec{r})}$$

$$-\rho_{xy}(\vec{r}) = \rho_{yx}(\vec{r}) = \frac{\mu_n(\vec{r})\, B(\vec{r})}{\sigma_n(\vec{r})} = \frac{B(\vec{r})}{en(\vec{r})}. \tag{1.14}$$

The diagonal resistivity (e.g., $\rho_{xx} = E_x/J_x|_{J_y=0}$) is called the *longitudinal resistivity*, and the off-diagonal resistivity (e.g., $\rho_{yx} = E_y/J_x\big|_{J_y=0}$) is called the *Hall resistivity*. The latter is linearly proportional to the magnetic flux density and the proportionality constant depends only on the electron concentration $n(\vec{r})$. Hence, measurement of this proportionality constant, called the *Hall constant*, will allow one to determine the average carrier concentration in the sample. If there are two types of carriers in a sample, namely both electrons and holes, then the situation becomes a little more complicated, but as long as there is only one type of carrier, or there are two types but one is much more numerous than the other, this simple derivation holds. In the latter case, it holds for the majority carriers.

Hall measurement is a routine way of measuring majority carrier concentration in a sample. The experiment is carried out by measuring the voltage V_y between the transverse edges of a sample with a voltmeter, while applying a magnetic field in the z-direction and passing a current in the x-direction using a current source. Because an ideal voltmeter has infinite impedance, it should draw no current so that $J_y = 0$. The electric field E_y is estimated from the relation $E_y = V_y/W$, where W is the sample's width. The current density J_x is measured with an ammeter. All this yields the Hall resistivity $\rho_{yx} = E_y/J_x\big|_{J_y=0}$. By measuring this quantity under various flux densities, one can obtain the Hall constant and from it the carrier concentration as long as the sample is unipolar (only one type of charge carrier is present) or bipolar (both types of charge carrier are present) but one type is the majority carrier.

Note that we can define a conductivity tensor $[\sigma(\vec{r})]$ as

$$\begin{bmatrix} \sigma_{xx}(\vec{r}) & \sigma_{xy}(\vec{r}) \\ \sigma_{yx}(\vec{r}) & \sigma_{yy}(\vec{r}) \end{bmatrix} \begin{bmatrix} E_x(\vec{r}) \\ E_y(\vec{r}) \end{bmatrix} = \begin{bmatrix} J_x(\vec{r}) \\ J_y(\vec{r}) \end{bmatrix}. \tag{1.15}$$

Since, we should have $[\rho(\vec{r})]^{-1} = [\sigma(\vec{r})]$, we will obtain that

$$\begin{bmatrix} \sigma_{xx}(\vec{r}) & \sigma_{xy}(\vec{r}) \\ \sigma_{yx}(\vec{r}) & \sigma_{yy}(\vec{r}) \end{bmatrix} = \begin{bmatrix} \frac{\rho_{yy}(\vec{r})}{\rho_{xx}(\vec{r})\rho_{yy}(\vec{r}) - \rho_{xy}(\vec{r})\rho_{yx}(\vec{r})} & -\frac{\rho_{yx}(\vec{r})}{\rho_{xx}(\vec{r})\rho_{yy}(\vec{r}) - \rho_{xy}(\vec{r})\rho_{yx}(\vec{r})} \\ -\frac{\rho_{xy}(\vec{r})}{\rho_{xx}(\vec{r})\rho_{yy}(\vec{r}) - \rho_{xy}(\vec{r})\rho_{yx}(\vec{r})} & \frac{\rho_{xx}(\vec{r})}{\rho_{xx}(\vec{r})\rho_{yy}(\vec{r}) - \rho_{xy}(\vec{r})\rho_{yx}(\vec{r})} \end{bmatrix}. \tag{1.16}$$

Hence, $\sigma_{xx}(\vec{r}) \neq 1/\rho_{xx}(\vec{r})$, unless the Hall resistances vanish!

1.4 Diffusion Current

Ohm's law relates only the *drift current* density to the electric field in a solid. It makes no allowance for the diffusion current which does not depend on the electric field in any case. Ohm's law could never have accounted for diffusion since it is based on the Drude model which deals only with a *single* electron. As mentioned earlier, diffusion requires multiple electrons to be present. It is caused by a concentration gradient (electrons diffuse from a region with higher concentration to a region with lower concentration) and a single electron does not define a concentration. One needs many electrons to form a "concentration." Therefore, diffusion is a phenomenon that cannot be captured within the single electron picture and will not emerge from the Drude model.

We have mentioned earlier that diffusion causes particles to move away from a region of high concentration to a region of low concentration. This happens due to Brownian motion of particles that causes them to bounce off one another and execute a random walk. There will be more frequent collisions in the region of high concentration, which will cause the average particle to move toward the low concentration region, as shown in Fig. 1.4a. This will happen regardless of whether

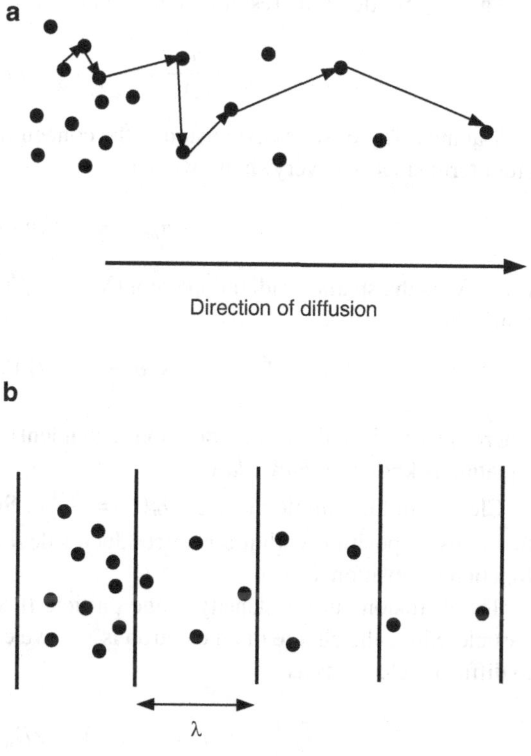

Fig. 1.4 (**a**) Particles, whether or not they are charged, diffuse from a region of higher concentration to a region of lower concentration through a process of random walk. The random walk takes place since the particles collide with each other and with other scatterers present within a solid; (**b**) A region is compartmentalized into bins of width λ which is the mean free path. Within each bin, the concentration is assumed to be uniform so that the probability of a particle diffusing to the *left* is equal to the probability of diffusing to the *right*. The concentration, however, varies from one bin to the next

the particles are charged or uncharged, but if the particles are charged, then the net motion will result in an electric current. We call this current the "diffusion current."

In order to derive an expression for the diffusion current, let us consider the situation in Fig. 1.4b where a sample with nonuniform particle distribution is partitioned into a number of compartments of equal width λ. Here, λ is the mean free path, or the average distance a particle travels before suffering a collision. Although λ could be different in different compartments, we will ignore that difference in the ensuing zeroth-order analysis.

We will assume that within any compartment, the concentration is approximately uniform (although it varies from one compartment to another), so that a particle has equal probability of moving to the left and to the right. In that case, the number of electrons moving from the m-th compartment to the $m+1$-th compartment in time τ is $\frac{1}{2}n_m\lambda A$, where n_m is the volume concentration in the m-th compartment, τ is the mean free time between collisions, and A is the cross-sectional area of the sample. Similarly, the number of electrons moving from the $m+1$-th compartment to the m-th compartment in time τ is $\frac{1}{2}n_{m+1}\lambda A$. The net number crossing the interface between the two compartments in time τ is

$$n_{\text{net}} = \frac{1}{2}\lambda A[n_m - n_{m+1}]. \tag{1.17}$$

The magnitude of the resulting particle flux is

$$\mathcal{F} = \frac{n_{\text{net}}}{A\tau} = \frac{\lambda}{2\tau}[n_m - n_{m+1}]. \tag{1.18}$$

Making a Taylor series expansion of the concentration and retaining only the first order term since λ is very small, we get

$$n_m - n_{m+1} = -\vec{\nabla}_{\text{r}} n \cdot \vec{\lambda}, \tag{1.19}$$

where $\vec{\nabla}_{\text{r}}$ is the spatial gradient operator ($\vec{\nabla}_{\text{r}} = (\partial/\partial x)\hat{x} + (\partial/\partial y)\hat{y} + (\partial/\partial z)\hat{z}$). This results in

$$\vec{\mathcal{F}} = -\frac{\lambda^2}{2\tau}\vec{\nabla}_{\text{r}} n = -D_n(\vec{r})\vec{\nabla}_{\text{r}} n(\vec{r}), \tag{1.20}$$

where $D_n(\vec{r})$ is called the (position-dependent) *diffusion coefficient*. The last equation is known as *Fick's law*.

Clearly, in this simple picture, $D_n(\vec{r}) = \frac{\lambda^2(\vec{r})}{2\tau(\vec{r})}$. Since both λ and τ are generally functions of position within a semiconductor device in steady-state, D_n will be a function of position as well.

The diffusion current density is the particle flux multiplied by the charge of a particle. Since the charge of an electron is $-e$, we can write the current density due to diffusing electrons as

$$\vec{J}_n^{\text{diffusion}}(\vec{r}) = -e\vec{\mathcal{F}}(\vec{r}) = eD_n(\vec{r})\vec{\nabla}_{\text{r}} n(\vec{r}), \tag{1.21}$$

1.5 Electron and Hole Currents

So far, we have talked mostly about electrons. Let us now turn our attention to holes. Electrons and holes have negative and positive charges, respectively. Therefore, electrons drift in the direction opposite to the electric field and holes drift in the same direction as the electric field. Consequently, electrons and holes always drift in opposite directions in an electric field. However, since conventional current is viewed as the flow of *positive* charge, the drift current due to electrons flows in the direction *opposite* to the direction in which the electrons are drifting, whereas the drift current due to holes flows in the same direction as that in which holes are drifting. Therefore, even though electrons and holes drift in opposite directions, their associated drift currents flow in the *same* direction, namely the direction of the electric field. Hence, the electron drift current density and hole drift current density always have the same sign and they *add*; they do not subtract. We can write the electron and hole (steady state) drift currents as

$$\begin{aligned} \vec{J}_n(\vec{r})|_{\text{drift}} &= n(\vec{r})e\mu_n(\vec{r})\vec{\mathcal{E}}(\vec{r}) \\ \vec{J}_p(\vec{r})|_{\text{drift}} &= p(\vec{r})e\mu_p(\vec{r})\vec{\mathcal{E}}(\vec{r}), \end{aligned} \tag{1.22}$$

where n refers to electrons and p to holes. Note that the drift current densities have the same sign, which is the sign of the electric field. In our convention, the quantity e is always positive; the charge of an electron is $-e$ and the charge of a hole is $+e$.

Even though electrons and holes always *drift* in opposite directions, they need not *diffuse* in opposite directions as well. The direction of diffusion is determined by the direction of the concentration gradient which is a vector quantity. Particles, regardless of whether they are electrons or holes or even charged or uncharged, always diffuse from a region of higher concentration to one of lower concentration. Therefore, they always diffuse in the direction *opposite* to the concentration gradient because in going from a region of higher concentration to a region of lower concentration, the gradient is negative. Remembering that the direction of conventional current is defined as the direction of flow of positive charges, we get

$$\begin{aligned} \vec{J}_n(\vec{r})|_{\text{diffusion}} &= eD_n(\vec{r})\vec{\nabla}_{\text{r}}n(\vec{r}) \\ \vec{J}_p(\vec{r})|_{\text{diffusion}} &= -eD_p(\vec{r})\vec{\nabla}_{\text{r}}p(\vec{r}). \end{aligned} \tag{1.23}$$

The total steady-state current density for each type of carrier is obtained by adding the drift and diffusion current densities *vectorially*. This will yield:

$$\begin{aligned} \vec{J}_n(\vec{r})|_{\text{steady-state}} &= n(\vec{r})e\mu_n(\vec{r})\vec{\mathcal{E}}(\vec{r}) + eD_n(\vec{r})\vec{\nabla}_{\text{r}}n(\vec{r}) \\ \vec{J}_p(\vec{r})|_{\text{steady-state}} &= p(\vec{r})e\mu_p(\vec{r})\vec{\mathcal{E}}(\vec{r}) - eD_p(\vec{r})\vec{\nabla}_{\text{r}}p(\vec{r}). \end{aligned} \tag{1.24}$$

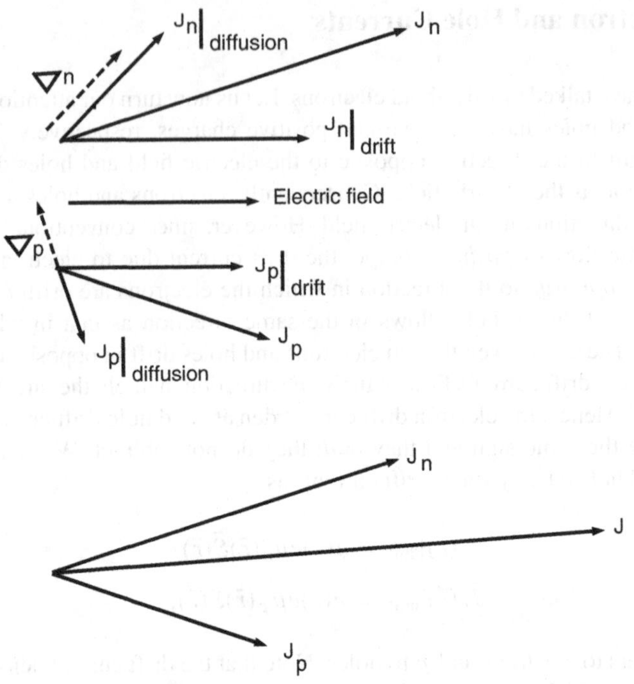

Fig. 1.5 The *arrows* show the directions of the different components of the steady-state electron and hole currents at any given location $\vec{r}$. Note that the drift components are always in the direction of the local electric field. The electron diffusion current component is in the direction of the concentration gradient for electrons while the hole diffusion current component is directed opposite to the direction of the hole concentration gradient

Finally, the total steady-state current due to both electrons and holes is given by the vector sum of the electron and hole currents:

$$\vec{J}|_{\text{steady-state}} = \vec{J}_n(\vec{r})|_{\text{steady-state}} + \vec{J}_p(\vec{r})|_{\text{steady-state}}. \tag{1.25}$$

In Fig. 1.5, we show pictorially the different *steady-state* current components that can arise in a semiconductor.

1.6 Conservation of Charge and the Continuity Equation

Both electrons and holes, being charged particles, obey the principle of conservation of charge which mandates that charge cannot be created or destroyed *arbitrarily*. This principle leads to the *continuity equation*, which we derive below for *steady-state* situations.

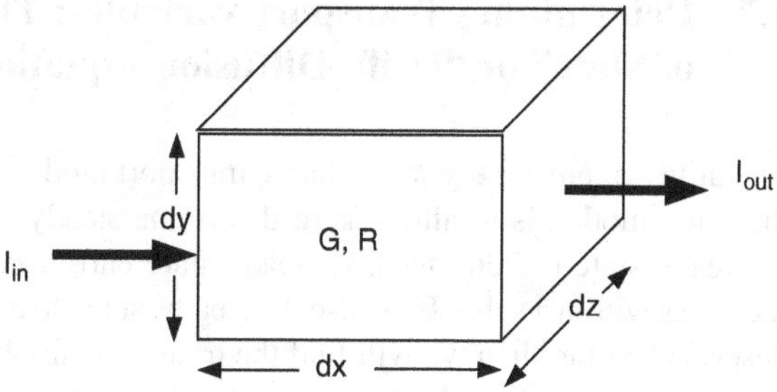

Fig. 1.6 An incremental volume in space where current is entering and exiting

Consider an incremental region in space with dimensions dx, dy, and dz as shown in Fig. 1.6. Let current flow in the x-direction, with I_{in} being the current entering the box and I_{out} the current exiting.

Therefore, we get that

$$I_{in} = J_{in} dy dz = \frac{dq_{in}}{dt}$$
$$I_{out} = J_{out} dy dz = \frac{dq_{out}}{dt}, \tag{1.26}$$

where q_{in} and q_{out} are the charge entering and exiting.

In steady state situations, conservation of charge dictates that

$$q_{out} - q_{in} = q_{gained} - q_{lost} = \rho_{gained} dx dy dz - \rho_{lost} dx dy dz, \tag{1.27}$$

where q_{gained} is the charge gained within the box due to generation processes such as light shining on a semiconductor and creating electron-hole pairs, while q_{lost} is the charge lost due to electron-hole recombination. Here, ρ is the charge density.

Combining the last two equations, we obtain

$$dJ = J_{out} - J_{in} = \frac{\left(\rho_{gained} - \rho_{lost}\right) dx}{dt} = \frac{\left(\rho_{gained} - \rho_{lost}\right) dx}{dt}, \tag{1.28}$$

which yields

$$\frac{dJ}{dx} = qG - qR, \tag{1.29}$$

where G is the particle generation rate, R is the particle recombination rate, and q is the charge of a particle. Remembering that electrons are negatively charged so that $q = -e$ for them, and holes are positively charged, so that $q = e$ for them, we get the following continuity equations for electrons and holes in *steady state* (in three dimensions):

$$\vec{\nabla}_r \cdot \vec{J}_n + eG_n(\vec{r}) - eR_n(\vec{r}) = 0 \quad \texttt{(for electrons)}$$
$$\vec{\nabla}_r \cdot \vec{J}_p - eG_p(\vec{r}) + eR_p(\vec{r}) = 0 \quad \texttt{(for holes)}. \tag{1.30}$$

1.7 Determining Transport Variables: The "Equations of State" or "Drift–Diffusion Equations"

The utility of any steady-state charge transport model, such as the steady-state drift–diffusion model, is to allow us to determine steady-state *transport variables* such as steady-state current density, steady-state carrier concentration, etc. at different locations within a solid. If we use the steady-state drift–diffusion model that we have described so far, then we will find the transport variables by solving three different sets of equations, namely the current equation (one for electrons and one for holes), the continuity equation (again, one for electrons and one for holes) and the Poisson equation. There are a total of five unknowns: $\vec{J}_n(\vec{r}), \vec{J}_p(\vec{r}), n(\vec{r}), p(\vec{r}), \vec{\mathcal{E}}(\vec{r})$, so that we need five equations. These five equations are:

$$\begin{aligned}
\vec{\nabla}_r \cdot [\epsilon(\vec{r})\vec{\mathcal{E}}(\vec{r})] &= e[p(\vec{r}) - n(\vec{r})] \quad \texttt{(Poisson equation)} \\
\vec{\nabla}_r \cdot \vec{J}_n + eG_n(\vec{r}) - eR_n(\vec{r}) &= 0 \quad \texttt{(electron continuity equation)} \\
\vec{J}_n(\vec{r}) &= n(\vec{r})e\mu_n(\vec{r})\vec{\mathcal{E}}(\vec{r}) + eD_n(\vec{r})\vec{\nabla}_r n(\vec{r}) \quad \texttt{(electron current equation)} \\
\vec{\nabla}_r \cdot \vec{J}_p - eG_p(\vec{r}) + eR_p(\vec{r}) &= 0 \quad \texttt{(hole continuity equation)} \\
\vec{J}_p(\vec{r}) &= p(\vec{r})e\mu_p(\vec{r})\vec{\mathcal{E}}(\vec{r}) - eD_p(\vec{r})\vec{\nabla}_r p(\vec{r}) \quad \texttt{(hole current equation).}
\end{aligned} \tag{1.31}$$

Here, $G_n(G_p)$ and $R_n(R_p)$ are the generation and recombination rates of electrons (holes). The last five equations are sometimes referred to as the "equations of state," or simply, "drift–diffusion equations." They form the basis of the drift–diffusion model of steady-state charge transport [5].

The Poisson equation is simply a restatement of the famous Gauss' law of electrostatics. It can be rewritten as

$$\vec{\nabla}_r \cdot [\epsilon(\vec{r})\vec{\nabla}_r V(\vec{r})] = -\rho = -e[p(\vec{r}) - n(\vec{r})], \tag{1.32}$$

where the potential $V(\vec{r})$ is related to the electric field $\vec{\mathcal{E}}(\vec{r})$ as

$$\vec{\mathcal{E}}(\vec{r}) = -\vec{\nabla}_r V(\vec{r}). \tag{1.33}$$

Equation (1.32) is solved subject to the boundary conditions that specify the potential $V(\vec{r})$ at two ends of a device. The unknowns $p(\vec{r})$ and $n(\vec{r})$ are found by solving (1.31), (1.33), and (1.32) simultaneously or iteratively. In the iterative approach, one starts with a guess for the potential $V(\vec{r})$ everywhere within the device, then extracts the electric field from this potential using (1.33), uses this field in (1.31) to find the carrier concentrations, and finally uses those concentrations in (1.32) to come up with a more refined guess for $V(\vec{r})$. The process is repeated until satisfactory convergence is obtained, i.e., until we reach a situation when $\left|V_m(\vec{r}) - V_{m-1}(\vec{r})\right| \leq \epsilon$, where ϵ is a small voltage, $V_m(\vec{r})$ is the voltage found in the m-th iteration and $V_{m-1}(\vec{r})$ is the voltage found in the previous iteration. This method of finding the transport variables within a solid state device is sometimes called *self-consistent device analysis*.

1.8 Generation and Recombination Processes

In the continuity equations within (1.31), we have two terms G and R associated with generation and recombination rates of charge carriers. Electron-hole pairs can be generated within a solid by external agents such as light shining on the solid, thermal fluctuations, electron bombardment, etc. Additionally, internal processes such as impact ionization can also generate electron-hole pairs. The number of pairs generated per unit time by such processes is the generation rate G. On the other hand, electron-hole pairs can be annihilated when an electron recombines with a hole while giving off light or heat, or via Auger recombination. The number of pairs recombining per unit time is the recombination rate R. In both recombination and generation, net charge is conserved since an electron and a hole have equal but opposite charges. This is consistent with charge conservation. Note that whenever an electron is generated, a hole is generated as well, and whenever an electron is annihilated, a hole is annihilated with it. Therefore, it is obvious that $G_n(\vec{r},t) = G_p(\vec{r},t)$.

In light induced generation—also known as "radiative generation"—light is absorbed in a semiconductor to provide the energy to break a covalent bond between atoms and release an electron which becomes free. This process is viewed as a photon being absorbed to excite an electron from the valence band to the conduction band, leaving behind a hole in the valence band. It is depicted in Fig. 1.7a.

In thermally induced generation, an electron in the valence band absorbs a phonon associated with thermal lattice vibrations and gets excited to a trap level in the bandgap. Since phonons typically have much less energy than photons, direct transfer from the valence band to the conduction band is usually not possible unless the semiconductor has a very small bandgap. The trapped electron may then absorb another phonon to get to the conduction band. This multiphonon process can excite electrons from the valence band to the conduction band in multiple steps and cause electron–hole pair generation. It is shown in Fig. 1.7b.

Electron–hole pairs can also be generated due to internal processes such as impact ionization. An electron with large kinetic energy collides with an atom, breaks a bond and knocks off an extra electron leaving behind a hole. In the energy band diagram, this process is viewed as an electron high in the conduction band (large kinetic energy) falling down in the conduction band and the released energy is absorbed to excite an electron from the valence band into the conduction band leaving behind a hole in the valence band. It is depicted in Fig. 1.7c.

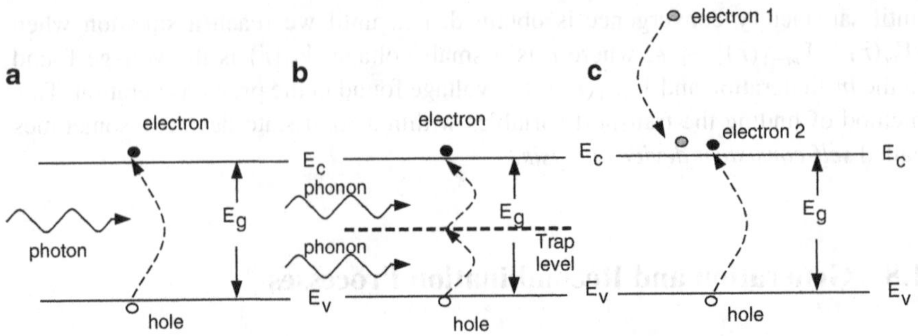

Fig. 1.7 Energy band depictions of generation processes: (**a**) The process of radiative generation in a semiconductor. A photon is absorbed to excite an electron from the valence band to the conduction band, leaving behind a hole in the valence band. The conduction band edge is denoted by E_{c0} and the valence band edge by E_{v0}. The bandgap is E_g. (**b**) The process of nonradiative generation through the intercession of traps. An electron in the valence band absorbs phonons to reach up to a trap level in the bandgap, leaving behind a hole in the valence band. It then absorbs more phonons to ultimately reach the conduction band. In the process, an electron–hole pair is generated. (**c**) Process of nonradiative generation via impact ionization. An energetic electron gives up some of its energy, falling down in the conduction band. The released energy is used to excite a second electron from the valence band to the conduction band, leaving behind a hole in the valence band. In physical terms, a photon (in process "a") or a phonon (in process "b"), will provide enough energy to an electron in the outermost (valence) shell of an atom to get free from the atom (ionization). This is depicted as excitation of an electron from the valence to the conduction band, leaving behind a hole in the valence band. In impact ionization, an energetic electron traveling through a semiconductor collides with an electron in the outermost shell of an atom and knocks it free. The first electron loses some of its kinetic energy, but the second electron gets free from its parent atom. This process is depicted as the first electron falling down in the conduction band and the second electron getting excited from the valence to the conduction band

Recombination involves an electron and a hole annihilating each other, resulting in the loss of both the electron and the hole. In the energy band diagram, this process is viewed as an electron from the conduction band falling down into the valence band (in one or multiple steps), releasing energy, and annihilating a hole in the valence band. The released energy can be given off as light (a photon) or heat (phonons). The former is called *radiative recombination* and the latter *nonradiative recombination*. The radiative processes can be accomplished in a single hop, in which case the energy of the photon emitted will be equal to or larger than the bandgap energy. If there are bandgap states, i.e., allowed energy levels in the bandgap of the semiconductor caused by impurities and crystal defects, then it is possible that the falling electron makes multiple hops, with pit stops at the bandgap states. Photons may or may not be emitted during any of these hops, but since the energy difference between the initial and final state of the electron at the beginning and end of a hop is less than the bandgap energy, the energy of the emitted photon (if one is emitted) will be less than the bandgap energy. However, the probability of a photon being emitted while transitioning through bandgap states is usually very small. It is much more likely that phonons are emitted while transitioning through bandgap states. Since the energy of a phonon in a semiconductor is typically much less than the bandgap energy, a phonon cannot be usually emitted in the single hop

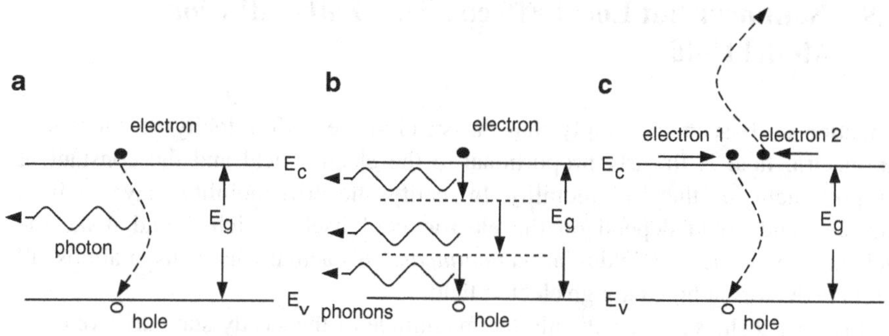

Fig. 1.8 Energy band depictions of recombination processes: (**a**) The process of radiative recombination in a semiconductor. An electron falls down from the conduction band into the valence band, releasing a photon, and annihilating itself as well as a hole in the valence band; (**b**) The process of nonradiative recombination in a semiconductor. An electron falls down from the conduction band to the valence band, making stops at bandgap states (shown by *dashed lines*), and emitting phonons at each step. Ultimately, the electron reaches the valence band where it annihilates itself and a hole; (**c**) Auger recombination process. Two electrons collide where one gives up its energy to the other. The electron which gains energy goes higher up in the conduction band, while the electron which loses energy falls down into the valence band and annihilates a hole. In physical terms, recombination is the process of an atom capturing a free electron and making it quasi-bound. The captured electron settles down into an outer shell of the atom, thereby annihilating a vacancy (hole) there

process. Thus, the presence of bandgap states caused by defects and impurities helps the nonradiative (phonon emission) processes, while their absence inhibits them. Semiconductor lasers that emit light through the radiative recombination process would prefer that all recombination processes are radiative and none are nonradiative. In fact, the internal quantum efficiency of such a laser is defined as the ratio of the radiative recombination rate to the total (radiative + nonradiative) recombination rate. Hence, in order to make this efficiency high, the nonradiative processes must be suppressed. This requires eliminating the bandgap states by using very pure and perfect semiconductor materials.

Finally, there is one other important process of recombination known as Auger recombination. Here, two electrons in the conduction band collide. One gains energy from the collision and goes high up in the conduction band, while the other loses energy and drops down into the valence band where it recombines with a hole. In the collision process, the total energy is conserved, i.e., the energy gained by one electron is equal to that lost by the other. Clearly, this process is the converse of impact ionization. Since no light is given off, it is also a form of nonradiative recombination. Radiative, nonradiative, and Auger recombination processes are depicted in Fig. 1.8a–c, respectively.

Radiative processes—both recombination and generation—can take place efficiently only in so-called "direct-gap" semiconductors like GaAs. They will not happen efficiently in "indirect-gap semiconductors" like Silicon and Germanium. We will see the reason for that in Chap. 6.

1.9 Nonlinear but Local Effects: The Drift–Diffusion Model Holds

Equation (1.3) seems to imply that the steady-state drift velocity of carriers in an electric field is linearly proportional to the electric field and the constant of proportionality is the drift mobility. In reality, the drift mobility may not be a constant and could depend on the electric field itself, which would make the velocity versus electric field relation *nonlinear*. This actually happens in almost all semiconductors at high enough electric field.

Figure 1.9 shows schematically the magnitude of the steady-state drift velocity versus electric field characteristic in two well-known semiconductors—silicon and GaAs. In both cases, the steady-state drift-velocity is actually proportional to the electric field at low fields, so that the drift mobility is indeed a field-independent constant at low field strengths. However, in the case of silicon, the drift velocity saturates at high field strengths, so that it is no longer proportional to the field and the mobility is no longer a field-independent constant. If we still define mobility as the ratio of the drift velocity to the electric field, then it becomes progressively smaller with increasing electric field strength, showing that it is a field-dependent quantity. Consequently, mobility is not a material parameter, since it is not a constant for a given material. Rather, it is a "transport parameter" indicating how mobile a charge carrier is in a semiconductor. If the mobility becomes zero, then the drift velocity becomes zero at any electric field, meaning that the carrier is completely immobile and no amount of force or electric field will make it move. Recall that mobility is proportional to the mean time between collisions and inversely proportional to the effective mass. Obviously, a particle with infinite effective mass will be immobile and a particle that experiences scattering so frequent that the mean time between collisions is zero, will be also effectively immobile. Hence, the term "mobility" for the quantity μ in (1.3) is completely apt.

In the case of GaAs, the drift velocity peaks at a critical electric field and then actually drops, resulting in a negative slope, or a negative differential mobility. This feature is the basis of the well-known Ridley–Watkins–Hilsum effect or Gunn effect

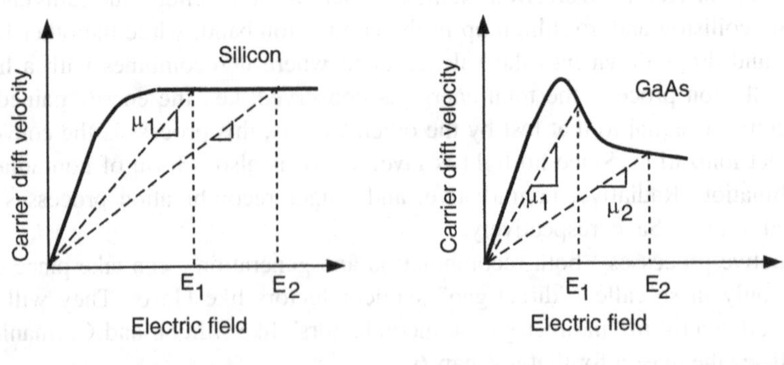

Fig. 1.9 Schematic depiction of the steady-state carrier drift velocity versus electric field characteristics in Silicon and GaAs

[1–3] first observed in GaAs, which is exploited in microwave generators. Here too, the drift mobility will be a function of the electric field strength, since the velocity versus electric-field relation is not linear.

Because of the nonlinearity of the velocity-field characteristics in Fig. 1.9, one might wonder if the drift–diffusion model can hold up and handle this kind of nonlinearity. The answer is "yes." All we have to do is to replace $\mu(\vec{r})$ with $\mu(\vec{\mathcal{E}}(\vec{r}))$, i.e., the mobility is made a function of the local electric field. This will allow us to model nonlinear effects correctly. We simply have to extract the dependence of the mobility on the local electric field using the type of characteristic as shown in Fig. 1.9, and use that in determining $\mu(\vec{\mathcal{E}}(\vec{r}))$. As long as the drift mobility *is a function of local electric field alone*, our drift–diffusion model holds up.

1.10 Nonlocal Effects: Invalidity of Drift–Diffusion Model

The drift–diffusion model will unfortunately crumble if there are *nonlocal* effects so that the mobility cannot be expressed as a function of local parameters (such as local electric field, or local electron energy, etc.) alone. One example of nonlocal effects (in time) is "velocity overshoot" where the velocity of an electron in response to an electric field exhibits transient behavior. After the electric field is switched on (at time $t = 0$), the average velocity of the electron ensemble subjected to the electric field [8] increases rapidly, reaches a peak in time, and then subsides gradually to settle down to the steady-state value as shown in Fig. 1.10. This effect cannot be captured within the drift–diffusion model since the velocity is no longer a *unique function of the local electric field.* The final steady-state velocity may be a unique function, but the transient velocity is not. In very short devices, the electron may traverse the entire device without ever reaching steady-state. In such a device, the velocity is never a unique function of the local electric field. Consequently, (1.31) (and the entire drift–diffusion model) will fail in such a device.

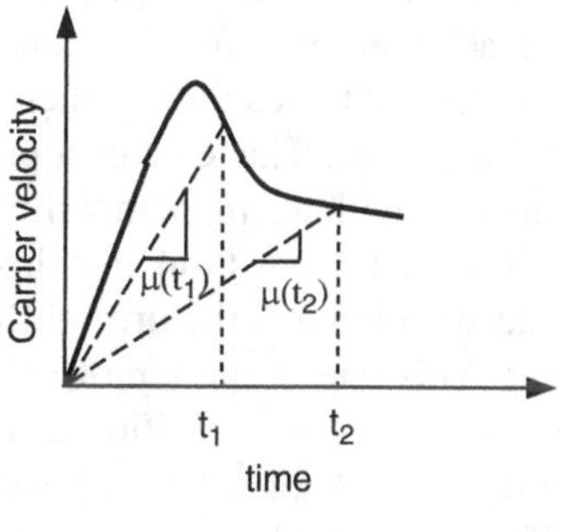

Fig. 1.10 Velocity overshoot effect. The applied electric field is constant and turned on at time $t = 0$. The velocity begins to rise with time, reaches a peak, and then drops, ultimately settling to a constant value

[8]We call this the "average velocity of the electron ensemble" as opposed to "drift velocity" since it is not steady-state.

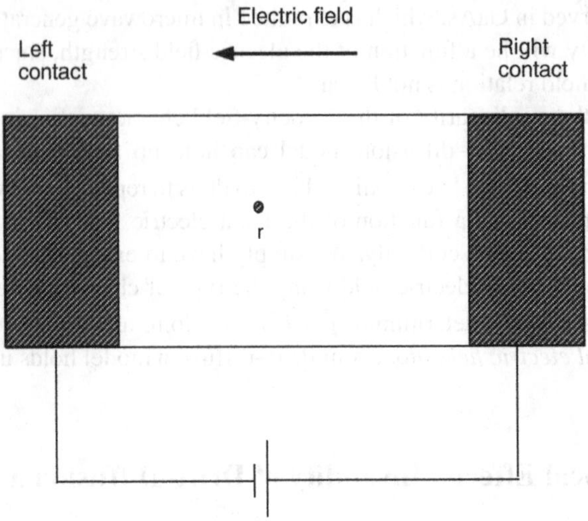

Fig. 1.11 A resistor where electrons are injected from the left contact with zero velocity. The injected electrons encounter an electric field after entering the resistor (due to the battery), which propels them toward the right contact. The electrons may suffer velocity overshoot. The mobility at location **r** is not well-defined (it is certainly not uniquely prescribed by the electric field at that location) since it depends on how long it took for electrons to get there. This time depends not only on the electric field at location **r** but also on electric fields everywhere to the left of **r** since they will determine the electron's acceleration and hence the time it takes to reach **r**. If there is scattering within the device, then the time to reach **r**—and hence the mobility at **r**—also depends on the scattering history! Therefore, the mobility cannot be expressed in terms of local variables alone

In order to understand the problem caused by velocity overshoot, consider the device shown in Fig. 1.11 which is subjected to a potential gradient (or electrical bias) caused by a battery. The electrons were injected from the left contact with, say, zero velocity and encountered the electric field after entering the device. At location $\vec{r}$, we do not have a unique mobility $\mu(\vec{r})$ determined by the electric field at location $\vec{r}$, since the mobility depends on how long it took *on the average* for an electron to reach that location. If it took a time t_1, then the mobility is $\mu(t_1)$ (see Fig. 1.10), whereas if it took a time t_2, then the mobility is $\mu(t_2)$. There is not a unique mobility at $\vec{r}$ anymore! This creates a problem for the drift–diffusion equations which can only deal with a *unique* mobility $\mu(\vec{r})$. As a result, the drift–diffusion model fails.

The reader may think that this happens because we are dealing with a nonsteady-state situation and the drift–diffusion model was derived for steady-state transport only. This line of thinking would be wrong. First, the drift–diffusion model works in nonsteady-state situation as well and we will derive the nonsteady-state version in the next chapter. Second, we will show another example now where the drift–diffusion model breaks down in steady-state situation as well.

Figure 1.12 shows the approximate conduction band profile within an n^+-p^+–n heterojunction bipolar transistor (HBT) where the emitter is composed of AlGaAs

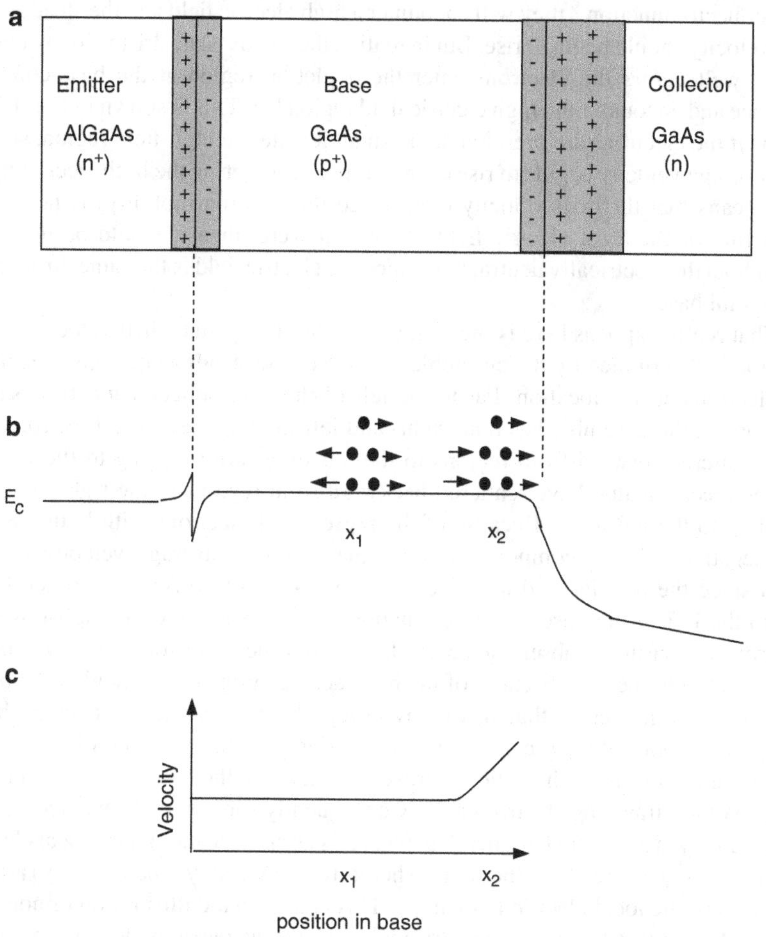

Fig. 1.12 (**a**) An n^+-p^+-n heterojunction bipolar transistor where the emitter is composed of AlGaAs and the base and collector are composed of GaAs. The emitter and base are heavily doped. The depletion regions are shown as positively and negatively charged regions. (**b**) The approximate conduction band profile with the electron ensemble velocity distribution (indicated by *arrows*) shown at two different locations in the base. (**c**) Qualitative behavior of the steady-state ensemble average velocity versus position within the base. Note that the velocity begins to rise even before the electrons encounter the high-field region, resulting in an apparent "anticipatory effect"

and the collector and base are composed of GaAs. The transistor is biased in the active region, i.e., the emitter-base junction is forward biased and the base-collector junction is reverse biased. Electrons are injected into the base from the emitter. The base is heavily doped so that the electric field in the base is nearly zero (and constant) as indicated by the fact that the conduction band profile within the base is nearly flat. We do expect that once the electrons enter the depletion region at the base-collector junction,[9] they will encounter a high electric field and the steady-state drift velocity should begin to rise. But in reality, the steady-state drift velocity begins to rise well *before* the electrons enter the depletion region at the base-collector interface and encounter the high electric field region [4]. This is shown in Fig. 1.12c. It is as if the electrons are prescient and can anticipate the high-field region, so that their average velocity begins to rise even before encountering the high electric field! This means that the drift velocity (and hence the electron mobility) is no longer a function of the local electric field, since if it were, then it would be the same throughout the electrically neutral base since the electric field is the same throughout the neutral base.

What really happens here is the following. The steady-state drift velocity at any location is determined by the ensemble average of the steady-state velocities of all the electrons at that location. Far to the left of the base-collector junction, say at location x_1, the ensemble has both right- and left-moving electrons. Electrons that have scattered forward from regions to the left of x_1 are traveling to the right at x_1, and electrons that have scattered backward from regions to the right of x_1 are traveling to the left at x_1. Because of the presence of electrons with both positive and negative velocity components at x_1, the ensemble average velocity here is small since the positive and negative components tend to offset each other. Now, just to the left of the base-collector junction, say at location x_2, the left-traveling electrons are virtually absent since electrons from the collector region will never travel back into the base because of the huge accelerating electric field in the base-collector depletion region that inexorably sweeps them forward, away from the base. Viewed differently, there is a huge potential barrier at the base-collector junction that prevents electrons from the collector traveling back into the base and increasing the fraction of left traveling electrons at x_2. Consequently, the ensemble average steady-state velocity at x_2 is much higher than what it is at x_1, even though the local electric fields at x_1 and x_2 are about the same. Therefore, the velocity (and mobility) are not functions of the local electric field alone. This is a dramatic illustration of nonlocal effects. The drift velocity begins to rise when electrons reach the location x_2, even though the electric field at x_2 is still very weak. This is the basis of the "anticipatory effect."

Needless to say, such an effect cannot be captured within the standard drift–diffusion model. The drift–diffusion model fails here since assumption 2 is violated. There is not enough backscattering near the collector edge within the base to allow the electron ensemble to reach local equilibrium with the surroundings. The inevitable aftermath is nonlocal effects that invalidate the drift–diffusion model.

The final issue to address before we conclude this chapter is how does one go about determining the *transport parameters* mobility and diffusion coefficient for both electrons and holes at different locations within a piece of solid. These quantities appear in the drift–diffusion equations, i.e., (1.31), and must be known before

[9] Because of the much heavier doping in the base compared to the collector, the depletion region resides mostly in the collector and hence starts almost at the base-collector metallurgical junction.

we can solve (1.31) to find the steady-state transport variables $J_n(\vec{r})$, $J_p(\vec{r})$, $n(\vec{r})$, and $p(\vec{r})$. As long as there are *no nonlocal effects*, we can always find the mobility from the velocity field relationships of the type shown in Fig. 1.9. The diffusion coefficient can then be related to the mobility by the so-called *Einstein relation* discussed next. This relation is at least approximately valid as long as the current density is small, quasi equilibrium conditions prevail, and the semiconductor is nondegenerate. Therefore, we can determine the transport parameters if no nonlocal effect shows up. However, all bets are off if nonlocal effects take hold. In that case, the transport parameters cannot be expressed as functions of the local electric field, or any other variable that depends only on the local coordinate, and the entire drift–diffusion model fails. In fact, the mobility and diffusion coefficient then become functionals of a certain function known as the distribution function $f(\vec{r}, \vec{k}, t)$, which is the probability of finding an electron at a position $\vec{r}$, with a wavevector $\vec{k}$, at an instant of time t. The distribution function can be found by solving the *Boltzmann Transport equation (BTE)* which is the subject of the next chapter.

Special Topic 2: The Einstein relation

The Einstein relation relates the diffusion coefficient to the drift mobility in a *nondegenerate* semiconductor under *global equilibrium* conditions. When global equilibrium exists, no electron or hole current is flowing through the semiconductor. However, if some electron or hole current is flowing but it is small, then we can say that quasi-equilibrium condition prevails. In that case, the Einstein relation will be approximately valid. We will derive this relation here.

A nondegenerate semiconductor is one where the Fermi level is at least $3k_BT$ energy below the conduction band edge and $3k_BT$ energy above the valence band edge (k_B is the Boltzmann constant and T is the absolute ambient temperature). In such a semiconductor, the steady-state electron concentration is related to the Fermi level μ_F[10] as

$$n(\vec{r}) = N_c e^{\frac{\mu_F - E_{c0}(\vec{r})}{k_BT}}, \tag{1.34}$$

where N_c is the effective density of states in the conduction band and E_{c0} is the conduction band edge. We will derive this relation later in Chap. 4. The quantity N_c depends on the effective mass of the electron and temperature, which we assume to

---

[10]The Fermi level μ_F is defined in the following way: the probability that an electron in a solid under global equilibrium has *total* energy (kinetic + potential) equal to the Fermi level μ_F is exactly one-half or 50%. This follows from Fermi–Dirac statistics which dictates that the probability of finding an electron with total energy E under global equilibrium is $1/\{\exp[(E - \mu_F)] + 1\}$.

be spatially invariant. We will also show in Chap. 2 that if the effective mass and temperature are spatially invariant, then the Fermi level μ_F does not vary in space as long as the semiconductor is in global equilibrium.

Using this result in the electron current equation in (1.31), and setting $J_n = 0$ for global equilibrium, we get

$$0 = N_\mathrm{c}\mathrm{e}^{\frac{\mu_\mathrm{F}-E_{\mathrm{c}0}(\vec{r})}{k_\mathrm{B}T}} e\mu_n(\vec{r})\vec{\mathcal{E}}(\vec{r}) + eD_n(\vec{r})\vec{\nabla}_\mathrm{r}\left[N_\mathrm{c}\mathrm{e}^{\frac{\mu_\mathrm{F}-E_{\mathrm{c}0}(\vec{r})}{k_\mathrm{B}T}}\right], \tag{1.35}$$

which simplifies to

$$0 = \mu_n(\vec{r})\vec{\mathcal{E}}(\vec{r}) - D_n(\vec{r})\frac{1}{k_\mathrm{B}T}\vec{\nabla}_\mathrm{r}E_{\mathrm{c}0}(\vec{r}). \tag{1.36}$$

Now, the electric field is the negative of the potential gradient, so that $\vec{\nabla}_\mathrm{r}E_{\mathrm{c}0}(\vec{r}) = e\vec{\mathcal{E}}(\vec{r})$. Therefore, we immediately get

$$D_n(\vec{r}) = \frac{k_\mathrm{B}T}{e}\mu_n(\vec{r}). \tag{1.37}$$

Similarly, it can be shown that

$$D_p(\vec{r}) = \frac{k_\mathrm{B}T}{e}\mu_p(\vec{r}). \tag{1.38}$$

The above relations between the local diffusion coefficient and mobility are collectively known as the *Einstein relation*.

Special Topic 3: Shockley–Read–Hall recombination

Whenever the concentrations of electrons and holes in a semiconductor exceed the local equilibrium concentrations $n_0(\vec{r},t)$ and $p_0(\vec{r},t)$, the excess electrons and holes recombine radiatively and/or nonradiatively to restore local equilibrium conditions. Similarly, whenever the concentrations are less than the local equilibrium concentrations, electrons and holes are generated (nonradiatively, if there are no light sources present) until local equilibrium is restored.

Shockley, Read, and Hall developed an expression for the recombination rate of electrons and holes, which stated

$$\begin{aligned} R_n(\vec{r},t) &= R_p(\vec{r},t) \\ &= \frac{n(\vec{r},t)p(\vec{r},t) - n_0(\vec{r},t)p_0(\vec{r},t)}{\tau_{p0}(\vec{r},t)[n(\vec{r},t) + n_i(\vec{r},t)] + \tau_{n0}(\vec{r},t)[(p(\vec{r},t) + n_i(\vec{r},t)]}, \end{aligned} \tag{1.39}$$

where n_i is the intrinsic carrier concentration in the semiconductor, n and p are the electron and hole concentrations, and τ_{p0}, τ_{n0} are lifetimes associated with recombination and generation processes.

We can write

$$n(\vec{r}) = n_0(\vec{r}) + \Delta n(\vec{r})$$
$$p(\vec{r}) = p_0(\vec{r}) + \Delta p(\vec{r}), \tag{1.40}$$

which reduces the last equation to

$$R_n(\vec{r},t) = R_p(\vec{r},t)$$
$$= \frac{\Delta n(\vec{r},t)\Delta p(\vec{r},t) + \Delta n(\vec{r},t)(n_0(\vec{r},t) + p_0(\vec{r},t))}{\tau_p(\vec{r},t)(n_0(\vec{r},t) + n_i(\vec{r},t) + \Delta n(\vec{r},t)) + \tau_n(\vec{r},t)(p_0(\vec{r},t) + n_i(\vec{r},t) + \Delta n(\vec{r},t))}, \tag{1.41}$$

where we have used the fact that because electrons and holes always generate or recombine in pairs (charge neutrality), $\Delta n(\vec{r},t) = \Delta p(\vec{r},t)$. Furthermore, the deviations from local equilibrium are assumed to be small so that we can neglect product terms $\Delta n \Delta p$. Finally, defining a quantity

$$\tau_R(\vec{r},t) = \frac{\tau_p(\vec{r},t)(n_0(\vec{r},t) + n_i(\vec{r},t)) + \tau_n(\vec{r},t)(p_0(\vec{r},t) + n_i(\vec{r},t))}{n_0(\vec{r},t) + p_0(\vec{r},t)} \tag{1.42}$$

and recognizing that as long as we are not too far off equilibrium (known as low level injection conditions), $\Delta n(\vec{r}) = \Delta p(\vec{r},t) \ll n_i(\vec{r},t)$, we get

$$R_n(\vec{r},t) = R_p(\vec{r},t) = \frac{\Delta n(\vec{r},t)}{\tau_R(\vec{r},t)} = \frac{\Delta p(\vec{r},t)}{\tau_R(\vec{r},t)}. \tag{1.43}$$

The above equation is the expression for the Shockley–Reed–Hall recombination rate.

1.11 Summary

In this chapter, we learned about the drift–diffusion model. The important points to remember are:

1. The drift–diffusion model is predicated on two basic assumptions about the nature of charge transport in a solid, namely that (1) charge carriers are classical particles obeying Newton's laws, and (2) they experience forces due to external electric and/or magnetic fields as well as scattering. The latter tends

to restore local equilibrium. A carrier scatters many times before the electric and/or magnetic fields change significantly, so that a carrier is always at local equilibrium with its surroundings.
2. The single particle drift model is the Drude model of conduction that yields Ohm's law.
3. Electron and hole *drift* current densities at any location within a solid always point in the same direction, which is the direction of the local electric field.
4. Electron *diffusion* current density at any location points in the direction of the local electron concentration gradient, while the hole diffusion current density points opposite to the direction of the local hole concentration gradient.
5. Steady-state transport variables, like current densities $\vec{J}_p(\vec{r})$, $\vec{J}_n(\vec{r})$ and carrier concentrations $p(\vec{r})$, $n(\vec{r})$, at any location in space, are found by simultaneously solving the continuity equations, current equations, and Poisson equation (1.31). These equations are collectively known as the "equations of state." In order to solve them, we must know the transport parameters mobility $[\mu_p(\vec{r}), \mu_n(\vec{r})]$ and diffusion coefficient $[D_p(\vec{r}), D_n(\vec{r})]$ everywhere in the device. In the absence of nonlocal effects (which would have invalidated the drift–diffusion model in any case), the mobility can be found from the known drift velocity versus electric field characteristics in the material, and the diffusion coefficient can be related to the mobility by the Einstein relation.
6. The drift–diffusion model can handle *local* nonlinear effects, such as velocity saturation or negative differential mobility, by defining transport parameters—mobilities and diffusion coefficients—as functions of the local electric field.
7. The drift–diffusion model breaks down in the presence of nonlocal effects that arise when carriers fail to reach local equilibrium with their surroundings. This can happen in devices with rapid compositional or doping variations (recall the device in Fig. 1.12) which make the electric field in the device change so rapidly in space that a charge carrier does not have the opportunity to scatter many times before the field changes significantly. This prevents the charge carrier from equilibrating with the local environment, so that its behavior is no longer determined uniquely by its local surroundings. Nonlocal effects can also arise in ultrasmall devices where transport is quasi-ballistic and velocity overshoot can occur.

Problems

Problem 1.1. Newton's second law states

$$\frac{\mathrm{d}\vec{p}(\vec{r},t)}{\mathrm{d}t} = \vec{F}(\vec{r},t). \tag{1.44}$$

where $\vec{p}$ is a particle's momentum and the force $\vec{F}$ is usually defined as

$$\vec{F}(\vec{r},t) = -\vec{\nabla}_r U(\vec{r},t), \tag{1.45}$$

where $\vec{\nabla}_r$ is the spatial gradient and $U(\vec{r},t)$ is the potential energy .

Hamiltonian mechanics treats $\vec{p}$ and $\vec{r}$ as independent variables and postulates

$$\frac{d\vec{r}}{dt} = \vec{\nabla}_p H(\vec{r}(t), \vec{p}(t))\Big|_{\vec{r}=\texttt{constant}}$$
$$\frac{d\vec{p}}{dt} = -\vec{\nabla}_r H(\vec{r}(t), \vec{p}(t))\Big|_{\vec{p}=\texttt{constant}}, \tag{1.46}$$

where $\vec{\nabla}_p$ is the momentum gradient and H is the Hamiltonian which is the sum of kinetic energy T and potential energy U.

Consider the situation when the mass of the particle *varies in space*. If we define momentum as $\vec{p}(\vec{r},t) = m(\vec{r},t)\vec{v}(\vec{r},t)$ and $T(\vec{r},t) = p^2(\vec{r},t)/2m(\vec{r},t)$, show that the first of Hamilton's equation in (1.46) is satisfied. Show also that the second of Hamilton's equations, (1.44) and (1.45) cannot all be satisfied. Which of these three equations need to be modified in the case of spatially varying mass so that all three equations can be satisfied simultaneously? Note that if mass did not vary in space, then $\vec{\nabla}_r T|_{\vec{p}=\texttt{constant}} = 0$, but otherwise it is nonzero.

Problem 1.2. In this chapter, we saw that current is carried in a solid by electrons and holes that drift and diffuse. Such current is called "conduction current" since it is caused by the flow (or conduction) of actual charged particles. There can be another type of current in a solid (or even vacuum), called "displacement current," which does not require flow of particles. It is caused by the time variation of the displacement vector which is the product of the dielectric constant ϵ and the electric field $\vec{\mathcal{E}}$ in a medium. Both types of current can give rise to a magnetic field according to the well-known Ampere's law.

Maxwell's equation provides a mathematical statement of Ampere's law. The magnetic field intensity H arising from conduction and displacement currents are related to those currents by the relation

$$\vec{\nabla}_r \times \vec{H} = \vec{J} + \frac{\partial(\epsilon\vec{\mathcal{E}})}{\partial t},$$

where the first term in the right-hand side is the conduction current, and the second term is the displacement current.

Using the above equation and Poisson equation, derive the *nonsteady-state* continuity equation

$$\vec{\nabla}_r \cdot \vec{J} + \frac{\partial \rho}{\partial t} = 0,$$

in the absence of any recombination or generation. Here, ρ is the charge density. We will derive this equation in the next chapter from the BTE. This equation immediately shows that in the absence of any recombination and generation, the steady-state conduction current density must be *spatially invariant.*

Problem 1.3. Consider a situation when minority carrier electrons are injected into a piece of p-type semiconductor where the electric field is weak. Assuming that the equilibrium minority carrier concentration n_0 does not vary with position or time, show that the excess minority carrier concentration $\Delta n(\vec{r},t)$ obeys the equation

$$\frac{\partial \Delta n(\vec{r},t)}{\partial t} \approx D_n \nabla_{\mathrm{r}}^2 \Delta n(\vec{r},t) + G_n(\vec{r},t) - \frac{\Delta n(\vec{r},t)}{\tau_R(\vec{r},t)}$$

if we assume that the diffusion coefficient is spatially invariant and the recombination rate is given by the Shockley–Read–Hall expression.

Hint: Use the equation in Problem 1.1 and add the generation and recombination terms on the right-hand-side of that equation. Then express the conduction current in terms of its drift and diffusion components.

The above equation is called the *minority carrier diffusion equation.* Can this equation be applied to find the minority carrier concentration at the edge of the depletion region of a diode?

Problem 1.4. Consider a one-dimensional n-type semiconductor structure of length L stretching from coordinate $x = 0$ to $x = L$. There is no recombination or generation inside this structure since the semiconductor is maintained in the dark and the nonradiative lifetime is extremely long. Minority carrier holes are injected from the end $x = 0$ and the electric field inside the semiconductor is uniform and equal to $\mathcal{E}$. Assume that the current through the semiconductor is small enough that Einstein relation is approximately valid. Assume also that the hole mobility is spatially invariant. Show that the steady-state minority hole concentration is given by

$$p(x) = \frac{p(L)}{2\exp(e\mathcal{E}L/2k_{\mathrm{B}}T)\sinh(e\mathcal{E}L/2k_{\mathrm{B}}T)}\left[\exp\left(\frac{e\mathcal{E}}{k_{\mathrm{B}}T}x\right) - 1\right].$$

if the hole concentration at $x = 0$ is zero.

Problem 1.5. Under steady-state conditions, is the ensemble average velocity always the same as the drift velocity?

Solution. No. The drift velocity is the velocity imparted by the electric field. However, carriers can have a net velocity due to diffusion when carriers migrate from a region of high concentration to a region of low concentration. This "diffusion velocity" contributes to the ensemble average velocity but has nothing to do with any electric field. Hence, the ensemble average velocity need not be equal to the drift velocity.

Problem 1.6. Consider a bar of nondegenerate semiconductor doped with spatially varying donor concentration $N_D(x) = N_D(0)e^{-x/L}$. Assume that $N_D(x) \gg n_i$, where n_i is the intrinsic carrier concentration in the semiconductor. Assume that all the donors are ionized. Show that under equilibrium, there will be a homogeneous built-in electric field within the semiconductor. Find the strength of this electric field at room temperature if $L = 1\,\mu$m. Does this electric field satisfy the Poisson equation?

Solution. Since all the donors are ionized,

$$n(x) = N_D(x) = N_D(0)\exp[-x/L].$$

The law of mass action (students should be familiar with this law; otherwise, see [6]) mandates that at equilibrium,

$$n(x)p(x) = n_i^2.$$

Hence, $p(x) = n_i^2/n(x) \ll n_i \approx 0$ since $n(x) = N_D(x) \gg n_i$.

At equilibrium, $J_n(x) = 0$. Therefore, we can write

$$0 = e\mu_n(x)n(x)\mathcal{E}(x) + eD_n(x)\frac{dn(x)}{dx},$$

where $\mathcal{E}(x)$ is the electric field in the semiconductor.

Using Einstein relation, we can write

$$0 = e\mu_n(x)N_D(0)\exp[-x/L]\mathcal{E}(x) + e\frac{k_BT}{e}\mu_n(x)\left[-\frac{1}{L}\right]N_D(0)\exp[-x/L],$$

which yields

$$\mathcal{E}(x) = \frac{k_BT}{eL}.$$

Note that the electric field turns out to be spatially invariant (independent of x); hence, it is homogeneous.

At room temperature, $k_BT/e = 26\,$mV. Therefore, the strength of this field is 26 kV/m everywhere since $L = 1\,\mu$m.

The Poisson equation states

$$\epsilon\frac{d\mathcal{E}(x)}{dx} = e[N_D(x) + p(x) - n(x)] = e\left[N_D(x) + \frac{n_i^2}{n(x)} - n(x)\right] \approx 0.$$

Since the electric field is homogeneous, its spatial derivative vanishes and Poisson's equation is satisfied.

Problem 1.7. Consider an infinite one-dimensional silicon bar whose central section extending from $x = -L$ to $x = L$ is selectively illuminated with light that

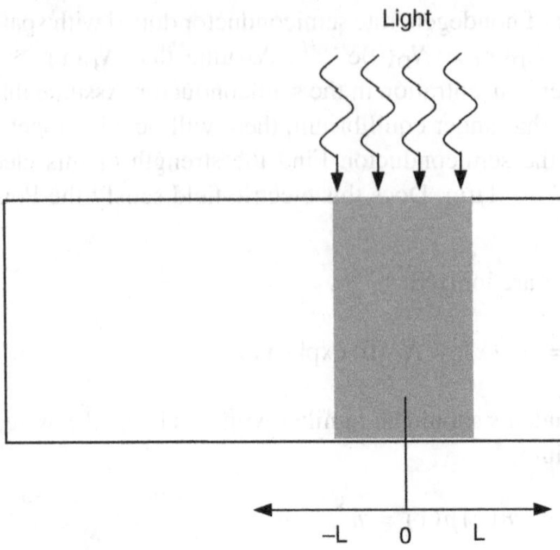

Fig. 1.13 A uniformly doped infinite semiconductor bar whose central region extending from $x = -L$ to $x = L$ is illuminated. There is no electric field inside the semiconductor

generates G electron-hole pairs per unit volume per unit time. The semiconductor is uniformly doped with Boron at a level of 10^{18} cm^{-3}. There is no electric field inside the semiconductor since it is uniformly doped and no battery is connected across it. Refer to Fig. 1.13 for this problem. (Hint: You should use the minority carrier diffusion equation derived in Problem 1.2 for this exercise.)

1. If the semiconductor were not uniformly doped, would an electric field inevitably exist inside the semiconductor at equilibrium? Explain.
2. Explain why the excess minority carrier concentration generated by light and its first derivative must be continuous everywhere.
3. If the semiconductor is maintained at steady-state, derive an equation for the minority carrier concentration everywhere.

Problem 1.8. Consider a nonuniformly doped semiconductor doped with donors. The majority carrier concentration is approximately equal to the donor concentration everywhere.

Apply Poisson equation to derive an expression for the intrinsic level profile $[E_i(x)$ versus $x]$ at equilibrium assuming that the semiconductor is nondegenerate, which allows one to use the equation

$$p(x) = n_i \exp\left[\frac{E_i(x) - \mu_F}{k_B T}\right],$$

where n_i is the intrinsic carrier concentration and μ_F is the Fermi level.

The intrinsic level is approximately half-way between the conduction and valence band edges and is given by $E_i \approx \frac{E_{c0}+E_{v0}}{2}$. The electric field inside the semiconductor is related to the intrinsic level as $\mathcal{E}(x) = \frac{1}{e}\frac{\mathrm{d}E_i(x)}{\mathrm{d}x}$.

Show that if the intrinsic level intersects the Fermi level at $x = 0$ and at that point the electric field is zero, then the intrinsic level profile in the vicinity of that region can be written as

$$E_i(x) = \mu_\mathrm{F} + k_\mathrm{B}T\left[1 - \cosh\left(\frac{x}{L_D}\right)\right],$$

where $L_D = \sqrt{\frac{\epsilon_r\epsilon_0 k_\mathrm{B}T}{e^2 n_i}}$ is called the intrinsic Debye length.

Problem 1.9. Consider an n-type semiconductor bar that is at steady state but not at equilibrium since a small current is flowing through it. If the majority carrier concentration varies linearly in space, then show that the built-in electric field $\mathcal{E}(x)$ in the semiconductor obeys the equation (in one dimension):

$$\frac{\mathrm{d}n(x)}{\mathrm{d}x}\mathcal{E}(x) + n(x)\frac{\mathrm{d}\mathcal{E}(x)}{\mathrm{d}x} = 0.$$

References

1. B. K. Ridley and T. B. Watkins, "The possibility of negative resistance effect in semiconductors", Proc. Phys. Soc. Lond., **78**, 293 (1961).
2. C. Hilsum, "Transferred electron amplifiers and oscillators", Proc. IRE, **50**, 185 (1962).
3. J. B. Gunn, "Microwave oscillation of current in III-V semiconductors", Solid State Commun., **1**, 88 (1963).
4. S. Bandyopadhyay, M. E. Klausmeier-Brown, C. M. Maziar, S. Datta and M. S. Lundstrom, "A rigorous technique to couple Monte Carlo and drift–diffusion models for computationally efficient device simulation", IEEE Trans. Elec. Dev., **34**, 392 (1987).
5. S. M. Sze, *Physics of Semiconductor Devices*, 2nd. edition, (John Wiley & Sons, New York, 1981).
6. R. F. Pierret, *Semiconductor Device Fundamentals* (Addison Wesley Longman, Reading, Massachusetts, 1996).

Chapter 2
Boltzmann Transport: Beyond the Drift–Diffusion Model

2.1 Transport in the Presence of Nonlocal Effects: The Carrier Distribution Function

In the previous chapter, we mentioned that when nonlocal effects are present, transport parameters—such as mobility and diffusion coefficient—become functionals of a quantity known as the carrier *distribution function*. This quantity $f(\vec{r}, \vec{k}, t)$ is the probability of finding an electron at a position $\vec{r}$, with a wavevector $\vec{k}$ at an instant of time t.[1]

The distribution function $f(\vec{r}, \vec{k}, t)$ is a function in 7-dimensional phase space (three coordinates of $\vec{r}$, three coordinates of $\vec{k}$, and one coordinate of t). Being a "probability," it is always positive indefinite (i.e., its value is never negative). In order to determine this quantity (which we must determine if we wish to find transport parameters in the presence of nonlocal effects), we have to solve an equation known as the *BTE* which we discuss in this chapter. Once the distribution function is found from the BTE, we can find any transport parameter as a function of position $\vec{r}$ and time t from the distribution function.

At a given location $\vec{r}$ in a solid and at a given instant of time t, the distribution function $f(\vec{r}, \vec{k}, t)$ will tell us the probability of finding an electron with a wavevector $\vec{k}$, or velocity $\vec{v}$ which is a single-valued function of $\vec{k}$. If there are N electrons in that location at that instant of time, it will tell us how many of them have wavevectors between $\vec{k}_1 - \Delta\vec{k}_1$ and $\vec{k}_1 + \Delta\vec{k}_1$, how many have wavevectors between

[1]The electron's wavevector may seem to be a quantum-mechanical attribute since only a wave has a wavevector and the electron behaves as a wave only in the domain of quantum mechanics. The Boltzmann transport formalism is however almost entirely classical. The wavevector is uniquely related to the electron's velocity which may appear to be more of a classical attribute. We could have defined the distribution function as $f(\vec{r}, \vec{v}, t)$, but is more common to define it as $f(\vec{r}, \vec{k}, t)$ since it is easy to count electron states in wavevector space. This will become clearer when we discuss the concept of density of states in a later chapter.

S. Bandyopadhyay, *Physics of Nanostructured Solid State Devices*,
DOI 10.1007/978-1-4614-1141-3_2, © Springer Science+Business Media, LLC 2012

$\vec{k}_2 - \Delta\vec{k}_2$ and $\vec{k}_2 + \Delta\vec{k}_2$, etc. In the example of Fig. 1.12, it would have told us that there are very few electrons with negative wavevectors at x_2 (or that the probability of finding an electron with negative wavevector is very small there). Therefore, it would have correctly predicted the anticipatory rise in the ensemble average velocity at $x = x_2$. In other words, it would have *correctly described nonlocal effects*.

Now, we said that the transport parameters are functionals of the distribution function and therefore we can determine them if we know the distribution function. The procedure for doing this is straightforward but unfortunately, it is cumbersome, as we shall see later. However, the utility of transport parameters is ultimately in finding transport variables from (1.31). What if it were possible to find the transport variables directly from the distribution function, without having to first evaluate the transport parameters and then finding the transport variables from (1.31)? This is not only possible, but preferred since it is direct. The transport variables can be found directly from the distribution function since they are nothing but different "moments" of the distribution function. For example,

$$
\begin{aligned}
n(\vec{r},t) &= \frac{1}{\Omega}\sum_{\vec{k}} f(\vec{r},\vec{k},t) = \frac{1}{4\pi^3}\int d^3\vec{k}\; f(\vec{r},\vec{k},t) \;\; \texttt{(0-th moment)} \\
\vec{J}_n(\vec{r},t) &= \frac{-e}{\Omega}\sum_{\vec{k}} \vec{v}(\vec{k}) f(\vec{r},\vec{k},t) = \frac{-e}{4\pi^3}\int d^3\vec{k}\; \vec{v}(\vec{k}) f(\vec{r},\vec{k},t) \;\; \texttt{(1st. moment)} \\
\epsilon_n(\vec{r},t) &= \frac{1}{\Omega}\sum_{\vec{k}} E(\vec{k}) f(\vec{r},\vec{k},t) = \frac{1}{4\pi^3}\int d^3\vec{k} E(\vec{k}) f(\vec{r},\vec{k},t) \;\; \texttt{(2nd. moment)}
\end{aligned}
\tag{2.1}
$$

and so on. Here, $\epsilon_n(\vec{r},t)$ is the electron energy density at location $\vec{r}$ at time t, $E(\vec{k})$ is the energy of an electron with wavevector $\vec{k}$, $\vec{v}(\vec{k})$ is the velocity of an electron with wavevector $\vec{k}$, and Ω is a normalizing volume, which does not matter in the end since the right-hand sides are independent of this quantity.

This direct approach appears to be far better than finding the transport parameters from the distribution function and then solving (1.31) to find the transport variables. However, there is one caveat. Since the distribution function is a 7-dimensional function, the BTE is not easy to solve. It is usually solved numerically or with MC simulation. Both techniques are computationally demanding and require considerable resources. Therefore, it is sometimes more efficient to solve the BTE *only in selected regions* of a device where we suspect that nonlocal effects will prevail, instead of solving it globally. From these solutions, we can extract the transport parameters in the suspect regions, while the transport parameters in the remaining regions are found from local field-dependent models (such as from the velocity-field characteristic of Fig. 1.9). Finally, we can solve the drift–diffusion equation (1.31) globally using the transport parameters found in every region of the device. This kind of an approach, although seemingly indirect and tortuous, is actually computationally more efficient in many cases and is used in commercial device simulation packages such as MINIMOS that analyze charge transport in solid state devices.

2.2 Boltzmann Transport Equation for the Carrier Distribution Function

We are now ready to introduce the BTE. It is really nothing but a statement about the conservation of particles (or charge). Since it is difficult for us humans to mentally visualize a 7-dimensional phase space, let us consider a simpler situation where there is only one spatial dimension x, one wavevector dimension k_x, and time t. For the purpose of this discussion, we have mentally shrunk the 7-dimensional phase space to a 3-dimensional phase space because it is easy to visualize.

Now consider a rectangular "box" located at (x, k_x, t) which has dimensions Δx, Δk_x, and Δt, as shown in Fig. 2.1. The number of electrons within this box is proportional to the probability of finding an electron at position x, with wavevector k_x at time t, or, in other words, it is proportional to $f(x, k_x, t)$. The rate at which the number of electrons within this box is changing must be equal to the difference of the rate at which electrons are scattering into this box from the outside and the rate at which electrons are scattering out of this box to the outside. Electrons cannot spontaneously appear or disappear within this box because of the conservation of particles or conservation of charge. Therefore, we can write:

$$\frac{\mathrm{d}f(x,k_x,t)}{\mathrm{d}t} = \left.\frac{\mathrm{d}f(x,k_x,t)}{\mathrm{d}t}\right|_{\text{scatter-in}} - \left.\frac{\mathrm{d}f(x,k_x,t)}{\mathrm{d}t}\right|_{\text{scatter-out}} = S_{\text{in}} - S_{\text{out}}. \quad (2.2)$$

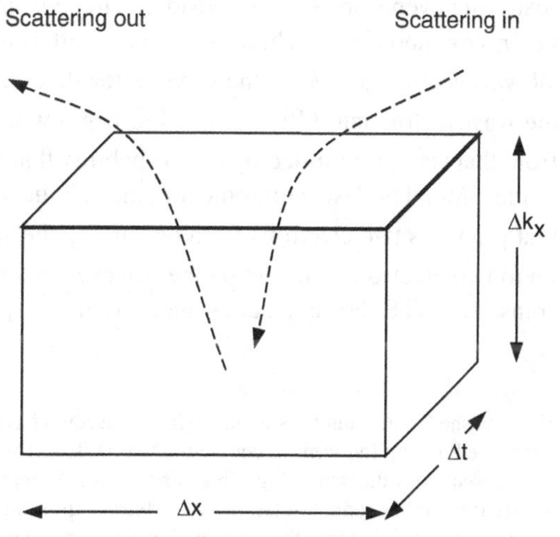

Fig. 2.1 A rectangular box in three-dimensional phase space located at (x, k_x, t). The box has dimensions of Δx, Δk_x, and Δt. The number of electrons in this box is proportional to $f(x, k_x, t)$. Because of charge conservation, the only way the number of electrons within this box can change is if electrons scatter into this box from outside to increase the number, or scatter from this box to the outside to decrease the number

Generalizing back to seven dimensions and converting the total derivative in the left-hand side to partial derivatives, we get

$$\frac{\mathrm{d}f(\vec{r},\vec{k},t)}{\mathrm{d}t} = \frac{\partial f(\vec{r},\vec{k},t)}{\partial t} + \frac{\partial f(\vec{r},\vec{k},t)}{\partial \vec{r}} \cdot \frac{\mathrm{d}\vec{r}}{\mathrm{d}t} + \frac{\partial f(\vec{r},\vec{k},t)}{\partial \vec{k}} \cdot \frac{\mathrm{d}\vec{k}}{\mathrm{d}t} = S_{\mathrm{in}} - S_{\mathrm{out}}, \tag{2.3}$$

where S_{in} and S_{out} are the in- and out-scattering rates.

Now, Newton's second law tells us that the force on a particle in a crystal is equal to the rate of change of crystal momentum $\hbar\vec{k}$:

$$\vec{F}(\vec{r},t) = \frac{\mathrm{d}(\hbar\vec{k})}{\mathrm{d}t}. \tag{2.4}$$

Furthermore, by definition,

$$\frac{\mathrm{d}\vec{r}}{\mathrm{d}t} = \vec{v}(\vec{k}), \tag{2.5}$$

where $\vec{v}$ is the velocity. Using these two results in (2.3), we get:

$$\frac{\partial f(\vec{r},\vec{k},t)}{\partial t} + \vec{v}(\vec{k}) \cdot \vec{\nabla}_{\mathrm{r}} f(\vec{r},\vec{k},t) + \frac{\vec{F}(\vec{r},t)}{\hbar} \cdot \vec{\nabla}_{\mathrm{k}} f(\vec{r},\vec{k},t) = S_{\mathrm{in}} - S_{\mathrm{out}}, \tag{2.6}$$

where $\vec{\nabla}_{\mathrm{r}}$ represents gradient in real space and $\vec{\nabla}_{\mathrm{k}}$ represents gradient in wavevector space.

We are almost there, but we have one last step to complete. We need to find more explicit expressions for S_{in} and S_{out}. The former is the rate of scattering into the "box" from outside the "box," i.e., from all other wavevector states $\vec{k}'$, from all other sites $\vec{r}'$, and from all other times t'. However, in the Boltzmann picture, scattering events are always considered to be *instantaneous in both time and space*, i.e., $\vec{r} = \vec{r}'$ and $t = t'$. Therefore, the in-scattering rate is the rate of scattering from all wavevector states $\vec{k}'$ to the wavevector state $\vec{k}$, multiplied by the probability that the wavevector state $\vec{k}'$ is occupied to begin with so that an electron can scatter in from that state, multiplied by the probability that the wavevector state $\vec{k}$ is empty to scatter into. The last requirement comes about from the Pauli Exclusion Principle that prohibits two electrons from occupying the same wavevector state $\vec{k}$. In reality up to two electrons can occupy the wavevector state $\vec{k}$, but they must have opposite spins. The BTE, however, does not account for spin.[2] Hence,

[2]If we wished to account for spin in the BTE, we could have written one BTE for each spin and then introduced a coupling term to couple the two BTE-s. The coupling term would have been a term to represent spin-flip scattering. This is an approach implicit in the drift–diffusion model of spin transport but it has some limitations. An arbitrary spin can be written as the *coherent* superposition of two antiparallel spins. Spin precession about a magnetic field is equivalent to continuously

$$S_{\text{in}} = \sum_{\vec{k}'} S\left(\vec{k}',\vec{k}\right) f\left(\vec{r},\vec{k}',t\right)\left[1 - f\left(\vec{r},\vec{k},t\right)\right]$$

$$S_{\text{out}} = \sum_{\vec{k}'} S\left(\vec{k},\vec{k}'\right) f\left(\vec{r},\vec{k},t\right)\left[1 - f\left(\vec{r},\vec{k}',t\right)\right], \tag{2.7}$$

where $S(\vec{k}_1,\vec{k}_2)$ is the rate at which a particle scatters from a wavevector state $\vec{k}_1$ to a wavevector state $\vec{k}_2$.

Therefore, the final form of the BTE is

$$\begin{aligned}\frac{\partial f\left(\vec{r},\vec{k},t\right)}{\partial t} &+ \vec{v}\left(\vec{k}\right)\cdot\vec{\nabla}_{\mathrm{r}} f\left(\vec{r},\vec{k},t\right) + \frac{\vec{F}\left(\vec{r},t\right)}{\hbar}\cdot\vec{\nabla}_{\mathrm{k}} f\left(\vec{r},\vec{k},t\right)\\ &= \sum_{\vec{k}'} S\left(\vec{k}',\vec{k}\right) f\left(\vec{r},\vec{k}',t\right)\left[1 - f\left(\vec{r},\vec{k},t\right)\right]\\ &- \sum_{\vec{k}'} S\left(\vec{k},\vec{k}'\right) f\left(\vec{r},\vec{k},t\right)\left[1 - f\left(\vec{r},\vec{k}',t\right)\right].\end{aligned} \tag{2.8}$$

Note that the left-hand-side contains derivatives in wavevector space whereas the right-hand-side contains summations (or equivalently integrals) in wavevector space. Hence, the BTE is an *integro-differential equation*.

2.2.1 *Limitations of the BTE*

At the outset, let us point out that the BTE is purely classical and does not account for any quantum-mechanical effect. It cannot. Note that the distribution function $f(\vec{r},\vec{k},t)$ is a function of both $\vec{r}$ and $\vec{k}$. This is forbidden by the laws of *quantum mechanics* since $\vec{r}$ and $\vec{k}$ are Fourier transform pairs and obey the Heisenberg Uncertainty relation

$$\Delta\vec{r}\cdot\Delta\vec{k} \geq \frac{\hbar}{2}. \tag{2.9}$$

That means both a position $\vec{r}$ and a wavevector $\vec{k}$ cannot be defined at the *same time* with precision, according to quantum mechanics. In other words, a function such as $f(\vec{r},\vec{k},t)$, which is simultaneously a function of both $\vec{r}$ and $\vec{k}$, cannot

changing the phase relationship between the two components coherently. This coherence aspect cannot be included satisfactorily in such an approach, which is why the coupled BTE method is not an accurate way to model spin transport.

even exist according to orthodox quantum mechanics. Therefore, the BTE can never model quantum-mechanical effects. Furthermore, quantum mechanics allows interference between electron waves. Destructive interference would involve two waves canceling each other out to produce a "null' or "zero." BTE can never model such an effect since the distribution function is never negative. To produce a zero, a negative quantity must cancel a positive quantity. That is not possible if the quantity in question can never be negative.

Quantum mechanical effects arise when charge carriers cease to behave as classical particles like billiard balls, and instead behave as waves propagating through a device in accordance with the Schrödinger equation. Starting in the mid-1980s, a branch of physics, known as "mesoscopic physics," emerged that deals with such situations in ultrasmall metal and semiconductor structures [1–4]. These structures are so small that an electron can traverse them without suffering a single phonon collision or collision with another electron. Such *inelastic* collisions would have randomized the electron's phase and restored its particle nature, but if they are infrequent or absent, the wave nature is preserved. Charge transport in such structures cannot be captured by the BTE, but will require strictly quantum-mechanical formulations. Many such formulations exist and some are discussed in Chap. 8.

2.2.2 Limits of Validity of the BTE

The Heisenberg Uncertainty Principle sets limits on the validity of the BTE. This Principle states

$$\Delta E \Delta t \geq \frac{\hbar}{2}, \tag{2.10}$$

where ΔE is the uncertainty in the carrier's energy and Δt is the associated uncertainty in time. Therefore, the carrier must remain in a state for a sufficiently long time to have a well-defined energy E (or wavevector k, since the latter is related to the energy). If we assume that the electron remains in a fixed energy state between two successive inelastic collisions (which always change the electron's energy), then we can set $\Delta t = \tau_{\text{inelastic}}$, which is the mean time between inelastic collisions. Now, if we assume that the electron energy level is thermally broadened to the effect that $\Delta E = k_B T$ [5], then we require that

$$\tau_{\text{inelastic}} \gg \frac{\hbar}{2k_B T}. \tag{2.11}$$

At room temperature, this means that the inelastic scattering time should far exceed 12.5 fs for the BTE to be valid ($\hbar/(2k_B T) = 12.5$ fs at room temperature). In some metals with high electron concentration, the electron–electron scattering (an inelastic scattering process) can be frequent enough that the above inequality is not fulfilled. In that case, the BTE is not valid.

Note that if the inelastic scattering time is too long, the BTE may be invalidated because the wave nature of electrons will be manifested. On the other hand, if it is too short, then the BTE may be invalidated again because of the Uncertainty Principle. Thus, there is a range of the inelastic scattering time over which the BTE is meaningful.

2.3 The Generalized Moment Equation or Hydrodynamic Balance Equation

In the last chapter, we stated that there is a transient (or time-varying) version of the equations of state, even though we derived (1.31) only for steady-state situations. In this section, we will show how to generalize (1.31) to the transient case. In fact, we will derive a general equation from the BTE which will yield the continuity equation and the current equation of (1.31) as mere special cases. This general equation will allow us to derive not just (1.31), but an equation for any arbitrary transport variable, that is valid for both steady-state and nonsteady-state situations. In order to accomplish this, we will first define a general *moment* of the distribution function and show that this moment obeys an equation that is variously known as the "generalized moment equation," or "hydrodynamic balance equation" because of its similarity with the equations of hydrodynamics. Once we obtain this equation, we will show how the current and continuity equations of the preceding chapter can be derived as corollaries. Moreover, this new equation will allow us to extend the equations of state to nonsteady-state or time-varying situations. Clearly, since the drift–diffusion equations can be derived from the BTE, but not the other way around, it is obvious that the Boltzmann model is more advanced and powerful than the drift–diffusion model.

Let us define the m-th moment of the distribution function as the quantity

$$G_m(\vec{r},t) = \sum_{\vec{k}} g(k^m) f(\vec{r},\vec{k},t), \tag{2.12}$$

where $g(k^m)$ is a scalar, vector, or tensor function of k^m, the m-th power of the wavevector.

Multiplying both sides of the BTE in (2.8) with $g(k^m)$ and summing over all $\vec{k}$, we get

$$\sum_{\vec{k}} g(k^m) \frac{\partial f(\vec{r},\vec{k},t)}{\partial t} + \sum_{\vec{k}} g(k^m) \vec{v}(\vec{k}) \cdot \vec{\nabla}_{\mathrm{r}} f(\vec{r},\vec{k},t)$$
$$+ \sum_{\vec{k}} g(k^m) \frac{\vec{F}(\vec{r},t)}{\hbar} \cdot \vec{\nabla}_{\mathrm{k}} f(\vec{r},\vec{k},t)$$

$$= \sum_{\vec{k}} g(k^m) \left\{ \sum_{\vec{k}'} S(\vec{k}',\vec{k}) f(\vec{r},\vec{k}',t)[1 - f(\vec{r},\vec{k},t)] \right.$$
$$\left. - \sum_{\vec{k}'} S(\vec{k},\vec{k}') f(\vec{r},\vec{k},t)[1 - f(\vec{r},\vec{k}',t)] \right\}. \quad (2.13)$$

Since $g(k^m)$ is a function of only wavevector and not time, the first term on the left-hand-side of (2.13) can be simplified to

$$\sum_{\vec{k}} g(k^m) \frac{\partial f(\vec{r},\vec{k},t)}{\partial t} = \sum_{\vec{k}} \frac{\partial [g(k^m) f(\vec{r},\vec{k},t)]}{\partial t}$$
$$= \frac{\partial \left\{ \sum_{\vec{k}} [g(k^m) f(\vec{r},\vec{k},t)] \right\}}{\partial t}$$
$$= \frac{\partial G_m(\vec{r},t)}{\partial t} \quad (2.14)$$

To evaluate the second term on the left-hand-side of (2.13), we need to use the vector identity

$$\vec{\nabla} \cdot (\vec{V} S) = S \vec{\nabla} \cdot \vec{V} + \vec{V} \cdot \vec{\nabla} S, \quad (2.15)$$

where $\vec{V}$ is a vector and S is a scalar.

The second term becomes

$$\sum_{\vec{k}} g(k^m) \vec{v}(\vec{k}) \cdot \vec{\nabla}_{\mathrm{r}} f(\vec{r},\vec{k},t) = \sum_{\vec{k}} g(k^m) \left\{ \vec{\nabla}_{\mathrm{r}} \cdot \left[\vec{v}(\vec{k}) f(\vec{r},\vec{k},t) \right] \right.$$
$$\left. - f(\vec{r},\vec{k},t) \vec{\nabla}_{\mathrm{r}} \cdot \vec{v}(\vec{k}) \right\}$$
$$= \left\{ \vec{\nabla}_{\mathrm{r}} \cdot \sum_{\vec{k}} g(k^m) \vec{v}(\vec{k}) f(\vec{r},\vec{k},t) \right\}$$
$$= \vec{\nabla}_{\mathrm{r}} \cdot \vec{J}_{Gm}(\vec{r},t), \quad (2.16)$$

where we have used the fact that since $\vec{r}$, $\vec{k}$, and t are independent variables, neither $\vec{v}(\vec{k})$ nor $g(k^m)$ will be functions of $\vec{r}$ *unless the quantities that relate* $\vec{v}(\vec{k})$ *or* $g(k^m)$ *with k change in space or time.* The latter can happen, for example, if the material has a spatially varying bandstructure that makes the effective mass spatially variant. In that case, $\vec{v}(\vec{k})$ will depend on *vecr*, but we will *not* be considering such situations here. The quantity $\vec{J}_{Gm}(\vec{r},t)$ is called "flux of the moment $G_m(\vec{r},t)$". Note that since

$\vec{v}(\vec{k})$ is roughly proportional to k (in metals or semiconductors with approximately parabolic band structures), this flux is like the $(m+1)$-th moment of the distribution function.

Using the vector identity in (2.15), we can write the third term on the left-hand side of (2.13) as

$$\sum_{\vec{k}} g(k^m)\frac{\vec{F}(\vec{r},t)}{\hbar}\cdot\vec{\nabla}_{\mathrm{k}} f(\vec{r},\vec{k},t) = \sum_{\vec{k}}\left\{\vec{\nabla}_{\mathrm{k}}\cdot\left[g(k^m)\frac{\vec{F}(\vec{r},t)}{\hbar}f(\vec{r},\vec{k},t)\right]\right.$$
$$\left. - f\,(\vec{r},\vec{k},t)\vec{\nabla}_{\mathrm{k}}\cdot\left[g(k^m)\frac{\vec{F}(\vec{r},t)}{\hbar}\right]\right\}. \quad (2.17)$$

The first term on the right-hand-side in the last equation can be written as

$$\sum_{\vec{k}}\vec{\nabla}_{\mathrm{k}}\cdot\left[g(k^m)\frac{\vec{F}(\vec{r},t)}{\hbar}f(\vec{r},\vec{k},t)\right] \propto \int_{-\infty}^{\infty}\frac{\partial\left[g(k^m)\frac{\vec{F}(\vec{r},t)}{\hbar}f(\vec{r},\vec{k},t)\right]}{\partial\vec{k}}\mathrm{d}\vec{k}$$
$$= \frac{\vec{F}(\vec{r},t)}{\hbar}\{g(\infty)f(\vec{r},\infty,t) - g(-\infty)f(\vec{r},-\infty,t)\}$$
$$= 0, \quad (2.18)$$

This term vanishes since the distribution function is bounded and must vanish when $\vec{k} = \pm\infty$.[3]

Using (2.15) yet again, the second term on the right-hand-side of (2.17) can be written as

$$\sum_{\vec{k}} f(\vec{r},\vec{k},t)\left\{g(k^m)\vec{\nabla}_{\mathrm{k}}\cdot\left(\frac{\vec{F}(\vec{r},t)}{\hbar}\right) + \frac{\vec{F}(\vec{r},t)}{\hbar}\cdot\vec{\nabla}_{\mathrm{k}}[g(k^m)]\right\}$$
$$= \frac{\vec{F}(\vec{r},t)}{\hbar}\cdot\sum_{\vec{k}}[\vec{\nabla}_{\mathrm{k}}g(k^m)]f(\vec{r},\vec{k},t) = \frac{\vec{F}(\vec{r},t)}{\hbar}\cdot G'_m(\vec{r},t), \quad (2.19)$$

where we have made use of the fact that the force $\vec{F}(\vec{r},t)$ is not a function of wavevector and hence its divergence in wavevector space is zero. The quantity $G'_m(\vec{r},t)$ is called the "wavevector gradient of the moment $G_m(\vec{r},t)$." Note that it is like the $(m-1)$-th moment of the distribution function. Therefore, the third term on the left-hand-side of (2.13) will be $-\frac{\vec{F}(\vec{r},t)}{\hbar}\cdot G'_m(\vec{r},t)$.

[3]The quantity $g(k^m)$ is a transport variable, like electron velocity or electron energy. Hence, it must always remain finite.

The right-hand-side of (2.13) is

$$\begin{aligned}
&\sum_{\vec{k}} g(k^m)\left\{\sum_{\vec{k}'} S(\vec{k}',\vec{k}) f(\vec{r},\vec{k}',t)[1-f(\vec{r},\vec{k},t)]\right.\\
&\qquad\left.-\sum_{\vec{k}'} S(\vec{k},\vec{k}') f(\vec{r},\vec{k},t)[1-f(\vec{r},\vec{k}',t)]\right\}\\
&=\sum_{\vec{k},\vec{k}'} g(k^m)\left\{S(\vec{k}',\vec{k}) f(\vec{r},\vec{k}',t)[1-f(\vec{r},\vec{k},t)]\right.\\
&\qquad\left.-S(\vec{k},\vec{k}') f(\vec{r},\vec{k},t)[1-f(\vec{r},\vec{k}',t)]\right\}.
\end{aligned} \tag{2.20}$$

Since $\vec{k}$ and $\vec{k}'$ are dummy variables running from $-\infty$ to ∞, we can interchange the order of summation over the first term only, i.e., interchange the variables in the first term, and write the right-hand-side of (2.13) as

$$\begin{aligned}
&\sum_{\vec{k},\vec{k}'} g(k'^m)\left\{S(\vec{k},\vec{k}') f(\vec{r},\vec{k},t)[1-f(\vec{r},\vec{k}',t)]\right.\\
&\qquad\left.-\sum_{\vec{k},\vec{k}'} g(k^m) S(\vec{k},\vec{k}') f(\vec{r},\vec{k},t)[1-f(\vec{r},\vec{k}',t)]\right\}\\
&=\sum_{\vec{k},\vec{k}'} g(k^m)\left\{S(\vec{k},\vec{k}') f(\vec{r},\vec{k},t)[1-f(\vec{r},\vec{k}',t)]\frac{g(k'^m)-g(k^m)}{g(k^m)}\right\}\\
&=-\sum_{\vec{k}} g(k^m) f(\vec{r},\vec{k},t)\frac{1}{\tau_{G_m}(\vec{r},\vec{k},t)},
\end{aligned} \tag{2.21}$$

where

$$\frac{1}{\tau_{G_m}(\vec{r},\vec{k},t)}=\sum_{\vec{k}'} S(\vec{k},\vec{k}')\frac{g(k^m)-g(k'^m)}{g(k^m)}[1-f(\vec{r},\vec{k}',t)]. \tag{2.22}$$

Let us now define a quantity

$$\begin{aligned}
\frac{1}{\langle\tau_{G_m}\rangle(\vec{r},t)}&=\frac{\sum_{\vec{k}} g(k^m) f(\vec{r},\vec{k},t)\frac{1}{\tau_{G_m}(\vec{r},\vec{k},t)}}{\sum_{\vec{k}} g(k^m) f(\vec{r},\vec{k},t)}\\
&=\frac{\sum_{\vec{k}} g(k^m) f(\vec{r},\vec{k},t)\frac{1}{\tau_{G_m}(\vec{r},\vec{k},t)}}{G_m(\vec{r},t)}.
\end{aligned} \tag{2.23}$$

Therefore, the right-hand-side of (2.13) can be written as $-\frac{G_m(\vec{r},t)}{\langle \tau_{G_m} \rangle (\vec{r},t)}$.

Collecting all the terms of (2.13), we finally get our generalized moment equation or hydrodynamic balance equation:

$$\frac{\partial G_m(\vec{r},t)}{\partial t} + \vec{\nabla}_{\mathrm{r}} \cdot \vec{J}_{G_m}(\vec{r},t) + \frac{G_m(\vec{r},t)}{\langle \tau_{G_m} \rangle (\vec{r},t)} = \frac{\vec{F}(\vec{r},t)}{\hbar} \cdot G'_m(\vec{r},t). \tag{2.24}$$

We remind the reader that we derived this equation assuming spatially invariant bandstructure. Hence, this will not exactly hold in a heterostructure where the bandstructure (and hence effective mass of the carrier) will be changing in space.

2.4 Deriving the Drift–Diffusion Equations from the Generalized Moment Equation

2.4.1 The Zero-th Moment

Let us choose the function $g(k^0)$ as

$$g(k^0) = \frac{1}{\Omega}, \tag{2.25}$$

so that the generalized zero-th moment becomes

$$G_0(\vec{r},t) = \sum_{\vec{k}} g(k^0) f(\vec{r},\vec{k},t) = \frac{1}{\Omega} \sum_{\vec{k}} f(\vec{r},\vec{k},t) = n(\vec{r},t), \tag{2.26}$$

which is the carrier concentration.

The flux of this moment becomes

$$\vec{J}_{G_0}(\vec{r},t) = \sum_{\vec{k}} g(k^0) \vec{v}(\vec{k}) f(\vec{r},\vec{k},t) = \frac{1}{\Omega} \sum_{\vec{k}} \vec{v}(\vec{k}) f(\vec{r},\vec{k},t) = -\frac{1}{e} \vec{J}_n(\vec{r},t). \tag{2.27}$$

and the relaxation rate of the generalized zero-th moment becomes

$$\begin{aligned} \frac{1}{\tau_{G_0}(\vec{r},\vec{k},t)} &= \sum_{\vec{k}'} S(\vec{k},\vec{k}') \frac{g(k^0) - g(k'^0)}{g(k^0)} [1 - f(\vec{r},\vec{k}',t)] \\ &= \sum_{\vec{k}'} S(\vec{k},\vec{k}') \frac{1/\Omega - 1/\Omega}{1/\Omega} [1 - f(\vec{r},\vec{k}',t)] \\ &= 0. \end{aligned} \tag{2.28}$$

Therefore,

$$\frac{1}{\langle \tau_{G_0} \rangle (\vec{r},t)} = \frac{\sum_{\vec{k}} g(k^0) f(\vec{r},\vec{k},t) \frac{1}{\tau_{G_0}(\vec{r},\vec{k},t)}}{G_0(\vec{r},t)} = 0. \tag{2.29}$$

The wavevector-gradient of the generalized zero-th moment $G_0'(\vec{r},t)$ is also zero since $\vec{\nabla}_k g(k^0) = \vec{\nabla}_k (1/\Omega) = 0$.

Substituting these results in (2.24), we get the equation

$$\frac{\partial n(\vec{r},t)}{\partial t} - \frac{1}{e} \vec{\nabla}_r \cdot \vec{J}_n(\vec{r},t) = 0. \tag{2.30}$$

Compare the above equation with the continuity equation of the preceding chapter, which was

$$-\frac{1}{e} \vec{\nabla}_r \cdot \vec{J}_n(\vec{r},t) = G_n(\vec{r},t) - R_n(\vec{r},t). \tag{2.31}$$

We can see that we have gained a new term $\frac{\partial n(\vec{r},t)}{\partial t}$ that did not appear in Chap. 1 because it is a time-dependent term that vanishes in steady-state transport. We have also lost two terms, namely the generation and recombination terms. That too is not surprising since generation and recombination require two types of carriers—electrons and holes. Only an electron can recombine with a hole, and they always generate in pairs because charge must be conserved. Therefore, generation-recombination is a concept of bipolar transport, *not* unipolar transport. A single BTE can describe a single type of carrier, either electron or hole, but not both. In other words, it can describe unipolar transport only. Hence, the recombination-generation terms do not show up. In order to make them appear, we will have to write down two coupled BTE-s, one for the conduction band and one for the valence band, and add scattering terms that would scatter an electron from the conduction band to the valence band resulting in recombination, or scatter an electron from the valence band to the conduction band, resulting in generation. These "inter-band" scattering terms will couple the two equations. Since we have not gone that route, we must add the recombination/generation terms by hand. Thereupon, the continuity equation becomes

$$\frac{\partial n(\vec{r},t)}{\partial t} - \frac{1}{e} \vec{\nabla}_r \cdot \vec{J}_n(\vec{r},t) = G_n(\vec{r},t) - R_n(\vec{r},t). \tag{2.32}$$

which can also be written in a compact form as

$$\frac{\partial \rho(\vec{r},t)}{\partial t} + \vec{\nabla}_r \cdot \vec{J}(\vec{r},t) = G(\vec{r},t) - R(\vec{r},t), \tag{2.33}$$

where ρ is the charge density.

The above equation is the general continuity equation that is valid in all cases. The reader can easily verify that it is valid for both electrons and holes

2.4.2 The First Moment

Let

$$g(k^1) = -\frac{e}{\Omega}\vec{v}(\vec{k}) = -\frac{e}{\Omega}v_i(\vec{k}), \tag{2.34}$$

where we have used tensor notation. The quantity v_i is a velocity vector directed along the i-th direction.

The generalized first moment is

$$G_1(\vec{r},t) = \sum_{\vec{k}} g(k^1)f(\vec{r},\vec{k},t) = -\frac{e}{\Omega}\sum_{\vec{k}} \vec{v}(\vec{k})f(\vec{r},\vec{k},t) = \vec{J}_n(\vec{r},t) = J_i(\vec{r},t). \tag{2.35}$$

The flux of this moment is

$$\begin{aligned} J_{G_1}(\vec{r},t) &= \sum_{\vec{k}} g(k^1)\vec{v}(\vec{k})f(\vec{r},\vec{k},t) \\ &= -\frac{e}{\Omega}\sum_{\vec{k}} \vec{v}_i(\vec{k})\vec{v}_j(\vec{k})f(\vec{r},\vec{k},t) \\ &= -\frac{e}{\Omega}\sum_{\vec{k}} \frac{2\hat{u}_{ij}(\vec{r},\vec{k},t)}{m_n^*}f(\vec{r},\vec{k},t) \\ &= -\frac{2e}{m_n^*}n(\vec{r},t)u_{ij}(\vec{r},t), \end{aligned} \tag{2.36}$$

where

$$\hat{u}_{ij}(\vec{r},\vec{k},t) = \frac{1}{2}m_n^*\vec{v}_i(\vec{k})\vec{v}_j(\vec{k}). \tag{2.37}$$

The quantities $u_{ij}(\vec{r},t)$ and $\hat{u}_{ij}(\vec{r},\vec{k},t)$ are second-ranked tensors or "dyadics." They are like kinetic energy since m_n^* here is the effective mass of the carrier. Note that we have assumed m_n^* to be independent of space and time consistent with our assumption of spatially invariant bandstructure when we derived the generalized moment equation.

Next, the wavevector gradient of the generalized first moment is

$$G_1'(\vec{r},t) = \sum_{\vec{k}} f(\vec{r},\vec{k},t)\vec{\nabla}_k[g(k^1)] = -\frac{e}{\Omega}\sum_{\vec{k}} f(\vec{r},\vec{k},t)\vec{\nabla}_k\vec{v}(\vec{k}). \tag{2.38}$$

Now, since $\vec{v}(\vec{k})$ is considered to be linearly proportional to $\vec{k}$, the quantity $\vec{\nabla}_k\vec{v}(\vec{k})$ must be a constant independent of $\vec{k}$. If we assume a parabolic energy dispersion relation, then $\vec{v}(\vec{k}) = \hbar\vec{k}/m_n^*$. In that case,

$$G_1'(\vec{r},t) = -\frac{e\hbar}{m_n^*}\frac{1}{\Omega}\sum_{\vec{k}} f(\vec{r},\vec{k},t) = -\frac{e\hbar}{m_n^*}n(\vec{r},t). \tag{2.39}$$

Finally, the relaxation rate is

$$\begin{aligned}\frac{1}{\tau_{G_1}(\vec{r},\vec{k},t)} &= \sum_{\vec{k}} S(\vec{k},\vec{k}')\frac{g(k^1)-g(k'^1)}{g(k^1)}[1-f(\vec{r},\vec{k}',t)] \\ &= \sum_{\vec{k}} S(\vec{k},\vec{k}')\frac{\vec{v}(\vec{k})-\vec{v}(\vec{k}')}{\vec{v}(\vec{k})}[1-f(\vec{r},\vec{k}',t)] \\ &= \frac{1}{\tau_{\mathrm{m}}(\vec{r},\vec{k},t)}\end{aligned} \tag{2.40}$$

so that

$$\begin{aligned}\frac{1}{\langle\tau_{G_1}(\vec{r},t)\rangle} &= \frac{\sum_{\vec{k}} g(k^1)f(\vec{r},\vec{k},t)\frac{1}{\tau_{G_1}(\vec{r},\vec{k},t)}}{G_1(\vec{r},t)} \\ &= \frac{\sum_{\vec{k}} \vec{v}(\vec{k})f(\vec{r},\vec{k},t)\frac{1}{\tau_m(\vec{r},\vec{k},t)}}{\sum_{\vec{k}} \vec{v}(\vec{k})f(\vec{r},\vec{k},t)} \\ &= \frac{1}{\langle\tau_{\mathrm{m}}\rangle(\vec{r},t)}.\end{aligned} \tag{2.41}$$

We will call the quantity $\langle\tau_{\mathrm{m}}\rangle(\vec{r},t)$ the momentum relaxation time or velocity relaxation time since it is associated with relaxation of momentum or velocity.

Substituting the above terms in the hydrodynamic balance equation (2.24), we get

$$\frac{\partial\vec{J}_n(\vec{r},t)}{\partial t}+\vec{\nabla}_{\mathrm{r}}\cdot\left[-\frac{2e}{m_n^*}n(\vec{r},t)u_{ij}(\vec{r},t)\right]+\frac{\vec{J}_n(\vec{r},t)}{\langle\tau_{\mathrm{m}}\rangle(\vec{r},t)} = -\frac{e\vec{F}(\vec{r},t)}{m_n^*}n(\vec{r},t), \tag{2.42}$$

Since the quantity $u_{ij}(\vec{r},t)$ is like kinetic energy, let us write

$$u_{ij}(\vec{r},t) = \frac{1}{2}kT_{ij}(\vec{r},t) = \frac{1}{2}e\mathrm{V}_{ij}(\vec{r},t), \tag{2.43}$$

where T_{ij} is a tensor "electron temperature" and V_{ij} is a tensor potential. Using tensor calculus,

$$\vec{\nabla}_{\mathrm{r}} \cdot [e\mathrm{V}_{ij}(\vec{r},t)] = e\sum_{j=1}^{3} \frac{\partial \mathrm{V}_{ij}(\vec{r},t)}{\partial x_j} = -eE_i(\vec{r},t) = -e\vec{E}_n(\vec{r},t), \tag{2.44}$$

where $\vec{E}_n$ is an effective electric field arising from the divergence of the electron temperature or electron energy. This field would be identically zero if the electron temperature does not vary in space.

Using this result in (2.42), we get

$$\frac{\partial \vec{J}_n(\vec{r},t)}{\partial t} + \frac{e^2}{m_n^*} n(\vec{r},t)\vec{E}_n(\vec{r},t) - \frac{ek_{\mathrm{B}}T_{ij}(\vec{r},t)}{m_n^*}\vec{\nabla}_{\mathrm{r}} n(\vec{r},t) + \frac{\vec{J}_n(\vec{r},t)}{\langle \tau_{\mathrm{m}} \rangle (\vec{r},t)}$$

$$= \frac{e^2}{m_n^*} n(\vec{r},t)\vec{E}(\vec{r},t). \tag{2.45}$$

Multiplying the above equation throughout by $\langle \tau_{\mathrm{m}} \rangle (\vec{r},t)$ and defining *electron mobility* as

$$\mu_n(\vec{r},t) = \frac{e\,\langle \tau_{\mathrm{m}} \rangle (\vec{r},t)}{m_n^*}, \tag{2.46}$$

based on what we found in Chap. 1, we get

$$\langle \tau_{\mathrm{m}} \rangle (\vec{r},t)\frac{\partial \vec{J}_n(\vec{r},t)}{\partial t} + \vec{J}_n(\vec{r},t) = en(\vec{r},t)\mu_n(\vec{r},t)\left[\vec{E}(\vec{r},t) - \vec{E}_n(\vec{r},t)\right]$$

$$+ e\frac{k_{\mathrm{B}}T_{ij}(\vec{r},t)}{e}\mu_n(\vec{r},t)\vec{\nabla}_{\mathrm{r}} n(\vec{r},t). \tag{2.47}$$

Finally, defining the *electron diffusion coefficient* as (note its similarity with the Einstein relation)

$$D_n(\vec{r},t) = \frac{kT_{ij}(\vec{r},t)}{e}\mu_n(\vec{r},t), \tag{2.48}$$

we get

$$\langle \tau_{\mathrm{m}} \rangle (\vec{r},t)\frac{\partial \vec{J}_n(\vec{r},t)}{\partial t} + \vec{J}_n(\vec{r},t) = en(\vec{r},t)\mu_n(\vec{r},t)\left[\vec{E}(\vec{r},t) - \vec{E}_n(\vec{r},t)\right]$$

$$+ eD_n(\vec{r},t)\vec{\nabla}_{\mathrm{r}} n(\vec{r},t). \tag{2.49}$$

The last equation is the electron current density equation derived from the BTE. Comparing it with the electron current density equation derived in the preceding chapter, we find that we have obtained three new terms. The first of these, $\frac{\partial \vec{J}_n(\vec{r},t)}{\partial t}$, is the time-varying term which would not have appeared in the steady state analysis of Chap. 1, and the second of these is the effective electric field $\vec{E}_n$ arising from spatial variation of the electron energy or temperature T_{ij}.

The *steady-state* electron current density equation will become

$$\vec{J}_n(\vec{r},t) = en(\vec{r},t)\mu_n(\vec{r},t)[\vec{E}(\vec{r},t) - \vec{E}_n(\vec{r},t)] + eD_n(\vec{r},t)\vec{\nabla}_r n(\vec{r},t), \quad (2.50)$$

which is the familiar current equation with one exception. The exception is the term involving the energy gradient field $\vec{E}_n(\vec{r},t)$, which is usually ignored in low field transport where the electron energy varies little in space. However, it cannot be ignored in high-field transport where the electron energy can vary quite rapidly in space giving rise to a large energy-gradient field.

2.5 Too Many Unknowns

When we derived the electron current density equation rigorously from the BTE, we ended up introducing a new transport variable which is the tensor temperature $T_{ij}(\vec{r},t)$ (or, equivalently, the energy gradient field $\vec{E}_n(\vec{r},t)$). So, now we have a total of four transport variables—$J_n(\vec{r},t)$, $n(\vec{r},t)$, $\vec{E}(\vec{r},t)$, $T_{ij}(\vec{r},t)$—and yet only three equations to work with, namely the new electron current density equation (2.49), the new continuity equation (2.30) and the Poisson equation. Thus, we have one more unknown than the number of equations. We can try to overcome this problem by deriving an equation for the second moment of the distribution function which will be the electron energy density related to the tensor temperature $T_{ij}(\vec{r},t)$. Unfortunately, each time we derive a new hydrodynamic balance equation for a new generalized moment we will invariably end up introducing yet another new transport variable. For example, in the equation for the second moment which we have not derived here, that new transport variable will be the flux of the energy density. Thus, we can continue to add more hydrodynamic equations, but in the end we will always have one more unknown than the number of equations. In bipolar transport, we will have seven transport variables—$J_n(\vec{r},t)$, $J_p(\vec{r},t)$, $n(\vec{r},t)$, $p(\vec{r},t)$, $\vec{E}(\vec{r},t)$, $T^n_{ij}(\vec{r},t)$, $T^p_{ij}(\vec{r},t)$—and five equations which are current density equations for electrons and holes (2 equations), continuity equations for electrons and holes (2 equations) and the Poisson equation. Thus, we will have two more unknowns than the number of equations.

In order to close the hierarchy of equations (so that we have as many equations as unknowns), we must introduce a completely ad-hoc assumption. For example, we may assume (without any justifiable basis) that

$$T^n_{ij} = T^p_{ij} = T \quad (2.51)$$

everywhere, where T is some temperature (not necessarily the ambient temperature). This will make $\vec{E}_n = \vec{E}_p = 0$ everywhere and eliminate our problem because now we will have just as many unknowns as equations. But, obviously this is an ad-hoc measure that has no scientific basis. Yet, we must do this in order to be able to solve the hydrodynamic equations. This is the primary shortcoming of the hydrodynamic approach.

We do not have to live with this shortcoming if we evaluate the distribution function $f(\vec{r}, \vec{k}, t)$ and find the transport variables directly from (2.1). However, as mentioned before, solving the BTE is not easy and therefore $f(\vec{r}, \vec{k}, t)$ is not easily found everywhere within a device. We may be able to find it in some selected regions of a device without expending too much effort, but finding it everywhere within a large device becomes nearly intractable. The most common method of solving the BTE is MC simulation which is discussed later in this chapter. This method becomes particularly inefficient in regions of a device where the electric field is low (or, even worse, retards the forward motion of the carrier). Fortunately, in those regions, nonlocal effects are most unlikely to occur so that we can find the mobility from the steady-state velocity versus field characteristics. That is indeed a very fortuitous happenstance.

The advantage of the hydrodynamic equation approach is that we never have to find $f(\vec{r}, \vec{k}, t)$. Therefore, we can bypass a computationally challenging step, but the disadvantage is that we will ultimately have to make an ad-hoc assumption to close the hierarchy of equations, which may make the final results dubious.

2.6 Finding the Mobility and Diffusion Coefficient from the Carrier Distribution Function

We mentioned earlier that mobility and diffusion coefficient are functionals of the distribution function $f(\vec{r}, \vec{k}, t)$. Here, we provide the explicit steps for finding them from the distribution function.

As defined earlier,

$$\mu(\vec{r}, t) = \frac{e \langle \tau_m \rangle (\vec{r}, t)}{m*}, \tag{2.52}$$

where

$$\frac{1}{\langle \tau_m \rangle (\vec{r}, t)} = \frac{\sum_{\vec{k}} \vec{v}(\vec{k}) f(\vec{r}, \vec{k}, t) \frac{1}{\tau_m(\vec{r}, \vec{k}, t)}}{\sum_{\vec{k}} \vec{v}(\vec{k}) f(\vec{r}, \vec{k}, t)} \tag{2.53}$$

and

$$\frac{1}{\tau_m(\vec{r}, \vec{k}, t)} = \sum_{\vec{k}'} S(\vec{k}, \vec{k}') \frac{\vec{v}(\vec{k}) - \vec{v}(\vec{k}')}{\vec{v}(\vec{k})} [1 - f(\vec{r}, \vec{k}', t)]. \tag{2.54}$$

The diffusion coefficient is defined as

$$D(\vec{r},t) = \frac{kT_{ij}(\vec{r},t)}{e}\mu(\vec{r},t) = \frac{m^*}{2e}\frac{\sum_{\vec{k}} \vec{v}_i(\vec{k})\vec{v}_j(\vec{k}) f(\vec{r},\vec{k},t)}{\sum_{\vec{k}} f(\vec{r},\vec{k},t)}\mu(\vec{r},t). \quad (2.55)$$

The above equations tell us how to find the transport parameters—mobility and diffusion coefficient—from the distribution function $f(\vec{r},\vec{k},t)$ and the scattering rate $S(\vec{k},\vec{k}')$. Finding the latter quantity requires quantum mechanics and it is usually evaluated from a quantum mechanical prescription known as *Fermi's Golden Rule*, which we will discuss in Chap. 5. One should note that mobility and diffusion coefficient found in this fashion have nonlocal effects incorporated in them implicitly since the distribution function at a location $\vec{r}$ is influenced by scattering events that occurred at a different location $\vec{r}'$ and caused electrons to drift and diffuse down to the location $\vec{r}$. In other words, the transport parameters are no longer just a function of the local electric field and therefore may be able to reproduce some nonlocal effects.

Special Topic 1: The variational principle of transport

We state a very important principle here without proof. The so-called *variational principle of transport* [6] states that a first order change in the distribution function will cause only a second order change in a transport parameter such as mobility. Therefore, the transport parameters, which are functionals of the distribution function, are ultimately not particularly sensitive to the distribution function.

Special Topic 2: Principle of detailed balance: a relationship between $S(\vec{k},\vec{k}')$ and $S(\vec{k}',\vec{k})$

Without deriving specific expressions for the scattering rates $S(\vec{k},\vec{k}')$ and $S(\vec{k}',\vec{k})$, we can still derive a universal relation between them, which is sometimes referred to as the Principle of Detailed Balance.

When global equilibrium exists, the distribution function will not change due to collisions, i.e.,

$$\left.\frac{\mathrm{d}f(\vec{r},\vec{k},t)}{\mathrm{d}t}\right|_{\text{collisions}} = S_{\text{in}} - S_{\text{out}} = 0. \quad (2.56)$$

To understand this, recall the box in Fig. 2.1. The number of particles inside this box should become constant under global equilibrium and should not change, so that the rate at which particles are scattering in should be equal to the rate at which they are scattering out. This leads to the above equation.

Therefore,

$$\sum_{\vec{k}'} \Big\{ S(\vec{k}',\vec{k}) f_0(\vec{r},\vec{k}',t)[1 - f_0(\vec{r},\vec{k},t)] - S(\vec{k},\vec{k}') f_0(\vec{r},\vec{k},t)[1 - f_0(\vec{r},\vec{k}',t)] \Big\} = 0, \tag{2.57}$$

where the subscript "0" on the distribution function implies that it is the (global) *equilibrium* distribution function.

Since the scattering rates are positive indefinite and can never assume negative values, this implies that

$$S(\vec{k}',\vec{k}) f_0(\vec{r},\vec{k}',t)[1 - f_0(\vec{r},\vec{k},t)] = S(\vec{k},\vec{k}') f_0(\vec{r},\vec{k},t)[1 - f_0(\vec{r},\vec{k}',t)], \tag{2.58}$$

which, in turn, yields

$$S(\vec{k}',\vec{k}) \left\{ \frac{1}{f_0(\vec{r},\vec{k},t)} - 1 \right\} = S(\vec{k},\vec{k}') \left\{ \frac{1}{f_0(\vec{r},\vec{k}',t)} - 1 \right\}. \tag{2.59}$$

Since the distribution function at global equilibrium is always the Fermi–Dirac function, it is given by

$$f_0(\vec{r},\vec{k},t) = \frac{1}{e^{\frac{E(\vec{r},\vec{k},t)+E_{c0}(\vec{r},t)-\mu_F}{k_B T}} + 1}, \tag{2.60}$$

so that the last equation can be recast as

$$S(\vec{k}',\vec{k}) e^{\frac{E(\vec{r},\vec{k},t)+E_{c0}(\vec{r},t)-\mu_F}{k_B T}} = S(\vec{k},\vec{k}') e^{\frac{E(\vec{r},\vec{k}',t)+E_{c0}(\vec{r},t)-\mu_F}{k_B T}}, \tag{2.61}$$

or

$$S(\vec{k}',\vec{k}) = S(\vec{k},\vec{k}') e^{\frac{E(\vec{r},\vec{k}',t)-E(\vec{r},\vec{k},t)}{k_B T}}. \tag{2.62}$$

The above relation between the scattering rates is valid even out of equilibrium since the scattering rates themselves do *not* generally depend on the distribution function, and therefore do not depend on whether global equilibrium conditions prevail or not. This relation is often called the *Principle of Detailed Balance* [7]. Note that the Principle of Detailed Balance implies that it is *more probable* to scatter from a higher energy state to a lower energy state. This is a manifestation of the general principle of thermodynamics that all systems tend to lower their energy by dissipation.

Special Topic 3: Electrostatic potential versus chemical potential

Students of device physics and device engineering are very familiar with the notion that under global equilibrium conditions, the Fermi level is spatially invariant, i.e., $\vec{\nabla}_r \mu_F = 0$. This is among the first things one is taught in undergraduate device physics courses. We even used this result in the previous chapter to derive the Einstein relation. Here, we prove this result explicitly from the BTE.

Under global equilibrium conditions, the number of particles in the box in Fig. 2.1 cannot change, so that $\left.\frac{\mathrm{d}f(\vec{r},\vec{k},t)}{\mathrm{d}t}\right|_{\text{collisions}} \equiv 0$. In other words, the right-hand-side of the BTE vanishes. That means, the-left-hand-side should vanish as well under global equilibrium.

Now, global equilibrium implies steady-state, so that the distribution function will be independent of time and the BTE will read

$$\vec{v}(\vec{k}) \cdot \vec{\nabla}_r f_{\text{eq}}(\vec{r},\vec{k}) + \frac{\vec{F}(\vec{r})}{\hbar} \cdot \vec{\nabla}_k f_{\text{eq}}(\vec{r},\vec{k}) = 0, \tag{2.63}$$

where the subscript "eq" represents global equilibrium.

There are two ways in which the above equation can be satisfied: (1) transport is spatially homogeneous so that the distribution function does not vary in space and there is no electric field anywhere, or (2) transport is spatially varying and there is an electric field. The latter case is the nontrivial one and we encountered it in Problems 1.5 and 1.6 of the previous chapter. This situation occurs, for example, in the depletion region of a p–n junction diode whose energy band diagram under equilibrium is shown in Fig. 2.2. There is an electric field in the depletion region because the conduction/valence band edge varies spatially. This causes a drift current, but the distribution function (and hence the electron and hole concentrations) vary spatially within the depletion region which causes a diffusion current as well. Under global equilibrium conditions, the two currents are equal and opposite, so that the total current is exactly zero and global equilibrium exists. This is an example of the Case 2 situation.

Now, let us define a quantity $\theta(\vec{r},\vec{k})$ such that

$$\theta \equiv \theta(\vec{r},\vec{k}) = \frac{E_{c0}(\vec{r}) + E(\vec{r},\vec{k}) - \mu_F(\vec{r})}{k_B T(\vec{r})}, \tag{2.64}$$

where $E_{c0}(\vec{r})$ is the electron's potential energy, $E(\vec{r},\vec{k})$ is the electron's kinetic energy which would vary in space if the electron's effective mass varies spatially,[4]

[4]Readers having some familiarity with device physics will recognize $E_{c0}(\vec{r})$ as the conduction band edge.

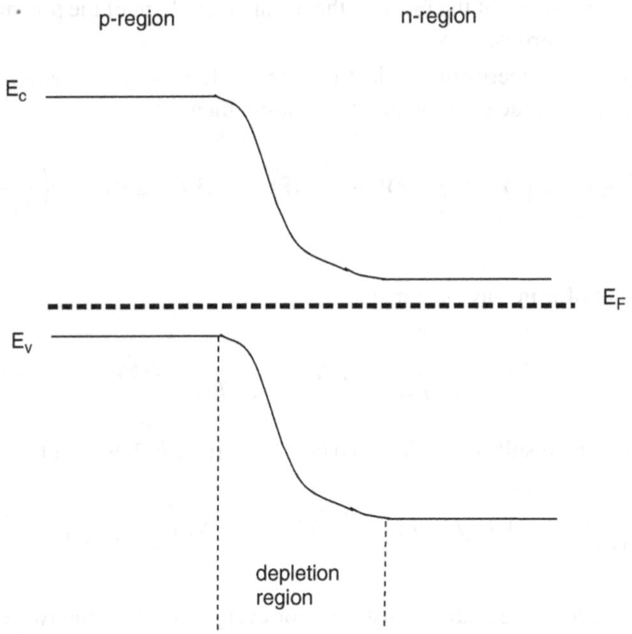

Fig. 2.2 Energy band diagram of a p-n junction diode under global equilibrium conditions. Note that there is an electrostatic potential gradient (or an electric field) in the depletion region but the Fermi level is "flat" (or spatially invariant), so that there is no chemical potential gradient. Hence there is no current flow and the diode is maintained in global equilibrium

and $\mu_F(\vec{r})$ is the Fermi level. Let us assume that it could vary in space and test this assumption. For the sake of generality, we also assume that the electron temperature varies in space.

Using the chain rule of differentiation, (2.63) can be written as

$$\frac{\mathrm{d}f(\vec{r},\vec{k})}{\mathrm{d}\theta}\left[\vec{v}(\vec{k})\cdot\vec{\nabla}_r\theta + \frac{\vec{F}(\vec{r})}{\hbar}\cdot\vec{\nabla}_k\theta\right] = 0. \tag{2.65}$$

Next, we find the gradients of θ in both real space and wavevector space:

$$\begin{aligned}\vec{\nabla}_r\theta &= \frac{1}{k_B T(\vec{r})}[\vec{\nabla}_r E_{c0}(\vec{r}) + \vec{\nabla}_r E(\vec{r},\vec{k}) - \vec{\nabla}_r\mu_F(\vec{r})] \\ &\quad + [E_{c0}(\vec{r}) + E(\vec{r},\vec{k}) - \mu_F(\vec{r})]\vec{\nabla}_r\left(\frac{1}{k_B T(\vec{r})}\right) \\ &= \frac{1}{k_B T(\vec{r})}[-\vec{F}(\vec{r}) + \vec{\nabla}_r E(\vec{r},\vec{k}) - \vec{\nabla}_r\mu_F(\vec{r})] \\ &\quad + [E_{c0}(\vec{r}) + E(\vec{r},\vec{k}) - \mu_F(\vec{r})]\vec{\nabla}_r\left(\frac{1}{k_B T(\vec{r})}\right),\end{aligned} \tag{2.66}$$

where we have made use of the fact that the negative gradient of the potential energy is the force on the particle.

Now, if the kinetic energy of the electron $E(\vec{r}, \vec{k})$ does not vary spatially (e.g., due to spatial variation of the electron effective mass), then

$$\vec{\nabla}_{\rm r}\theta = \frac{1}{k_{\rm B}T(\vec{r})}[-\vec{F}(\vec{r}) - \vec{\nabla}_{\rm r}\mu_{\rm F}(\vec{r})] + [E_{\rm c0}(\vec{r}) + E(\vec{k}) - \mu_{\rm F}(\vec{r})]\vec{\nabla}_{\rm r}\left(\frac{1}{k_{\rm B}T(\vec{r})}\right). \tag{2.67}$$

The gradient of θ in wavevector space is

$$\vec{\nabla}_{\rm k}\theta = \frac{1}{k_{\rm B}T(\vec{r})}\vec{\nabla}_{\rm k}E(\vec{k}) = \frac{1}{k_{\rm B}T(\vec{r})}\hbar\vec{v}(\vec{k}). \tag{2.68}$$

Substituting the results of the last two equations in (2.65), we get that

$$-\frac{1}{k_{\rm B}T(\vec{r})}\vec{\nabla}_{\rm r}\mu_{\rm F}(\vec{r}) + [E_{\rm c0}(\vec{r}) + E(\vec{k}) - \mu_{\rm F}(\vec{r})]\vec{\nabla}_{\rm r}\left(\frac{1}{k_{\rm B}T(\vec{r})}\right) = 0. \tag{2.69}$$

Since the preceding equation must hold for every $\vec{k}$, each of the two terms must vanish individually. Therefore,

$$\begin{aligned}\vec{\nabla}_{\rm r}\mu_{\rm F}(\vec{r}) &= 0\\ \vec{\nabla}_{\rm r}T(\vec{r}) &= 0.\end{aligned} \tag{2.70}$$

The last equation shows that under global equilibrium conditions (when no current flows), the Fermi level is spatially invariant and there can be no thermal gradients. Of course, all this is strictly true as long as the effective mass is spatially invariant. The term "Fermi level" is only meaningful in global equilibrium since only then it is a *constant* independent of position within a solid. If current flows, then equilibrium is disturbed and the Fermi level is not spatially invariant any more. In that case, the concept of Fermi level is no longer meaningful. Instead, we can define a "quasi-Fermi level" which replaces the Fermi level. The defining relation for the quasi-Fermi level is proved in Problem 2.2 which shows that the *total* current density (including both drift and diffusion components) in a solid is proportional to the gradient of the quasi Fermi-level.

All of this has profound implications. Most of us, steeped in the culture of Ohm's law that is taught in high-school physics classes, subconsciously tend to think of current as being entirely due to drift and almost never think of the diffusion component. In that mindset, current appears to be caused by an electric field, or an *electrostatic potential* gradient. Problem 2.2 shows that it is instead caused by a Fermi level gradient. In the physics and chemistry literature, the Fermi level is typically referred to as "chemical potential." It is important to realize that current is *not* caused by electrostatic potential gradients, but rather by chemical potential

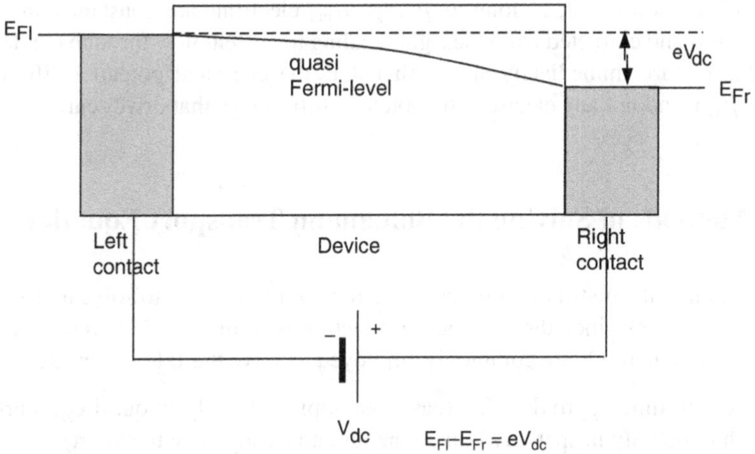

Fig. 2.3 A two-terminal device has two contacts that are in local thermodynamic equilibrium. The electron distribution function in each contact is Fermi–Dirac defined by a local Fermi level or chemical potential. A battery delivering a voltage V_{dc} connected between the two contacts will make the Fermi levels different such that $\mu_{Fl} - \mu_{Fr} = eV_{dc}$. The left contact will try to raise the quasi Fermi level in the device to μ_{Fl} by injecting carriers from the left, while the right contact will try to lower it to μ_{Fr} by extracting carriers from the right. As long as $V_{dc} \neq 0$, so that $\mu_{Fl} \neq \mu_{Fr}$, carriers will be constantly injected from the left and extracted from the right resulting in a current flow

gradients. The obvious example of this is the p-n junction diode in equilibrium shown in Fig. 2.2. There is clearly an electrostatic potential gradient in the depletion region, but no chemical potential gradient. Hence, no current flows.

A two-terminal device, such as a resistor, is a piece of material connected to two contacts (see Fig. 2.3). It is very difficult to drive the contacts out of local thermodynamic equilibrium since they are enormous reservoirs containing numerous electrons. As a result, the distribution function in each contact is Fermi–Dirac and prescribed by a local Fermi level or chemical potential. We can write the distribution functions in the left and right contacts (see Fig. 2.3) as

$$
\begin{aligned}
f_{\text{left}}(E) &= \frac{1}{e^{\frac{E-\mu_{Fl}}{k_B T}} + 1} \\
f_{\text{right}}(E) &= \frac{1}{e^{\frac{E-\mu_{Fr}}{k_B T}} + 1}
\end{aligned}
\tag{2.71}
$$

where μ_{Fl} and μ_{Fr} are the local chemical potentials.

Let us assume that $\mu_{Fl} > \mu_{Fr}$. In that case, the left contact will try to raise the chemical potential within the device to μ_{Fl} and the right contact will try to lower it to μ_{Fr}. Thus, the left contact will inject electrons into the device to raise the quasi Fermi-level, and the right contact will extract those electrons to lower the quasi

Fermi-level. As a result, as long as $\mu_{\mathrm{Fl}} \neq \mu_{\mathrm{Fr}}$, electrons are constantly injected from the left and extracted from the right, leading to current flow through the device [8]. This picture immediately shows that it is the chemical potential difference ($\mu_{\mathrm{Fl}} \neq \mu_{\mathrm{Fr}}$), and not any electrostatic potential difference, that drives current flow.

2.7 Methods of Solving the Boltzmann Transport Equation

In order to find the distribution function from the BTE we have to solve it. That can be a daunting task since the distribution function is a function of seven variables. Four different methods are commonly employed to solve the BTE. They are

- Relaxation time approximation (easy, but applicable only in quasi-equilibrium, and that too only in spatially homogeneous and steady-state transport)
- Series expansion method (harder)
- MC simulation (numerical and even harder)
- Numerical methods (hardest)

The last method is computationally most challenging because of the infamous "curse of dimensionality." Remember that the distribution function is a function of seven variables—three space coordinates, three wavevector coordinates, and one time coordinate. Therefore, even if we were to discretize the distribution function by taking just 100 grid points in each coordinate (a relatively coarse mesh), we will still be dealing with matrices of dimension $10^{14} \times 10^{14}$, which will burden even the most powerful supercomputers. We will not discuss this technique here, but conceptually it is straightforward.

We will not discuss the series expansion technique either since this is not a standard method. This method involves expanding the distribution function in a complete orthonormal set and then finding the coefficients of expansion which will finally yield the entire distribution function. For an example of this, see [9]. A related method is called the variational method of solution which, too, is nonstandard and omitted from discussion. Therefore, we will briefly discuss the other two techniques, spending most of our time on the relaxation time approximation, which we discuss next.

2.8 Relaxation Time Approximation Method

This is a particularly simple method of solving the BTE to find an analytical solution for the distribution function when transport is: (1) space- and time-invariant, and (2) the deviation of the distribution function from its equilibrium value is small.

The physical assumption that undergirds the relaxation time approximation is that if any perturbation makes the distribution function deviate from its equilibrium value, then scattering forces will tend to restore the distribution function to its equilibrium value *exponentially* in time with a *wavevector-dependent* time constant, i.e.,

$$f(\vec{k},t) = f_0(\vec{k})\left[1 - e^{-\frac{t}{\tau(\vec{k})}}\right], \tag{2.72}$$

where $f(\vec{k},t)$ is the perturbed distribution function and $f_0(\vec{k})$ is the equilibrium distribution function (normally Fermi–Dirac).

From the last equation, we get

$$\left.\frac{\mathrm{d}f(\vec{k},t)}{\mathrm{d}t}\right|_{\text{collisions}} = S_{\text{in}} - S_{\text{out}} = \frac{f_0(\vec{k})}{\tau(\vec{k})}e^{-\frac{t}{\tau(\vec{k})}} = \frac{f_0(\vec{k}) - f(\vec{k},t)}{\tau(\vec{k})}. \tag{2.73}$$

Equation (2.73) is obtained by simply differentiating the preceding one with respect to time.

Let us now consider space- and time-invariant transport so that the distribution function becomes a function of only wavevector. The terms in the BTE involving the spatial and temporal gradients will vanish, and the equation will reduce to

$$\frac{\vec{F}}{\hbar} \cdot \vec{\nabla}_{\mathrm{k}} f(\vec{k}) = S_{\text{in}} - S_{\text{out}} = \frac{f_0(\vec{k}) - f(\vec{k})}{\tau(\vec{k})}. \tag{2.74}$$

Using the chain rule of differentiation, the above equation can be written as

$$\vec{F} \cdot \left[\frac{1}{\hbar}\vec{\nabla}_{\mathrm{k}} E(\vec{k})\right] \frac{\partial f(\vec{k})}{\partial E(\vec{k})} = \frac{f_0(\vec{k}) - f(\vec{k})}{\tau(\vec{k})}. \tag{2.75}$$

The term within the square brackets is the carrier velocity $\vec{v}(\vec{k})$. Next, defining $\Delta f(\vec{k}) = f(\vec{k}) - f_0(\vec{k})$, we get

$$\vec{F} \cdot \vec{v}(\vec{k}) \left\{\frac{\partial f_0(\vec{k})}{\partial E(\vec{k})} + \frac{\partial \Delta f(\vec{k})}{\partial E(\vec{k})}\right\} = \frac{f_0(\vec{k}) - f(\vec{k})}{\tau(\vec{k})}. \tag{2.76}$$

Because the distribution function is assumed to deviate little from the global equilibrium value, $\Delta f(\vec{k}) \ll f_0(\vec{k})$ so that we can neglect the second term within the curly brackets in comparison with the first. Consequently, the solution of the BTE is

$$f(\vec{k}) = f_0(\vec{k}) - \tau(\vec{k})\vec{F} \cdot \vec{v}(\vec{k}) \frac{\partial f_0(\vec{k})}{\partial E(\vec{k})}. \tag{2.77}$$

Finally, recognizing that the equilibrium distribution function is the Fermi factor in (2.60), we get

$$f(\vec{k}) = \frac{1}{e^{\frac{E(\vec{k})+E_{c0}-\mu_F}{k_B T}}+1} + \tau(\vec{k})\vec{F}\cdot\vec{v}(\vec{k})\frac{1}{4k_B T}\mathrm{sech}^2\left[\frac{E(\vec{k})+E_{c0}-\mu_F}{2k_B T}\right]. \quad (2.78)$$

The above solution is the so-called relaxation-time-approximation solution which is valid for quasi-equilibrium (when the distribution function is close to its global equilibrium value) and transport is both spatially homogeneous and time-invariant.

2.9 When Can We Define a Constant Relaxation Time?

In order for the relaxation time approximation method to be meaningful, it is imperative that one is able to define a constant relaxation time that is independent of the distribution function. Let us examine under what circumstances that is possible.

Earlier, we had shown that the momentum relaxation time $\langle \tau_m \rangle$ is a functional of the distribution function. Therefore, one would expect that the relaxation time $\tau(\vec{k})$ too would depend on the distribution function. If that were true, then it would invalidate the entire solution in (2.78) which was based on the assumption of a constant relaxation time that depends only on the wavevector $\vec{k}$. Therefore, we need to identify the situations in which $\tau(\vec{k})$ becomes independent of the distribution function. Only in those situations will the relaxation time method (and the resulting solution in (2.78)) will have any validity.

Here, we identify the situations when this happens. The relaxation time $\tau(\vec{k})$ becomes independent of the distribution function in two cases: Either

1. The scattering processes are *elastic*, or
2. The scattering processes are isotropic and also the carrier population is nondegenerate, i.e., the carrier density is small, so that the probability of finding a carrier with wavevector $\vec{k}$ at any position $\vec{r}$ at time t is much smaller than unity.

In this section, we will prove the above two results.

First, let us derive an expression for the relaxation time $\tau(\vec{k})$ that we used in (2.73). From this equation, we obtain

$$\begin{aligned} S_{in} - S_{out} &= \sum_{\vec{k}'} S(\vec{k}',\vec{k}) f(\vec{k}')[1 - f(\vec{k})] - \sum_{\vec{k}'} S(\vec{k},\vec{k}') f(\vec{k})[1 - f(\vec{k}')] \\ &= \frac{f_0(\vec{k}) - f(\vec{k})}{\tau(\vec{k})} = -\frac{\Delta f(\vec{k})}{\tau(\vec{k})}. \end{aligned} \quad (2.79)$$

Using a short-hand notation $f = f(\vec{k})$, $f' = f(\vec{k}')$, $\Delta f(\vec{k}) = \Delta f$ and $\Delta f(\vec{k}') = \Delta f'$, we can rewrite the above equation as

$$\sum_{\vec{k}'} \left\{ S(\vec{k}',\vec{k}) f'[1-f] - S(\vec{k},\vec{k}') f[1-f'] \right\} = -\frac{\Delta f}{\tau(\vec{k})}. \tag{2.80}$$

Noting that $f = f_0 + \Delta f$ and $f' = f_0' + \Delta f'$, the last equation is recast as

$$\sum_{\vec{k}'} \left\{ S(\vec{k}',\vec{k})(f_0' + \Delta f')[1 - f_0 - \Delta f] - S(\vec{k},\vec{k}')(f_0 + \Delta f)[1 - f_0' - \Delta f'] \right\}$$
$$= -\frac{\Delta f}{\tau(\vec{k})}. \tag{2.81}$$

Regrouping the terms, we get

$$\sum_{\vec{k}'} \left\{ S(\vec{k}',\vec{k}) f_0'(1-f_0) - S(\vec{k},\vec{k}') f_0(1-f_0') \right\}$$
$$-\sum_{\vec{k}'} \left\{ S(\vec{k}',\vec{k})[f_0'\Delta f - \Delta f'(1-f_0)] - S(\vec{k},\vec{k}')[f_0\Delta f' - \Delta f(1-f_0')] \right\}$$
$$= -\frac{\Delta f}{\tau(\vec{k})}, \tag{2.82}$$

where we have neglected terms that are second order in the deviation, such as $\Delta f \Delta f'$. We can do that as long as the deviations are small, which is what we assumed in the first place.

Because of the equality in (2.58) (Principle of Detailed Balance), the first sum vanishes, leaving us with

$$\sum_{\vec{k}'} \left\{ S(\vec{k}',\vec{k})[f_0'\Delta f - \Delta f'(1-f_0)] - S(\vec{k}',\vec{k}) \frac{f_0'(1-f_0)}{f_0(1-f_0')} [f_0\Delta f' - \Delta f(1-f_0')] \right\}$$
$$= \frac{\Delta f}{\tau(\vec{k})}, \tag{2.83}$$

where we have again used (2.58).

Simplifying, we obtain

$$\sum_{\vec{k}'} S(\vec{k}',\vec{k}) \left\{ f_0'\Delta f - \Delta f'(1-f_0) - \frac{f_0'(1-f_0)}{(1-f_0')}\Delta f' + \frac{f_0'}{f_0}\Delta f - f_0'\Delta f \right\}$$
$$= \frac{\Delta f}{\tau(\vec{k})}, \tag{2.84}$$

or

$$\sum_{\vec{k}'} S(\vec{k}',\vec{k})\left\{\Delta f'(1-f_0)\left(1+\frac{f_0'}{1-f_0'}\right)-\frac{f_0'}{f_0}\Delta f\right\}=-\frac{\Delta f}{\tau(\vec{k})}, \tag{2.85}$$

or

$$\frac{1}{\tau(\vec{k})}=\sum_{\vec{k}'} S(\vec{k}',\vec{k})\left\{\frac{f_0'}{f_0}-\frac{\Delta f'[1-f_0])}{\Delta f[1-f_0']}\right\}. \tag{2.86}$$

The above is the expression for the relaxation time $\tau(\vec{k})$ used in the relaxation time approximation method. Clearly, this time depends on the distribution function, in general. In the following two subsections, we will show that this dependence drops out when scattering is either elastic or isotropic in a nondegenerate population. However, first, we need to derive a slightly different form of (2.86).

Using the relaxation time solution in (2.78), we get

$$\begin{aligned}
\Delta f &= f(\vec{k})-f_0(\vec{k})=\tau(\vec{k})\vec{F}\cdot\vec{v}(\vec{k})\frac{1}{4k_\mathrm{B}T}\mathrm{sech}^2\left(\frac{E(\vec{k})+E_{c0}(\vec{r},t)-\mu_\mathrm{F}}{2k_\mathrm{B}T}\right)\\
&= \Xi(k)\cos\theta\\
\Delta f' &= f(\vec{k}')-f_0(\vec{k}')=\tau(\vec{k}')\vec{F}\cdot\vec{v}(\vec{k}')\frac{1}{4k_\mathrm{B}T}\mathrm{sech}^2\left(\frac{E(\vec{k}')+E_{c0}(\vec{r},t)-\mu_\mathrm{F}}{2k_\mathrm{B}T}\right)\\
&= \Xi(k')\cos\theta',
\end{aligned} \tag{2.87}$$

where

$$\begin{aligned}
\Xi(k) &= \tau(k)|\vec{F}||\vec{v}(\vec{k})|\frac{1}{4k_\mathrm{B}T}\mathrm{sech}^2\left(\frac{E(k)+E_{c0}(\vec{r},t)-\mu_\mathrm{F}}{2k_\mathrm{B}T}\right)\\
\Xi(k') &= \tau(k')|\vec{F}||\vec{v}(\vec{k}')|\frac{1}{4k_\mathrm{B}T}\mathrm{sech}^2\left(\frac{E(k')+E_{c0}(\vec{r},t)-\mu_\mathrm{F}}{2k_\mathrm{B}T}\right)
\end{aligned} \tag{2.88}$$

and θ is the angle between the direction of the force $\vec{F}$ and the direction of $\vec{v}(\vec{k})$ while θ' is the angle between the direction of the force $\vec{F}$ and the direction of $\vec{v}\left(\vec{k}'\right)$. Note that since the τ-s depend only the magnitude of the wavevectors, $\Xi(k)$ depends only on the magnitude of k and $\Xi(k')$ only on the magnitude of k'.

Substituting these results in (2.86), we get

$$\frac{1}{\tau(\vec{k})}=\sum_{\vec{k}'} S(\vec{k}',\vec{k})\left\{\frac{f_0'}{f_0}-\frac{\Xi(k')\cos\theta'[1-f_0]}{\Xi(k)\cos\theta[1-f_0']}\right\}. \tag{2.89}$$

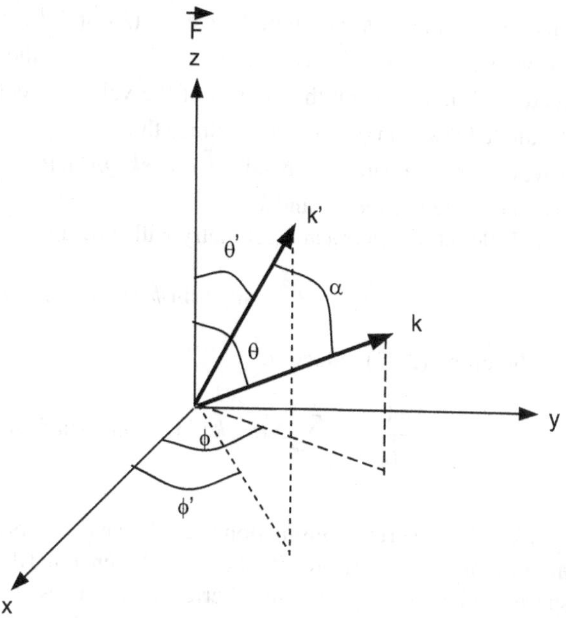

Fig. 2.4 A spherical coordinate system shown with the z-axis aligned along the direction of the force acting on the carrier

2.9.1 *Elastic Scattering*

Elastic scattering events are energy conserving, i.e., the energy of the carrier (electron or hole) is the same before and after scattering. Let us assume that the carrier's wavevector before scattering was $\vec{k}$ and after scattering becomes $\vec{k}'$. Elasticity implies

$$E(\vec{k}) = E(\vec{k}'), \tag{2.90}$$

and since the carrier energy depends only on the magnitude of the wavevector,

$$|\vec{k}| = |\vec{k}'|. \tag{2.91}$$

Therefore, for elastic scattering, $\Xi(k) = \Xi(k')$. Also, since the equilibrium distribution function depends only on the magnitude (and not the direction) of the wavevector, $f_0(k) = f_0(k')$. Consequently, in the case of elastic scattering, (2.89) reduces to

$$\frac{1}{\tau(\vec{k})} = \sum_{\vec{k}'} S(\vec{k}', \vec{k}) \left\{ 1 - \frac{\cos\theta'}{\cos\theta} \right\}. \tag{2.92}$$

To proceed further, we need to derive a relation between $\cos\theta$ and $\cos\theta'$. For this purpose, refer to Fig. 2.4 where we show a spherical coordinate frame assuming that the z-axis is aligned along the direction of the force $\vec{F}$ acting on a carrier.

Since the angle θ is the angle between $\vec{v}(\vec{k})$ and $\vec{F}$, it will be the polar angle of the velocity vector $\vec{v}(\vec{k})$. Similarly, the angle θ' is the polar angle of the velocity vector $\vec{v}(\vec{k}')$. The azimuthal angles of the velocity vectors are ϕ and ϕ'. Let α be the angle between $\vec{v}(\vec{k})$ and $\vec{v}(\vec{k}')$. Since the velocity is generally proportional to the wavevector (in a parabolic band $\vec{v}(\vec{k}) = \hbar\vec{k}/m^*$), they are collinear and hence α is also the angle between $\vec{k}$ and $\vec{k}'$.

A little bit of algebra and geometry will show that

$$\cos\theta' = \sin\theta\sin\phi'\sin\alpha + \cos\theta\cos\alpha. \tag{2.93}$$

Therefore, (2.92) reduces to

$$\frac{1}{\tau(\vec{k})} = \sum_{\vec{k}'} S(\vec{k}',\vec{k})\left\{1 - \tan\theta\sin\phi'\sin\alpha - \cos\alpha\right\}. \tag{2.94}$$

We will convert the summation over all wavevectors $\vec{k}'$ to an integral in $\vec{k}'$ by first multiplying the summand with the three-dimensional density of states in wavevector space[5] and then integrating in spherical coordinates. This is the standard method of converting summations to integrals and this leads to

$$\begin{aligned}\frac{1}{\tau(\vec{k})} = \frac{\Omega}{4\pi^3}\int_{-\infty}^{\infty}\int_0^{2\pi}\int_{-1}^{1} k'^2\mathrm{d}k'\mathrm{d}\phi'\mathrm{d}(\cos\theta')S(\vec{k}',\vec{k})\\ \times\left\{1 - \tan\theta\sin\phi'\sin\alpha - \cos\alpha\right\}.\end{aligned} \tag{2.95}$$

Since θ' is uniquely related with α, we can replace the variable θ' in the above integral with α. This yields

$$\begin{aligned}\frac{1}{\tau(\vec{k})} = \frac{\Omega}{4\pi^3}\int_{-\infty}^{\infty}\int_0^{2\pi}\int_{-1}^{1} k'^2\mathrm{d}k'\mathrm{d}\phi'\mathrm{d}(\cos\alpha)S(\vec{k}',\vec{k})\\ \times\left\{1 - \tan\theta\sin\phi'\sin\alpha - \cos\alpha\right\}.\end{aligned} \tag{2.96}$$

Finally, noting that scattering rates for anisotropic scattering mechanisms may depend only on the angle between the initial and final wavevectors, i.e., only on α, but not on the azimuthal angle ϕ', we find that the second term within the curly brackets integrates to zero because of the integration over ϕ'. Therefore, converting the integral back to a sum, we find that

$$\frac{1}{\tau(\vec{k})} = \sum_{\vec{k}'} S(\vec{k}',\vec{k})[1 - \cos\alpha]. \tag{2.97}$$

[5]We will see more of the density of states in Chap. 6.

This relaxation time is clearly independent of the distribution function since the latter does not appear anywhere. Therefore, in the case of elastic scattering, the relaxation time approximation method is valid.

Compare now the last equation with (2.40). Note that since we are discussing elastic scattering, $|\vec{v}(k)| = |\vec{v}(k')|$. Furthermore, if we assume that the carrier population is nondegenerate, then the probability of finding an electron at any given position, with any given wavevector, at any given instant of time is much smaller than unity, so that $f(\vec{k} \ll 1)$. Therefore, the relaxation time in the last equation is nothing but the momentum relaxation time $\tau_m(k)$ for a nondegenerate carrier population. In other words, the appropriate relaxation time to use in the relaxation time approximation, when all scattering mechanisms are elastic and the carrier population is nondegenerate, is the momentum relaxation time if .

2.9.2 Isotropic Scattering in a Nondegenerate Carrier Population

In this case, $|k| \neq |k'|$, so that $\Xi(k) \neq \Xi(k')$ and $f_0(k) \neq f_0(k')$. Therefore, from (2.89), we obtain

$$\frac{1}{\tau(\vec{k})} = \sum_{\vec{k}'} S(\vec{k}', \vec{k}) \left\{ \frac{f_0'}{f_0} - \frac{\Xi(k')[1 - f_0]}{\Xi(k)[1 - f_0']} [\tan\theta \sin\phi' \sin\alpha + \cos\alpha] \right\}. \quad (2.98)$$

Converting the summation to an integral, we get

$$\frac{1}{\tau(\vec{k})} = \frac{\Omega}{4\pi^3} \int_{-\infty}^{\infty} \int_0^{2\pi} \int_{-1}^{1} k'^2 \mathrm{d}k' \mathrm{d}\phi' \mathrm{d}(\cos\alpha) \Big[S(\vec{k}', \vec{k}) \times \left\{ \frac{f_0'}{f_0} - \frac{\Xi(k')[1 - f_0]}{\Xi(k)[1 - f_0']} [1 - \tan\theta \sin\phi' \sin\alpha - \cos\alpha] \right\} \Big]. \quad (2.99)$$

Since the scattering mechanism is isotropic, $S(\vec{k}', \vec{k})$ no longer can depend on α. Therefore, the second term within the curly brackets integrates to zero and we get

$$\frac{1}{\tau(\vec{k})} = \sum_{\vec{k}'} S(\vec{k}', \vec{k}) \frac{f_0'}{f_0} = \sum_{\vec{k}'} S(\vec{k}, \vec{k}') \frac{1 - f_0'}{1 - f_0}, \quad (2.100)$$

where we have used (2.58) to arrive at the last equality.

Now, in a nondegenerate carrier population, $f_0 \ll 1$ and $f_0' \ll 1$. Therefore, when scattering mechanisms are isotropic and the carrier population in nondegenerate,

$$\frac{1}{\tau(\vec{k})} \approx \sum_{\vec{k}'} S(\vec{k}, \vec{k}'). \quad (2.101)$$

The last result shows that once again $\tau(\vec{k})$ is independent of the distribution function. Therefore, the relaxation time approximation holds for isotropic scattering in a nondegenerate carrier population as well. However, in this case, the relaxation time is not the momentum relaxation time, but the so-called mean time between collisions defined by (2.101).

2.10 Low Field Steady-State Conductivity in a Homogeneous Nondegenerate Electron Gas

The relaxation time solution for the distribution function was derived for quasi-equilibrium conditions when the distribution function deviates little from its global equilibrium value. Therefore, we can trust it only when the electric field driving transport is weak and does not drive the distribution function far out of equilibrium. Furthermore, the solution is valid only in steady-state and spatially homogeneous transport. Using this solution, we will derive an expression for the low field steady-state conductivity of a homogeneous nondegenerate electron gas.

From (2.1), the expression for the steady-state current density in a spatially homogeneous medium is

$$\vec{J}_n = \frac{-e}{\Omega}\sum_{\vec{k}}\vec{v}(\vec{k})f(\vec{k}) = \frac{-e}{\Omega}\sum_{\vec{k}}\vec{v}(\vec{k})\left[f_0(\vec{k}) - \tau(\vec{k})\vec{F}\cdot\vec{v}(\vec{k})\frac{\partial f_0(\vec{k})}{\partial E(\vec{k})}\right], \quad (2.102)$$

where we have used the relaxation-time-approximation solution for the distribution function.

The summation over the first term within the square bracket must be zero since there can be no current flowing under global equilibrium. Mathematically, $f_0(\vec{k})$ is an even function in $\vec{k}$ and $\vec{v}(\vec{k})$ is an odd function in $\vec{k}$. Therefore, this term should sum to zero when summed over all $\vec{k}$ from $-\infty$ to $+\infty$. As a result,

$$\vec{J}_n = \frac{e}{\Omega}\sum_{\vec{k}}\vec{v}(\vec{k})\tau(\vec{k})\vec{F}\cdot\vec{v}(\vec{k})\frac{\partial f_0(\vec{k})}{\partial E(\vec{k})}. \quad (2.103)$$

In an electric field $\vec{\mathcal{E}}$, $\vec{F} = -e\vec{\mathcal{E}}$. Consequently,

$$\vec{J}_n = \frac{-e^2}{\Omega}\sum_{\vec{k}}\vec{v}(\vec{k})\tau(\vec{k})\vec{\mathcal{E}}\cdot\vec{v}(\vec{k})\frac{\partial f_0(\vec{k})}{\partial E(\vec{k})}. \quad (2.104)$$

Now

$$\vec{\mathcal{E}}\cdot\vec{v} = \mathcal{E}_x v_x + \mathcal{E}_y v_y + \mathcal{E}_z v_z = \sum_{j=1}^{3}\mathcal{E}_j v_j \quad (2.105)$$

In tensor notation, we will write the right-hand-side of the above equation simply as $\mathcal{E}_j v_j$ since repeated indices always imply summation over that index. Moreover, a vector is always written with a single index. Using this notation, (2.104) is written as

$$J_i = \frac{-e^2}{\Omega} \sum_{\vec{k}} v_i(\vec{k}) \tau(\vec{k}) \mathcal{E}_j v_j(\vec{k}) \frac{\partial f_0(\vec{k})}{\partial E(\vec{k})}. \tag{2.106}$$

The above equation shows immediately that if we make the relaxation time approximation and assume spatially homogeneous, steady-state, quasi-equilibrium transport, then the current density turns out to be proportional to the electric field, which is essentially Ohm's law. Since we assumed spatial homogeneity, we will not have any diffusion current since that is proportional to the spatial gradient of the carrier concentration. Hence, we will only have the drift current, and we have seen earlier from the Drude model that the drift current density is proportional to the electric field and yields Ohm's law. Therefore, the relaxation time approximation for spatially homogeneous, steady-state, quasi-equilibrium transport yields what the Drude model for a single particle yields. Everything is consistent.

Transport where the current density is *linearly* proportional to the electric field is sometimes referred to as *linear response transport*. Therefore, the relaxation time approximation for spatially homogeneous, steady-state, quasi-equilibrium transport results in linear response. The converse, however, is not true, i.e., linear response does not necessarily imply that the relaxation time approximation is valid.

We will define the electron conductivity tensor as

$$\sigma_{ij} = \frac{J_i}{\mathcal{E}_j} \tag{2.107}$$

which is the ratio of the electron current density flowing in the i-th direction in response to an electric field applied in the j-th direction. It is a second-ranked tensor (also known as a dyadic) and has 9 components:

$$\sigma = \begin{pmatrix} \sigma_{xx} & \sigma_{xy} & \sigma_{xz} \\ \sigma_{yx} & \sigma_{yy} & \sigma_{yz} \\ \sigma_{zx} & \sigma_{zy} & \sigma_{zz} \end{pmatrix}. \tag{2.108}$$

From (2.106), the expression for the conductivity tensor is

$$\begin{aligned} \sigma_{ij} &= -\frac{e^2}{\Omega} \sum_{\vec{k}} \left[v_i(\vec{k}) v_j(\vec{k}) \tau(\vec{k}) \frac{\partial f_0(\vec{k})}{\partial E(\vec{k})} \right] \\ &= -\frac{e^2}{\Omega} \frac{\Omega}{4\pi^3} \int d^3\vec{k} \left[v_i(\vec{k}) v_j(\vec{k}) \tau(\vec{k}) \frac{\partial f_0(\vec{k})}{\partial E(\vec{k})} \right], \end{aligned} \tag{2.109}$$

where once again we have converted the summation to an integral using the density of states.

We will now make two assumptions:

- The energy dispersion relation for the electrons is parabolic, i.e.,

$$
\begin{aligned}
E(k) &= \frac{\hbar^2 k^2}{2m^*}, \quad \texttt{so that} \\
\vec{v}_i(\vec{k}) &= \frac{1}{\hbar}\frac{\partial E(k)}{\partial k_i} = \frac{\hbar k_i}{m^*} \\
\vec{v}_j(\vec{k}) &= \frac{1}{\hbar}\frac{\partial E(k)}{\partial k_j} = \frac{\hbar \vec{k}_j}{m^*}.
\end{aligned}
\tag{2.110}
$$

- The carrier population in nondegenerate so that Fermi–Dirac statistics can be approximated by Boltzmann statistics. Therefore,

$$
\begin{aligned}
f_0(\vec{k}) &= \mathrm{e}^{\frac{\mu_F - E(\vec{k}) - E_{c0}}{k_B T}} \\
\frac{\partial f_0(\vec{k})}{\partial E(\vec{k})} &= -\frac{1}{k_B T}\mathrm{e}^{\frac{\mu_F - E(\vec{k}) - E_{c0}}{k_B T}}.
\end{aligned}
\tag{2.111}
$$

Substituting these results in (2.109), we get

$$
\begin{aligned}
\sigma_{ij} &= \frac{e^2}{4\pi^3 k_B T}\int_0^\infty \int_0^{2\pi} \int_{-1}^{1} \mathrm{d}k\mathrm{d}\phi\mathrm{d}(\cos\theta) k^2 \left(\frac{\hbar}{m^*}\right)^2 k_i k_j \tau(\vec{k})\mathrm{e}^{\frac{\mu_F - E_{c0} - \frac{\hbar^2 k^2}{2m^*}}{k_B T}} \\
&= \frac{e^2 \hbar^2}{m^{*2} 4\pi^3 k_B T}\int_0^\infty \int_0^{2\pi} \int_{-1}^{1} \mathrm{d}k\mathrm{d}\phi\mathrm{d}(\cos\theta) k^4 \frac{k_i k_j}{k^2}\tau(\vec{k})\mathrm{e}^{\frac{\mu_F - E_{c0} - \frac{\hbar^2 k^2}{2m^*}}{k_B T}} \\
&= \frac{e^2 \hbar^2}{m^{*2} 4\pi^3 k_B T}\mathrm{e}^{\frac{\mu_F - E_{c0}}{k_B T}}\int_0^\infty \int_0^{2\pi} \int_{-1}^{1} \mathrm{d}k\mathrm{d}\phi\mathrm{d}(\cos\theta) k^4 \frac{k_i k_j}{k^2}\tau(\vec{k})\mathrm{e}^{\frac{-\frac{\hbar^2 k^2}{2m^*}}{k_B T}},
\end{aligned}
\tag{2.112}
$$

where we have used the fact that $E(\vec{k}) = E_{c0} + \frac{\hbar^2 k^2}{2m^*}$.

Next, let us assume a specific functional form for $\tau(\vec{k})$, which is valid for many different scattering mechanisms:

$$
\tau(\vec{k}) = \tau_0 \left(\frac{\hbar^2 k^2}{2m^* k_B T}\right)^s. \tag{2.113}
$$

Substituting this form in the last equation, we get

$$
\begin{aligned}
\sigma_{ij} &= \frac{e^2 \hbar^2}{m^{*2} 4\pi^3 k_B T}\mathrm{e}^{\frac{\mu_F - E_{c0}}{k_B T}}\tau_0 \\
&\times \int_0^\infty \int_0^{2\pi} \int_{-1}^{1} \mathrm{d}k\mathrm{d}\phi\mathrm{d}(\cos\theta) k^4 \frac{k_i k_j}{k^2}\left(\frac{\hbar^2 k^2}{2m^* k_B T}\right)^s \mathrm{e}^{\frac{-\frac{\hbar^2 k^2}{2m^*}}{k_B T}}.
\end{aligned}
\tag{2.114}
$$

Let us now define a new variable β such that

$$\beta = \frac{\hbar^2 k^2}{2m^* k_\mathrm{B} T}. \tag{2.115}$$

Therefore,

$$\begin{aligned} k &= \frac{\sqrt{2m^* k_\mathrm{B} T}}{\hbar}\sqrt{\beta} \\ \mathrm{d}k &= \frac{1}{2}\frac{\sqrt{2m^* k_\mathrm{B} T}}{\hbar}\frac{1}{\sqrt{\beta}}\mathrm{d}\beta. \end{aligned} \tag{2.116}$$

The expression for the tensor conductivity can then be written as

$$\begin{aligned} \sigma_{ij} &= \frac{e^2\hbar^2}{m^{*2}4\pi^3 k_\mathrm{B} T}\mathrm{e}^{\frac{\mu_\mathrm{F}-E_\mathrm{c0}}{k_\mathrm{B}T}}\tau_0 \\ &\times \int_0^\infty \int_0^{2\pi} \int_{-1}^{1} \mathrm{d}\phi \mathrm{d}(\cos\theta) \left(\frac{2m^* k_\mathrm{B} T}{\hbar^2}\right)^{5/2} \frac{k_i k_j}{k^2}\frac{1}{2}\beta^{(3/2+s)}\mathrm{e}^{-\beta}\mathrm{d}\beta \\ &= \frac{e^2}{4\pi^3 m^*}\left(\frac{2m^* k_\mathrm{B} T}{\hbar^2}\right)^{3/2} \mathrm{e}^{\frac{\mu_\mathrm{F}-E_\mathrm{c0}}{k_\mathrm{B}T}}\tau_0 \\ &\times \int_0^\infty \int_0^{2\pi} \int_{-1}^{1} \mathrm{d}\phi \mathrm{d}(\cos\theta)\frac{k_i k_j}{k^2}\beta^{(3/2+s)}\mathrm{e}^{-\beta}\mathrm{d}\beta. \end{aligned} \tag{2.117}$$

Now, referring to Fig. 2.4, which shows the spherical coordinate frame, we see that

$$\begin{aligned} k_x &= k\sin\theta\cos\phi \\ k_y &= k\sin\theta\sin\phi \\ k_z &= k\cos\theta, \end{aligned} \tag{2.118}$$

where θ and ϕ are the polar and azimuthal angles, respectively.

It is obvious that the ratio $k_i k_j/k^2$ depends only on θ and ϕ and not on k. Therefore, we can write the conductivity tensor as

$$\begin{aligned} \sigma_{ij} &= \frac{e^2}{4\pi^3 m^*}\left(\frac{2m^* k_\mathrm{B} T}{\hbar^2}\right)^{3/2} \mathrm{e}^{\frac{\mu_\mathrm{F}-E_\mathrm{c0}}{k_\mathrm{B}T}}\tau_0 \\ &\times \int_0^\infty \beta^{(3/2+s)}\mathrm{e}^{-\beta}\mathrm{d}\beta \int_0^{2\pi}\int_{-1}^{1} \mathrm{d}\phi\mathrm{d}(\cos\theta)\frac{k_i k_j}{k^2}. \end{aligned} \tag{2.119}$$

It is easy to see from (2.118) that

$$\int_0^{2\pi}\int_{-1}^{1} \mathrm{d}\phi \mathrm{d}(\cos\theta)\frac{k_i k_j}{k^2} = \frac{4\pi}{3}\delta_{i,j}, \tag{2.120}$$

where $\delta_{i,j}$ is the Kronecker delta. Moreover,

$$\int_0^{\infty} \beta^{(3/2+s)}\mathrm{e}^{-\beta}\mathrm{d}\beta = \Gamma\left(\frac{5}{2}+s\right), \tag{2.121}$$

where Γ is the gamma function.

Therefore, the expression for the conductivity tensor can be written as

$$\begin{aligned}\sigma_{ij} &= \frac{e^2}{4\pi^3 m^*}\left(\frac{2m^* k_\mathrm{B}T}{\hbar^2}\right)^{3/2} \mathrm{e}^{\frac{\mu_\mathrm{F}-E_{\mathrm{c}0}}{k_\mathrm{B}T}}\tau_0\Gamma\left(\frac{5}{2}+s\right)\frac{4\pi}{3}\delta_{i,j} \\ &= \frac{e^2}{\pi^{3/2}m^*}\left[2\left(\frac{2\pi m^* k_\mathrm{B}T}{h^2}\right)^{3/2} \mathrm{e}^{\frac{\mu_\mathrm{F}-E_{\mathrm{c}0}}{k_\mathrm{B}T}}\right]\tau_0\Gamma\left(\frac{5}{2}+s\right)\frac{4\pi}{3}\delta_{i,j}. \end{aligned} \tag{2.122}$$

In the above equation, the term within the square brackets can be written as $N_\mathrm{c}\mathrm{e}^{\frac{\mu_\mathrm{F}-E_{\mathrm{c}0}}{k_\mathrm{B}T}}$ where N_c is the effective density of states in the conduction band [10]. Since the electron concentration is given by

$$n = N_\mathrm{c}\mathrm{e}^{\frac{\mu_\mathrm{F}-E_{\mathrm{c}0}}{k_\mathrm{B}T}}, \tag{2.123}$$

we can write the conductivity tensor as

$$\sigma_{ij} = \frac{e^2}{\pi^{3/2}m^*}n\tau_0\Gamma\left(\frac{5}{2}+s\right)\frac{4\pi}{3}\delta_{i,j} \tag{2.124}$$

Finally, noting that $\Gamma(5/2) = (3/4)\sqrt{\pi}$, we can write the final expression for the conductivity tensor as

$$\sigma_{ij} = ne\frac{e\tau_0}{m^*}\frac{\Gamma\left(\frac{5}{2}+s\right)}{\Gamma\left(\frac{5}{2}\right)}\delta_{i,j} = ne\mu\delta_{i,j}, \tag{2.125}$$

where the mobility μ is given by

$$\mu = \frac{e\tau_0}{m^*}\frac{\Gamma\left(\frac{5}{2}+s\right)}{\Gamma\left(\frac{5}{2}\right)}. \tag{2.126}$$

If we write an effective momentum relaxation time as

$$\langle \tau_{\mathrm{m}} \rangle = \tau_0 \frac{\Gamma\left(\frac{5}{2} + s\right)}{\Gamma\left(\frac{5}{2}\right)}, \tag{2.127}$$

then $\mu = e \langle \tau_{\mathrm{m}} \rangle / m^*$, which is consistent with the result in Chap. 1 and the Drude model.

The above expression shows that the conductivity tensor is "diagonal" and all the diagonal terms are equal, i.e.,

$$\sigma_{ij} = \begin{pmatrix} \sigma & 0 & 0 \\ 0 & \sigma & 0 \\ 0 & 0 & \sigma \end{pmatrix}, \tag{2.128}$$

where $\sigma = ne\mu$ and the expression for μ is given by (2.126).

2.11 The Monte Carlo Simulation Method for Evaluating the Distribution Function

The most popular method for evaluating the distribution function from the BTE is the MC simulation method [7, 11]. In this method, the BTE is not directly solved, but the distribution function $f(\vec{r}, \vec{k}, t)$ is evaluated by simulating electron trajectories in the presence of an external electric field and collisions. This is achieved by generating random numbers. The effect of the external electric field on the trajectories is deterministic and found from Newton's second law, but the effect of collisions is probabilistic.

In MC simulation, random numbers uniformly distributed between 0 and 1 are used to determine:

1. The time between successive collisions, or the so-called "free-flight time"
2. The nature of the collision at the end of the free-flight time
3. The trajectory and the electron state after the collision

There are two types of MC simulation: single particle (where the trajectories of a single electron are simulated) and multi-particle (where the trajectories of many particles are simulated). Obviously, the latter is more time consuming and computationally demanding, but its advantage is that it can yield the time-dependent distribution function $f(\vec{r}, \vec{k}, t)$, while the former can yield only the steady-state distribution function $f(\vec{r}, \vec{k})$. Here, we describe the multi-particle method.

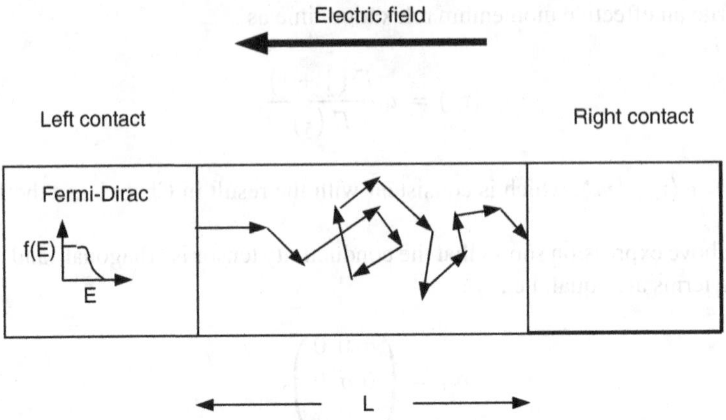

Fig. 2.5 In Monte Carlo simulation, an electron whose energy is picked from the Fermi–Dirac distribution in the left contact, is injected into the device. The injected electron is accelerated by the electric field during the free flight and then suffers a collision. The free flight duration, the nature of the collision, and the state of the wavevector after collision are all determined by random numbers. After executing random walk motion (where the electron is accelerated during free flight and the wavevector changes randomly after collision), the electron finally reaches the right contact where it is immediately absorbed and exits the device. When this happens, a new electron is injected at the left contact and the process is repeated. Statistics of electron variables (e.g., wavevector, energy, velocity) are collected at fixed time intervals Δt. After the trajectories of a sufficient number of electrons have been simulated, the gathered statistics is used to compute the electron distribution function

2.11.1 Initial Conditions in the Simulation

In the multi-particle simulation, the first electron is injected into the device from the left contact as indicated in Fig. 2.5. Electrons in the left contact are assumed to be in local thermodynamic equilibrium, i.e., the distribution function in the left contact is Fermi–Dirac distribution characterized by the local Fermi level:

$$f_{\text{left-contact}}(\vec{r},\vec{k},t) = \frac{1}{e^{\frac{E_{\text{cl}}+\epsilon(k)-\mu_{\text{Fl}}}{k_{\text{B}}T}}+1}, \tag{2.129}$$

where E_{cl} is the conduction band edge energy in the device at the interface with the left contact and μ_{Fl} is the Fermi level in the left contact. The entering electron's kinetic energy $\epsilon(k)$ is picked from this distribution. The kinetic energy with which the electron enters the device is found from the conservation of energy:

$$\epsilon_{\text{entering}} = \epsilon(k) + E_{\text{cl}} - E_{\text{c}}^{\text{valley}}, \tag{2.130}$$

where $E_{\text{c}}^{\text{valley}}$ is the energy at the bottom of a particular valley in the conduction band of the device. Thus, $E_{\text{c}}^{\text{valley}} - E_{\text{cl}}$ is the conduction band offset between the valley and the left contact. Semiconductors typically have multiple valleys in the conduction band and their bottoms have different energies.[6] Let us say that there are m valleys for which the energy $\epsilon_{\text{entering}}$ turns out to be positive. Then all these valleys can be populated and as a first level of approximation, we can assume that there is equal probability of the electron going into any of these valleys. Once an electron enters a valley, it need not stay there forever. It can later transfer to a different valley by suffering intervalley scattering.

The magnitude of the wavevector k_0 with which the electron enters the device is found from the dispersion relation $\epsilon_{\text{entering}} = \epsilon(k_0)$ for the initial valley. This relation may not be parabolic, but must be specified so that k_0 can be determined. A frequently used dispersion relation is

$$\frac{\hbar^2 k^2}{2m^*_{\text{valley}}} = \epsilon(1 + \alpha_{\text{valley}}\epsilon), \tag{2.131}$$

where m^*_{valley} is the effective mass and α_{valley} is the so-called nonparabolicity factor in the conduction band valley occupied by the electron.

Once the magnitude of k_0 is determined as above, the azimuthal angle is chosen as uniformly distributed between 0 and π (not 2π since electrons that have a negative component of the wavevector along the direction of current flow do not enter the device to begin with) and the polar angle is also chosen as uniformly distributed between 0 and π. If we call these angles ϕ_0 and θ_0, respectively, then

$$\begin{aligned} \phi_0 &= \pi r_1 \\ \theta_0 &= \pi r_2. \end{aligned} \tag{2.132}$$

where r_1 and r_2 are two random numbers uniformly distributed in the interval [0, 1].

2.11.2 Determination of the Duration of Free Flight

The next task is to determine the duration of free flight, i.e., the time to the next collision. Let the probability that the free flight lasts for a time between t and $t + \mathrm{d}t$ be $P(t)\mathrm{d}t$. The simulator generates a third random number r_3 uniformly distributed in the interval [0, 1]. If collisions are true random events, then we can relate r_3 to the free-flight time t_c as

$$r_3 = \frac{\int_0^{t_c} P(t)\mathrm{d}t}{\int_0^{\infty} P(t)\mathrm{d}t}. \tag{2.133}$$

---

[6] GaAs, Si, and Ge all have three conduction band valleys known as the Γ, L, and X valleys, all of which can be occupied by electrons at sufficiently high electric fields. The lowest valley in GaAs is the Γ, followed by L and X. In Si, the ordering is X, L, and Γ (in ascending order) and in Ge, the ordering in L, X, and Γ.

Let $S(k(t))$ be the scattering probability per unit time (or the scattering rate) for an electron with wavevector $k(t)$ at time t.[7] Then the number of electrons suffering a collision between time t and $t + \mathrm{d}t$ is $n(t)S(k(t))\mathrm{d}t$ where $n(t)$ is the number of electrons that have not suffered a prior collision at time t. Since the number $n(t)$ is reduced by the number that suffered a collision between time t and $t + \mathrm{d}t$, we get

$$-\,\mathrm{d}n(t) = n(t)Sk(t))\mathrm{d}t, \tag{2.134}$$

or

$$\frac{\mathrm{d}n(t)}{n(t)} = -S(k(t))\mathrm{d}t. \tag{2.135}$$

Integrating,

$$ln\left(\frac{n(t)}{n(0)}\right) = -\int_0^t S(k(t'))\mathrm{d}t', \tag{2.136}$$

which yields

$$n(t) = n(0)\mathrm{e}^{-\int_0^t S(k(t'))\mathrm{d}t'}. \tag{2.137}$$

Since we called $n(t)$ the number of electrons that have not suffered a prior collision up to time t, it is clearly the number that have not suffered a collision between time $t = 0$ and $t = t$. The probability that an electron which suffered a collision at time $t = 0$ has not suffered another one till time t is $n(t)/n(0)$, which according to the last equation is

$$\frac{n(t)}{n(0)} = \mathrm{e}^{-\int_0^t S(k(t'))\mathrm{d}t'}. \tag{2.138}$$

Let us now calculate the probability that an electron which suffered a collision at time $t = 0$ suffered the next one precisely between time t and $t + \mathrm{d}t$. This is the product of the probability of not suffering a collision between time $t = 0$ and $t = t$ and the probability of suffering a collision between time t and $t + \mathrm{d}t$. This quantity is also the probability that the free flight lasts for a duration between t and $t + \mathrm{d}t$, which is $P(t)\mathrm{d}t$. Therefore,

$$P(t)\mathrm{d}t = S(k(t))\mathrm{e}^{-\int_0^t S(k(t'))\mathrm{d}t'}\mathrm{d}t. \tag{2.139}$$

As a result, (2.133) can be recast as

$$r_3 = \frac{\int_0^{t_c} P(t)\mathrm{d}t}{\int_0^{\infty} P(t)\mathrm{d}t} = \frac{\int_0^{t_c} S(k(t))\mathrm{e}^{-\int_0^t S(k(t'))\mathrm{d}t'}\mathrm{d}t}{\int_0^{\infty} S(k(t))\mathrm{e}^{-\int_0^t S(k(t'))\mathrm{d}t'}\mathrm{d}t}. \tag{2.140}$$

[7] This rate depends only on the magnitude of $\vec{k}$ and is independent of the direction.

Once the random number r_3 is generated by the simulator, the above equation is solved to find the free flight duration t_c. Note that since the wavevector changes with time because of the external electric field and collisions, the complicated double integral on the right-hand-side has to be evaluated dynamically during the simulation. In order to evaluate the double integral, we need to know how to find $S(k(t))$. It is the total scattering rate at any instant t and is given by

$$S(k(t)) = \sum_{i,\vec{k}'} S_i(\vec{k}(t), \vec{k}'), \tag{2.141}$$

where $S_i(\vec{k}(t), \vec{k}')$ is the scattering rate for an electron in the initial wavevector state $\vec{k}$ scattering into the final wavevector state $\vec{k}'$ via the i-th type of scattering mechanism.

Solving (2.140) is too complicated and computationally demanding. Rees developed a method of generating the free flight time more easily [12]. He defined a constant Γ that is larger than the largest value of $S(k(t))$ to be encountered at any time during the simulation.

$$\Gamma > S(k(t))|_{\text{maximum}}. \tag{2.142}$$

Since $S(k(t))$ tends to increase monotonically with increasing electron energy (or, equivalently, the magnitude of the electron wavevector), we need to estimate what is the maximum possible energy an electron can acquire during the simulation. Once we estimate that, we can determine $S(k(t))|_{\text{maximum}}$ and pick an appropriate value for Γ that satisfies the last equation. Rees showed that the free flight time is then given by

$$t_c = -\frac{1}{\Gamma} ln(r_3), \tag{2.143}$$

provided we introduce a new fictitious scattering mechanism called "self-scattering" that has a rate of $\Gamma - S(k(t))$ at any time t. If a self-scattering event takes place, then the wavevector of the electron does not change at all, i.e.,

$$\vec{k}_{\text{final}} = \vec{k}_{\text{initial}} \quad (\texttt{for self-scattering}). \tag{2.144}$$

The method due to Rees considerably reduces the computational burden in MC simulation, but it is unfortunately not applicable in quasi one-dimensional structures (quantum wires). Many scattering mechanisms (e.g., electron scattering due to acoustic phonons) are proportional to the density of final states, and the latter quantity diverges at subband edges in quasi one-dimensional systems. Therefore, Γ is infinity in quasi one-dimensional structures, which invalidates the Rees procedure of self-scattering. That is why MC simulation of carrier transport in quantum wires is especially demanding because there one must solve (2.140) rigorously to find the free flight time. This is usually done by rejection techniques which is often quite time consuming.

2.11.3 Updating the Electron Trajectory at the End of Free Flight and Beginning of the Collision Event

The electron trajectories, i.e., the different electron variables such as wavevector, position $\vec{r}$, energy E, velocity $\vec{v}$, etc. are updated at the end of the free flight following Newton's second law:

$$-e\vec{\mathcal{E}} = \frac{\mathrm{d}(\hbar\vec{k})}{\mathrm{d}t}, \tag{2.145}$$

which yields

$$\begin{aligned}
\vec{k}(t+t_c) &= \vec{k}(t) - \frac{e}{\hbar}\int_t^{t+t_c} \vec{\mathcal{E}}(t')\mathrm{d}t' \\
E(t+t_c) &= E(\vec{k}(t+t_c)) \\
\vec{v}(t+t_c) &= \vec{v}(\vec{k}(t+t_c)) \\
\vec{r}(t+t_c) &= \vec{r}(t) - \frac{E(t+t_c)-E(t)}{e\vec{\mathcal{E}}}
\end{aligned} \tag{2.146}$$

The updated wavevector, i.e., the wavevector at time $t + t_c$, is the wavevector at the end of free flight and at the beginning of the collision event that ends the free flight.

2.11.4 Updating the Electron State After the Collision Event

In order to determine the electron wavevector (and all other electron variables such as energy and velocity) after the collision event, we need to first determine what type of scattering took place. Let the total number of possible scattering types (including "self-scattering") that the electron can encounter be m. We generate a random number r_4 uniformly distributed between 0 and 1. If

$$\frac{\sum_{i=1}^{n-1} S_i(k(t))}{\sum_{i=1}^{m} S_i(k(t))} \le r_4 \le \frac{\sum_{i=1}^{n} S_i(k(t))}{\sum_{i=1}^{m} S_i(k(t))}, \tag{2.147}$$

then we determine that the n-th type of scattering took place. Basically, we partition the domain of scattering rate (normalized to Γ) into different bins for different types of scattering mechanisms, such that the width of each bin is proportional to the scattering rate $S_i(k(t))$ for that type. This is schematically shown in Fig. 2.6. If the generated random number r_4 falls in the n-th bin, then we determine that the n-th mechanism was operative. This modality guarantees that the frequency of the n-th scattering mechanism is indeed proportional to scattering rate associated with that mechanism, as it should be. Note that since the scattering rate depends on the

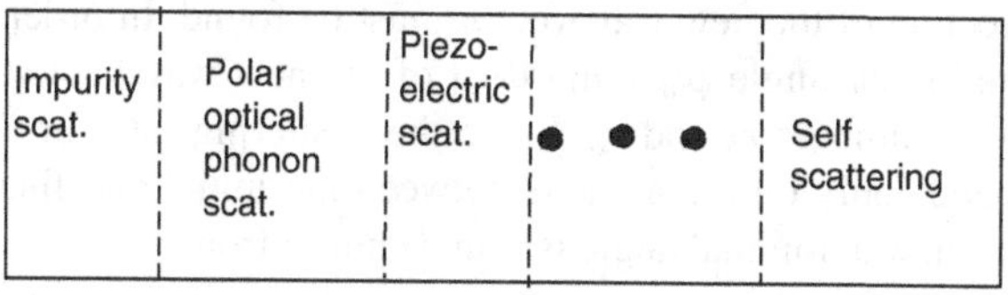

Fig. 2.6 The domain of the scattering rate is broken up into m bins, where m is the total number of scattering mechanisms, including self-scattering. The width of the i-th bin is $S_i(\vec{k}(t))/\Gamma$; hence, it is always proportional to the probability of the i-th scattering mechanism. If the generated random number r_4 (uniformly distributed between 0 and 1) falls in the n-th bin, then we determine that the n-th type of scattering occurred. Note that this method guarantees that the frequency of every scattering mechanism is proportional to its scattering probability so that more probable events occur more frequently, as they should. Since the width of any bin changes with time t, this partitioning is dynamic and has to be updated whenever a scattering event occurs during the simulation

magnitude of the wavevector, which changes with time, this partitioning is *dynamic*. That is why, MC simulation is usually computationally burdensome.

Once we have determined the type of scattering mechanism, we need to update the electron wavevector after scattering. That is, we need to find the magnitude and direction of the new wavevector. If the scattering type that occurred was elastic (such as impurity scattering), then the electron energy (and hence the magnitude of the wavevector) is not changed by the collision, so that we have

$$|k_{\text{final}}| = |k_{\text{initial}}| \tag{2.148}$$

If the scattering type was inelastic (such as phonon emission or absorption) but no intervalley transfer took place, then

$$E(k_{\text{final}}) = E(k_{\text{initial}}) \pm \hbar\omega_{\text{phonon}}, \tag{2.149}$$

where $\hbar\omega_{\text{phonon}}$ is the phonon energy. The plus sign is for absorption and the minus sign for emission of a phonon during the scattering process. Once the final energy is determined, the magnitude of the final wavevector is found from the energy dispersion relation in (2.131). If the scattering type involved a valley transfer (i.e., the electron makes a transition from one conduction band valley to another), then

$$E(k_{\text{final}}) = E(k_{\text{initial}}) \pm \hbar\omega_{\text{int-phonon}} \pm \Delta, \tag{2.150}$$

where Δ is the energy separation between the valley bottoms. The plus sign (before Δ) applies when the final valley's bottom is lower in energy than the initial valley's bottom and the minus sign applies when it is higher in energy. Such transfers are mediated by intervalley phonons with large wavevectors whose energy is $\hbar\omega_{\text{int-phonon}}$ and once again the plus sign applies when a phonon is absorbed and the minus sign applies when a phonon is emitted. Once the new energy is found, the magnitude of the new wavevector is determined from the dispersion relation in (2.131).

Next, the direction of the new wavevector must be found. In order to determine the azimuthal and polar angle ϕ_{final} and θ_{final} of the new wavevector, we generate two more random numbers r_5 and r_6. Since the scattering rate for any scattering mechanism depends only on the angle α between the initial and final wavevector (see Fig. 2.4), the new azimuthal angle is found simply from

$$\phi_{\text{final}} = 2\pi r_5. \tag{2.151}$$

The new polar angle is found from

$$\cos\theta_{\text{final}} = \cos\alpha\cos\theta_{\text{initial}} - \sin\alpha\sin\theta_{\text{initial}}\cos\phi_{\text{initial}}, \tag{2.152}$$

where the only unknown α is found by solving the equation

$$r_6 = \frac{\int_{-1}^{\cos\alpha} S(\vec{k}_{\text{initial}}, \vec{k}_{\text{final}})\mathrm{d}(\cos\alpha)}{\int_{-1}^{1} S(\vec{k}_{\text{initial}}, \vec{k}_{\text{final}})\mathrm{d}(\cos\alpha)}. \tag{2.153}$$

Once the new polar and azimuthal angles are found, the new wavevector is completely determined.

2.11.5 *Terminating a Trajectory*

The electron trajectories are sampled at fixed time intervals of Δt, where $\Delta t \ll t_c$. At every sampling instant, we gather information about the position, wavevector, velocity, energy, etc. of the sampled electron. If at any sampling instant, the position of the electron is found to be outside the device, i.e., the electron has reached or exited at the right contact, we terminate the trajectory and inject the next electron from the left contact. We repeat this process until we have exhausted the pool of electrons whose trajectories we intended to simulate. This pool typically will have several tens of thousands of electrons. The larger the pool, the more reliable will the statistics be.

2.11.6 *Collecting Statistics and Evaluating the Distribution Function*

From the sampled data obtained during the s-th sampling, we produce histograms of the electron's wavevector components at any location $\vec{r}$ at the time $t = t_s = s\Delta t$. We specify the location by partitioning each of the dimensions—the length, width and thickness—of the device into a number of bins of equal size. Let the location

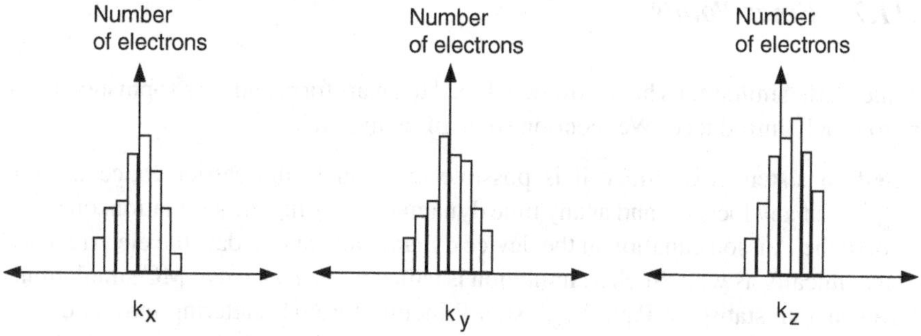

Fig. 2.7 Histograms of the electron's wavevector components at any location $\vec{r}(t_{\text{av}})$, where t_{av} is the average time for the electrons to reach that location after being injected from the left contact

designated by $\vec{r}_{i,j,k}$ denote a position within the i-th bin along the length, j-th bin along the thickness and k-th bin along the width. The histogram is produced from all the electrons that happen to be located within that bin at the sampling instant $t_s = s\Delta t$. This histogram may look like the data in Fig. 2.7, and if the number of electrons plotted along the vertical axis is normalized to the total number of electrons in the pool, then it is the distribution function $f(\vec{r}_{i,j,k}, \vec{k}, t_s)$. We can also plot histograms of the energy distribution $f(\vec{r}_{i,j,k}, E, t_s)$, or velocity distribution $f(\vec{r}_{i,j,k}, \vec{v}, t_s)$, etc. since we gather statistics about energy and velocity as well. From these distributions, we can calculate the ensemble averaged velocity or energy or electron number (or concentration) at any location, which yields the transport variables $n(\vec{r}, t)$, $\vec{J}_n(\vec{r}, t)$, etc. It is also possible to evaluate the transport parameters $\mu_{ij}(\vec{r}, t)$ and $D_{ij}(\vec{r}, t)$. The method for extracting these parameters from MC simulation has been described in [13]. Those quantities were shown to be tensors and given by:

$$\mu_{ij}(\vec{r}, t) = \frac{\langle v_i \rangle(\vec{r}, t)}{\mathcal{E}_j(\vec{r}, t) + \frac{2}{en(\vec{r},t)} \frac{\partial \langle n u_{ij} \rangle(\vec{r},t)}{\partial x_j}}$$

$$D_{ij}(\vec{r}, t) = \mu_{ik}(\vec{r}, t) \frac{2 \langle u_{kj} \rangle(\vec{r}, t)}{e}, \tag{2.154}$$

where $\langle v_i \rangle(\vec{r}, t)$ is the ensemble averaged electron velocity component in the i-th direction, $\mathcal{E}_j(\vec{r}, t)$ is the electric field in the j-th direction, and $\langle u_{ij} \rangle(\vec{r}, t)$ is the ensemble averaged tensor energy $u_{ij} = (1/2)m^* v_i v_j$—all evaluated at location $\vec{r}$ and at time t. The ensemble averaging is carried out over all the electrons at location $\vec{r} = \vec{r}_{i,j,k}$ at time $t = s\Delta t$. The quantities in the right-hand-side of (2.154) are found directly from the results of MC simulation. This provides a method for computing the mobility and diffusion coefficient at any arbitrary location $\vec{r}$ and at any arbitrary time t in a device using MC simulation.

2.11.7 Finer Points

Monte Carlo simulations have now developed to an art form and new sophistications are routinely introduced. We mention some of them here:

- Self-consistent MC: Since it is possible to evaluate the carrier concentration $n(\vec{r}, t)$ at any location and at any time dynamically during the simulation, one can solve the Poisson equation in the device dynamically and update the electric field dynamically as well. Such a simulation is called a *self-consistent* MC simulation.
- Degenerate statistics: Pauli Exclusion Principle forbids scattering into an occupied state. Once a scattering event occurs at a location $\vec{r}$ and at time t, and the final wavevector state $\vec{k}_{\text{final}}$ is determined, one generates a random number r, uniformly distributed in the interval [0, 1], and compares it with the distribution function $f(\vec{r}, \vec{k}_{\text{final}}, t)$ at the location $\vec{r}$, at time t. This distribution is dynamically computed from the histograms sampled. If $r \leq f(\vec{r}, \vec{k}_{\text{final}}, t)$, then a decision is made that the scattering event is blocked by Pauli Exclusion Principle and the event is ignored. This strategy accounts for the Pauli Exclusion Principle and is particularly important when the carrier population is degenerate so that the probability of a state being occupied is high.
- Simulations in quantum wires: The density of states of electrons has singularities at the subband edge energies in quasi one-dimensional structures (or "quantum wires") where the electron motion is constrained in two of the three dimensions. This is discussed in Chap. 4. Scattering rates for many scattering mechanisms are proportional to the density of the final state that the electron scatters into, which could be infinity in a quantum wire, if the final state is at a subband bottom. Therefore, the Rees parameter Γ will be infinity and the Rees self-scattering technique cannot be applied. In that case, (2.140) is solved numerically using rejection techniques. This is time consuming. As a result, MC simulation in quasi one-dimensional structures is usually computationally burdensome.
- Full-band MC: The energy-dispersion relation of an electron is required in MC simulations to relate the electron energy and velocity to the wavevector (see (2.131)) and also for calculating the scattering rates $S(k(t))$ since it is proportional to the density of final states. In the most sophisticated simulators, the band structure of the material (in which electron transport is taking place) is solved to find the accurate energy dispersion relation to replace the empirical relation in (2.131). The density of states as a function of energy is then found from this relation. Such an approach is called *full-band MC* [14].
- Hot phonon effects: The rates $S_i(\vec{k}, \vec{k}')$ of electron scattering events that involve phonon absorption or stimulated emission are proportional to the phonon occupation probability (see Chap. 5). If the phonons are in thermodynamic equilibrium, then the occupation probability (i.e., the probability of finding a phonon of a particular energy) is given by Bose–Einstein statistics. However, if numerous phonons are emitted or absorbed, then the phonon population may be driven out of equilibrium so that the occupation probability is no longer determined by Bose–Einstein statistics. Phonons that are out of thermodynamic equilibrium

are called "hot phonons." Hot phonon effects are incorporated by keeping track of the phonons that are emitted and absorbed during the simulation, and then filling up the phase space for phonons with these phonons. This will change the phonon occupation probability and alter the scattering rates dynamically.

- In MC simulations, scattering events are considered instantaneous in both time and space since that is the fundamental assumption in BTE. However, in reality, scatterings take place over a finite duration. In very high electric fields, the electron can be accelerated substantially *while* it is scattering. This is known as *intracollisional field effect.* A related effect is *collision retardation* which inhibits scattering into certain states. These effects are beyond the scope of this introductory textbook, but detailed discussions can be found in many references such as [15–18]. Such effects can be accounted for in MC simulations as well using numerical rejection techniques (see, for example, [17, 18]).

2.12 Summary

In this chapter, we learned about Boltzmann transport. We learned that

1. The distribution function $f(\vec{r},\vec{k},t)$ is the probability of finding an electron at position $\vec{r}$, with wavevector $\vec{k}$ at an instant of time t. This quantity is positive indefinite.
2. Transport variables like carrier concentration, current density, etc. are different moments of the distribution function. Hence, these variables can be found directly if we can find the distribution function.
3. The distribution function is found by solving the BTE which describes its evolution in time, real space, and wavevector space. This equation is an equation of classical mechanics and cannot describe quantum-mechanical effects that arise when electrons suffer no "phase-breaking" inelastic scattering. It is also inconsistent with the Heisenberg Uncertainty Principle of quantum mechanics.
4. The generalized moment equations, or hydrodynamic balance equations, describe the evolution in time and space of any arbitrary moment of the distribution function, i.e., any arbitrary transport variable. However, in dealing with these equations in unipolar transport, we always end up with one more unknown transport variable than the number of equations. Therefore, we have to arbitrarily close the hierarchy of equations by making an ad-hoc assumption. This is the disadvantage of the hydrodynamic equation method. The advantage is that we can find any arbitrary transport variable without having to find the distribution function which is not easy to evaluate.
5. Transport parameters, such as mobility and diffusion coefficient are functionals of the distribution function and we can find them from the latter if we know the scattering rate $S(\vec{k},\vec{k}')$ associated with scattering of an electron from a wavevector state $\vec{k}$ to a wavevector state $\vec{k}'$. This rate is found from a quantum-mechanical prescription, e.g., Fermi's Golden Rule. This Rule is discussed in Chap. 5.

6. The two popular methods of solving the BTE are the (1) relaxation time approximation for steady-state, spatially homogeneous, quasi-equilibrium transport, and (2) MC simulation. The former is easy but has limited validity, while the latter is more difficult but has a much wider range of validity.
7. The relaxation time approximation is applicable only if the scattering mechanisms are (i) elastic, or (ii) isotropic and the carrier population is nondegenerate.
8. The relaxation time approximation yields Ohm's law and an analytical expression for the conductivity.
9. MC simulation simulates electron trajectories with the help of random numbers. These trajectories are sampled at fixed time intervals to gather statistics about electron variables such as concentration, velocity, energy, etc. as a function of wavevector. The statistics also yield the distribution function $f(\vec{r},\vec{k},t)$.

Problems

Problem 2.1. Consider a semiconductor like Silicon which has an ellipsoidal conduction band valley [10] so that the energy dispersion relation in the valley can be written as

$$E_c(k) = E_{c0} + \frac{\hbar^2 k_x^2}{2m_1} + \frac{\hbar^2 k_y^2}{2m_2} + \frac{\hbar^2 k_z^2}{2m_3}.$$

Note that the kinetic energy $E(k)$ of the electron is $E_c(k) - E_{c0}$. Assume that $\tau(\vec{k}) = \tau_0$.

Starting from the general expression for the low-field steady-state tensor conductivity in (2.109), show that for a nondegenerate electron gas

$$\sigma_{zz} = ne\frac{e\tau_0}{m_3}.$$

You may find the following integrals helpful:

$$\int_{-\infty}^{\infty} \mathrm{d}x \mathrm{e}^{-\alpha x^2} = \sqrt{\frac{\pi}{\alpha}}$$

$$\int_{-\infty}^{\infty} \mathrm{d}x x^2 \mathrm{e}^{-\alpha x^2} = \frac{1}{2\alpha}\sqrt{\frac{\pi}{\alpha}}$$

Remember that for a nondegenerate population,

$$n = 2\left(\frac{2\pi m_d k_B T}{h^2}\right)^{3/2} \mathrm{e}^{(\mu_F - E_{c0})/k_B T},$$

where the "density of states effective mass" is $m_d = (m_1 m_2 m_3)^{1/3}$.

Problem 2.2. The equilibrium electron concentration in a nondegenerate semiconductor maintained at a uniform temperature is [10]

$$n(\vec{r}) = N_c(\vec{r})e^{\frac{\mu_F - E_{c0}(\vec{r})}{k_B T}},$$

where E_{c0} is the conduction band edge, μ_F is the Fermi level, and N_c is the effective density of states in the conduction band. Under *quasi-equilibrium* steady-state condition when a small current flows, the above expression changes to

$$n(\vec{r}) = N_c(\vec{r})e^{\frac{F(\vec{r}) - E_{c0}(\vec{r})}{k_B T}},$$

where $F(\vec{r})$ is called the quasi Fermi level.

Using the fact that the electric field is related to the gradient of the conduction band edge according to

$$e\vec{\mathcal{E}}(\vec{r}) = \vec{\nabla}_r E_{c0}(\vec{r}),$$

show that the total steady-state electron current density in the drift–diffusion model can be written as

$$\vec{J}_n = n(\vec{r})\mu_n(\vec{r})\vec{\nabla}_r[F(\vec{r})]. \tag{2.155}$$

State any assumption that you make.

Problem 2.3. Equation (2.155) is approximately valid for a degenerate semiconductor as well. Using this equation show that under quasi-equilibrium conditions, the following relation holds for the spatial gradient of the quasi-Fermi level:

$$\vec{\nabla}_r F(\vec{r}) = \vec{\nabla}_r E_{c0}(\vec{r}) + \frac{k_B T \vec{\nabla}_r n(\vec{r})}{(1/\Omega)\sum_{\vec{k}} f_0(\vec{r},\vec{k})[1 - f_0(\vec{r},\vec{k})]}.$$

Use the above relation in (2.155) to derive the familiar steady-state current density equation

$$\vec{J}_n = n(\vec{r})\mu_n(\vec{r})\vec{\mathcal{E}}(\vec{r}) + eD_n(\vec{r})\vec{\nabla}_r n(\vec{r}).$$

where the relation between the diffusion coefficient and mobility for the degenerate semiconductor is

$$D_n(\vec{r}) = \frac{k_B T}{e}\mu_n(\vec{r})\frac{\sum_{\vec{k}} f_0(\vec{r},\vec{k})}{\sum_{\vec{k}} f_0(\vec{r},\vec{k})[1 - f_0(\vec{r},\vec{k})]}.$$

Note that this equation reduces to the Einstein relation for a nondegenerate semiconductor because in the latter, $f_0(\vec{r},\vec{k}) \ll 1$.

Problem 2.4. This problem and some of the ensuing ones in this chapter are inspired by [7].

Let us define two *real* quantities $\Phi(\vec{r},\vec{k},t)$ and $\Phi(\vec{r},\vec{k}',t)$ as follows:

$$\Delta f = f(\vec{r},\vec{k},t) - f_0(\vec{r},\vec{k},t) = -\Phi(\vec{r},\vec{k},t)\frac{\partial f_0(\vec{r},\vec{k},t)}{\partial E(\vec{k})}$$

$$\Delta f' = f(\vec{r},\vec{k}',t) - f_0(\vec{r},\vec{k}',t) = -\Phi(\vec{r},\vec{k}',t)\frac{\partial f_0(\vec{r},\vec{k}',t)}{\partial E(\vec{k}')},$$

where

$$f_0(\vec{r},\vec{k},t) = \frac{1}{\mathrm{e}^{\frac{E(\vec{k})+E_{c0}(\vec{r},t)-\mu_F}{k_B T}}+1}.$$

Show that under quasi-equilibrium conditions

$$\left.\frac{\mathrm{d}f(\vec{r},\vec{k},t)}{\mathrm{d}t}\right|_{\text{collisions}}$$
$$= \frac{1}{k_B T}\sum_{\vec{k}'} S(\vec{k}',\vec{k}) f_0(\vec{r},\vec{k}',t)[1 - f_0(\vec{r},\vec{k},t)]\{\Phi(\vec{r},\vec{k}',t) - \Phi(\vec{r},\vec{k},t)\}$$
$$= \frac{1}{k_B T}\sum_{\vec{k}'} S(\vec{k},\vec{k}') f_0(\vec{r},\vec{k},t)[1 - f_0(\vec{r},\vec{k}',t)]\{\Phi(\vec{r},\vec{k}',t) - \Phi(\vec{r},\vec{k},t)\} \tag{2.156}$$

Solution. By definition

$$\left.\frac{\mathrm{d}f(\vec{r},\vec{k},t)}{\mathrm{d}t}\right|_{\text{collisions}}$$
$$= \sum_{k'}\Big\{S(\vec{k}',\vec{k}) f(\vec{r},\vec{k}',t)[1 - f(\vec{r},\vec{k},t)]$$
$$- S(\vec{k},\vec{k}') f(\vec{r},\vec{k},t)[1 - f(\vec{r},\vec{k}',t)]\Big\}$$
$$= \sum_{k'}\Big\{S(\vec{k}',\vec{k})\,(f_0' + \Delta f')\,[1 - f_0 - \Delta f] - S(\vec{k},\vec{k}')(f_0 + \Delta f)[1 - f_0' - \Delta f']\Big\}$$
$$= \sum_{k'}\Big\{S(\vec{k}',\vec{k})[\Delta f'(1 - f_0) - \Delta f f_0'] - S(\vec{k},\vec{k}')[\Delta f(1 - f_0') - \Delta f' f_0]\Big\}, \tag{2.157}$$

where we have made use of the Principle of Detailed Balance and neglected second order terms involving deviation of the distribution function from equilibrium. Such deviations are small when quasi-equilibrium conditions prevail. We have also used the shorthand notation $f_0 = f_0(\vec{r},\vec{k},t)$ and $f_0' = f_0(\vec{r},\vec{k}',t)$.

Note that

$$\frac{\partial f_0}{\partial E(k)} = -\frac{1}{k_B T} f_0(1 - f_0)$$

$$\frac{\partial f_0'}{\partial E(k)} = -\frac{1}{k_B T} f_0'(1 - f_0')$$

Using the last two relations in the defining equations given in the problem statement, we get

$$\Delta f = \frac{1}{k_B T} \Phi f_0(1 - f_0)$$

$$\Delta f' = \frac{1}{k_B T} \Phi' f_0'(1 - f_0')$$

Using these relations in (2.157), we obtain

$$\left.\frac{\mathrm{d} f(\vec{r},\vec{k},t)}{\mathrm{d}t}\right|_{\text{collisions}}$$

$$= \sum_{k'} \left\{ S(\vec{k}',\vec{k}) \left[\frac{1}{k_B T} f_0'(1 - f_0')(1 - f_0)\Phi' - \frac{1}{k_B T} f_0(1 - f_0) f_0' \Phi \right] \right.$$

$$\left. -S(\vec{k},\vec{k}') \left[\frac{1}{k_B T} f_0(1 - f_0)(1 - f_0')\Phi - \frac{1}{k_B T} f_0' \left(1 - f_0'\right) f_0 \Phi' \right] \right\}$$

$$= \sum_{k'} \left\{ S(\vec{k},\vec{k}') \left[\frac{1}{k_B T} f_0(1 - f_0')^2 \Phi' - \frac{1}{k_B T} f_0^2 \left(1 - f_0'\right) \Phi \right] \right.$$

$$\left. -S(\vec{k},\vec{k}') \left[\frac{1}{k_B T} f_0(1 - f_0)(1 - f_0')\Phi - \frac{1}{k_B T} f_0'(1 - f_0') f_0 \Phi' \right] \right\}$$

using the Principle of Detailed Balance

$$= \frac{1}{k_B T} \sum_{k'} S(\vec{k},\vec{k}') f_0(1 - f_0') \left[(1 - f_0')\Phi' - f_0\Phi - (1 - f_0)\Phi + f_0'\Phi' \right]$$

$$= \frac{1}{k_B T} \sum_{k'} S(\vec{k},\vec{k}') f_0(1 - f_0')[\Phi' - \Phi]$$

$$= \frac{1}{k_B T} \sum_{k'} S(\vec{k},\vec{k}') f_0(\vec{r},\vec{k},t)[1 - f_0(\vec{r},\vec{k}',t)][\Phi(\vec{r},\vec{k}',t) - \Phi(\vec{r},\vec{k},t)]$$

$$= \frac{1}{k_B T} \sum_{k'} S(\vec{k}',\vec{k}) f_0(\vec{r},\vec{k}',t)[1 - f_0(\vec{r},\vec{k},t)][\Phi(\vec{r},\vec{k}',t) - \Phi(\vec{r},\vec{k},t)],$$

where we have used the Principle of Detailed Balance once again to arrive at the last equality. This proves the result in the problem statement.

Problem 2.5. Using (2.86) derived for the relaxation time approximation in the case of steady-state spatially-invariant transport, and remembering that $\left.\frac{\mathrm{d}f(\vec{k})}{\mathrm{d}t}\right|_{\text{collisions}} = -\Delta f(\vec{k})/\tau(\vec{k})$, rederive (2.156). This is a special case. Equation (2.156) is valid for space- and time-varying transport as well, since we showed that in Problem 2.4. Here, we show that it is certainly valid for the special case when transport is space- and time-invariant.

Problem 2.6. Define an operator L given by

$$\begin{aligned}
L\Phi(\vec{r},\vec{k},t) &= \frac{\mathrm{d}f(\vec{r},\vec{k},t)}{\mathrm{d}t}|_{\text{collisions}} \\
&= \frac{1}{k_{\mathrm{B}}T}\sum_{\vec{k}'} S(\vec{k},\vec{k}')f_0(\vec{r},\vec{k},t)\left[1 - f_0(\vec{r},\vec{k}',t)\right] \\
&\quad \times\{\Phi(\vec{r},\vec{k},t) - \Phi(\vec{r},\vec{k}',t)\} \\
&= \frac{\Omega}{4\pi^3 k_{\mathrm{B}}T}\int d^3\vec{k}' S(\vec{k},\vec{k}')f_0(\vec{r},\vec{k},t)[1 - f_0(\vec{r},\vec{k}',t)] \\
&\quad \times\{\Phi(\vec{r},\vec{k},t) - \Phi(\vec{r},\vec{k}',t)\}.
\end{aligned}$$

Show that this operator is Hermitian (or self-adjoint), i.e., for two arbitrary functions $A(\vec{r},\vec{k},t)$ and $B(\vec{r},\vec{k},t)$, the following relation is always satisfied:

$$\int d^3\vec{k} A(\vec{r},\vec{k},t)LB(\vec{r},\vec{k},t) = \int d^3\vec{k} B(\vec{r},\vec{k},t)LA(\vec{r},\vec{k},t).$$

Solution. Using the defining relation for the operator L, we get

$$\begin{aligned}
LB(\vec{r},\vec{k},t) &= \frac{\Omega}{4\pi^3 k_{\mathrm{B}}T}\int d^3\vec{k}' S(\vec{k},\vec{k}')f_0(\vec{r},\vec{k},t)[1 - f_0(\vec{r},\vec{k}',t)] \\
&\quad \times\{B(\vec{r},\vec{k},t) - B(\vec{r},\vec{k}',t)\}
\end{aligned}$$

Therefore,

$$\begin{aligned}
&\int d^3\vec{k} A(\vec{r},\vec{k},t)LB(\vec{r},\vec{k},t) \\
&= \frac{\Omega}{4\pi^3 k_{\mathrm{B}}T}\int d^3\vec{k} A(\vec{r},\vec{k},t)\int d^3\vec{k}' S(\vec{k},\vec{k}')f_0(\vec{r},\vec{k},t)[1 - f_0(\vec{r},\vec{k}',t)] \\
&\quad \times\{B(\vec{r},\vec{k},t) - B(\vec{r},\vec{k}',t)\} \\
&= \frac{\Omega}{4\pi^3 k_{\mathrm{B}}T}\int\int d^3\vec{k} d^3\vec{k}' A(\vec{r},\vec{k},t)S(\vec{k},\vec{k}')f_0(\vec{r},\vec{k},t)[1 - f_0(\vec{r},\vec{k}',t)] \\
&\quad \times\{B(\vec{r},\vec{k},t) - B(\vec{r},\vec{k}',t)\}
\end{aligned}$$

$$= \frac{\Omega}{4\pi^3 k_B T} \int\int d^3\vec{k} d^3\vec{k}' S(\vec{k},\vec{k}') f_0(\vec{r},\vec{k},t)[1 - f_0(\vec{r},\vec{k}',t)]$$
$$\times \{A(\vec{r},\vec{k},t)B(\vec{r},\vec{k},t) - A(\vec{r},\vec{k},t)B(\vec{r},\vec{k}',t)\}$$

Since $\vec{k}$ and $\vec{k}'$ are dummy variables of integration, we will interchange them in the *second term only* and obtain

$$\int d^3\vec{k} A(\vec{r},\vec{k},t) L B(\vec{r},\vec{k},t)$$
$$= \frac{\Omega}{4\pi^3 k_B T} \int\int d^3\vec{k} d^3\vec{k}' S(\vec{k},\vec{k}') f_0(\vec{r},\vec{k},t)[1 - f_0(\vec{r},\vec{k}',t)]$$
$$\times \{A(\vec{r},\vec{k},t)B(\vec{r},\vec{k},t) - A(\vec{r},\vec{k}',t)B(\vec{r},\vec{k},t)\}$$
$$= \frac{\Omega}{4\pi^3 k_B T} \int d^3\vec{k} B(\vec{r},\vec{k},t) \int d^3\vec{k}' S(\vec{k},\vec{k}') f_0(\vec{r},\vec{k},t)[1 - f_0(\vec{r},\vec{k}',t)]$$
$$\times \{A(\vec{r},\vec{k},t) - A(\vec{r},\vec{k}',t)\}$$
$$= \int d^3\vec{k} B(\vec{r},\vec{k},t) L A(\vec{r},\vec{k},t).$$

This proves the hermiticity. One can also prove the same result in a different way by interchanging the dummy variables of integration in both terms, instead of just the second term, and then using the Principle of Detailed Balance. This is left to the reader.

Problem 2.7. Show that for a real function $A(\vec{r},\vec{k},t)$, the quantity $\int d^3\vec{k} A(\vec{r},\vec{k},t) LA(\vec{r},\vec{k},t)$ is positive indefinite (or nonnegative).

Problem 2.8. Let λ be an eigenvalue of the operator L, i.e.,

$$L\Psi(\vec{r},\vec{k},t) = \lambda\Psi(\vec{r},\vec{k},t),$$

where the eigenfunction Ψ is real. Show that λ is real and nonnegative.

Solution. Since the operator L is Hermitian, its eigenvalue λ must be real (well-known property of all Hermitian operators; see the next chapter). We merely have to show that this real value is positive indefinite.

$$\int d^3\vec{k} \Psi(\vec{r},\vec{k},t) L\Psi(\vec{r},\vec{k},t) = \lambda \int d^3\vec{k} \Psi(\vec{r},\vec{k},t)\Psi(\vec{r},\vec{k},t)$$
$$= \lambda \int d^3\vec{k} [\Psi(\vec{r},\vec{k},t)]^2.$$

According to the statement of Problem 2.7, the left-hand-side is positive indefinite, so that the right-hand-side must also be the same, i.e.,

$$\lambda \int d^3\vec{k}[\Psi(\vec{r},\vec{k},t)]^2 \geq 0.$$

But, for real Ψ,

$$\int d^3\vec{k}[\Psi(\vec{r},\vec{k},t)]^2 \geq 0.$$

Therefore,

$$\lambda \geq 0.$$

Problem 2.9. Show using the relaxation time approximation that the function $\Phi(\vec{r},\vec{k},t)$ defined in Problem 2.4 is generally *not* an eigenfunction of the operator L.

Problem 2.10. Show that the relaxation time τ in the relaxation time approximation is nonnegative *if it is independent of the wavevector* $\vec{k}$. This shows that when the distribution function is disturbed from its equilibrium value, the fluctuation continuously dies down with time and does not grow, leading to instabilities. However, note that the relaxation time approximation is valid under spatially homogeneous transport and hence it is entirely possible for the relaxation time to have negative values in spatially inhomogeneous transport. This will allow fluctuations to grow, temporarily, but they must ultimately die down if equilibrium is restored.

Solution. According to the relaxation time approximation,

$$\left.\frac{\mathrm{d}f}{\mathrm{d}t}\right|_{\text{collisions}} = -\frac{\Delta f(\vec{r},\vec{k},t)}{\tau(\vec{k})}.$$

Using the definition of $\Phi(\vec{r},\vec{k},t)$ given in Problem 2.4,

$$\Delta f(\vec{r},\vec{k},t) = -\Phi(\vec{r},\vec{k},t)\frac{\partial f_0(\vec{r},\vec{k},t)}{\partial E(\vec{k})}$$

we obtain

$$\left.\frac{\mathrm{d}f}{\mathrm{d}t}\right|_{\text{collisions}} = \left[\frac{1}{\tau(\vec{k})}\frac{\partial f_0(\vec{r},\vec{k},t)}{\partial E(\vec{k})}\right]\Phi(\vec{r},\vec{k},t).$$

Finally, using the definition of the operator L given in Problem 2.6, we get

$$L\Phi(\vec{r},\vec{k},t) = -\left[\frac{1}{\tau(\vec{k})}\frac{\partial f_0(\vec{r},\vec{k},t)}{\partial E(\vec{k})}\right]\Phi(\vec{r},\vec{k},t).$$

From the last equation, we get that in case $\tau(\vec{k}) = \tau_0$ (independent of wavevector), then

$$\int d^3\vec{k}\Phi(\vec{r},\vec{k},t)L\Phi(\vec{r},\vec{k},t) = \frac{1}{\tau_0}\int d^3\vec{k}\left[-\frac{\partial f_0(\vec{r},\vec{k},t)}{\partial E(\vec{k})}\right]\Phi^2(\vec{r},\vec{k},t). \quad (2.158)$$

Note that

$$-\frac{\partial f_0(\vec{r},\vec{k},t)}{\partial E(\vec{k})} = \frac{1}{4k_BT}\text{sech}^2\left[\frac{E(\vec{k}) + E_{c0}(\vec{r},t) - \mu_F}{k_BT}\right] \geq 0,$$

Additionally, note from the defining statement in Problem 2.4 that $\Phi(\vec{r},\vec{k},t)$ is real, so that

$$\Phi^2(\vec{r},\vec{k},t) \geq 0 \, for \, all \, \vec{k}.$$

Thus, the integral on the right-hand-side of (2.158) is positive indefinite. But Problem 2.7 showed that the left-hand-side is also positive indefinite. Therefore, τ_0 must be positive indefinite as well, i.e.,

$$\tau_0 \geq 0.$$

We made two assumptions to derive this result:

1. Relaxation time approximation is valid, which also implies that transport is spatially homogeneous since the relaxation time approximation is not valid otherwise. The variable $\vec{r}$ should have been omitted in the problem.
2. Wavevector independence of the relaxation time.

If either of these assumptions is violated, the "relaxation time" need not be positive definite.

References

1. Y. Imry, *Introduction ot Mesoscopic Physics* (Oxford University Press, Oxford, 2002).
2. C. W. J. Beenakker and H. van Houten, "Quantum transport in semiconductor nanostructures", Chapter 1 in *Solid State Physics*, Eds. H. Ehrenreich and D. Turnbull, Vol. 44, (Academic Press. San Diego, 1991).
3. M. Cahay and S. Bandyopadhyay, "Semiconductor quantum devices", in *Advances in Electronics and Electron Physics*, Ed. P. W. Hawkes, Vol. 89, (Academic Press, San Diego, 1994).
4. S. Datta, *Electronic Transport in Mesoscopic Systems*, (Cambridge University Press, Cambridge, 1995).
5. M. Lundstrom, *Fundamentals of Carrier Transport*, Modular Series on Solid State Devices, Eds. G. W. Neudeck and R. F. Pierret (Addison-Wesley, Reading, 1990)
6. J. M. Ziman, *Electrons and Phonons*, (Clarendon Press, Oxford, 1960).

7. B. R. Nag, *Electron Transport in Compound Semiconductors*, (Springer-Verlag, New York, 1980).
8. S. Datta, "Nanoelectronic devices: A unified view", The Oxford Handbook on Nanoscience and Nanotechnology: Frontiers and Advances, Vol. I, Eds. A. V. Narlikar and Y. Y. Fu, Chapter 1.
9. S. Bandyopadhyay, C. M. Maziar, S. Datta and M. S. Lundstrom, "An analytical technique for calculating high-field transport parameters in semiconductors", J. Appl. Phys., **60**, 278 (1986).
10. S. M. Sze, *Physics of Semiconductor Devices*, 2nd. edition, (John Wiley & Sons, New York, 1981).
11. C. Jacobini and L. Reggiani, "The Monte Carlo method for the solution of charge transport in semiconductors with applications in covalent materials", Rev. Mod. Phys., **55**, 645 (1983).
12. H. D. Rees, "Calculation of distribution functions by exploiting stability of steady state", J. Phys. Chem. Solid., **30**, 643 (1969).
13. S. Bandyopadhyay, M. E. Klausmeier-Brown, C. M. Maziar, S. Datta and M. S. Lundstrom, "A rigorous technique to couple Monte Carlo and drift–diffusion models for computationally efficient device simulation", IEEE Trans. Elec. Dev., **34**, 392 (1987).
14. *Monte Carlo Device Simulation: Full Band and Beyond*, Ed. K. Hess, (Kluwer Academic Publishers, Boston, 1991).
15. J. R. Barker and D. K. Ferry, "Self scattering path variable formulation of high-field time-dependent quantum kinetic equations for semiconductor transport in the finite collision duration regime", Phys. Rev. Lett., **42**, 1779 (1979).
16. P. Lipavsky, F. S. Khan, A. Kalvova and J. W. Wilkins, "High field transport in semiconductors 2: Collision duration time", Phys. Rev. B, **43**, 6650 (1991).
17. D. K. Ferry, A. M. Kriman, H. Hida and S. Yamaguchi, "Collision retardation and its role in femtosecond-laser excitation of semiconductor plasmas", Phys. Rev. Lett., **67**, 633 (1991).
18. N. Telang and S. Bandyopadhyay, "Effects of collision retardation of hot electron transport in a two dimensional electron gas", Phys. Rev. B, **47**, 9900 (1993).

Chapter 3
Some Essential Elements of Quantum Mechanics

3.1 Introduction

In the previous two chapters, we treated electrons as particles that obey Newton's laws of classical mechanics. In this chapter, we will treat them as waves (quantum-mechanical entities) and revisit some essential concepts of quantum mechanics. This will allow us to treat effects arising from electron wave interference in solids, as well as device phenomena that are essentially quantum-mechanical in origin. Many of the concepts in this chapter should be already very familiar to the reader if he/she has had an advanced undergraduate class in quantum mechanics.

3.2 The Schrödinger Picture

An electron anywhere in the universe is described by a "wavefunction" $\psi(\vec{r},t)$ whose squared magnitude $|\psi(\vec{r},t)|^2$ is the probability of finding that electron at any position $\vec{r}$ at an instant of time t. In that sense, $|\psi(\vec{r},t)|^2$ has some similarities with the Boltzmann distribution function $f(\vec{r},\vec{k},t)$. Both are positive indefinite quantities, but the bare wavefunction $\psi(\vec{r},t)$ need not be positive indefinite. It is generally a complex quantity. Hence, adding (or superposing) two wavefunctions can produce a null, which is what is needed to represent the phenomenon of destructive interference. Thus, the wavefunction approach can do what the Boltzmann approach could not, namely represent the phenomenon of destructive interference.

The wavefunction of an electron has a wealth of information in it since if we know it, then we can determine the "expectation value" of any physical property of the electron (such as its position, momentum, energy, current density, etc.). If we make a measurement of that physical variable, we will not get the same answer every time because of the *probabilistic* nature of quantum mechanics.

S. Bandyopadhyay, *Physics of Nanostructured Solid State Devices*,
DOI 10.1007/978-1-4614-1141-3_3,

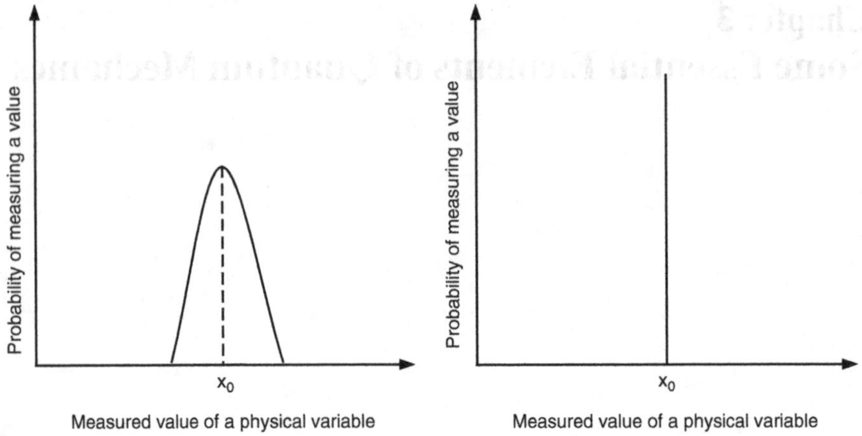

Fig. 3.1 The *left figure* shows the probability distribution of measured values of a physical quantity according to quantum mechanics. The expectation value is the average of the measured values and is denoted by x_0. The *right figure* shows the probability distribution according to classical mechanics. The classical distribution has zero width and is located at x_0

Hence, we cannot say what we will measure in any *one* given measurement. However, we can tell what will be the *average* value that we will measure if we make several measurements. The results could have a probability distribution like that shown in Fig. 3.1, but the average value (or expectation value) is deterministic. Classical mechanics, however, is completely deterministic, so that if we make repeated measurements of the same variable and that variable did not change between measurements, then we will measure exactly the same value *every time*. Hence, the "classical" probability distribution will be a delta function of unit amplitude, while the "quantum-mechanical" probability distribution will have a nonzero width. What is important, however, is that the classical value will be exactly equal to average quantum-mechanical value *in the appropriate limit*. This is known as the *Correspondence Principle*, which was introduced by Niels Bohr phenomenologically in 1923 to reconcile quantum theory with classical theory.

3.2.1 Calculating Expectation Values

Any physical variable like current density, velocity, momentum, position, etc. will have a linear operator O associated with it. If we want to determine the expectation value of that variable, then we will compute the quantity

$$\langle O \rangle = \frac{\int_{-\infty}^{\infty} d^3\vec{r}\,\psi^*(\vec{r},t) O \psi(\vec{r},t)}{\int_{-\infty}^{\infty} d^3\vec{r}\,\psi^*(\vec{r},t)\psi(\vec{r},t)}, \tag{3.1}$$

where the asterisk represents complex conjugate. Dirac introduced a shorthand notation (called the "bra-ket" notation) to represent the integrals in the above equation. In this notation $\langle\psi(\vec{r},t)|$ is called the "bra" and represents the complex conjugate of the wavefunction, while $|\psi(\vec{r},t)\rangle$ is called the "ket" and represents the wavefunction. The integral $\int_{-\infty}^{\infty} d^3\vec{r}\psi^*(\vec{r},t)O\psi(\vec{r},t)$ will be written as $\langle\psi(\vec{r},t)|O|\psi(\vec{r},t)\rangle$. Therefore, in the Dirac bra-ket notation, the last equation will be written as

$$\langle O\rangle = \frac{\langle\psi(\vec{r},t)|O|\psi(\vec{r},t)\rangle}{\langle\psi(\vec{r},t)|\psi(\vec{r},t)\rangle}, \tag{3.2}$$

Thus, it is obvious that in order to calculate expectation values of physical quantities, we need to know two things: (1) the wavefunction $\psi(\vec{r},t)$ and (2) the operator O for the particular physical variable that we seek.

3.3 The Wavefunction and the Schrödinger Equation

The wavefunction $\psi(\vec{r},t)$ is found by solving the celebrated *Schrödinger equation.* This equation is basically a statement of the conservation of energy:

$$\texttt{Kinetic Energy} + \texttt{Potential Energy} = \texttt{Total Energy} \tag{3.3}$$

The kinetic energy is equal to $(1/2)\vec{p}\cdot\vec{v}$ and $\vec{v} = \vec{p}/m_0$, where $\vec{p}$ is the electron's momentum, $\vec{v}$ is its velocity, and m_0 is its mass. Therefore, in operator terms, the conservation of energy can be written as

$$\left[\vec{p}_{\text{op}}\cdot\frac{\vec{p}_{\text{op}}}{2m_0} + V_{\text{op}}\right]\psi(\vec{r},t) = E_{\text{op}}\psi(\vec{r},t), \tag{3.4}$$

where p_{op} is the operator for momentum, V_{op} is the operator for potential energy and E_{op} is the operator for total energy. The operator for momentum was derived from the famous wave–particle duality equation of Louis DeBroglie who stipulated that the momentum p of an electron that has a dual wave and particle character, is given by

$$p = \hbar k = h/\lambda, \tag{3.5}$$

where h is Planck's constant, $\hbar = h/(2\pi)$, and k is the electron's wavevector which is related to the electron's wavelength λ as $k = 2\pi/\lambda$. The quantities k and λ are sometimes referred to as the DeBroglie wavevector and DeBroglie wavelength of a particle, respectively.

We will show later that the wavefunction of a free electron is the plane wave

$$\psi_{\text{free electron}} = \frac{1}{\sqrt{\Omega}}\mathrm{e}^{\mathrm{i}(\vec{k}\cdot\vec{r}-\omega t)}, \tag{3.6}$$

where Ω is a normalizing volume and the angular frequency ω is related to the electron's energy by Planck's famous equation

$$E = \hbar\omega. \tag{3.7}$$

If we define the momentum operator as $\vec{p}_{op} = -i\hbar\vec{\nabla}_r$, then for a free electron, the expectation value of the momentum is (see (3.1))

$$\langle p \rangle = \frac{(1/\Omega)\int_{-\infty}^{\infty} d^3\vec{r} e^{-i(\vec{k}\cdot\vec{r}-\omega t)}(-i\hbar\vec{\nabla}_r)e^{i(\vec{k}\cdot\vec{r}-\omega t)}}{(1/\Omega)\int_{-\infty}^{\infty} d^3\vec{r} e^{-i(\vec{k}\cdot\vec{r}-\omega t)}e^{i(\vec{k}\cdot\vec{r}-\omega t)}} = \hbar\vec{k}, \tag{3.8}$$

which agrees with (3.5). Hence, the appropriate choice for the momentum operator is $-i\hbar\vec{\nabla}_r$.

The appropriate choice for the total energy operator is $i\hbar\frac{\partial}{\partial t}$ since if we use the free electron wavefunction, then the expectation value of the total energy is

$$\langle E \rangle = \frac{(1/\Omega)\int_{-\infty}^{\infty} d^3\vec{r} e^{-i\left(\vec{k}\cdot\vec{r}-\omega t\right)}\left(i\hbar\frac{\partial}{\partial t}\right)e^{i\left(\vec{k}\cdot\vec{r}-\omega t\right)}}{(1/\Omega)\int_{-\infty}^{\infty} d^3\vec{r} e^{-i\left(\vec{k}\cdot\vec{r}-\omega t\right)}e^{i\left(\vec{k}\cdot\vec{r}-\omega t\right)}} = \hbar\omega, \tag{3.9}$$

which agrees with the Planck equation (3.7).

Using these operators in (3.4) and realizing that the potential energy operator is a multiplicative operator, we get the final form of the *Schrödinger equation* as

$$\left[-\hbar^2\vec{\nabla}_r\cdot\left(\frac{\vec{\nabla}_r}{2m_0(\vec{r})}\right)+V(\vec{r},t)\right]\psi(\vec{r},t) = H_{op}\psi(\vec{r},t) = i\hbar\frac{\partial\psi(\vec{r},t)}{\partial t}$$

$$\text{if the mass is spatially varying}$$

$$\left[-\frac{\hbar^2}{2m_0}\nabla_r^2+V(\vec{r},t)\right]\psi(\vec{r},t) = H_{op}\psi(\vec{r},t) = i\hbar\frac{\partial\psi(\vec{r},t)}{\partial t}$$

$$\text{if the mass is spatially invariant.} \tag{3.10}$$

where the operator H_{op} (called the Hamiltonian) is the operator for kinetic plus potential energy.

The wavefunction $\psi(\vec{r},t)$ is found by solving the above equation.

Boundary conditions on the wavefunction: Note immediately that:

- The wavefunction must be *continuous* in space since otherwise its gradient will blow up resulting in infinite momentum wherever the wavefunction is discontinuous. Infinite momentum is not physical (it also violates the principle of relativity) and therefore not allowed.
- The ratio of the spatial gradient of the wavefunction to the mass must also be continuous in space, since otherwise the kinetic energy term will blow up resulting

in infinite kinetic energy wherever the ratio becomes discontinuous. Of course, if the mass is spatially invariant, then the spatial gradient of the wavefunction must be continuous. In that case, the wavefunction and its first derivative in space are always continuous. The only exception occurs at boundaries of infinite potential barriers. There, since the potential energy is infinite, the kinetic energy is allowed to blow up and the ratio of the gradient of the wavefunction to the mass need not be continuous in space.

3.3.1 Schrödinger Equation in a Time-Independent Potential

If the potential energy is time independent, then the Schrödinger equation (3.10) reduces to

$$i\hbar\frac{\partial\psi(\vec{r},t)}{\partial t} = H_{\text{op}}\psi(\vec{r},t) = \left[-\frac{\hbar^2}{2m_0}\nabla_{\text{r}}^2 + V\left(\vec{r}\right)\right]\psi(\vec{r},t), \tag{3.11}$$

provided the mass is spatially invariant.

In the above equation, there are no "mixed terms", i.e., terms that depend on *both* time and space. Hence, we can write a product solution for the wavefunction that is the product of a space-dependent part and a time-dependent part, i.e.,

$$\psi(\vec{r},t) = T(t)\phi(\vec{r}). \tag{3.12}$$

Substituting this in (3.11) and then dividing throughout with the wavefunction $\psi(\vec{r},t)$, we get

$$i\hbar\frac{1}{T(t)}\frac{\partial T(t)}{\partial t} = -\frac{\hbar^2}{2m_0}\frac{1}{\phi(\vec{r})}\nabla_{\text{r}}^2\phi(\vec{r}) + V(\vec{r}). \tag{3.13}$$

Now, the left-hand-side does not depend on spatial coordinate $\vec{r}$ but depends on time t, while the right-hand-side does not depend on time t but depends on the spatial coordinate $\vec{r}$. The only way they can be equal to each other is if they actually depended *neither* on time t, *nor* on spatial coordinate $\vec{r}$. Therefore, each side is a constant. Setting this constant equal to energy E (since each side has the dimension of energy), we get

$$i\hbar\frac{\partial T(t)}{\partial t} = ET(t)$$

$$-\frac{\hbar^2}{2m}\nabla_{\text{r}}^2\phi(\vec{r}) + V(\vec{r})\phi(\vec{r}) = E\phi(\vec{r}),\ or$$

$$H_{\text{op}}\phi(\vec{r}) = E\phi(\vec{r}) \tag{3.14}$$

The first equation above has the solution

$$T(t) = T(0)\mathrm{e}^{-\mathrm{i}Et/\hbar} = T(0)\mathrm{e}^{-\mathrm{i}\omega t}. \tag{3.15}$$

The second equation can be written as

$$\nabla_{\mathrm{r}}^2\phi(\vec{r}) + k^2(\vec{r})\phi(\vec{r}) = 0, \tag{3.16}$$

where

$$k(\vec{r}) = \sqrt{2m_0\left(E - V(\vec{r})\right)/\hbar}. \tag{3.17}$$

The solution of (3.16) is

$$\phi(\vec{r}) = \frac{1}{\sqrt{\Omega}}\mathrm{e}^{\mathrm{i}\vec{k}(\vec{r})\cdot\vec{r}}, \tag{3.18}$$

where Ω is some normalizing volume such that $\int_\Omega d^3\vec{r}\,\left|\phi(\vec{r})\right|^2 = 1$. This last condition simply implies the probability of finding the electron within the volume Ω is 100%.

Substituting these results in (3.12), we get

$$\psi(\vec{r},t) = \frac{1}{\sqrt{\Omega}}\mathrm{e}^{\mathrm{i}(\vec{k}(\vec{r})\cdot\vec{r}-\omega t)}, \tag{3.19}$$

where we have posited the initial condition $T(0) = 1$ to ensure that at any instant of time t, $\int_\Omega d^3\vec{r}|\psi(\vec{r},t)|^2 = 1$.

3.3.2 *Wavefunction of a Free Electron*

By definition, a free particle is one that is not subjected to any force. Since force is the negative gradient of potential energy ($\vec{F} = -\vec{\nabla}V(\vec{r})$),[1] the potential energy V of a free particle will be spatially invariant. If we are considering steady state situations, then the potential energy will be time invariant as well. In other words, it is a constant. Now, since potential energy is always undefined to the extent of an

[1]Such a force is a *conservative* (or irrotational) force whose curl vanishes: $\vec{\nabla}_{\mathrm{r}} \times \vec{F} = -\vec{\nabla}_{\mathrm{r}} \times \vec{\nabla}_{\mathrm{r}}V(\vec{r}) = 0$. There can be nonconservative forces which may not be expressed as the gradient of a scalar potential. A familiar example of a nonconservative force is friction which dissipates energy. For a conservative force, the work done to move a particle around a closed contour C is zero: $W = \oint_C \vec{F}\cdot\vec{r} = 0$, which will of course not hold true if the force is dissipative. Dissipation cannot be handled in the Schrödinger equation in a straightforward way. This discussion is however deferred until Chap. 8.

arbitrary constant, we can assume any *constant* potential energy to be identically zero. Thus, for a free particle, the wavevector given in (3.17) will be spatially invariant and will be given by $k = \sqrt{2m_0E}/\hbar$ since $V = 0$. Thus, the wavefunction of a free particle is indeed given by (3.6).

3.3.3 Electron Wave and Electron Waveguide

The one-dimensional version of the time-independent Schrödinger equation in (3.16) is

$$\frac{d^2\phi(z)}{dz^2} + k^2(z)\phi(z) = 0. \tag{3.20}$$

Compare this with the one-dimensional version of the Maxwell equation for the electric field in an electromagnetic wave:

$$\frac{d^2\mathcal{E}(z)}{dz^2} + \overline{k}^2(z)\mathcal{E}(z) = 0, \tag{3.21}$$

where $\overline{k}(z) = \eta(z)\omega_0/c$. Here, $\eta(z)$ is the spatially varying refractive index, ω_0 is the angular frequency of the (monochromatic) electromagnetic wave, and c is the speed of light in vacuum.

The mathematical similarity between (3.20) and (3.21) is intriguing. It is also physically meaningful, since $|\phi(z)|^2$ is the probability of finding an electron at the location z and $|\mathcal{E}(z)|^2$ is proportional to the wave intensity or the probability of finding a photon (in the electromagnetic wave) at location z. It is amazing that the amplitudes of an electron wave and an electromagnetic wave obey mathematically identical equations! Needless to say, this has profound implications. It tells us that the physics of electron propagation through a medium with time-independent (but spatially varying) potential is analogous to the physics of electromagnetic wave propagation through a medium with time-independent (but spatially varying) refractive index! Indeed, at low temperatures, when time-dependent perturbations on an electron in a solid (such as electron–phonon or electron–electron interactions) are weak, the physics of electron transport will be very similar to the physics of electromagnetic wave propagation in a static medium. Electromagnetic waves (like light waves or microwaves) are known to interfere and produce interference fringes. Therefore, electrons in a disordered solid with spatially varying (but not time-varying) potential should also produce interference effects. Such effects should be, and have been, observed in ultrasmall metallic and semiconductor structures at low temperatures because of the absence of inelastic electron scattering events which are time-varying perturbations. Since these structures can propagate electron waves coherently and therefore produce interference effects, they are sometimes referred to as *electron waveguides*.

The similarity works both ways. The Maxwell equation in (3.21) has sometimes been referred to as the Schrödinger equation for a *single* photon. Why a single (not multiple) photons? That is because only for a single photon it is always possible to define an electric field $\mathcal{E}$. For multi-photon ensembles, the average field will be zero if the ensemble forms an *incoherent state* (like photons from a light bulb where the photons are in a Bose–Einstein distribution). The average value will be nonzero only if the photons form a *coherent state* (like photons from a laser where the photons are in a Poisson distribution). Further discussion of this concept is beyond the scope of this book, but can be found in textbooks dealing with many body physics. Suffice it to say that we can always associate an electric field with a single photon; hence, Maxwell's equation is aptly viewed as the Schrödinger equation for a single photon.

3.4 How to Find the Quantum-Mechanical Operator for a Physical Quantity

We had stated earlier that two things are needed to determine the expectation value of any physical quantity: (1) the wavefunction which is found by solving the Schrödinger equation, and (2) the quantum-mechanical operator for the physical quantity. We showed how the operators for momentum and energy were found, but how do we find the operator for any arbitrary quantity like, say, current density? The usual prescription for that is to relate the physical quantity to other quantities for which the quantum-mechanical operators are known, and then use the same relationship to come up with the unknown operator. Once this is done, we must check for two features: first, the expectation value of the operator that we come up with must correspond to the classical value in accordance with the Correspondence Principle, and second, we must make sure that the operator is Hermitian since all quantum-mechanical operators have to be Hermitian. That latter requirement comes about from the fact that the expectation value of any physical quantity must be *real*, and not imaginary or complex. We cannot measure an imaginary or complex current density, or charge density, or position, or any other such thing! Note that if the wavefunction is an eigenfunction of the operator, then the expectation value of that operator is nothing but its eigenvalue. Since we cannot have a complex or imaginary expectation value of a physical quantity, we cannot have a complex or imaginary eigenvalue of the operator either. Only Hermitian operators are guaranteed to have real eigenvalues (see Problem 3.1) and hence only Hermitian operators can make legitimate quantum-mechanical operators. Conversely, all quantum-mechanical operators must be Hermitian.

Since examples are better than precepts, we will exemplify this method of finding quantum-mechanical operators of physical quantities by finding the operators for charge density and current density due to a *single electron* at a location $\vec{r}_0$ at an instant of time t. The charge density must be proportional to the probability of finding the electron at $\vec{r}_0$ at the instant t since charge density is proportional to electron density. Therefore, we insist that the expectation value of the charge density operator $\langle \rho_{op} \rangle (\vec{r}_0, t)$ should be

$$\langle \rho_{op} \rangle (\vec{r}_0, t) = -e|\psi(\vec{r}_0, t)|^2, \tag{3.22}$$

as long as $\int_{-\infty}^{\infty} d^3\vec{r} \psi^*(\vec{r}, t)\psi(\vec{r}, t) = \int_{-\infty}^{\infty} d^3\vec{r}|\psi(\vec{r}, t)|^2 = 1$, so that the probability of finding the electron somewhere in the universe is unity. The last condition—namely, normalization of the wavefunction—is needed to interpret $|\psi(\vec{r}_0, t)|^2$ as the probability of finding the electron at location $\vec{r}_0$ at time t.

Equation (3.22) actually follows from the Correspondence Principle since the expectation value is assumed to be equal to the classical value. Now, we need to find the operator ρ_{op} that satisfies the relation

$$\langle \rho_{op} \rangle (\vec{r}_0, t) = \frac{\int_{-\infty}^{\infty} d^3\vec{r} \psi^*(\vec{r}, t)\rho_{op}\psi(\vec{r}, t)}{\int_{-\infty}^{\infty} d^3\vec{r} \psi^*(\vec{r}, t)\psi(\vec{r}, t)} = \int_{-\infty}^{\infty} d^3\vec{r} \psi^*(\vec{r}, t)\rho_{op}\psi(\vec{r}, t)$$
$$= -e|\psi(\vec{r}_0, t)|^2. \tag{3.23}$$

A little reflection shows that $\rho_{op} = -e\delta(\vec{r} - \vec{r}_0)$ (where $\delta(\vec{r} - \vec{r}_0)$ is the Dirac delta) since it satisfies the preceding equation. The final check is to make sure that this operator is Hermitian, i.e., it obeys the condition

$$\int_{-\infty}^{\infty} d^3\vec{r} \psi^*(\vec{r}, t)\rho_{op}\psi(\vec{r}, t) = \int_{-\infty}^{\infty} d^3\vec{r} [\rho_{op}\psi(\vec{r}, t)]^* \psi(\vec{r}, t). \tag{3.24}$$

Since $\delta(\vec{r} - \vec{r}_0) = [\delta(\vec{r} - \vec{r}_0)]^*$, the above equation is obviously satisfied. Therefore, the charge density operator for electrons is indeed $\rho_{op} = -e\delta(\vec{r} - \vec{r}_0)$.

To find the current density operator $\vec{J}_n^{op}$, we use the following analogy: Since classically $\vec{J}_n(\vec{r}, t) = \rho(\vec{r}, t)v(\vec{r}, t)$, we will start with the guess that

$$\vec{J}_n^{op} = \rho_{op} \times \frac{\vec{p}_{op}}{m_0} = -e\delta(\vec{r} - \vec{r}_0) \times \frac{-i\hbar\vec{\nabla}_r}{m_0}|_{\vec{r}=\vec{r}_0}. \tag{3.25}$$

The problem now is that operators generally do not commute and hence there is a difference between $e\delta(\vec{r}) \times \frac{i\hbar\vec{\nabla}_r}{m_0}$ and $\frac{i\hbar\vec{\nabla}_r}{m_0} \times e\delta(\vec{r})$. The velocity and charge density operators indeed do not commute (see Problem 3.2), so that we are in a quandary to choose between the two. At first glance, both appear equally legitimate. The quandary is resolved easily since it turns out that neither is actually Hermitian and therefore neither is legitimate. What is Hermitian is the "average" or symmetric combination of the two (see Problem 3.3), and hence we choose

$$\vec{J}_n^{\rm op} = \frac{1}{2}\left[\rho_{\rm op} \times \frac{\vec{p}_{\rm op}}{m_0} + \frac{\vec{p}_{\rm op}}{m_0} \times \rho_{\rm op}\right]$$
$$= \frac{1}{2}\left[e\delta(\vec{r}-\vec{r}_0) \times \left.\frac{i\hbar\vec{\nabla}_{\rm r}}{m_0}\right|_{\vec{r}=\vec{r}_0} + \left.\frac{i\hbar\vec{\nabla}_{\rm r}}{m_0}\right|_{\vec{r}=\vec{r}_0} e\delta(\vec{r}-\vec{r}_0)\right]. \quad (3.26)$$

We can now proceed to find the expectation value of the current density operator.

$$\langle \vec{J}_n^{\rm op}\langle(\vec{r}_0,t) = \frac{\int_{-\infty}^{\infty} d^3\vec{r}\psi^*(\vec{r},t)\vec{J}_n^{\rm op}\psi(\vec{r},t)}{\int_{-\infty}^{\infty} d^3\vec{r}\psi^*(\vec{r},t)\psi(\vec{r},t)} = \int_{-\infty}^{\infty} d^3\vec{r}\psi^*(\vec{r},t)\vec{J}_n^{\rm op}\psi(\vec{r},t)$$
$$= \frac{1}{2}\int_{-\infty}^{\infty} d^3\vec{r}\psi^*(\vec{r},t)e\delta(\vec{r}-\vec{r}_0)\frac{i\hbar\vec{\nabla}_{\rm r}}{m_0}\psi(\vec{r},t)$$
$$+\frac{1}{2}\int_{-\infty}^{\infty} d^3\vec{r}\psi^*(\vec{r},t)\frac{i\hbar\vec{\nabla}_{\rm r}}{m_0}e\delta(\vec{r}-\vec{r}_0)\psi(\vec{r},t). \quad (3.27)$$

Now use the Hermiticity property of the momentum operator that guarantees

$$\int_{-\infty}^{\infty} d^3\vec{r}\psi^*\left(\frac{-i\hbar\vec{\nabla}_{\rm r}}{m_0}\right)[e\delta(\vec{r}-\vec{r}_0)\psi] = \int_{-\infty}^{\infty} d^3\vec{r}\left[\frac{-i\hbar\vec{\nabla}_{\rm r}}{m_0}\psi\right]^* e\delta(\vec{r}-\vec{r}_0)\psi. \quad (3.28)$$

Using the last relation in (3.27), we get that the expectation value of current density operator (which is the current density at location $\vec{r}_0$ at time t because of the Correspondence Principle) is

$$\langle \vec{J}_n^{\rm op}\rangle(\vec{r}_0,t) = \frac{-ie\hbar}{2m_0}\left[\psi(\vec{r}_0,t)(\vec{\nabla}_{\rm r}\psi(\vec{r},t)|_{\vec{r}=\vec{r}_0})^* - \psi^*(\vec{r}_0,t)(\vec{\nabla}_{\rm r}\psi(\vec{r},t)|_{\vec{r}=\vec{r}_0})\right]$$
$$= \frac{e\hbar}{m_0}{\rm Im}\left[\psi(\vec{r}_0,t)(\vec{\nabla}_{\rm r}\psi(\vec{r},t)|_{\vec{r}=\vec{r}_0})^*\right], \quad (3.29)$$

where "Im" stands for the "imaginary part." Note that the expectation value of the current density operator is completely *real* just as the expectation value of any quantum-mechanical operator should be.

The above two examples illustrate the usual procedure for finding the quantum-mechanical operator for any physical variable. There are three basic steps:

1. First, relate the physical variable whose operator is being sought to other variables whose quantum-mechanical operators are known.
2. Next, assume that the operators obey the same relation between them as the physical variables do. This will yield the first guess for the operator that we seek.
3. Finally, ensure that the operator thus obtained is Hermitian. If not, then try symmetric combinations as we did in the case of the current density operator.

The above approach works in most cases.

Special Topic 1: Common potential well problems encountered in nanostructured solid state devices

An electron in any solid structure finds itself in a *potential well* since the structure's boundaries act as potential barriers which prevent the electron from escaping. Confinement of an electron within a potential well causes its energy to become discretized so that only specific energies are allowed. The energy difference between any two allowed states of the electron increases with decreasing size of the potential well. In the case of nanostructures (when the potential well has nanometer scale dimensions), this energy difference usually exceeds the thermal energy $k_B T$, so that thermal smearing of the energy levels cannot mask the energy discretization. We will show that "energy discretization" or "energy quantization" is a consequence of the electron's motion being restricted within the boundaries of the potential well.

The exact values of an electron's energy in an allowed state is determined by material properties such as the electron's effective mass in the material, and also the shape and size of the potential well. The two shapes that one typically encounters in modern solid state devices are "rectangular" and "parabolic." We will explain them shortly.

Since any solid structure will be three dimensional, an electron's motion can be confined along either one, two, or all three dimensions. A structure that restricts the electron's motion along only one dimension, but keeps it free along the other two, is called a *quantum well* or quasi two-dimensional structure. A structure that restricts the motion along two dimensions, but keeps it free along one is called a *quantum wire* or quasi one-dimensional structure, while a structure that restricts the motion along all three dimensions is called a *quantum dot* or quasi zero-dimensional structure. They are pictorially depicted in Fig. 3.2.

Assume that the structures shown in Fig. 3.2 are placed in vacuum which presents an infinite potential barrier at the surfaces of the structures. If we are to plot the potential profile along any restricted dimension,—say, the z-axis—then the profile might look like that in Fig. 3.3a. Such a shape is called a rectangular shape and the corresponding potential well is called a rectangular potential well. The potential energy $V(z)$ in a rectangular well is given by

$$V(z) = \texttt{constant} \quad \texttt{if}\ 0 \leq z \leq W_z$$
$$V(z) = \infty \quad \texttt{otherwise.}$$

A quantum well having this kind of (rectangular) potential profile along the direction of confinement can be realized by sandwiching a thin layer of a narrow gap semiconductor (e.g., 100 Å of InSb) between two wide gap semiconductors (e.g., CdTe). This situation is shown in Fig. 3.4. The potential barrier for electrons will be the conduction band offset and the potential barrier for holes will be the valence band offset between CdTe and InSb. Of course the potential barriers will not be really infinite, but if they are large enough (~1 eV or more), we can consider

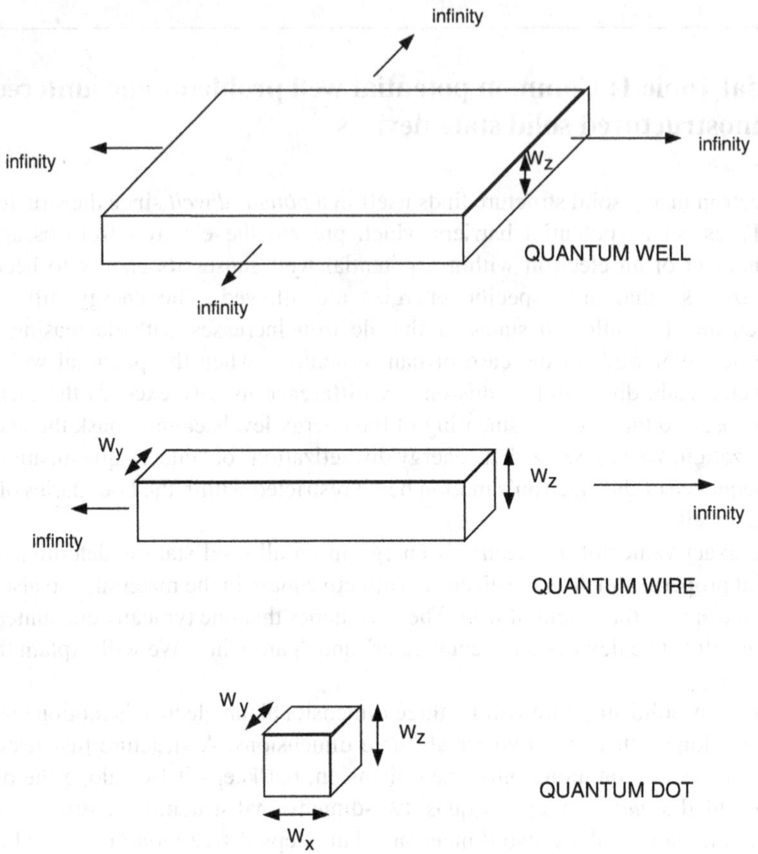

Fig. 3.2 Quantum well, quantum wire and quantum dot structures

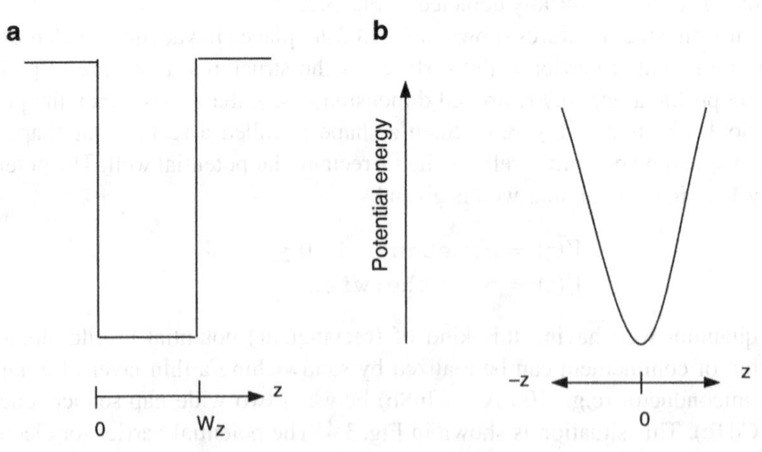

Fig. 3.3 Potential energy profile in (**a**) a rectangular quantum well, and (**b**) a parabolic quantum well

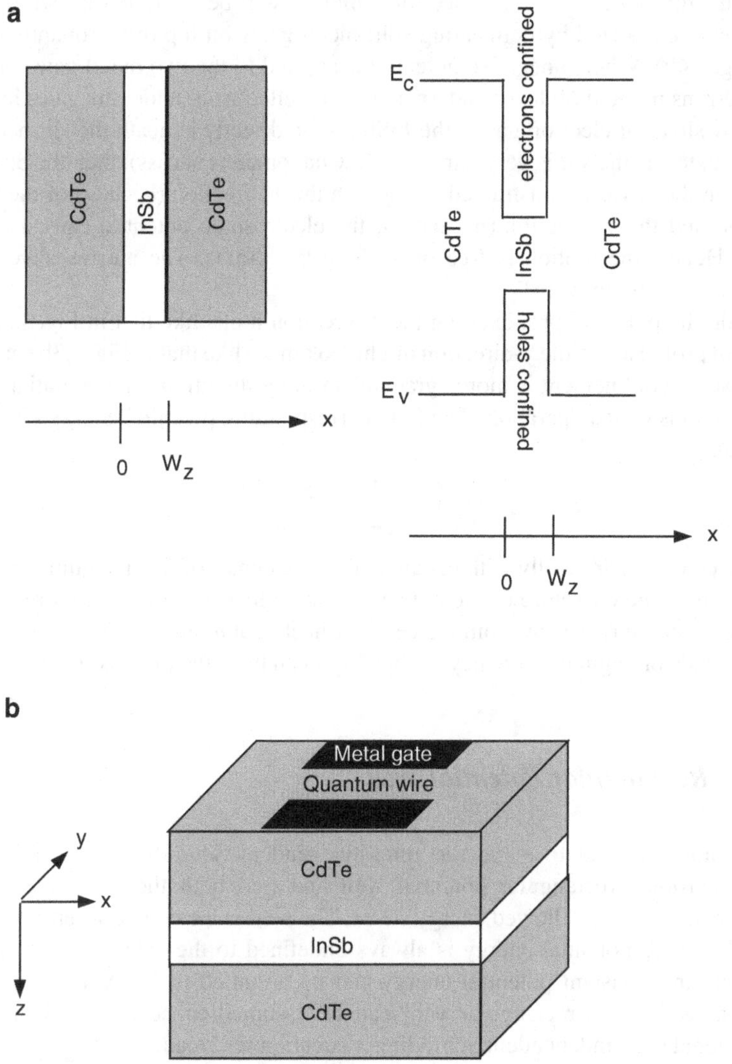

Fig. 3.4 (**a**) A quantum well is realized by sandwiching a narrow gap semiconductor like InSb between two widegap semiconductors like CdTe. (**b**) A quantum wire is fabricated by delineating a split metal gate on top of a quantum well, and then applying a negative potential to the gates to deplete electrons from underneath. This leaves behind a narrow sliver of electron pool underneath the slit in the gate, resulting in the formation of a quantum wire

them to be nearly infinite. The conduction band offset between InSb and CdTe is quite large, so that we can consider the electrons in the InSb layer to be confined in a potential well with nearly infinite barriers. In this structure, the holes are also quantum-confined (see Fig. 3.4a).

A quantum wire, where the electron's motion will be restricted in two dimensions, can be realized by delineating split metal gates on top of the quantum well (see Fig. 3.4b). When a negative potential is applied to the two metal gates on top, the electrons in the InSb layer underneath are repelled from under the gates leaving a narrow sliver of electron gas in the InSb region directly beneath the slit between the two gates. If the slit is very narrow (a few nanometers across), then the electron motion in the sliver is constrained along both the z-direction (because of the CdTe barriers) and the y-direction (because of the electrostatic potential caused by the gates). Hence, the motion is free only along the x-axis. The narrow sliver then becomes a "quantum wire."

While the potential profile along the z-direction looks like that in Fig. 3.3a, the potential profile along the y-direction might look more like that in Fig. 3.3b since the electrostatic confinement is more "gradual" in the y-direction. The potential shape in Fig. 3.3b is called "parabolic" and the corresponding potential energy $V(y)$ can be written as

$$V(y) = \frac{1}{2} m^* \omega^2 y^2,$$

since it varies quadratically with distance from the center of the quantum wire. The quantity ω is the curvature of the potential. The higher its value, the more steeply will the potential rise away from the center. Check that ω must have the dimension of frequency or angular frequency so that $V(y)$ can have the dimension of energy.

3.4.1 Rectangular Potential Well

Let us now proceed to solve the time-independent Schrödinger equation in a one-dimensional rectangular potential well and find both the time-independent wavefunction and the allowed energy states. The well is located between $z = 0$ and $z = W_z$. Since potential energy is always undefined to the extent of an arbitrary constant, any constant potential energy can be assumed to be 0. Therefore, the potential within the rectangular well can be assumed to be zero and the one-dimensional time-independent Schrödinger equation will read:

$$-\frac{\hbar^2}{2m^*} \frac{\partial^2 \Psi(z)}{\partial z^2} = E\Psi(z), \tag{3.30}$$

where m^* is the electron's effective mass in the well, Ψ is the time-independent wavefunction, and the confinement is assumed to exist along the z-axis as shown in Fig. 3.3a.

The above equation can be recast as

$$\frac{\partial^2 \Psi(z)}{\partial z^2} + k^2 \Psi(z) = 0, \tag{3.31}$$

where

$$k = \frac{\sqrt{2m^* E}}{\hbar}. \tag{3.32}$$

This is very reminiscent of the free electron and we note that (3.31) is identical with the Schrödinger equation for a free electron. Therefore, we may be tempted to write the wavefunction as that of a free electron given in (3.6). However, note that the electron is *not* free; it is confined within the well. Therefore, its wavefunction should not be that of a free electron's. It cannot venture outside the well and that restriction is expressed via the *boundary conditions* on the wavefunction. The solution of a differential equation is not independent of the boundary conditions and the same equation can have very different solutions for different boundary conditions. It is the boundary conditions that make a difference and will result in a solution for the wavefunction that is very different from that in (3.6). This is an example of how important boundary conditions are. In fact, we will show now that energy discretization (also known as energy quantization) is purely a consequence of the boundary conditions in a potential well.

We can write the general solution of the above second order linear differential equation with constant coefficients as

$$\Psi(z) = A\sin(kz) + B\cos(kz), \tag{3.33}$$

where A and B are two constants to be determined from the boundary conditions.

Let us now proceed to establish the correct boundary conditions. Since the potential barriers at $z = 0$ and $z = W_z$ are infinite (see Fig. 3.3a), the electron cannot be found outside the well. Hence, the probability $|\Psi(z)|^2$ will be zero for $z < 0$ and $z > W_z$. But since the wavefunction must be continuous at $z = 0$ and W_z, we conclude that the probability must vanish at $z = 0$ and $z = W_z$ as well, so that

$$\Psi(0) = \Psi(W_z) = 0. \tag{3.34}$$

Applying the first boundary condition [$\Psi(0) = 0$] to the solution in (3.33), we get that $\Psi(z) = A\sin(kz)$, and then applying the second boundary condition [$\Psi(W_z) = 0$], we get that either $A = 0$, or $kW_z = n\pi$, where n is an integer. The former condition makes the wavefunction vanish everywhere and is a trivial solution, while the latter is the nontrivial solution and yields the energy quantization condition:

$$k = \frac{\sqrt{2m^* E}}{\hbar} = \frac{n\pi}{W_z}$$

$$\Rightarrow E = E_n = \frac{\hbar^2}{2m^*}\left(\frac{n\pi}{W_z}\right)^2. \tag{3.35}$$

The above result shows that only discrete energy states are allowed in a rectangular potential well, corresponding to $n = 1, 2, 3, \ldots$, etc. These states are called "subbands." Note that the formation of subbands, or energy quantization, is entirely a consequence of boundary conditions.

We will normalize the wavefunction to represent the fact that the probability of finding it *somewhere* within the well is exactly unity. Therefore,

$$\int_0^{W_z} |\Psi(z)|^2 \mathrm{d}z = A^2 \int_0^{W_z} \sin^2\left(\frac{n\pi z}{W_z}\right) \mathrm{d}z = 1, \tag{3.36}$$

which yields that $A^2 = 2/W_z$. Hence, the solution for the complete wavefunction is

$$\Psi(z) = \sqrt{\frac{2}{W_z}} \sin\left(\frac{n\pi z}{W_z}\right). \tag{3.37}$$

3.4.2 *Parabolic Well*

Let us now solve the time-independent Schrödinger equation in a parabolic well, such as along the y-direction in Fig. 3.4b. The equation in this case is

$$-\frac{\hbar^2}{2m^*}\frac{\partial^2 \Psi(y)}{\partial y^2} + \frac{1}{2}m^*\omega^2 y^2 = E\Psi(y). \tag{3.38}$$

We define a quantity $\xi = \alpha y$, so that the Schrödinger equation can be written as

$$\frac{\partial^2 \Psi(\xi)}{\partial \xi^2} + (\zeta - \xi^2)\Psi(\xi) = 0, \tag{3.39}$$

where

$$\alpha = \left(\frac{m^*\omega}{\hbar}\right)^{1/2}$$

$$\zeta = \frac{2E}{\hbar\omega}. \tag{3.40}$$

The solution of this equation is [1]

$$\Phi(y) = \left(\frac{\alpha}{\pi^{1/2}2^n n!}\right)^{1/2} \mathcal{H}_n(\alpha y) \mathrm{e}^{-(1/2)\alpha^2 y^2}, \tag{3.41}$$

where $\mathcal{H}_n$ is the Hermite polynomial of the n-th order.

Since the potential energy becomes infinity at $y = \pm\infty$, the wavefunction must vanish there. Application of this boundary condition yields that $\zeta = 2n + 1$, where n is an integer. This leads to the energy quantization result:

$$E = E_n = \left(n + \frac{1}{2}\right)\hbar\omega. \tag{3.42}$$

Thus, a parabolic well too has discrete subbands corresponding to $n = 1, 2, 3, \ldots$ etc. The Schrödinger equation for a parabolic potential is sometimes referred to as the Schrödinger equation for a simple harmonic motion oscillator since the parabolic potential energy is reminiscent of the potential energy of a simple harmonic motion oscillator.

Comparison of (3.35) and (3.42) shows that the ladder of allowed energy states (subbands) increases *quadratically* with the index n in a rectangular well, but *linearly* with the index n in a parabolic well. Hence, the energy states are unequally spaced in energy in a rectangular well (the spacing continues to increase with increasing n) but equally spaced in energy in a parabolic well. Note also that the energy of the state with $n = 0$ is zero in the rectangular well, but it is nonzero ($= \frac{1}{2}\hbar\omega$) in a parabolic well. The latter energy is often called the zero-point energy (or "quantum noise energy"). Therefore, what this result shows is that a simple harmonic oscillator that does not oscillate ($n = 0$) still has some energy! This is one of the surprising results of quantum mechanics.

3.5 Perturbation Theory

In quantum mechanics, we are sometimes faced with a situation where we know the wavefunction and allowed energy states of a particle in a potential, but then a perturbation changes the potential landscape and alters the wavefunction and the energy states. In the new potential landscape, the wavefunction and energy states are unknown and we have to find them. Solving the Schrödinger equation in the new landscape may be very difficult and may not even be analytically possible. In that case, we can approximately find the new wavefunction and new energy states from knowledge of the old wavefunction and energy states as long as the perturbation is relatively weak and the change in the potential landscape is not drastic. The method that allows us to accomplish this task is known as *perturbation theory*.

Perturbation theory is widely applied in different areas of solid state physics, including calculation of bandstructure in crystals, operation of wavefunction-engineering devices (see Problem 3.9), and calculation of electron scattering rates associated with such phenomena as electron–phonon, electron–impurity, and electron–electron interaction. It is therefore an extremely important recipe and will be repeatedly used in this textbook in different scenarios.

The idea behind perturbation theory is simply this: Suppose we know the solution of the time-independent Schrödinger Equation

$$H_0(\vec{r})\phi_n(\vec{r}) = E_n\phi_n(\vec{r}), \tag{3.43}$$

i.e., we know the allowed eigenenergies E_n and the corresponding wavefunctions (or eigenfunctions) $\phi_n(\vec{r})$, can we then find the new wavefunction $\Psi(\vec{r}, t)$ and the new eigenenergy E when the Hamiltonian changes according to

$$H \to H_0(\vec{r}) + H'(\vec{r}, t),$$

where $H'(\vec{r}, t)$ is the Hamiltonian associated with the perturbation. In general, $H'(\vec{r}, t)$ is both time and space dependent. If it is time independent, then the new Hamiltonian H will be time-independent as well and the new wavefunction will obey the time-independent Schrödinger equation, but if $H'(\vec{r}, t)$ is time-dependent, then the new Hamiltonian will be time-dependent and the new wavefunction will be time-dependent and obey the time-dependent Schrödinger equation. The method of ascertaining the new wavefunction and the new energy in the former case is known as *time-independent perturbation theory*, while in the latter case, it is the *time-dependent perturbation theory*.

3.5.1 Time-Independent Perturbation Theory

When the perturbation is time independent, our task is to solve the time-independent Schrödinger equation

$$[H_0(\vec{r}) + H'(\vec{r})]\Psi(\vec{r}) = E\Psi(\vec{r}), \tag{3.44}$$

to find the new energy E and the new wavefunction $\Psi(\vec{r})$.

Now, there is a mathematical theorem which stipulates that any analytical function can be written as a weighted sum of functions that form a complete orthonormal set. The new wavefunction, being an analytical function, can therefore be written as

$$\Psi(\vec{r}) = \sum_n C_n w_n(\vec{r}), \tag{3.45}$$

where the constants C_n are the coefficients of expansion (or "weights") and $w_n(\vec{r})$ are functions that form a complete orthonormal set, i.e., they satisfy the condition $\int_0^\infty d^3\vec{r} w_n^*(\vec{r}) w_m(\vec{r}) = \delta_{m,n}$, where $\delta_{m,n}$ is the Kronecker delta. The orthonormal functions w_n are called "basis functions." The space spanned by a denumerably infinite number of orthonormal functions is called Hilbert space, so that the basis functions can be viewed as coordinate axes in a (denumerably) infinite dimensional

Hilbert space. We can, of course, choose any arbitrary orthonormal set as basis functions, e.g., sine and cosine functions, but we will choose the unperturbed wavefunctions $\phi_n(\vec{r})$ as basis functions. The reason for this specific choice will become clear later. We therefore write:

$$\Psi(\vec{r}) = \sum_n C_n \phi_n(\vec{r}). \tag{3.46}$$

The unperturbed wavefunctions $\phi_n(\vec{r})$ are guaranteed to form a complete orthonormal set because they are eigenfunctions of the *Hermitian* operator $H_0(\vec{r})$ and eigenfunctions of any Hermitian operator (corresponding to distinct eigenvalues) are mutually orthogonal (see Problem 3.3). If the unperturbed wavefunctions have been normalized, then it is easy to show that

$$\int_0^\infty d^3\vec{r}\phi_n^*(\vec{r})\phi_m(\vec{r}) = \delta_{m,n}, \tag{3.47}$$

so that they indeed form a complete orthonormal set.

Substituting (3.46) in (3.44), we get

$$\sum_n C_n H_0(\vec{r})\phi_n(\vec{r}) + \sum_n C_n H'(\vec{r})\phi_n(\vec{r}) = E\sum_n C_n\phi_n(\vec{r}), \tag{3.48}$$

where we have used the fact that the coefficients C_n are constants independent of both time and space.

The reason for choosing the unperturbed wavefunctions as basis functions for expansion will now become clear. Because of this choice, we can write the first term as $\sum_n C_n E_n \phi_n(\vec{r})$ since the basis functions *obey* the unperturbed Schrödinger equation (3.43). Therefore, we get:

$$\sum_n C_n E_n\phi_n(\vec{r}) + \sum_n C_n H'(\vec{r})\phi_n(\vec{r}) = E\sum_n C_n\phi_n(\vec{r}), \tag{3.49}$$

Premultiplying the last equation throughout by $\phi_m^*(\vec{r})$ and integrating over all space, we obtain

$$\sum_n C_n E_n \int_0^\infty d^3\vec{r}\phi_m^*(\vec{r})\phi_n(\vec{r}) + \sum_n C_n \int_0^\infty d^3\vec{r}\phi_m^*(\vec{r})H'(\vec{r})\phi_n(\vec{r})$$
$$= E\sum_n C_n \int_0^\infty d^3\vec{r}\phi_m^*(\vec{r})\phi_n(\vec{r}). \tag{3.50}$$

Next, using the orthonormality property in (3.47), we get

$$\sum_n C_n E_n \delta_{m,n} + \sum_n C_n H'_{mn} = E\sum_n C_n \delta_{m,n}, \tag{3.51}$$

where

$$H'_{mn} = \int_0^\infty d^3\vec{r}\phi_m^*(\vec{r})H'(\vec{r})\phi_n(\vec{r}). \tag{3.52}$$

The quantity H'_{mn} is usually referred to as a "matrix element."

The Kronecker delta in (3.51) removes the summation, so that we end up with

$$C_m E_m + \sum_n C_n H'_{mn} = EC_m. \tag{3.53}$$

The above equation is a set of coupled equations for the coefficient C_m, which can be written as

$$\begin{aligned}
C_1 E_1 + \sum_n C_n H'_{1n} &= EC_1 \quad (m = 1) \\
C_2 E_2 + \sum_n C_n H'_{2n} &= EC_2 \quad (m = 2) \\
&\cdots \\
&\cdots \\
C_p E_p + \sum_n C_n H'_{pn} &= EC_p \quad (m = p) \\
&\cdots \\
&\cdots \\
C_n E_n + \sum_n C_n H'_{nn} &= EC_n \quad (m = n).
\end{aligned} \tag{3.54}$$

Such coupled equations can be written in a matrix form:

$$\begin{bmatrix} E_1 + H'_{11} & H'_{12} & \cdots\cdots\cdots & H'_{1n} \\ H'_{21} & E_2 + H'_{22} & \cdots\cdots\cdots & H'_{2n} \\ \cdots & \cdots & \cdots\cdots\cdots & \cdots \\ \cdots & \cdots & \cdots\cdots\cdots & \cdots \\ \cdots & \cdots & \cdots\cdots\cdots & \cdots \\ H'_{n1} & H'_{n2} & \cdots\cdots\cdots & E_n + H'_{nn} \end{bmatrix} \begin{bmatrix} C_1 \\ C_2 \\ \cdots \\ \cdots \\ \cdots \\ C_n \end{bmatrix} = E \begin{bmatrix} C_1 \\ C_2 \\ \cdots \\ \cdots \\ \cdots \\ C_n \end{bmatrix}. \tag{3.55}$$

Note that since H'_{mn} appears as elements of a matrix, it is commonly referred to as a "matrix element."

The last equation shows that the new eigenenergy E is the eigenvalue of the square matrix on the left-hand-side and for each eigenvalue we can find the corresponding eigenfunction which will give us the coefficients $C_1, C_2, \ldots, C_n$.

Finally, substituting these coefficients in (3.46), we can get the new wavefunction $\Psi(\vec{r})$. It is easy to see that if the new wavefunction $\Psi(\vec{r})$ is normalized, then

$$\sum_n |C_n|^2 = 1. \tag{3.56}$$

Let us find the first perturbed eigenstate, i.e., the first energy eigenvalue E and the corresponding wavefunction. As long as the perturbation is *weak*, we can find an *approximate* analytical answer. The answer is

$$\begin{aligned} E^{(1)} &\approx E_1 + H'_{11} + \sum_{p=2}^{n} \frac{|H'_{1p}|^2}{E_1 + H'_{11} - E_p - H'_{1p}} \\ C^{(1)}_{n,n\neq 1} &\approx \frac{H'_{n1}}{E_1 + H'_{11} - E_n - H'_{1n}} C_1, \end{aligned} \tag{3.57}$$

where the superscript "(1)" denotes that this is the first eigenstate of the perturbed system. These results are valid if: (1) the perturbation is weak, and (2) the first unperturbed energy state E_1 is not degenerate with any other unperturbed energy state. We will prove this answer for a simple 2×2 matrix later since that can be handled analytically (larger matrices will require a computer).

Substituting the last result in (3.46), the wavefunction for the first eigenstate of the perturbed system will be

$$\begin{aligned} \Psi^{(1)}(\vec{r}) &= C_1 \left[\phi_1(\vec{r}) + \frac{H'_{21}}{E_1 + H'_{11} - E_2 - H'_{12}} \phi_2(\vec{r}) + \dots\dots\dots \right. \\ &\quad \left. + \frac{H'_{n1}}{E_1 + H'_{11} - E_n - H'_{1n}} \phi_n(\vec{r}) \right] \\ &= C_1 \left[\phi_1(\vec{r}) + \sum_{p=2}^{n} \frac{H'_{p1}}{E_1 + H'_{11} - E_p - H'_{1p}} \phi_p(\vec{r}) \right]. \end{aligned} \tag{3.58}$$

The only remaining unknown is C_1 which is found from the normalization condition in (3.56):

$$\begin{aligned} |C_1|^2 &\left[1 + \left| \frac{H'_{12}}{E_1 + H'_{11} - E_2 - H'_{12}} \right|^2 + \dots\dots\dots \right. \\ &\quad \left. + \left| \frac{H'_{n1}}{E_1 + H'_{11} - E_n - H'_{1n}} \right|^2 \right] = 1. \end{aligned} \tag{3.59}$$

This gives the magnitude of C_1, but not its phase. It does not matter since the absolute phase of the wavefunction is not something we can measure in any case. It is always undefined to the extent of an arbitrary constant.

3.5.1.1 The 2x2 Matrix Case: An Analytical Result

If the square matrix on the left-hand-side of (3.55) were a simple 2×2 matrix, then we would have

$$\begin{bmatrix} E_1 + H'_{11} - E & H'_{12} \\ H'_{21} & E_2 + H'_{22} - E \end{bmatrix} \begin{bmatrix} C_1 \\ C_2 \end{bmatrix} = 0. \tag{3.60}$$

In order to have nontrivial solutions for C_1, C_2, the determinant of the matrix must vanish, which yields

$$\left(E_1 + H'_{11} - E\right)\left(E_2 + H'_{22} - E\right) - H'_{12}H'_{21} = 0. \tag{3.61}$$

Since H' is a Hermitian operator, $H'_{12} = H'^{\,*}_{21}$ (where the asterisk denotes complex conjugate), so that $H'_{12}H'_{21} = |H'_{12}|^2$. Therefore, the last equation becomes a quadratic equation

$$E^2 - (E_1 + H'_{11} + E_2 + H'_{22})E + [(E_1 + H'_{11})(E_2 + H'_{22}) - |H'_{12}|^2] = 0. \tag{3.62}$$

Assume now that $E_1 \neq E_2$ and $H'_{11} \neq H'_{22}$. In that case, the two unperturbed energy states E_1 and E_2 are nondegenerate and the resulting perturbation theory is called *nondegenerate* perturbation theory. The solution for the last equation is

$$E = \frac{E_1 + H'_{11} + E_2 + H'_{22}}{2} \pm \frac{E_1 + H'_{11} - E_2 - H'_{22}}{2} \times \sqrt{1 + \frac{4|H'_{12}|^2}{(E_1 + H'_{11} - E_2 - H'_{22})^2}}. \tag{3.63}$$

Making a binomial expansion of the terms under the radical, we get

$$E = \frac{E_1 + H'_{11} + E_2 + H'_{22}}{2} \pm \frac{E_1 + H'_{11} - E_2 - H'_{22}}{2} \times \left[1 + \frac{2|H'_{12}|^2}{(E_1 + H'_{11} - E_2 - H'_{22})^2} + \texttt{higher order terms.}\right]. \tag{3.64}$$

Now if the perturbation had been *weak* so that $|H'_{12}| \ll |E_1 + H'_{11} - E_2 - H'_{22}|$, we can retain only the first term in the binomial expansion and neglect the higher order terms which are progressively smaller. The resulting theory is known as *first-order* nondegenerate perturbation theory. This theory then yields the solution for the new (perturbed) energy as

$$E \approx \begin{cases} E_1 + H'_{11} + \frac{|H'_{12}|^2}{E_1 + H'_{11} - E_2 - H'_{22}} \\ E_2 + H'_{22} - \frac{|H'_{12}|^2}{E_1 + H'_{11} - E_2 - H'_{22}} \end{cases} \tag{3.65}$$

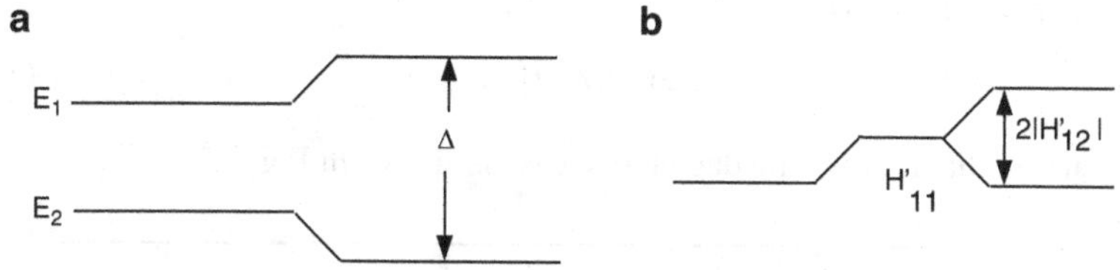

Fig. 3.5 Energy splitting of states caused by time-independent perturbation: (**a**) nondegenerate case, and (**b**) degenerate case

Generalizing the first solution to the case of n-states instead of just two states, we get the result in (3.57).

Let us find the eigenfunction corresponding to the first eigenvalue in the preceding equation. It is easy to show that the solution is

$$C_2 = \frac{H'_{21}}{E_1 + H'_{11} - E_2 - H'_{22}} C_1, \tag{3.66}$$

which, when generalized to n states instead of just two states, yields the result in (3.57).

From (3.65), we see that the energy splitting between the two states in this case is

$$\Delta = (E_2 - E_1) + (H'_{22} - H'_{11}) + 2\frac{|H'_{12}|^2}{E_2 + H'_{22} - E_1 - H'_{11}} \tag{3.67}$$

The energy splitting due to perturbation is shown schematically in Fig. 3.5a.

Special Topic 2: Degenerate perturbation theory

In the case when the unperturbed states E_1 and E_2 are degenerate, i.e., $E_1 = E_2$, (3.60) becomes

$$\begin{bmatrix} E_1 + H'_{11} - E & H'_{12} \\ H'_{21} & E_1 + H'_{11} - E \end{bmatrix} \begin{bmatrix} C_1 \\ C_2 \end{bmatrix} = 0. \tag{3.68}$$

This equation can be solved exactly for the energies of the perturbed states. These solutions are

$$E = E_1 + H'_{11} \pm |H'_{12}|. \tag{3.69}$$

The coefficients C_1 and C_2 are related as $C_2 = \pm e^{i\theta} C_1$, where θ is the phase of the generally complex quantity H'_{21}.

The energy splitting is

$$\Delta = 2|H'_{12}|. \tag{3.70}$$

The energy diagram for the degenerate case is shown in Fig. 3.5b.

3.5.2 *Time-Dependent Perturbation Theory*

Consider the situation when the perturbation is both time and space dependent. In that case, our task will be to solve the time-dependent Schrödinger equation

$$i\hbar \frac{\partial \Psi(\vec{r},t)}{\partial t} = [H_0(\vec{r}) + H'(\vec{r},t)]\Psi(\vec{r},t). \tag{3.71}$$

We will once again write the perturbed wavefunction as a series expansion involving the unperturbed states as basis functions, but this time the coefficients of expansion will be time dependent. Furthermore, we will include the time-dependent phase of the unperturbed wavefunctions. Hence, (3.46) will be modified as

$$\Psi(\vec{r},t) = \sum_n C_n(t)\phi_n(\vec{r})e^{-iE_nt/\hbar}. \tag{3.72}$$

Substituting (3.72) in (3.71), we get

$$i\hbar \sum_n \frac{\partial C_n(t)}{\partial t}\phi_n(\vec{r})e^{-iE_nt/\hbar} = \sum_n C_n(t)[H_0(\vec{r}) - E_n]\phi_n(\vec{r})e^{-iE_nt/\hbar} + \sum_n H'(\vec{r},t)C_n(t)\phi_n(\vec{r})e^{-iE_nt/\hbar}. \tag{3.73}$$

The first term on the right-hand-side is obviously zero since $\phi_n(\vec{r})e^{-iE_nt/\hbar}$ is the unperturbed solution of the Schrödinger equation. Next, premultiplying both sides of the above equation once again with $\phi_m^*(\vec{r})e^{iE_mt/\hbar}$, integrating over all space and using the orthonormality condition in (3.47), we obtain

$$i\hbar \frac{\partial C_m(t)}{\partial t} = \sum_n C_n(t)H'_{mn}(t)e^{i\frac{E_m-E_n}{\hbar}t}, \tag{3.74}$$

where

$$H'_{mn}(t) = \int_0^\infty d^3\vec{r}\phi_m^*(\vec{r})H'(\vec{r},t)\phi_n(\vec{r}). \tag{3.75}$$

Unlike in the time-independent case, the matrix element $H'_{mn}(t)$ is now a function of time since the perturbation is time-dependent.

Equation (3.74) is a collection of coupled differential equations for the coefficients and can be written in a matrix form

$$
i\hbar\frac{\partial}{\partial t}\begin{bmatrix} C_1(t) \\ C_2(t) \\ \cdots \\ \cdots \\ \cdots \\ C_n(t) \end{bmatrix}
= \begin{bmatrix} H'_{11}(t) & H'_{12}(t)e^{i\frac{E_1-E_2}{\hbar}t} & \cdots\cdots\cdots & H'_{1n}(t)e^{i\frac{E_1-E_n}{\hbar}t} \\ H'_{21}(t)e^{i\frac{E_2-E_1}{\hbar}t} & H'_{22}(t) & \cdots\cdots\cdots & H'_{2n}(t)e^{i\frac{E_2-E_n}{\hbar}t} \\ \cdots & & & \\ \cdots & & & \\ \cdots & & & \\ H'_{n1}(t)e^{i\frac{E_n-E_1}{\hbar}t} & H'_{n2}(t)e^{i\frac{E_n-E_2}{\hbar}t} & \cdots\cdots\cdots & H'_{nn} \end{bmatrix}\begin{bmatrix} C_1(t) \\ C_2(t) \\ \cdots \\ \cdots \\ \cdots \\ C_n(t) \end{bmatrix}. \tag{3.76}
$$

The solution of the above equation is

$$
\begin{bmatrix} C_1(t) \\ C_2(t) \\ \cdots \\ \cdots \\ \cdots \\ C_n(t) \end{bmatrix} = e^{-i[\mathcal{H}]t/\hbar}\begin{bmatrix} C_1(0) \\ C_2(0) \\ \cdots \\ \cdots \\ \cdots \\ C_n(0) \end{bmatrix}, \tag{3.77}
$$

where $[\mathcal{H}]$ is the square matrix in the right-hand-side of (3.76). Obviously, in order to find the perturbed wavefunction given in (3.72), we will have to know the initial condition, i.e., what state the electron initially was in before the perturbation was turned on. If the electron was in the m-th eigenstate at time $t = 0$ when the perturbation was turned on, then $C_m(0) = 1$ and all other coefficients at time $t = 0$ were zero. This will allow us to find the coefficients at any time t and correspondingly find the perturbed wavefunction at any time t from (3.72).

Now, in the case of time-independent perturbation theory, we found not only the perturbed wavefunction, but also the perturbed energy eigenstate. So, what about the perturbed energy when the perturbation is time dependent? Actually it makes no sense to talk about the energy any more since the energy is no longer conserved when the perturbation is time dependent. For example, if the perturbing agent is a phonon, it can be absorbed to increase the electron's energy by the phonon energy, or a phonon could be emitted (stimulated emission) so that the electron's energy drops by an amount equal to the phonon energy. Since the energy is no longer conserved, we say that energy is no longer a "good" quantum number.

In Sect. 3.3.1, we showed that only when the entire Hamiltonian is time independent, we can write the wavefunction as the product of a time-dependent part and a space-dependent part. The variable separation solution then yields a constant energy eigenvalue. But now, looking at (3.72), we see immediately that the perturbed wavefunction under a time-dependent perturbation can no longer be written as the product of a time-dependent part and a space-dependent part. Each term in the summation can be written as such a product, but the entire sum cannot be written as such a product. Hence, we will not get a constant eigenvalue.

Time-dependent perturbations change the energy state of the electron due to absorption or emission of the perturbing agent. In Chap. 5, we will make use of the time-dependent perturbation theory to derive Fermi's Golden Rule that provides a prescription for calculating electron scattering rates associated with scattering of an electron due to time-dependent perturbations.

Special Topic 3: Exponentiating a matrix

In (3.77), the matrix $e^{-i[\mathcal{H}]t/\hbar}$ is called a unitary transformation since it is a unitary matrix. It is unitary since $\mathcal{H}$ is a Hermitian operator. In order to find this matrix, first evaluate the different eigenvalues of $[\mathcal{H}]$ and call them $\Lambda_1, \Lambda_2, \cdots, \Lambda_n$, and let their corresponding eigenvectors be

$$\begin{bmatrix} \xi_1^{(1)} \\ \xi_1^{(2)} \\ \cdots \\ \cdots \\ \cdots \\ \xi_1^{(n)} \end{bmatrix}, \begin{bmatrix} \xi_2^{(1)} \\ \xi_2^{(2)} \\ \cdots \\ \cdots \\ \cdots \\ \xi_2^{(n)} \end{bmatrix}, \begin{bmatrix} \cdots \\ \cdots \\ \cdots \\ \cdots \\ \cdots \\ \cdots \end{bmatrix}, \begin{bmatrix} \cdots \\ \cdots \\ \cdots \\ \cdots \\ \cdots \\ \cdots \end{bmatrix}, \begin{bmatrix} \cdots \\ \cdots \\ \cdots \\ \cdots \\ \cdots \\ \cdots \end{bmatrix}, \begin{bmatrix} \xi_n^{(1)} \\ \xi_n^{(2)} \\ \cdots \\ \cdots \\ \cdots \\ \xi_n^{(n)} \end{bmatrix}.$$

Now, define a matrix M whose columns are the eigenvectors, i.e.,

$$[M] = \begin{bmatrix} \xi_1^{(1)} & \xi_2^{(1)} & \cdots\cdots\cdots & \xi_n^{(1)} \\ \xi_1^{(2)} & \xi_2^{(2)} & \cdots\cdots\cdots & \xi_n^{(2)} \\ \cdots & \cdots & \cdots\cdots\cdots & \cdots \\ \cdots & \cdots & \cdots\cdots\cdots & \cdots \\ \cdots & \cdots & \cdots\cdots\cdots & \cdots \\ \xi_1^{(n)} & \xi_2^{(n)} & \cdots\cdots\cdots & \xi_n^{(n)} \end{bmatrix}$$

and another diagonal matrix $[\Lambda]$ as

$$[\Lambda] = \begin{bmatrix} e^{-i\Lambda_1 t/\hbar} & 0 & 0 & 0\ 0 & 0 \\ 0 & e^{-i\Lambda_2 t/\hbar} & 0 & 0\ 0 & 0 \\ 0 & 0 & e^{-i\Lambda_3 t/\hbar} & \ldots\ldots & \ldots \\ \ldots & \ldots & \ldots & \ldots\ldots & \ldots \\ \ldots & \ldots & \ldots & \ldots\ldots & \ldots \\ 0 & 0 & 0 & 0\ 0 & e^{-i\Lambda_n t/\hbar} \end{bmatrix}.$$

The matrix

$$e^{-i[\mathcal{H}]t/\hbar} = [M][\Lambda][M]^{-1}. \tag{3.78}$$

The last equation tells us how to exponentiate a matrix.

3.6 Wavefunction Engineering: Device Applications of Perturbation Theory

Perturbation theory, which we have just discussed, showed us that by applying an external perturbation to an electron (or a hole) in a solid state structure, we can change its wavefunction. This has found applications in a number of electronic and optical devices. We will discuss two of them here.

3.6.1 The Velocity Modulation Transistor

The velocity modulation transistor (VMT), proposed in the early 1980s [2, 3], is an excellent example of the application of perturbation theory to a real solid state device problem. This device is an analog of the traditional metal oxide semiconductor field effect transistor (MOSFET). A normal n-channel MOSFET works as follows. An electrostatic potential applied to the gate terminal changes the carrier concentration in the channel (see Fig. 3.6). For example, if we apply a positive potential to the gate, then it draws electrons into the channel from the source contact by electrostatic attraction, thereby increasing the channel conductance since the channel conductivity $\sigma_n = en\mu_n$ (where n is the electron concentration in the channel and μ_n is the electron mobility). This, in turn, will increase the source-to-drain current at a fixed source-to-drain bias, which realizes transistor action. All ordinary n-channel MOSFETs work on the basis of changing n with a gate potential.

We can, however, also increase or decrease the mobility μ_n with a gate potential. That too will change the channel conductance and realize transistor action. This is

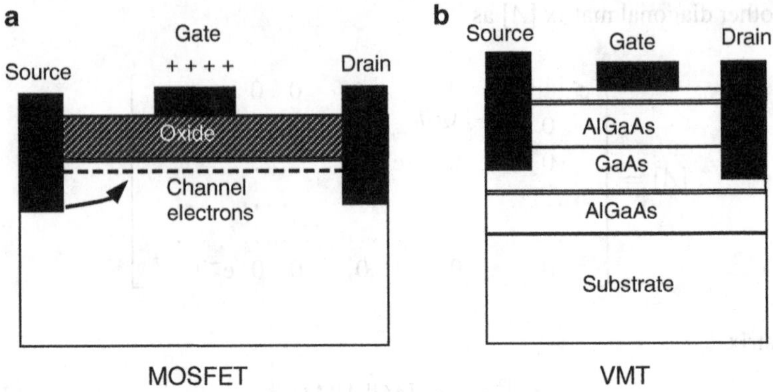

Fig. 3.6 (**a**) A traditional n-channel metal-oxide-semiconductor-field-effect-transistor (MOSFET), (**b**) a Velocity Modulation Transistor (VMT)

the operating principle of the VMT.[2] The typical VMT configuration is shown in Fig. 3.6. It consists of a thin layer of, say, GaAs sandwiched between two AlGaAs layers. The GaAs layer acts as a rectangular potential well (it is called a "quantum well") since it has a smaller bandgap than AlGaAs. When a negative electrostatic potential is applied to the gate, it exerts an electric field in a direction perpendicular to the heterointerfaces, which tilts the conduction band diagram (see Fig. 3.8 in Problem 3.9). The electron wavefunction is skewed toward the lower GaAs/AlGaAs interface and the electrons now reside closer to this interface. There, they suffer severe scattering from the rough interface which decreases electron mobility. That, in turn, decreases the channel conductance, thereby realizing transistor action. The theory of this device has been discussed in [2, 3].

In order to calculate the change in electron mobility induced by the gate potential, we will treat the latter as a time-independent perturbation and apply time-independent perturbation theory to calculate the new wavefunction. From this wavefunction, we can compute the new scattering rate and hence the new mobility. This is a direct application of perturbation theory to a real-life device problem.

Changing mobility instead of carrier concentration is preferable from considerations of switching speed, i.e., how fast the conductance can be changed by the gate voltage. The disadvantage of changing n is that it takes too long. Typically, the current will respond in a timescale no shorter than the *transit time* through the channel, which is the average time it takes for electrons to travel from the source to the drain. This is because additional electrons have to flow in from the source and populate the entire channel up to the drain, and that takes at least as long as the transit time. The transit time depends on the channel length and the average electron velocity (it is the ratio of the two).

[2]Since changing the mobility changes the drift velocity of carriers in a constant electric field, this device is called a "velocity modulation transistor."

Changing μ_n however does not take that long. The wavefunction can be skewed very fast—on a timescale much faster than the transit time through the channel. Nevertheless, the mobility cannot change any faster than the momentum relaxation time (why?). Since the transit time is typically much longer than the momentum relaxation time (at least in transistors that existed in the 1980s), the VMT can be faster than the traditional MOSFET. However, the disadvantage of the VMT is that we can change n by several orders of magnitude with a reasonable gate voltage ($\sim$1 V) but we certainly cannot change μ_n by several orders of magnitude. Changing n can easily result in a conductance ON/OFF ratio of $\sim 10^6$, but changing μ_n may not be able to produce a conductance ON/OFF ratio of even 10. This is the reason why VMTs did not ultimately succeed as mainstream transistors against MOSFETs, but the device still serves as an useful example of perturbation theory at work.

3.6.2 The Electro-Optic Modulator Based on the Quantum-Confined Stark Effect

Perturbation theory can be applied to find the device characteristics of optical modulators based on the so-called *quantum-confined dc Stark effect*. Problem 3.9 deals with direct application of the time-independent perturbation theory to this device. Consider a quantum well in which carriers are photoexcited from the heavy hole band to the conduction band (by a light source), and then the photoexcited electrons and holes recombine to produce photons or light. This process is known as photoluminescence (PL). The energy of the emitted photons is the energy gap between the lowest electronic subband and the highest heavy-hole subband. If a dc electric field is applied transverse to the well interface, then the wavefunctions of electrons and holes are skewed in opposite directions because these particles are oppositely charged (see Fig. 3.8), and the intensity of the emitted light decreases since the intensity is proportional to the square of the overlap between the electron and hole wavefunctions. This overlap decreases when the wavefunctions are skewed. At the same time, the effective bandgap (energy difference between the lowest electron state and highest heavy hole state) decreases, causing a redshift in the PL spectrum. Thus, by applying a transverse electric field, we can change both the intensity and the frequency of the emitted light. In fact, if we apply an ac transverse electric field, we can modulate both the intensity and the frequency of the emitted light with the electric field's frequency, resulting in both amplitude modulation (AM) and frequency modulation (FM). The theory of this device has been discussed in [4].

The above two examples show how device functionality can be implemented by "wavefunction engineering," which is the art of changing an electron's wavefunction with an external perturbation, such as a voltage. These devices best exemplify the application of perturbation theory to real-life solid-state systems. Perturbation theory also has other applications, such as in calculation of bandstructure in a crystal and computation of electron scattering rates. We will see examples of them in the next two chapters.

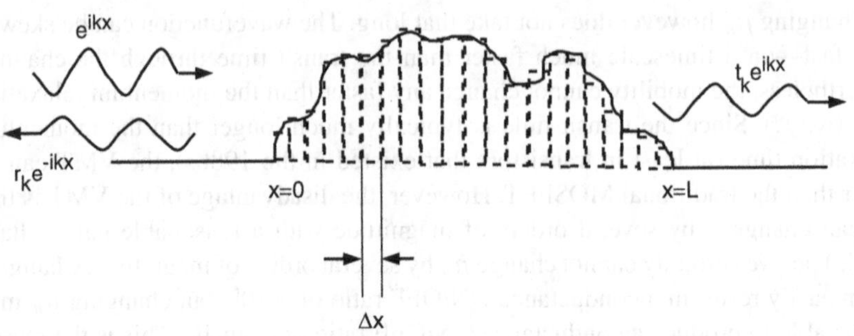

Fig. 3.7 An arbitrary one-dimensional potential profile approximated as a series of steps

---

Special Topic 4: A simple technique to solve the one-dimensional time-independent Schrödinger equation

We would like to solve the one-dimensional time-independent Schrödinger equation in an arbitrary potential profile $V(x)$ as shown in Fig. 3.7. The potential profile is approximated by a series of steps of uniform width Δx. The step height V_n of the n-th segment is made equal to the average potential within that segment, i.e., $\int_{x_n}^{x_n+1} \mathrm{d}x\, V(x) = V_n \Delta x$.

The time-independent Schrödinger equation within the n-th segment is then written as

$$-\frac{\hbar^2}{2m^*}\frac{\mathrm{d}^2\phi(x)}{\mathrm{d}x^2} + (V_n - E)\phi(x) = 0. \tag{3.79}$$

which is recast as

$$\frac{\mathrm{d}^2\phi(x)}{\mathrm{d}x^2} + k_n^2\phi(x) = 0, \tag{3.80}$$

where $k_n = \sqrt{2m^*(E - V_n)}/\hbar$.

We can break up the last equation into two equations and write them in a matrix form:

$$\frac{\mathrm{d}}{\mathrm{d}x}\begin{bmatrix}\phi(x)\\ \phi'(x)\end{bmatrix} = \begin{bmatrix}0 & 1\\ -k_n^2 & 0\end{bmatrix}\begin{bmatrix}\phi(x)\\ \phi'(x)\end{bmatrix}, \tag{3.81}$$

where $\phi'(x) = \frac{\mathrm{d}\phi(x)}{\mathrm{d}x}$.

Consider the segment located between x and $x + \Delta x$. We can relate the wavefunction at the right edge of the segment to that at the left edge according to

$$\begin{bmatrix} \phi(x+\Delta x) \\ \phi'(x+\Delta x) \end{bmatrix} = \mathrm{e}^{[A]\Delta x} \begin{bmatrix} \phi(x) \\ \phi'(x) \end{bmatrix}, \tag{3.82}$$

where the matrix $[A]$ is the 2×2 matrix in (3.81).

We now have to exponentiate the matrix $[A\Delta x]$. For this, we need to first find the eigenvalues of this matrix and the corresponding eigenvectors.

The two eigenvalues are $\Lambda_1 = \mathrm{i}k_n\Delta x$ and $\Lambda_2 = -\mathrm{i}k_n\Delta x$, while the corresponding eigenvectors are:

$$\lambda_1 = \begin{bmatrix} -\mathrm{i} \\ k_n \end{bmatrix}$$

$$\lambda_2 = \begin{bmatrix} \mathrm{i} \\ k_n \end{bmatrix}. \tag{3.83}$$

Therefore, (see Special Topic 3)

$$\mathrm{e}^{[A]\Delta x} = \begin{bmatrix} -\mathrm{i} & \mathrm{i} \\ k_n & k_n \end{bmatrix} \begin{bmatrix} \mathrm{e}^{\mathrm{i}k_n\Delta x} & 0 \\ 0 & \mathrm{e}^{-\mathrm{i}k_n\Delta x} \end{bmatrix} \begin{bmatrix} -\mathrm{i} & \mathrm{i} \\ k_n & k_n \end{bmatrix}^{-1}. \tag{3.84}$$

The last matrix above can be easily evaluated:

$$\begin{bmatrix} -\mathrm{i} & \mathrm{i} \\ k_n & k_n \end{bmatrix}^{-1} = \begin{bmatrix} \mathrm{i}/2 & 1/2k_n \\ -\mathrm{i}/2 & 1/2k_n \end{bmatrix}. \tag{3.85}$$

From (3.82), we now get

$$\begin{bmatrix} \phi(x+\Delta x) \\ \phi'(x+\Delta x) \end{bmatrix} = \begin{bmatrix} \cos(k_n\Delta x) & \sin(k_n\Delta x)/k_n \\ -k_n\sin(k_n\Delta x) & \cos(k_n\Delta x) \end{bmatrix} \begin{bmatrix} \phi(x) \\ \phi'(x) \end{bmatrix}, \tag{3.86}$$

Consider now a plane wave with wavevector k impinging on the potential profile from the left as shown in Fig. 3.7. It will be partially reflected and partially transmitted. If there are no inelastic processes within the region, then the energy with which the electron will be transmitted will be the same energy with which it was incident on the region, so that the wavevector of the transmitted wave will be k', where $E(k') - E(k) = E_{\mathrm{cl}} - E_{\mathrm{cr}}$. Here, $E(k)$ is the energy of an electron with wavevector k, E_{cl} is the conduction band level to the left of the region, and E_{cr} is the conduction band level to the right of the region. The wavefunction on the left of the region ($x \leq 0$) is the superposition of the incident and transmitted waves and hence will be $\mathrm{e}^{\mathrm{i}kx} + r_k\mathrm{e}^{-\mathrm{i}kx}$, while the wavefunction on the right of the region ($x \geq L$) will be $t_k\mathrm{e}^{\mathrm{i}k'(x-L)}$, where r_k and t_k are the (wavevector-dependent) reflection and transmission coefficients, respectively, of an electron impinging on the region with wavevector k.

The wavefunction at the right edge of the first segment to the right of $x = 0$ will be $\phi(\Delta x)$ which will obey (see (3.86))

$$\begin{bmatrix} \phi(\Delta x) \\ \phi'(\Delta x) \end{bmatrix} = \begin{bmatrix} \cos(k_1 \Delta x) & \sin(k_1 \Delta x)/k_1 \\ -k_1 \sin(k_1 \Delta x) & \cos(k_1 \Delta x) \end{bmatrix} \begin{bmatrix} \phi(0) \\ \phi'(0) \end{bmatrix}$$
$$= \begin{bmatrix} \cos(k_1 \Delta x) & \sin(k_1 \Delta x)/k_1 \\ -k_1 \sin(k_1 \Delta x) & \cos(k_1 \Delta x) \end{bmatrix} \begin{bmatrix} 1 + r_k \\ ik - ikr_k \end{bmatrix}, \tag{3.87}$$

since $\phi(0) = 1 + r_k$ and $\phi'(0) = ik - ikr_k$.

Since the wavefunction and its first derivative are continuous in space, the wavefunction to the right of the second segment ($\phi(2\Delta x)$) obeys the equation

$$\begin{bmatrix} \phi(2\Delta x) \\ \phi'(2\Delta x) \end{bmatrix} = \begin{bmatrix} \cos(k_2 \Delta x) & \sin(k_2 \Delta x)/k_2 \\ -k_2 \sin(k_2 \Delta x) & \cos(k_2 \Delta x) \end{bmatrix} \begin{bmatrix} \phi(\Delta x) \\ \phi'(\Delta x) \end{bmatrix}$$
$$= \begin{bmatrix} \cos(k_2 \Delta x) & \sin(k_2 \Delta x)/k_2 \\ -k_2 \sin(k_2 \Delta x) & \cos(k_2 \Delta x) \end{bmatrix}$$
$$\times \begin{bmatrix} \cos(k_1 \Delta x) & \sin(k_1 \Delta x)/k_1 \\ -k_1 \sin(k_1 \Delta x) & \cos(k_1 \Delta x) \end{bmatrix} \begin{bmatrix} 1 + r_k \\ ik - ikr_k \end{bmatrix}. \tag{3.88}$$

Continuing in this fashion, we can cascade (or multiply) the matrices all the way from $x = 0$ to $x = L$ and obtain

$$\begin{bmatrix} \phi(L) \\ \phi'(L) \end{bmatrix} = \begin{bmatrix} t_k \\ ik' t_k \end{bmatrix} = \begin{bmatrix} \cos(k_M \Delta x) & \sin(k_M \Delta x)/k_M \\ -k_M \sin(k_M \Delta x) & \cos(k_M \Delta x) \end{bmatrix} \times \cdots\cdots\cdots$$
$$\times \begin{bmatrix} \cos(k_1 \Delta x) & \sin(k_1 \Delta x)/k_1 \\ -k_1 \sin(k_1 \Delta x) & \cos(k_1 \Delta x) \end{bmatrix} \begin{bmatrix} 1 + r_k \\ ik - ikr_k \end{bmatrix}, \tag{3.89}$$

where we have assumed that there are M segments of width Δx between $x = 0$ and $x = L$. The reader can easily grasp that the segments do not actually have to be of equal width. They could very easily be of different widths (a nonuniform mesh) and the process still works.

Note that we have to cascade the matrices from right to left. Their ordering matters since, in general, matrices do not commute. But now we have two equations and two unknowns t_k and r_k in (3.89). We can solve these two equations to find the transmission and reflection coefficients t_k and r_k. We show how to do this in Special Topic 5, which follows immediately.

Now suppose that we want to know the wavefunction at any arbitrary location x' which happens to be located within a segment that is N segments from the left edge at $x = 0$. It is easy to see that

$$\begin{bmatrix} \phi(x') \\ \phi'(x') \end{bmatrix} = \begin{bmatrix} \cos(k_N \Delta x) & \sin(k_N \Delta x)/k_N \\ -k_N \sin(k_N \Delta x) & \cos(k_N \Delta x) \end{bmatrix} \times \ldots\ldots\ldots$$

$$\times \begin{bmatrix} \cos(k_1 \Delta x) & \sin(k_1 \Delta x)/k_1 \\ -k_1 \sin(k_1 \Delta x) & \cos(k_1 \Delta x) \end{bmatrix} \begin{bmatrix} 1 + r_k \\ ik - ik r_k \end{bmatrix}. \tag{3.90}$$

Since we have already determined t_k and r_k, we can now completely find the wavefunction $\phi(x')$ from the last equation. This provides a simple, yet elegant, technique to solve the one-dimensional time-independent Schrödinger equation in any arbitrary potential profile and find the wavefunction at any arbitrary location. The accuracy of the method increases with increasing fine-graining of the mesh, i.e., with decreasing step width Δx.

Special Topic 5: Finding the transmission and reflection amplitudes

Equation (3.89) provides the prescription for finding the transmission amplitude t_k through any arbitrary section of length L. We rewrite this equation as

$$t_k \begin{bmatrix} 1 \\ ik' \end{bmatrix} = [W] \begin{bmatrix} 1 + r_k \\ ik[1 - r_k] \end{bmatrix}, \tag{3.91}$$

where $[W]$ is the total transfer matrix of the section of length L obtained by cascading (or multiplying) the transfer matrices of the constituent sections in the proper order as indicated in (3.89).

Eliminating r_k from the above equation, we get

$$t_k = \frac{2ik(W_{11}W_{22} - W_{12}W_{21})}{i(k'W_{11} + kW_{22}) + (kk'W_{12} - W_{21})}, \tag{3.92}$$

where W_{pq} is the (p, q)-th element of the matrix $[W]$.

It is easy to verify that the matrix $[W]$ is unimodular, i.e., its determinant is always unity [5]. Therefore, the term within the parenthesis in the numerator is 1. Hence,

$$t_k = \frac{2ik}{(ik'W_{11} + ikW_{22}) + (kk'W_{12} - W_{21})}. \tag{3.93}$$

Similarly,

$$r_k = \frac{(ikW_{22} - ik'W_{11}) + (kk'W_{12} + W_{21})}{(ik'W_{11} + ikW_{22}) + (kk'W_{12} - W_{21})}. \tag{3.94}$$

Using the unimodularity of $[W]$, it is easy to show that $\frac{k'}{k}|t_k|^2 + |r_k|^2 = 1$. This guarantees current conservation (Kirchoff's current law) as we will see later in Chap. 8 when we discuss current scattering matrices. The reader may be uncomfortable with the thought that $|t_k|^2 + |r_k|^2 \neq 1$ and worry that it violates conservation of probability since it appears that the transmission and reflection probabilities are not adding up to unity. One should note that the transmission and reflection probabilities are not calculated at the same point in space; the reflection probability is calculated at the left edge and the transmission probability at the right edge. Hence, they need not add up to unity. Actually, what happens is that the current entering the region is proportional to $1 - |r_k|^2$ and the current exiting is proportional to $\frac{k'}{k}|t_k|^2$, so that the identity correctly enforces the requirement that current in is equal to current out.

3.6.2.1 Matrices for Evanescent States

By repeating the procedure used to derive (3.86), the reader can easily show that if the electron wave is evanescent in any section, i.e., if its energy is less than the potential ($E < V_n$), then for that section, (3.86) will be replaced with

$$\begin{bmatrix} \phi(x + \Delta x) \\ \phi'(x + \Delta x) \end{bmatrix} = \begin{bmatrix} \cosh(\kappa_n \Delta x) & \sinh(\kappa_n \Delta x)/\kappa_n \\ \kappa_n \sinh(\kappa_n \Delta x) & \cosh(\kappa_n \Delta x) \end{bmatrix} \begin{bmatrix} \phi(x) \\ \phi'(x) \end{bmatrix}, \tag{3.95}$$

where $\kappa_n = \sqrt{2m^*(V_n - E)}/\hbar = ik_n$. The 2×2 matrix above is also unimodular and its elements are real. The transmission and reflection coefficients of an electron impinging on such a section with wavevector k are still given by (3.93) and (3.94) with the elements of the $[W]$ matrix replaced with the corresponding elements of the matrix above.

3.7 Multi-Electron Schrödinger Equation

So far everything that we have discussed in connection with the Schrödinger equation pertains to the *one-electron* or *single-particle* Schrödinger equation that describes a single isolated electron. However, in any solid, there are numerous electrons which interact with each other, so that the one-electron equation does not appear to be describing reality. Strictly speaking, we should always be dealing with the multi-electron Schrdinger equation in any real situation, but this is enormously complex and therefore avoided in device physics. What we will show in this section is that as long as we solve the one-electron Schrödinger equation simultaneously (or self-consistently) with the Poisson equation, we can account for the effect of other electrons (and any surrounding fixed charges) reasonably well. This is not always true, but is true in most cases of interest. By self-consistent solution, we mean the following: the Schrödinger equation has a potential term in the Hamiltonian. That potential must satisfy the Poisson equation. The latter equation has an electron charge density term which is proportional to the square of the wavefunction's modulus. That wavefunction must satisfy the Schrödinger equation. Thus, the Schrödinger and the Poisson equations have to be mutually consistent. In other words, we have to solve the Schrödinger and Poisson equations simultaneously to find the two variables—potential and wavefunction. If we do this, the resulting wavefunction will have taken into account the effect of all other electrons on the test electron *approximately*.

If the approximate result is not adequate and we need to be exact and accurate (which we almost never have to be), then we must solve N-coupled Schrödinger equations for all the electrons in a system. This is computationally prohibitive, as we illustrate below with just the two-electron problem. It is difficult enough to solve the 2-electron problem, and one can only imagine how difficult it will be to solve the N-electron problem, when $N \gg 2$.

The 2-electron time-independent Schrödinger equation is

$$-\frac{\hbar^2}{2m_0}\nabla^2_{r_1} - \frac{\hbar^2}{2m_0}\nabla^2_{r_2} + V(\vec{r}_1) + V(\vec{r}_2) + \frac{e^2}{4\pi\epsilon|\vec{r}_1 - \vec{r}_2|} = E\psi(\vec{r}_1, \vec{r}_2), \quad (3.96)$$

where $\psi(\vec{r}_1, \vec{r}_2)$ is the 2-electron wavefunction, $\vec{r}_1$ is the coordinate of the first electron, and $\vec{r}_2$ is the coordinate of the second. The interaction between the two electrons is Coulombic (which gives rise to the last term in the left-hand-side) and ϵ is the dielectric constant of the medium hosting the electrons.

We can write the Hamiltonian in the above equation as

$$H(\vec{r}_1, \vec{r}_2) = H_1(\vec{r}_1) + H_2(\vec{r}_2) + H'(\vec{r}_1, \vec{r}_2), \quad (3.97)$$

where

$$\begin{aligned} H_1(\vec{r}_1) &= -\frac{\hbar^2}{2m_0}\nabla^2_{r_1} + V(\vec{r}_1) \\ H_2(\vec{r}_2) &= -\frac{\hbar^2}{2m_0}\nabla^2_{r_2} + V(\vec{r}_2) \\ H'(\vec{r}_1, \vec{r}_2) &= \frac{e^2}{4\pi\epsilon|\vec{r}_1 - \vec{r}_2|}. \end{aligned} \quad (3.98)$$

Now concentrate on the Hamiltonian $H_0 = H_1 + H_2$ (which is the Hamiltonian without the Coulomb interaction term) and let its eigenfunction be $\Phi(\vec{r}_1, \vec{r}_2)$. That means $\Phi(\vec{r}_1, \vec{r}_2)$ satisfies the equation

$$[H_1(\vec{r}_1) + H_2(\vec{r}_2)]\Phi(\vec{r}_1, \vec{r}_2) = \varepsilon\Phi(\vec{r}_1, \vec{r}_2), \quad (3.99)$$

where ε is the energy of the 2-electron system without the Coulomb interaction.

Since $H_1(\vec{r}_1)$ depends only on $\vec{r}_1$, $H_2(\vec{r}_2)$ depends only on $\vec{r}_2$, and there are no mixed terms in H_0 that depend on both $\vec{r}_1$ and $\vec{r}_2$, we can write $\Phi(\vec{r}_1, \vec{r}_2)$ as a product:

$$\Phi(\vec{r}_1, \vec{r}_2) = \phi_m(\vec{r}_1)\phi_n(\vec{r}_2), \tag{3.100}$$

where it is easy to show that $\phi_m(\vec{r}_1)$ and $\phi_n(\vec{r}_2)$ obey the equations

$$\begin{aligned} H_1\phi_m(\vec{r}_1) &= \varepsilon_m\phi_m(\vec{r}_1) \\ H_2\phi_m(\vec{r}_2) &= \varepsilon_n\phi_n(\vec{r}_2) \\ \varepsilon &= \varepsilon_m + \varepsilon_n. \end{aligned} \tag{3.101}$$

We can think of H_0 $(= H_1 + H_2)$ as the unperturbed Hamiltonian, H' as the perturbation, and apply perturbation theory to calculate $\psi(\vec{r}_1, \vec{r}_2)$ and E, but there is a *subtlety* that needs to be addressed before proceeding further. Since electrons are Fermi particles, they obey the so-called *Symmetry Principle* which states that the *total* wavefunction of two electrons must be anti-symmetric under exchange of the two electrons. What we mean by the "total" wavefunction is the wavefunction that includes both the orbital part and the spin part. So far, we have ignored the fact that the electron has a quantum-mechanical property called "spin" that needs to be accounted for in the wavefunction. Roughly speaking—and again this is not accurate—we can think of the total wavefunction as the tensor product of the orbital part (which we are dealing with) and the spin part:

$$\texttt{Total wavefunction} = \texttt{Orbital part} \bigotimes \texttt{Spin part.}$$

If the spins of the two electrons are mutually anti-parallel, then they form the so-called "singlet" state which is considered anti-symmetric in spin space. In that case, the orbital part of the 2-electron wavefunction must be symmetric to make the total wavefunction anti-symmetric. On the other hand, if the spins of the electrons are mutually parallel, then they form the "triplet" state, which is considered symmetric in spin space. In that case, the orbital part of the 2-electron wavefunction must be anti-symmetric.

Looking at (3.100), it is obvious that if $m = n$, then (the orbital part of) the 2-electron wavefunction must be symmetric and given by

$$\Phi_S(\vec{r}_1, \vec{r}_2) = \phi_m(\vec{r}_1)\phi_m(\vec{r}_2), \tag{3.102}$$

which of course mandates that the spins will be anti-parallel and the electrons will form the singlet state. Note that this is in conformity with Pauli Exclusion Principle which states that no two electrons whose wavefunctions have a nonzero overlap can have all their quantum states identical. Therefore, if they are in the same orbital state ϕ_m, then they must have anti-parallel spins.

However, if $m \neq n$, then we have two possibilities. The orbital part could be either symmetric and given by

$$\Phi_S(\vec{r}_1, \vec{r}_2) = \frac{1}{\sqrt{2}}[\phi_m(\vec{r}_1)\phi_n(\vec{r}_2) + \phi_n(\vec{r}_1)\phi_m(\vec{r}_2)], \tag{3.103}$$

or anti-symmetric and given by

$$\Phi_A(\vec{r}_1, \vec{r}_2) = \frac{1}{\sqrt{2}}[\phi_m(\vec{r}_1)\phi_n(\vec{r}_2) - \phi_n(\vec{r}_1)\phi_m(\vec{r}_2)]. \tag{3.104}$$

In the former case, the spins must be anti-parallel (singlet) and in the latter case they must be parallel (triplet) to satisfy the Symmetry Principle. The anti-symmetric orbital wavefunction can be written as a so-called *Slater determinant*:

$$\Phi_A(\vec{r}_1, \vec{r}_2) = \frac{1}{\sqrt{2}} \det \begin{vmatrix} \phi_m(\vec{r}_1) & \phi_m(\vec{r}_2) \\ \phi_n(\vec{r}_1) & \phi_n(\vec{r}_2) \end{vmatrix}. \tag{3.105}$$

The advantage of writing the wavefunction as a Slater determinant is that it can be extended to any arbitrary number of electrons (we state this without proof). Hence, for an N-electron system, the many-body wavefunction could be written as

$$\Phi_A(\vec{r}_1, \vec{r}_2, \ldots \vec{r}_N) = \frac{1}{\sqrt{N}} \det \begin{vmatrix} \phi_1(\vec{r}_1) & \phi_1(\vec{r}_2) & \ldots & \phi_1(\vec{r}_N) \\ \phi_2(\vec{r}_1) & \phi_2(\vec{r}_2) & \ldots & \phi_2(\vec{r}_N) \\ \ldots & \ldots & \ldots & \ldots \\ \phi_N(\vec{r}_1) & \phi_N(\vec{r}_2) & \ldots & \phi_N(\vec{r}_N) \end{vmatrix}. \tag{3.106}$$

Note from (3.104), that if we put $\vec{r}_1 = \vec{r}_2$, then the wavefunction vanishes entirely! This means that if the two electrons have parallel spins, they will never be found at the same location in space. Once again, this is in conformity with the Pauli Exclusion Principle. The Symmetry Principle has the Pauli Exclusion Principle built in it. The fact that electrons of parallel spins avoid each other implies that electron motion in a solid is *correlated* in a way to make this happen, and the wavefunction in (3.104) contains this correlation effect.

It is easy to verify that both $\Phi_S(\vec{r}_1, \vec{r}_2)$ and $\Phi_A(\vec{r}_1, \vec{r}_2)$ satisfy the 2-electron "unperturbed" Schrödinger equation without the Coulomb interaction term, as given in (3.99). Our normal tendency will then be to consider the Coulomb interaction term as a perturbation, $\Phi_S(\vec{r}_1, \vec{r}_2)$ and $\Phi_A(\vec{r}_1, \vec{r}_2)$ as unperturbed wavefunctions, and use them as basis functions in perturbation theory to determine $\psi(\vec{r}_1, \vec{r}_2)$ and E. However, this is an enormously complicated enterprise since any attempt to apply (3.57) and (3.58) will involve nightmarish bookkeeping. The summation over p in these equations will be a double summation over m and n, we will have to account for symmetric and anti-symmetric orbital parts of the wavefunctions, and take spin into account explicitly. Therefore, perturbation theory is not an elegant or practical approach for this problem. There are more convenient and elegant formalisms that deal with many-particle Schrödinger equation, which are outside the scope of this textbook.

3.7.1 *Hartree and Exchange Potentials*

We can calculate the expectation value of the Coulomb repulsion energy between the two electrons *approximately* using the unperturbed wavefunctions $\Phi_S(\vec{r}_1, \vec{r}_2)$ and $\Phi_A(\vec{r}_1, \vec{r}_2)$, provided that the actual perturbed wavefunctions are not too different from the unperturbed ones. This works only if the perturbation (Coulomb interaction) is weak, which unfortunately it may not be. Nonetheless, this is the best we can do without getting too complicated. Within this approach, the Coulomb energy K_{mm} for two electrons in the same orbital state ϕ_m is given by

$$K_{mm} = \langle \phi_m(\vec{r}_1)\phi_m(\vec{r}_2)| \frac{e^2}{4\pi\epsilon|\vec{r}_1 - \vec{r}_2|} |\phi_m(\vec{r}_1)\phi_m(\vec{r}_2)\rangle$$
$$= \int d^3\vec{r}_1 \int d^3\vec{r}_2 \frac{e^2 \left|\phi_m(\vec{r}_1)\right|^2 \left|\phi_m(\vec{r}_2)\right|^2}{4\pi\epsilon \left|\vec{r}_1 - \vec{r}_2\right|}, \tag{3.107}$$

where we have omitted the spin part of the wavefunction.

When $m \neq n$, the Coulomb repulsion energy will have two different values depending on whether we choose the symmetric orbital wavefunction $\Phi_S(\vec{r}_1, \vec{r}_2)$ [singlet state] or the anti-symmetric orbital wavefunction $\Phi_A(\vec{r}_1, \vec{r}_2)$ [triplet state] to calculate the expectation value of the Coulomb interaction.

3.7.1.1 Singlet State; Symmetric Orbital Wavefunction

In this case,

$$K_{mn} = \left\langle \Phi_S(\vec{r}_1, \vec{r}_2) \left| \frac{e^2}{4\pi\epsilon\left|\vec{r}_1 - \vec{r}_2\right|} \right| \Phi_S(\vec{r}_1, \vec{r}_2) \right\rangle$$
$$= \frac{1}{2}\int d^3\vec{r}_1 \int d^3\vec{r}_2 \left\{ \frac{e^2}{4\pi\epsilon\left|\vec{r}_1 - \vec{r}_2\right|} \times \left[\phi_m^*(\vec{r}_1)\phi_n^*(\vec{r}_2) + \phi_n^*(\vec{r}_1)\phi_m^*(\vec{r}_2)\right] \right.$$
$$\left. \times \left[\phi_m(\vec{r}_1)\phi_n(\vec{r}_2) + \phi_n(\vec{r}_1)\phi_m(\vec{r}_2)\right] \right\}$$
$$= \kappa_{mn} + J_{mn}, \tag{3.108}$$

where

$$\kappa_{mn} = \int d^3\vec{r}_1 \int d^3\vec{r}_2 \frac{e^2 \left|\phi_m(\vec{r}_1)\right|^2 \left|\phi_n(\vec{r}_2)\right|^2}{4\pi\epsilon\left|\vec{r}_1 - \vec{r}_2\right|}$$
$$J_{mn} = \int d^3\vec{r}_1 \int d^3\vec{r}_2 \frac{e^2 \phi_m^*(\vec{r}_1)\phi_n^*(\vec{r}_2)\phi_n(\vec{r}_1)\phi_m^*(\vec{r}_2)}{4\pi\epsilon\left|\vec{r}_1 - \vec{r}_2\right|}. \tag{3.109}$$

In deriving the above, we used the fact that since $\vec{r}_1$ and $\vec{r}_2$ are dummy variables of integration, we can interchange them within the integral.

The term κ_{mn} is called the Hartree term, or Hartree potential, and the term J_{mn} is called the *exchange* interaction, or exchange potential. The latter's origin can be traced back to the Symmetry Principle which comes about because electron motion is correlated. If we define two quantities

$$\begin{aligned}\rho_{mn}(\vec{r}_1) &= e\phi_m^*(\vec{r}_1)\phi_n(\vec{r}_1)\\ \rho_{nm}(\vec{r}_2) &= e\phi_n^*(\vec{r}_2)\phi_m(\vec{r}_2),\end{aligned} \tag{3.110}$$

then the Hartree and exchange terms can be written as

$$\begin{aligned}\kappa_{mn} &= \int d^3\vec{r}_1 \int d^3\vec{r}_2 \frac{\rho_{mm}(\vec{r}_1)\rho_{nn}(\vec{r}_2)}{4\pi\epsilon|\vec{r}_1 - \vec{r}_2|}\\ J_{mn} &= \int d^3\vec{r}_1 \int d^3\vec{r}_2 \frac{\rho_{mn}(\vec{r}_1)\rho_{nm}(\vec{r}_2)}{4\pi\epsilon|\vec{r}_1 - \vec{r}_2|}.\end{aligned} \tag{3.111}$$

3.7.1.2 Triplet State; Anti-Symmetric Orbital Wavefunction

In this case,

$$\begin{aligned}K_{mn} &= \left\langle \Phi_A(\vec{r}_1, \vec{r}_2) \,|\frac{e^2}{4\pi\epsilon|\vec{r}_1 - \vec{r}_2|}|\Phi_A(\vec{r}_1, \vec{r}_2)\right\rangle\\ &= \frac{1}{2}\int d^3\vec{r}_1 \int d^3\vec{r}_2 \left\{ \frac{e^2}{4\pi\epsilon\,|\vec{r}_1 - \vec{r}_2|} \times [\phi_m^*(\vec{r}_1)\phi_n^*(\vec{r}_2) - \phi_n^*(\vec{r}_1)\phi_m^*(\vec{r}_2)]\right.\\ &\quad \left. \times [\phi_m(\vec{r}_1)\phi_n(\vec{r}_2) - \phi_n(\vec{r}_1)\phi_m(\vec{r}_2)] \right\}\\ &= \kappa_{mn} - J_{mn}.\end{aligned} \tag{3.112}$$

The energy difference between the triplet state and the singlet state is therefore $-2J_{mn}$. If J_{mn} is positive, then the triplet state is at a lower energy than the singlet state. If J_{mn} is negative, then the opposite is true. When two electrons are bound to a common attractive potential, like the nucleus of a helium atom, J_{mn} is usually positive, and the triplet state is at a lower energy than the singlet state. In the context of the helium atom, this state is called orthohelium, which is at a lower energy than the singlet state known as parahelium. On the other hand, if the two electrons are bound to two different potentials, such as in the hydrogen molecule, or two adjacent quantum dots, then J_{mn} is usually negative and the singlet state is at the lower energy.

In a many- (more than two) electron ensemble, the Hartree term is written as

$$\kappa_{mn} = \frac{1}{2}\sum_{i,j}\int d^3\vec{r}_i \int d^3\vec{r}_j \frac{\rho_{mm}(\vec{r}_i)\rho_{nn}(\vec{r}_j)}{4\pi\epsilon|\vec{r}_i - \vec{r}_j|}, \tag{3.113}$$

where the factor of one-half is there to account for the fact that every electron is counted twice in the summation. The last equation is the standard expression for the Coulomb energy due to interaction between two charge densities $\rho(\vec{r}_i)$ and $\rho(\vec{r}_j)$.

The above term can also be written as

$$\kappa_{mn} = \frac{1}{2}\int d^3\vec{r}_i \rho_{mm}(\vec{r}_i) e V_{sc}(\vec{r}_i), \tag{3.114}$$

where

$$V_{sc} = \frac{1}{e}\sum_{i,j}\int d^3\vec{r}_j \frac{\rho_{nn}(\vec{r}_j)}{4\pi\epsilon|\vec{r}_i - \vec{r}_j|}, \tag{3.115}$$

which is a solution of the Poisson equation $-\nabla^2 V_{sc} = \rho/\epsilon$ in a homogeneous medium. Therefore, the Hartree term (but not the exchange term) which is due to the Coulomb interaction of all other electrons with the test electron, can be incorporated in the single-particle Schrödinger equation for the test electron by adding the self-consistent potential that comes out of the Poisson equation! This is a remarkable result and simplifies matters considerably since now we can include the effects of many electrons in the single-particle Schrödinger equation if we solve it simultaneously (or self-consistently) with the Poisson equation. The only thing that we will miss is the influence of exchange and correlation. The latter can be accounted for separately by using empirical expressions for the exchange energy. One such expression is

$$J = -\frac{e^2}{4\pi\epsilon} C\,[n(x, y, z)]^{1/3}, \tag{3.116}$$

where C is a constant of the order of unity. This expression is known as the Slater approximation for the exchange energy.

3.7.2 Self-Consistent Schrödinger–Poisson Solution

3.7.2.1 Structure with Confinement in Three Dimensions, or Quantum Dot

In a structure with quantum confinement in all three dimensions, e.g., a quantum dot, we have to solve the following effective mass Schrödinger and Poisson equations simultaneously or self-consistently subject to appropriate boundary conditions:

$$\left[-\frac{\hbar^2}{2m^*}\left(\frac{\partial^2}{\partial x^2}+\frac{\partial^2}{\partial y^2}+\frac{\partial^2}{\partial z^2}\right)+V_{\text{sc}}(x,y,z)\right]\psi_{m,n,p}(x,y,z)=E_{m,n,p}\psi_{m,n,p}(x,y,z)$$

$$\texttt{Schrodinger equation} \qquad (3.117)$$

$$-\frac{\partial}{\partial x}\left(\epsilon(x,y,z)\frac{\partial V_{\text{sc}}(x,y,z)}{\partial x}\right)-\frac{\partial}{\partial y}\left(\epsilon(x,y,z)\frac{\partial V_{\text{sc}}(x,y,z)}{\partial y}\right)-\frac{\partial}{\partial z}\left(\epsilon(x,y,z)\frac{\partial V_{\text{sc}}(x,y,z)}{\partial z}\right)$$
$$=en(x,y,z)+\rho_{\text{fixed}}=e\sum_{m,n,p}|\psi_{m,n,p}(x,y,z)|^2 f(E_{m,n,p}-\mu_{\text{F}})+\rho_{\text{fixed}}$$

$$\texttt{Poisson equation} \qquad (3.118)$$

where m, n, p are integers denoting the subband indices for quantization along x-, y-, and z-directions, $n(x,y,z)$ is the electron concentration, ρ_{fixed} is the fixed background charge concentration (such as due to ionized dopant atoms), μ_{F} is the Fermi level, $f(E)$ is the occupation probability of an electron of energy E (given by the Fermi–Dirac factor at equilibrium), and $E_{m,n,p}$ is the energy of the subband with indices m, n, p. The subband energies $E_{m,n,p}$ are found from boundary conditions on the wavefunction. There are two unknowns—$\psi_{m,n,p}(x,y,z)$ and $V_{\text{sc}}(x,y,z)$—which are found from the above two equations.

3.7.2.2 Structure with Confinement in Two Dimensions, or Quantum Wire

In a structure with quantum confinement in two dimensions, e.g., a quantum wire with its axis along the x-direction, we have to solve the following Schrödinger and Poisson equations simultaneously or self-consistently subject to appropriate boundary conditions:

$$\left[-\frac{\hbar^2}{2m^*}\left(\frac{\partial^2}{\partial x^2}+\frac{\partial^2}{\partial y^2}+\frac{\partial^2}{\partial z^2}\right)+V_{\text{sc}}(x,y,z)\right]\psi_{m,n,k_x}(x,y,z)=E_{m,n,k_x}\psi_{m,n,k_x}(x,y,z)$$

$$\texttt{Schrodinger equation} \qquad (3.119)$$

$$-\frac{\partial}{\partial x}\left(\epsilon(x,y,z)\frac{\partial V_{\text{sc}}(x,y,z)}{\partial x}\right)-\frac{\partial}{\partial y}\left(\epsilon(x,y,z)\frac{\partial V_{\text{sc}}(x,y,z)}{\partial y}\right)-\frac{\partial}{\partial z}\left(\epsilon(x,y,z)\frac{\partial V_{\text{sc}}(x,y,z)}{\partial z}\right)$$

$$= en(x,y,z) + \rho_{\text{fixed}} = e \sum_{m,n,k_x} |\psi_{m,n,k_x}(x,y,z)|^2 f(E_{m,n} + E(k_x) - \mu_F) + \rho_{\text{fixed}}$$

$$= e\frac{2}{\pi}\sum_{m,n}\int_0^\infty \mathrm{d}k_x |\psi_{m,n,k_x}(x,y,z)|^2 f(E_{m,n} + E(k_x) - \mu_F) + \rho_{\text{fixed}}$$

`Poisson equation` (3.120)

where k_x is the wavevector along the unconfined direction and $E(k_x)$ is the energy of an electron with wavevector k_x associated with motion in the x-direction. Note that in this case, confinement along the y- and z-directions give rise to the transverse subband indices m and n. Since k_x is a continuous variable, unlike the discrete variables m and n, we have converted the summation over k_x to an integral. The procedure for doing this is to multiply the quantity to be summed by the "density of states" and then integrate over the wavevector. The density of states concept will be introduced in the next chapter. Suffice it to say that the density of states in one dimension is simply $2/\pi$, which explains why this factor appears in the last equation.

3.7.2.3 Structure with Confinement in One Dimension, or Quantum Well

In a structure with quantum confinement in one dimension, e.g., a quantum well in the x–y plane, we have to solve the following Schrödinger and Poisson equations simultaneously or self-consistently subject to appropriate boundary conditions:

$$\left[-\frac{\hbar^2}{2m^*}\left(\frac{\partial^2}{\partial x^2}+\frac{\partial^2}{\partial y^2}+\frac{\partial^2}{\partial z^2}\right)+V_{\text{sc}}(x,y,z)\right]\psi_{m,k_x,k_y}(x,y,z) = E_{m,k_x,k_y}\psi_{m,k_x,k_y}(x,y,z)$$

`Schrodinger equation` (3.121)

$$-\frac{\partial}{\partial x}\left(\epsilon(x,y,z)\frac{\partial V_{\text{sc}}(x,y,z)}{\partial x}\right) - \frac{\partial}{\partial y}\left(\epsilon(x,y,z)\frac{\partial V_{\text{sc}}(x,y,z)}{\partial y}\right)$$
$$-\frac{\partial}{\partial z}\left(\epsilon(x,y,z)\frac{\partial V_{\text{sc}}(x,y,z)}{\partial z}\right)$$

$$= en(x,y,z) = e^2 \sum_{m,k_x,k_y} |\psi_{m,k_x,k_y}(x,y,z)|^2 f(E_m + E(k_x,k_y) - \mu_F) + \rho_{\text{fixed}}$$

$$= e\frac{1}{2\pi^2}\sum_m \int_0^\infty \int_0^{2\pi} k\mathrm{d}k\mathrm{d}\theta |\psi_{m,k}(x,y,z)|^2 f(E_{m,n} + E(k) - \mu_F) + \rho_{\text{fixed}}$$

$$= \frac{e}{\pi} \sum_m \int_0^\infty k\mathrm{d}k |\psi_{m,k}(x, y, z)|^2 f(E_{m,n} + E(k) - \mu_\mathrm{F}) + \rho_\mathrm{fixed}$$

`Poisson equation` (3.122)

where $k^2 = k_x^2 + k_y^2$, and $E(k)$ is the energy of an electron with wavevector k in the plane of the quantum well. Note that in this case, confinement along the z-direction gives rise to the transverse subband index m. In two dimensions, the density of states in wavevector space is $1/2\pi^2$.

3.8 Summary

In this chapter, we learned the following important concepts and recipes:

1. The electron is described by a wavefunction $\Psi(\vec{r}, t)$, which is a function of space and time, whose squared magnitude $|\Psi(\vec{r}, t)|^2$ gives us the probability of finding the electron at position $\vec{r}$ at an instant of time t.
2. The expectation value of any quantum-mechanical operator, which tells us the average value we will measure if we conducted repeated experiments to measure the corresponding physical quantity, is given by the integral $\frac{\langle \Psi(\vec{r},t)|O|\Psi(\vec{r},t)\rangle}{\langle \Psi(\vec{r},t)|\Psi(\vec{r},t)\rangle}$ in Dirac's bra-ket notation.
3. All quantum-mechanical operators are Hermitian. A quantum-mechanical operator for any physical variable can be found by relating the variable to other variables with known operators, relating the operators in the same way, ensuring that the final operator is Hermitian and finally checking that the expectation value of the operator yields a result that corresponds with the classical result.
4. The physics of an electron wave propagating through a disordered medium is similar to the physics of an electromagnetic wave propagating through an inhomogeneous medium with spatially varying refractive index.
5. An electron (or hole) in a potential well has discrete energy eigenstates. This is known as energy quantization.
6. Perturbation theory gives us the recipe to find the perturbed wavefunction if we know the unperturbed energy eigenstates and the corresponding wavefunctions.
7. Perturbation theory has direct applications in many areas such as bandstructure calculation, wavefunction engineering devices, and calculation of electron scattering rates.
8. The Symmetry Principle that governs multi-electron wavefunction gives rise to the exchange energy. This Principle states that the overall wavefunction of two fermions (i.e., the orbital and spin parts together) must be anti-symmetric under exchange of the fermions.
9. As long as we can ignore exchange, the Coulomb forces due to all surrounding electrons on a test electron in a solid can be accounted for by solving the one-electron Schrödinger and Poisson equations simultaneously or self-consistently.

Problems

Problem 3.1. Show that the eigenvalues of Hermitian operators are real.

Problem 3.2. Show that the operators for velocity and charge density do not commute.

Problem 3.3. Show that neither the operator $\frac{i\hbar}{m^*}\vec{\nabla}_r e\delta(\vec{r})$, nor the operator $e\delta(\vec{r})\frac{i\hbar}{m^*}\vec{\nabla}_r$, is Hermitian, but their symmetric combination is. To show this, use the definition of hermiticity, i.e., a Hermitian operator obeys the identity

$$\int_{-\infty}^{\infty} d^3\vec{r}\Phi^*(\vec{r})H\Psi(\vec{r}) = \int_{-\infty}^{\infty} d^3\vec{r}[H\Phi(\vec{r})]^*\Psi(\vec{r}).$$

Problem 3.4. Show that the eigenfunctions of a Hermitian operator corresponding to *distinct* eigenvalues are mutually orthogonal.

Problem 3.5. Sturm–Liouville equation: Equation (3.72) can be written as

$$\Psi(\vec{r},t) = \sum_n R_n(t)\phi_n(\vec{r}),$$

where $R_n(t) = C_n(t)\mathrm{e}^{-\mathrm{i}E_nt/\hbar}$. Substitution of this in (3.71) yields

$$\mathrm{i}\hbar\frac{\partial R_m(t)}{\partial t} = \sum_j R_j(t)H_{mj}(t),$$

where $H_{mj}(t) = \int_0^{\infty} d^3\vec{r}\phi_m^*(\vec{r})H(\vec{r},t)\phi_j(\vec{r}) = \langle\phi_m|H(\vec{r},t)|\phi_j\rangle$ in Dirac's bra-ket notation.

Show that

$$\mathrm{i}\hbar\frac{\partial\left[R_n^*(t)R_m(t)\right]}{\partial t} = \sum_j [R_n^*(t)R_j(t)H_{mj}(t) - R_j^*(t)R_m(t)H_{jn}(t)].$$

Define a "density matrix" $[\rho(t)]$ whose elements are

$$\rho_{mn}(t) = R_m(t)R_n^*(t).$$

Show that

$$\mathrm{i}\hbar\frac{\partial[\rho(t)]}{\partial t} = [H(t)\rho(t) - \rho(t)H(t)] = [H(t),\rho(t)]. \tag{3.123}$$

The last equation is called the *Sturm–Liouville* equation.

Problem 3.6. *Ehrenfest Theorem*: Show that the expectation value of any time-dependent operator A obeys the equation

$$i\hbar\frac{\partial\langle A\rangle}{\partial t} = \langle[A, H]\rangle + i\hbar\left\langle\frac{\partial A}{\partial t}\right\rangle, \tag{3.124}$$

where H is the Hamiltonian and the angular brackets $\langle\ldots\rangle$ denote expectation values.

If the operator is time independent, only the last term in the right-hand-side will vanish. However, the other term (commutator) will not vanish unless A commutes with H. How can you explain the oddity that even though the operator is time independent, its expectation value may not be time independent?

An example of this is the momentum operator $-i\hbar\vec{\nabla}$ which is time independent. However, it does not commute with the Hamiltonian if the potential energy is position dependent, so that the expectation value of the momentum will change with time. Note that a position-dependent potential energy gives rise to a force on the particle (force is negative gradient of potential energy) and this force will accelerate or decelerate the particle, causing its momentum to change with time.

Problem 3.7. Consider a region of space where there is an incident electron wave and a reflected electron wave. The wavefunction is given by the coherent superposition of these two waves:

$$\psi(z) = e^{\nu z} + Re^{-\nu z},$$

where R is the complex reflection coefficient and ν is the complex propagation constant given by

$$\nu = \kappa + ik,$$

where κ and k are real.

1. Find an expression for the expectation values of the charge and current density.
2. Assume $\kappa = 0$, i.e., the wave is purely propagating and has no evanescent component. In that case, show that in the absence of reflection (i.e., $R = 0$), the charge density is spatially uniform, but not otherwise. Explain the significance of this result.
3. Show that if the wave is purely propagating, then the current density is the difference between the incident current density and the reflected current density.
4. If the wave is purely evanescent (as in the case of tunneling through a potential barrier), i.e., when $k = 0$, show that the current density is zero if the reflection coefficient R is real. Therefore, a nonzero tunneling current requires an imaginary or complex reflection coefficient.

Problem 3.8. Starting from the time-dependent Schrödinger equation, show that the expectation values of the electron and current densities obey the familiar continuity equation as long as the Hamiltonian is hermitian, i.e.,

$$\frac{\partial\langle n\rangle}{\partial t} = \frac{1}{e}\vec{\nabla}\cdot\langle\vec{J}_n\rangle.$$

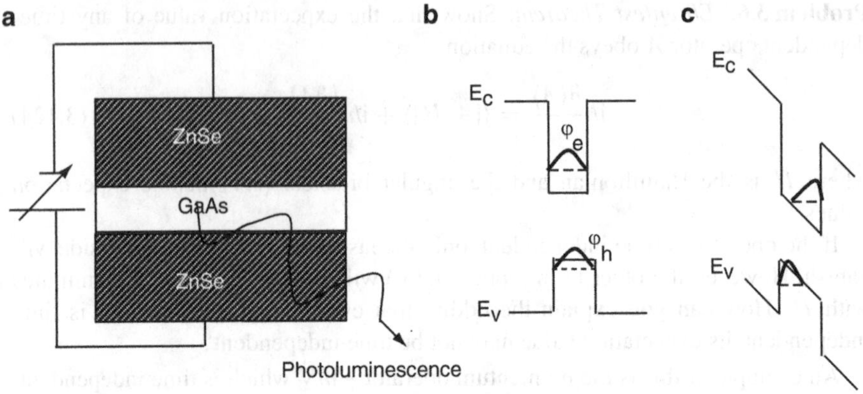

Fig. 3.8 Quantum-confined dc Stark Effect. (**a**) A heterostructure consisting of a GaAs quantum well sandwiched between ZnSe barriers. The battery is used to apply an electric field perpendicular to the heterointerfaces. (**b**) the conduction band (E_{c0}) and valence band (E_{v0}) profile along the direction perpendicular to the heterointerfaces in the absence of any applied electric field. The unperturbed wavefunctions ψ_e and ψ_h are shown, and (**c**) the band diagrams in the presence of the electric field. The perturbed (skewed) wavefunctions are shown

Hint: You may find the vector identity in (2.15) useful.

Show also that the wavefunction and its complex conjugate obey the same Schrödinger equation with time reversed, as long as the Hamiltonian is hermitian. This property is known as *time reversal symmetry*.

Problem 3.9. Refer to Fig. 3.8 for this problem.

Consider a 10 nm layer of GaAs sandwiched between two ZnSe layers. The bulk bandgap of ZnSe and GaAs are 2.83 and 1.42 eV so that for all practical purposes, we can consider the conduction and valence band offsets to be large enough that the potential barriers for both electrons and holes in the GaAs layer are effectively infinite. Suppose that the temperature and carrier concentration in GaAs are low enough that only the lowest electron and heavy hole subband in GaAs are occupied. The effective masses of electrons and heavy holes in GaAs are 0.067 and 0.45 times the free electron mass, respectively.

Assume that an electric field of 100 kV/cm is applied transverse to the heterointerfaces which tilts the energy band diagram as shown in Fig. 3.8c. Answer the following questions:

1. Use first-order, time-independent, nondegenerate perturbation theory to come up with an expression for the skewed wavefunctions of electrons and heavy holes when the electric field is turned on. Remember that the unperturbed wavefunctions for both electrons and holes (in the absence of the electric field) are particle-in-a-box states given by

$$\psi_e(x) = \psi_h(x) = \sqrt{\frac{2}{W}} \sin\left(\frac{n\pi x}{W}\right),$$

where W is the width of the GaAs layer (quantum well). When the field is turned on, the wavefunctions of electrons and holes are skewed in opposite directions since the field exerts oppositely directed forces on the electrons and holes. The perturbed wavefunctions are very different for electrons and holes. Note that the electric field $\mathcal{E}$ causes a perturbation $-e\mathcal{E}x$ on the electrons and $+e\mathcal{E}x$ on the holes; i.e., $H' = \pm e\mathcal{E}x$.

2. Which wavefunction is skewed more—the electron or the heavy hole? Why?
3. The device emits light when electrons and holes recombine to produce photons. We can shine an ultraviolet lamp on the quantum well which excites carriers from the heavy hole subband to the electron subband and the photoexcited carriers subsequently recombine to emit light. This phenomenon is known as PL. For light polarized in the plane of the quantum well, the PL intensity I depends on the overlap between the electron and heavy hole wavefunctions:

$$I \propto \left| \int_0^W \psi_e^*(x)\psi_h(x)\mathrm{d}x \right|^2 .$$

   Find the percentage decrease in the PL intensity when then electric field is turned on. Provide a numerical answer.
4. Not only will the PL be partially quenched by the electric field, but the PL peak frequency (i.e., the frequency of the emitted photons) will be redshifted. Calculate the redshift in the photon energy and provide a numerical answer. For this part, first calculate the change in the lowest subband energies of the electrons and heavy holes from perturbation theory. The redshift in photon energy is the sum of the changes in the electron and hole subband energies.
5. If H is a Hamiltonian and ϕ is a wavefunction, then is it always true that

$$H\phi = E\phi?$$

   This question was posed to 89 advanced undergraduates and more than 200 first-year physics graduate students in 7 US universities (see C. Singh, M. Belloni and W. Christian, "Improving students' understanding of quantum mechanics", Physics Today, August 2006, p. 43). Only 29% of the respondents gave the supposedly correct answer. However, according to this author, even those 29% did not give the completely correct answer.

Problem 3.10. Prove that the factors $1/\sqrt{2}$ appearing in (3.103) and (3.104) are needed to normalize $\Phi_S(\vec{r}_1, \vec{r}_2)$ and $\Phi_A(\vec{r}_1, \vec{r}_2)$. In other words, show that

$$\int\int d^3\vec{r}_1 d^3\vec{r}_2 \Phi_S(\vec{r}_1, \vec{r}_2)|^2 = \int\int d^3\vec{r}_1 d^3\vec{r}_2 \Phi_A(\vec{r}_1, \vec{r}_2)|^2 = 1,$$

provided the individual ϕ-s are normalized, i.e.,

$$\int d^3\vec{r}_1 |\phi_m(\vec{r}_1)|^2 = \int d^3\vec{r}_2 |\phi_m(\vec{r}_2)|^2 = 1.$$

Problem 3.11. Show that each of the transmission matrices in (3.89) have the eigenvalues

$$\lambda_1 = \lambda_2^{-1} = \exp(i\theta) = \frac{Tr[W]}{2} + \sqrt{\left(\frac{Tr[W]}{2}\right)^2 - 1},$$

where $\theta = k_n \Delta x$ and $[W]$ is the transmission matrix.

Problem 3.12. Consider a periodic structure, like a finite superlattice, consisting of N identical sections. Clearly, if $[W]$ is the transmission matrix describing propagation of an electron wave through a single section, then the total transmission matrix describing propagation through all sections will be

$$[W]_{\text{tot}} = [W]^N.$$

Show that the matrix $[W]_{\text{tot}}$ can be expressed as

$$[W]_{\text{tot}} = [W]\frac{\sin(N\theta)}{\sin\theta} - [I]\frac{\sin[(N-1)\theta]}{\sin\theta},$$

where $[I]$ is the 2×2 identity matrix and θ is defined in the preceding problem.

Also show that the matrix $[W]$ is unimodular even for evanescent states, i.e., if $k = \pm i\kappa$.

Problem 3.13. Coupling and crosstalk between interconnects in ultra-large-scale-integrated (ULSI) circuits

In modern ULSI circuit chips, interconnects are so densely packed that two interconnects can be extremely close to each other, which makes it very possible for electrons in one interconnect to tunnel into an adjacent one, causing coupling and crosstalk. This corrupts data processed in the chip. In this problem, we will study this issue to assess how serious this problem can be by calculating the coupling coefficient between two very narrow and closely spaced interconnects that we will treat as electron waveguides. The approach used here are similar to (but not identical with) perturbation theory.

Problem statement: Consider two linear interconnects of rectangular cross section as shown in Fig. 3.9. The electron wavefunctions leak out of the interconnects and overlap in space, so that an electron can laterally tunnel from one interconnect to another, causing coupling between them. Calculate the coupling coefficient which will tell us how much of the signal from one interconnect leaks into the other over a given distance.

Solution. The time-independent Schrödinger equation governing an electron in the coupled waveguide pair is

$$\left\{-\frac{\hbar^2}{2m^*}\left[\frac{\partial^2}{\partial x^2} + \frac{\partial^2}{\partial y^2} + \frac{\partial^2}{\partial z^2}\right] + H'(y,z)\right\}\psi(x,y,z) = E\psi(x,y,z),$$

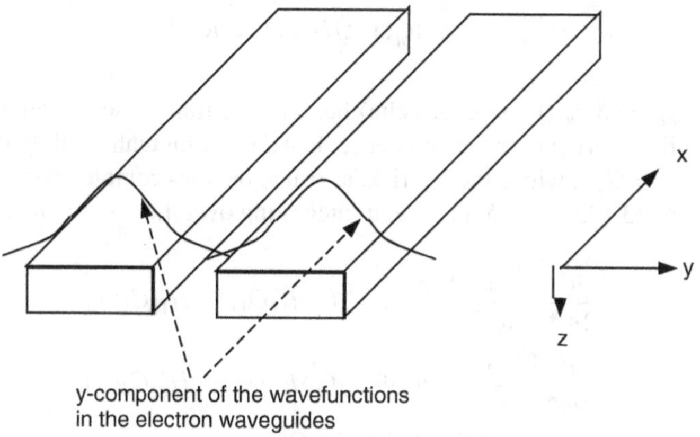

Fig. 3.9 Two interconnects modeled as rectangular electron waveguides. The wavefunctions leak out and overlap in space as shown, causing tunneling of electrons between the interconnects. Current flows along the length of the interconnects, i.e., along the x-direction, but tunneling will cause some current to flow along the y-direction as well

where H' is the perturbation in either interconnect caused by its neighbor. We assume that H' (causing coupling) is constant along the lengths of the interconnects, i.e., it does not depend on the coordinate x.

We will write the wavefunction as

$$\psi(x, y, z) = C_1(x)\phi_1(y, z) + C_2(x)\phi_2(y, z),$$

where $\phi_n(y, z)$ is the transverse component of the wavefunction in the n-th waveguide ($n = 1, 2$). Bear in mind that when the electron is completely in the first waveguide, $C_1 = 1$ and $C_2 = 0$; similarly, when the electron is completely in the second waveguide, $C_2 = 1$ and $C_1 = 0$.

Substituting the last equation in the Schrödinger equation, we get

$$\begin{aligned}
&-\frac{\hbar^2}{2m^*}\left\{\frac{\partial^2 C_1(x)}{\partial x^2} + C_1(x)\left[\frac{\partial^2}{\partial y^2} + \frac{\partial^2}{\partial z^2}\right]\right\}\phi_1(y, z) + H'C_1(x)\phi_1(y, z)\\
&-\frac{\hbar^2}{2m^*}\left\{\frac{\partial^2 C_2(x)}{\partial x^2} + C_2(x)\left[\frac{\partial^2}{\partial y^2} + \frac{\partial^2}{\partial z^2}\right]\right\}\phi_2(y, z) + H'C_2(x)\phi_2(y, z)\\
&= E\left[C_1(x)\phi_1(y, z) + C_2(x)\phi_2(y, z)\right].
\end{aligned} \tag{3.125}$$

Let us define the following quantities:

$$O_{mn} = \int_0^\infty \phi_m^*(y, z)\phi_n(y, z)\mathrm{d}y\mathrm{d}z$$

$$H'_{mn} = \int_0^\infty \phi_m^*(y,z) H' \phi_n(y,z) \mathrm{d}y\mathrm{d}z.$$

Note that $O_{mn} \neq \delta_{m,n}$ (Kronecker delta) because the transverse wavefunctions in two waveguides overlap in space and hence they are *not* mutually orthogonal. Note also that $O_{mn} = O_{nm}^*$, where the asterisk, as usual, denotes complex conjugate.

Multiplying (3.125) with $\phi_1^*(y,z)$ and integrating over the $y-z$ plane, we get

$$\begin{aligned}
&-\frac{\hbar^2}{2m^*}\frac{\partial^2 C_1(x)}{\partial x^2} O_{11} + C_1(x)E_1 O_{11} + H'_{11}C_1(x) \\
&-\frac{\hbar^2}{2m^*}\frac{\partial^2 C_2(x)}{\partial x^2} O_{12} + C_2(x)E_2 O_{12} + H'_{12}C_2(x) \\
&= EC_1(x)O_{11} + EC_2(x)O_{12},
\end{aligned} \tag{3.126}$$

where

$$-\frac{\hbar^2}{2m^*}\left[\frac{\partial^2}{\partial y^2} + \frac{\partial^2}{\partial z^2}\right]\phi_n(y,z) = E_n \phi_n(y,z).$$

Similarly, multiplying (3.125) with $\phi_2^*(y,z)$ and integrating over the $y-z$ plane, we get

$$\begin{aligned}
&-\frac{\hbar^2}{2m^*}\frac{\partial^2 C_2(x)}{\partial x^2} O_{22} + C_2(x)E_2 O_{22} + H'_{22}C_2(x) \\
&-\frac{\hbar^2}{2m^*}\frac{\partial^2 C_1(x)}{\partial x^2} O_{21} + C_1(x)E_1 O_{21} + H'_{21}C_1(x) \\
&= EC_2(x)O_{22} + EC_1(x)O_{21}.
\end{aligned} \tag{3.127}$$

Defining two other quantities

$$\begin{aligned}
k_n^2 &= 2m^*(E - E_n)/\hbar^2 \\
\kappa_{mn}^2 &= \frac{2m^*}{\hbar^2} H'_{mn},
\end{aligned}$$

we get from (3.126) and (3.127)

$$\begin{aligned}
\frac{\partial^2 C_1(x)}{\partial x^2} O_{11} + \frac{\partial^2 C_2(x)}{\partial x^2} O_{12} &= k_1^2 C_1(x) O_{11} + k_2^2 C_2(x) O_{12} \\
&\quad + \kappa_{11}^2 C_1(x) + \kappa_{12}^2 C_2(x) \\
\frac{\partial^2 C_1(x)}{\partial x^2} O_{21} + \frac{\partial^2 C_2(x)}{\partial x^2} O_{22} &= k_1^2 C_1(x) O_{21} + k_2^2 C_2(x) O_{22} \\
&\quad + \kappa_{21}^2 C_1(x) + \kappa_{22}^2 C_2(x).
\end{aligned} \tag{3.128}$$

We now define yet another variable as follows:

$$D_n(x) = C_n(x)\mathrm{e}^{-\mathrm{i}k_n x}. \tag{3.129}$$

Using this in (3.128), we get

$$\begin{aligned}
&\frac{\partial^2 D_1(x)}{\partial x^2}\mathrm{e}^{\mathrm{i}k_1 x}O_{11} + 2ik_1\frac{\partial D_1(x)}{\partial x}\mathrm{e}^{\mathrm{i}k_1 x}O_{11} \\
&+\frac{\partial^2 D_2(x)}{\partial x^2}\mathrm{e}^{\mathrm{i}k_2 x}O_{12} + 2ik_2\frac{\partial D_2(x)}{\partial x}\mathrm{e}^{\mathrm{i}k_2 x}O_{12} \\
&= \kappa_{11}^2 D_1(x)\mathrm{e}^{\mathrm{i}k_1 x} + \kappa_{12}^2 D_2(x)\mathrm{e}^{\mathrm{i}k_2 x} \\
&\frac{\partial^2 D_2(x)}{\partial x^2}\mathrm{e}^{\mathrm{i}k_2 x}O_{22} + 2ik_2\frac{\partial D_2(x)}{\partial x}\mathrm{e}^{\mathrm{i}k_2 x}O_{22} \\
&+\frac{\partial^2 D_1(x)}{\partial x^2}\mathrm{e}^{\mathrm{i}k_1 x}O_{21} + 2ik_1\frac{\partial D_1(x)}{\partial x}\mathrm{e}^{\mathrm{i}k_1 x}O_{21} \\
&= \kappa_{22}^2 D_2(x)\mathrm{e}^{\mathrm{i}k_2 x} + \kappa_{21}^2 D_1(x)\mathrm{e}^{\mathrm{i}k_1 x}.
\end{aligned} \tag{3.130}$$

The last set of equations are often referred to as *coupled mode equations*. They can be written in a matrix form as

$$[\mathbf{A}]\frac{\partial^2}{\partial x^2}[\mathbf{D}] + 2i[\mathbf{B}]\frac{\partial}{\partial x}[\mathbf{D}] = [\mathbf{K}][\mathbf{D}],$$

where $[\mathbf{D}]$ is a 2×1 matrix whose elements are D_1 and D_2, $[\mathbf{A}]$ is a 2×2 matrix whose elements are $A_{mn} = O_{mn}\mathrm{e}^{\mathrm{i}k_n x}$, $[\mathbf{B}]$ is a 2×2 matrix whose elements are $B_{mn} = k_n O_{mn}\mathrm{e}^{\mathrm{i}k_n x}$ and $[\mathbf{K}]$ is a 2×2 matrix whose elements are given by $K_{mn} = \kappa_{mn}^2\mathrm{e}^{\mathrm{i}k_n x}$.

The quantity k_n is a wavevector that can be written as $2\pi/\lambda_n$, where λ_n will be of the order of the DeBroglie wavelength in the interconnect. This will be on the order of a few nanometers to few tens of nanometers. Over a length scale of a few tens of nanometers, we do not expect significant coupling to occur between the two interconnects (we justify this later). Hence, we can always neglect the second derivative terms in (3.130) in comparison with the first derivative terms. This yields

$$\begin{aligned}
\frac{\partial D_1}{\partial x} &= -\mathrm{i}\Delta_1 D_1 - \mathrm{i}\Omega_{12}D_2\mathrm{e}^{-\mathrm{i}[k_1-k_2]x} \\
\frac{\partial D_2}{\partial x} &= -\mathrm{i}\Delta_2 D_2 - \mathrm{i}\Omega_{21}D_1\mathrm{e}^{-\mathrm{i}[k_2-k_1]x},
\end{aligned} \tag{3.131}$$

where

$$\Delta_1 = \frac{\kappa_{11}^2 O_{22} - \kappa_{21}^2 O_{12}}{2k_1(O_{11}O_{22} - |O_{12}|^2)}$$

$$\Delta_2 = \frac{\kappa_{22}^2 O_{11} - \kappa_{12}^2 O_{21}}{2k_2(O_{11}O_{22} - |O_{12}|^2)}$$

$$\Omega_{12} = \frac{\kappa_{12}^2 O_{22} - \kappa_{22}^2 O_{12}}{2k_1(O_{11}O_{22} - |O_{12}|^2)}$$

$$\Omega_{21} = \frac{\kappa_{21}^2 O_{11} - \kappa_{11}^2 O_{21}}{2k_2(O_{11}O_{22} - |O_{12}|^2)}.$$

We now make another transformation of variables

$$\begin{aligned} D_1 &= R_1 e^{-i\Delta_1 x} \\ D_2 &= R_2 e^{-i\Delta_2 x} \\ k_1 &= k_1' + \Delta_1 \\ k_2 &= k_2' + \Delta_2. \end{aligned} \tag{3.132}$$

This reduces (3.131) to

$$\begin{aligned} \frac{\partial R_1}{\partial x} &= -i\Omega_{12} R_2 e^{i[k_2' - k_1']x} \\ \frac{\partial R_2}{\partial x} &= -i\Omega_{21} R_1 e^{i[k_1' - k_2']x}. \end{aligned} \tag{3.133}$$

The above two coupled differential equations can be decoupled in the following way. First, differentiate the top relation in (3.133) with respect to x to obtain:

$$\frac{\partial^2 R_1}{\partial x^2} = -i\Omega_{12} \frac{\partial R_2}{\partial x} e^{i[k_2' - k_1']x} + \Omega_{12}[k_2' - k_1'] R_2 e^{i[k_2' - k_1']x}.$$

Next, use the bottom relation in (3.133) to replace the first derivative, which will yield

$$\frac{\partial^2 R_1}{\partial x^2} = (-i\Omega_{12})(-i\Omega_{21}) R_1 + \Omega_{12}[k_2' - k_1'] R_2 e^{i[k_2' - k_1']x}.$$

Now, use the top relation in (3.133) again in the last equation, which will result in the equation below that involves only R_1 and not R_2.

$$\frac{\partial^2 R_1}{\partial x^2} + i(k_1' - k_2') \frac{\partial R_1}{\partial x} + \Omega_{12}\Omega_{21} R_1 = 0. \tag{3.134}$$

The solution of the last equation is

$$R_1(x) = e^{i\delta x}[P e^{i\nu x} + Q e^{-i\nu x}] \tag{3.135}$$

where P and Q are constants, $2\delta = k_2' - k_1'$ and $\nu = \sqrt{\kappa^2 + \delta^2}$ with $\kappa = \Omega_{12} = \Omega_{21}$. The last equality follows from the fact that the two interconnects are assumed to be identical. The reader can easily find a solution for $R_2(x)$ following the same procedure.

Finally, using (3.129) and (3.132), we get

$$C_1(x) = \mathrm{e}^{ik_0x}[P\mathrm{e}^{\mathrm{i}\nu x} + Q\mathrm{e}^{-\mathrm{i}\nu x}] \tag{3.136}$$

where $k_0 = k_2' + k_1'$.

Similarly,

$$C_2(x) = -\mathrm{e}^{ik_0x}[P'\mathrm{e}^{\mathrm{i}\nu x} + Q'\mathrm{e}^{-\mathrm{i}\nu x}]. \tag{3.137}$$

where $P' = P\frac{k^+}{\kappa}$ and $Q' = Q\frac{k^-}{\kappa}$ with $k^+ = 2\delta + 2\nu$ and $k^- = 2\delta - 2\nu$.

To evaluate the constants P and Q, we need to apply boundary conditions. Let

$$\begin{aligned} C_1(x=0) &= P + Q = A_1^+ \\ C_2(x=0) &= -\frac{k^+P + k^-Q}{\kappa} = A_2^+. \end{aligned} \tag{3.138}$$

This tells us

$$\begin{aligned} P &= -\frac{k^-}{2\nu}A_1^+ - \frac{\kappa}{2\nu}A_2^- \\ Q &= \frac{k^+}{2\nu}A_1^+ + \frac{\kappa}{2\nu}A_2^-. \end{aligned} \tag{3.139}$$

Now let

$$\begin{aligned} C_1(x=L) &= B_1^+ \\ C_2(x=L) &= B_2^+. \end{aligned} \tag{3.140}$$

Using (3.135)–(3.139), we get

$$\begin{aligned} B_1^+ &= \mathrm{e}^{\mathrm{i}k_0L}\left[\left(\cos(\nu L) + \frac{\mathrm{i}\delta}{\nu}\sin(\nu L)\right)A_1^+ + \left(-\frac{\mathrm{i}\kappa}{\nu}\sin(\nu L)\right)A_2^+\right] \\ B_2^+ &= \mathrm{e}^{\mathrm{i}k_0L}\left[\left(\cos(\nu L) - \frac{\mathrm{i}\delta}{\nu}\sin(\nu L)\right)A_2^+ + \left(-\frac{\mathrm{i}\kappa}{\nu}\sin(\nu L)\right)A_1^+\right]. \end{aligned} \tag{3.141}$$

This equation can be written in a matrix form as

$$\begin{pmatrix} B_1{}^+ \\ B_2{}^+ \end{pmatrix} = \begin{pmatrix} a & c \\ b & d \end{pmatrix} \cdot \begin{pmatrix} A_1{}^+ \\ A_2{}^+ \end{pmatrix}, \tag{3.142}$$

where

$$a = e^{ik_0L}\left(\cos(\nu L) + \frac{i\delta}{\nu}\sin(\nu L)\right)$$

$$b = c = e^{ik_0L}\left(-\frac{i\kappa}{\nu}\sin(\nu L)\right)$$

$$d = e^{ik_0L}\left(\cos(\nu L) - \frac{i\delta}{\nu}\sin(\nu L)\right).$$

Note that the 2×2 matrix in (3.142) is in the form of a transmission matrix. The matrix element b (= c) is an indication of the coupling from one interconnect to another. It is the fraction of the signal in interconnect 1 that gets coupled to interconnect 2 after a distance L. Substituting back the values of κ and ν, we find that the quantity b is given by

$$|b| = |c| = \frac{\kappa}{\sqrt{\kappa^2 + \delta^2}}\sin\left[\sqrt{\kappa^2 + \delta^2}L\right].$$

It is easy to show that this coupling is maximum if $\delta = 0$ as long as $\tan(\kappa L) > \kappa L$ or as long as $\kappa L \leq \pi/2$. To have $\delta = 0$ would require that the two interconnects be identical and carry the same signal frequency. Note that when δ is zero it is possible for 100 % of the signal in one interconnect to get coupled to the other and this happens at a distance

$$L_{100\%} = \frac{\pi}{2\kappa}.$$

In the case of $\delta = 0$, the coupling over a distance L is simply given by

$$|b| = |c| = \sin(\kappa L)$$

Therefore, the fraction of the signal power from one interconnect that is coupled into another is given by

$$|b|^2 = |c|^2 = \sin^2(\kappa L)$$

The last relation is the sought after relation for the coupling coefficient. Note that this relation indicates that signal power alternately flows along interconnect 1 and interconnect 2 because of the sinusoidal dependence on length L. As long as the mutual coupling does not vary along the length of the interconnects (as we have assumed), the signal will alternate between the two waveguides, and this is a typical characteristic of coupled mode behavior.

The coupling coefficient $|b|$ can be calculated exactly if we know the nature of the perturbation Hamiltonian H' and the y-components of the wavefunctions in the two waveguides, which will allow us to calculate κ. It has been calculated in [6] for a specific set of conditions. There, it was found that significant coupling occurs only over lengths of few millimeters, which justifies our previous assumption that negligible coupling occurs over distance scales of tens of nanometers. Interconnects that are several millimeters long are actually not that unusual in chips with areas of several square centimeters that are found today.

References

1. L. I. Schiff, *Quantum Mechanics*, 3rd. edition (McGraw Hill, New York, 1968).
2. H. Sakaki, "The velocity modulation transistor (VMT): A new field effect transistor concept", Jpn. J. Appl. Phys., Part 2, **21**, L381 (1982).
3. C. Hamaguchi, K. Miyatsuji and H. Hihara, "Proposal for a single quantum well transistor (SQWT): Self consistent calculation of 2-D electrons in a quantum well with external voltage", Jpn. J. Appl. Phys., Part 2, **23**, L132 (1984).
4. D. A. B. Miller, D. S. Chemla, T. C. Damen, A. C. Gossard, W. Weigmann, T. H. Wood and C. A. Burrus, "Band edge electroabsorption in quantum well structures - The quantum confined Stark effect", Phys. Rev. Lett., **53**, 2173 (1984).
5. D. J. Vezetti and M. Cahay, "Transmission resonance in finite repeated structures", J. Phys. D., **19**, L53 (1986).
6. S. Bandyopadhyay, "Coupling and crosstalk between high speed interconnects in ultra large sale integrated circuits", IEEE J. Quant. Elec., **28**, 1554 (1992).

Chapter 4
Band Structures of Crystalline Solids

4.1 Introduction

In the last chapter, we saw a real-life application of time-independent perturbation theory in explaining the working of an electro-optic modulator based on the quantum-confined Stark effect. In this chapter, we will see yet another—perhaps a more important—application of this theory in calculating the energy dispersion relations of electrons in a crystal. The energy dispersion relation, also referred to as "band structure," yields the relation between the energy and wavevector of an electron. It is important in understanding both optical and electronic properties of crystalline semiconductors, such as why certain semiconductors like GaAs can emit light easily and certain others like Si cannot, or why certain semiconductors like InAs have high electron mobilities while others like GaN do not. However, before we delve into bandstructure calculations based on time-independent perturbation theory, let us first understand what a "crystal" is.

All solids can be broadly classified into three distinct types: (1) crystalline, (2) polycrystalline, and (3) amorphous. In a *crystalline* solid, all the atoms are periodically arranged throughout the entire material, although the distance between the nearest neighbors (which is the so-called "lattice period") may be different along different directions since a crystal is generally anisotropic. In a *polycrystalline* material, the atoms are periodically arranged only within finite-sized regions called "grains." While there is perfect spatial ordering within every grain, if we examine the lattice period in any fixed direction, it will vary between different grains, since different grains are oriented differently. Neighboring grains are separated from each other by "grain boundaries." Thus, a polycrystalline material has short-range order, but no long-range order unlike a pure crystal. *Amorphous* materials, on the other hand, exhibit no order at all and the atoms are found in random locations. The three different types are portrayed in Fig. 4.1.

In a perfectly crystalline material, we can find a well-defined relationship between the energy and the wavevector of an electron. This relationship is called the energy dispersion relation, or the $E - k$ relation, or the "bandstructure." In

DOI 10.1007/978-1-4614-1141-3_4,

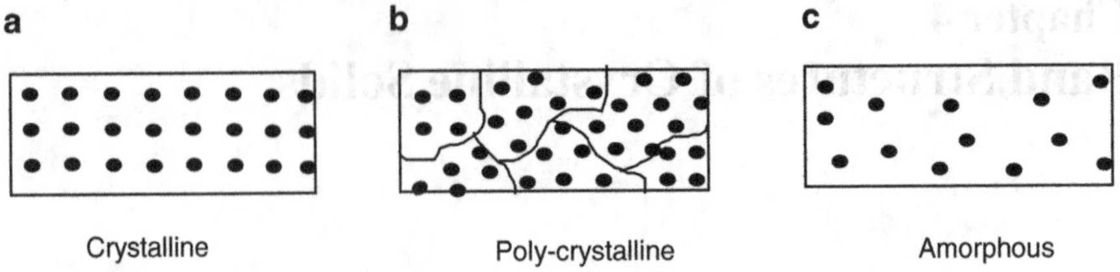

Fig. 4.1 Three different classes of solids: (**a**) crystalline, (**b**) polycrystalline, and (**c**) amorphous. The *dark circles* represent individual atoms. In (**b**), the different grains are shown

order to derive the bandstructure, we will assume that at high enough temperatures, the atoms ionize to release electrons from the outermost shells, and the released electrons constitute the mobile charge carriers that carry current in the crystal. We are seeking the $E-k$ relation of one such electron. Since the ionized atoms are electrically charged and immobile, they form a perfectly periodic potential landscape for the mobile electron. Such a system is sometimes referred to as "jellium" since the electron gas will have the consistency of jelly. If the gas is displaced from its equilibrium position by some force, there will be a restoring force acting on it due to the electrostatic attraction between the electrons and the positively charged ions, which will endow the jellium with some viscosity. In order to find the $E-k$ relation of an electron in this jellium, we will neglect interactions between different electrons, or between electrons and holes, and consider only the interaction between the electron and the *rigid* background of ions. The background is considered rigid since we neglect lattice vibrations (phonons) that will invariably be present at any nonzero temperature. Thus, the $E-k$ relation that we will derive is strictly valid at absolute zero of temperature.

This raises a conundrum. At absolute zero of temperature, the atoms will not have thermal energy to ionize and hence there will be no jellium formation! One way to overcome this dichotomy is to assume that the temperature was initially high enough to produce a jellium and then the system was gradually cooled down to absolute zero. However, the ionic potentials were shielded by the free electron gas which prevented recapture of the electrons by their parent donors when the temperature fell to 0 K. This allows the jellium to exist at 0 K. It is a convenient mental tool and merely allows us to think of bandstructure without worrying about phonons.

In this chapter, we will discuss four different methods for calculating the bandstructure, or E-k relation, in a crystalline solid at effectively 0 K. They are: (1) *the nearly free electron (NFE) method*, (2) *the orthogonalized plane wave expansion method*, (3) *the TBA method* and (4) *the* $\vec{k} \cdot \vec{p}$ *perturbation method*. All, except the third, draw upon timeindependent, first order perturbation theory discussed in the previous chapter. The reader should understand that these four methods are not the only methods available, nor are they necessarily the most popular methods. The intention here is not to discuss band structure calculation for the sake of it, but rather to exemplify time-independent perturbation theory through applications to band structure calculations. Once a reader grasps the essential elements of the four methods discussed here, other methods of calculating band structure will be relatively transparent. This is the philosophy throughout this textbook; the idea is to teach principles and not recipes.

Knowing the bandstructure in a crystalline solid is extremely useful and important in many contexts. We need it to calculate many physical properties, such as optical absorption spectra discussed in Chap. 6, the electron (or hole) density of states which determines the electron-phonon or electron-impurity scattering rates discussed in the next chapter, and several other parameters, such as equilibrium electron and hole concentrations in a crystal. We also need it in full band MC simulations of Boltzmann transport discussed in Chap. 2. Therefore, it is vital in many different applications.

4.2 The Schrödinger Equation in a Crystal

The starting point for calculating the relation between the energy and the wavevector of an electron in a crystal is the time-independent Schrödinger equation that describes an electron in a periodically modulated potential. The potential that an electron sees inside a crystal will be periodic in space since the atoms (or ions) are periodically arranged in space. Since the lattice constant (or the "period") is different in different directions within a crystal, the potential term in the Schrdinger equation will be direction dependent. This will make the bandstructure anisotropic, meaning that the $E - k$ relation will be different along different crystallographic directions.

Solution of the time-independent Schrödinger equation will give us the wavefunction and the corresponding energy. Both will depend on the electron's wavevector, and the relation between the energy and the wavevector will be the bandstructure relation that we are after. Thus, the bandstructure along a particular crystallographic direction in a crystal is obtained by solving the following Schrödinger equation:

$$-\frac{\hbar^2}{2m_0}\nabla_{\mathrm{r}}^2\psi(\vec{r}) + V_{\mathrm{L}}(\vec{r})\psi(\vec{r}) = E\psi(\vec{r}), \tag{4.1}$$

where $\psi(\vec{r})$ is the wavefunction, E is the energy, and $V_{\mathrm{L}}(\vec{r})$ is the periodic crystal potential such that

$$V_{\mathrm{L}}(\vec{r}) = V_{\mathrm{L}}(\vec{r} + n\vec{a}), \tag{4.2}$$

where $\vec{a}$ is the lattice period along any given direction in the crystal and n is an integer. The last equation is a result of the periodicity of the lattice potential, i.e., the potential repeats itself after integral multiples of the lattice constant $\vec{a}$.

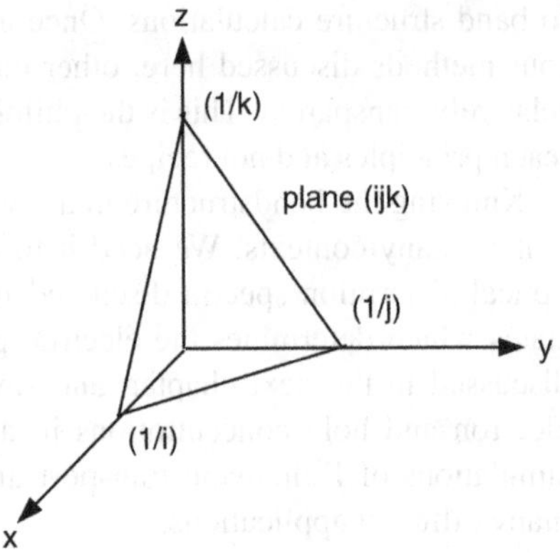

Fig. 4.2 The Miller system for designating a crystallographic plane

4.2.1 Crystallographic Directions and Miller Indices

In a crystal, different crystallographic directions are denoted by three Miller indices $[ijk]$. The direction denoted by $[ijk]$ is perpendicular to the plane denoted by the Miller indices (ijk). Readers who are not already familiar with the Miller index system can find good descriptions in many undergraduate textbooks on solid state physics, but here we will provide a very short introduction. Let us first adopt a Cartesian coordinate system to describe directions within a crystal. Such a system is shown in Fig. 4.2. Consider next an arbitrary plane whose intercepts with the x, y, and z axes are a, b, and c (in some arbitrary units), respectively. The three Miller indices describing this plane will be chosen as follows. The first Miller index i is proportional to the reciprocal of the intercept the plane makes with the x-axis. Hence, we can choose $i = 1/a$. The second index j will be proportional to the reciprocal of the intercept with the y-axis ($j = 1/b$) and the last index k is proportional to the reciprocal of the intercept with the z-axis ($k = 1/c$). Hence, the Miller index for this plane will be $(ijk) = (1/a, 1/b, 1/c)$. Obviously, the x-y plane will have Miller indices (001), the plane that makes equal intercepts with the x-, y-, and z-axes will have Miller indices (111), and so on. The plane that makes intercepts of $-1/l$ with the x-axis, $1/m$ with the y-axis, and $-1/n$ with the z-axis will have Miller indices $(\bar{l}m\bar{n})$, where the "overline" or "bar" stands for negative quantities. The direction that is perpendicular to the plane with Miller indices (ijk) is denoted by the Miller indices $[ijk]$. Note that while the indices for planes are enclosed within curved braces, the indices for directions are enclosed with square braces.

Since a crystal's bandstructure is different along different crystallographic directions, bandstructure plots (E versus k) typically carry a label indicating the Miller indices of the crystallographic direction along which the bandstructure is calculated. In this chapter, we discuss four different methods of finding the bandstructure and the electron wavefunction in a crystal. In each case, we will solve the full three-dimensional Schrödinger equation given in (4.2), which will yield the E versus k relation along any arbitrary direction.

4.3 The Nearly Free Electron Method for Calculating Bandstructure

In some atoms like sodium, the outermost electron is very loosely bound to the nucleus and is "nearly free." In a solid made of such atoms, we can treat the lattice potential $V_L(\vec{r})$ in (4.2) as a weak time-independent perturbation and apply first order time-independent nondegenerate perturbation theory to find the two quantities of interest: the wavefunction $\psi(\vec{r})$ and the energy E. The unperturbed Schrödinger equation in this case is obviously the equation describing a free electron, since, if we remove the crystal potential, the Schrödinger equation becomes that of a free electron. Consequently, the unperturbed wavefunctions are free electron wavefunctions or plane wave states given by (3.6). Now, in order to come up with a complete orthonormal set that we will use as basis functions for expanding the perturbed wavefunction, we need to find a set of plane wave states that are mutually orthogonal. For this purpose, we will take the normalizing volume Ω to be the volume of a unit lattice and define the set of unperturbed wavefunctions as

$$\phi_n(\vec{r}) = \frac{1}{\sqrt{a^3}} \mathrm{e}^{\mathrm{i}(\vec{k}+n\vec{G})\cdot\vec{r}}, \tag{4.3}$$

where a is the lattice constant so that a^3 is the volume of the unit lattice, and $\vec{G} = 2\pi/\vec{a}$ is the *reciprocal lattice vector*. Note that

$$\begin{aligned} \int_0^{a^3} d^3\vec{r}\phi_m^*(\vec{r})\phi_n(\vec{r}) &= \frac{1}{a^3}\int_0^{a^3} d^3\vec{r}\mathrm{e}^{-\mathrm{i}(\vec{k}+m\vec{G})\cdot\vec{r}}\mathrm{e}^{\mathrm{i}(\vec{k}+n\vec{G})\cdot\vec{r}} \\ &= \frac{1}{a^3}\int_0^{a^3} d^3\vec{r}\mathrm{e}^{\mathrm{i}(n-m)(2\pi/\vec{a})\cdot\vec{r}} = \delta_{m,n}, \end{aligned} \tag{4.4}$$

where $\delta_{m,n}$ is the Kronecker delta. Hence, we find that the wavefunctions in (4.3) indeed form a complete orthonormal set.

Following standard time-independent nondegenerate perturbation theory that we discussed in the previous chapter, we expand the perturbed wavefunction in a complete orthonormal set using the wavefunctions in (4.3) as basis functions. This yields:

$$\begin{aligned} \psi^{\mathrm{NFE}}(\vec{r}) &= \sum_{p=0}^{n} C_p \frac{1}{\sqrt{a^3}} \mathrm{e}^{\mathrm{i}(\vec{k}+p\vec{G})\cdot\vec{r}} \\ &= \frac{1}{\sqrt{a^3}}\left[\mathrm{e}^{\mathrm{i}\vec{k}\cdot\vec{r}} + C_1\mathrm{e}^{\mathrm{i}(\vec{k}+\vec{G})\cdot\vec{r}} + \cdots + C_n\mathrm{e}^{\mathrm{i}(\vec{k}+n\vec{G})\cdot\vec{r}}\right], \end{aligned} \tag{4.5}$$

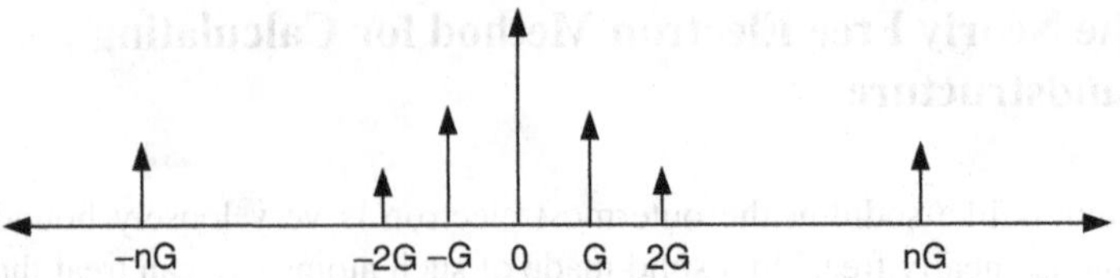

Fig. 4.3 Fourier components of the spatially periodic lattice potential at integral multiples of the reciprocal lattice vector

where we have assumed that $C_0 = 1$, and the superscript *NFE* stands for the nearly-free-electron model that we are discussing here.

Recall now the results of time-independent first order nondegenerate perturbation theory. We get:

$$C_p = \frac{H'_{0p}}{E_0 + H'_{00} - E_p - H'_{pp}} C_0 = \frac{H'_{0p}}{E_0 + H'_{00} - E_p - H'_{pp}}, \tag{4.6}$$

where

$$\begin{aligned}
H'_{0p} &= \langle \phi_0 | H' | \phi_p \rangle = \frac{1}{a^3} \int_0^{a^3} d^3\vec{r} e^{-i\vec{k}\cdot\vec{r}} V_L(\vec{r}) e^{i(\vec{k}+p\vec{G})\cdot\vec{r}} \\
&= \frac{1}{(2\pi/G)^3} \int_0^{(2\pi/G)^3} d^3\vec{r} e^{ip\vec{G}\cdot\vec{r}} V_L(\vec{r}) = \mathcal{V}_{p\vec{G}} \\
H'_{pp} &= \frac{1}{(2\pi/G)^3} \int_0^{(2\pi/G)^3} d^3\vec{r} e^{-i(\vec{k}+p\vec{G})\cdot\vec{r}} V_L(\vec{r}) e^{i(\vec{k}+p\vec{G})\cdot\vec{r}} \\
&= \frac{1}{(2\pi/G)^3} \int_0^{(2\pi/G)^3} d^3\vec{r} V_L(\vec{r}) \\
&= V_0 = \texttt{dc component, independent of p} \\
E_p &= \frac{1}{(2\pi/G)^3} \int_0^{(2\pi/G)^3} d^3\vec{r} e^{-i(\vec{k}+p\vec{G})\cdot\vec{r}} \left[-\frac{\hbar^2}{2m_0} \nabla_r^2 \right] e^{i(\vec{k}+p\vec{G})\cdot\vec{r}} \\
&= \frac{\hbar^2 |\vec{k} + p\vec{G}|^2}{2m_0}.
\end{aligned} \tag{4.7}$$

The first equation tells us that $H'_{0p} = \mathcal{V}_{p\vec{G}}$, which is the *Fourier component* of the spatially periodic lattice potential at $\vec{q} = p\vec{G}$. These Fourier components are pictorially depicted in Fig. 4.3.

Using the above results, we get

$$C_p = \frac{\mathcal{V}_{p\vec{G}}}{E_0 + H'_{00} - E_p - H'_{pp}} = \frac{\mathcal{V}_{p\vec{G}}}{\frac{\hbar^2|\vec{k}|^2}{2m_0} - \frac{\hbar^2|\vec{k}+p\vec{G}|^2}{2m_0}}. \tag{4.8}$$

Finally, using the results of time-independent nondegenerate first order perturbation theory, we obtain

$$\begin{aligned}\psi^{\mathrm{NFE}}(\vec{r}) &= \frac{1}{\sqrt{a^3}}\mathrm{e}^{\mathrm{i}\vec{k}\cdot\vec{r}} + \sum_{p=1}^{n} \frac{\mathcal{V}_{p\vec{G}}}{E_0 + H'_{00} - E_p - H'_{pp}} \frac{1}{\sqrt{(2\pi/G)^3}}\mathrm{e}^{\mathrm{i}(\vec{k}+p\vec{G})\cdot\vec{r}} \\ &= \frac{1}{\sqrt{(2\pi/G)^3}}\mathrm{e}^{\mathrm{i}\vec{k}\cdot\vec{r}} + \sum_{p=1}^{n} \frac{\mathcal{V}_{p\vec{G}}}{\frac{\hbar^2|\vec{k}|^2}{2m_0} - \frac{\hbar^2|\vec{k}+p\vec{G}|^2}{2m_0}} \frac{1}{\sqrt{(2\pi/G)^3}}\mathrm{e}^{\mathrm{i}(\vec{k}+p\vec{G})\cdot\vec{r}} \\ E &= E_0 + H'_{00} + \sum_{p=1}^{n} \frac{|\mathcal{V}_{p\vec{G}}|^2}{\frac{\hbar^2|\vec{k}|^2}{2m_0} - \frac{\hbar^2|\vec{k}+p\vec{G}|^2}{2m_0}} \\ &= \frac{\hbar^2 k^2}{2m_0} + V_0 + \sum_{p=1}^{n} \frac{|\mathcal{V}_{p\vec{G}}|^2}{\frac{\hbar^2|\vec{k}|^2}{2m_0} - \frac{\hbar^2|\vec{k}+p\vec{G}|^2}{2m_0}}\end{aligned} \tag{4.9}$$

The last equation gives the energy E versus the wavevector $\vec{k}$ relation and *therefore the bandstructure*. However, this equation is *not* valid at $\vec{k} = p\vec{G}/2$ because at those values, the denominator of the p-th term in the summation may vanish, making that term diverge. What happens is that when $\vec{k} = p\vec{G}/2$, the plane wave (or free electron) states with wavevectors $\vec{k}$ and $\vec{k} + p\vec{G}$ are *degenerate* in energy and hence we *cannot* apply nondegenerate perturbation theory, as we have done so far. But this is easily handled since we know exactly what happens at points of degeneracy (see Special Topic 2 in Chap. 3). Energy gaps open up at these wavevectors, whose values will be $2|H'_{0p}| = 2\mathcal{V}_{p\vec{G}}$.

In Fig. 4.4 [left panel], we have plotted the energy versus wavevector relation (or bandstructure) of an arbitrary crystal calculated using the NFE method. Note that indeed energy gaps appear at wavevectors $\vec{k} = p\vec{G}/2$. The region between $\vec{k} = -\vec{G}/2$ and $\vec{G}/2$ is called the first *Brillouin zone*, while the regions between $\vec{k} = -\vec{G}$ and $-\vec{G}/2$ together with the regions between $\vec{k} = \vec{G}/2$ and $\vec{G}$ make up the second Brillouin zone, and so on. This plot is called the *extended zone representation* of the bandstructure. If we translate the bandstructure curves along the wavevector in either direction (positive or negative $\vec{k}$) by multiples of the reciprocal lattice vector $\vec{G}$, then we can map every zone on to the first Brillouin zone and obtain the dispersion relation shown in Fig. 4.4 [right panel]. This is called the *reduced zone representation* of the bandstructure.

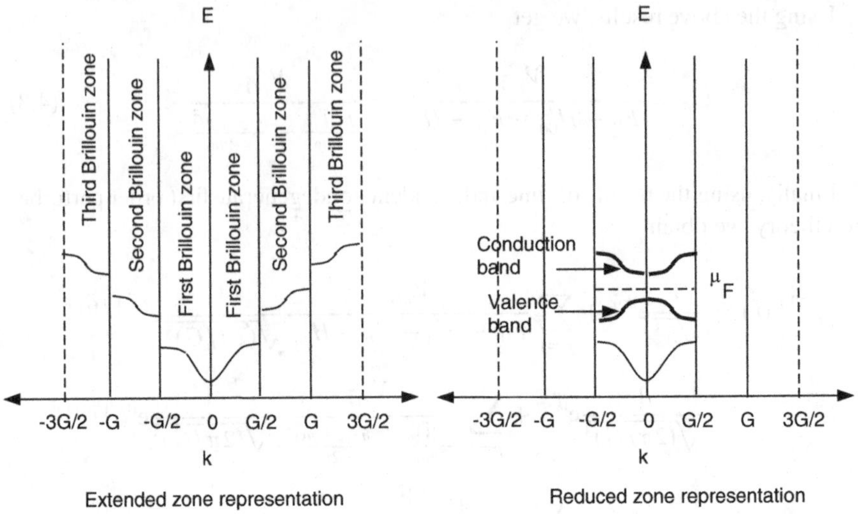

Fig. 4.4 The bandstructure calculated from the nearly free electron model. The *left panel* shows the extended zone representation and the *right panel* shows the reduced zone representation. In the *right panel*, we indicate the conduction and valence bands depending on the placement of the Fermi level μ_F

4.3.1 Why Do Energy Gaps Appear in the Bandstructure?

By solving the Schrödinger equation in a periodic potential, we found that energy gaps open up at wavevectors $\vec{k} = p\vec{G}/2$, which are called the *Brillouin zone edges* or *zone boundaries*. We will now hint at a physical reason as to why these gaps open up in the bandstructure. They are entirely a consequence of the periodic lattice potential. An electron wave suffers distributed reflection from the periodically placed atoms in the crystal. Readers familiar with electromagnetic wave theory know that whenever a wave propagates through a periodic array of scatterers (e.g., a transmission line), "stop bands" and "pass bands" appear, i.e., certain bands of frequencies can transmit through the array of scatterers and certain other bands cannot. The former types constitute the "pass bands" and the latter the "stop bands". The wave is evanescent in the stop bands and does not propagate, while it does propagate in the pass bands. The same physics is at work here. Note that energy and frequency are synonymous in this context since they are related according to $E = hf$, where E is energy, h is Planck's constant, and f is frequency. There is no real solution for the wavevector $\vec{k}$ in the energy gaps, which means that electron waves with those energies have imaginary or complex wavevectors, which is another way of saying that they are evanescent. Since the wave nature of electrons is a quantum-mechanical attribute, the bandstructure is obviously of quantum-mechanical origin.

4.3.2 Designation of Conduction and Valence Bands

Note from the reduced zone representation that there are denumerably infinite number of bands in a crystal. The band that is just above the Fermi level will become the "conduction band" and the band just below this level will become the "valence band," as shown in the right panel of Fig. 4.4. Therefore, the conduction and valence bands are determined by the placement of the Fermi level.[1] Because of Fermi–Dirac statistics, at any reasonable temperature, bands below the valence band will be completely filled with electrons, meaning every k-state in those bands will contain two electrons of opposite spins, while bands above the conduction band will be completely empty. Only the conduction and the valence bands will be partially filled. Since neither completely filled bands, nor completely empty bands, can carry any current, it is only the conduction and valence bands that contribute to current in a crystal and therefore are important for transport. They are also the only bands that matter for optical properties which are usually determined by electron transitions from the conduction band to the valence band and vice versa. Transitions involving bands above the conduction band or below the valence band will involve photons of very high energies (e.g., x-rays) that are not usually encountered.

One last question is why do we call this method the *nearly free* electron method. We call it that because we applied *first order* perturbation theory to calculate the bandstructure and that theory is applicable only when the perturbation is relatively *weak*. If the perturbation is weak, then the electron must be "nearly free" because if the perturbation (i.e., the attractive lattice potential) were absent, then the electron would have been completely free. This implies that the NFE method is most suitable for a material like sodium which has only one electron in the outermost shell that barely feels the heavily screened electrostatic attraction from the nucleus. On the other hand, an electron in an element like arsenic is strongly bound to the nucleus and is not nearly free. Hence, the nearly free electron method will not be suitable for arsenic, but it will be very suitable for sodium.

4.4 The Orthogonalized Plane Wave Expansion Method

In the NFE method, we use plane waves (wavefunctions of free electrons) as basis functions to expand the wavefunction. The question we wish to ask now is can we still use the plane waves as basis functions if the electron is not nearly free, but

[1]The placement of the Fermi level is determined by the number of electrons released by each atom in the crystal. We will fill up every available wavevector state in every band starting from the bottom, putting two electrons of opposite spins in each state in accordance with the Pauli Exclusion Principle. When we exhaust all the electrons in the crystal, the last wavevector state that we filled up in the last band becomes the Fermi wavevector and the corresponding energy becomes the Fermi energy at 0 K. A little reflection should convince the reader that this is consistent with Fermi–Dirac statistics at absolute zero of temperature.

instead quite tightly bound to the ions, as in arsenic? The answer is YES as long as we use the exact (3.55) and not the approximate solution in (3.57) to find the energy eigenvalues E and the eigenfunctions $[C_1, C_2, \ldots C_n]$ which will yield the wavefunction. Note that while (3.57) is valid only for a weak perturbation, (3.55) is valid for perturbation of any arbitrary strength. Therefore, we can use plane waves as basis functions to find the bandstructure in any arbitrary material as long as we use the exact (3.55) and not resort to the approximate (3.57). The only problem then is a computational one. Equation (3.55) represents N-coupled equations and therefore we have to deal with an $N \times N$ matrix. The computational burden to find the eigenvalues and eigenfunctions of an $N \times N$ matrix increases rapidly as N becomes larger. On the other hand, the accuracy of the calculated bandstructure improves with increasing N. How large N has to be in order to yield results with acceptable accuracy depends on the strength of the perturbation; the stronger it is, the larger will N have to be. To understand this, note that N is also the number terms in the summation in (4.5) where we are trying to replicate the wavefunction of the electron by adding up free electron wavefunctions. If the electron is not free to begin with, then we will obviously need numerous terms in the summation to successfully replicate the wavefunction. In other words, N will be reasonably small in sodium, but very large in a material like arsenic, where the electron is quite strongly bound to the background ions, and the lattice potential (or perturbation) is strong.

A clever strategy to overcome the computational challenge in a material like arsenic was suggested by Herring in 1940 [1]. He realized that the actual wavefunction in such a material will deviate most strongly from the plane wave state of free electrons right at the core of the ions where the electrostatic attraction is strongest. There, the wavefunction will resemble the core state (bound electron) wavefunctions rather than the free electron wavefunction. Therefore, the best approach is to use as basis functions for expansion in (4.5) not free electron wavefunctions, but instead wavefunctions that have a *mixed* character—partly free and partly bound. In other words, these wavefunctions will be superpositions of plane waves and core state wavefunctions. They are written as

$$\phi_n^{\text{OPW}}(\vec{r}) = \frac{1}{\sqrt{a^3}} e^{i(\vec{k}+n\vec{G})\cdot\vec{r}} - \sum_{i,j} \beta_{i,j} |\zeta_i(\vec{r} - \vec{R}_j)\rangle, \tag{4.10}$$

where $|\zeta_i(\vec{r} - \vec{R}_j)\rangle$ are the core state wavefunctions. They are, in fact, the wavefunction of an electron orbiting around an atom located at location $\vec{R}_j$ in the i-th atomic orbital state ($i = 2s, 3p, 5d$, etc.). In other words, $|\zeta_i(\vec{r} - \vec{R}_j)\rangle$ are the solutions of the Schrödinger equation

$$-\frac{\hbar^2}{2m_0} \nabla_r^2 \zeta_i(\vec{r} - \vec{R}_j) + V_a(\vec{r} - \vec{R}_j)\zeta_i(\vec{r} - \vec{R}_j) = E_{i,j}\zeta_i(\vec{r} - \vec{R}_j), \tag{4.11}$$

where $E_{i,j}$ is the energy of an electron in the i-th orbital state of an atom located at $\vec{R}_j$. The atomic potential is $V_a(\vec{r} - \vec{R}_j)$. Since the core state wavefunctions are strongly localized over their parent ions, they will not even see the potential due to other ions. Hence, we might as well write

$$-\frac{\hbar^2}{2m_0}\nabla_{\mathrm{r}}^2\zeta_i(\vec{r}-\vec{R}_j)+V_{\mathrm{L}}(\vec{r})\zeta_i(\vec{r}-\vec{R}_j)=E_{i,j}\zeta_i(\vec{r}-\vec{R}_j). \tag{4.12}$$

The quantity $\beta_{i,j}$ is a weighting coefficient in the linear expansion. Compare (4.10) with (4.3) to appreciate the difference between the NFE method and the OPW expansion method.

The perturbed wavefunction is now written as a linear superposition of the basis functions:

$$\psi^{\mathrm{OPW}}(\vec{r})=\sum_{p=0}^{n}C_p\phi_p^{\mathrm{OPW}}(\vec{r}), \tag{4.13}$$

which is very different from (4.5).

The next order of business is to find the coefficients $\beta_{i,j}$. Herring showed that the actual wavefunction $\psi(\vec{r})$ will contain many oscillations where the ions are located (i.e., at the core of the ions) and we can account for them quite well if we make $\phi_n^{\mathrm{OPW}}(\vec{r})$ orthogonal to the core states $|\zeta_i(\vec{r}-\vec{R}_j)\rangle$. Hence,

$$\langle\zeta_p(\vec{r}-\vec{R}_q)|\phi_n^{\mathrm{OPW}}(\vec{r})\rangle=0. \tag{4.14}$$

The last equation can be recast as

$$\left\langle\zeta_p(\vec{r}-\vec{R}_q)\left|\frac{1}{\sqrt{a^3}}\mathrm{e}^{\mathrm{i}(\vec{k}+n\vec{G})\cdot\vec{r}}\right.\right\rangle-\sum_{i,j}\beta_{i,j}\langle\zeta_p(\vec{r}-\vec{R}_q)|\zeta_i(\vec{r}-\vec{R}_j)\rangle=0. \tag{4.15}$$

Since the core state wavefunctions of adjacent ions do not overlap in space (these wavefunctions are strongly localized over their parent ions), it is obvious that

$$\langle\zeta_p(\vec{r}-\vec{R}_q)|\zeta_i(\vec{r}-\vec{R}_j)\rangle=\delta_{p,i}\delta_{q,j}, \tag{4.16}$$

where the delta-s are Kronecker deltas. The first Kronecker delta comes about because the core state wavefunctions in an atom are orthonormal (since they are eigenfunctions of the atom's Hamiltonian which is a Hermitian operator), and the second Kronecker delta comes about because core state wavefunctions of neighboring atoms have no overlap in space. Using this result in (4.15), we obtain

$$\left\langle\zeta_p(\vec{r}-\vec{R}_q)\left|\frac{1}{\sqrt{a^3}}\mathrm{e}^{\mathrm{i}(\vec{k}+n\vec{G})\cdot\vec{r}}\right.\right\rangle-\beta_{p,q}=0, \tag{4.17}$$

or

$$\beta_{i,j}=\left\langle\zeta_i(\vec{r}-\vec{R}_j)\left|\frac{1}{\sqrt{a^3}}\mathrm{e}^{\mathrm{i}(\vec{k}+n\vec{G})\cdot\vec{r}}\right.\right\rangle. \tag{4.18}$$

Using this result in (4.10), we get

$$\begin{aligned}\phi_n^{\rm OPW}(\vec{r}) &= \frac{1}{\sqrt{a^3}}\mathrm{e}^{\mathrm{i}(\vec{k}+n\vec{G})\cdot\vec{r}} - \sum_{i,j}|\zeta_i(\vec{r}-\vec{R}_j)\rangle\left\langle\zeta_i(\vec{r}-\vec{R}_j)\left|\frac{1}{\sqrt{a^3}}\mathrm{e}^{\mathrm{i}(\vec{k}+n\vec{G})\cdot\vec{r}}\right.\right\rangle \\ &= (1-\mathcal{P})\frac{1}{\sqrt{a^3}}\mathrm{e}^{\mathrm{i}(\vec{k}+n\vec{G})\cdot\vec{r}} \\ &= (1-\mathcal{P})\frac{1}{\sqrt{(2\pi/G)^3}}\mathrm{e}^{\mathrm{i}(\vec{k}+n\vec{G})\cdot\vec{r}},\end{aligned} \tag{4.19}$$

where $\mathcal{P}$ is called the "projection operator" and is given by

$$\begin{aligned}\mathcal{P} &= \sum_{i,j}|\zeta_i(\vec{r}-\vec{R}_j)\rangle\langle\zeta_i(\vec{r}-\vec{R}_j)| \\ &= \sum_{i,j}\zeta_i(\vec{r}-\vec{R}_j)\int_0^\infty d^3\vec{r}'\zeta_i(\vec{r}'-\vec{R}_j).\end{aligned} \tag{4.20}$$

The last expression immediately shows that $\mathcal{P}$ is zero outside the core since the core state wavefunctions vanish outside the core of the ion.

Finally, (4.13) can be written as

$$\psi^{\rm OPW}(\vec{r}) = (1-\mathcal{P})\sum_{p=0}^{n} C_p\frac{1}{\sqrt{(2\pi/G)^3}}\mathrm{e}^{\mathrm{i}(\vec{k}+p\vec{G})\cdot\vec{r}}, \tag{4.21}$$

so that the only difference with (4.5) is the factor $(1-\mathcal{P})$.

We can now find the bandstructure and the wavefunction in two ways. The first is referred to as the *direct method*, and the second as the *pseudopotential method*. In the direct method, we will find the energy eigenvalues from (3.55) after evaluating the matrix elements using the wavefunctions in (4.19). That is,

$$\begin{aligned}E_m &= \left\langle\phi_m^{\rm OPW}\left|-\frac{\hbar^2}{2m_0}\nabla_{\rm r}^2\right|\phi_m^{\rm OPW}\right\rangle \\ H'_{mn} &= \left\langle\phi_m^{\rm OPW}\right| V_{\rm L}(\vec{r})\left|\phi_n^{\rm OPW}\right\rangle.\end{aligned} \tag{4.22}$$

We will also find the eigenfunctions $[C_1, C_2, \ldots . C_n]$ from (3.55) and then use that in (4.21) to find the wavefunction. We will still need a computer to do all this, but fortunately, the size of the matrix N in (4.21) is now a lot smaller which relieves the computational burden significantly. The reader should realize that N is going to be much smaller simply because $\phi_n^{\rm OPW}$ is a better basis function than plane waves and is a lot closer to the actual wavefunction than plane waves, so that we will need far fewer terms in the summation in (4.13) to replicate the actual wavefunction on the left-hand-side of this equation.

4.4.1 The Pseudopotential Method

Let

$$\xi(\vec{r}) = \sum_{p=0}^{n} C_p \frac{1}{\sqrt{a^3}} \mathrm{e}^{\mathrm{i}(\vec{k}+p\vec{G})\cdot\vec{r}}, \tag{4.23}$$

so that

$$\psi^{\mathrm{OPW}}(\vec{r}) = (1 - \mathcal{P})\xi(\vec{r}). \tag{4.24}$$

Substituting the last result in the Schrödinger equation (4.1), we get

$$-\frac{\hbar^2}{2m_0}\nabla_{\mathrm{r}}^2\xi(\vec{r}) + V_{\mathrm{L}}(\vec{r})\xi(\vec{r}) - \left[-\frac{\hbar^2}{2m_0}\nabla_{\mathrm{r}}^2 + V_{\mathrm{L}}(\vec{r})\right]\mathcal{P}\xi(\vec{r}) + E\mathcal{P}\xi(\vec{r}) = E\xi(\vec{r}). \tag{4.25}$$

The last three terms on the left-hand-side are combined and called the *pseudopotential U*, so that the Schrödinger now equation becomes

$$-\frac{\hbar^2}{2m_0}\nabla_{\mathrm{r}}^2\xi(\vec{r}) + U\xi(\vec{r}) = E\xi(\vec{r}). \tag{4.26}$$

From (4.12) and (4.20), we can write

$$\left[-\frac{\hbar^2}{2m_0}\nabla_{\mathrm{r}}^2 + V_{\mathrm{L}}(\vec{r})\right]\mathcal{P} = \sum_{i,j} E_{i,j}|\zeta_i(\vec{r} - \vec{R}_j)\rangle\langle\zeta_i(\vec{r} - \vec{R}_j)|, \tag{4.27}$$

so that the pseudopotential can be conveniently written as

$$U = V_{\mathrm{L}}(\vec{r}) + \sum_{i,j}(E - E_{i,j})|\zeta_i(\vec{r} - \vec{R}_j)\rangle\langle\zeta_i(\vec{r} - \vec{R}_j)|. \tag{4.28}$$

The last equation is often referred to as the *pseudopotential equation*. Note that because of the presence of the projection operator, the pseudopotential U is *nonlocal* unlike the potential $V_{\mathrm{L}}(\vec{r})$ which depends only on the local coordinate $\vec{r}$. This complicates matters, but the redeeming feature is that the pseudopotential U is much weaker than the actual crystal potential $V_{\mathrm{L}}(\vec{r})$. This is because the crystal potential is negative (attractive) while the second term on the right-hand-side of (4.28) is clearly positive since it is the difference of the total energy and core state energy, while the projection operator is also essentially positive. Therefore, the positive and negative terms in the right-hand-side of (4.28) tend to cancel each other, which makes U a small quantity. This is sometimes referred to as the *cancellation theorem*. The result of all this is that we can solve (4.26) using first order nondegenerate perturbation theory to find the energy E and the wavefunction ξ *because* U is a weak perturbation while $V_{\mathrm{L}}(\vec{r})$ was not. Once we have found ξ, we can use (4.24) to find the actual wavefunction $\psi^{\mathrm{OPW}}(\vec{r})$. This yields the complete bandstructure and the wavefunction in a material where the electron is strongly bound to the ions.

4.5 The Tight-Binding Approximation Method

Consider a material where the electron is tightly bound to its parent ion so that its wavefunction is essentially the core state wavefunction. It is possible, however, for the electron to tunnel or "hop" to the next ion elastically (i.e., without absorbing or emitting any form of energy) or inelastically (where energy is absorbed/emitted in some form). Therefore, we can represent the electron wavefunction as a linear superposition of core state wavefunctions at different atomic (or ionic) sites. This is known as *linear combination of atomic orbitals* (LCAO):

$$\psi^{\text{TBA}}(\vec{r}) = \sum_{i,j} C_{ij}\zeta_i(\vec{r} - \vec{R}_j), \tag{4.29}$$

where $\zeta_i(\vec{r} - \vec{R}_j)$ is the wavefunction of an electron bound in the i-th orbital state of an atom located at the j-th atomic site in the crystal, and C_{ij} is the coefficient of expansion. We are now in a position to make a comparison between the wavefunctions used in the three approaches we have discussed so far. They are shown in Table 4.1.

In the TBA method, we assume that electrons are so tightly bound to their parent ions that their wavefunctions are strongly localized over the parent ions. As a result, there can be overlap between the wavefunctions of electrons in nearest neighbor ions *only* (see Fig. 4.5) so that the following condition holds:

$$\langle\zeta_i(\vec{r} - \vec{R}_j)|V_{\text{L}}(\vec{r})|\zeta_p(\vec{r} - \vec{R}_q)\rangle = \delta_{i,p}\xi, \tag{4.30}$$

where the delta is a Kronecker delta and

$$\xi = \begin{cases} A_p \ \texttt{if q = j} \\ t_p \ \texttt{if q = j} \pm \texttt{1} \\ 0 \ \texttt{otherwise} \end{cases}. \tag{4.31}$$

Substituting (4.29) in the time-independent Schrödinger equation in a crystal, we obtain

$$H\psi^{\text{TBA}}(\vec{r}) = \left[-\frac{\hbar^2}{2m}\nabla_{\text{r}}^2 + V_{\text{L}}(\vec{r})\right]\psi^{\text{TBA}}(\vec{r}) = E\psi^{\text{TBA}}(\vec{r})$$

and then following the usual prescription of multiplying from the left with the bra $\langle\zeta_i(\vec{r} - \vec{R}_j)|$, integrating over all volume, and making using of the conditions in

Table 4.1 Comparison between the wavefunctions used in the nearly free electron method (NFE), the OPW expansion method (OPW), and the TBA method

NFE	OPW	TBA
$\sum_{m=0}^{n} C_m \frac{1}{\sqrt{a^3}} e^{i(\vec{k}+m\vec{G})\cdot\vec{r}}$	$\sum_{m=0}^{n} C_m (1-P)\frac{1}{\sqrt{a^3}} e^{i(\vec{k}+m\vec{G})\cdot\vec{r}}$	$\sum_{i,j} C_{ij}\zeta_i(\vec{r} - \vec{R}_j)$

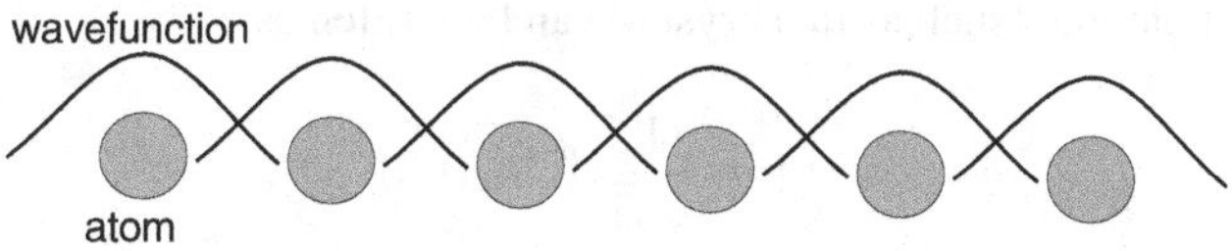

Fig. 4.5 The wavefunctions of electrons bound to nearest neighbor electrons have nonzero overlap

(4.31), we get the matrix equation

$$\begin{bmatrix} \epsilon_p & t_p & 0 & \ldots\ldots & 0 \\ t_p & \epsilon_p & t_p & 0 \ldots & 0 \\ 0 & t_p & \epsilon_p & t_p \ldots & 0 \\ \ldots & \ldots & \ldots & \ldots & \ldots \\ \ldots & \ldots & \ldots & \ldots & \ldots \\ 0 & \ldots\ldots & 0 & t_p & \epsilon_p \end{bmatrix} \begin{bmatrix} C_{p,1} \\ C_{p,2} \\ \ldots \\ \ldots \\ C_{p,n-1} \\ C_{p,n} \end{bmatrix} = E \begin{bmatrix} C_{p,1} \\ C_{p,2} \\ \ldots \\ \ldots \\ C_{p,n-1} \\ C_{p,n} \end{bmatrix}, \tag{4.32}$$

where

$$\begin{aligned} \epsilon_{p,q} &= \left\langle \zeta_p(\vec{r} - \vec{R}_q) \middle| -\frac{\hbar^2}{2m}\nabla_{\mathrm{r}}^2 + V_{\mathrm{L}}(\vec{r}) \middle| \zeta_p(\vec{r} - \vec{R}_q) \right\rangle \\ &= \left\langle \zeta_p(\vec{r} - \vec{R}_q) \middle| -\frac{\hbar^2}{2m}\nabla_{\mathrm{r}}^2 \middle| \zeta_p(\vec{r} - \vec{R}_q) \right\rangle + A_p \\ t_{p,q} &= \left\langle \zeta_p(\vec{r} - \vec{R}_{q-1}) \left| V_{\mathrm{L}}(\vec{r}) \right| \zeta_p(\vec{r} - \vec{R}_q) \right\rangle \\ &= \left\langle \zeta_p(\vec{r} - \vec{R}_q) | V_{\mathrm{L}}(\vec{r}) | \zeta_p(\vec{r} - \vec{R}_{q+1}) \right\rangle. \end{aligned} \tag{4.33}$$

If all the atoms (or ions) are identical, then we can drop the subscript q and write $\epsilon_p \equiv \epsilon_{p,q}$ and $t_p \equiv t_{p,q}$.

The matrix element ϵ_p is called the "on-site energy" (for an electron in the p-th orbital) and t_p is called the "tunneling or hopping matrix element" (again, for an electron in the p-th orbital). Note that t_p will be zero if there is no overlap between the wavefunctions of nearest neighbor atoms, in which case an electron cannot tunnel or hop between nearest neighbors. Now, considering any arbitrary row in the matrix in (4.32) (note that the matrix is a band matrix with a band of nonzero elements about the diagonal), we get the equation

$$\begin{aligned} t_p C_{p,m-1} + \epsilon_p C_{p,m} + t_p C_{p,m+1} &= E C_{p,m}, \text{ or} \\ t_p \left(\frac{C_{p,m-1}}{C_{p,m}} + \frac{C_{p,m+1}}{C_{p,m}} \right) &= E - \epsilon_p. \end{aligned} \tag{4.34}$$

In order to find a solution for the coefficient $C_{i,n}$ in the above equation, we invoke the well-known *Bloch theorem* [2] which states that the wavefunction of an electron in a periodic potential (such as in a crystal) can be written as

$$\psi(\vec{r}) = \frac{1}{\sqrt{\Omega}} \mathrm{e}^{\mathrm{i}\vec{k}\cdot\vec{r}} u_{\vec{k}}(\vec{r}), \tag{4.35}$$

where Ω is a normalizing volume such that $\frac{1}{\Omega}\int_{\Omega} d^3\vec{r} \mathrm{e}^{\mathrm{i}(\vec{k}-\vec{k}')\cdot\vec{r}}$ is equal to the Kronecker delta $\delta_{\vec{k},\vec{k}'}$, and the function $u_{\vec{k}}(\vec{r})$, known as the *Bloch function*, is a periodic function that has the same period as the lattice potential $V_{\mathrm{L}}(\vec{r})$. In other words, it obeys the relation

$$u_{\vec{k}}(\vec{r}) = u_{\vec{k}}(\vec{r} + \vec{a}), \tag{4.36}$$

where $\vec{a}$ is the lattice period or lattice constant. From (4.35) and (4.36), it is easy to show that the wavefunctions at locations $\vec{r}$ and $\vec{r} + \vec{a}$ are related according to

$$\psi(\vec{r} + \vec{a}) = \mathrm{e}^{\mathrm{i}\vec{k}\cdot\vec{a}} \psi(\vec{r}) \tag{4.37}$$

Since the tight-binding wavefunction, like all legitimate wavefunctions in a crystal, must obey the Bloch equation, it follows that

$$C_{p,m+1}\zeta_i(\vec{r} - \vec{R}_{m+1}) = \mathrm{e}^{\mathrm{i}\vec{k}\cdot\vec{a}} C_{p,m}\zeta_i(\vec{r} - \vec{R}_m), \tag{4.38}$$

because $\vec{R}_{m+1} - \vec{R}_m = \vec{a}$.

Furthermore, since all atoms are identical in a crystal (we are assuming an elemental crystal here; atoms in a compound semiconductor like GaAs can be different since they can be either Ga atoms or As atoms), it stands to reason that $\zeta_p(\vec{r} - \vec{R}_{m+1}) = \zeta_p(\vec{r} - \vec{R}_m)$. Therefore,

$$\frac{C_{p,m+1}}{C_{p,m}} = \frac{C_{p,m}}{C_{p,m-1}} = \mathrm{e}^{\mathrm{i}\vec{k}\cdot\vec{a}} \tag{4.39}$$

Using the last relation in (4.38), we get

$$t_p(\mathrm{e}^{-\mathrm{i}\vec{k}\cdot\vec{a}} + \mathrm{e}^{\mathrm{i}\vec{k}\cdot\vec{a}}) = E - \epsilon_p, \text{ or}$$
$$2t_p \cos(\vec{k}\cdot\vec{a}) = E - \epsilon_p. \tag{4.40}$$

The last equation yields the *energy dispersion relation* or the sought after $E - k$ relation in a crystal:

$$E = \epsilon_p + 2t_p \cos(\vec{k}\cdot\vec{a}). \tag{4.41}$$

Approximate plots of (4.41) are shown in Fig. 4.6.

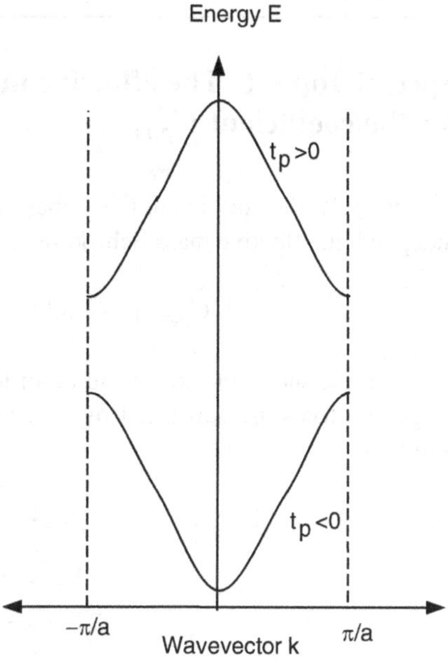

Fig. 4.6 Dispersion relations found from the tight-binding approximation

From (4.29) and (4.38), we get that the tight-binding wavefunction is

$$\psi^{\mathrm{TBA}}(\vec{r}) = \sum_{i,n} \mathrm{e}^{in\vec{k}\cdot\vec{a}} \zeta_i(\vec{r} - \vec{R}_n). \tag{4.42}$$

The TBA method has proved to be very useful in heterostructures composed of two or more materials. It also highlights the relation between the effective mass of an electron in a crystal and the ability of the electron to transit between neighboring atoms by tunneling and hopping. If we use the TBA dispersion relation in (4.41), then the (group) velocity v and effective mass m^* of an electron in the crystal turn out to be

$$\begin{aligned} v &= \frac{1}{\hbar}\frac{\partial E}{\partial k} = -2t_p\vec{a}\sin(\vec{k}\cdot\vec{a}) \\ m^* &= \frac{\hbar^2}{\partial^2 E/\partial^2 k} = -\frac{\hbar^2 \sec(\vec{k}\cdot\vec{a})}{2t_p a^2} \end{aligned} \tag{4.43}$$

The last relation shows that if the tunneling or hopping matrix element t_p is zero, then an electron cannot transit between neighboring sites and must remain static and completely localized over an atom. Such an electron should have no velocity and its effective mass should be infinity corresponding to infinite "inertia." This is exactly what we find. Therefore, the TBA approximation is firmly grounded on a very physical basis.

Special Topic 1: The effective mass Schrödinger equation for the coefficient $C_{p,m}$

Interestingly, the coefficient $C_{p,m}$ obeys an equation that is reminiscent of the time-independent effective mass Schrödinger equation. From (4.38), we get

$$t_p C_{p,m-1} + \epsilon_p C_{p,m} + t_p C_{p,m+1} = E C_{p,m}. \tag{4.44}$$

Since the index m indexes an atom located at the m-th lattice site, and nearest neighbor atoms are separated by a distance equal to the lattice period a, we can write

$$\begin{aligned} C_{p,m-1} &= C(x-a) \\ C_{p,m} &= C(x) \\ C_{p,m+1} &= C(x+a). \end{aligned} \tag{4.45}$$

Using a Taylor series expansion, we rewrite (4.38) as

$$\begin{aligned} &t_p \left[C(x) - \frac{\partial C}{\partial x} a + \frac{1}{2} \frac{\partial^2 C}{\partial^2 x} a^2 + \ldots \right] + \epsilon_p C(x) \\ &+ t_p \left[C(x) + \frac{\partial C}{\partial x} a + \frac{1}{2} \frac{\partial^2 C}{\partial^2 x} a^2 + \ldots \right] = E C(x). \end{aligned} \tag{4.46}$$

Retaining only up to second order terms (since the lattice period a is typically small), the above equation reduces to

$$t_p a^2 \frac{\partial^2 C(x)}{\partial x^2} + (\epsilon_p + 2t_p) C(x) = E C(x). \tag{4.47}$$

Now, near $k = 0$, $t_p a^2 = -\hbar^2/2m^*$. Therefore, we get

$$-\frac{\hbar^2}{2m^*} C(x) + (\epsilon_p + 2t_p) C(x) = E C(x) \tag{4.48}$$

which is mathematically similar to the time-independent effective mass Schrödinger equation with the coefficient $C_{p,m}$ playing the role of the wavefunction and $(\epsilon_p + 2t_p)$ playing the role of the potential energy.

4.6 The $\vec{k} \cdot \vec{p}$ Perturbation Method

The $\vec{k} \cdot \vec{p}$ perturbation theory for calculating bandstructures is a very powerful approach that has been particularly successful for narrow gap semiconductors like InSb [3]. This technique makes explicit use of the Bloch Theorem, as well as (4.35) and (4.36).

The time-independent Schrödinger equation for an electron in a crystal is given by (4.1). If we substitute (4.35) for the wavefunction in this equation, we obtain

$$-\frac{\hbar^2}{2m_0}\nabla_{\mathrm{r}}^2\left[\mathrm{e}^{\mathrm{i}\vec{k}\cdot\vec{r}}u_{\vec{k}}(\vec{r})\right] + V_{\mathrm{L}}(\vec{r})\left[\mathrm{e}^{\mathrm{i}\vec{k}\cdot\vec{r}}u_{\vec{k}}(\vec{r})\right] = E\left[\mathrm{e}^{\mathrm{i}\vec{k}\cdot\vec{r}}u_{\vec{k}}(\vec{r})\right]. \tag{4.49}$$

Note that

$$\begin{aligned}
\vec{\nabla}_{\mathrm{r}}\left[\mathrm{e}^{\mathrm{i}\vec{k}\cdot\vec{r}}u_{\vec{k}}(\vec{r})\right] &= \left[\mathrm{i}\vec{k}u_{\vec{k}}(\vec{r}) + \vec{\nabla}_{\mathrm{r}}u_{\vec{k}}(\vec{r})\right]\mathrm{e}^{\mathrm{i}\vec{k}\cdot\vec{r}} \\
\nabla_{\mathrm{r}}^2\left[\mathrm{e}^{\mathrm{i}\vec{k}\cdot\vec{r}}u_{\vec{k}}(\vec{r})\right] &= \vec{\nabla}_{\mathrm{r}}\cdot\vec{\nabla}_{\mathrm{r}}\left[\mathrm{e}^{\mathrm{i}\vec{k}\cdot\vec{r}}u_{\vec{k}}(\vec{r})\right] \\
&= \vec{\nabla}_{\mathrm{r}}\cdot\left[\mathrm{i}\vec{k}u_{\vec{k}}(\vec{r})\,\mathrm{e}^{\mathrm{i}\vec{k}\cdot\vec{r}}\right] + \vec{\nabla}_{\mathrm{r}}\cdot\left[\vec{\nabla}_{\mathrm{r}}u_{\vec{k}}(\vec{r})\,\mathrm{e}^{\mathrm{i}\vec{k}\cdot\vec{r}}\right].
\end{aligned} \tag{4.50}$$

We will now make use of the familiar vector identity in (2.15). Treating the term $\mathrm{i}\vec{k}$ as a vector and the term $u_{\vec{k}}(\vec{r})\mathrm{e}^{\mathrm{i}\vec{k}\cdot\vec{r}}$ as a scalar, we obtain:

$$\begin{aligned}
\vec{\nabla}_{\mathrm{r}}\cdot\left[\mathrm{i}\vec{k}u_{\vec{k}}(\vec{r})\,\mathrm{e}^{\mathrm{i}\vec{k}\cdot\vec{r}}\right] &= \mathrm{i}\vec{k}\cdot\vec{\nabla}_{\mathrm{r}}\left[u_{\vec{k}}(\vec{r})\,\mathrm{e}^{\mathrm{i}\vec{k}\cdot\vec{r}}\right] + u_{\vec{k}}(\vec{r})\,\mathrm{e}^{\mathrm{i}\vec{k}\cdot\vec{r}}\,\underbrace{\vec{\nabla}_{\mathrm{r}}\cdot\left(\mathrm{i}\vec{k}\right)} \\
&= \mathrm{i}\vec{k}\cdot\left[\mathrm{i}\vec{k}u_{\vec{k}}(\vec{r}) + \vec{\nabla}_{\mathrm{r}}u_{\vec{k}}(\vec{r})\right]\mathrm{e}^{\mathrm{i}\vec{k}\cdot\vec{r}} \\
&= \left[-k^2u_{\vec{k}}(\vec{r}) + \mathrm{i}\vec{k}\cdot\vec{\nabla}_{\mathrm{r}}u_{\vec{k}}(\vec{r})\right]\mathrm{e}^{\mathrm{i}\vec{k}\cdot\vec{r}} \\
\vec{\nabla}_{\mathrm{r}}\cdot\left[\vec{\nabla}_{\mathrm{r}}u_{\vec{k}}(\vec{r})\,\mathrm{e}^{\mathrm{i}\vec{k}\cdot\vec{r}}\right] &= \vec{\nabla}_{\mathrm{r}}u_{\vec{k}}(\vec{r})\cdot\vec{\nabla}_{\mathrm{r}}\left(\mathrm{e}^{\mathrm{i}\vec{k}\cdot\vec{r}}\right) + \mathrm{e}^{\mathrm{i}\vec{k}\cdot\vec{r}}\vec{\nabla}_{\mathrm{r}}\cdot\vec{\nabla}_{\mathrm{r}}u_{\vec{k}}(\vec{r}) \\
&= \vec{\nabla}_{\mathrm{r}}u_{\vec{k}}(\vec{r})\cdot\mathrm{i}\vec{k}\mathrm{e}^{\mathrm{i}\vec{k}\cdot\vec{r}} + \mathrm{e}^{\mathrm{i}\vec{k}\cdot\vec{r}}\nabla_{\mathrm{r}}^2u_{\vec{k}}(\vec{r}) \\
&= \left[\mathrm{i}\vec{k}\cdot\vec{\nabla}_{\mathrm{r}}u_{\vec{k}}(\vec{r}) + \nabla_{\mathrm{r}}^2u_{\vec{k}}(\vec{r})\right]\mathrm{e}^{\mathrm{i}\vec{k}\cdot\vec{r}}.
\end{aligned} \tag{4.51}$$

Note that the term with the underbrace vanishes since the wavevector is not a function of position.

Adding the two terms on the left-hand side in the last equation and substituting in (4.50), we get that

$$\nabla_{\mathrm{r}}^2\left[\mathrm{e}^{\mathrm{i}\vec{k}\cdot\vec{r}}u_{\vec{k}}(\vec{r})\right] = \left[-k^2u_{\vec{k}}(\vec{r}) + 2i\vec{k}\cdot\vec{\nabla}_{\mathrm{r}}u_{\vec{k}}(\vec{r}) + \nabla_{\mathrm{r}}^2u_{\vec{k}}(\vec{r})\right]\mathrm{e}^{\mathrm{i}\vec{k}\cdot\vec{r}}. \tag{4.52}$$

Using the last result in (4.49), we obtain

$$\frac{\hbar^2}{2m_0}\left[k^2u_{\vec{k}}(\vec{r}) - 2i\vec{k}\cdot\vec{\nabla}_{\rm r}u_{\vec{k}}(\vec{r}) - \nabla_{\rm r}^2u_{\vec{k}}(\vec{r})\right]{\rm e}^{{\rm i}\vec{k}\cdot\vec{r}} + V_{\rm L}(\vec{r})u_{\vec{k}}(\vec{r}){\rm e}^{{\rm i}\vec{k}\cdot\vec{r}} = Eu_{\vec{k}}(\vec{r}){\rm e}^{{\rm i}\vec{k}\cdot\vec{r}}, \tag{4.53}$$

which can be simplified to

$$-\frac{\hbar^2}{2m_0}\nabla_{\rm r}^2u_{\vec{k}}(\vec{r}) - \frac{\hbar^2}{m_0}{\rm i}\vec{k}\cdot\vec{\nabla}_{\rm r}u_{\vec{k}}(\vec{r}) + V_{\rm L}(\vec{r})u_{\vec{k}}(\vec{r}) = \left(E - \frac{\hbar^2k^2}{2m_0}\right)u_{\vec{k}}(\vec{r}), \tag{4.54}$$

or

$$-\frac{\hbar^2}{2m_0}\nabla_{\rm r}^2u_{\vec{k}}(\vec{r}) + \frac{\hbar}{m_0}\vec{k}\cdot(-{\rm i}\hbar\vec{\nabla}_{\rm r})u_{\vec{k}}(\vec{r}) + V_{\rm L}(\vec{r})u_{\vec{k}}(\vec{r}) = \epsilon_k u_{\vec{k}}(\vec{r}), \tag{4.55}$$

where

$$\epsilon_k = E - \frac{\hbar^2k^2}{2m_0} \tag{4.56}$$

Noting that the momentum operator $\vec{p}_{\rm op} \equiv \vec{p} = -{\rm i}\hbar\vec{\nabla}_{\rm r}$, the last equation can be recast as

$$\left[-\frac{\hbar^2}{2m_0}\nabla_{\rm r}^2 + V_{\rm L}(\vec{r}) + \frac{\hbar}{m_0}\vec{k}\cdot\vec{p}\right]u_{\vec{k}}(\vec{r}) = \epsilon_k u_{\vec{k}}(\vec{r}). \tag{4.57}$$

We now treat the term $-\frac{\hbar^2}{2m_0}\nabla_{\rm r}^2 + V_{\rm L}(\vec{r})$ as the unperturbed Hamiltonian H_0 and the term $\frac{\hbar}{m_0}\vec{k}\cdot\vec{p}$ as the perturbation term H'. Hence the name $\vec{k}\cdot\vec{p}$ *perturbation theory*. The above equation now becomes the familiar equation:

$$[H_0 + H']u_{\vec{k}}(\vec{r}) = \epsilon_k u_{\vec{k}}(\vec{r}). \tag{4.58}$$

Next, we apply normal time-independent first-order perturbation theory to find the Bloch function $u_{\vec{k}}(\vec{r})$ and the energy ϵ_k. Once we have found the former, we can find the entire wavefunction from (4.35) and from the latter, we can find the energy dispersion ($E - k$) relation using (4.56).

4.6.0.1 Orthogonality of the Bloch Functions in Different Bands at the Same Wavevector

It should be obvious that since the Bloch functions $u_{\vec{k},m}(\vec{r})$ in different bands *at the same wavevector* $\vec{k}$ are nondegenerate solutions of the same Hamiltonian $[H_0 + H']$, which is a hermitian operator, they must be orthonormal functions, i.e.,

$$\int_0^\infty d^3\vec{r}u_{\vec{k},m}^*(\vec{r})u_{\vec{k},n}^*(\vec{r}) = \delta_{m,n}, \tag{4.59}$$

where $\delta_{m,n}$ is a Kronecker delta. We will have occasion to use this property in Chap. 6 where we discuss optical properties of semiconductor crystals.

4.6.1 Application of Perturbation Theory to Calculate Bandstructure

Note that the perturbation term vanishes at zero wavevector ($\vec{k} = 0$). Hence, the zero-wavevector Bloch functions $u_0(\vec{r})$ are obviously the unperturbed states, which must obey the equation

$$H_0 u_0(\vec{r}) = \epsilon_0 u_0(\vec{r}). \tag{4.60}$$

Now, from (4.56), it is obvious that the energy ϵ_k is the total energy E minus the free (particle) energy $\hbar^2 k^2/2m_0$, and is therefore the core energy. Consequently, ϵ_0 must be the core energy of a particle that has no wavevector ($k = 0$) and hence no free energy.[2] Hence, it must be the energy of a completely bound state (or atomic orbital state). In other words, since the core energy of the unperturbed states (with zero wavevector) are the bound orbital state energies, the unperturbed states *must be bound orbital states* in an atom. Therefore,

$$u_0(\vec{r}) = |S\rangle, |P_x\rangle, |P_y\rangle, |P_z\rangle, \texttt{ etc.} \tag{4.61}$$

where $|S\rangle$ is the s-orbital state, $|P_x\rangle, |P_y\rangle, |P_z\rangle$ are the p-orbital states, etc. Normal perturbation theory will then tell us that

$$\begin{aligned} u_{\vec{k}}(\vec{r}) &= C_1 u_0^1 + C_2 u_0^2 + C_3 u_0^3 + \cdots \\ &= C_1|S\rangle + C_2|P_x\rangle + C_3|P_y\rangle + C_4|P_z\rangle + \cdots . \end{aligned} \tag{4.62}$$

We display the typical bandstructure of a crystal (in the reduced Brillouin zone) in Fig. 4.7. Note that the Bloch functions in the different bands at the zone center (i.e., at $\vec{k} = 0$) are the quantities $u_0^n(\vec{r})$, and these are the wavefunctions of bound orbital states.

From (4.62), it then becomes obvious that the Bloch function in any band at nonzero wavevector is a mixture of the Bloch functions at the zone center ($k = 0$) of all the bands. Therefore, every band perturbs the states in any given band at nonzero wavevectors. As a result, it is important to include the effects of all the other bands on the wavefunction at any nonzero k-state in any band, no matter how remote the other bands are in energy. This has important consequences, as we will see later.

[2]Physically, an electron in a core state is completely bound to the atom or ion and therefore has no resultant momentum. Hence $\hbar k = 0$, which makes $k = 0$.

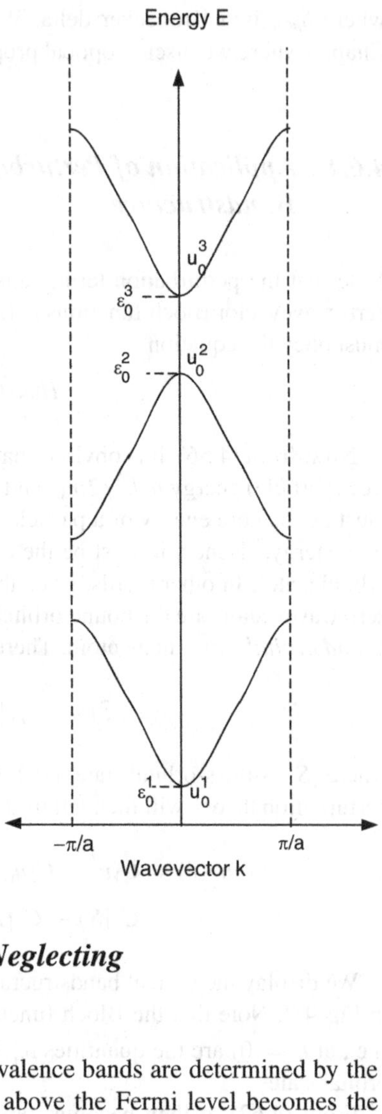

Fig. 4.7 The Bloch functions at the zone center are the atomic orbital states and the Bloch functions at nonzero wavevector in any band is a mixture of the Bloch functions at the zone centers of all bands

4.6.2 A Simple Two-Band Theory: Neglecting the Remote Bands

We have seen before that the conduction and valence bands are determined by the placement of the Fermi level. The band just above the Fermi level becomes the conduction band and the band just below becomes the valence band. Let us now calculate the wavefunctions and energy dispersion relations in the conduction and valence bands by considering only the perturbation of one band on the other while neglecting all other remote bands. This will lead to interesting consequences.

Since we are considering only two bands, (4.62) will have only two terms. Therefore, the Bloch function for any k-state in the conduction band will be

$$u^{c}_{\vec{k}}(\vec{r}) = C_1 u^{c}_0(\vec{r}) + C_2 u^{v}_0(\vec{r}), \tag{4.63}$$

where the superscripts c and v refer to conduction and valence bands, respectively.

Using standard time-independent first order perturbation theory that we discussed in the previous chapter, we obtain

$$\frac{C_2}{C_1} = \frac{H'_{12}}{E_1 + H'_{11} - E_2 - H'_{22}} = \frac{\frac{\hbar}{m_0}\langle u_0^{\mathrm{c}}(\vec{r})|\vec{k}\cdot\vec{p}|u_0^{\mathrm{v}}(\vec{r})\rangle}{\epsilon_0^{\mathrm{c}} - \epsilon_0^{\mathrm{v}} + H'_{11} - H'_{22}}, \tag{4.64}$$

where ϵ_0^{c} and ϵ_0^{v} are the unperturbed energy states which must be the energies in the conduction and valence bands at the Brillouin zone center (see Fig. 4.7), i.e., at $\vec{k} = 0$, since the $\vec{k} \cdot \vec{p}$ perturbation vanishes when $\vec{k} = 0$.

Let us now attempt to evaluate the self energy terms H'_{11} and H'_{22}.

$$H'_{11} = \frac{\hbar}{m_0}\langle u_0^{\mathrm{c}}(\vec{r})|\vec{k}\cdot\vec{p}|u_0^{\mathrm{c}}(\vec{r})\rangle$$
$$H'_{22} = \frac{\hbar}{m_0}\langle u_0^{\mathrm{v}}(\vec{r})|\vec{k}\cdot\vec{p}|u_0^{\mathrm{v}}(\vec{r})\rangle. \tag{4.65}$$

We offer two different arguments as to why these terms should be zero. First, recognize that $u_0^{\mathrm{c}}(\vec{r})$ and $u_0^{\mathrm{v}}(\vec{r})$ are wavefunctions of atomic orbital states that are purely real, while the momentum operator $\vec{p}$ is purely imaginary. This would have made the self-energy terms imaginary had they not been zero. Any imaginary self-energy would be entirely unphysical since there is no such thing as imaginary energy. Therefore, these terms must vanish. Second, consider the simpler one-dimensional case:

$$\begin{aligned}
H'_{11} &= \frac{\hbar}{m_0}\langle u_0^{\mathrm{c}}(x)|k_x p_x|u_0^{\mathrm{c}}(x)\rangle \\
&= -\mathrm{i}\hbar k_x \frac{\hbar}{m_0}\int_{-\infty}^{\infty} u_0^{\mathrm{c}}(x)\left[\frac{\mathrm{d}u_0^{\mathrm{c}}(x)}{\mathrm{d}x}\right]\mathrm{d}x \\
&= -\mathrm{i}\frac{\hbar^2 k_x}{m_0}\left[(u_0^{\mathrm{c}}(\infty))^2 - (u_0^{\mathrm{c}}(-\infty))^2\right] + \mathrm{i}\frac{\hbar^2 k_x}{m_0}\int_{-\infty}^{\infty} u_0^{\mathrm{c}}(x)\left[\frac{\mathrm{d}u_0^{\mathrm{c}}(x)}{\mathrm{d}x}\right]\mathrm{d}x, \\
&= -\mathrm{i}\frac{\hbar^2 k_x}{m_0}\left[(u_0^{\mathrm{c}}(\infty))^2 - (u_0^{\mathrm{c}}(-\infty))^2\right] - H'_{11}
\end{aligned} \tag{4.66}$$

where we have carried out integration by parts. Since any atomic orbital wavefunction must vanish at $\pm\infty$, the first term on the right-hand side of the equation above reduces to zero, leaving us with

$$H'_{11} = -H'_{11}, \text{ or}$$
$$H'_{11} = 0. \tag{4.67}$$

Similarly, $H'_{22} = 0$. Note that at $\vec{k} = 0$, the conduction band energy is $E_{\mathrm{c}}(0) = \epsilon_0^{\mathrm{c}} + \hbar^2(0)^2/2m_0 = \epsilon_0^{\mathrm{c}} = E_{\mathrm{c}0}$ and likewise the valence band energy is $E_v(0) = \epsilon_0^v = E_{v0}$, where E_{c0} and E_{v0} are the energies at the conduction band edge (bottom of the conduction band) and valence band edge (top of the valence band) in a *direct-gap* semiconductor where the band edges occur at the Brillouin zone center (i.e., at $\vec{k} = 0$). Therefore, from (4.64), we obtain

$$\frac{C_2}{C_1} = \frac{\frac{\hbar}{m_0}\langle u_0^c(\vec{r})|\vec{k}\cdot\vec{p}|u_0^v(\vec{r})\rangle}{E_{c0} - E_{v0}} = \frac{\frac{\hbar}{m_0}\langle u_0^c(\vec{r})|\vec{k}\cdot\vec{p}|u_0^v(\vec{r})\rangle}{E_g}, \tag{4.68}$$

where, by definition, the bandgap $E_g = E_{c0} - E_{v0}$.

Application of standard time-independent first-order perturbation theory also yields:

$$\epsilon_k^c = \epsilon_0^c + H'_{11} + \frac{|H'_{12}|^2}{\epsilon_0^c + H'_{11} - \epsilon_0^v - H'_{22}} = E_{c0} + \frac{\frac{\hbar^2}{m_0^2}\left|\langle u_0^c(\vec{r})|\vec{k}\cdot\vec{p}|u_0^v(\vec{r})\rangle\right|^2}{E_g}. \tag{4.69}$$

Finally, using (4.56), we obtain

$$E_c(k) = E_{c0} + \frac{\hbar^2 k^2}{2m_0} + \frac{\frac{\hbar^2}{m_0^2}\left|\langle u_0^c(\vec{r})|\vec{k}\cdot\vec{p}|u_0^v(\vec{r})\rangle\right|^2}{E_g}. \tag{4.70}$$

In direct-gap semiconductors, where the conduction and valence band extrema occur at the Brillouin zone center, the conduction band states $u_0^c(\vec{r})$ tend to be s-type orbitals ($|S>$) and the valence band states $u_0^v(\vec{r})$ are usually p-type orbitals. There are *three* p-type orbitals, namely $|P_x\rangle$, $|P_y\angle$, and $|P_z\rangle$. Electrons in the valence band are weakly bound to their parent nuclei and experience the electric field due to the latter. This causes fairly strong spin–orbit interaction in the valence band which results in admixture between the three p-orbital states. Note that atomic spin–orbit interaction will require the orbital quantum number l to be nonzero, but this requirement is satisfied by p-type orbitals since their $l = \pm 1$. The result is that we end up with three valence bands that we call the *heavy-hole band*, the *light-hole band* and the *split-off band*. There is, however, only one conduction band since there is only one s-type orbital, and there is no intrinsic spin–orbit interaction in this band since s-type orbitals have orbital quantum number $l = 0$. In indirect-gap semiconductors like silicon, the conduction band edge is off-zone center where $\vec{k} \neq 0$. There, the orbitals need not be s-type and hence there can be intrinsic spin–orbit interaction in the conduction band as well.

From the last equation, we can get the dispersion relation in the conduction band of a direct-gap semiconductor along any wavevector direction. For example, along the x-direction, we will have

$$\begin{aligned} E_c(k_x) &= E_{c0} + \frac{\hbar^2 k_x^2}{2m_0} + \frac{\hbar^2 k_x^2 |\langle S|p_x|P_x\rangle|^2}{m_0^2 E_g} \\ &= E_{c0} + \frac{\hbar^2 k_x^2}{2m_0}\left[1 + \frac{2\Upsilon^2}{m_0 E_g}\right], \end{aligned} \tag{4.71}$$

where $\Upsilon = |\langle S|p_x|P_x\rangle|$. In this simple two-band model, $|\langle S|p_x|P_x\rangle| = |\langle S|p_y|P_y\rangle| = |\langle S|p_z|P_z\rangle| = \Upsilon$, so that there is no anisotropy of bandstructure along the principal crystallographic directions belonging to the $\langle 100\rangle >$ family. The quantity Υ is called the "momentum matrix element."

The dispersion relation in the valence band is

$$\epsilon_k^{\mathrm{v}} = \epsilon_0^{\mathrm{v}} + H'_{22} + \frac{|H'_{12}|^2}{\epsilon_0^{\mathrm{v}} + H'_{22} - \epsilon_0^{\mathrm{c}} - H'_{11}} = E_{\mathrm{v}0} + \frac{\frac{\hbar^2}{m_0^2}|\langle u_0^{\mathrm{v}}(\vec{r})|\vec{k}\cdot\vec{p}|u_0^{\mathrm{c}}(\vec{r})\rangle|^2}{-E_{\mathrm{g}}}. \quad (4.72)$$

Consequently, the valence band dispersion relation along the $\langle 100\rangle$ direction will be

$$\begin{aligned} E_{\mathrm{v}}(k_x) &= E_{\mathrm{v}0} + \frac{\hbar^2 k_x^2}{2m_0} - \frac{\hbar^2 k_x^2 |\langle P_x|p_x|S\rangle|^2}{m_0^2 E_{\mathrm{g}}} \\ &= E_{\mathrm{v}0} + \frac{\hbar^2 k_x^2}{2m_0}\left[1 - \frac{2\Upsilon^2}{m_0 E_{\mathrm{g}}}\right], \end{aligned} \quad (4.73)$$

since $|\langle P_x|p_x|S\rangle| = |\langle S|p_x|P_x\rangle|$ because of the hermiticity of the p_x operator.

We can represent the dispersion relation in either band as

$$E_{\mathrm{c,v}}(k_x) = E_{\mathrm{c0,v0}} + \frac{\hbar^2 k_x^2}{2m_0}\left[1 \pm \frac{2\Upsilon^2}{m_0 E_{\mathrm{g}}}\right], \quad (4.74)$$

where the plus sign applies to the conduction band and the minus sign to the valence band. The quantity $\frac{2\Upsilon^2}{m_0 E_{\mathrm{g}}}$ is obviously the deviation from "parabolicity" since without this term, the dispersion relation in either band would have been parabolic. Note that narrower gap materials (smaller E_{g}) are more nonparabolic because of this term. Note also that if we write the dispersion relation in the usual form

$$E(k_x) = E_0 + \frac{\hbar^2 k_x^2}{2m^*}, \quad (4.75)$$

where m^* is the effective mass, then

$$m^* = \begin{cases} \frac{m_0}{1+\frac{2\Upsilon^2}{m_0 E_{\mathrm{g}}}} & \texttt{(electrons)} \\ \frac{m_0}{1-\frac{2\Upsilon^2}{m_0 E_{\mathrm{g}}}} & \texttt{(holes)} \end{cases} . \quad (4.76)$$

Therefore, holes have a larger effective mass than electrons. Equally important, narrower gap materials (smaller E_{g}) have a smaller effective mass at zero wavevector (the effective masses are wavevector-dependent since Υ could depend on wavevector). In this simple two-band theory, the effective mass is isotropic, i.e., it is the same along x-, y-, and z-directions (i.e., any direction in the $\langle 100\rangle$ family), since the bandstructure is isotropic.

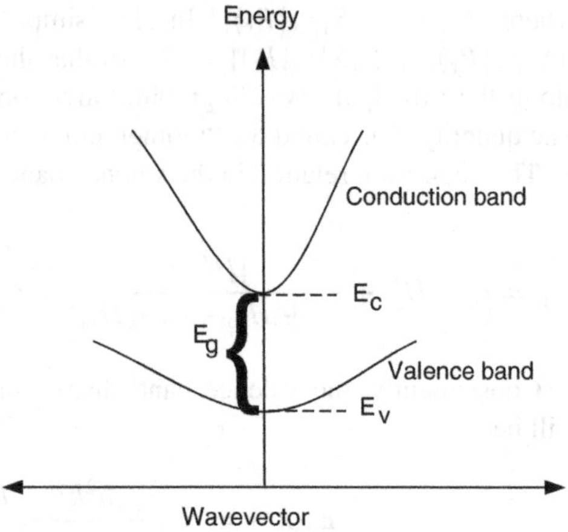

Fig. 4.8 The dispersion relations of the conduction and valence bands obtained from the simple two-band $\vec{k} \cdot \vec{p}$ perturbation theory

Limitation of the 2-band theory: Since we applied first-order perturbation theory, we tacitly assumed that the perturbation is weak so that $\frac{2\Upsilon^2}{m_0 E_g} \ll 1$. In that case, if we plot the energy dispersion relations of the conduction and valence bands (E versus k) from (4.74), then the plots will look like in Fig. 4.8.

The obvious problem with Fig. 4.8 (and therefore (4.74)) is that the valence band curves up the wrong way! Instead of bending downward with increasing wavevector, as it should, it bends upward. This is of course wrong and is the result of neglecting the influence of the remote bands in the two-band model. Therefore, a simple two-band model is not only quantitatively inaccurate, it is even qualitatively incorrect. Consequently, the remote bands have a very nontrivial consequence on the dispersion relations.

4.6.3 The Effect of Spin–Orbit Interaction on Bandstructure

A second problem with the dispersion relations in Fig. 4.8 is that there is only one valence band instead of the three we expect (heavy-hole, light-hole, and split-off). This is a consequence of neglecting spin–orbit interaction. It is this interaction that lifts the degeneracy between the three valence bands at nonzero wavevector and makes them distinct. Without this interaction, the three valence bands will be degenerate and hence indistinguishable from each other.

Spin–orbit interaction is a phenomenon of relativistic origin. When a charged particle, like an electron, is moving in an electric field (that arises from the Coulomb potential of the charged ions that the electrons see in the background), a magnetic field that did not exist in the laboratory frame, appears in the electron's rest frame as a result of Lorentz transformation. The flux density of this magnetic field is given by

$$\vec{B}_{SOI} = \frac{\vec{\mathcal{E}} \times \vec{v}}{2c^2\sqrt{1 - v^2/c^2}}, \tag{4.77}$$

where $\vec{\mathcal{E}}$ is the electric field associated with the ionic potential, $\vec{v}$ is the electron velocity, and c is the speed of light in vacuum. This magnetic field interacts with the magnetic moment $\vec{\mu}_e$ of the electron due to its spin. The resulting interaction is spin–orbit interaction and its strength is given by

$$E_{so} = -\vec{\mu}_e \cdot \vec{B}_{SOI}. \tag{4.78}$$

Landé had shown that $= \vec{\mu}_e = -g\mu_B\vec{s}$, where g is the Landé g-factor of the crystal, μ_B is the Bohr magneton ($\mu_B = e\hbar/2m_0$; e is the electronic charge), and $\vec{s}$ is the normalized spin operator which is equal to $(1/2)[\vec{\sigma}]$, with $[\vec{\sigma}]$ being the Pauli spin matrix which we discuss later in Chap. 7. It is a 2×2 unitary matrix and is the quantum mechanical operator for a nonrelativistic electron's spin [4].[3] Therefore, the operator for spin–orbit interaction is

$$[H_{so}] = -\frac{g}{2}\frac{e\hbar}{2m_0}[\vec{\sigma}] \cdot \frac{\vec{\mathcal{E}} \times \vec{v}}{2c^2\sqrt{1 - v^2/c^2}}. \tag{4.79}$$

Noting that in a crystal, the effective electric field seen by an electron is $\vec{\mathcal{E}} = -\vec{\nabla}_r V_L(\vec{r})$ and taking note of the fact that for a nonrelativistic electron $v \ll c$, we get

$$[H_{so}] = -\frac{ge\hbar}{8m_0^2c^2}(\vec{\nabla}_r V_L(\vec{r}) \times \vec{p}) \cdot [\vec{\sigma}]. \tag{4.80}$$

This operator is a 2×2 matrix since the Pauli spin matrix is a 2×2 matrix.

In the presence of spin–orbit interaction, one must account for the electron's quantum mechanical spin explicitly. We can do this by replacing the Schrödinger equation describing a "spinless" electron with the *Pauli equation* [4] which deals with a 2×1 component "spinor" $[\phi(\vec{r}, t)]$, instead of a scalar wavefunction $\psi(\vec{r}, t)$. Just as $\psi(\vec{r}, t)$ will tell us the probability of finding an electron at a location $\vec{r}$ at time t, the spinor will give us information about the spin orientation at any given location $\vec{r}$ at time t.

The time-independent Pauli equation in a crystal will read

$$\left[\left(-\frac{\hbar^2}{2m_0}\nabla_r^2 + V_L(\vec{r})\right)[\mathbf{I}] - \frac{ge\hbar}{8m_0^2c^2}(\vec{\nabla}_r V_L(\vec{r}) \times \vec{p}) \cdot [\vec{\sigma}]\right][\phi(\vec{r})] = E[\phi(\vec{r})], \tag{4.81}$$

where

$$[\phi(\vec{r})] = \begin{bmatrix} \phi_1(\vec{r}) \\ \phi_2(\vec{r}) \end{bmatrix}.$$

[3]Things are much more complicated for a relativistic electron moving with speeds comparable to that of light in vacuum. There, the electron gets coupled with the corresponding positron (antimatter) and the spin operator for the coupled system is a 4×4 Dirac matrix. Discussion of all this is of course outside the scope of this book.

Here, [**I**] is the 2×2 identity matrix, and the vector $[\vec{\sigma}]$ is given by

$$[\vec{\sigma}] = [\sigma_x]\hat{x} + [\sigma_y]\hat{y} + [\sigma_z]\hat{z}.$$

where $\hat{n}$ is the unit vector along the n-direction, and the $[\sigma_n]$ are 2×2 matrices given by

$$[\sigma_x] = \begin{bmatrix} 0 & 1 \\ 1 & 0 \end{bmatrix}, [\sigma_y] = \begin{bmatrix} 0 & -\mathrm{i} \\ \mathrm{i} & 0 \end{bmatrix}, [\sigma_z] = \begin{bmatrix} 1 & 0 \\ 0 & -1 \end{bmatrix} \tag{4.82}$$

The Bloch theorem can be extended to the spinor as well. In a crystal, which has a periodic potential, the solution of the time-independent Pauli equation can be written as

$$[\phi(\vec{r})] = \mathrm{e}^{\mathrm{i}\vec{k}\cdot\vec{r}}[v_{\vec{k}}(\vec{r})], \tag{4.83}$$

where

$$[v_{\vec{k}}(\vec{r})] = \begin{bmatrix} a_{\vec{k}}(\vec{r}) \\ b_{\vec{k}}(\vec{r}) \end{bmatrix}. \tag{4.84}$$

The Bloch spinor has the same property as the Bloch function, i.e., it has the same periodicity as that of the lattice. In other words, $[v_{\vec{k}}(\vec{r})] = [v_{\vec{k}}(\vec{r} + n\vec{a})]$, where n is an integer and $\vec{a}$ is the lattice period in the direction of $\vec{r}$.

Substitution of the Bloch expression for $[\phi(\vec{r})]$ in the Pauli equation (4.81) results in

$$\begin{aligned}
&\left[\left(-\frac{\hbar^2}{2m_0}\nabla_{\mathrm{r}}^2 + V_{\mathrm{L}}(\vec{r})\right)[\mathbf{I}] - \frac{ge\hbar}{8m_0^2c^2}\left(\vec{\nabla}_{\mathrm{r}}V_{\mathrm{L}}(\vec{r}) \times \vec{p}\right)\cdot[\vec{\sigma}]\right. \\
&\left. + \hbar\vec{k}\cdot\left\{\frac{p}{m_0} - \frac{ge\hbar}{8m_0^2c^2}\left([\vec{\sigma}] \times \vec{\nabla}_{\mathrm{r}}V_{\mathrm{L}}(\vec{r})\right)\right\}\right][v_{\vec{k}}(\vec{r})] = \left\{E - \frac{\hbar^2k^2}{2m_0}\right\}[v_{\vec{k}}(\vec{r})] \\
&= \epsilon_k[v_{\vec{k}}(\vec{r})].
\end{aligned} \tag{4.85}$$

Since this equation contains 2×2 matrices, the eigenenergies ϵ_k will have two values for every wavevector k, which, in general, will be distinct, i.e., *not* degenerate. The corresponding eigenspinors will be denoted by

$$\begin{aligned}
\left[v_{\vec{k}}^{\uparrow}(\vec{r})\right] &= \begin{bmatrix} a_{\vec{k}}^{\uparrow}(\vec{r}) \\ b_{\vec{k}}^{\uparrow}(\vec{r}) \end{bmatrix} \quad \texttt{(upspin state)} \\
\left[v_{\vec{k}}^{\downarrow}(\vec{r})\right] &= \begin{bmatrix} a_{\vec{k}}^{\downarrow}(\vec{r}) \\ b_{\vec{k}}^{\downarrow}(\vec{r}) \end{bmatrix} \quad \texttt{(downspin state).}
\end{aligned} \tag{4.86}$$

These two eigenspinors are orthogonal in Hilbert space, i.e.,

$$\left[\left(a_{\vec{k}}^{\uparrow}\right)^* \left(b_{\vec{k}}^{\uparrow}\right)^*\right]\begin{bmatrix} a_{\vec{k}}^{\downarrow}(\vec{r}) \\ b_{\vec{k}}^{\downarrow}(\vec{r}) \end{bmatrix} = 0, \tag{4.87}$$

since they are eigenspinors of a Hamiltonian which is a Hermitian operator. The fact that they are orthogonal in Hilbert space means that the spin polarizations represented by these two eigenspinors are *anti-parallel*. Therefore, we can call them the "upspin" state and the "downspin" state.

The nondegeneracy of the eigenenergies at every value of $\vec{k}$ leads to "spin-splitting", i.e., at any wavevector, there are two possible values of the energy *in any band* — one for "upspin" electrons and one for "downspin" electrons. Although not obvious, it also leads to three distinct valence bands: the heavy-hole band, the light-hole band, and the split-off band.

4.6.4 The 8-band $\vec{k} \cdot \vec{p}$ Perturbation Theory

Spin–orbit interaction lifts the spin degeneracy of every band and breaks up the valence band into three distinct bands. Therefore, at any wavevector, there will be a total of 8 bands, instead of just 2, even if we neglect all remote bands and focus only on the conduction and valence bands. These eight bands are: the upspin conduction band, the downspin conduction band, the upspin heavy hole band, the downspin heavy hole band, the upspin light hole band, the downspin light hole band, the upspin split-off band, and the downspin split-off band. We will now deal with these 8 conduction and valence bands, while still continuing to neglect the effect of all remote bands.

Let the Bloch functions (spinors) in these 8 bands at the zone center ($\vec{k} = 0$) be $[v^{\uparrow}_{c0}(\vec{r})]$, $[v^{\downarrow}_{c0}(\vec{r})]$, $[v^{\uparrow}_{hh0}(\vec{r})]$, $[v^{\downarrow}_{hh0}(\vec{r})]$, $[v^{\uparrow}_{lh0}(\vec{r})]$, $[v^{\downarrow}_{lh0}(\vec{r})]$, $[v^{\uparrow}_{so0}(\vec{r})]$, and $[v^{\uparrow}_{so0}(\vec{r})]$, respectively. Note that all these spinors are mutually orthogonal since they are eigenspinors of a Hamiltonian which is a Hermitian operator.

Equation (4.85) can be written as

$$[H][v_{\vec{k}}(\vec{r})] = \epsilon_k [v_{\vec{k}}(\vec{r})], \tag{4.88}$$

where the Hamiltonian $[H]$ now contains all the spin–orbit interaction terms and is a 2×2 matrix. Following standard perturbation theory, we can write the spinor $[v_{\vec{k}}(\vec{r})]$ as

$$\begin{aligned}
[v_{\vec{k}}(\vec{r})] &= C_1 \left[v^{\uparrow}_{c0}(\vec{r})\right] + C_2 \left[v^{\downarrow}_{c0}(\vec{r})\right] + C_3 \left[v^{\uparrow}_{hh0}(\vec{r})\right] + C_4 \left[v^{\downarrow}_{hh0}(\vec{r})\right] \\
&\quad + C_5 \left[v^{\uparrow}_{lh0}(\vec{r})\right] + C_6 \left[v^{\downarrow}_{lh0}(\vec{r})\right] + C_7 \left[v^{\uparrow}_{so0}(\vec{r})\right] + C_8 \left[v^{\uparrow}_{so0}(\vec{r})\right] \\
&= C_1[v_1] + C_2[v_2] + C_3[v_3] + C_1[v_4] \\
&\quad + C_5[v_5] + C_6[v_6] + C_7[v_7] + C_8[v_8].
\end{aligned} \tag{4.89}$$

Substituting the last result in (4.88) and defining the matrix element H_{mn} as

$$H_{mn} = [v_m]^{\dagger}[H][v_n],$$

where the superscript "dagger" represents Hermitian conjugate, we obtain the eigenequation

$$\begin{bmatrix} H_{11} & H_{12} & H_{13} & H_{14} & H_{15} & H_{16} & H_{17} & H_{18} \\ H_{21} & H_{22} & H_{23} & H_{24} & H_{25} & H_{26} & H_{27} & H_{28} \\ H_{31} & H_{32} & H_{33} & H_{34} & H_{35} & H_{36} & H_{37} & H_{38} \\ H_{41} & H_{42} & H_{43} & H_{44} & H_{45} & H_{46} & H_{47} & H_{48} \\ H_{51} & H_{52} & H_{53} & H_{54} & H_{55} & H_{56} & H_{57} & H_{58} \\ H_{61} & H_{62} & H_{63} & H_{64} & H_{65} & H_{66} & H_{67} & H_{68} \\ H_{71} & H_{72} & H_{73} & H_{74} & H_{75} & H_{76} & H_{77} & H_{78} \\ H_{81} & H_{82} & H_{83} & H_{84} & H_{85} & H_{86} & H_{87} & H_{88} \end{bmatrix} \begin{bmatrix} C_1 \\ C_2 \\ C_3 \\ C_4 \\ C_5 \\ C_6 \\ C_7 \\ C_8 \end{bmatrix} = \epsilon_k \begin{bmatrix} C_1 \\ C_2 \\ C_3 \\ C_4 \\ C_5 \\ C_6 \\ C_7 \\ C_8 \end{bmatrix}. \tag{4.90}$$

There are 8 eigenvalues of this matrix for every value of $\vec{k}$. These eight eigenvalues give the energies in the 8 bands—upspin conduction band, downspin conduction band, upspin heavy-hole band, downspin heavy-hole band, upspin light-hole band, downspin light-hole band, upspin split-off band, and downspin split-off band—at the corresponding value of the wavevector $\vec{k}$.

The basis functions in the 8-band model in the presence of spin–orbit interaction: We mentioned earlier that there is no spin–orbit interaction in a conduction band whose minimum is in the zone center since the orbital states are s-type. Therefore, the conduction band states are pure s-states $|S>$. There is, however, spin–orbit interaction in the valence band where the orbital states are p-type. Therefore, the valence band states are not pure $|P_x\rangle$, $|P_y\rangle$ or $|P_z\rangle$, but mixtures of them. Accordingly, the zone-center Bloch spinors in the conduction, heavy-hole, light-hole and split-off bands are given by [5]

$$\left[v_{c0}^{\uparrow}(\vec{r})\right] = \begin{bmatrix} |S> \\ 0 \end{bmatrix}$$

$$\left[v_{c0}^{\downarrow}(\vec{r})\right] = \begin{bmatrix} 0 \\ |S> \end{bmatrix}$$

$$\left[v_{hh0}^{\uparrow}(\vec{r})\right] = \begin{bmatrix} \frac{1}{\sqrt{2}}\left(|P_x\rangle + \mathrm{i}|P_y\rangle\right) \\ 0 \end{bmatrix} \propto \begin{bmatrix} \frac{1}{\sqrt{2}}(x + \mathrm{i}y) \\ 0 \end{bmatrix}$$

$$\left[v_{hh0}^{\downarrow}(\vec{r})\right] = \begin{bmatrix} 0 \\ \frac{i}{\sqrt{2}}\left(|P_x\rangle - \mathrm{i}|P_y\rangle\right) \end{bmatrix} \propto \begin{bmatrix} 0 \\ \frac{\mathrm{i}}{\sqrt{2}}(x - \mathrm{i}y) \end{bmatrix}$$

$$\left[v_{lh0}^{\uparrow}(\vec{r})\right] = \begin{bmatrix} -i\sqrt{\frac{2}{3}}|P_z\rangle \\ \frac{\mathrm{i}}{\sqrt{6}}\left(|P_x\rangle + \mathrm{i}|P_y\rangle\right) \end{bmatrix} \propto \begin{bmatrix} -\mathrm{i}\sqrt{\frac{2}{3}}z \\ \frac{\mathrm{i}}{\sqrt{6}}(x + \mathrm{i}y) \end{bmatrix}$$

$$
\left[v^{\downarrow}_{lh0}(\vec{r})\right] = \begin{bmatrix} \frac{1}{\sqrt{6}}\left(|P_x\rangle - \mathrm{i}|P_y\rangle\right) \\ \sqrt{\frac{2}{3}}|P_z\rangle \end{bmatrix} \propto \begin{bmatrix} \frac{1}{\sqrt{6}}(x - \mathrm{i}y) \\ \sqrt{\frac{2}{3}}z \end{bmatrix}
$$

$$
\left[v^{\uparrow}_{so0}(\vec{r})\right] = \begin{bmatrix} \frac{1}{\sqrt{3}}|P_z\rangle \\ \frac{1}{\sqrt{3}}\left(|P_x\rangle + \mathrm{i}|P_y\rangle\right) \end{bmatrix} \propto \begin{bmatrix} \frac{1}{\sqrt{3}}z \\ \frac{1}{\sqrt{3}}(x + \mathrm{i}y) \end{bmatrix}
$$

$$
\left[v^{\downarrow}_{so0}(\vec{r})\right] = \begin{bmatrix} -\frac{\mathrm{i}}{\sqrt{3}}\left(|P_x\rangle - \mathrm{i}|P_y\rangle\right) \\ \frac{\mathrm{i}}{\sqrt{3}}|P_z\rangle \end{bmatrix} \propto \begin{bmatrix} -\frac{\mathrm{i}}{\sqrt{3}}(x - \mathrm{i}y) \\ \frac{\mathrm{i}}{\sqrt{3}}z \end{bmatrix}. \tag{4.91}
$$

The conduction and heavy-hole band share the same spin quantization axis. That is, spins in the upspin conduction band and upspin heavy-hole band are parallel. The same is true of spins in the downspin conduction band and downspin heavy hole band. Therefore, it is impossible to excite (such as by shining light) an electron from the downspin heavy-hole band to the upspin conduction band, or from the upspin heavy hole band to the downspin conduction band since the spins of the initial and final states are anti-parallel or orthogonal in Hilbert space. Such transitions are "spin-blocked" and forbidden. The light hole band, however, has a spin quantization axis that is at an angle with the spin quantization axis of the conduction and heavy hole bands. Therefore, strictly speaking, we should not call the light hole bands "upspin" and "downspin" bands if we call the conduction and heavy hole bands upspin and downspin bands. Because spins in neither of the two light hole bands are anti-parallel to spins in either of the two conduction bands, it is possible to photoexcite an electron from either light hole band to either conduction band because they are not spin blocked.

The 8×8 Hamiltonian in (4.90) written in the basis of the spinors $[v_n]$ has the following elements:

$$
H_{8\times 8}(\vec{k}) =
\begin{bmatrix}
E_{c0} + \frac{\hbar^2 k^2}{2m_0} & 0 & 0 & 0 & 0 & 0 & 0 & 0 \\
0 & E_{c0} + \frac{\hbar^2 k^2}{2m_0} & 0 & 0 & 0 & 0 & 0 & 0 \\
0 & 0 & Q_1 & R & \Delta_{so}/3 & 0 & 0 & S_1 \\
0 & 0 & R^* & Q_2 & -\mathrm{i}\Delta_{so}/3 & 0 & 0 & S_2 \\
0 & 0 & \Delta_{so}/3 & \mathrm{i}\Delta_{so}/3 & Q_3 & S_1 & S_2 & 0 \\
0 & 0 & 0 & 0 & S_1 & Q_1 & R^* & \Delta_{so}/3 \\
0 & 0 & 0 & 0 & S_2 & R & Q_2 & \mathrm{i}\Delta_{so}/3 \\
0 & 0 & S_1 & S_2 & 0 & \Delta_{so}/3 & -\mathrm{i}\Delta_{so}/3 & Q_3
\end{bmatrix},
\tag{4.92}
$$

where the asterisk represents complex conjugate, Δ_{so} is the spin–orbit splitting energy, and

$$\begin{aligned}
Q_1 &= E_{v0} - Ak_x^2 - B(k_y^2 + k_z^2) \\
Q_2 &= E_{v0} - Ak_y^2 - B(k_z^2 + k_x^2) \\
Q_3 &= E_{v0} - Ak_z^2 - B(k_x^2 + k_y^2) \\
R &= -Ck_xk_y - \mathrm{i}\Delta_{so}/3 \\
S_1 &= -Ck_zk_x \\
S_2 &= -Ck_yk_z.
\end{aligned} \tag{4.93}$$

The coefficients A, B, and C are material parameters that are called *Kip–Kittel–Dresselhaus parameters*. Note from the Hamiltonian matrix in (4.92) that the conduction band is completely decoupled from the three valence bands, but there is coupling between the three valence bands themselves. This 8×8 matrix is sometimes referred to as the *Kane matrix*. It is easy to verify that this matrix is Hermitian, i.e., $H_{mn} = H_{nm}^*$, as it should be.

The procedure for finding the energy dispersion relations of the eight bands is the following. First evaluate the eight eigenvalues of the matrix $H_{8\times8}(\vec{k})$ for a given wavevector $\vec{k}$. These are the energies in the eight bands at that $\vec{k}$-value. Repeat this procedure for other values of $\vec{k}$ to find the complete $E - k$ relation, or dispersion relation, for the eight bands.

4.6.5 The 6×6 Hamiltonian: Neglecting the Split-Off Valence Band

In materials with large spin–orbit splitting energy Δ_{so} that is larger than the bandgap, the split-off band is far below the other three bands in energy. Since it is so remote from the other bands, it is sometimes ignored and a 6×6 Hamiltonian is constructed in the basis of the zone-center spinors $[v_n(\vec{r})]$ in the conduction, light- and heavy-hole bands. This Hamiltonian is [6]:

$$H_{6\times6}(\vec{k}) =$$

$$\begin{bmatrix}
E_{c0} + \frac{\hbar^2k^2}{2m_0} & 0 & U(k_x + \mathrm{i}k_y) & 0 & -\mathrm{i}\sqrt{\frac{4}{3}}Uk_z & \frac{1}{\sqrt{3}}U(k_x - \mathrm{i}k_y) \\
0 & E_{c0} + \frac{\hbar^2k^2}{2m_0} & 0 & U(k_y + \mathrm{i}k_x) & \frac{1}{\sqrt{3}}U(k_y + \mathrm{i}k_x) & \sqrt{\frac{4}{3}}Uk_z \\
U(k_x - \mathrm{i}k_y) & 0 & X & 0 & M & N \\
0 & U(k_y - \mathrm{i}k_x) & 0 & X & M^* & -N^* \\
\mathrm{i}\sqrt{\frac{4}{3}}Uk_z & \frac{1}{\sqrt{3}}U(k_y - \mathrm{i}k_x) & M^* & N & W & 0 \\
\frac{1}{\sqrt{3}}U(k_x + \mathrm{i}k_y) & \sqrt{\frac{4}{3}}Uk_z & N^* & -M & 0 & W
\end{bmatrix}, \tag{4.94}$$

where

$$
\begin{aligned}
U &= \frac{1}{\sqrt{2}} \frac{\hbar}{m_0} \Upsilon \\
X &= E_{v0} - \frac{\hbar^2}{2m_0}\left[(\gamma_1 + \gamma_2)(k_x^2 + k_y^2) + (\gamma_1 - 2\gamma_2)k_z^2\right] \\
W &= E_{v0} - \frac{\hbar^2}{2m_0}\left[(\gamma_1 - \gamma_2)(k_x^2 + k_y^2) + (\gamma_1 + 2\gamma_2)k_z^2\right] \\
M &= \frac{\hbar^2}{2m_0}\left[2\sqrt{3}\gamma_3 k_z(k_x + \mathrm{i}k_y)\right] \\
N &= \frac{\hbar^2}{2m_0}\left[\sqrt{3}\gamma_2(k_y^2 - k_x^2) + \mathrm{i}2\sqrt{3}\gamma_3 k_x k_y\right],
\end{aligned}
\tag{4.95}
$$

and the three γ-s are called *Lüttinger parameters*. The Lüttinger parameters are related to Kip–Kittel–Dresselhaus parameters that appeared in the 8×8 Hamiltonian. Note that the effect of neglecting the split-off band is that the conduction band is no longer completely decoupled from the heavy- and light-hole valence bands. There is now coupling between them.

The wavefunctions for the electrons, light-holes, and heavy-holes at $k = 0$ are still given by the expressions in (4.91), if we use the 6×6 Hamiltonian [6].

Once again, the procedure for finding the dispersion relations of the conduction, light-hole and heavy hole bands is to find the six eigenvalues of the matrix $H_{6\times 6}(\vec{k})$ for a given $\vec{k}$ and then repeat this process for other values of $\vec{k}$ to find the complete $E - k$ relations in the six bands.

An important question to ask now is will the methods that we have described so far, i.e., the 8×8 Hamiltonian and the simpler 6×6 Hamiltonian, yield the correct energy-dispersion relations in a crystal. The answer is actually no, since we have neglected the remote bands. Those bands have a very nontrivial effect on the dispersion relations of the conduction and valence bands. We illustrate that in the next subsection where we neglect spin–orbit interaction (and hence spin-splitting of the bands), but attempt to include the effect of remote bands.

4.6.6 The 4×4 Hamiltonian: Neglecting Spin–Orbit Interaction

If we neglect spin–orbit interaction and consider only the conduction, heavy-hole, light-hole and split-off bands, then we will be dealing with a 4×4 Hamiltonian since the spin degeneracy would not have been lifted. We now consider two cases:

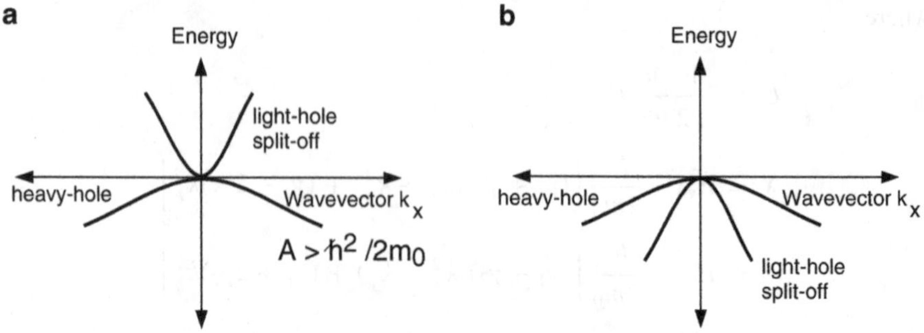

Fig. 4.9 The energy dispersion relation derived from the 4×4 Hamiltonian that ignores spin–orbit coupling. (**a**) effect of remote bands neglected, (**b**) effect of remote bands included

Case I. Neglecting remote bands:

When the effect of the remote bands is ignored, the Hamiltonian is

$$H_{4\times4}(\vec{k}) = \begin{bmatrix} E_{c0} + \frac{\hbar^2k^2}{2m_0} & 0 & 0 & 0 \\ 0 & E_{v0} + \frac{\hbar^2k^2}{2m_0} - Ak_x^2 & -Ak_xk_y & -Ak_xk_z \\ 0 & -Ak_yk_x & E_{v0} + \frac{\hbar^2k^2}{2m_0} - Ak_y^2 & -Ak_yk_z \\ 0 & -Ak_zk_x & -Ak_zk_y & E_{v0} + \frac{\hbar^2k^2}{2m_0} - Ak_z^2 \end{bmatrix}. \tag{4.96}$$

Case II. Including the effect of remote bands:

When the effect of the remote bands is taken into account, the Hamiltonian becomes

$$H_{4\times4}(\vec{k}) = \begin{bmatrix} E_{c0} + \frac{\hbar^2k^2}{2m_0} & 0 & 0 & 0 \\ 0 & E_{v0} - Ak_x^2 - B(k_y^2 + k_z^2) & -Ck_xk_y & -Ck_xk_z \\ 0 & -Ck_yk_x & E_{v0} - Ak_y^2 - B(k_x^2 + k_z^2) & -Ck_yk_z \\ 0 & -Ck_zk_x & -Ck_zk_y & E_{v0} - Ak_z^2 - B(k_x^2 + k_y^2) \end{bmatrix} \tag{4.97}$$

Interestingly, if we evaluate the energy dispersion relations from (4.96) along the [100] crystallographic direction, i.e., finding the four eigenvalues of the 4×4 matrix for different values of k_x and then plotting the $E - k_x$ relations, we will get the absurd result for the three valence bands shown in Fig. 4.9a. The heavy-hole dispersion will curve the right way if $A > \hbar^2/2m_0$, but the light-hole and split-off bands will be degenerate (at all wavevectors) and curve the wrong way. All three bands will curve the wrong way if $A < \hbar^2/2m_0$. But if we plot the same relations from (4.97), where the effect of remote bands has been included, all three valence bands will go the right way, but once again, the light-hole and split-off bands will be degenerate (see Problem 4.4). This situation is shown in Fig. 4.9b. The fact that the light-hole and split-off bands are degenerate at all wavevectors should not come as a surprise since we have ignored spin–orbit interaction which makes the bands nondegenerate. Normally, we would have expected all three bands to be degenerate in the absence of spin–orbit interaction, but what makes the heavy-hole band distinct from the other two is the coupling between the three valence bands. It is this coupling that makes the lower right 3×3 block in the matrix (representing the three valence bands) in (4.96) and (4.97) have nonzero off-diagonal terms. Note that the conduction band is still decoupled from the valence band since the off-diagonal terms in the first row and first column are all zero.

Special Topic 2: More on spin–orbit interaction in a crystal

Spin–orbit interaction clearly has a nontrivial effect in the valence band of a crystal since it resolves this band into three distinct bands—the heavy-hole, the light-hole, and the split-off. In the conduction band, however, there is no such dramatic effect since the atomic orbital states are s-type with orbital angular momentum quantum number $l = 0$. Such states do not experience *atomic* spin–orbit interaction and therefore the conduction band is not split into multiple bands. In contrast, the valence band states are p-type with $l = \pm 1$, so that these states do experience atomic spin–orbit interaction which causes this band to split up into three distinct branches.

The existence of spin–orbit interaction in an atom can also be posited from Biot–Savart's law applied to an orbiting electron [4] and it can be shown that the energy of this interaction is approximately

$$E_{\mathrm{so}} = -\vec{\mu}_{\mathrm{e}} \cdot \vec{B}_{\mathrm{SOI}} \approx \frac{Zge^2\hbar^2}{8\pi\epsilon(m^*)^2c^2r^3}\vec{l} \cdot \vec{s}, \tag{4.98}$$

where Z is the atomic number of the nucleus, g is the so-called Landé g-factor or gyromagnetic ratio characterizing the electron's surrounding (or the medium the electron is in), m^* is the effective mass of the electron, ϵ is the dielectric constant of the medium, r is the radius of the atomic orbital, $\vec{l}$ is the orbital quantum number operator associated with the orbit, and $\vec{s}$ is the spin quantum number operator.

If the orbit is s-type, then $\vec{l} = 0$ and the *atomic* spin–orbit interaction vanishes. This is the reason why there is no atomic or *intrinsic* spin–orbit interaction in a conduction band whose minimum is at the zone center, because $\vec{l} = 0$. However, there still can be weak, *extrinsic* spin–orbit interaction in the conduction band that accrues from more long-range (macroscopic) electric fields in a crystal. The macroscopic electric fields could originate from bend-bending due to space charges, externally applied electrostatic potentials, or simply lack of inversion symmetry along any given direction in a crystal. Externally applied electric fields and space charge fields break structural inversion symmetry and cause what is known as Rashba spin–orbit interaction [7], while crystallographic asymmetry (which occurs in compound semiconductors like GaAs, but not in elemental semiconductors like

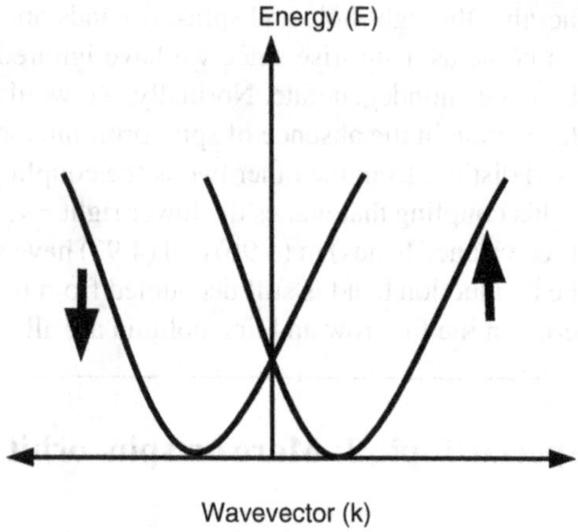

Fig. 4.10 The energy-dispersion relations in the conduction band in the presence of Rashba and Dresselhaus interactions. The bands for orthogonal spin states are horizontally displaced from each other so that (4.99) is satisfied. Note that the bands cross (and hence are degenerate) at $k = 0$, since the Rashba and Dresselhaus interactions vanish at $k = 0$

Si) breaks bulk inversion symmetry and causes what is known as Dresselhaus spin–orbit interaction [8]. These two *extrinsic* spin–orbit interactions can occur in the conduction band of a crystal and lift the spin degeneracy at nonzero wavevectors. The result is that "upspin" and "downspin" bands will be horizontally displaced from each other as shown is Fig. 4.10. The consequence of this is that

$$
\begin{aligned}
E(k\uparrow) &\neq E(-k\uparrow)\\
E(k\downarrow) &\neq E(-k\downarrow)\\
E(k\uparrow) &= E(-k\downarrow)\\
E(k\downarrow) &= E(-k\uparrow). \qquad (4.99)
\end{aligned}
$$

This band splitting is usually quite small because the Rashba and Dresselhaus interactions are normally weak in crystals. However, they can have a few observable effects that are not discussed here since they are outside the scope of this book.

Special Topic 3: Satellite valleys

Normally, the $E-k$ plot for any band in a crystal in any crystallographic direction is nearly symmetric about $k = 0$, unless there is very strong spin–orbit interaction making $E(+k) \neq E(-k)$. Therefore, it is customary not to plot the $E-k$ relation for both positive and negative values of k in any given direction since that will be superfluous. We will usually plot the $E-k$ relation for positive k values along one chosen direction and for negative k values in another direction in order to provide as much information as possible in the same plot. A representative plot for a hypothetical semiconductor is shown in Fig. 4.11 where we have plotted the $E-k$ relation of the conduction band along the [100] direction for positive k values and [111] direction for negative k values in the Brillouin zone. Note that there are two minima in either crystallographic direction; one at the zone center and another

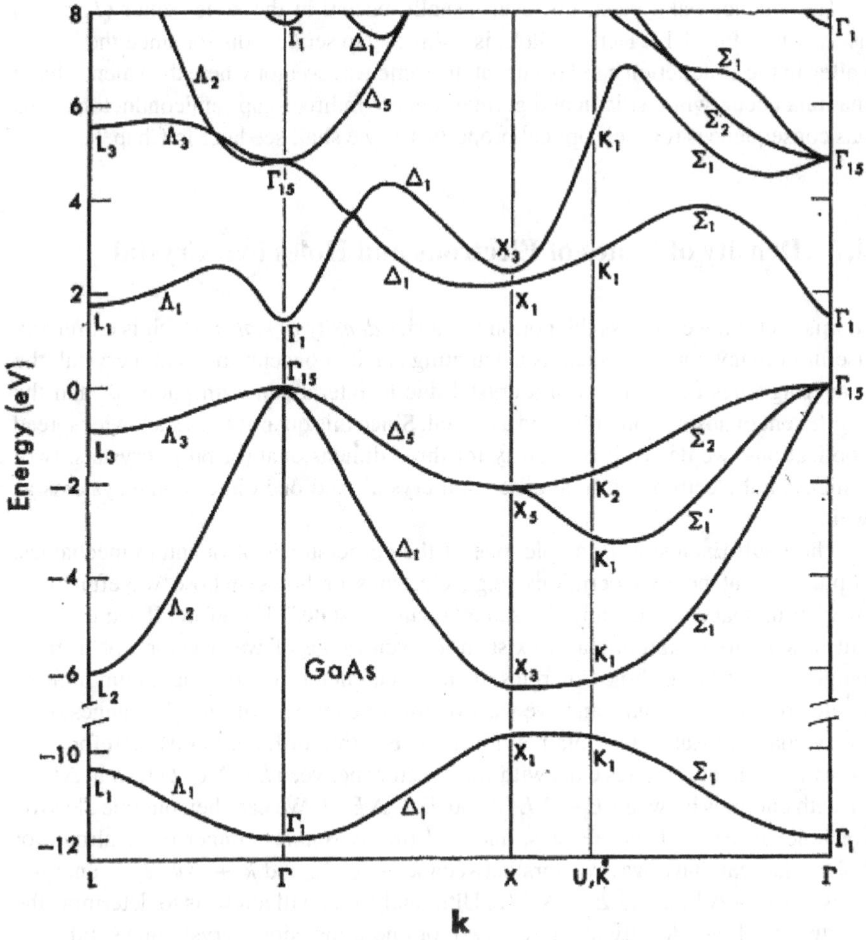

Fig. 4.11 The energy-dispersion $(E - k)$ plot for GaAs the [100] and [111] directions. Reproduced with permission from John P. Walter and Marvin L. Cohen, "Calculation of the reflectivity, modulated reflectivity and band structure of GaAs, GaP, ZnSe and ZnS", Physical Review, 183, 763 (1969). Copyright American Physical Society, 1969. See *http : //prola.aps.org/abstract/PR/v183/i3/p763_1*

at the zone edge. The lowest minimum is called the primary valley and the others are called *satellite valleys*. Whether the zone edge minimum is the lowest valley or the zone center minimum is the lowest valley depends on the semiconductor.

The valley occurring at the zone center is referred to as the Γ-valley, while the valley occurring at the zone edge in the [100] direction is referred to as the X-valley and the one occurring at the zone edge in the [111] direction is referred to as the L-valley. In GaAs, the Γ-valley is the lowest valley; in silicon, the X-valley is the lowest valley and in germanium, the L-valley is the lowest valley. The nomenclature adopted here—Γ, X, L, etc. have their origins in group theory and are not discussed here.

The valence band peak, however, usually occurs in the zone center (Γ-point) as shown in Fig. 4.11. Hence, GaAs is a direct-gap semiconductor since the lowest valley in the conduction band occurs at the same wavevector where the valence band maxima occur, while silicon and germanium are indirect-gap semiconductors. This has consequences for their optical properties as we shall see later in Chap. 6.

4.7 Density of States of Electrons and Holes in a Crystal

In this section, we discuss the notion of carrier *density of states* which is extremely useful in many contexts, such as calculating carrier concentrations in a crystal, the scattering rates of electrons in a crystal due to interaction with phonons, and the coefficient of absorption of light in a crystal. Since this quantity has such widespread applications, we derive it rigorously for three-dimensional (or bulk) crystals, two-dimensional electron and hole gases in a crystal, and one dimensional systems as well.

The Pauli Exclusion Principle, one of the cornerstones of quantum mechanics, stipulates that no two Fermions (e.g., electrons or holes) whose wavefunctions overlap in space can occupy the same "quantum state." Therefore, if we can find out how many quantum states exist in a given range of wavevectors, or a given range of electron (or hole) energies, then we can determine the maximum number of electrons or holes that can have wavevectors or energies within those ranges since every quantum state can contain at most one electron or hole. Let us therefore find how many quantum states exist with wavevectors between $k - \Delta k/2$ and $k + \Delta k/2$, or with energies between $E - \Delta E/2$ and $E + \Delta E/2$. We can then put one electron (or hole) in each of these states, and find the *maximum* number of electrons (or holes) that can have wavevectors between $k - \Delta k/2$ and $k + \Delta k/2$, or energies between $E - \Delta E/2$ and $E + \Delta E/2$. Ultimately, this will allow us to determine the electron (or hole) density in a three- two- or one-dimensional crystalline solid.

A quantum state in a crystal is labeled by two quantities: the wavevector and spin. Each quantum state has a unique wavevector and one of two orthogonal spin polarizations that we will call "upspin" and "downspin." Therefore, the number of quantum states is simply twice the number of wavevector states.

We will first find out how many allowed wavevector states exist within a given range of wavevectors in a crystal. To do this, we need to find out: (1) if the wavevector states are uniformly distributed in wavevector space, and if so (2) what the separation between neighboring states is. If the distribution is uniform so that the separation between neighboring states is constant, then we can find the number of wavevector states in a range of wavevectors by simply dividing that range with the separation.

In order to answer the two questions posed above, we appeal to the Bloch Theorem. According to this theorem, the electron (or hole) wavefunction at any arbitrary location $\vec{r}$ in a crystal obeys the relation

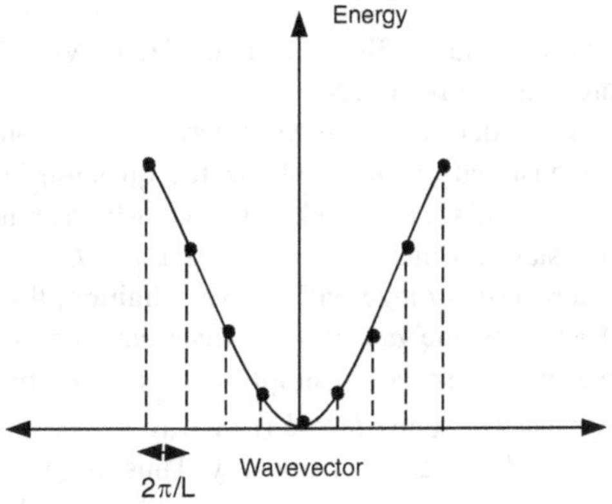

Fig. 4.12 Allowed states in the bandstructure of a one-dimensional crystal of length L

$$\psi_{\vec{k}}(\vec{r}+n\vec{a}) = \mathrm{e}^{\mathrm{i}\vec{k}\cdot(\vec{r}+n\vec{a})}u_{\vec{k}}(\vec{r}+n\vec{a}) = \mathrm{e}^{\mathrm{i}\vec{k}\cdot(\vec{r}+n\vec{a})}u_{\vec{k}}(\vec{r}) = \mathrm{e}^{\mathrm{i}\vec{k}\cdot n\vec{a}}\mathrm{e}^{\mathrm{i}\vec{k}\cdot\vec{r}}u_{\vec{k}}(\vec{r})$$
$$= \mathrm{e}^{\mathrm{i}\vec{k}\cdot n\vec{a}}\psi_{\vec{k}}(\vec{r}), \tag{4.100}$$

where n is an integer and $\vec{a}$ is the lattice period.

The periodicity of the crystal mandates that the wavefunction has to repeat every lattice period, so that

$$\psi_{\vec{k}}(\vec{r}+n\vec{a}) = \psi_{\vec{k}}(\vec{r}). \tag{4.101}$$

This condition is often referred to as "periodic boundary condition."

From the last two equations, we infer that

$$\mathrm{e}^{\mathrm{i}\vec{k}\cdot n\vec{a}} = 1, \text{ or}$$
$$\vec{k}\cdot n\vec{a} = 2n\pi, \text{ or}$$
$$\vec{k}\cdot\vec{a} = 2\pi. \tag{4.102}$$

Consider now a one-dimensional crystal of length L that has m atoms in it. The distance between neighboring atoms in this crystal will be L/m, so that the lattice period $a = L/m$. (4.102) therefore tells us that the magnitudes of the allowed wavevectors in the crystal will obey the relation

$$k_{\text{allowed}} = 2\pi/|\vec{a}| = 2m\pi/L. \tag{4.103}$$

Thus, there is a discrete set of allowed wavevector states in a crystal which are separated from their nearest neighbors by $2\pi/L$. Since this separation is independent of k, we have answered the two questions that we posed at the beginning of this section: the separation between allowed wavevector states is constant and therefore the states are distributed uniformly in wavevector space. Moreover, the separation is exactly $2\pi/L$.

Figure 4.12 shows schematically what the actual allowed wavevector states are in the bandstructure of a one-dimensional crystal. If the crystal is very large, then

$L \to \infty$ and the separation is vanishingly small so that the allowed states form a near-continuum. But in a small-sized crystal, where L is small, the allowed states are reasonably discrete.

Consider now a range of wavevectors from $k - \Delta k/2$ to $k + \Delta k/2$ in a one-dimensional crystal, and ask the question how many states are in this range of wavevectors. That number is obviously the range divided by the separation between the states, which is $\Delta k/(2\pi/L) = L\Delta k/(2\pi)$. The "density" of wavevector states in this range will be, by definition, this number divided by the range itself. Therefore, the *density of wavevector states* in this one-dimensional crystal (in wavevector space) is simply $L/(2\pi)$. In a polydimensional crystal, the density of wavevector states ($\mathcal{D}^d(k)$) in wavevector space will therefore simply be $\left(\frac{L}{2\pi}\right)^d$, where d is the dimensionality. Thus, in one-, two-, or three-dimensional crystals, the density of states in wavevector space will be:

$$\mathcal{D}^{3-D}(k) = \left(\frac{L}{2\pi}\right)^3 = \frac{\Omega}{8\pi^3} \quad \texttt{in three dimensions}$$

$$\mathcal{D}^{2-D}(k) = \left(\frac{L}{2\pi}\right)^2 = \frac{A}{4\pi^2} \quad \texttt{in two dimensions}$$

$$\mathcal{D}^{1-D}(k) = \left(\frac{L}{2\pi}\right) = \frac{L}{2\pi} \quad \texttt{in one dimension,} \tag{4.104}$$

where A is the area of a two-dimensional crystal and Ω is the volume of a three-dimensional crystal.

There is one caveat however. If we use spherical coordinates in wavevector space, then in three dimensions, any wavevector will be specified by the magnitude k, polar angle θ, and azimuthal angle ϕ as shown in Fig. 4.13. In two-dimensions, it will be the magnitude k and the angle φ. In one dimension, the wavevector will have a magnitude k and a sign (positive or negative direction). When we integrate over all wavevectors in *one dimension*, we should be integrating from $k = -\infty$ to $k = +\infty$. But, if we want to integrate only over the magnitude (as we should do since the density of wavevector states was calculated by considering only the magnitude of k), we should incorporate the sign by *multiplying the density of wavevector states by a factor of* 2. Therefore, we modify the last relation in the preceding equation and write the correct density of wavevector states in one-dimension as

$$\mathcal{D}^{1-D}(k) = 2\left(\frac{L}{2\pi}\right) = \frac{L}{\pi} \quad \texttt{in one dimension.} \tag{4.105}$$

In the following discussion, we will concentrate on electrons, but everything that applies to electrons also applies to holes. We will start by calculating the number of electrons in a crystal that has wavevectors between $k - \Delta k/2$ and $k + \Delta k/2$. For this purpose, recall that every wavevector state can accommodate up to 2 electrons of opposite spins, so that the number of quantum states in every wavevector state is

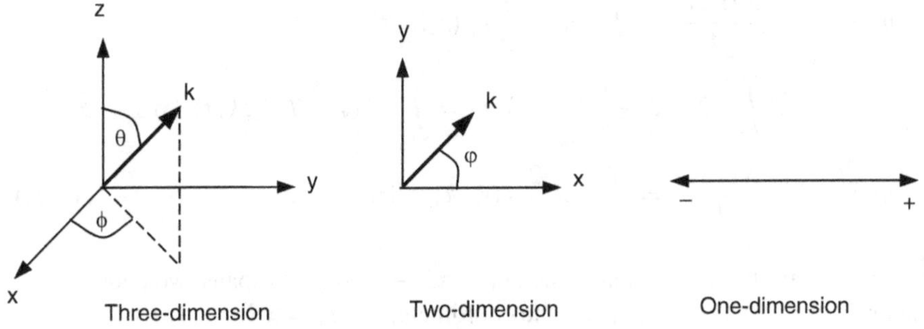

Fig. 4.13 Wavevector in three, two and one dimension

exactly 2. Therefore, the density of *quantum states* in wavevector space is obtained by multiplying the density of wavevector states by 2, i.e.,

$$D_q^{3-D}(k) = 2\mathcal{D}^{3-D}(k) = 2\left(\frac{L}{2\pi}\right)^3 = \frac{\Omega}{4\pi^3} \quad \texttt{in three dimensions}$$

$$D_q^{2-D}(k) = 2\mathcal{D}^{2-D}(k) = 2\left(\frac{L}{2\pi}\right)^2 = \frac{A}{2\pi^2} \quad \texttt{in two dimensions}$$

$$D_q^{1-D}(k) = 2\mathcal{D}^{1-D}(k) = 2\left(\frac{L}{\pi}\right) = \frac{2L}{\pi} \quad \texttt{in one dimension.} \tag{4.106}$$

Since Pauli Exclusion Principle guarantees that each quantum state can be occupied by either 0 or 1 electron, the number of electrons at a position $\vec{r}$ in a d-dimensional crystal possessing wavevectors between $k - \Delta k/2$ and $k + \Delta k/2$ at time t is therefore $D_q^d(k)$ times Δk times the probability that the wavevector state is actually occupied by an electron at position $\vec{r}$ at the instant of time t. Calling the latter probability $f(\vec{r}, k, t)$ (recall the Boltzmann distribution function of Chap. 2), we get that the total number of electrons at position $\vec{r}$ and at time t having *any* wavevector magnitude between zero and infinity is

$$N(\vec{r}, t) = \int_0^\infty d^d\vec{k} D_q^d(k) f(\vec{r}, k, t). \tag{4.107}$$

in a d-dimensional crystal.

The electron density in a crystal will be then given by

$$n(\vec{r}_3, t) = \frac{N(\vec{r}_3, t)}{\Omega} = \int_0^\infty d^3\vec{k} \frac{1}{4\pi^3} f(\vec{r}_3, k, t)$$

$$= \int_0^\infty 4\pi k^2 \mathrm{d}k \frac{1}{4\pi^3} f(\vec{r}_3, k, t) = \int_0^\infty k^2 \mathrm{d}k \frac{1}{\pi^2} f(\vec{r}_3, k, t) \texttt{ in 3 - D}$$

$$n_s(\vec{r}_2,t) = \frac{N(\vec{r}_2,t)}{A} = \int_0^\infty d^2\vec{k}\frac{1}{2\pi^2}f(\vec{r}_2,k,t)$$
$$= \int_0^\infty 2\pi k\mathrm{d}k\frac{1}{2\pi^2}f(\vec{r}_2,k,t) = \int_0^\infty k\mathrm{d}k\frac{1}{\pi}f(\vec{r}_2,k,t) \text{ in } 2-\mathrm{D}$$
$$n_l(\vec{r}_1,t) = \frac{N(\vec{r}_1,t)}{L} = \int_0^\infty \mathrm{d}k\frac{2}{\pi}f(\vec{r}_1,k,t) \text{ in } 1-\mathrm{D}, \tag{4.108}$$

where $\vec{r}_3$ spans three coordinates, i.e., $\vec{r}_3 = x\hat{x} + y\hat{y} + z\hat{z}$, $\vec{r}_2$ spans two coordinates, i.e., $\vec{r}_2 = x\hat{x} + y\hat{y}$, and $\vec{r}_1$ spans one coordinate, i.e., $\vec{r}_1 = x\hat{x}$.

4.7.1 Equilibrium Electron Concentrations at Absolute Zero of Temperature

In a population of electrons that are at thermodynamic equilibrium, the occupation probability (or Boltzmann distribution function) is given by the Fermi–Dirac function that has the form

$$f(\vec{r},k,t) = \frac{1}{e^{\frac{E(k)+E_{c0}(\vec{r},t)-\mu_F}{k_B T}}+1}, \tag{4.109}$$

where $E(k)$ is the kinetic energy of an electron having a wavevector of magnitude k, $E_{c0}(\vec{r},t)$ is the time- and position-dependent conduction band edge which is the electron's potential energy, and μ_F is of course the Fermi level. At 0 K temperature, this function has the form

$$f_{[T=0K]}(\vec{r},k,t) = \Theta(k_F(\vec{r},k,t) - k), \tag{4.110}$$

where Θ is the Heaviside function or unit step function and $k_F(\vec{r},t)$ is the Fermi wavevector such that an electron with this wavevector has the kinetic energy $\mu_F - E_{c0}(\vec{r},t)$ in the crystal. In other words, $E(k_F) = \mu_F - E_{c0}(\vec{r},t)$, where $E(k_F)$ is the kinetic energy of an electron with wavevector k_F and is found from the $E-k$ relation or bandstructure. Using this form in (4.108), we get

$$n(\vec{r}_3,t) == \int_0^{k_F(\vec{r}_3,t)} k^2\mathrm{d}k\frac{1}{\pi^2} = \frac{k_F^3(\vec{r}_3,t)}{3\pi^2} \text{ in } 3-\mathrm{D}$$
$$n_s(\vec{r}_2,t) = \int_0^{k_F(\vec{r}_2,t)} k\mathrm{d}k\frac{1}{\pi} = \frac{k_F^2(\vec{r}_2,t)}{2\pi} \text{ in } 2-\mathrm{D}$$
$$n_l(\vec{r}_1,t) = \int_0^{k_F(\vec{r}_1,t)} \mathrm{d}k\frac{2}{\pi} = \frac{2k_F(\vec{r}_1,t)}{\pi} \text{ in } 1-\mathrm{D}. \tag{4.111}$$

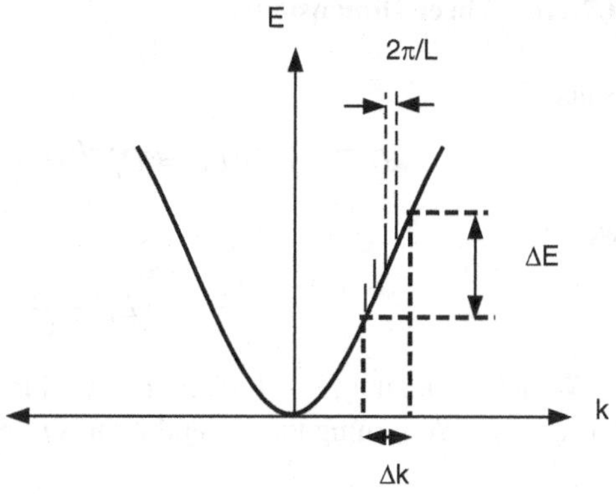

Fig. 4.14 The number of states in the wavevector range Δk and the corresponding energy range ΔE are the same

The quantity $E(k_F) = \mu_F - E_{c0}(\vec{r}, t)$ is called the *Fermi energy* E_F. Clearly, it can vary in space and time since $E_{c0}(\vec{r}, t)$ varies in space and time.[4] If we know the Fermi energy in the crystal at any position at any time, we can find the Fermi wavevector from the bandstructure (or E-k relation) and thus determine the electron densities at that position and that instant of time. Conversely, the Fermi wavevector and Fermi energy in a crystal are determined by the electron densities. This brings us to the next sub-topic, which is density of (quantum) states in energy space, as opposed to wavevector space.

4.7.2 *Density of States as a Function of Kinetic Energy*

Since we know the density of states in wavevector space, we will be able to find the density of states in energy space easily if we know the E-k relation or bandstructure (the energy we are talking of here is the kinetic energy). All we have to do is use the knowledge that the number of states in the wavevector interval between $k - \Delta k/2$ and $k + \Delta k/2$ is the same as the number of states in the interval between $E(k - \Delta k/2)$ and $E(k + \Delta k/2)$ as shown in Fig. 4.14. Therefore, we can write

$$D_q^d(E_k)\Delta E_k = D_q^d(k)\Delta k, \tag{4.112}$$

where $D_q^d(E_k)$ is the density of (quantum) states in a d-dimensional crystal in (kinetic) energy space.

In the limit when the range becomes infinitesimally small, we get

$$D_q^d(E_k)\mathrm{d}E_k = D_q^d(k)d^d\vec{k}. \tag{4.113}$$

We will now work out the density of states in all three dimensions.

[4]Note that μ_F cannot vary in space under global equilibrium, but E_F can.

4.7.2.1 Three Dimensions

Since

$$D_q^{3-D}(E_k)\mathrm{d}E_k = D_q^{3-D}(k)d^3\vec{k} = \frac{\Omega}{4\pi^3}4\pi k^2\mathrm{d}k, \tag{4.114}$$

we get

$$D_q^{3-D}(E_k)\frac{dE_k}{\mathrm{d}k} = \frac{\Omega}{\pi^2}k^2 \tag{4.115}$$

We now need the $E_k - k$ relation to evaluate the derivative $\frac{\mathrm{d}E_k}{\mathrm{d}k}$ and express k in terms of E_k. Assuming that this relation is *parabolic* and given by

$$E_k = \frac{\hbar^2k^2}{2m^*}, \tag{4.116}$$

we get that $\mathrm{d}E_k/\mathrm{d}k = \hbar^2k/m^* = \hbar v$, where v is the velocity of an electron with wavevector k or kinetic energy E_k. After substituting this derivative in (4.115), we find

$$D_q^{3-D}(E_k) = \frac{\Omega m^*}{\hbar^2\pi^2}k = 4\pi\Omega\left(\frac{2m^*}{h^2}\right)^{3/2}\sqrt{E_k}. \tag{4.117}$$

The above result yields the density of states in energy space, in three dimensions. Note that it depends on the kinetic energy E_k and increases as the square root of kinetic energy.

4.7.2.2 Two Dimensions

Similarly, in two-dimensional systems,

$$D_q^{2-D}(E_k)\frac{\mathrm{d}E_k}{\mathrm{d}k} = \frac{A}{\pi}k, \text{ or}$$
$$D_q^{2-D}(E_k)\frac{\mathrm{d}E_k}{\mathrm{d}k^2} = \frac{A}{2\pi}. \tag{4.118}$$

Once again, assuming a parabolic band and using (4.116), we obtain

$$D_q^{2-D}(E_k) = A\frac{m^*}{\pi\hbar^2}. \tag{4.119}$$

Note that the density of states in two dimensions is independent of kinetic energy, but *only because of the assumption of the parabolic E-k relation*. Had the dispersion relation not been parabolic, the density of states would not have been independent of kinetic energy.

4.7.2.3 One Dimension

Finally, in one dimension,

$$D_q^{1-D}(E_k)\frac{dE_k}{dk} = \frac{2L}{\pi}. \tag{4.120}$$

The assumption of a parabolic band and use of (4.116) then yields

$$D_q^{1-D}(E_k) = \frac{2Lm^*}{\pi\hbar^2 k} = \frac{\sqrt{2m^*}L}{\pi\hbar\sqrt{E_k}}. \tag{4.121}$$

Note the striking difference between the density of states in wavevector space and in kinetic energy space. The former is independent of wavevector in every dimension, but the latter is generally *not* independent of kinetic energy. Only in two dimensions it is independent of kinetic energy if the energy dispersion relation is parabolic. In three dimensions, it is directly proportional to the square root of kinetic energy and in one dimension, it is inversely proportional to the square root of kinetic energy—again only if the energy dispersion relation is parabolic. Also note that in a parabolic band, the density of states in kinetic energy space is always proportional to $(m^*)^d/2$ in every dimension, where d is the dimensionality. Therefore, there is a nice pattern to all the dependences.

We conclude this section by reminding the reader that everything we have said about electrons here also applies equally to holes. Whenever we are dealing with electrons, we will use the effective mass of electrons and when we are dealing with holes we will use their effective mass. Since heavy holes, light holes, and split-off holes have different effective masses, they will all have different density of states in energy space.

4.7.2.4 Anisotropic Effective Mass

In many semiconductors, such as Si and Ge, the conduction band bottom does not occur at $k = 0$, but may occur at a nonzero value of the wavevector. In Si, it occurs at the Brillouin zone edge in the [100] crystallographic direction (also known as the X-point) and in Ge, it occurs at the Brillouin zone edge in the [111] direction (also known as the L-point, in the language of group theory). The electron's effective mass at the band bottom in such semiconductors is often *anisotropic*, i.e., it is very different along the longitudinal direction (say, the [100] direction in the case of Si) and the two transverse directions ([010] and [001] for Si). We can denote the effective mass along the longitudinal direction as m_l^* and that along the transverse direction as m_t^*. The question then is which effective mass shall we consider when calculating the density of states near the band bottom in silicon? The correct answer is that we shall consider the so-called density of states-effective-mass m_{de}^* which is given by $m_{de}^* = (m_l^* m_t^* m_t^*)^{1/3}$ [9]. In Si, $m_l^* = 0.98m_0$, and $m_t^* = 0.19m_0$ [10]. Therefore, the density of states-effective-mass is $0.328m_0$, where m_0 is the free electron mass.

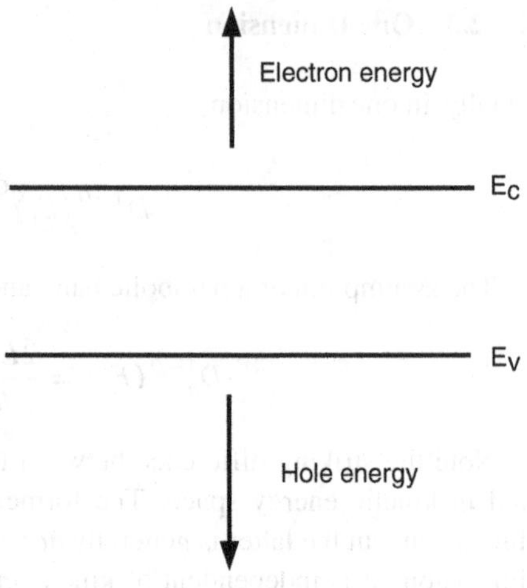

Fig. 4.15 The energy band diagram of a crystal in real space at a given instant of time. Energy of electrons is measured from the conduction band bottom $E_{c0}(\vec{r}, t)$ and extends upward in the energy band diagram. The energy of holes is measured from the valence band top $E_{v0}(\vec{r}, t)$ and extends downward in the same energy band diagram

4.7.3 Calculating the Electron and Hole Concentrations in a Crystal Using the Density of States in Kinetic Energy Space

We can use the density of states in kinetic energy space to calculate the electron or hole concentrations in a three-, two-, or one-dimensional crystal, assuming that the energy dispersion relation is parabolic. We must remember, however, that when we measure a hole's kinetic energy, we measure it from the valence band top $E_{v0}(\vec{r}, t)$ and extend downward as shown in Fig. 4.15. In the case of electrons, we measure kinetic energy from the conduction band bottom $E_{c0}(\vec{r}, t)$ and extend upward.

Note that the total energy of a particle is the sum of kinetic and potential energy, i.e., $E = E_k + E_{c0}$ in the conduction band and $E = E_{v0} - E_k$ in the valence band in a three-dimensional crystal. In one- and two-dimensions, the conduction and valence bands are quantized into subband levels because of electron confinement in one- or two-dimensions. Hence, $E = E_k + E_i$ in the conduction band and $E = E_j - E_k$ in the valence band, where E_m is the minimum energy in the m-th subband. Consequently, the expressions for the electron and hole concentrations will be

$$\begin{aligned} n(\vec{r}_3, t) &= \frac{1}{\Omega} \int_0^{\infty} D_q^{3-D}(E_k) f_e(E) \mathrm{d}E_k \\ &= 4\pi \left(\frac{2m_e^*(\vec{r}_3, t)}{h^2} \right)^{3/2} \int_0^{\infty} \sqrt{E_k} f_e(E_k + E_{c0}(\vec{r}_3, t)) \mathrm{d}E_k \; (\text{in } 3-\text{D}) \end{aligned} \tag{4.122}$$

$$p(\vec{r}_3,t) = \frac{1}{\Omega}\int_{-\infty}^{0} D_q^{3-D}(E_k) f_h(E) dE_k$$
$$= 4\pi\left(\frac{2m_h^*(\vec{r}_3,t)}{h^2}\right)^{3/2}\int_{-\infty}^{0}\sqrt{E_k} f_h(E_{v0}(\vec{r}_3,t)-E_k)dE_k \ (\texttt{in 3}-\texttt{D}) \tag{4.123}$$

$$n_s(\vec{r}_2,t) = \sum_{i=1}^{\infty}\frac{1}{A}\int_{0}^{\infty} D_q^{2-D}(E_k) f_e(E) dE_k$$
$$= \sum_{i=1}^{\infty}\frac{m_e^*(\vec{r}_2,t)}{\pi\hbar^2}\int_{0}^{\infty} f_e(E_k+E_i(\vec{r}_2,t))dE_k \ (\texttt{in 2}-\texttt{D}) \tag{4.124}$$

$$p_s(\vec{r}_2,t) = \sum_{j=1}^{\infty}\frac{1}{A}\int_{-\infty}^{0} D_q^{2-D}(E_k) f_h(E) dE_k$$
$$= \sum_{j=1}^{\infty}\frac{m_h^*(\vec{r}_2,t)}{\pi\hbar^2}\int_{-\infty}^{0} f_h(E_j(\vec{r}_2,t)-E_k)dE_k \ (\texttt{in 2}-\texttt{D}) \tag{4.125}$$

$$n_l(\vec{r}_1,t) = \sum_{i=1}^{\infty}\frac{1}{L}\int_{0}^{\infty} D_q^{1-D}(E_k) f_e(E) dE_k$$
$$= \sum_{i=1}^{\infty}\frac{\sqrt{2m_e^*(\vec{r}_1,t)}}{\pi\hbar}\int_{0}^{\infty}\frac{1}{\sqrt{E_k}} f_e(E_k+E_i(\vec{r}_1,t))dE_k \ (\texttt{in 1}-\texttt{D}) \tag{4.126}$$

$$p_l(\vec{r}_1,t) = \sum_{j=1}^{\infty}\frac{1}{L}\int_{-\infty}^{0} D_q^{1-D}(E_k) f_h(E) dE_k$$
$$= \sum_{j=1}^{\infty}\frac{\sqrt{2m_h^*(\vec{r}_1,t)}}{\pi\hbar}\int_{-\infty}^{0}\frac{1}{\sqrt{E_k}} f_h(E_j(\vec{r}_1,t)-E_k)dE_k \ (\texttt{in 1}-\texttt{D}), \tag{4.127}$$

where $m_e^*(\vec{r},t)$ is the effective mass of electrons at position $\vec{r}$ at time t, $m_h^*(\vec{r},t)$ is the effective mass of holes, $f_e(\eta)$ is the probability of an electron occupying a state in the conduction band with total energy η and $f_h(\nu)$ is the probability of a hole occupying a state in the valence band with total energy ν.

Since the probability of finding a hole with energy ν is the probability of *not* finding an electron at that state, $f_h(\nu) = 1 - f_e(\nu)$. At equilibrium, these occupation probabilities are given by

$$f_e^{\text{equil}}(\eta) = \frac{1}{e^{\frac{\eta-\mu_F}{k_B T}} + 1}$$
$$f_h^{\text{equil}}(\nu) = 1 - f_e^{\text{equil}}(\nu) = 1 - \frac{1}{e^{\frac{\nu-\mu_F}{k_B T}} + 1} = \frac{1}{e^{\frac{\mu_F-\nu}{k_B T}} + 1}. \tag{4.128}$$

4.7.3.1 The Effective Two-Dimensional Density of States

Consider a quantum well and we wish to find the electron sheet concentration $n_s(\vec{r}, t)$ in this quasi two-dimensional structure. From (4.124), we find that the sheet concentration is the sum of the contributions due to each subband in the quantum well and will be given by

$$\begin{aligned} n_s(\vec{r}_2, t) &= \sum_{i=1}^{\infty} \frac{m_e^*(\vec{r}_2, t)}{\pi\hbar^2} \int_0^{\infty} f_e(E_k + E_i(\vec{r}_2, t))\mathrm{d}E_k \\ &= \sum_{i=1}^{\infty} \frac{m_e^*(\vec{r}_2, t)}{\pi\hbar^2} \int_{E_i(\vec{r}_2,t)}^{\infty} f_e(E)\mathrm{d}E = \frac{1}{A}\sum_{i=1}^{\infty} \int_{E_i(\vec{r}_2,t)}^{\infty} D_q^{2-D}(E) f_e(E)\mathrm{d}E. \end{aligned} \tag{4.129}$$

Let us say that we wish to write the above equation in a more simple form such as

$$n_s(\vec{r}_2, t) = \frac{1}{A}\int_0^{\infty} D_{\text{eff}}^{2-D}(E, \vec{r}, t) f_e(E)\mathrm{d}E, \tag{4.130}$$

where $D_{\text{eff}}^{2-D}(E, \vec{r}_2, t)$ is the *effective* two-dimensional density of states in *total* energy space at position $\vec{r}$ and at time t. What will this effective density of states be? Equating the two expressions for n_s in the last two equations, we obtain that

$$D_{\text{eff}}^{2-D}(E, \vec{r}_2, t) = A\frac{m_e^*(\vec{r}_2, t)}{\pi\hbar^2}\sum_{i=1}^{\infty} \Theta(E - E_i(\vec{r}_2, t)). \tag{4.131}$$

This expression is plotted as a function of total energy E for a given $\vec{r}$ and t in Fig. 4.16. It has a step-like structure. Note that while the density of states $D_q^{2-D}(E_k)$ in *kinetic* energy space is independent of energy, the density of states $D_{\text{eff}}^{2-D}(E, \vec{r}_2, t)$ in *total* energy space is clearly not.

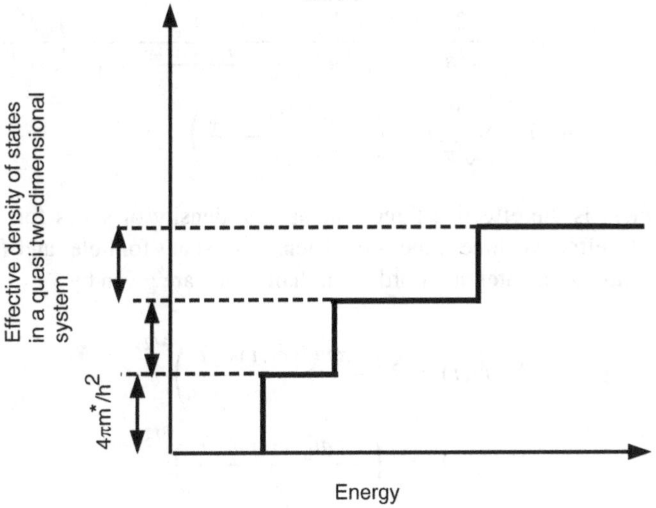

Fig. 4.16 The effective density of states in a quasi two-dimensional structure as a function of energy

4.7.4 *Expressions for the Carrier Concentrations at Nonzero Temperatures*

Compact analytical expressions for electron and hole concentrations cannot be obtained in all cases at nonzero temperatures.

4.7.4.1 3-D Systems

Using (4.122), we get that for three-dimensional systems

$$
\begin{aligned}
n(\vec{r}_3,t) &= 4\pi\left(\frac{2m_\mathrm{e}^*(\vec{r}_3,t)}{h^2}\right)^{3/2}\int_0^\infty \sqrt{E_\mathrm{k}}\, f_\mathrm{e}\left(E_\mathrm{k}+E_\mathrm{c0}(\vec{r}_3,t)\right)\mathrm{d}E_\mathrm{k}\\
&= N_\mathrm{c}(\vec{r}_3,t)\frac{2}{\sqrt{\pi}}\int_0^\infty \sqrt{\frac{E_\mathrm{k}}{k_\mathrm{B}T}}\frac{1}{\mathrm{e}^{\frac{E_\mathrm{k}}{k_\mathrm{B}T}-\frac{\mu_\mathrm{F}-E_\mathrm{c0}(\vec{r}_3,t)}{k_\mathrm{B}T}}+1}d\left(\frac{E_\mathrm{k}}{k_\mathrm{B}T}\right)\\
&= N_\mathrm{c}(\vec{r}_3,t)\frac{2}{\sqrt{\pi}}F_{1/2}\left(\frac{\mu_\mathrm{F}-E_\mathrm{c0}(\vec{r}_3,t)}{k_\mathrm{B}T}\right)
\end{aligned}
$$

$$
p(\vec{r}_3,t) = 4\pi\left(\frac{2m_\mathrm{h}^*(\vec{r}_3,t)}{h^2}\right)^{3/2}\int_0^\infty \sqrt{E_\mathrm{k}}\, f_\mathrm{h}\left(E_\mathrm{v0}(\vec{r}_3,t)-E_\mathrm{k}\right)\mathrm{d}E_\mathrm{k}
$$

$$= N_{\rm v}(\vec{r}_3,t)\frac{2}{\sqrt{\pi}}\int_0^\infty \sqrt{\frac{E_{\rm k}}{k_{\rm B}T}}\frac{1}{{\rm e}^{\frac{E_{\rm k}}{k_{\rm B}T}-\frac{E_{\rm v0}(\vec{r}_3,t)-\mu_{\rm F}}{k_{\rm B}T}}+1}d\left(\frac{E_{\rm k}}{k_{\rm B}T}\right)$$

$$= N_{\rm v}(\vec{r}_3,t)\frac{2}{\sqrt{\pi}}F_{1/2}\left(\frac{E_{\rm v0}(\vec{r}_3,t)-\mu_{\rm F}}{k_{\rm B}T}\right), \tag{4.132}$$

where $N_{\rm c}(\vec{r}_3,t)$ is the effective three-dimensional density of states for electrons, $N_{\rm v}(\vec{r}_3,t)$ is the effective three-dimensional density of states for holes and $F_{1/2}$ is the so-called Fermi–Dirac integral of order one-half. They are given by

$$N_{\rm c}(\vec{r}_3,t) = 2\left(\frac{2\pi m_{\rm e}^*(\vec{r}_3,t)k_{\rm B}T}{h^2}\right)^{3/2}$$

$$N_{\rm v}(\vec{r}_3,t) = 2\left(\frac{2\pi m_{\rm h}^*(\vec{r}_3,t)k_{\rm B}T}{h^2}\right)^{3/2}$$

$$F_{1/2}(a) = \int_0^\infty \sqrt{x}\frac{1}{{\rm e}^{x-a}+1}{\rm d}x. \tag{4.133}$$

In a nondegenerate semiconductor, $\frac{\mu_{\rm F}-E_{\rm c0}(\vec{r},t)}{k_{\rm B}T} \ll 0$ and $\frac{E_{\rm v0}(\vec{r},t)-\mu_{\rm F}}{k_{\rm B}T} \ll 0$. When a is a large negative quantity, the Fermi–Dirac integral $F_{1/2}(a) \rightarrow \sqrt{\pi}{\rm e}^a/2$. Therefore, for a nondegenerate electron or hole population, we get the following simple expressions for the carrier concentration in bulk (3-D) systems:

$$n(\vec{r}_3,t) = N_{\rm c}(\vec{r}_3,t)\exp\left(\frac{\mu_{\rm F}-E_{\rm c0}(\vec{r}_3,t)}{k_{\rm B}T}\right)$$

$$p(\vec{r}_3,t) = N_{\rm v}(\vec{r}_3,t)\exp\left(\frac{E_{\rm v0}(\vec{r}_3,t)-\mu_{\rm F}}{k_{\rm B}T}\right). \tag{4.134}$$

4.7.4.2 2-D Systems

For quasi two-dimensional systems, we can derive exact analytical expressions for the sheet carrier concentrations. Once again, using (4.124), we get

$$n_s(\vec{r}_2,t) = \frac{m_{\rm e}^*(\vec{r}_2,t)}{\pi\hbar^2}\sum_{i=1}^{\infty}\int_0^\infty f_{\rm e}(E_{\rm k}+E_i(\vec{r}_2,t)){\rm d}E_{\rm k}$$

$$= \frac{m_{\rm e}^*(\vec{r}_2,t)}{\pi\hbar^2}\sum_{i=1}^{\infty}\int_0^\infty \frac{1}{{\rm e}^{\frac{E_{\rm k}+E_i(\vec{r}_2,t)-\mu_{\rm F}}{k_{\rm B}T}}+1}{\rm d}E_{\rm k}$$

$$= \frac{m_{\mathrm{e}}^{*}(\vec{r}_2,t)k_{\mathrm{B}}T}{\pi\hbar^2}\sum_{i=1}^{\infty} ln\left(\mathrm{e}^{\frac{\mu_{\mathrm{F}}-E_i(\vec{r}_2,t)}{k_{\mathrm{B}}T}}+1\right)$$

$$= \frac{m_{\mathrm{e}}^{*}(\vec{r}_2,t)k_{\mathrm{B}}T}{\pi\hbar^2} ln\left[\prod_{i=1}^{\infty}\left(\mathrm{e}^{\frac{\mu_{\mathrm{F}}-E_i(\vec{r}_2,t)}{k_{\mathrm{B}}T}}+1\right)\right] \quad (4.135)$$

$$p_s(\vec{r}_2,t) = \frac{m_{\mathrm{h}}^{*}(\vec{r}_2,t)}{\pi\hbar^2}\sum_{j=1}^{\infty}\int_0^{\infty} f_{\mathrm{h}}(E_j(\vec{r}_2,t)-E_{\mathrm{k}})\mathrm{d}E_{\mathrm{k}}$$

$$= \frac{m_{\mathrm{h}}^{*}(\vec{r}_2,t)}{\pi\hbar^2}\sum_{j=1}^{\infty}\int_0^{\infty} \frac{1}{\mathrm{e}^{\frac{E_j(\vec{r}_2,t)-E_{\mathrm{k}}-\mu_{\mathrm{F}}}{k_{\mathrm{B}}T}}+1} dE_{\mathrm{k}}$$

$$= \frac{m_{\mathrm{h}}^{*}(\vec{r}_2,t)k_{\mathrm{B}}T}{\pi\hbar^2}\sum_{j=1}^{\infty} ln\left(\mathrm{e}^{\frac{E_j(\vec{r}_2,t)-\mu_{\mathrm{F}}}{k_{\mathrm{B}}T}}+1\right)$$

$$= \frac{m_{\mathrm{h}}^{*}(\vec{r}_2,t)k_{\mathrm{B}}T}{\pi\hbar^2} ln\left[\prod_{j=1}^{\infty}\left(\mathrm{e}^{\frac{E_j(\vec{r}_2,t)-\mu_{\mathrm{F}}}{k_{\mathrm{B}}T}}+1\right)\right]. \quad (4.136)$$

4.7.4.3 1-D systems

For quasi one-dimensional systems, we cannot derive compact analytical expressions for the carrier concentrations since the integrals are improper integrals owing to the presence of a singularity in the density of states at $E_{\mathrm{k}} = 0$. Such a singularity is called a van-Hove singularity.

4.8 Density of States of Phonons and Photons

Succinctly put, a "phonon" is a quantum of lattice vibration. At any finite temperature, the atoms in a crystal vibrate about a mean position. This vibration propagates through the entire crystal like a wave and carries heat and sound, thereby helping thermal and acoustic conduction. The atoms can vibrate in different ways: if at any instant of time, every atom is moving in the same direction, then the associated phonon mode is called an *acoustic phonon* mode,[5] whereas if neighboring atoms are moving in opposite directions, then the associated mode is called an *optical phonon mode*. These two modes are pictorially depicted in Fig. 4.17. Being a wave,

[5]These modes mostly carry sound and hence the name "acoustic".

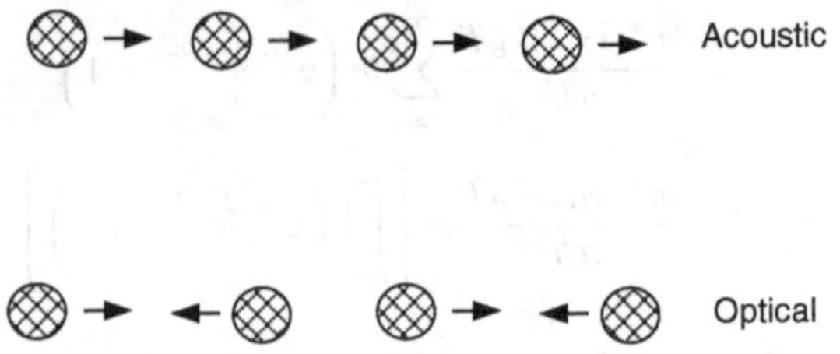

Fig. 4.17 A snapshot in time. Atoms moving in the same direction at any given instant of time constitutes an acoustic phonon mode, while neighboring atoms moving in opposite directions constitutes an optical phonon mode

Fig. 4.18 Approximate dispersion relations of acoustic and optical phonons within the first Brillouin zone of a crystal

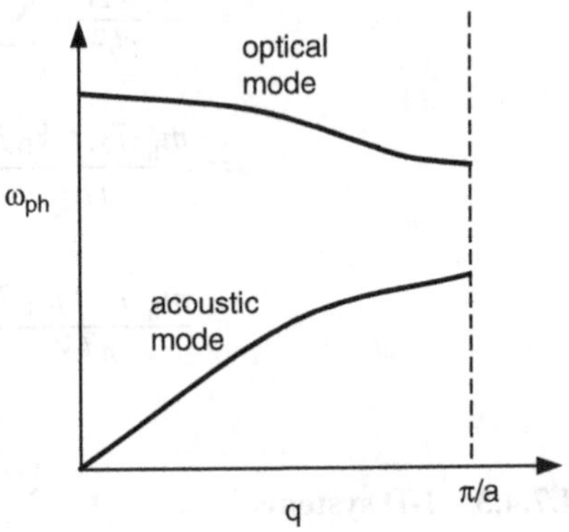

the phonon has a dispersion relation that relates the wave's frequency ω_{ph} with the wavevector q. The approximate dispersion relations of acoustic and optical phonon modes in a crystal are shown in Fig. 4.18. Note that at small wavevectors q, the acoustic phonon's dispersion relation is approximately linear so that the phase velocity ω_{ph}/q and the group velocity $\partial\omega_{\text{ph}}/\partial q$ are nearly equal. The optical phonon's dispersion relation, however, is nearly "flat," so that the phonon frequency is nearly independent of the wavevector. As a result, the optical phonon's group velocity is nearly zero. This makes sense since the atoms are moving in opposite directions and hence their net displacement is small.

At small wavevectors, the optical phonon frequency is quite a bit higher than the acoustic phonon frequency. We can rationalize this qualitatively in terms of a classical analog. The vibrations of atoms in a crystal lattice giving rise to phonons can be viewed as the motion of two masses connected by a spring (imagine a spring between two neighboring atoms in Fig. 4.17). Balancing the force on the mass of an atom with the force due to the spring, we can write

$$M\frac{d^2u}{dt^2} = -Ku^2, \tag{4.137}$$

where K is the spring constant, M is the atomic mass, and u is the displacement of the atom from its equilibrium position. The solution of this equation is that the displacement varies sinusoidally with time as $u \sim e^{i\omega_{ph}t}$, where ω_{ph} is the frequency of the oscillation (or frequency of the phonon). Substituting this solution in the previous equation, we get

$$\omega_{ph} = \sqrt{K/M}. \tag{4.138}$$

The above phenomenological relation shows that the phonon frequency increases with the spring constant or stiffness of the spring. An acoustic mode with small wavevectors corresponds to atoms moving in the same direction with small velocity so that the spring connecting two neighboring atoms is barely stretched and the spring does not appear stiff at all. Such a mode will therefore have a small ω_{ph} which agrees with Fig. 4.18. On the other hand, an optical mode will involve two neighbors moving in opposite directions, which pulls on the spring and makes it appear stiff. Hence, optical phonons will have higher frequency, which is what we see in Fig. 4.18.

In a "polar" material like GaAs, which consists of two different types of atoms (Ga and As), any displacement of an atom from its mean position causes local charge polarization since the two different atoms have different numbers of electrons. Therefore, the phonons in a polar material cause an oscillating electric field to build up within the crystal. Such phonons are called *polar* phonons. In contrast, an elemental semiconductor like Si has only one kind of atom and is therefore "nonpolar" in the sense that atomic vibrations do not cause charge polarization within the crystal. Hence, the phonons in a nonpolar material are nonpolar phonons.

Additionally, a phonon wave has a *polarization* just as most waves do. If the displacement of the vibrating atoms in the crystal is in the same direction in which the phonon wave propagates, the polarization is "longitudinal," whereas if the displacement is transverse to the direction in which the wave propagates, the polarization is "transverse." The velocity of an acoustic wave traveling through a crystal typically depends on its polarization. Sometimes a wave has mixed polarizations or other types of complex and exotic polarizations, particularly in quantum-confined structures like a quantum wire. Those are beyond the scope of this book, but are discussed in [11]. Suffice it to say then that a phonon has three different characteristics: (1) mode (acoustic or optical), (2) nature (polar or nonpolar), and (3) polarization (longitudinal or transverse). A phonon in a crystal will therefore fall under one of eight different categories characterized by the two types of mode, two types of nature, and two types of polarization. The phonon family is pictorially depicted in Fig. 4.19.

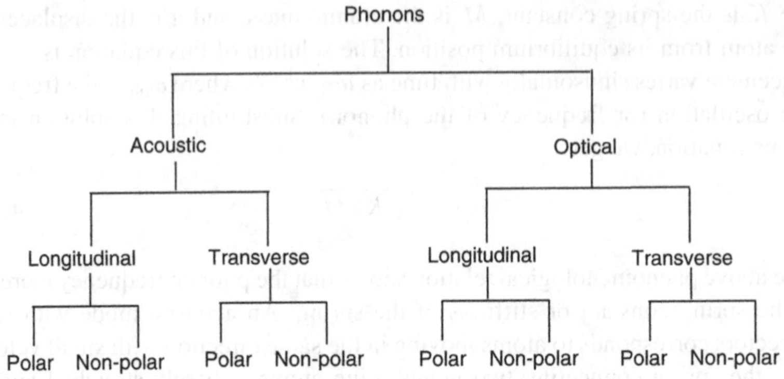

Fig. 4.19 The phonon family

4.8.1 Density of Modes or Density of States of Phonons

A phonon in a solid behaves as a plane wave and hence must obey the plane wave equation

$$\nabla_r^2 u(x,y,z,t) + \frac{1}{v_{\text{ph}}^2}\frac{\partial^2 u(x,y,z,t)}{\partial t^2} = 0, \tag{4.139}$$

where v_{ph} is the wave velocity of the phonon.

Now consider the fact that the phonon is confined within a crystal of volume L^3. We will think of the crystal as a cube of edge L. Because of this confinement, the amplitude $u(x,y,z,t)$ must vanish at $x = 0,\ L$; $y = 0,\ L$ and $z = 0,\ L$. With these boundary conditions, a solution of the wave equation is

$$u(x,y,z,t) = A\sin\left(\frac{n_x\pi x}{L}\right)\sin\left(\frac{n_y\pi y}{L}\right)\sin\left(\frac{n_z\pi z}{L}\right)\sin\left(qv_{\text{ph}}t\right), \tag{4.140}$$

where A is a constant, q is the phonon's wavevector and n_x, n_y, and n_z are integers.

Substitution of this solution into the wave equation immediately yields the relation

$$\left(\frac{n_x\pi}{L}\right)^2 + \left(\frac{n_y\pi}{L}\right)^2 + \left(\frac{n_z\pi}{L}\right)^2 = q^2. \tag{4.141}$$

In order to find how many phonon modes are allowed within the crystal, we need to evaluate how many modes can meet the above condition. This is equivalent to counting all possible combinations of n_x, n_y, and n_z that satisfy (4.141). The number of combinations can be viewed as the volume of a three-dimensional grid of the values of n in n-space as shown in Fig. 4.20. This is the so-called Rayleigh scheme of counting modes [12]. In spherical coordinates, this volume is $(4/3)\pi(n_x^2 + n_y^2 + n_z^2)^{3/2}$. However, we have to make two corrections to this volume. First, in using the sphere in n-space, we have used both positive and negative values of n, while the wave equation solution admits only positive values.

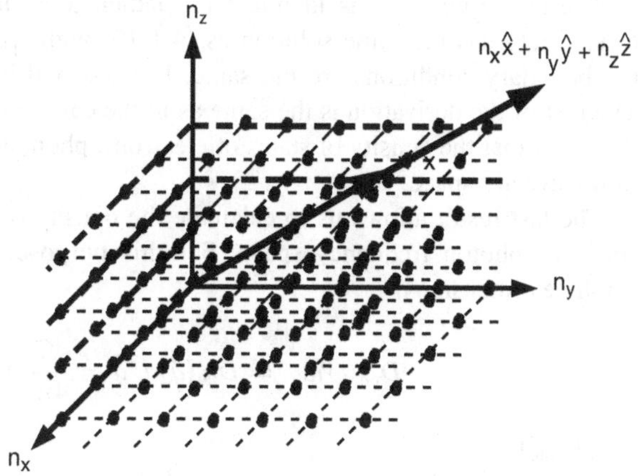

Fig. 4.20 Rayleigh scheme of counting modes

Therefore the number of combinations (or, equivalently, the number of allowed phonon modes) is $(1/2)^3$ or 1/8-th of the sphere's volume and is therefore equal to $(1/6)\pi(n_x^2+n_y^2+n_z^2)^{3/2}$. Second, a plane wave can be polarized in two mutually perpendicular directions (the longitudinal and transverse phonon modes). Therefore, we have to multiply the reduced volume by two to account for that. Hence, the number of phonon modes will be $(1/3)\pi(n_x^2+n_y^2+n_z^2)^{3/2}$. Using (4.141), this number can be written as $(1/3)q^3L^3/\pi^2$. Now the volume that this number of modes occupies in wavevector space or q-space is $(4/3)\pi q^3$. Hence, the density of modes in q-space is $(1/3)q^3L^3/\pi^2$ divided by $(4/3)\pi q^3$ which is $L^3/(4\pi^3)$ or $V/4\pi^3$, which is reminiscent of the density of (electron) states in wavevector space. If we think of phonon modes as phonon states, then the density of states of phonons in wavevector space is equal to the density of states of electrons in wavevector space.

4.8.2 Density of States of Photons

A photon is a quantum of electromagnetic wave just as a phonon is a quantum of acoustic wave. Even without a detailed derivation, we can see easily that the density of modes or density of states of photons in wavevector space will be the same as that of phonons. This follows from the fact that an electromagnetic wave obeys the Maxwell's equation

$$\nabla_r^2\mathcal{E}(x,y,z,t)+\frac{\eta^2}{c^2}\frac{\partial^2\mathcal{E}(x,y,z,t)}{\partial t^2}=0, \tag{4.142}$$

where $\mathcal{E}$ is the electric field in the wave, c is the wave velocity of the photon (i.e., the velocity of the electromagnetic wave) and η is the refractive index of the medium in which the electromagnetic wave is traveling.

The above equation is identical in mathematical form to (4.139). Therefore, $\mathcal{E}(x, y, z, t)$ has the same solution as (4.140) (with v_{ph} replaced by c/η) because the boundary conditions are the same. Here, q will be the photon's wavevector. The rest of the derivation is the same as in the case of phonons. In the end, we get the result that the density of states of electrons, phonons, and photons are the same in wavevector space.

The last result allows us to calculate the density of state of photons in energy space or photon frequency space. For this purpose, we use the usual relation (in three dimensions)

$$D_{\text{ph}}(\omega)\text{d}\omega = D_{\text{ph}}(\vec{q})d^3\vec{q} = \frac{\Omega}{4\pi^3}4\pi q^2\text{d}q, \tag{4.143}$$

which yields

$$D_{\text{ph}}(\omega) = \Omega\frac{q^2}{\pi^2}\frac{\text{d}q}{\text{d}\omega} = \Omega\frac{\omega^2}{\pi^2}\frac{\eta^3}{c^3}, \tag{4.144}$$

where we have assumed that the dispersion relation of photons is $\omega = (c/\eta)q$.

4.9 Summary

In this chapter, we learned four different techniques to calculate an electron's wavefunction and the bandstructure, i.e., energy versus wavevector (E–k) relations, in a crystalline solid. The important points to remember are:

1. In the NFE approach, the electron wavefunction is written as a linear superposition of free electron wavefunctions, i.e., expanded in the basis of free-electron wavefunctions, which are plane waves. Time-independent, nondegenerate, first order perturbation theory is applied to calculate the coefficients of expansion, which then yields the complete wavefunction in the crystal as well as the energy versus wavevector relation. This method is most suitable for crystals like sodium where the electron is loosely bound to the parent ion and is therefore nearly free.
2. The opening up of bandgaps at Brillouin zone edges is explained by invoking *degenerate* perturbation theory, which also allows us to calculate the bandgap energies.
3. In the OPW expansion method, the electron's wavefunction is written as a mixture of free electron wavefunctions and atomic orbital wavefunctions. Once again, time-independent, nondegenerate, first order perturbation theory is applied to calculate the complete wavefunction and the bandstructure. This can be achieved in two ways: (1) the direct method, and (2) the pseudopotential method.
4. In the TBA, the electron wavefunction is written as a LCAO wavefunctions. The coefficients of expansion are not found from perturbation theory but by assuming that electrons can tunnel or hop between nearest neighbor atomic sites only, and then applying the Bloch Theorem to derive a relation between the coefficients.

This approach also directly yields the energy dispersion relation. From the latter, we can derive analytical expressions for the electron velocity and effective mass in terms of the tunneling matrix element, which is a measure of how easily an electron can tunnel between neighboring sites. This is the only nonperturbative approach to calculation of bandstructure that we have discussed.

5. In the $\vec{k} \cdot \vec{p}$ perturbation method, we substitute the Bloch expression for the electron wavefunction in the Schrödinger equation. This yields a Schrödinger-like equation for the Bloch function, where the scalar product $\vec{k} \cdot \vec{p}$ appears as a perturbation term. Time-independent, nondegenerate, first-order perturbation theory is applied to find the Bloch function and the energy dispersion relations. This technique is very illustrative since it shows clearly the consequences of neglecting spin–orbit interaction for holes and the perturbation caused by remote bands.
6. The density of states can be expressed in either the wavevector space or in energy space. The advantage of the former is that in wavevector space, the density of states of one-, two-, or three-dimensional structures is independent of wavevector. In energy space, the density of states of two-dimensional structures is independent of energy, but this is not true in the case of three- or one-dimensional structures.
7. The density of states allows us to calculate carrier concentrations in two- or three-dimensional systems at any arbitrary temperature. In the case of a one-dimensional structure, we cannot derive compact analytical expressions for the carrier concentration because the density of states has a singularity at the zero of energy. Such a singularity is known as a van Hove singularity.
8. The Fermi energy is determined by the carrier concentration and vice versa.
9. The density of states of electrons, phonons, and photons are the same in wavevector space.

Problems

Problem 4.1. The Fourier transform of the unscreened Coulomb potential $V(\vec{r}) = \frac{e^2}{4\pi\epsilon|\vec{r}|}$ is given by

$$V_{\vec{k}'-\vec{k}} = \frac{1}{\Omega}\int_{-\infty}^{\infty} V(\vec{r})d^3\vec{r} = \frac{1}{\Omega}\frac{e^2}{4\pi\epsilon\left|\vec{k}-\vec{k}'\right|^2},$$

where ϵ is the dielectric constant of the crystal.

Assuming that the lattice potential in a particular crystal is unscreened Coulombic, apply the NFE technique to calculate the value of the lowest energy gap opening up at the so-called L-point in the first Brillouin zone, i.e., at $\vec{k} = [\pi/a, \pi/a, \pi/a]$, where a is the lattice period. The normalizing volume Ω is equal to the volume of the unit lattice, i.e., $V = a^3$. For this problem, assume the following values: $a = 0.5$ nm and $\epsilon = 12.9 \times 8.854 \times 10^{-12}$ Farads/meter. Do you expect materials with larger dielectric constant to have smaller or larger bandgaps? Explain.

Problem 4.2. Using the Bloch Theorem, first derive the following relation for the electron wavefunction in a crystal:

$$\psi(\vec{r} + n\vec{a}) = e^{i\vec{k}\cdot n\vec{a}}\psi(\vec{r}),$$

where $\vec{a}$ is the lattice constant or lattice period.

Show that the OPW basis function $\phi_n^{\mathrm{OPW}}(\vec{r})$ obeys the above relation and therefore is a legitimate choice to represent an electron's wavefunction in a crystal.

Problem 4.3. Bloch–Zener oscillations: According to the simple TBA formalism that we considered in this chapter, the energy dispersion relation in a single band can be expressed as

$$E(k) = \epsilon - 2t\cos(ka), \tag{4.145}$$

where t is the tunneling matrix element and a is the lattice constant or lattice period.

Consider an electron in a crystal which is subjected to an external (uniform) electric field $\vec{\mathcal{E}}$. Ignore scattering forces. The field exerts a force on the electron, which accelerates the particle according to Newton's second law:

$$\frac{d(\hbar k)}{dt} = -e\vec{\mathcal{E}}. \tag{4.146}$$

Using (4.145) and (4.146), show that the *velocity* of the electron will oscillate in time with an angular frequency given by

$$\omega = \frac{e|\vec{\mathcal{E}}|a}{\hbar}. \tag{4.147}$$

Hint: To show this, you will have to show that the electron's velocity v in the band obeys the equation of a simple-harmonic-motion oscillator:

$$\frac{d^2v}{dt^2} + \left(\frac{e|\vec{\mathcal{E}}|a}{\hbar}\right)^2 v = 0.$$

Physically explain the cause of this oscillation which is called Bloch–Zener oscillation. Find the angular frequency in (4.147) if the electric field is 100 kV/cm and the lattice constant is 0.5 nm. It is awfully hard to observe these oscillations in practice since scattering forces (which we ignored) tend to dephase these oscillations.

Problem 4.4. Justify the plots in Fig. 4.9 using the 4×4 Hamiltonians in (4.96) and (4.97).

Problem 4.5. Consider the 4×4 Hamiltonian in (4.97) which includes the effect of remote bands, but ignores spin–orbit interaction.

(a) We wish to find the $E - k$ relation of the three valence bands along the [100] and [111] crystallographic (Miller) directions. In the former case, $k_x = k, k_y = k_z = 0$, and in the latter case, $k_x = k_y = k_z = k/\sqrt{3}$. Find the $E - k$ relations and show that two of the bands are degenerate.
(b) The effective mass of holes along the [100] direction was determined experimentally (from cyclotron resonance experiments) and was found to be $0.3m_0$ in the doubly degenerate bands and $0.2m_0$ in the other band (m_0 is the free electron mass). Also, the effective mass of the doubly degenerate band along the [111] direction was determined to be $0.5m_0$. Armed with this knowledge, determine the effective mass of holes in the three bands along the [110] crystallographic direction. Are any two bands degenerate (at nonzero wavevector values) along the [110] direction?

Problem 4.6. We wish to determine the bandstructure of a hypothetical crystal where the effect of remote bands can be neglected for small wavevectors, so that it is adequately described by the 8×8 Kane matrix.

Plot the $E - k$ relations for small k-values along the [100] direction. For our hypothetical material, the Lüttinger parameters are $\gamma_1 = 7.65, \gamma_2 = 2.41, \gamma_3 = 3.28$ and the spin–orbit splitting energy Δ_{so} is 0.34 eV.

Concentrate on the lower 6×6 block in the Kane matrix which describes the valence band. You will find that for the [100] crystallographic direction, this block decouples into two 3×3 blocks, which means that the bands are spin-degenerate (spin–orbit coupling does not lift the spin degeneracy). The Kip–Kittel–Dresselhaus parameters are related to the Lüttinger parameters as

$$A = \frac{\hbar^2}{2m_0}\gamma_1$$

$$B = \frac{\hbar^2}{m_0}\gamma_2$$

$$\sqrt{C^2 + 3B^2} = \sqrt{3}\frac{\hbar^2}{m_0}\gamma_3.$$

(a) Find the dispersion relations of the three valence bands in the [100] direction. (Hint: One of the eigenvalues of the 3×3 block is $E_{v0} - Bk^2 + \Delta_{so}/3$.)
(b) You will find that one of the three bands has a parabolic dispersion relation, while the other two do not. Is the parabolic band the heavy-hole, light-hole, or split-off band?
(c) Calculate the effective masses of the holes (for motion along the [100] direction) around $k = 0$.

Problem 4.7. Derive expressions for the three-dimensional density of states as functions of energy in the three valence bands that you derived in the previous problem. Express your answer in terms of the Kip–Kittel–Dresselhaus parameters.

Problem 4.8. (a) Consider the dispersion relation in (2.131). Find an expression for the two-dimensional density of states as function of electron kinetic energy E. Is this independent of kinetic energy?

(b) Consider a quasi two-dimensional electron gas (2-DEG) where the Fermi level is well below the bulk conduction band edge so that the Boltzmann approximation to Fermi–Dirac statistics is valid. Show that the equilibrium sheet concentration of electrons in the 2-DEG at any temperature T is

$$n_s = \frac{m^* k_B T}{\pi \hbar^2} \sum_n \mathrm{e}^{\frac{\mu_F - E_n}{k_B T}} \left[1 + 2\alpha \left(E_n + k_B T\right)\right],$$

where m^* is the effective mass, $k_B T$ is the thermal energy, μ_F is the Fermi level and E_n is the energy at the bottom of the n-th subband.

(c) By what percentage would you underestimate the equilibrium sheet electron concentration at room temperature (300 K) in a 100 Å wide GaAs quantum well if you did not account for nonparabolicity? Assume that the effective mass is $0.067 m_0$, $\alpha = 0.6$/eV and the Fermi level is $k_B T$ energy below the bulk conduction band edge.

Problem 4.9. (a) Consider a one-dimensional lattice maintained at 0 K temperature. Assume that only the lowest band is filled with electrons since the Fermi level is below the second lowest band. Show that if there are two electrons per atom, then the energy at the Brillouin zone edge is the Fermi energy E_F. In other words, the Fermi level μ_F intersects the $E - k$ curve at the Brillouin zone edge.

(b) Will this crystal be an insulator or a metal? An insulator can carry no current when a voltage is applied across it, while a metal will carry current. A completely filled band cannot carry any current since for every electron with a wavevector $+k$, there is one with wavevector $-k$ and the currents due to these two states cancel.

(c) Assume that the bandstructure has been calculated using the TBA formalism and there is only one electron per atom. Derive an expression for the $E - k$ relation in terms of the lattice period a and the Fermi energy E_F. Assume also that the energy is zero when the wavevector is zero. Show that the one-dimensional density of states in kinetic energy space is given by

$$D^{1-d}(E_k) = \frac{2L}{\pi E_F a \sqrt{1 - (E_k/E_F - 1)^2}},$$

where L is the crystal's length.

(d) At what energies do van-Hove singularities occur, i.e., the one-dimensional density of state diverges?

Problem 4.10. The heavy-hole, light-hole, and split-off band effective masses in a hypothetical semiconductor are $0.2 m_0$, $0.06 m_0$, and $0.1 m_0$. The valence band top occurs at $k = 0$, where the light- and heavy-hole bands are degenerate and the split-

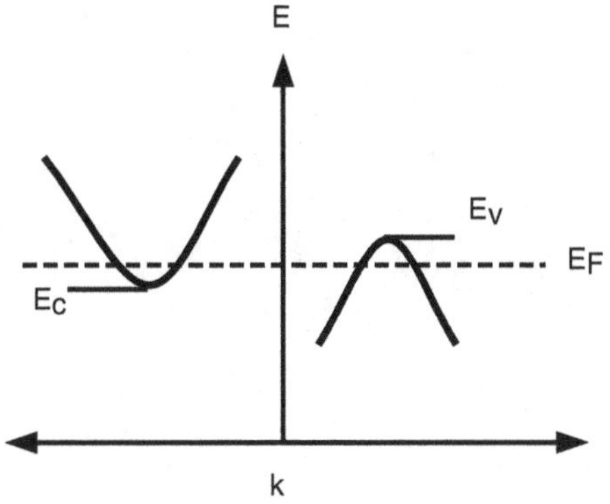

Fig. 4.21 The energy dispersion diagram of an intrinsic indirect-gap semi-metal

off band top also occurs at $k = 0$, but is located is 90 meV below the valence band top. Calculate the relative equilibrium populations of the three types of carriers at room temperature.

Problem 4.11. Consider an *indirect-gap* semi-metal where the conduction band bottom and valence band top occur at different wavevectors. Assume that the semi-metal is intrinsic so that the electron and hole concentrations are equal. Find the location of the Fermi level μ_F at 0 K in terms of the conduction band edge E_{c0}, valence band edge E_{v0}, and the effective masses m_e^* and m_h^* of electrons and holes. Refer to Fig. 4.21 for this problem.

References

1. C. Herring, "A new method for calculating wavefunctions in crystals", Phys. Rev., **57**, 1169 (1940).
2. F. Bloch, Z. Physik, **52**, 555 (1928).
3. E. O. Kane, "Band structure of Indium Antimonide", J. Phys. Chem. Solids, **1**, 249 (1957).
4. S. Bandyopadhyay and M. Cahay, *Introduction to Spintronics*, (CRC Press, Boca Raton, 2008).
5. S. Datta, *Quantum Phenomena*, Eds. R. F. Pierret and G. W. Neudeck, (Addison-Wesley, Reading, Massachusetts, 1989).
6. J. M. Luttinger and W. Kohn, "Motion of electrons and holes in perturbed periodic fields", Phys. Rev., **97**, 869 (1955).
7. Yu. A. Bychkov and E. I. Rashba, "Oscillatory effects and the magnetic susceptibility of carriers in inversion layers", J. Phys. C, **17**, 6039 (1984).
8. G. Dresselhaus, "Spin orbit coupling effects in zinc-blende structures", Phys. Rev., **100**, 580 (1955).
9. R. A. Smith, *Semiconductors*, 2nd. ed., (Cambridge University Press, London, 1979).
10. S. M. Sze, *Physics of Semiconductor Devices*, 2nd. ed., (John Wiley & Sons, New York, 1981).
11. N. Nishiguchi, "Resonant acoustic phonon modes in a quantum wire", Phys. Rev. B, **52**, 5279 (1995).
12. F. K. Richtmyer, E. H. Kennard and J. N. Cooper, *Introduction to Modern Physics*, 6th edition, (McGraw Hill, New York, 1969).

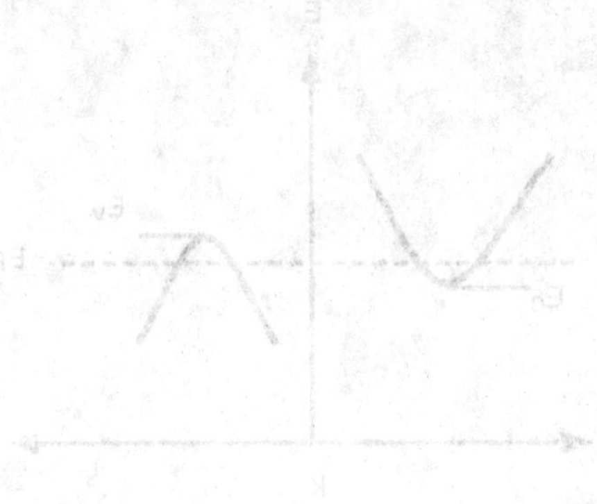

Chapter 5
Carrier Scattering in Solids

5.1 Charge Carrier Scattering

An electron (or a hole) moving through any solid encounters random scatterers that can change its energy and/or its momentum. A scatterer could be a charged impurity in a crystal lattice (e.g., an ionized donor or acceptor atom), another electron or hole, a phonon which is associated with thermal vibrations of the lattice at a finite temperature, or anything else that can take the charged carrier from one wavevector state to another. Since the wavevector changes (in either magnitude or direction or both), the momentum will change. Additionally, the kinetic energy will change if the wavevector's magnitude changes. If the kinetic energy does change, then the electron will either absorb energy from the scatterer or emit energy. We will be interested in finding the rates with which such scattering events or interactions occur since they will determine the momentum and energy relaxation rates. Since the electron's wavevector state will change from $\vec{k}$ to $\vec{k}'$ as a result of scattering, the associated scattering rate will be denoted as $S(\vec{k}, \vec{k}')$. We encountered this quantity in Chap. 2 in the context of the Boltzmann Transport Equation.

5.2 Fermi's Golden Rule for Calculating Carrier Scattering Rates

The prescription for calculating the scattering rate $S(\vec{k}, \vec{k}')$ is known as *Fermi's Golden Rule* [1, 2]. This rule provides an approximate analytical expression for the scattering rate and is derived from time-dependent perturbation theory which we derived in Chap. 3. Since Fermi's Golden Rule tells us the rate of transition of an electron from one wavevector state to another, it has applications beyond scattering theory. For example, it can tell us the rate at which electrons transition from a wavevector state in the valence band of a crystal to another wavevector state in the conduction band upon absorption of a photon from a light source.

S. Bandyopadhyay, *Physics of Nanostructured Solid State Devices*,
DOI 10.1007/978-1-4614-1141-3_5,

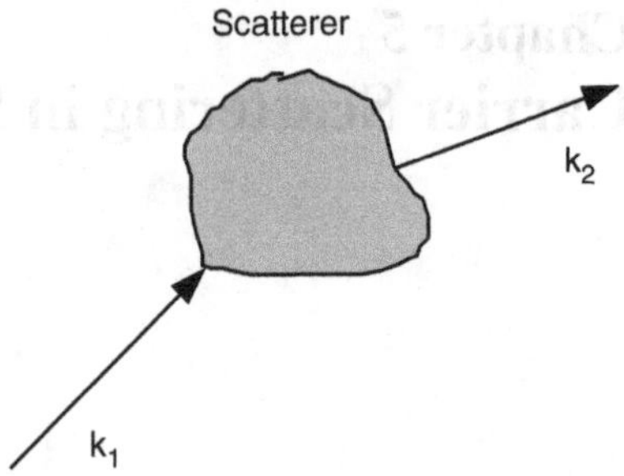

Fig. 5.1 An electron with wavevector $\vec{k}_1$ at time $t = 0$ interacts with a time-dependent scatterer and emerges with wavevector $\vec{k}_2$ at time $t = t$

Therefore, it can be used to calculate the absorption coefficient of light in a solid as well. We will see more of that in Chap. 6, but our immediate task is to derive an expression for $S(\vec{k}, \vec{k}')$ using time-dependent perturbation theory. The expression that we will derive is Fermi's Golden Rule.

5.2.1 Derivation of Fermi's Golden Rule

Consider an electron in a crystal with initial wavevector $\vec{k}_1$ at time $t = 0$ that interacts with a scatterer as shown in Fig. 5.1 and emerges with a wavevector $\vec{k}_2$ at time $t = t$. In general, the scatterer will be a time-dependent scatterer (e.g., a phonon or another electron passing by) and therefore it will act as a time-dependent perturbation. Prior to encountering the scatterer, at time $t = 0$, the electron was obviously in an eigenstate of the crystal with wavevector $\vec{k}_1$. Hence, its wavefunction was $\frac{1}{\sqrt{\Omega}} e^{i\vec{k}_1 \cdot \vec{r}} u_{\vec{k}_1}(\vec{r})$ in accordance with the Bloch Theorem. After interacting with the scatterer which acts as a time-dependent perturbation, the electron's wavefunction will change. The final wavefunction will be a superposition of all allowed wavevector states $\vec{k}$. Therefore, the wavefunction at time $t = t$ can be written as (recall time-dependent perturbation theory of Chap. 3)

$$\psi(\vec{r}, t) = \frac{1}{\sqrt{\Omega}} \sum_{\vec{k}} C_{\vec{k}}(t) e^{i\vec{k} \cdot \vec{r}} u_{\vec{k}}(\vec{r}) e^{-iE_{\vec{k}} t/\hbar}, \tag{5.1}$$

where $E_{\vec{k}}$ is the energy in wavevector state $\vec{k}$ (found from the $E - k$ relation in the crystal) and the coefficient $C_{\vec{k}}(t)$ will depend on time because of the time-dependent nature of the scatterer. We will find the latter using standard time-dependent perturbation theory of Chap. 3.

After the scattering event is complete at time $t = t$, the probability of finding the electron emerging in the wavevector state $\vec{k}_2$ is, by definition,

$$P(\vec{k}_2) = |C_{\vec{k}_2}(t)|^2. \tag{5.2}$$

Therefore, the scattering rate $S(\vec{k}_1, \vec{k}_2)$, which is nothing but the probability of an electron initially in state $\vec{k}_1$ ending up in state $\vec{k}_2$ per unit time, is

$$S(\vec{k}_1, \vec{k}_2) = \lim_{t \to \infty} \frac{P(\vec{k}_2)}{t} = \lim_{t \to \infty} \frac{|C_{\vec{k}_2}(t)|^2}{t}. \tag{5.3}$$

Note that the physical significance of allowing the time t to go to infinity in the previous equation is that the collision duration is viewed as infinitely long. Therefore, what we will derive—namely Fermi's Golden Rule—is strictly valid for prolonged interactions and *not* instantaneous ones. Consequently, Fermi's Golden Rule should never be applicable in Boltzmann Transport formalism because there scattering events are viewed as instantaneous in both time and space. Yet, in spite of this contradiction, the result of Fermi's Golden Rule is routinely used to calculate scattering rates in Boltzmann transport and in many other situations. To go beyond Fermi's Golden Rule requires considerable effort which is seldom made.

There are, however, many inherent contradictions in Fermi's Golden Rule. If the collision duration is infinite and the electron is in an electric or magnetic field when it scatters, then its energy should change immensely *while* it is scattering because the electric and/or magnetic field will accelerate or decelerate it. This effect is not included in Fermi's Golden Rule which, as we shall see, does not allow the electron's energy to change during scattering events except via emission or absorption of scattering entities such as phonons. The infinite collision duration is nevertheless consistent with the fact that the electron's energy before and after scattering are precisely defined (anything else would have violated the Heisenberg Uncertainty Principle). On the other hand, the infinite collision duration also seems to imply that the mean time between collisions must be infinite since otherwise two scattering events have to overlap in time. But an infinite mean time will make the scattering rate identically zero! Thus, there are many contradictions. Despite all these bothersome conundrums, Fermi's Golden Rule has become the most widely used recipe to calculate scattering rates since it provides an elegant and simple analytical expression to calculate the latter.

Going back to standard time-dependent perturbation theory, we note from (3.76) that the coefficients $C_{\vec{k}_1}(t), C_{\vec{k}_2}(t), \ldots C_{\vec{k}_n}(t)$ must obey the equation

$$i\hbar \frac{\partial}{\partial t} \begin{bmatrix} C_{\vec{k}_1}(t) \\ C_{\vec{k}_2}(t) \\ \cdots \\ \cdots \\ \cdots \\ C_{\vec{k}_n}(t) \end{bmatrix} = \begin{bmatrix} H'_{\vec{k}_1,\vec{k}_1}(t) & H'_{\vec{k}_1,\vec{k}_2}(t)e^{i\frac{E_{\vec{k}_1}-E_{\vec{k}_2}}{\hbar}t} & \cdots & H'_{\vec{k}_1,\vec{k}_n}(t)e^{i\frac{E_{\vec{k}_1}-E_{\vec{k}_n}}{\hbar}t} \\ H'_{\vec{k}_2,\vec{k}_1}(t)e^{i\frac{E_{\vec{k}_2}-E_{\vec{k}_1}}{\hbar}t} & H'_{\vec{k}_2,\vec{k}_2}(t) & \cdots & H'_{\vec{k}_2,\vec{k}_n}(t)e^{i\frac{E_{\vec{k}_2}-E_{\vec{k}_n}}{\hbar}t} \\ \cdots & & & \\ \cdots & & & \\ \cdots & & & \\ H'_{\vec{k}_n,\vec{k}_1}(t)e^{i\frac{E_{\vec{k}_n}-E_{\vec{k}_1}}{\hbar}t} & H'_{\vec{k}_n,\vec{k}_2}(t)e^{i\frac{E_{\vec{k}_n}-E_{\vec{k}_2}}{\hbar}t} & \cdots & H'_{\vec{k}_n,\vec{k}_n} \end{bmatrix} \begin{bmatrix} C_{\vec{k}_1}(t) \\ C_{\vec{k}_2}(t) \\ \cdots \\ \cdots \\ \cdots \\ C_{\vec{k}_n}(t) \end{bmatrix}. \tag{5.4}$$

Hence, this equation must be solved to find the coefficients $C_{\vec{k}}(t)$ that yield the expression for the scattering rate.

Born approximation: We will now have to make an approximation known as the *Born approximation*. We will assume that since the scattering potential is weak, it perturbs the electron wavefunction very slightly. As a result, the perturbed wavefunction is only marginally different from the unperturbed wavefunction. Consequently, $C_{\vec{k}_1}(t) \approx 1$ at all times t and $C_{\vec{k}_n}(t) \approx 0$ if $n \neq 1$. This is known as the Born approximation. It allows us to ignore terms in every column of the above matrix except the first, so that we can write

$$i\hbar \frac{\partial C_{\vec{k}_2}(t)}{\partial t} \approx H'_{\vec{k}_2,\vec{k}_1}(t) e^{i\frac{E_{\vec{k}_2}-E_{\vec{k}_1}}{\hbar}t}, \tag{5.5}$$

where, from (3.75), the matrix element is

$$H'_{\vec{k}_2,\vec{k}_1}(t) = \frac{1}{\Omega}\int_0^\infty d^3\vec{r} e^{-i\vec{k}_2\cdot\vec{r}} u^*_{\vec{k}_2}(\vec{r}) V_s(\vec{r},t) e^{i\vec{k}_1\cdot\vec{r}} u_{\vec{k}_1}(\vec{r}), \tag{5.6}$$

with $V_s(\vec{r},t)$ being perturbation term in the Hamiltonian. This term is the space- and time-dependent scattering potential, which is the potential energy of the scatterer (e.g., phonons, impurities, etc.) responsible for the scattering event.

From (5.5), we obtain

$$C_{\vec{k}_2}(t) \approx \frac{1}{i\hbar}\int_0^t H'_{\vec{k}_2,\vec{k}_1}(t') e^{i\frac{E_{\vec{k}_2}-E_{\vec{k}_1}}{\hbar}t'} dt' + C_{\vec{k}_2}(0). \tag{5.7}$$

The last term on the right-hand side in the above equation is exactly zero since at time $t = 0$, the electron was in state $\vec{k}_1$. Therefore, the probability of finding it in state $\vec{k}_2$ at time $t = 0$, which is $|C_{\vec{k}_2}(0)|^2$, is exactly zero. As a result,

$$C_{\vec{k}_2}(t) \approx \frac{1}{i\hbar}\int_0^t H'_{\vec{k}_2,\vec{k}_1}(t') e^{i\frac{E_{\vec{k}_2}-E_{\vec{k}_1}}{\hbar}t'} dt'. \tag{5.8}$$

Next, we will consider time-dependent scattering potentials that vary *periodically* in time and space. Such a potential can be expanded in a Fourier series and viewed as the superposition of forward and backward traveling plane waves. Accordingly, we will write the scattering potential as

$$V_s(\vec{r},t) = V_s^+(\vec{r},t) + V_s^-(\vec{r},t) = \sum_{\omega,\vec{q},\hat{p}} V^+_{\omega,\vec{q},\hat{p}} e^{i(\omega t - \vec{q}\cdot\vec{r})} + \sum_{\omega,\vec{q},\hat{p}} V^-_{\omega,\vec{q},\hat{p}} e^{-i(\omega t - \vec{q}\cdot\vec{r})}, \tag{5.9}$$

where $\hat{p}$ is the polarization of the wave, ω is the angular frequency, and $\vec{q}$ is the wavevector.

The first term in the right-hand side is the forward traveling component and the last term is the backward traveling component. The quantity $V^+_{\omega,\vec{q},\hat{p}}$ is the Fourier component of the scattering potential at frequency ω and wavevector $\vec{q}$ in the forward traveling wave of polarization $\hat{p}$ and $V^-_{\omega,\vec{q}}$ is the Fourier component of the scattering potential at frequency ω and wavevector $\vec{q}$ in the backward traveling wave of polarization $\hat{p}$.

Let us now focus on one particular polarization $\hat{p}$ and one particular Fourier component of the scattering potential with frequency ω and wavevector $\vec{q}$. The perturbation Hamiltonian due to this component alone is (from (5.6))

$$\begin{aligned}
H'_{\omega,\vec{q},\hat{p},\vec{k}_2,\vec{k}_1}(t') &= \frac{1}{\Omega}\int_0^\infty d^3\vec{r}\mathrm{e}^{-\mathrm{i}\vec{k}_2\cdot\vec{r}}u^*_{\vec{k}_2}(\vec{r})V^+_{\omega,\vec{q},\hat{p}}\mathrm{e}^{\mathrm{i}(\omega t'-\vec{q}\cdot\vec{r})}\mathrm{e}^{\mathrm{i}\vec{k}_1\cdot\vec{r}}u_{\vec{k}_1}(\vec{r}) \\
&\quad + \frac{1}{\Omega}\int_0^\infty d^3\vec{r}\mathrm{e}^{-\mathrm{i}\vec{k}_2\cdot\vec{r}}u^*_{\vec{k}_2}(\vec{r})V^-_{\omega,\vec{q},\hat{p}}\mathrm{e}^{-\mathrm{i}(\omega t'-\vec{q}\cdot\vec{r})}\mathrm{e}^{\mathrm{i}\vec{k}_1\cdot\vec{r}}u_{\vec{k}_1}(\vec{r}) \\
&= M^+_{\omega,\vec{q},\hat{p},\vec{k}_2,\vec{k}_1}\mathrm{e}^{\mathrm{i}\omega t'} + M^-_{\omega,\vec{q},\hat{p},\vec{k}_2,\vec{k}_1}\mathrm{e}^{-\mathrm{i}\omega t'},
\end{aligned} \tag{5.10}$$

where

$$M^\pm_{\omega,\vec{q},\hat{p},\vec{k}_2,\vec{k}_1} = \frac{1}{\Omega}\int_0^\infty d^3\vec{r}\mathrm{e}^{-\mathrm{i}\vec{k}_2\cdot\vec{r}}u^*_{\vec{k}_2}(\vec{r})V^\pm_{\omega,\vec{q},\hat{p}}\mathrm{e}^{\mp\mathrm{i}\vec{q}\cdot\vec{r}}\mathrm{e}^{\mathrm{i}\vec{k}_1\cdot\vec{r}}u_{\vec{k}_1}(\vec{r}). \tag{5.11}$$

The quantity $V^\pm_{\omega,\vec{q},\hat{p}}$ is actually an operator and should be treated like one. For some types of time-dependent scatterers like phonons, it is a simple multiplicative operator, while for some other types, it may not be. In the case of photons, it is a differential operator. This can make a serious difference and has profound physical consequences (see Special Topic 2 in Chap. 6). However, if we assume $V^\pm_{\omega,\vec{q},\hat{p}}$ to be a simple multiplicative operator, then we can simplify the previous equation to

$$\begin{aligned}
M^\pm_{\omega,\vec{q},\hat{p},\vec{k}_2,\vec{k}_1} &= \frac{V^\pm_{\omega,\vec{q},\hat{p}}}{\Omega}\int_0^\infty d^3\vec{r}\mathrm{e}^{-\mathrm{i}\vec{k}_2\cdot\vec{r}}u^*_{\vec{k}_2}(\vec{r})\mathrm{e}^{\mp\mathrm{i}\vec{q}\cdot\vec{r}}\mathrm{e}^{\mathrm{i}\vec{k}_1\cdot\vec{r}}u_{\vec{k}_1}(\vec{r}) \\
&\approx \frac{V^\pm_{\omega,\vec{q},\hat{p}}}{\Omega}\int_0^\infty d^3\vec{r}\mathrm{e}^{-\mathrm{i}\vec{k}_2\cdot\vec{r}}\mathrm{e}^{\mp\mathrm{i}\vec{q}\cdot\vec{r}}\mathrm{e}^{\mathrm{i}\vec{k}_1\cdot\vec{r}} \ \texttt{[for parabolic bands]} \\
&= V^\pm_{\omega,\vec{q},\hat{p}}\delta_{\vec{k}_2,\vec{k}_1\mp\vec{q}},
\end{aligned} \tag{5.12}$$

where the "delta" is a Kronecker delta.

The approximation for parabolic bands (with parabolic $E-k$ relation) is based on the fact that electron wavefunctions in parabolic bands are plane waves just like that of a free electron. In other words, the Bloch functions $u_{\vec{k}}(\vec{r})$ become exactly unity. Note that the Kronecker delta simply represents momentum conservation in the scattering process ($\hbar\vec{k}_2 = \hbar\vec{k}_1 \pm \hbar\vec{q}$).

In very nonparabolic bands, the Bloch functions $u_{\vec{k}}(\vec{r})$ cannot be assumed to be unity when calculating $M^\pm_{\omega,\vec{q},\vec{k}_2,\vec{k}_1}$. Additionally, they play an important role in *interband* scattering, i.e., scattering between two different bands in the Brillouin zone, regardless of whether the band is parabolic or not. The Bloch functions of two states in two different bands at the same wavevector in the reduced Brillouin zone are orthogonal. In the extended zone representation, these wavevectors will differ by integer multiples of the reciprocal lattice vector $\vec{G}$. Scattering between states in different bands are called *Umklapp* processes and they are rare since they involve a very large momentum change ($\hbar\vec{q} > \pi/\vec{a}$; $\vec{a}$ =lattice constant). However, it is clear that the matrix element for Umklapp scattering from one band to another will vanish if the initial and final state wavevectors differ by integral multiples of $\vec{G}$ since the corresponding Bloch functions will be orthogonal. Therefore, such Umklapp processes are forbidden. This is an important fact that we will miss if we completely ignore the Bloch functions. For intraband scattering events, however, ignoring the Bloch functions will not have any serious consequence as long as the bands are at least nearly parabolic.

Using (5.10) in (5.8), we obtain

$$C_{\vec{k}_2}(t) = C^{+}_{\vec{k}_2}(t) + C^{-}_{\vec{k}_2}(t), \tag{5.13}$$

where

$$C^{\pm}_{\vec{k}_2}(t) \approx \frac{1}{\mathrm{i}\hbar} M^{\pm}_{\omega,\vec{q},\vec{p},\vec{k}_2,\vec{k}_1} \frac{\mathrm{e}^{\mathrm{i}\frac{E_{\vec{k}_2}-E_{\vec{k}_1}\pm\hbar\omega}{\hbar}t} - 1}{\left(E_{\vec{k}_2} - E_{\vec{k}_1} \pm \hbar\omega\right)/\hbar} = \frac{1}{\mathrm{i}\hbar} M^{\pm}_{\omega,\vec{q},\vec{p},\vec{k}_2,\vec{k}_1} \mathrm{e}^{\mathrm{i}\theta^{\pm}t/2} \frac{\sin(\theta^{\pm}t/2)}{\theta^{\pm}t/2} t, \tag{5.14}$$

and

$$\theta^{\pm} = \frac{E_{\vec{k}_2} - E_{\vec{k}_1} \pm \hbar\omega}{\hbar} \tag{5.15}$$

Substituting (5.13) and (5.14) in (5.3), we obtain

$$S(\omega,\vec{q},\vec{p},\vec{k}_1,\vec{k}_2) = \lim_{t\to\infty} \frac{P(\vec{k}_2)}{t} = \lim_{t\to\infty} \frac{|C_{\vec{k}_2}(t)|^2}{t}$$

$$= \lim_{t\to\infty} \left\{ \frac{\left|M^{+}_{\omega,\vec{q},\vec{p},\vec{k}_2,\vec{k}_1}\right|^2}{t\hbar^2} \left[\frac{\sin(\theta^{+}t/2)}{\theta^{+}t/2}\right]^2 t^2 + \frac{\left|M^{-}_{\omega,\vec{q},\vec{p},\vec{k}_2,\vec{k}_1}\right|^2}{t\hbar^2} \left[\frac{\sin(\theta^{-}t/2)}{\theta^{-}t/2}\right]^2 t^2 \right\}$$

$$= +\texttt{product terms}. \tag{5.16}$$

As $t \to \infty$, the term within the square brackets in the above equation becomes sharply peaked at $\theta^{\pm} = 0$ and begins to resemble a delta-function centered at

$\theta^{\pm} = 0$. It can therefore be approximated as $A\delta(\theta^{\pm})$, where the strength A is found from the relation

$$\int_{-\infty}^{\infty} \left[\frac{\sin(\theta^{\pm}t/2)}{\theta^{\pm}t/2}\right]^2 \mathrm{d}\theta^{\pm} = \frac{2\pi}{t} \approx A \int_{-\infty}^{\infty} \delta(\theta^{\pm})\mathrm{d}\theta^{\pm} = A. \tag{5.17}$$

This yields that $A = \frac{2\pi}{t}$. Therefore,

$$\lim_{t\to\infty} \left[\frac{\sin(\theta^{\pm}t/2)}{\theta^{\pm}t/2}\right]^2 = \frac{2\pi}{t}\delta(\theta^{\pm}). \tag{5.18}$$

Substituting the last relation in (5.16), we finally get the scattering rate as

$$\begin{aligned} S(\omega, \vec{q}, \hat{p}, \vec{k}_1, \vec{k}_2) &\approx \frac{2\pi}{\hbar^2}\left\{\left|M^{+}_{\omega,\vec{q},\hat{p},\vec{k}_2,\vec{k}_1}\right|^2 \delta(\theta^{+}) + \left|M^{-}_{\omega,\vec{q},\hat{p},\vec{k}_2,\vec{k}_1}\right|^2 \delta(\theta^{-}) \right. \\ &\quad \left. + \, \texttt{product terms}\right\} \\ &= \frac{2\pi}{\hbar}\left|M^{+}_{\omega,\vec{q},\hat{p},\vec{k}_2,\vec{k}_1}\right|^2 \delta(E_{\vec{k}_2} - E_{\vec{k}_1} + \hbar\omega) + \frac{2\pi}{\hbar}\left|M^{-}_{\omega,\vec{q},\hat{p},\vec{k}_2,\vec{k}_1}\right|^2 \\ &\quad \times \delta(E_{\vec{k}_2} - E_{\vec{k}_1} - \hbar\omega) + \texttt{product terms}. \end{aligned} \tag{5.19}$$

In deriving the above, we used the fact that $\delta(ax) = \frac{1}{|a|}\delta(x)$.

Now, the product terms are comparatively negligible since the first term of the product would have been peaked around $E_{\vec{k}_2} = E_{\vec{k}_1} + \hbar\omega$, while the second would have been peaked around a different energy $E_{\vec{k}_2} = E_{\vec{k}_1} - \hbar\omega$. Therefore, the product is negligible as long as $\hbar\omega \neq 0$. Consequently, we can write

$$S(\omega, \vec{q}, \hat{p}, \vec{k}_1, \vec{k}_2) \approx S^{+}(\omega, \vec{q}, \hat{p}, \vec{k}_1, \vec{k}_2) + S^{-}(\omega, \vec{q}, \hat{p}, \vec{k}_1, \vec{k}_2), \tag{5.20}$$

where

$$S^{\pm}(\omega, \vec{q}, \hat{p}, \vec{k}_1, \vec{k}_2) = \frac{2\pi}{\hbar}\left|M^{\pm}_{\omega,\vec{q},\hat{p},\vec{k}_2,\vec{k}_1}\right|^2 \delta(E_{\vec{k}_2} - E_{\vec{k}_1} \pm \hbar\omega). \tag{5.21}$$

In a parabolic band, we can use (5.12) to write the above as

$$S^{\pm}(\omega, \vec{q}, \hat{p}, \vec{k}_1, \vec{k}_2) = \frac{2\pi}{\hbar}\left|V^{\pm}_{\omega,\vec{q},\hat{p}}\right|^2 \delta_{\vec{k}_2,\vec{k}_1\mp\vec{q}}\,\delta(E_{\vec{k}_2} - E_{\vec{k}_1} \pm \hbar\omega). \tag{5.22}$$

Because of the Dirac delta and the Kronecker delta in the preceding equation, the rate $S^{+}(\omega, \vec{q}, \hat{p}, \vec{k}_1, \vec{k}_2)$ in (5.20) is nonzero only when $E_{\vec{k}_2} = E_{\vec{k}_1} - \hbar\omega$ *and* $\vec{k}_2 = \vec{k}_1 - \vec{q}$, while the rate $S^{-}(\omega, \vec{q}, \hat{p}, \vec{k}_1, \vec{k}_2)$ is nonzero only when $E_{\vec{k}_2} = E_{\vec{k}_1} + \hbar\omega$

and $\vec{k}_2 = \vec{k}_1 + \vec{q}$. In the first case, the final energy is the initial energy *minus* the energy of the scatterer (e.g., phonon) and the final momentum is the initial momentum *minus* the momentum of the scatterer. In the second case, the final energy is the initial energy *plus* the energy of the scatterer and the final momentum is the initial momentum *plus* the momentum of the scatterer. Energy and momentum conservations imply that the first case corresponds to *emission* of a scatterer of energy $\hbar\omega$ and momentum $\hbar\vec{q}$ (since the electron's energy and momentum decrease by these amounts after scattering), while the second case corresponds to *absorption* of the scatterer (since the electron's energy and momentum increase by these amounts after scattering). Thus, a time varying scattering potential can cause scattering by either emission or absorption. The total scattering rate, given in (5.20), is the *sum* of the emission and the absorption rates.

Equation (5.21) is the final expression that we sought for the scattering rate (either emission or absorption). This expression is referred to as *Fermi's Golden Rule*. It tells us what the rate of transition from wavevector state $\vec{k}_1$ to wavevector state $\vec{k}_2$ is due to any type of perturbation (phonons, photons, etc.). Note also that Fermi's Golden Rule automatically enforces energy conservation by means of the Dirac delta function, while the matrix element incorporates momentum conservation by means of the Kronecker delta function. The Dirac delta, however, implies that energy states before and after scattering are precisely defined. This, in turn, requires the collision duration to be infinite in order to satisfy Heisenberg's Uncertainty Principle.

The energy conserving Dirac delta function in Fermi's Golden Rule tells us something else which is important. If the scattering potential $V_s(\vec{r}, t)$ does *not* depend on time, then according to (5.9), $\omega = 0$. In that case, $E_{\vec{k}_2} = E_{\vec{k}_1}$, which means that the scattering will be *elastic*. Thus, time-*independent* scatterers, which do not change with time (e.g., a crystallographic defect), can only cause elastic scattering where the electron's energy before and after scattering remains the same. Inelastic scattering events invariably require a time-dependent scatterer.

5.3 Dressed Electron States and Second Quantization Operators

In (5.6), we used the "bare" electron states in a crystal $|\Psi_{\text{bare}}\rangle = \frac{1}{\sqrt{\Omega}} e^{i\vec{k}\cdot\vec{r}} u_{\vec{k}}(\vec{r})$ to calculate the perturbation term $H'_{\vec{k}_2,\vec{k}_1}(t)$. However, we now know that this perturbation causes emission and absorption of entities with energy $\hbar\omega$. Emission creates these entities and absorption annihilates them. This immediately tells us that these entities must be bosons and not fermions, since fermions cannot be created or destroyed at will, at least in low-energy physics.

Because of the presence of these bosons, we should use "dressed" electron states instead of the bare electron states to calculate $H'_{\vec{k}_2,\vec{k}_1}(t)$. Dressed states are coupled electron-boson states describing an electron dressed with bosons. When we use them, we automatically account for the possibility that transitions between such states can cause emission or absorption of a boson. Thus, we should use dressed states $|\Psi_{\text{dressed}}\rangle = \frac{1}{\sqrt{\Omega}}\mathrm{e}^{\mathrm{i}\vec{k}\cdot\vec{r}}u_{\vec{k}}(\vec{r})|N_{\omega_1,\vec{q}_1,\hat{p}_1}, N_{\omega_2,\vec{q}_2,\hat{p}_2}, N_{\omega_3,\vec{q}_3,\hat{p}_3},\ldots\rangle$ which represents an electron state dressed with $N_{\omega_1,\vec{q}_1,\hat{p}_1}$ number of bosons (of frequency ω_1, wavevector $\vec{q}_1$, and polarization $\hat{p}_1$), $N_{\omega_2,\vec{q}_2,\hat{p}_2}$ number of bosons (of frequency ω_2, wavevector $\vec{q}_2$ and polarization $\hat{p}_2$), etc. The quantity $N_{\omega,\vec{q},\hat{p}}$ is the effective number of bosons in the mode with frequency ω, wavevector $\vec{q}$ and polarization $\hat{p}$. This number, however, does not have to be an integer; it can be a fraction or a mixed number. It is quite possible that one-half of a boson, or three and three-fifths of a boson, occupies a mode. This number is a measure of the likelihood that a mode is occupied. Obviously if there are 3.6 bosons in mode A and 0.5 bosons in mode B, then there is a 7.2 times higher probability for a boson to occupy mode A than mode B.

Recall that the scattering potential in (5.9) is actually an *operator* where the first term will lead to emission of bosons and the second term will lead to absorption of bosons. Hence, if we want to write it properly, then we should not write it as a mere multiplicative constant, but instead write it as

$$V_{\mathrm{s}}(\vec{r},t) = \sum_{\omega,\vec{q},\hat{p}} V^{+}_{\omega,\vec{q},\hat{p}} a^{\dagger}_{\omega,\vec{q},\hat{p}} \mathrm{e}^{\mathrm{i}(\omega t-\vec{q}\cdot\vec{r})} + \sum_{\omega,\vec{q},\hat{p}} V^{-}_{\omega,\vec{q},\hat{p}} a_{\omega,\vec{q},\hat{p}} \mathrm{e}^{-\mathrm{i}(\omega t-\vec{q}\cdot\vec{r})}, \tag{5.23}$$

where the operator $a^{\dagger}_{\omega,\vec{q},\hat{p}}$ acting on a dressed state will "create" a boson of frequency ω, wavevector $\vec{q}$ and polarization $\hat{p}$ because of emission, while the operator $a_{\omega,\vec{q},\hat{p}}$ will "annihilate" a boson of frequency ω, wavevector $\vec{q}$, and polarization $\hat{p}$ because of absorption. These operators are called *creation* and *annihilation* operators, respectively. They are also called Bose operators or second quantization operators since they are associated with quantization of the boson field.

When these operators operate on a dressed state, they will create or annihilate a boson in that state. This is expressed via the relations:

$$\begin{aligned} &a^{\dagger}_{\omega_n,\vec{q}_n,\hat{p}_n} \frac{1}{\sqrt{\Omega}}\mathrm{e}^{\mathrm{i}\vec{k}\cdot\vec{r}}u_{\vec{k}}(\vec{r})|N_{\omega_1,\vec{q}_1,\hat{p}_1}, N_{\omega_2,\vec{q}_2,\hat{p}_2},\ldots,N_{\omega_n,\vec{q}_n,\hat{p}_n},\ldots\rangle \\ &\quad= \sqrt{N_{\omega_n,\vec{q}_n,\hat{p}_n}+1}\frac{1}{\sqrt{\Omega}}\mathrm{e}^{\mathrm{i}\vec{k}\cdot\vec{r}}u_{\vec{k}}(\vec{r})|N_{\omega_1,\vec{q}_1,\hat{p}_1}, N_{\omega_2,\vec{q}_2,\hat{p}_2},\ldots,N_{\omega_n,\vec{q}_n,\hat{p}_n}+1,\ldots\rangle \end{aligned} \tag{5.24}$$

$$\begin{aligned} &a_{\omega_n,\vec{q}_n,\hat{p}_n} \frac{1}{\sqrt{\Omega}}\mathrm{e}^{\mathrm{i}\vec{k}\cdot\vec{r}}u_{\vec{k}}(\vec{r})|N_{\omega_1,\vec{q}_1,\hat{p}_1}, N_{\omega_2,\vec{q}_2,\hat{p}_2},\ldots,N_{\omega_n,\vec{q}_n,\hat{p}_n},\ldots\rangle \\ &\quad= \sqrt{N_{\omega_n,\vec{q}_n,\hat{p}_n}}\frac{1}{\sqrt{\Omega}}\mathrm{e}^{\mathrm{i}\vec{k}\cdot\vec{r}}u_{\vec{k}}(\vec{r})|N_{\omega_1,\vec{q}_1,\hat{p}_1}, N_{\omega_2,\vec{q}_2,\hat{p}_2},\ldots,N_{\omega_n,\vec{q}_n,\hat{p}_n}-1,\ldots\rangle. \end{aligned} \tag{5.25}$$

The above two equations show that the number of bosons (of the right frequency ω, wavevector $\vec{q}$, and polarization $\hat{p}$) increases or decreases by one in the dressed state when we operate with creation and annihilation operators, respectively. Note from the above equations that the dressed state $|\Psi_{\text{dressed}}\rangle = \frac{1}{\sqrt{\Omega}} e^{i\vec{k}\cdot\vec{r}} u_{\vec{k}}(\vec{r}) |N_{\omega_1,\vec{q}_1,\hat{p}_1}, \ldots, N_{\omega_n,\vec{q}_n,\hat{p}_n}, \ldots\rangle$ is *not* an eigenstate of either the creation operator $a^\dagger$ or the annihilation operator a. Also, neither is $\sqrt{N_{\omega_n,\vec{q}_n,\hat{p}_n}}$ an eigenvalue of the annihilation operator $a_{\omega_n,\vec{q}_n,\hat{p}_n}$, nor is $\sqrt{N_{\omega_n,\vec{q}_n,\hat{p}_n}+1}$ an eigenvalue of the creation operator $a^\dagger_{\omega_n,\vec{q}_n,\hat{p}_n}$. It can be shown that these two operators are indeed hermitian conjugates of each other. Since $a^\dagger \neq a$, clearly these operators are *not* hermitian. Second quantized operators do not have to be hermitian since their eigenvalues, if any, do not represent the expectation value of any observable.

We can write (5.25) in short-hand notation as

$$a^\dagger_{\omega_n,\vec{q}_n,\hat{p}_n} |\ldots, N_{\omega_n,\vec{q}_n,\hat{p}_n}, \ldots\rangle = \sqrt{N_{\omega_n,\vec{q}_n,\hat{p}_n}+1}\, |\ldots, N_{\omega_n,\vec{q}_n,\hat{p}_n}+1, \ldots\rangle$$
$$a_{\omega_n,\vec{q}_n,\hat{p}_n} |\ldots, N_{\omega_n,\vec{q}_n,\hat{p}_n}, \ldots\rangle = \sqrt{N_{\omega_n,\vec{q}_n,\hat{p}_n}}\, |\ldots, N_{\omega_n,\vec{q}_n,\hat{p}_n}-1, \ldots\rangle, \tag{5.26}$$

where $|\ldots, N_{\omega_n,\vec{q}_n,\hat{p}_n}, \ldots\rangle \equiv |\Psi_{\text{dressed}}\rangle = \frac{1}{\sqrt{\Omega}} e^{i\vec{k}\cdot\vec{r}} u_{\vec{k}}(\vec{r}) |N_{\omega_1,\vec{q}_1,\hat{p}_1}, \ldots, N_{\omega_n,\vec{q}_n,\hat{p}_n}, \ldots\rangle$.

Applying the above operation rules twice, we get

$$a^\dagger_{\omega_n,\vec{q}_n,\hat{p}_n} a_{\omega_n,\vec{q}_n,\hat{p}_n} |\ldots, N_{\omega_n,\vec{q}_n,\hat{p}_n}, \ldots\rangle = N_{\omega_n,\vec{q}_n,\hat{p}_n} |\ldots, N_{\omega_n,\vec{q}_n,\hat{p}_n}, \ldots\rangle, \text{ or}$$
$$a^\dagger_{\omega_n,\vec{q}_n,\hat{p}_n} a_{\omega_n,\vec{q}_n,\hat{p}_n} |\Psi_{\text{dressed}}\rangle = N_{\omega_n,\vec{q}_n,\hat{p}_n} |\Psi_{\text{dressed}}\rangle \tag{5.27}$$

In this case, the dressed state $|\Psi_{\text{dressed}}\rangle = |\ldots, N_{\omega_n,\vec{q}_n,\hat{p}_n}, \ldots\rangle$ is indeed an eigenstate of the product operator $a^\dagger_{\omega_n,\vec{q}_n,\hat{p}_n} a_{\omega_n,\vec{q}_n,\hat{p}_n}$, whose eigenvalue is $N_{\omega_n,\vec{q}_n,\hat{p}_n}$. Therefore, this product operator is called the "number operator" since $N_{\omega_n,\vec{q}_n,\hat{p}_n}$ is the *number* of bosons in the dressed state.

Using (5.26), we can easily show that

$$a_{\omega_n,\vec{q}_n,\hat{p}_n} a^\dagger_{\omega_n,\vec{q}_n,\hat{p}_n} - a^\dagger_{\omega_n,\vec{q}_n,\hat{p}_n} a_{\omega_n,\vec{q}_n,\hat{p}_n} = 1, \tag{5.28}$$

or, in commutator notation, $\left[a_{\omega_n,\vec{q}_n,\hat{p}_n}, a^\dagger_{\omega_n,\vec{q}_n,\hat{p}_n}\right] = 1$. This is the universal commutation rule for Bose operators and it shows that the creation and annihilation operators for bosons do not commute.

5.3.1 Matrix Elements with Dressed States

If we correctly use dressed states instead of bare electron states to calculate the matrix elements for the perturbation Hamiltonian, then (5.10) and (5.12) will be modified to

$$H'_{\omega,\vec{q},\hat{p},\vec{k}_2,\vec{k}_1}(t') = \frac{\sqrt{N_{\omega,\vec{q},\hat{p}}+1}}{\Omega}\int_0^{\infty} d^3\vec{r}\mathrm{e}^{-\mathrm{i}\vec{k}_2\cdot\vec{r}}u^*_{\vec{k}_2}(\vec{r})V^+_{\omega,\vec{q}}\mathrm{e}^{\mathrm{i}(\omega t'-\vec{q}\cdot\vec{r})}\mathrm{e}^{\mathrm{i}\vec{k}_1\cdot\vec{r}}u_{\vec{k}_1}(\vec{r})$$
$$+\frac{\sqrt{N_{\omega,\vec{q},\hat{p}}}}{\Omega}\int_0^{\infty} d^3\vec{r}\mathrm{e}^{-\mathrm{i}\vec{k}_2\cdot\vec{r}}u^*_{\vec{k}_2}(\vec{r})V^-_{\omega,\vec{q}}\mathrm{e}^{-\mathrm{i}(\omega t'-\vec{q}\cdot\vec{r})}\mathrm{e}^{\mathrm{i}\vec{k}_1\cdot\vec{r}}u_{\vec{k}_1}(\vec{r})$$
$$= M^+_{\omega,\vec{q},\hat{p},\vec{k}_2,\vec{k}_1}\mathrm{e}^{\mathrm{i}\omega t'} + M^-_{\omega,\vec{q},\hat{p},\vec{k}_2,\vec{k}_1}\mathrm{e}^{-\mathrm{i}\omega t'}, \tag{5.29}$$

where

$$M^{\pm}_{\omega,\vec{q},\hat{p},\vec{k}_2,\vec{k}_1} = \frac{V^{\pm}_{\omega,\vec{q},\hat{p}}}{\Omega}\sqrt{N_{\omega_n,\vec{q}_n,\hat{p}_n}+(1/2)\pm(1/2)}$$
$$\times\int_0^{\infty} d^3\vec{r}\mathrm{e}^{-\mathrm{i}\vec{k}_2\cdot\vec{r}}u^*_{\vec{k}_2}(\vec{r})\mathrm{e}^{\mp\mathrm{i}\vec{q}\cdot\vec{r}}\mathrm{e}^{\mathrm{i}\vec{k}_1\cdot\vec{r}}u_{\vec{k}_1}(\vec{r}). \tag{5.30}$$

For parabolic bands, the Bloch functions are unity, so that

$$M^{\pm}_{\omega,\vec{q},\hat{p},\vec{k}_2,\vec{k}_1} = \frac{V^{\pm}_{\omega,\vec{q},\hat{p}}}{\Omega}\sqrt{N_{\omega_n,\vec{q}_n,\hat{p}_n}+(1/2)\pm(1/2)}\int_0^{\infty} d^3\vec{r}\mathrm{e}^{-\mathrm{i}\vec{k}_2\cdot\vec{r}}\mathrm{e}^{\mp\mathrm{i}\vec{q}\cdot\vec{r}}\mathrm{e}^{\mathrm{i}\vec{k}_1\cdot\vec{r}}$$
$$= V^{\pm}_{\omega,\vec{q},\hat{p}}\sqrt{N_{\omega_n,\vec{q}_n,\hat{p}_n}+(1/2)\pm(1/2)}\delta_{\vec{k}_2,\vec{k}_1\mp\vec{q}}. \tag{5.31}$$

In deriving the above, we used the orthonormality property of the dressed states:

$$\langle\ldots,M_{\omega,\vec{q},\hat{p}},\ldots|\ldots,N_{\omega,\vec{q},\hat{p}},\ldots\rangle. = \delta_{M_{\omega,\vec{q},\hat{p}},N_{\omega,\vec{q},\hat{p}}}, \tag{5.32}$$

where the delta is a Kronecker delta.

Equation (5.22) should now be modified to

$$S^{\pm}(\omega,\vec{q},\hat{p},\vec{k}_1,\vec{k}_2) = \frac{2\pi}{\hbar}\left[N_{\omega_n,\vec{q}_n,\hat{p}_n}+(1/2)\pm(1/2)\right]$$
$$\times\left|V^{\pm}_{\omega,\vec{q},\hat{p}}\right|^2\delta_{\vec{k}_2,\vec{k}_1\mp\vec{q}}\delta\left(E_{\vec{k}_2}-E_{\vec{k}_1}\pm\hbar\omega\right). \tag{5.33}$$

5.4 Applying Fermi's Golden Rule to Calculate Phonon Scattering Rates

One of the major causes of electron scattering in a crystal, particularly at room temperature, is due to interactions of the electron with the oscillating host lattice. The oscillations cause a disturbance that propagates through a bulk crystal like a plane wave. A "phonon" is a quantum of this wave or lattice vibration. Hence, this type of scattering is simply called phonon scattering.

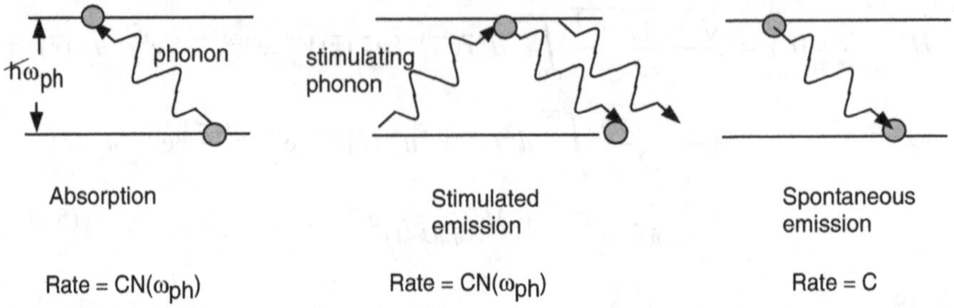

Fig. 5.2 The three different phonon scattering processes. Absorption and stimulated emission require the presence of phonons to absorb or stimulate emission. Hence their rates are proportional to the phonon occupation probability $N(\omega_{\mathrm{ph}})$. Spontaneous emission does not require the presence of other phonons

The displacement of an oscillating atom from its mean position is periodic in time and space. Hence, it can be expressed as a Fourier series:

$$u(\vec{r},t) = \sum_{\omega_{\mathrm{ph}},\vec{q}} \left[A^{+}_{\omega_{\mathrm{ph}},\vec{q}} \mathrm{e}^{\mathrm{i}(\omega_{\mathrm{ph}}t - \vec{q}\cdot\vec{r})} + A^{-}_{\omega_{\mathrm{ph}},\vec{q}} \mathrm{e}^{-\mathrm{i}(\omega_{\mathrm{ph}}t - \vec{q}\cdot\vec{r})} \right], \tag{5.34}$$

where ω_{ph} is phonon frequency and $\vec{q}$ is phonon wavevector.

In a parabolic band, we can use (5.33) to find the electron–phonon scattering rate (for both emission and absorption of a phonon of frequency ω_{ph}, wavevector $\vec{q}$ and polarization $\hat{p}$), as long as we can find the Fourier components $V^{\pm}_{\omega_{\mathrm{ph}},\vec{q},\hat{p}}$ of the scattering potential. Our next task is to find $V^{\pm}_{\omega_{\mathrm{ph}},\vec{q},\hat{p}}$, but before we do that, let us digress for a moment and discuss something else.

As we have already seen, phonon scattering will cause an electron to either absorb energy and momentum from the lattice vibration (*phonon absorption*), or give up some of its own energy and momentum to the lattice (*phonon emission*). The latter process (emission) can be of two types: *spontaneous* and *stimulated*. In the former type, an electron spontaneously emits a phonon and loses energy and momentum. In the latter type, a second phonon stimulates the electron to emit a phonon. Thus, there are really three different types of electron–phonon scattering processes—absorption, stimulated emission, and spontaneous emission. All three are pictorially depicted in Fig. 5.2.

From (5.33) it is clear that the absorption rate is proportional to $N_{\omega_{\mathrm{ph}},\vec{q},\hat{p}}$, while the emission rate is proportional to $N_{\omega_{\mathrm{ph}},\vec{q},\hat{p}} + 1$. Since absorption requires a phonon of frequency ω_{ph}, wavevector $\vec{q}$ and polarization $\hat{p}$ to be available for absorption, it stands to reason that the absorption rate will be proportional to the number of phonons available in that mode, i.e., the quantity $N_{\omega_{\mathrm{ph}},\vec{q},\hat{p}}$. Hence, this makes perfect sense.

Similarly, stimulated emission requires a phonon of frequency ω_{ph}, wavevector $\vec{q}$, and polarization $\hat{p}$ to be available to stimulate an electron to emit a phonon.

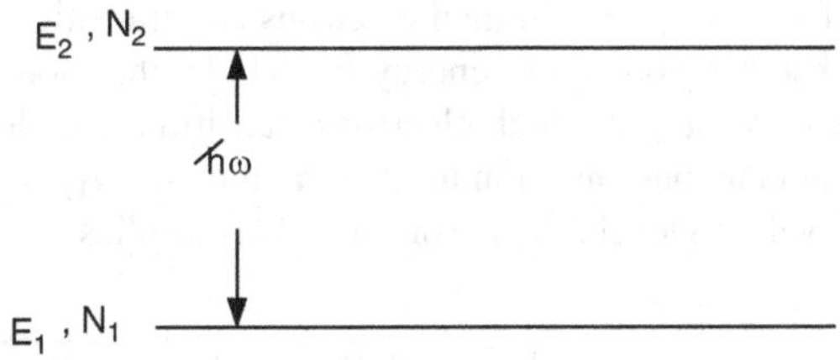

Fig. 5.3 A two-level system with energies E_1 and E_2 and carrier populations N_1 and N_2, respectively

Therefore, the stimulated emission rate should be proportional to $N_{\omega_{ph},\vec{q},\hat{p}}$ as well. That leaves spontaneous emission which does not require any phonon to be available. Hence, out of the total emission rate which is proportional to $N_{\omega_{ph},\vec{q},\hat{p}} + 1$, the stimulated rate must be proportional to $N_{\omega_{ph},\vec{q},\hat{p}}$ and the spontaneous rate must be proportional to 1. Next, we show that all this is consistent with thermodynamics.

5.4.1 Einstein's Thermodynamic Balance Argument

The quantity $N(\omega, \vec{q}, \hat{p})$ (sometimes referred to as the "boson occupation number") can be thought of as the number of bosons occupying the mode with frequency ω, wavevector $\vec{q}$ and polarization $\hat{p}$. If the boson system is in *thermal equilibrium*, then at any temperature T, this quantity is given by the Bose–Einstein factor:

$$N(\omega, \vec{q}, \hat{p})\big|_{\text{equilibrium}} = \frac{1}{e^{\frac{\hbar\omega}{k_B T}} - 1}. \tag{5.35}$$

Note that the equilibrium boson occupation number depends only on the frequency ω, or equivalently the energy of the particle $\hbar\omega$, and does not depend on the wavevector or polarization. Note also that unlike the Fermi–Dirac factor which involves a Fermi level μ_F, the Bose–Einstein factor does not involve a corresponding Bose level μ_B. This happens because fermions are conserved (they cannot be created or destroyed at will; at least in low energy physics), but bosons can be created or destroyed easily (as in phonon emission or phonon absorption). Only particles whose numbers are conserved have an energy like the Fermi energy, while particles whose numbers are not conserved cannot have any such energy. Note also that $N(\omega, \vec{q}, \hat{p})$ can vary between zero and infinity, unlike the Fermi–Dirac factor, which, being an occupation "probability," is bounded by 0 and 1.

Consider now the two-level system shown in Fig. 5.3 which consists of two energy levels E_2 and E_1. The electron population in level 1 is N_1 and that in level 2 is N_2. Each level's equilibrium occupation probability is assumed to be determined by Boltzmann's statistics so that

$$\frac{N_1}{N_2} = e^{\frac{E_2 - E_1}{k_B T}}. \tag{5.36}$$

If $E_2 - E_1 = \hbar\omega$, then electrons can transition between the levels by emitting or absorbing bosons of energy $\hbar\omega$. Under thermodynamic equilibrium conditions, the rate N_2/τ_{21} at which electrons transition from level 2 to level 1 by stimulated and spontaneous emission must equal the rate N_1/τ_{12} at which electrons transition from level 1 to level 2 by absorption. This implies

$$\frac{1/\tau_{21}}{1/\tau_{12}} = \frac{N_1}{N_2} = \mathrm{e}^{\frac{E_2-E_1}{k_\mathrm{B}T}} = \mathrm{e}^{\frac{\hbar\omega}{k_\mathrm{B}T}} = \frac{\frac{1}{\mathrm{e}^{\frac{\hbar\omega}{k_\mathrm{B}T}}-1}+1}{\frac{1}{\mathrm{e}^{\frac{\hbar\omega}{k_\mathrm{B}T}}-1}} = \frac{N(\omega)+1}{N(\omega)}. \tag{5.37}$$

Therefore, the total emission rate $(1/\tau_{21})$ must be proportional to $N(\omega)+1$ if the absorption rate $(1/\tau_{12})$ is proportional to $N(\omega)$. Since the emission has two components—stimulated and spontaneous—and the former is proportional to $N(\omega)$, the ratio of the stimulated and spontaneous rates (and hence the ratio of their squared matrix elements) must be proportional to $N(\omega)$.

Returning to (5.33), we see that the Kronecker delta function basically enforces momentum conservation so that

$$\begin{aligned} \hbar\vec{k}_\mathrm{final} &= \hbar\vec{k}_\mathrm{initial} - \hbar\vec{q}\ \texttt{(for emission)} \\ \hbar\vec{k}_\mathrm{final} &= \hbar\vec{k}_\mathrm{initial} + \hbar\vec{q}\ \texttt{(for absorption)}. \end{aligned} \tag{5.38}$$

All that is left for us to do now in order to evaluate phonon scattering rates is to find the quantity $V^{\pm}_{\omega,\vec{q},\hat{p}}$

5.4.1.1 Determination of $V^{\pm}_{\omega,\vec{q},\hat{p}}$

Let us now focus on just acoustic phonons and determine their $V^{\pm}_{\omega,\vec{q},\hat{p}}$. Acoustic phonons are quanta of acoustic waves in a crystal that carry sound and heat waves via lattice vibrations. They make the lattice constant $\vec{a}$ oscillate in time and space because the lattice is continuously expanding and shrinking. Recall from the previous chapter that the bandgap of a crystal depends on the lattice constant. Therefore, if the lattice constant is oscillating in time and space, the bandgap of the crystal will oscillate in time and space as shown in Fig. 5.4. As a result, an electron in the conduction band of a crystal will experience an oscillating electrostatic potential of amplitude $\Delta E_\mathrm{c}(\vec{r},t)/2$ which acts as a scattering potential. We can expand this potential in a Fourier series and write:

$$\Delta E_\mathrm{c}(\vec{r},t) = \sum_{\omega_\mathrm{ph},\vec{q}} \Delta E_\mathrm{c}^{+}(\omega_\mathrm{ph},\vec{q})\mathrm{e}^{\mathrm{i}(\omega_\mathrm{ph}t-\vec{q}\cdot\vec{r})} + \sum_{\omega_\mathrm{ph},\vec{q}} \Delta E_\mathrm{c}^{-}(\omega_\mathrm{ph},\vec{q})\mathrm{e}^{-\mathrm{i}(\omega_\mathrm{ph}t-\vec{q}\cdot\vec{r})}. \tag{5.39}$$

If the atoms are not displaced too far from their equilibrium positions, their vibrations about the equilibrium positions can be viewed as simple harmonic motion

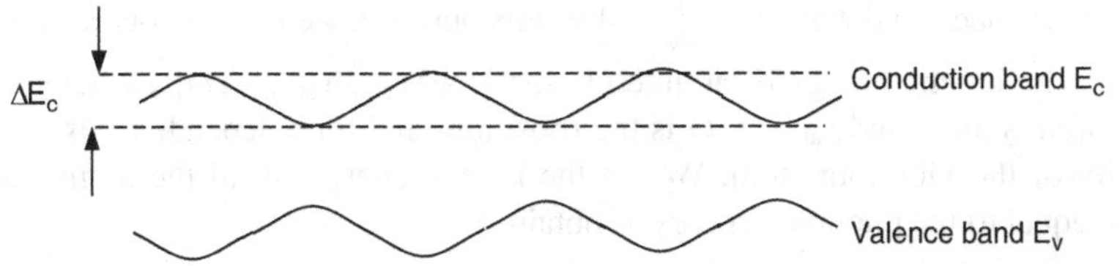

Fig. 5.4 The conduction and valence band edges in a crystal oscillate in time and space as the lattice constant oscillates because of phonons

resulting in longitudinally polarized and transversely polarized acoustic waves traveling through the crystal. In the Γ valley of a semiconductor, the transverse modes (or more precisely the shear strains associated with transverse modes) produce no energy modulation in the conduction band or valence band, although in L or X valleys, they can [3]. Accordingly, $\Delta E_c(\vec{r}, t) = 0$ for *transverse* acoustic phonons in the Γ valley. Therefore, if we just restrict ourselves to the Γ valley, then we can write:

$$\left|V^{\pm}_{\omega,\vec{q},\hat{p}}\right|^2 = \left|\frac{\Delta E^{\pm}_{c}(\omega_{ph}, \vec{q})}{2}\right|^2 \hat{p} \cdot \frac{\vec{q}}{|q|}, \tag{5.40}$$

where $\hat{p}$ is the polarization vector (of unit norm) of the phonon wave and $\vec{q}$ is the phonon wavevector. Note that the scalar product vanishes for transverse phonon modes (because $\hat{p}$ and $\vec{q}$ are mutually perpendicular) and becomes unity for longitudinal phonon modes.

5.4.1.2 Nonpolar Longitudinal Acoustic Phonons

For nonpolar acoustic phonons, ΔE_c is determined by the spatial rate of displacement of the atoms from their mean position and is given by [3]

$$\Delta E_c(\vec{r}, t) = D\vec{\nabla}_r \cdot \vec{u}(\vec{r}, t), \tag{5.41}$$

where D is the so-called *dilational deformation potential* and $\vec{u}$ is the displacement of the atoms from their mean position. Since we are dealing with *longitudinal* phonons, $\vec{q}$ and $\vec{u}$ are collinear. Therefore, using (5.34) for $\vec{u}$, we get from (5.39)

$$\Delta E^{\pm}_{c}(\omega_{ph}, \vec{q}) = \mp DiqA^{\pm}_{\omega_{ph},\vec{q}}. \tag{5.42}$$

As a result,

$$\left|V^{\pm}_{\omega,\vec{q},\hat{p}}\right|^2 = \left|\frac{\Delta E^{\pm}_{c}(\omega_{ph}, \vec{q})}{2}\right|^2 = \frac{D^2q^2\left(A^{\pm}_{\omega_{ph},\vec{q}}\right)^2}{4}. \tag{5.43}$$

Next, we need to determine $A^{\pm}_{\omega_{\rm ph},\vec{q}}$. For this purpose, we invoke conservation of energy. The kinetic energy of atomic motion is $\frac{1}{2}M_{\rm atom}|v_{\rm atom}(\vec{r},t)|^2$, where $M_{\rm atom}$ is the atom's mass and $v_{\rm atom}(\vec{r},t)$ is the (position- and time-dependent) oscillation velocity of the vibrating atom. We set the kinetic energy of all the atoms in the crystal equal to the phonon's energy to obtain

$$\sum_{\rm crystal}\frac{1}{2}M_{\rm atom}|v_{\rm atom}(\vec{r},t)|^2 = \frac{1}{2}\rho\Omega|v_{\rm atom}(\vec{r},t)|^2 = \hbar\omega_{\rm ph}, \tag{5.44}$$

where ρ is the mass density and Ω is the volume of the crystal. Now,

$$\begin{aligned} v_{\rm atom}(\vec{r},t) = \frac{{\rm d}u(\vec{r},t)}{{\rm d}t} &= {\rm i}\omega_{\rm ph}A^{+}_{\omega_{\rm ph},\vec{q}}{\rm e}^{{\rm i}(\omega_{\rm ph}t-\vec{q}\cdot\vec{r})}\,\texttt{(forward traveling wave)} \\ &= -{\rm i}\omega_{\rm ph}A^{-}_{\omega_{\rm ph},\vec{q}}{\rm e}^{-{\rm i}(\omega_{\rm ph}t-\vec{q}\cdot\vec{r})}\,\texttt{(backward traveling wave)}, \end{aligned} \tag{5.45}$$

which yields that for both forward and backward traveling waves

$$|v_{\rm atom}(\vec{r},t)|^2 = \omega_{\rm ph}^2\left(A^{\pm}_{\omega_{\rm ph},\vec{q}}\right)^2 \;\texttt{(independent of r and t)}. \tag{5.46}$$

Substitution of the last result in (5.44) yields

$$\left(A^{+}_{\omega_{\rm ph},\vec{q}}\right)^2 = \left(A^{-}_{\omega_{\rm ph},\vec{q}}\right)^2 = \frac{2\hbar}{\Omega\rho\omega_{\rm ph}}. \tag{5.47}$$

Finally, since acoustic phonons have a nearly linear dispersion relation at sufficiently low phonon energies and wavevectors, we can write that $\omega_{\rm ph} = v_{\rm l}q$, where $v_{\rm l}$ is the velocity of the longitudinal acoustic phonons, or the longitudinal velocity of sound propagation in the crystal (not to be confused with $v_{\rm atom}$). Normally, this velocity is given by $v_{\rm l} = \sqrt{\frac{C_{11}}{\rho}}$, where C_{11} is one of the elastic constants in the crystal [7]. From (5.40), (5.43) and (5.47), we get

$$\left|V^{\pm}_{\omega,\vec{q},\vec{p}}\right|^2 = \frac{D^2 q\hbar}{2\rho v_{\rm l}\Omega}. \tag{5.48}$$

For low-energy phonons such that $k_{\rm B}T \gg \hbar\omega_{\rm ph}$,

$$N(\omega_{\rm ph})\big|_{\rm equilibrium} = \frac{1}{{\rm e}^{\frac{\hbar\omega_{\rm ph}}{k_{\rm B}T}}-1} \approx \frac{k_{\rm B}T}{\hbar\omega_{\rm ph}}. \tag{5.49}$$

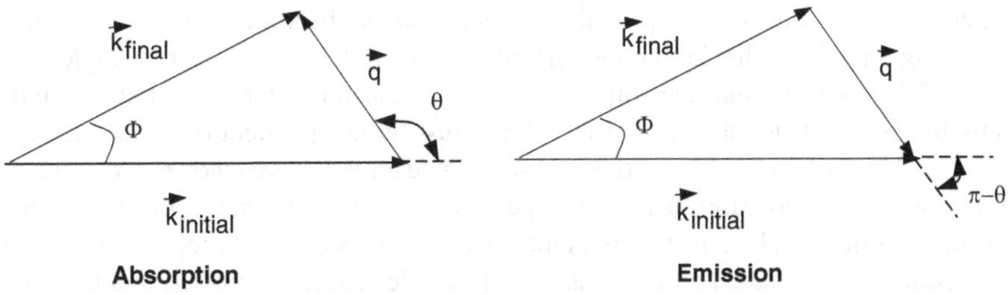

Fig. 5.5 Conservation of momentum in electron–phonon scattering process

This result is sometimes referred to as *equipartition of energy* since the total available thermal energy $k_B T$ is equally divided between $N(\omega_{ph})$ phonons such that each phonon's energy $\hbar\omega_{ph}$ is $k_B T/N(\omega_{ph})$.

Using this result and (5.48) in (5.33), we get that for low-energy longitudinal nonpolar acoustic phonons

$$
\begin{aligned}
S^{-}(\omega_{ph},\vec{q},\vec{k}_1,\vec{k}_2) &= \frac{2\pi}{\hbar}\frac{D^2 q\hbar}{2\rho v_1 \Omega}\frac{k_B T}{\hbar\omega_{ph}}\delta_{\vec{k}_2,\vec{k}_1+\vec{q}}\,\delta\left(E_{\vec{k}_2}-E_{\vec{k}_1}-\hbar\omega\right)\\
&= \frac{\pi D^2 k_B T}{\hbar\rho v_1^2 \Omega}\delta_{\vec{k}_2,\vec{k}_1+\vec{q}}\,\delta\left(E_{\vec{k}_2}-E_{\vec{k}_1}-\hbar\omega\right)\\
&\quad \texttt{for absorption}\\
S^{+}(\omega_{ph},\vec{q},\vec{k}_1,\vec{k}_2) &= \frac{2\pi}{\hbar}\frac{D^2 q\hbar}{2\rho v_1 \Omega}\frac{k_B T}{\hbar\omega_{ph}}\delta_{\vec{k}_2,\vec{k}_1-\vec{q}}\,\delta\left(E_{\vec{k}_2}-E_{\vec{k}_1}+\hbar\omega\right)\\
&= \frac{\pi D^2 k_B T}{\hbar\rho v_1^2 \Omega}\delta_{\vec{k}_2,\vec{k}_1-\vec{q}}\,\delta\left(E_{\vec{k}_2}-E_{\vec{k}_1}+\hbar\omega\right)\\
&\quad \texttt{for stimulated emission}\\
S^{+}(\omega_{ph},\vec{q},\vec{k}_1,\vec{k}_2) &= \frac{2\pi}{\hbar}\frac{D^2 q\hbar}{2\rho v_1 \Omega}\delta_{\vec{k}_2,\vec{k}_1-\vec{q}}\,\delta\left(E_{\vec{k}_2}-E_{\vec{k}_1}+\hbar\omega\right)\\
&= \frac{\pi D^2 q\hbar}{\hbar\rho v_1 \Omega}\delta_{\vec{k}_2,\vec{k}_1-\vec{q}}\,\delta\left(E_{\vec{k}_2}-E_{\vec{k}_1}+\hbar\omega\right)\\
&\quad \texttt{for spontaneous emission.}
\end{aligned}
\tag{5.50}
$$

5.4.1.3 Angular Dependence of Scattering Probability

A phonon scattering event will change a carrier's wavevector. In Fig. 5.5, we show the angle Φ between the initial and final wavevectors, i.e., the wavevectors before and after scattering. Conservation of momentum in the scattering process (represented by (5.38) leads to Fig. 5.5. Obviously, the angle of scattering Φ

depends on the wavevector $\vec{q}$ of the phonon that is absorbed or emitted in the scattering process. The larger the magnitude of $\vec{q}$, the larger is the angle Φ. From (5.50), we find that the squared matrix element for absorption or stimulated emission is independent of q and therefore must be independent of the scattering angle Φ. This means that absorption and stimulated emission are *isotropic scattering processes* whose probabilities are independent of the scattering angle. However, the squared matrix element for spontaneous emission is directly proportional to q. Consequently, stimulated emission is an anisotropic scattering process and there is a clear preference for scattering through larger angles. Such a scattering process is extremely efficient in relaxing the electron's momentum since it changes the direction of momentum by a lot. Therefore, spontaneous emission of nonpolar longitudinal acoustic phonons degrades an electron's mobility in a crystal very efficiently.

5.4.1.4 Polar Longitudinal Acoustic Phonons

Polar acoustic phonons are associated with time- and space-varying local charge polarization within a polar material (like GaAs) that occur due to lattice vibrations of atoms. Since the bond between two atoms is polar, displacement of the atoms from their mean positions causes charge build up in the crystal and an associated electric field. This phenomenon is the cause of piezoelectricity; hence, an electron scattering event mediated by longitudinal polar acoustic phonons is also often referred to as *piezoelectric scattering*.

The time- and space-varying charge polarization gives rise to an electric field which propagates as a wave through the crystal. The Fourier component of the longitudinal mode with frequency ω_{ph} and wavevector $\vec{q}$ can be written as

$$\vec{\mathcal{E}}_{\omega_{\mathrm{ph}},\vec{q}}(\vec{r},t) = \mathcal{E}^{+}_{\omega_{\mathrm{ph}},\vec{q}} \mathrm{e}^{\mathrm{i}\left(\omega_{\mathrm{ph}}t - \vec{q}\cdot\vec{r}\right)} \hat{q} + \mathcal{E}^{-}_{\omega_{\mathrm{ph}},\vec{q}} \mathrm{e}^{-\mathrm{i}\left(\omega_{\mathrm{ph}}t - \vec{q}\cdot\vec{r}\right)} \hat{q}, \tag{5.51}$$

where $\hat{q}$ is the unit vector in the direction of the wavevector.

The electric potential associated with the electric field in the direction of the wavevector is given by

$$V(\vec{r},t) = -\int \vec{\mathcal{E}}_{\omega_{\mathrm{ph}},\vec{q}}(\vec{r},t)\cdot \mathrm{d}\vec{r}. \tag{5.52}$$

The modulation in the conduction band edge as a result of this potential is simply

$$\Delta E_{\mathrm{c}}(\vec{r},t) = -eV(\vec{r},t). \tag{5.53}$$

Hence, from the last three equations and (5.39), we get

$$\Delta E_{\mathrm{c}}^{\pm}(\omega_{\mathrm{ph}},\vec{q}) = \pm e\frac{\mathcal{E}^{\pm}_{\omega_{\mathrm{ph}},\vec{q}}}{iq}. \tag{5.54}$$

Therefore, for longitudinal polar acoustic phonon interaction,

$$\left|V^{\pm}_{\omega,\vec{q},\hat{p}}\right|^2 = \left|\frac{\Delta E_c^{\pm}(\omega_{\text{ph}},\vec{q})}{2}\right|^2 = \frac{e^2\left|\mathcal{E}^{\pm}_{\omega_{\text{ph}},\vec{q}}\right|^2}{4q^2}, \tag{5.55}$$

so that, from (5.33), we get

$$S^-(\omega_{\text{ph}},\vec{q},\vec{k}_1,\vec{k}_2) = \frac{2\pi}{\hbar}N(\omega_{\text{ph}})\frac{e^2\left|\mathcal{E}^{-}_{\omega_{\text{ph}},\vec{q}}\right|^2}{4q^2}\delta_{\vec{k}_2,\vec{k}_1+\vec{q}}\,\delta\left(E_{\vec{k}_2}-E_{\vec{k}_1}-\hbar\omega\right)$$

for absorption

$$S^+(\omega_{\text{ph}},\vec{q},\vec{k}_1,\vec{k}_2) = \frac{2\pi}{\hbar}N(\omega_{\text{ph}})\frac{e^2\left|\mathcal{E}^{+}_{\omega_{\text{ph}},\vec{q}}\right|^2}{4q^2}\delta_{\vec{k}_2,\vec{k}_1-\vec{q}}\,\delta\left(E_{\vec{k}_2}-E_{\vec{k}_1}+\hbar\omega\right)$$

for stimulated emission

$$S^+(\omega_{\text{ph}},\vec{q},\vec{k}_1,\vec{k}_2) = \frac{2\pi}{\hbar}\frac{e^2\left|\mathcal{E}^{+}_{\omega_{\text{ph}},\vec{q}}\right|^2}{4q^2}\delta_{\vec{k}_2,\vec{k}_1-\vec{q}}\,\delta\left(E_{\vec{k}_2}-E_{\vec{k}_1}+\hbar\omega\right)$$

for spontaneous emission. (5.56)

For low acoustic phonon energies, when $k_BT \gg \hbar\omega_{\text{ph}}$, $N(\omega_{\text{ph}}) \approx k_BT/\left(\hbar\omega_{\text{ph}}\right) = k_BT/(\hbar v_l q)$ [equipartition of energy]. Therefore, the scattering rates for electron interaction with longitudinal polar acoustic phonons is

$$S^-(\omega_{\text{ph}},\vec{q},\vec{k}_1,\vec{k}_2) = \frac{\pi e^2\left|\mathcal{E}^{-}_{\omega_{\text{ph}},\vec{q}}\right|^2}{2\hbar v_l q^3}\delta_{\vec{k}_2,\vec{k}_1+\vec{q}}\,\delta\left(E_{\vec{k}_2}-E_{\vec{k}_1}-\hbar\omega\right)$$

for absorption

$$S^+(\omega_{\text{ph}},\vec{q},\vec{k}_1,\vec{k}_2) = \frac{\pi e^2\left|\mathcal{E}^{+}_{\omega_{\text{ph}},\vec{q}}\right|^2}{2\hbar v_l q^3}\delta_{\vec{k}_2,\vec{k}_1-\vec{q}}\,\delta\left(E_{\vec{k}_2}-E_{\vec{k}_1}+\hbar\omega\right)$$

for stimulated emission

$$S^+(\omega_{\text{ph}},\vec{q},\vec{k}_1,\vec{k}_2) = \frac{\pi e^2\left|\mathcal{E}^{+}_{\omega_{\text{ph}},\vec{q}}\right|^2}{2\hbar q^2}\delta_{\vec{k}_2,\vec{k}_1-\vec{q}}\,\delta\left(E_{\vec{k}_2}-E_{\vec{k}_1}+\hbar\omega\right)$$

for spontaneous emission. (5.57)

5.4.1.5 Angular Dependence

Note that the matrix elements for absorption and stimulated emission are inversely proportional to q^3 while the matrix element for spontaneous emission is inversely proportional to q^2. Recalling Fig. 5.5, it becomes obvious that electron-longitudinal polar acoustic phonon scattering is extremely anisotropic and there is a strong preference for scattering through small angles. Such scattering events will not change the direction of the electron's wavevector or momentum drastically since the angle Φ between the initial and final directions will be small. Therefore, scattering events mediated by longitudinal polar acoustic phonons—also known as piezoelectric scattering events—are not very effective in relaxing momentum and consequently will not degrade the electron mobility in a material significantly.

5.5 Calculation of Relaxation Rates

In MC simulations and in many other situations, we are interested in the total electron scattering rate due to one type of scattering (e.g., scattering due to longitudinal nonpolar acoustic phonons). This quantity is given by

$$\frac{1}{\tau(\omega_{\mathrm{ph}},\vec{k})} = \sum_{\vec{k}'} S(\omega_{\mathrm{ph}},\vec{k},\vec{k}'). \tag{5.58}$$

It depends on the initial wavevector and is found by summing $S(\omega_{\mathrm{ph}},\vec{k},\vec{k}')$ over all final wavevector states $\vec{k}'$.

We will evaluate this quantity for acoustic phonon scattering in three-, two-, and one-dimensional structures to illustrate several important properties of carrier scattering in solids. This exercise is very instructive since it brings out many interesting and intriguing features that have nontrivial consequences. We will not do this for every possible type of scattering mechanism since this is a textbook and not an encyclopedia. If the reader can understand the method for acoustic phonon scattering, he/she should be able to repeat the derivation for any other type of scattering mechanism.

5.5.1 Scattering Rates in 3-D (Bulk) Systems

In a bulk (or three-dimensional) system, the total scattering rate will be

$$\frac{1}{\tau^{3D}(\omega_{\mathrm{ph}},\vec{k})} = \sum_{\vec{k}'} S(\omega_{\mathrm{ph}},\vec{k},\vec{k}'). \tag{5.59}$$

where, once again, the upper sign stands for emission and the lower for absorption of a phonon. Recall that $\vec{k}'$ is the final state wavevector and $\vec{k}$ is the initial state wavevector. In 3-D, the summation is carried out over the magnitude, polar angle, and azimuthal angle of the final state wavevector (in spherical coordinates).

5.5.1.1 Nonpolar Longitudinal Acoustic Phonon Absorption

Using the matrix element for nonpolar longitudinal acoustic phonon absorption (NPLAPA), we find that the scattering rate for this type of interaction is

$$\left.\frac{1}{\tau^{3D}(\omega_{\text{ph}},\vec{k})}\right|_{\text{NPLAPA}} = \sum_{\vec{k}'} \frac{\pi D^2 k_B T}{\hbar \rho v_l^2 \Omega} \delta_{\vec{k}',\vec{k}+\vec{q}} \delta(E_{\vec{k}'} - E_{\vec{k}} - \hbar\omega_{\text{ph}}), \tag{5.60}$$

as long as the phonon energy is much smaller than the thermal energy $k_B T$ and the phonons are in equilibrium (so that equipartition of energy holds).

If we now assume that the energy dispersion relation of electrons in the material is parabolic, and that of acoustic phonons is linear ($\omega_{\text{ph}} = \hbar v_l q$), then

$$\begin{aligned} E_{\vec{k}'} &= \frac{\hbar^2 k'^2}{2m^*} \\ E_{\vec{k}} &= \frac{\hbar^2 k^2}{2m^*} \\ \hbar\omega_{\text{ph}} &= \hbar v_l q. \end{aligned} \tag{5.61}$$

In (5.60), we can absorb the Kronecker delta in the Dirac delta in the following way. The Kronecker delta yields (see (5.38))

$$\begin{aligned} &\vec{k}' = \vec{k} + \vec{q} \ \texttt{so that} \\ &k'^2 = \vec{k}' \cdot \vec{k}' = (\vec{k} + \vec{q}) \cdot (\vec{k} + \vec{q}) = k^2 + q^2 + 2kq\cos\theta, \end{aligned} \tag{5.62}$$

where θ is the angle between $\vec{k}$ and $\vec{q}$. We use this relation to rewrite the argument of the Dirac delta function as

$$\begin{aligned} E_{\vec{k}'} - E_{\vec{k}} - \hbar\omega_{\text{ph}} &= \frac{\hbar^2 k'^2}{2m^*} - \frac{\hbar^2 k^2}{2m^*} - \hbar v_l q \\ &= \frac{\hbar^2 q^2}{2m^*} + \frac{\hbar^2 kq\cos\theta}{m^*} - \hbar v_l q. \end{aligned} \tag{5.63}$$

Therefore,

$$\delta_{\vec{k}',\vec{k}+\vec{q}}\,\delta\left(E_{\vec{k}'} - E_{\vec{k}} - \hbar\omega_{\text{ph}}\right) = \delta\left(\frac{\hbar^2 q^2}{2m^*} + \frac{\hbar^2 kq\cos\theta}{m^*} - \hbar v_l q\right). \tag{5.64}$$

Using the last result in (5.60), we obtain

$$\left.\frac{1}{\tau^{3D}(\vec{k})}\right|_{\text{NPLAPA}} = \sum_{\vec{k}'} \frac{\pi D^2 k_B T}{\hbar \rho v_l^2 \Omega} \delta\left(\frac{\hbar^2 q^2}{2m^*} + \frac{\hbar^2 kq\cos\theta}{m^*} - \hbar v_l q\right). \tag{5.65}$$

The summation in the right-hand side is over the variable k', θ, and ϕ (where θ and ϕ are polar and azimuthal angles of $\vec{k}'$), but the quantity to be summed over does not explicitly contain k'. Instead, it contains the phonon wavevector q which is a unique function of k' for a given k (see (5.38)). Instead of performing a variable substitution and writing q in terms of k', we might just as well perform the summation over q instead of over k', since both k' and q are dummy variables for summation. This simplifies matters and we can write

$$\left.\frac{1}{\tau^{3D}\left(\omega_{\text{ph}},\vec{k}\right)}\right|_{\text{NPLAPA}} = \sum_{q,\theta,\phi} \frac{\pi D^2 k_B T}{\hbar \rho v_1^2 \Omega} \delta\left(\frac{\hbar^2 q^2}{2m^*} + \frac{\hbar^2 k q \cos\theta}{m^*} - \hbar v_1 q\right). \quad (5.66)$$

We can convert the summation to an integral in the usual way:

$$\begin{aligned}
\left.\frac{1}{\tau^{3D}(\omega_{\text{ph}},\vec{k})}\right|_{\text{NPLAPA}} &= \sum_{q,\theta,\phi} \frac{\pi D^2 k_B T}{\hbar \rho v_1^2 \Omega} \delta\left(\frac{\hbar^2 q^2}{2m^*} + \frac{\hbar^2 k q \cos\theta}{m^*} - \hbar v_1 q\right) \\
&= \int d^3\vec{q}\, D_{\text{ph}}^{3-D}(q) \frac{\pi D^2 k_B T}{\hbar \rho v_1^2 \Omega} \delta\left(\frac{\hbar^2 q^2}{2m^*} + \frac{\hbar^2 k q \cos\theta}{m^*} - \hbar v_1 q\right), \\
&= \int_0^\infty \int_{-1}^1 \int_0^{2\pi} q^2 \mathrm{d}q \mathrm{d}(\cos\theta) \mathrm{d}\phi D_{\text{ph}}^{3-D}(q) \frac{\pi D^2 k_B T}{\hbar \rho v_1^2 \Omega} \\
&\quad \times \delta\left(\frac{\hbar^2 q^2}{2m^*} + \frac{\hbar^2 k q \cos\theta}{m^*} - \hbar v_1 q\right), \quad (5.67)
\end{aligned}$$

where $D_{\text{ph}}^{3-D}(q)$ is the three-dimensional phonon density of states that we discussed in the previous chapter. However, there is one subtle point. In calculating the phonon density of states, we accounted for both possible phonon polarizations—longitudinal and transverse. Since only longitudinal phonons are considered here, the phonon density of states that we should use is $\Omega/(8\pi^3)$ and not $\Omega/(4\pi^3)$.

Our choice of the spherical coordinate system in three-dimensional space is arbitrary. Let us choose it in such a way that the polar angle θ is the angle between $\vec{k}$ and $\vec{q}$. (This corresponds to choosing the direction of the initial wavevector $\vec{k}$ as the direction of the polar axis). This yields

$$\begin{aligned}
\left.\frac{1}{\tau^{3D}(\omega_{\text{ph}},\vec{k})}\right|_{\text{NPLAPA}} &= \frac{D^2 k_B T}{8\hbar \pi^2 \rho v_1^2} \int_0^\infty \int_{-1}^1 \int_0^{2\pi} q^2 \mathrm{d}q \mathrm{d}(\cos\theta) \mathrm{d}\phi \\
&\quad \times \delta\left(\frac{\hbar^2 q^2}{2m^*} + \frac{\hbar^2 k q \cos\theta}{m^*} - \hbar v_1 q\right). \quad (5.68)
\end{aligned}$$

Next, we will use the following property of Dirac delta functions:

$$\delta(ax) = \frac{1}{|a|}\delta(x). \quad (5.69)$$

Furthermore, since the electron energy dispersion relation was assumed to be parabolic, the electron velocity v_e is related to the wavevector k as $v_e = \hbar k/m^*$. All this results in

$$\left.\frac{1}{\tau^{3D}\left(\omega_{ph}, \vec{k}\right)}\right|_{NPLAPA}$$
$$= \frac{D^2 k_B T}{8\hbar\pi^2\rho v_l^2}\int_0^\infty\int_{-1}^1\int_0^{2\pi} q^2 dq d(\cos\theta) d\phi \frac{m^*}{\hbar^2 kq}\delta\left(\cos\theta + \frac{q}{2k} - \frac{v_l}{v_e}\right)$$
$$= \frac{D^2 k_B T}{8\pi^2\rho v_l^2\hbar^2 v_e}\int_0^\infty\int_{-1}^1\int_0^{2\pi} q dq d(\cos\theta) d\phi\delta\left(\cos\theta + \frac{q}{2k} - \frac{v_l}{v_e}\right)$$
$$= \frac{D^2 k_B T}{4\pi\rho v_l^2\hbar^2 v_e}\int_0^\infty\int_{-1}^1 q dq d(\cos\theta)\delta\left(\cos\theta + \frac{q}{2k} - \frac{v_l}{v_e}\right). \tag{5.70}$$

Now, the integral $\int_a^b \delta(x-y)dx$ is unity *provided* y lies in the interval $[a,b]$; otherwise, it is zero. Therefore, $\left.\frac{1}{\tau^{3D}\left(\omega_{ph}, \vec{k}\right)}\right|_{NPLAPA}$ will be nonzero and the second integral over $\cos\theta$ will be unity as long as the quantity $\frac{v_l}{v_e} - \frac{q}{2k}$ lies in the interval $[-1, 1]$, i.e.,

$$-1 \le \frac{v_l}{v_e} - \frac{q}{2k} \le 1. \tag{5.71}$$

To check on the last condition, we will consider two cases since they will yield different results.

Case I: Electron is supersonic

In this case, the electron travels faster than sound and $v_e > v_l$. Equation (5.71) will then be satisfied as long as the maximum value of q is $2k\left(\frac{v_l}{v_e} + 1\right)$. The minimum value of q can be zero. Therefore, we have the condition:

$$q_{max} = 2k\left(\frac{v_l}{v_e} + 1\right)$$
$$q_{min} = 0. \tag{5.72}$$

The restriction on the maximum value accrues from the fact that the electron cannot scatter by an angle greater than 180°, i.e., Φ in Fig. 5.5 cannot exceed 180° (see Problem 5.1). Since the phonon energy is $\hbar v_l q$, this also means that the electron cannot absorb a phonon of arbitrary energy. The maximum energy of any phonon that it can absorb is $\hbar v_l q_{max} = 2\hbar k v_l\left(\frac{v_l}{v_e} + 1\right) = 2m^* v_l(v_l + v_e)$. Thus, a faster electron can absorb a more energetic phonon and become even more energetic and faster. If left unchecked, this could have caused a velocity runaway, but of course there are phonon emission and other processes that act as energy relaxation mechanisms and act as a check.

Case II: Electron is subsonic

In this case, the electron travels slower than sound and $v_e < v_l$. Equation (5.71) will then be satisfied as long as the maximum value of q is the same $2k\left(\frac{v_l}{v_e}+1\right)$ and the minimum value is $2k\left(\frac{v_l}{v_e}-1\right)$ Therefore, we have the condition:

$$q_{max} = 2k\left(\frac{v_l}{v_e}+1\right)$$
$$q_{min} = 2k\left(\frac{v_l}{v_e}-1\right). \quad (5.73)$$

The maximum value once again accrues from the condition that the electron cannot scatter by an angle larger than 180°. Because of this restriction, the electron cannot absorb a phonon of energy larger than $2m^*v_l(v_l+v_e)$. The minimum value does not have such a simple physical interpretation. Nonetheless, we see that a subsonic electron cannot absorb a phonon of energy any less than $2m^*v_l(v_l-v_e)$.

Going back to (5.70), the nonpolar acoustic phonon absorption rate of electrons is

$$\left.\frac{1}{\tau^{3D}(\omega_{ph},\vec{k})}\right|_{NPLAPA} = \frac{D^2k_BT}{4\pi\rho v_l^2\hbar^2 v_e}\left.\frac{q^2}{2}\right|_{q_{min}}^{q_{max}}$$
$$= \begin{cases} \frac{D^2k_BT(m^*)^2v_e}{2\pi\rho v_l^2\hbar^4}\left(\frac{v_l}{v_e}+1\right)^2 & \text{(supersonic electrons)} \\ \frac{2D^2k_BT(m^*)^2}{\pi\rho v_l\hbar^4}\left(\frac{v_l}{v_e}+1\right)^2 & \text{(subsonic electrons).} \end{cases} \quad (5.74)$$

Note from the above that the absorption rate for subsonic electrons does not depend on the electron's velocity or kinetic energy, but for supersonic electrons, there is a velocity dependence; faster electrons tend to scatter more. When the electron is extremely supersonic, i.e., $v_e \gg v_l$, then the absorption rate can be approximated as

$$\left.\frac{1}{\tau^{3D}(\omega_{ph},\vec{k})}\right|_{NPLAPA} = \frac{D^2k_BT(m^*)^2v_e}{2\pi\rho v_l^2\hbar^4}$$
$$= \frac{D^2k_BT(m^*)^{3/2}\sqrt{E}}{\sqrt{2}\pi\rho v_l^2\hbar^4}$$
$$= \frac{D^2k_BT\pi D^{3-d}(E)}{2\rho v_l^2\hbar}, \quad (5.75)$$

which is directly proportional to the density of states of the electrons.

5.5.1.2 Energy Relaxation Rate for Nonpolar Longitudinal Acoustic Phonon Absorption

From (2.22), the energy relaxation rate associated with any process is defined as

$$\frac{1}{\tau_E^{3D}(\vec{k})} \approx \sum_{\vec{k}'} S(\omega_{\text{ph}}, \vec{k}, \vec{k}') \frac{E_{\vec{k}} - E_{\vec{k}'}}{E_{\vec{k}}} \tag{5.76}$$

for a nondegenerate electron population, where $E_{\vec{k}}$ is the energy of an electron in the wavevector state $\vec{k}$.

In the phonon absorption process, $E_{\vec{k}} - E_{\vec{k}'} = -\hbar v_1 q$. Therefore, from (5.70), we immediately get that

$$\begin{aligned}\frac{1}{\tau_E^{3D}(\omega_{\text{ph}}, \vec{k})} &\approx -\frac{D^2 k_{\text{B}} T (m^*)^2}{2\pi\rho v_1 \hbar^4 k^3} \int_0^{\infty} \int_{-1}^{1} q^2 \text{d}q \text{d}(\cos\theta) \delta\left(\cos\theta + \frac{q}{2k} - \frac{v_1}{v_{\text{e}}}\right) \\ &= -\frac{D^2 k_{\text{B}} T (m^*)^2}{2\pi\rho v_1 \hbar^4 k^3} \left.\frac{q^3}{3}\right|_{q_{\min}}^{q_{\max}} .\end{aligned} \tag{5.77}$$

Consequently, the energy relaxation rate becomes

$$\frac{1}{\tau_E^{3D}(\omega_{\text{ph}}, \vec{k})} = \begin{matrix} -\frac{D^2 k_{\text{B}} T (m^*)^2}{3\pi\rho v_1 \hbar^4} \left(\frac{v_1}{v_{\text{e}}} + 1\right)^3 \; [\texttt{supersonic}] \\ -\frac{D^2 k_{\text{B}} T (m^*)^2}{3\pi\rho v_1 \hbar^4} \left[\left(\frac{v_1}{v_{\text{e}}} + 1\right)^3 - \left(\frac{v_1}{v_{\text{e}}} - 1\right)^3\right] [\texttt{subsonic}]. \end{matrix} \tag{5.78}$$

Note that the energy relaxation rate is negative since an electron *gains* energy—does not lose or relax energy—when it absorbs a phonon. The energy relaxation rate associated with an emission process will of course be positive.

5.5.1.3 Momentum Relaxation Rate for Nonpolar Longitudinal Acoustic Phonon Absorption

From (2.22), the momentum relaxation rate associated with nonpolar longitudinal acoustic phonon absorption is

$$\frac{1}{\tau_m^{3D}(\omega_{\text{ph}}, \vec{k})} \approx \sum_{\vec{k}'} S(\omega_{\text{ph}}, \vec{k}, \vec{k}') \frac{k - k'_k}{k} \tag{5.79}$$

for a nondegenerate electron population, where k'_k is the component of $\vec{k}'$ along the direction of $\vec{k}$. In the phonon absorption process, the quantity $k - k'_k = -q_k = -q\cos\theta$, where q_k is the component of $\vec{q}$ along the direction of $\vec{k}$. Equation (5.70) then yields that

$$\begin{aligned}\frac{1}{\tau_m^{3D}(\omega_{\rm ph},\vec{k})} &\approx -\frac{D^2 k_{\rm B} T m^*}{4\pi\rho v_{\rm l}^2\hbar^3 k^2}\int_0^\infty\int_{-1}^{1} q^2 {\rm d}q\cos\theta {\rm d}(\cos\theta)\delta\left(\cos\theta + \frac{q}{2k} - \frac{v_{\rm l}}{v_{\rm e}}\right)\\ &= \frac{D^2 k_{\rm B} T m^*}{4\pi\rho v_{\rm l}^2\hbar^3 k^2}\int_0^\infty {\rm d}q q^2\left[\frac{q}{2k} - \frac{v_{\rm l}}{v_{\rm e}}\right]\\ &= \frac{D^2 k_{\rm B} T m^*}{4\pi\rho v_{\rm l}^2\hbar^3 k^2}\left[\frac{q^4}{8k} - \frac{q^3}{3}\frac{v_{\rm l}}{v_{\rm e}}\right]\Bigg|_{q_{\rm min}}^{q_{\rm max}}.\end{aligned} \tag{5.80}$$

Therefore, the momentum relaxation rate becomes

$$\frac{1}{\tau_m^{3D}(\omega_{\rm ph},\vec{k})} = \begin{cases}\frac{D^2 k_{\rm B} T (m^*)^2 v_{\rm e}}{2\pi\rho v_{\rm l}^2\hbar^4}\left(\frac{v_{\rm l}}{v_{\rm e}}+1\right)^3\left[1-\frac{v_{\rm l}}{3v_{\rm e}}\right] & \text{[supersonic]}\\ \frac{8D^2 k_{\rm B} T (m^*)^2}{3\pi\rho v_{\rm l}\hbar^4}\left[\left(\frac{v_{\rm l}}{v_{\rm e}}\right)^2+1\right] & \text{[subsonic].}\end{cases} \tag{5.81}$$

Note that the momentum relaxation rate is always positive since phonon absorption invariably reduces the electron's momentum along the original direction.

5.5.1.4 Nonpolar Longitudinal Acoustic Phonon-Stimulated Emission

The total scattering rate in this case is

$$\frac{1}{\tau^{3D}(\omega_{\rm ph},\vec{k})}\Bigg|_{\rm NPLAPStE} = \sum_{\vec{k}'}\frac{\pi D^2 k_{\rm B} T}{\hbar\rho v_{\rm l}^2\Omega}\delta_{\vec{k}',\vec{k}-\vec{q}}\,\delta(E'_{\vec{k}} - E_{\vec{k}} + \hbar\omega_{\rm ph}), \tag{5.82}$$

as long as the phonon energy is much smaller than the thermal energy $k_{\rm B}T$ and the phonons are in equilibrium, so that equipartition of energy holds.

The Kronecker delta mandates that

$$\vec{k}' = \vec{k} - \vec{q}, \tag{5.83}$$

which results in

$$k'^2 = k^2 + q^2 - 2kq\cos\theta', \tag{5.84}$$

where (see Fig. 5.5), θ' is the polar angle if we take $\vec{k}$ as the direction of the polar axis as before. Note also that $\theta' = \pi - \theta$.

Assuming that the energy dispersion relation of electrons is parabolic and that of phonons is linear, the argument of the delta function in (5.82) can be written as $-\hbar^2 kq\cos\theta'/m^* + \hbar^2 q^2/(2m^*) + \hbar v_l q$. Therefore, (5.82) reduces to

$$\left.\frac{1}{\tau^{3D}(\omega_{ph},\vec{k})}\right|_{NPLAPStE} = \frac{D^2 k_B T}{8\hbar\pi^2\rho v_l^2}\int_0^\infty\int_{-1}^1\int_0^{2\pi} q^2 dq d(\cos\theta')d\phi \times\delta\left(\frac{\hbar^2 q^2}{2m^*} - \frac{\hbar^2 kq\cos\theta'}{m^*} + \hbar v_l q\right)$$
$$= \frac{D^2 k_B T}{8\hbar\pi^2\rho v_l^2}\int_0^\infty\int_{-1}^1\int_0^{2\pi} q^2 dq d(\cos\theta')d\phi \times\frac{m^*}{\hbar^2 kq}\delta\left(\cos\theta' - \frac{q}{2k} - \frac{v_l}{v_e}\right)$$
$$= \frac{D^2 k_B T}{4\pi\rho v_l^2\hbar^2 v_e}\int_0^\infty q dq\int_{-1}^1 d(\cos\theta')\delta\left(\cos\theta' - \frac{q}{2k} - \frac{v_l}{v_e}\right), \tag{5.85}$$

where we have used the fact that $\delta(x) = \delta(-x)$ and $\delta(ax) = \frac{1}{|a|}\delta(x)$. Once again, the integration over the delta function will be unity if $\frac{q}{2k} + \frac{v_l}{v_e}$ lies in the interval $[-1, 1]$. This leads to the condition

$$-1 \le \frac{q}{2k} + \frac{v_l}{v_e} \le 1, \tag{5.86}$$

which, in turn, determines the maximum and minimum values of the phonon wavevector q for supersonic and subsonic electrons.

Case I: Electron is supersonic

In this case, $v_e \ge v_l$ so that $q_{min} = 0$, but $q_{max} = 2k\left(1 - \frac{v_l}{v_e}\right)$, which means that an electron cannot emit a phonon of wavevector larger than $2k\left(1 - \frac{v_l}{v_e}\right)$ or energy larger than $2\hbar v_l k\left(1 - \frac{v_l}{v_e}\right) = 2m^* v_l\left(v_e - v_l\right)$.

Case II: Electron is subsonic

In this case, the condition in (5.86) can *never* be fulfilled. Hence, the integration over the delta function will vanish since $\frac{q}{2k} + \frac{v_l}{v_e}$ never lies in the interval $[-1, 1]$. In other words, *a subsonic electron can simply never emit a nonpolar acoustic phonon*. This is actually a familiar situation; only a supersonic jet can emit a sonic boom (acoustic waves) when flying at Mach 1 or greater, while a subsonic one cannot. The present situation is a microscopic analog of this macroscopic phenomenon.

We finally have

$$\left.\frac{1}{\tau^{3D}(\omega_{ph},\vec{k})}\right|_{NPLAPStE} = \begin{cases} \frac{D^2 k_B T (m^*)^2 v_e}{2\pi\rho v_l^2 \hbar^4}\left(1-\frac{v_l}{v_e}\right)^2 & \texttt{(supersonic)} \\ 0 & \texttt{(subsonic).} \end{cases} \tag{5.87}$$

Once again, note that faster electrons emit phonons more frequently, i.e., they have a higher stimulated emission rate.

5.5.1.5 Nonpolar Longitudinal Acoustic Phonon Spontaneous Emission

In this case,

$$S^+(\omega_{ph},\vec{q},\vec{k}_1,\vec{k}_2) = \frac{\pi e^2\left|\mathcal{E}^+_{\omega_{ph},\vec{q}}\right|^2}{2\hbar q^2}\delta_{\vec{k}_2,\vec{k}_1-\vec{q}}\delta(E_{\vec{k}_2}-E_{\vec{k}_1}+\hbar\omega) \tag{5.88}$$

so that proceeding as before, we obtain

$$\left.\frac{1}{\tau^{3D}(\omega_{ph},\vec{k})}\right|_{NPLAPSpE} = \frac{D^2}{4\pi\rho v_l \hbar v_e}\int_{q_{min}}^{q_{max}} q^2 dq \int_{-1}^{1} d(\cos\theta')\delta\left(\cos\theta' - \frac{q}{2k} - \frac{v_l}{v_e}\right). \tag{5.89}$$

where q_{max} and q_{min} are once again found from (5.86) in the case of supersonic and subsonic electrons. Obviously, (5.86) can never be satisfied by subsonic electrons which implies that they cannot spontaneously emit an acoustic phonon just as in the case of stimulated emission. Only supersonic electrons can spontaneously emit acoustic phonons.

For supersonic electrons, $q_{max} = 2k\left(1-\frac{v_l}{v_e}\right)$ and $q_{min} = 0$. Therefore, the double integration in (5.89) reduces to

$$\left.\frac{1}{\tau^{3D}(\omega_{ph},\vec{k})}\right|_{NPLAPSpE} = \frac{D^2}{4\pi\rho v_l \hbar v_e}\frac{q_{max}^3}{3}. \tag{5.90}$$

Combining both cases, we get

$$\left.\frac{1}{\tau^{3D}\left(\omega_{ph},\vec{k}\right)}\right|_{NPLAPSpE} = \begin{cases} \frac{2D^2(m^*)^3 v_e^2}{4\pi\rho v_l \hbar v_e}\left(1-\frac{v_l}{v_e}\right)^2 & \texttt{(supersonic)} \\ 0 & \texttt{(subsonic).} \end{cases} \tag{5.91}$$

5.5.2 Scattering Rates in 2-D (Quantum Well) Systems

Phonon scattering rates in quantum well systems have been evaluated by Price [8] and Ridley [9] using a number of approximations. Here, we present a more detailed treatment which results in a closed form expression for the scattering rate in some cases.

Consider a quantum well of width W_z. The electron motion is constrained in the z-direction, but free in the x-y plane. The wavefunction in the n-th subband of the well is written as

$$\psi_{n,\vec{k}_\parallel}\left(\vec{\rho}, z\right) = \frac{1}{\sqrt{W_z}}\frac{1}{\sqrt{A}}\phi_n(z)e^{i\vec{k}_\parallel\cdot\vec{\rho}}, \tag{5.92}$$

where $\vec{k}_\parallel$ is the electron's wavevector in the x-y plane and $\vec{\rho}$ is the radial position vector in that plane. The quantities W_z and A are normalizing width (in the z-direction) and transverse area (in the x-y plane), respectively.

Using (5.31), the matrix element is

$$\begin{aligned} M^{\pm}_{\vec{k}_{\parallel 2},n;\vec{k}_{\parallel 1},m} &= \frac{V^{\pm}_{\omega_{\text{ph}},\vec{q},\hat{p}}}{AW_z}\int_A\int_0^{W_z} d^2\vec{\rho}dz\phi_n^*(z)e^{-i\vec{k}_{\parallel 2}\cdot\vec{\rho}}e^{\mp i\left(\vec{q}_\parallel+q_z\hat{z}\right)\cdot\left(\vec{\rho}+z\hat{z}\right)}\phi_m(z)e^{i\vec{k}_{\parallel 1}\cdot\vec{\rho}} \\ &\quad\times\sqrt{N_{\omega_{\text{ph}},\vec{q},\hat{p}}+(1/2)\pm(1/2)} \\ &= \frac{V^{\pm}_{\omega_{\text{ph}},\vec{q},\hat{p}}}{AW_z}\int_A\int_0^{W_z} d^2\vec{\rho}dz\phi_n^*(z)e^{-i\vec{k}_{\parallel 2}\cdot\vec{\rho}}e^{\mp i\left(\vec{q}_\parallel\cdot\vec{\rho}+q_z z\right)}\phi_m(z)e^{i\vec{k}_{\parallel 1}\cdot\vec{\rho}} \\ &\quad\times\sqrt{N_{\omega_{\text{ph}},\vec{q},\hat{p}}+(1/2)\pm(1/2)}, \end{aligned} \tag{5.93}$$

where we assume that the electron is initially in the m-th subband, but ends up in the n-th subband after scattering. Here, $\vec{q}_\parallel$ is the phonon's wavevector in the x-y plane (plane of the quantum well) and q_z is the phonon's wavevector in the z-direction (transverse to the plane of the quantum well).

The process where $m = n$ is called *intra-subband scattering*, while the process where $m \neq n$ is called *inter-subband scattering*. As before, the upper sign is for emission and the lower for absorption.

Using (5.93), we get

$$\begin{aligned} \left|M^{\pm}_{\vec{k}_{\parallel 2},n;\vec{k}_{\parallel 1},m}\right|^2 &= \left|V^{\pm}_{\omega_{\text{ph}},\vec{q},\hat{p}}\right|^2\left(N(\omega_{\text{ph}})+\frac{1}{2}\pm\frac{1}{2}\right)\left|\frac{1}{W_z}\int_0^{W_z} dze^{\mp iq_z z}\phi_m^*(z)\phi_n(z)\right|^2 \\ &\quad\times\left|\frac{1}{A}\int_A d^2\vec{\rho}e^{i\left(k_{\parallel 1}-k_{\parallel 2}\mp q_\parallel\right)\cdot\vec{\rho}}\right|^2 \end{aligned}$$

$$= \left|V^{\pm}_{\omega_{\text{ph}},\vec{q},\vec{p}}\right|^2 \left(N(\omega_{\text{ph}}) + \frac{1}{2} \pm \frac{1}{2}\right) \left|\frac{1}{W_z}\int_0^{W_z} \mathrm{d}z e^{\mp i q_z z}\phi_m^*(z)\phi_n(z)\right|^2$$
$$\times\, \delta_{\vec{k}_{\parallel 2},\vec{k}_{\parallel 1}\mp\vec{q}_{\parallel}}$$
$$= \left|V^{\pm}_{\omega_{\text{ph}},\vec{q},\vec{P}}\right|^2 \left(N(\omega_{\text{ph}}) + \frac{1}{2} \pm \frac{1}{2}\right) \left|I_{mn,\mp}(q_z)\right|^2 \delta_{\vec{k}_{\parallel 2},\vec{k}_{\parallel 1}\mp\vec{q}_{\parallel}}, \tag{5.94}$$

where we have assumed that the phonon system is in equilibrium so that the phonon occupation number depends only on the phonon frequency (since it is the Bose–Einstein factor), and

$$I_{mn,\mp}(q_z) = \frac{1}{W_z}\int_0^{W_z} \mathrm{d}z e^{\mp i q_z z}\phi_m^*(z)\phi_n(z). \tag{5.95}$$

Note that the Kronecker delta in (5.94) implies momentum conservation in the x-y plane. There is no momentum conservation associated with motion in the z-direction since the electron's motion is constrained in that direction. An alternate way of viewing the absence of momentum conservation for z-directed motion is in terms of the Heisenberg uncertainty principle. Since the electron can be located anywhere in the well, the uncertainty in its position is W_z. The corresponding uncertainty in the z-component of the momentum will be greater than or equal to $\hbar/(2W_z)$, which is a large uncertainty since the well's width W_z is small. Hence, conservation of the z-component of the momentum is not meaningful. Furthermore, if we were to calculate the expectation value of the momentum in the z-direction, we will find that it is exactly zero, which simply reflects the fact that the electron has no free motion in the z-direction due to the confining potential of the quantum well. Since the electron never has any expectation value of momentum in the z-direction, it makes no sense to talk of conservation of the z-component of momentum.

Finally, note from (5.95) that $I_{mn,\mp}(q_z)$ is the Fourier transform of the quantity $\frac{1}{W_z}\phi_{mn}(z) = \frac{1}{W_z}\phi_m^*(z)\phi_n(z)$. Therefore, by Parceval's theorem, we have the following identity

$$\int_{-\infty}^{\infty} |I_{mn,\mp}(q_z)|^2 \,\mathrm{d}q_z = \frac{2\pi}{W_z^2}\int_0^{W_z} |\phi_{mn}(z)|^2 \,\mathrm{d}z = G_{mn}, \tag{5.96}$$

where G_{mn} is a constant and is sometimes called a "form factor." It has the unit of inverse length.

5.5.2.1 Nonpolar Longitudinal Acoustic Phonon Absorption

We wish to find an expression for the total scattering rate of an electron which is initially in the m-th subband and has a transverse wavevector $\vec{k}_{\parallel}$ in the x-y plane, ending up in the n-th subband after absorbing a nonpolar longitudinal acoustic phonon. This rate is given by (using Fermi's Golden Rule)

$$\left.\frac{1}{\tau^{2D}(\omega_{ph}, \vec{k}_{\parallel}, m, n)}\right|_{NPLAPA} = \sum_{\vec{k}'_{\parallel}, q_z} \left[\frac{\pi D^2 k_B T}{\hbar \rho v_l^2 V} |I_{mn,+}(q_z)|^2 \delta_{\vec{k}'_{\parallel}, \vec{k}_{\parallel}+\vec{q}_{\parallel}} \times \delta\left(E_{\vec{k}'_{\parallel},n} - E_{\vec{k}_{\parallel},m} - \hbar\omega_{ph}\right)\right], \tag{5.97}$$

as long as the phonon energy is much smaller than the thermal energy $k_B T$ and the phonons are in equilibrium. Here, we have used (5.48) to replace $V^{\pm}_{\omega_{ph},\vec{q},\hat{p}}$ and (5.49) to replace $N(\omega_{ph})$. As a reminder, ρ is the density of the well material, Ω is the normalizing volume, $\vec{k}'_{\parallel}$ is the final wavevector in the final n-th subband, and $E_{\vec{k}_{\parallel},m}$ is the energy of an electron in the initial m-th subband with a transverse wavevector $\vec{k}_{\parallel}$.

If the energy dispersion relation is parabolic, then

$$E_{\vec{k}_{\parallel},m} = \frac{\hbar^2 k_{\parallel}^2}{2m^*} + \epsilon_m, \tag{5.98}$$

where ϵ_m is the energy at the bottom of the m-th subband.

Following the usual prescription, we can convert the summation over the final state transverse wavevector $\vec{k}'_{\parallel}$ into an integral over the transverse component of the phonon wavevector, which will result in

$$\begin{aligned}\left.\frac{1}{\tau^{2D}(\omega_{ph}, \vec{k}_{\parallel}, m, n)}\right|_{NPLAPA} &= \int_0^{\infty}\int_{-\infty}^{\infty} d^2\vec{q}_{\parallel} dq_z \left[D_{ph}^{3-D}(q)\frac{\pi D^2 k_B T}{\hbar \rho v_l^2 \Omega} |I_{mn,+}(q_z)|^2 \right.\\ &\quad \left. \times \delta_{\vec{k}'_{\parallel}, \vec{k}_{\parallel}+\vec{q}_{\parallel}} \delta\left(E_{\vec{k}'_{\parallel},n} - E_{\vec{k}_{\parallel},m} - \hbar v_l \sqrt{q_z^2 + q_{\parallel}^2}\right)\right]\\ &= \frac{1}{8\pi^3}\frac{\pi D^2 k_B T}{\hbar \rho v_l^2}\int_0^{\infty}\int_{-\infty}^{\infty} dq_z d^2\vec{q}_{\parallel}\left[|I_{mn,+}(q_z)|^2\right.\\ &\quad \left. \times \delta\left(-\hbar v_l\sqrt{q_z^2+q_{\parallel}^2} - \epsilon_m + \epsilon_n + \frac{\hbar^2}{m^*}k_{\parallel}q_{\parallel}\cos\theta + \frac{\hbar^2 q_{\parallel}^2}{2m^*}\right)\right],\end{aligned} \tag{5.99}$$

where we have divided the phonon density of states $D_{ph}^{3-D}(q)$ by a factor of 2 to account for the fact that only longitudinal phonons interact with electrons while transverse phonons do not. We have also absorbed the Kronecker delta in the Dirac delta by incorporating momentum conservation in the argument of the Dirac delta function. In so doing, we have assumed that the dispersion relation in the x-y plane is parabolic and that the subband bottom energies are denoted by ϵ_m and ϵ_n. The angle between $\vec{k}_{\parallel}$ and $\vec{q}_{\parallel}$ is assumed to be θ.

Note that here we have treated the phonons as bulk (3-D) phonons since we are integrating over both q_z and $\vec{q}_\parallel$ and using the three dimensional phonon density of states. Quantum wells can cause phonon confinement in some cases which will make the 3-D treatment of phonons inappropriate. This, however, happens rarely. Near the end of this chapter, we have briefly discussed the aftermath of acoustic phonon confinement in the context of quasi one-dimensional (quantum wire) structures, where the effect of phonon confinement is much more pronounced than in the case of quantum well systems.

The preceding equation can be expanded as

$$\left.\frac{1}{\tau^{2D}(\omega_{\mathrm{ph}},\vec{k}_\parallel,m,n)}\right|_{\mathrm{NPLAPA}} = \frac{D^2 k_B T}{8\pi^2\hbar\rho v_1^2}\int_0^\infty\int_0^{2\pi}\int_{-\infty}^{\infty} q_\parallel \mathrm{d}q_\parallel \mathrm{d}\theta \mathrm{d}q_z \times\left[|I_{mn,+}(q_z)|^2\,\delta\left(-\hbar v_1\sqrt{q_z^2+q_\parallel^2}-\epsilon_m+\epsilon_n+\frac{\hbar^2}{m^*}k_\parallel q_\parallel\cos\theta+\frac{\hbar^2 q_\parallel^2}{2m^*}\right)\right]. \tag{5.100}$$

The above integration must be performed numerically. But we can attempt to obtain an analytical result if we make the approximation that $q_z \ll q_\parallel$, which then yields

$$\left.\frac{1}{\tau^{2D}\left(\omega_{\mathrm{ph}},\vec{k}_\parallel,m,n\right)}\right|_{\mathrm{NPLAPA}}^{q_z\ll q_\parallel} \approx \frac{D^2 k_B T}{8\pi^2\hbar\rho v_1^2}\left[\int_{-\infty}^{\infty}\mathrm{d}q_z\,|I_{mn,+}(q_z)|^2\right] \times\int_0^\infty\int_0^{2\pi}\mathrm{d}q_\parallel\mathrm{d}\theta\frac{1}{\hbar v_\parallel}\delta\left(\cos\theta+\frac{q_\parallel}{2k_\parallel}-\frac{v_1}{v_\parallel}-\frac{k_{mn}^2}{k_\parallel q_\parallel}\right)$$
$$= \frac{D^2 k_B T}{8\pi^2\hbar\rho v_1^2}G_{mn} \times\int_0^\infty\int_0^{2\pi}\mathrm{d}q_\parallel\mathrm{d}\theta\frac{1}{\hbar v_\parallel}\delta\left(\cos\theta+\frac{q_\parallel}{2k_\parallel}-\frac{v_1}{v_\parallel}-\frac{k_{mn}^2}{k_\parallel q_\parallel}\right) \tag{5.101}$$

where $k_{mn} = \frac{\sqrt{m^*(\epsilon_m-\epsilon_n)}}{\hbar}$ and $v_\parallel = (\hbar/m^*)k_\parallel$. Note that k_{mn} can be real or imaginary depending on whether $\epsilon_m > \epsilon_n$ or $\epsilon_m < \epsilon_n$, but k_{mn}^2 is always real. For intra-subband scattering, k_{mn} is, of course, identically zero.

Once again, we enforce the condition that in order for the scattering rate to be nonzero, the quantity $\frac{v_1}{v_\parallel}+\frac{k_{mn}^2}{k_\parallel q_\parallel}-\frac{q_\parallel}{2k_\parallel}$ must lie in the interval $[-1, 1]$. This yields the following results:

Case I: Supersonic electron

When $v_\parallel \geq v_l$, we get that

$$q_\parallel^{\max} = \sqrt{k_\parallel^2\left(1+\frac{v_l}{v_\parallel}\right)^2 + 2k_{mn}^2} + k_\parallel\left(1+\frac{v_l}{v_\parallel}\right)$$

$$q_\parallel^{\min} = \sqrt{k_\parallel^2\left(1-\frac{v_l}{v_\parallel}\right)^2 + 2k_{mn}^2} - k_\parallel\left(1-\frac{v_l}{v_\parallel}\right). \tag{5.102}$$

Note that for intra-subband scattering ($k_{mn} = 0$), $q_\parallel^{\max} = 2k_\parallel(1+\frac{v_l}{v_\parallel})$ and $q_\parallel^{\min} = 0$.

Case II: Subsonic electron

When $v_\parallel \leq v_l$, we get

$$q_\parallel^{\max} = \sqrt{k_\parallel^2\left(\frac{v_l}{v_\parallel}+1\right)^2 + 2k_{mn}^2} + k_\parallel\left(\frac{v_l}{v_\parallel}+1\right)$$

$$q_\parallel^{\min} = \sqrt{k_\parallel^2\left(\frac{v_l}{v_\parallel}-1\right)^2 + 2k_{mn}^2} + k_\parallel\left(\frac{v_l}{v_\parallel}-1\right). \tag{5.103}$$

For intra-subband scattering ($k_{mn} = 0$), $q_\parallel^{\max} = 2k_\parallel(\frac{v_l}{v_\parallel}+1)$ and $q_\parallel^{\min} = 2k_\parallel(\frac{v_l}{v_\parallel}-1)$.

The above conditions establish the maximum and minimum (transverse) wavevectors of phonons that can be absorbed. In order to evaluate the integrals in (5.101), we apply the following property of Dirac delta functions:

$$\delta(f(x)) = \sum_i \frac{\delta(x-x_i)}{|f'(x_i)|}, \tag{5.104}$$

where x_i-s are the zeros of the function $f(x)$ and $f'(x) = \frac{\mathrm{d}f(x)}{\mathrm{d}x}$. Therefore, (5.101) reduces to

$$\left.\frac{1}{\tau^{2D}(\omega_{\text{ph}}, \vec{k}_\parallel, m, n)}\right|_{\text{NPLAPA}}^{q_z \ll q_\parallel}$$

$$= \frac{D^2 k_{\mathrm{B}} T}{8\pi^2\hbar^2\rho v_l^2 v_\parallel} G_{mn} \int_{q_\parallel^{\min}}^{q_\parallel^{\max}} \int_0^{2\pi} \mathrm{d}q_\parallel \mathrm{d}\theta \sum_i \frac{\delta(\theta-\theta_i)}{|\sin\theta_i|}$$

$$= \frac{D^2 k_{\mathrm{B}} T}{8\pi^2\hbar^2\rho v_l^2 v_\parallel} G_{mn} \int_{q_\parallel^{\min}}^{q_\parallel^{\max}} \mathrm{d}q_\parallel \frac{1}{\sqrt{1-\left(\frac{v_l}{v_\parallel}+\frac{k_{mn}^2}{k_\parallel q_\parallel}-\frac{q_\parallel}{2k_\parallel}\right)^2}}, \tag{5.105}$$

where θ_i are roots of the function $\cos\theta + \frac{q_\parallel}{2k_\parallel} - \frac{v_l}{v_\parallel} - \frac{k_{mn}^2}{k_\parallel q_\parallel}$.

It is not possible to perform the last integral analytically to obtain a closed form solution. However, for *intra-subband* scattering ($k_{mn} = 0$), we can perform the integral analytically and obtain a closed form result which is

$$\left.\frac{1}{\tau^{2D}(\omega_{\mathrm{ph}}, \vec{k}_\parallel, m, m)}\right|_{\mathrm{NPLAPA}}^{q_z \ll q_\parallel} = \frac{D^2 k_B T}{8\pi^2\hbar^2\rho v_l^2 v_\parallel} G_{mm}\, 2k_\parallel \sin^{-1}\left(\frac{q_\parallel}{2k_\parallel} - \frac{v_l}{v_\parallel}\right)\Bigg|_{q_\parallel^{\min}}^{q_\parallel^{\max}}. \tag{5.106}$$

For supersonic electrons, the intra-subband scattering rate becomes

$$\begin{aligned}\left.\frac{1}{\tau^{2D}(\omega_{\mathrm{ph}}, \vec{k}_\parallel, m, m)}\right|_{\mathrm{NPLAPA}}^{q_z \ll q_\parallel} &= \frac{D^2 k_B T}{4\pi^2\hbar^2\rho v_l^2 v_\parallel} G_{mm} k_\parallel \left[\frac{\pi}{2} + \sin^{-1}\left(\frac{v_l}{v_\parallel}\right)\right] \\ &= \frac{D^2 m^* k_B T}{4\pi^2\hbar^3\rho v_l^2} G_{mm} \left[\frac{\pi}{2} + \sin^{-1}\left(\frac{v_l}{v_\parallel}\right)\right], \end{aligned} \tag{5.107}$$

which becomes approximately $\frac{D^2 m^* k_B T}{8\pi\hbar^3\rho v_l^2} G_{mm} = \frac{D^2 k_B T}{8\hbar\rho v_l^2} G_{mm} D_q^{2-d}$, if $v_l \ll v_\parallel$. Here, D_q^{2-d} is the two-dimensional density of states.

For subsonic electrons, the intra-subband scattering rate becomes

$$\left.\frac{1}{\tau^{2D}(\omega_{\mathrm{ph}}, \vec{k}_\parallel, m, m)}\right|_{\mathrm{NPLAPA}}^{q_z \ll q_\parallel} = \frac{D^2 m^* k_B T}{4\pi\hbar^3\rho v_l^2} G_{mm} = \frac{D^2 k_B T}{4\hbar\rho v_l^2} G_{mm} D_q^{2-d}. \tag{5.108}$$

We leave it to the readers to work out the results for spontaneous and stimulated emissions.

5.5.3 *Scattering Rates in 1-D (Quantum Wire) Systems*

Phonon scattering rates in quantum wire systems have been evaluated by a number of authors [11] using a number of approximations. Here, we present a calculation of the acoustic phonon absorption rate. Other rates can be calculated by the reader following the same prescription.

Consider a quantum wire of rectangular cross-section having width W_z and thickness W_y. The electron motion is free only along the x-direction. The wavefunction in the (m, n)-th subband of the well (each subband is indexed by two integers corresponding to confinement along the two transverse directions) is written as

$$\psi_{k,m,n} = \frac{1}{\sqrt{W_y}}\phi_m(z)\frac{1}{\sqrt{W_z}}\phi_n(y)\frac{1}{\sqrt{L}}\mathrm{e}^{ikx}, \tag{5.109}$$

where k is the electron's wavevector in the x direction and L is a normalizing length for the wavefunction component in that direction, i.e., $\frac{1}{L}\int_0^L \mathrm{e}^{\mathrm{i}(k-k')x}\mathrm{d}x = \delta_{k,k'}$.

Using (5.12), the matrix element for scattering is

$$M^{\pm}_{k,m,n;\,k',m',n'} = \frac{V^{\pm}_{\omega_{\mathrm{ph}},\vec{q},\hat{p}}}{W_x W_y L}\iiint \mathrm{d}x\mathrm{d}y\mathrm{d}z\phi^*_{m'}(z)\phi^*_{n'}(y)\mathrm{e}^{-\mathrm{i}k'x}(\vec{r})\phi_m(z)\phi_n(y)$$
$$\times\,\mathrm{e}^{\mathrm{i}kx}\mathrm{e}^{\mp\mathrm{i}(q_x x+q_y y+q_z z)}\sqrt{N_{\omega_{\mathrm{ph}},\vec{q},\hat{p}} + (1/2) \pm (1/2)}, \tag{5.110}$$

where we assume that the electron is initially in the (m,n)-th subband, but ends up in the (m',n')-th subband after scattering.

Once again, the process where $m = m'$ and $n = n'$ is called *intra-subband scattering*, while any other process will be *inter-subband scattering*. As always, the upper sign in the preceding equation is for emission and the lower for absorption.

Using (5.110), we get

$$\left|M^{\pm}_{k,m,n;\,k',m',n'}\right|^2 = \left|V^{\pm}_{\omega_{\mathrm{ph}},\vec{q},\hat{p}}\right|^2\left(N(\omega_{\mathrm{ph}})+\frac{1}{2}\pm\frac{1}{2}\right)\left|\frac{1}{W_z}\int_0^{W_z}\mathrm{d}z\mathrm{e}^{\mp\mathrm{i}q_z z}\phi^*_{m'}(z)\phi_m(z)\right|^2$$
$$\times\left|\frac{1}{W_y}\int_0^{W_y}\mathrm{d}y\mathrm{e}^{\mp\mathrm{i}q_y y}\phi^*_{n'}(y)\phi_n(y)\right|^2\left|\frac{1}{L}\int_0^{L}\mathrm{d}x\mathrm{e}^{\mathrm{i}(k-k'\mp q_x)x}\right|^2$$
$$= \left|V^{\pm}_{\omega_{\mathrm{ph}},\vec{q},\hat{p}}\right|^2\left(N(\omega_{\mathrm{ph}})+\frac{1}{2}\pm\frac{1}{2}\right)\left|\frac{1}{W_z}\int_0^{W_z}\mathrm{d}z\mathrm{e}^{\mp\mathrm{i}q_z z}\phi^*_{m'}(z)\phi_m(z)\right|^2$$
$$\times\left|\frac{1}{W_y}\int_0^{W_y}\mathrm{d}y\mathrm{e}^{\mp\mathrm{i}q_y y}\phi^*_{n'}(y)\phi_n(y)\right|^2\delta_{k',k\mp q_x}$$
$$= \left|V^{\pm}_{\omega_{\mathrm{ph}},\vec{q},\hat{p}}\right|^2\left(N(\omega_{\mathrm{ph}})+\frac{1}{2}\pm\frac{1}{2}\right)\left|I_{mm',\mp}(q_z)\right|^2\left|I_{nn',\mp}(q_y)\right|^2$$
$$\times\delta_{k',k\mp q_x}, \tag{5.111}$$

where

$$I_{pq,\mp}(q) = \int_0^{W_\eta} d\eta\mathrm{e}^{\mp\mathrm{i}q\eta}\phi^*_p(\eta)\phi_q(\eta). \tag{5.112}$$

Once again, Parceval's theorem dictates that

$$\int_{-\infty}^{\infty}|I_{pq,\mp}(q)|^2\mathrm{d}q = \frac{2\pi}{W_\eta^2}\int_0^{W_\eta}|\phi_{pq}(\eta)|^2\mathrm{d}\eta = G_{pq}, \tag{5.113}$$

where $\phi_{pq}(\eta) = \phi_p^*(\eta)\phi_q(\eta)$. Note that G_{pq} is a form factor that has dimensions of inverse length.

The Kronecker delta in (5.111) implies momentum conservation in the x-direction only. There is no momentum conservation associated with motion in the y- or z-direction since the electron's motion is constrained in those two directions. Because of the Heisenberg Uncertainty Principle, the uncertainty in the momentum in the y- and z-directions precludes conservation of y- or z-component of the momentum.

5.5.3.1 Nonpolar Longitudinal Acoustic Phonon Absorption

The total scattering rate for an electron in the (m, n)-th subband having a wavevector k in the x-direction, transitioning over to the (m', n')-th subband by absorbing a longitudinal acoustic phonon is given by Fermi's Golden Rule:

$$\left.\frac{1}{\tau^{1D}(\omega_{\text{ph}}, k;\ m, n;\ m', n')}\right|_{\text{NPLAPA}} = \sum_{k', q_y, q_z} \left[\frac{\pi D^2 k_B T}{\hbar \rho v_1^2 \Omega} \left| I_{m,m'}^{\mp}(q_z) \right|^2 \left| I_{n,n'}^{\mp}(q_y) \right|^2 \times \delta_{k', k \mp q_x} \delta(E_{\vec{k}'_{\|}, n} - E_{\vec{k}_{\|}, m} - \hbar\omega_{\text{ph}}) \right], \tag{5.114}$$

where we have made the usual assumption that the phonon energy is much smaller than the thermal energy $k_B T$ and the phonons are in equilibrium. We then convert the summation over final state wavevector k' into an integral over the x-component of the phonon wavevector which will result in

$$\begin{aligned}
\left.\frac{1}{\tau^{1D}(\omega_{\text{ph}}, k;\ m, n;\ m', n')}\right|_{\text{NPLAPA}} &= \frac{\pi D^2 k_B T}{\hbar \rho v_1^2 \Omega} \iiint \mathrm{d}q_x \mathrm{d}q_y \mathrm{d}q_z \Big[D_{\text{ph}}^{3-D}(q) \\
&\quad \times \left| I_{m,m'}^{\mp}(q_z) \right|^2 \left| I_{n,n'}^{\mp}(q_y) \right|^2 \delta_{k', k \mp q_x} \delta\left(E_{k', m', n'} - E_{k, m, n} - \hbar v_1 \sqrt{q_x^2 + q_y^2 + q_z^2} \right) \Big] \\
&= \frac{D^2 k_B T}{8\pi^2 \hbar \rho v_1^2} \int_{-\infty}^{\infty} \int_{-\infty}^{\infty} \int_{-\infty}^{\infty} \mathrm{d}q_x \mathrm{d}q_y \mathrm{d}q_z \left| I_{m,m'}^{\mp}(q_z) \right|^2 \left| I_{n,n'}^{\mp}(q_y) \right|^2 \\
&\quad \times \delta\left(\hbar v_1 \sqrt{q_x^2 + q_y^2 + q_z^2} + \epsilon_{m,n} - \epsilon_{m',n'} - \frac{\hbar^2}{m^*} k q_x - \frac{\hbar^2 q_x^2}{2m^*} \right).
\end{aligned} \tag{5.115}$$

Note that once again the phonons are treated as bulk (3-D) phonons since we are integrating over q_x, q_y, and q_z while using the three dimensional phonon density of states. However, we use only one-half of the phonon density of states since transverse phonons do not interact with electrons in the Γ valley.

The triple integration in the last equation must generally be performed numerically since no analytical result is possible. For a wide range of electron energies $q_x \ll \sqrt{q_y^2 + q_z^2}$ [11], but unfortunately that does not help in obtaining an analytical result. We can obtain an analytical solution in the opposite case when $q_x \gg \sqrt{q_y^2 + q_z^2}$. In that case,

$$\left.\frac{1}{\tau^{1D}(\omega_{\text{ph}}, k;\ m, n;\ m', n')}\right|_{\text{NPLAPA}}^{q_x \gg \sqrt{q_y^2+q_z^2}}$$
$$\approx \frac{D^2 k_B T}{8\pi^2 \hbar \rho v_1^2} \int_{-\infty}^{\infty} dq_y \left| I_{n,n'}^{\mp}(q_y) \right|^2 \int_{-\infty}^{\infty} \left| I_{m,m'}^{\mp}(q_z) \right|^2 dq_z$$
$$\times \int_{-\infty}^{\infty} dq_x \delta\left(\hbar v_1 |q_x| + \epsilon_{m,n} - \epsilon_{m',n'} - \frac{\hbar^2}{m^*} k q_x - \frac{\hbar^2 q_x^2}{2m^*} \right). \quad (5.116)$$

Now, $|q_x| = -q_x$ in the interval $[-\infty, 0]$, and $|q_x| = q_x$ in the interval $[0, \infty]$. Therefore,

$$\left.\frac{1}{\tau^{1D}(\omega_{\text{ph}}, k;\ m, n;\ m', n')}\right|_{\text{NPLAPA}}^{q_x \gg \sqrt{q_y^2+q_z^2}}$$
$$= \frac{D^2 k_B T}{8\pi^2 \hbar \rho v_1^2} G_{mm'} G_{nn'} \int_{-\infty}^{0} dq_x \frac{2m^*}{\hbar^2} \delta\left(q_x^2 + 2(m^*/\hbar)(v_e + v_1) q_x - k_{m,n:m',n'}^2 \right)$$
$$+ \frac{D^2 k_B T}{8\pi^2 \hbar \rho v_1^2} G_{mm'} G_{nn'} \int_{0}^{\infty} dq_x \frac{2m^*}{\hbar^2} \delta\left(q_x^2 + 2(m^*/\hbar)(v_e - v_1) q_x - k_{m,n:m',n'}^2 \right), \quad (5.117)$$

where $v_e = \frac{\hbar k_x}{m^*}$ is the electron velocity in the x-direction (assuming a parabolic band), and $\frac{\hbar^2 k_{m,n:m',n'}^2}{2m^*} = \epsilon_{m,n} - \epsilon_{m',n'}$.

Using (5.104), we can write the last equation as

$$\left.\frac{1}{\tau^{1D}(\omega_{\text{ph}}, k;\ m, n;\ m', n')}\right|_{\text{NPLAPA}} = \frac{D^2 m^* k_B T}{4\pi^2 \hbar^3 \rho v_1^2} G_{mm'} G_{nn'}$$
$$\times \left[\int_{-\infty}^{0} dq_x \sum_i \frac{\delta(q_x - q_x^i)}{2q_x^i + 2\frac{m^*}{\hbar}(v_e + v_1)} + \int_{0}^{\infty} dq_x \sum_i \frac{\delta(q_x - q_x^i)}{2q_x^i + 2\frac{m^*}{\hbar}(v_e - v_1)} \right] \quad (5.118)$$

where q_x^i is the i-th root of $q_x^2 + 2(m^*/\hbar)(v_e \pm v_1) q_x - k_{m,n:m',n'}^2$.

The final expression then becomes

$$\left.\frac{1}{\tau^{1D}(\omega_{\mathrm{ph}}, k;\ m, n;\ m', n')}\right|_{\mathrm{NPLAPA}} = \frac{D^2 m^* k_{\mathrm{B}} T}{4\pi^2 \hbar^3 \rho v_1^2} G_{mm'} G_{nn'}$$
$$\times \left[\frac{1}{2\sqrt{\left(\frac{m^*}{\hbar}\right)^2 (v_{\mathrm{e}} + v_1)^2 + k^2_{m,n:m',n'}}} + \frac{1}{2\sqrt{\left(\frac{m^*}{\hbar}\right)^2 (v_{\mathrm{e}} - v_1)^2 + k^2_{m,n:m',n'}}} \right]. \tag{5.119}$$

Note that the scattering rate is proportional to the joint electron–phonon density of states.

We leave it to the readers to work out similar expressions for the longitudinal nonpolar acoustic phonon emission rates.

5.6 Advanced Topics

5.6.1 *Phonon Confinement Effects*

So far in this chapter we had not considered the possibility that phonons, which are quanta of lattice vibrations and propagate like waves, can be confined within a small volume of space in a nanostructure much like electrons and holes. As always, confinement happens as a result of change in the material properties at boundaries. In the case of electron confinement, potential barriers at the boundaries of the structure, created by an abrupt change in material composition, reflect an electron wave back and forth and set up standing electron waves. Similarly, in the case of phonons, mechanical and electrical barriers at the boundaries of a structure can reflect acoustic and optical phonons (polar or nonpolar) back and forth and set up standing phonon waves. Just as electron confinement discretizes the electron's energy and therefore alters the energy dispersion relations of electrons, phonon confinement can alter the dispersion relation ω_{ph} versus q of phonons [10]. This will have an effect on the electron–phonon scattering rates since they usually depend on the joint electron–phonon density of states and the phonon density of states will be altered if the phonon dispersion relation changes. Additionally, the matrix element for scattering depends on the particle displacement vector $\vec{u}$, as we have seen earlier. When acoustic phonons are confined, they no longer behave as plane waves but behave as guided waves instead [4]. Hence, the expression for $\vec{u}$ given by (5.34) is no longer valid and different expressions hold. For example, consider a quantum wire of rectangular cross-section as shown in Fig. 5.6. Assume that the x-axis is along the length of the wire, the y-axis is along the width and the z-axis is along the thickness. The confined acoustic phonon modes in such a wire can

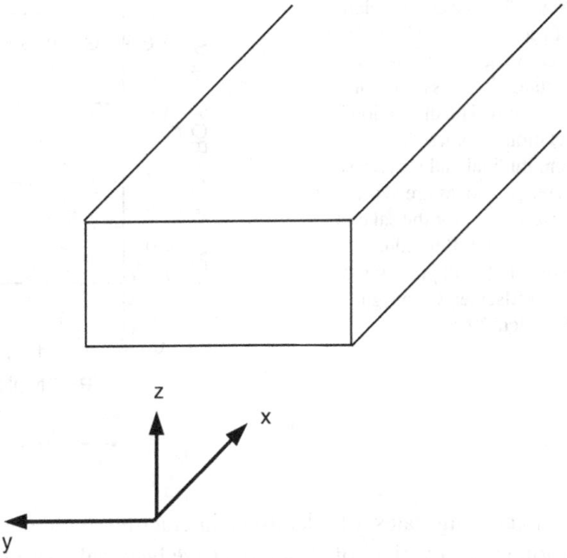

Fig. 5.6 A quantum wire with rectangular cross-section

be *approximately* decomposed into so-called "width-modes" and "thickness modes" [6] corresponding to phonon confinement along the two transverse directions. If we assume that the wire's axis is along the x-axis, then the expressions for the particle displacements along the x-, y-, and z-coordinates in the width modes are [5, 6, 17]:

$$\begin{aligned} u_x^w(x,y,z,t) &= A_w\alpha_w(q_x,z)\cos(hy)e^{iq_x(x-v_l(q_x,\omega_{ph})t)} \\ u_y^w(x,y,z,t) &= A_w\beta_w(q_x,z)\sin(hy)e^{iq_x(x-v_l(q_x,\omega_{ph})t)} \\ u_z^w(x,y,z,t) &= A_w\gamma_w(q_x,z)\cos(hy)e^{iq_x(x-v_l(q_x,\omega_{ph})t)}, \end{aligned} \tag{5.120}$$

and in the thickness modes, they are

$$\begin{aligned} u_x^t(x,y,z,t) &= A_t\alpha_t(q_x,y)\sin(\kappa z)e^{iq_x(x-v_l(q_x,\omega_{ph})t)} \\ u_y^t(x,y,z,t) &= A_t\beta_t(q_x,y)\cos(\kappa z)e^{iq_x(x-v_l(q_x,\omega_{ph})t)} \\ u_z^t(x,y,z,t) &= A_t\gamma_t(q_x,y)\cos(\kappa z)e^{iq_x(x-v_l(q_x,\omega_{ph})t)}, \end{aligned} \tag{5.121}$$

where q_x is the phonon's wavevector component in the x-direction, $v_l(q_x, \omega_{ph})$ is the longitudinal velocity of sound in a phonon branch, which depends on the phonon's wavevector and energy (or frequency), A_t and A_w are constants, $\alpha_t, \beta_t, \gamma_t, \alpha_w, \beta_w, \gamma_w$ are variables that depend on the phonon wavevector q_x, κ is another constant, and $h = (n + 1/2)\pi/W$, where W is the width of the wire and n is an integer. These expressions should be contrasted with (5.34) for unconfined acoustic phonon modes. Since the particle displacement enters the expression for the matrix element, it is obvious that the scattering rates will be modified if the phonons are confined (Fig. 5.7).

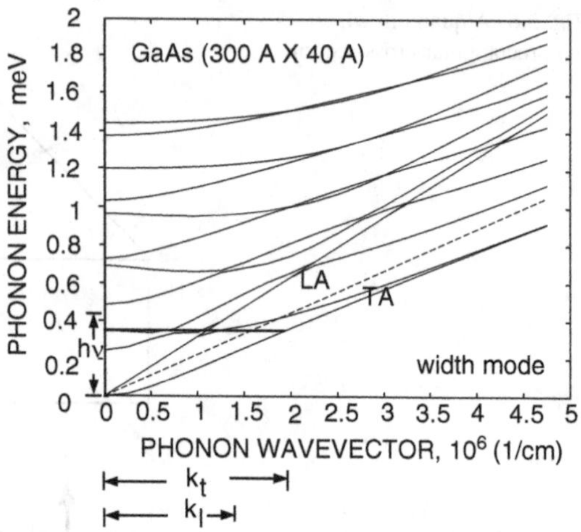

Fig. 5.7 Dispersion relations of acoustic "width" modes in a GaAs quantum wire of rectangular cross-section 3×40 nm. The dispersion relations of unconfined longitudinal and transverse acoustic modes are shown by *broken lines* for the sake of comparison. Reproduced from [10] with permission from Elsevier. Copyright Elsevier, 2000

Scattering rates of electrons interacting with confined acoustic and optical phonons in a variety of structures have been calculated by a number of authors, for example, [6, 12–17]. These calculations are fairly complex and the interested reader is referred to the cited work for the details. Here, we provide the expression for the scattering rate due to confined acoustic phonons in a quantum wire of rectangular cross-section derived (in the presence of an external magnetic field) in [6]:

$$\left.\frac{1}{\tau^{1D}(k;\ m,n)}\right|_{\text{confined acoustic phonons}}$$
$$= \frac{1}{\hbar}\sum_{i,j} D^{\mp}\left(E(k;\ m,n), E(k;\ m,n) \mp \omega_{\text{ph}}\left(q^{\mp}_{k,m,n,i,j}\right), q^{\mp}_{k,m,n,i,j}\right)$$
$$\times \left|M\left(E(k;\ m,n), E(k;\ m,n) \mp \omega_{\text{ph}}\left(q^{\mp}_{k,m,n,i,j}, i\right), q^{\mp}_{k,m,n,i,j}\right)\right|^2$$
$$\times \left[N\left(\omega_{\text{ph}}\left(q^{\mp}_{k,m,n,i,j}, i\right)\right) \mp 1/2 + 1/2\right], \tag{5.122}$$

where the upper sign is for emission and the lower for absorption, $\omega_{\text{ph}}(q,i)$ is the energy of an acoustic phonon with wavevector q in the i-th phonon branch, the quantity $q^{\mp}_{k,m,n,i,j}$ is the j-th solution for q in the equation $E(k;\ m,n) - E(k';\ m',n') \mp \hbar\omega_{\text{ph}}(q,i) = 0$ (we compute the solutions for all possible final state energies $E(k';\ m',n')$; they form a finite set as long as we view k' as a discrete variable separated by π/a with a being the lattice constant), and the matrix element $M\left(E(k;\ m,n), E(k;\ m,n) \mp \omega_{\text{ph}}(q^{\mp}_{k,m,n,i,j}, i), q^{\mp}_{k,m,n,i,j}\right)$ is different for thickness and width modes. The expression for the matrix elements are given in [6]. The quantity

$$D^{\mp}\left(E_1, E_2, q^{\mp}_{k,m,n,i,j}\right) = \frac{4}{|\partial(E_1, E_2 \mp \hbar\omega_{\text{ph}}(q,i)/\partial q|}\delta_{q,q^{\mp}_{k,m,n,i,j}}, \tag{5.123}$$

where the delta is a Kronecker delta, can be viewed as an effective one-dimensional joint electron–phonon density of states per unit length.

Not only is the electron-*confined*-phonon scattering rate very different from the electron-*unconfined*-phonon scattering rate, their energy dependences are also very different. The energy-averaged scattering rate is much higher when acoustic phonons are confined than when they are unconfined [5,6]. There may even be some experimental evidence of this fact [18]. Therefore, phonon confinement has a very dramatic effect on the electron–phonon scattering rates. Study of the interaction of confined electrons with confined phonons is very much an active area of research and the reader is encouraged to scour the literature for new and emerging results in this field.

5.6.2 Phonon Bottleneck Effects

While phonon confinement generally tends to increase the electron–phonon scattering rate (at least in the case of acoustic phonons), electron confinement may elicit the opposite effect. Consider, for example, a quantum dot where the electron energy is completely discretized since the electron motion is restricted in all three dimensions. Here, phonon emission or absorption can take place only between two discrete electron energy levels which must satisfy the energy conservation relation

$$E_{m,n,p} - E_{m',n',p'} = \pm\hbar\omega_{\text{ph}}, \tag{5.124}$$

where m, n, p are the subband indices of the initial state and m', n', p' of the final state, with the plus sign for emission and the minus for absorption. In a small enough quantum dot, the energy difference between two electron energy states (in the left-hand side of the above equation) will be large enough that the quantity on the right-hand side will have to be the energy of an optical phonon, since acoustic phonons have much smaller energies. Because of the "flat" dispersion relations of optical phonons shown in Fig. 4.18, their energy is essentially a material constant. Therefore, transitions between only very specific energy levels will be allowed in a quantum dot since (5.124) must be satisfied. The electron energy levels in a quantum dot are determined by the material, geometry, and size of the quantum dots. If the dot shape is rectangular parallelepiped, then the energy states are

$$E_{m,n,p} = \frac{\hbar^2}{2m^*}\left[\left(\frac{m\pi}{W_x}\right)^2 + \left(\frac{n\pi}{W_y}\right)^2 + \left(\frac{p\pi}{W_z}\right)^2\right], \tag{5.125}$$

where W_x, W_y, and W_z are the edge dimensions of the dot. If the effective mass of electrons in the material and the edge dimensions of the quantum dot are such that (5.124) cannot be satisfied by any pair of energy levels, optical phonon emission and absorption are blocked, leading to a severe suppression of electron-optical phonon interaction. The electron-optical phonon scattering rate then plummets. This is called the *phonon bottleneck effect*, first pointed out in [19], and requires electron confinement, but no phonon confinement. Here, the emission and absorption of optical phonons are blocked, but since acoustic phonons can have any energy (up to the Debye energy in a crystal) if unconfined, it is possible to have acoustic phonon emission and absorption between the levels, albeit these may be multi-phonon processes. In many semiconductors like GaAs, polar optical phonon emission rate far outweighs the acoustic phonon emission rate, so that the phonon bottleneck effect can dramatically suppress the overall emission rate.

Acoustic phonon confinement can exacerbate the bottleneck. For example, if these phonons are confined, then their dispersion relations will be such that phonons of arbitrary energies are no longer necessarily available. In that case, emission and absorption of acoustic phonons may be suppressed as well.

The phonon bottleneck effect has tangible outcomes. One outcome that was discussed in [19] is that it can decrease the luminescence efficiency of quantum dot light emitters. Another outcome, recently discussed, is that it can increase the dephasing time of an electron spin in a nanostructure when the latter is placed in a magnetic field [20]. Thus, the phonon bottleneck effect is an important phenomenon peculiar to nanostructures, and can have dramatic consequences.

5.7 Summary

1. Fermi's Golden Rule is a convenient recipe to derive the scattering rate $S(\vec{k},\vec{k}')$ of an electron that transitions from a wavevector state $\vec{k}$ to a wavevector state $\vec{k}'$ as a result of interaction with a scatterer. It relates the scattering rate to the squared matrix element which depends on the interaction Hamiltonian or the scattering potential. Since this "rule" is derived from first order time-dependent perturbation theory, it is applicable only if the scattering potential (or the interaction Hamiltonian) is relatively *weak*. Fermi's Golden Rule should not be applied for strong scattering potentials.
2. Fermi's Golden Rule is derived on the assumption that the electron interacts with a scatterer for a long time since we let the time of interaction $t \to \infty$. However, in the Boltzmann picture of transport discussed in Chap. 2, we assume that scattering events are instantaneous in time. Thus, there is an unavoidable dichotomy when Fermi's Golden Rule is applied to evaluate the scattering rate $S(\vec{k},\vec{k}')$ in the BTE.
3. Time-invariant scatterers can only cause elastic scattering. Inelastic scattering needs a time-varying scatterer (e.g., a phonon, which is a quantum of lattice vibration in time).
4. The total scattering rate $1/\tau(\vec{k})$ is found by summing $S(\vec{k},\vec{k}')$ over all final states, i.e., $1/\tau(\vec{k}) = \sum_{\vec{k}'} S(\vec{k},\vec{k}')$.

5. Energy and momentum are conserved in the scattering process, which sets restrictions on scattering (for example, a subsonic electron cannot emit a nonpolar acoustic phonon, or an electron with energy less than a longitudinal polar optical phonon cannot emit such a phonon—see Problem 5.3—etc.).
6. The total scattering rates $1/\tau(\vec{k})$ for an electron interacting with a given type of scatterer are very different in three-, two-, and one-dimensional systems.
7. Phonon confinement can dramatically affect the phonon dispersion relations and therefore the electron–phonon scattering rates.

Problems

Problem 5.1. Show that in the case of nonpolar acoustic phonon absorption, the restriction on the maximum value of the phonon wavevector comes about because an electron cannot scatter by an angle greater than 180°.

Solution. The maximum angle by which an electron can scatter, i.e., the maximum value of Φ, is obviously 180°. This corresponds to direct backscattering. Momentum conservation will then mandate that

$$q = k + k'.$$

Energy conservation mandates that

$$\frac{\hbar^2 k'^2}{2m^*} = \frac{\hbar^2 k^2}{2m^*} + \hbar v_l q.$$

Using the first relation in the second, we get

$$\frac{\hbar^2 (q-k)^2}{2m^*} = \frac{\hbar^2 k^2}{2m^*} + \hbar v_l q,$$

which immediately yields

$$\frac{\hbar q}{2m^*} = v_l + v_e.$$

or

$$q = 2k\left(\frac{v_l}{v_e} + 1\right),$$

where we have used the fact that $v_e = \hbar k/m^*$.

Problem 5.2. Consider a quantum wire where only the lowest electron subband is occupied by electrons and only the lowest confined acoustic phonon branch is occupied by phonons because the temperature is low. Assume that the electron's dispersion relation is parabolic and that of the phonon is linear. Show that the electron's wavevector states before and after scattering are symmetric about the wavevector $\pm\zeta = \pm m^* v_{ph}/\hbar$, where m^* is the electron's effective mass, and v_{ph} is the phonon velocity.

Solution. Energy conservation dictates

$$\frac{\hbar^2 k'^2}{2m^*} = \frac{\hbar^2 k^2}{2m^*} \pm \hbar v_{ph} q,$$

while momentum conservation dictates

$$k' = k \pm q,$$

where the upper sign is for absorption and the lower for emission. Here, the primed k refers to final state wavevector and the unprimed k to initial state wavevector. The phonon wavevector is q. From the last two equations, we obtain

$$\frac{k + k'}{2} = \pm\zeta = \pm\frac{m^* v_{ph}}{\hbar}.$$

Therefore, the electron wavevector states before (k) and after (k') scattering are *symmetric* about $\pm\zeta$. This situation is depicted in Fig. 5.8. The wavevector space can be broken up into three segments between $-\infty$ and $-\zeta$ (Region I), $-\zeta$ and 3ζ (Region II) and 3ζ to ∞ (Region III). Note that as a consequence of symmetric scattering, electrons can be scattered into Region III only from Region I and *never* from Region II. In other words, Region III is coupled to Region I, but not to Region II, by scattering.

Assume now that the equilibrium electron distribution was a Boltzmann or Fermi–Dirac distribution centered symmetrically about $k = 0$ and whose high velocity tail did not extend past $k = 3\zeta$. In that case, after an electric field is applied to break equilibrium, Region III can be populated only by electrons that were originally in Region I and scattered into Region III by the phonons they absorbed or emitted. If we apply a high electric field which depletes the electron population in Region I (which are traveling against the electric field), then it will also inevitably deplete the high velocity population in Region III, since that latter Region draws its electrons exclusively from Region I. Region III electrons are the largest contributors to the ensemble average velocity (or drift velocity) because they all have very large positive wavevectors and velocities. Therefore, if we increase the electric field too much, then instead of increasing, the drift velocity may actually decrease, resulting in a *negative differential mobility*. See [21] for further details.

Problem 5.3. The energy dispersion curve of optical phonons is relatively "flat" (see Fig. 4.18) so that we can write the dispersion relation as

$$\hbar\omega_{ph} = \hbar\omega_0,$$

where ω_0 is a constant.

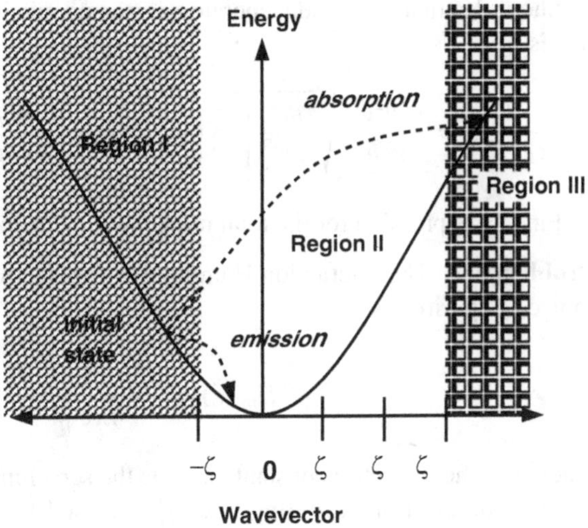

Fig. 5.8 Allowed scattering events which are symmetric about the wavevector $\pm\zeta$

The squared matrix element for electron interaction with longitudinal polar optical phonon is

$$\left|M^{\pm}_{\vec{k}_2,\vec{k}_1}\right|^2 = [N(\omega_0) + 1/2 \pm 1/2]\,\frac{e^2\hbar^2}{4m^*\Omega}\left(\frac{1}{\epsilon_\infty} - \frac{1}{\epsilon_0}\right)\frac{\hbar\omega_0}{E_\mathrm{k}},$$

where e is the electron charge, m^* is the electron effective mass, ϵ_∞ and ϵ_0 are the high- and low-frequency dielectric constants, Ω is the normalizing volume, and E_k is the energy of an electron with wavevector k (assume parabolic dispersion relation). The upper sign is for emission and the lower for absorption. This kind of interaction is often referred to as Frölich interaction.

Show using Fermi's Golden Rule that the total scattering rate (absorption + stimulated emission + spontaneous emission) in bulk material is given by

$$\frac{1}{\tau(E_\mathrm{k})} = \frac{e^2\hbar\omega_0}{4\pi\hbar^2}\sqrt{\frac{m^*}{2E_\mathrm{k}}}\left[N(\omega_0)\ln\left|\frac{\alpha+1}{\alpha-1}\right| + (N(\omega_0)+1)\ln\left|\frac{1+\beta}{1-\beta}\right|\right],$$

where

$$\alpha = \sqrt{1 + \hbar\omega_0/E_\mathrm{k}}$$
$$\beta = \sqrt{1 - \hbar\omega_0/E_\mathrm{k}}$$

Show that an electron cannot emit a longitudinal polar optical phonon unless its kinetic energy exceeds that of the phonon.

Show also that in a nondegenerate electron population, the total energy relaxation rate is

$$\frac{1}{\tau_E(E_k)} = \frac{e^2\hbar^2\omega_0^2}{4\pi\hbar^2}\sqrt{\frac{m^*}{2E_k^3}}\left[(N(\omega_0)+1)\ln\left|\frac{1+\beta}{1-\beta}\right| - N(\omega_0)\ln\left|\frac{\alpha+1}{\alpha-1}\right|\right],$$

Find an expression for the total momentum relaxation rate.

Problem 5.4. The interaction Hamiltonian for electron scattering due to screened ionized impurities is

$$H_{int}(\vec{r}) = \frac{e^2\exp[-\vec{\lambda}\cdot\vec{r}]}{4\pi\epsilon\left|\vec{r}\right|},$$

where ϵ is the dielectric constant and $\vec{\lambda}$ is the screening vector.

Find the electron scattering rate $\frac{1}{\tau(E_k)}$ caused by ionized impurities in a bulk crystal as a function of electron energy E_k. Treat the electrons as free particles. Find also the energy and momentum relaxation rates. Hint: The Fourier transform of the interaction Hamiltonian at wavevector $\left|\vec{k}-\vec{k}'\right|$ is $\frac{e^2}{4\pi\epsilon\left(|\vec{k}-\vec{k}'|^2+\lambda^2\right)}$.

Show that if the impurities are unscreened, i.e., $\vec{\lambda} \to 0$, then the scattering rate diverges. This is of course unphysical and exposes the limitation of Fermi's Golden Rule, or, equivalently, the Born approximation. The same problem arises in calculating the electron-acoustic phonon scattering rate in a quantum wire. Since this rate is proportional to the electron density of states, it diverges at electron energies corresponding to subband bottoms, where van-Hove singularities occur and the density of states diverges. Once again, this is a limitation of Fermi's Golden Rule.

Problem 5.5. Just by inspection of the rate $S(\vec{k},\vec{k}')$, determine if the associated scattering mechanism is elastic, isotropic, or both. Explain your answers.

$S(\vec{k},\vec{k}') \propto \frac{1}{\left|\vec{k}-\vec{k}'\right|}\delta(E_k - E_{k'})$.

$S(\vec{k},\vec{k}') \propto \Gamma\delta(E_k - E_{k'})$. Γ is a constant.

$S(\vec{k},\vec{k}') \propto \frac{1}{\left|\vec{k}-\vec{k}'\right|^2}\delta(E_k - E_{k'} + \hbar\omega)$.

References

1. E. Fermi, *Nuclear Physics*, (University of Chicago Press, Chicago, 1950).
2. P. A. M. Dirac, "The quantum theory of emission and abosrption of radiation", Proc. Royal Soc. (London) A, **114**, 243 (1927).
3. B. K. Ridley, *Quantum Processes in Semiconductors*, (Clarendon Press, Oxford, 1982).

4. N. Nishiguchi, "Resonant acoustic phonon modes in a quantum wire", Phys. Rev. B, **52**, 5279 (1995).
5. SeGi Yu, K. W. Kim, M. A. Stroscio, G. J. Iafrate and A. Ballato, "Electron-acoustic-phonon scattering rates in rectangular quantum wires", Phys. Rev. B., **50**, 1733 (1994).
6. A. Svizhenko, A. Balandin, S. Bandyopadhyay and M. A. Stroscio, "Electron interaction with confined acoustic phonons in quantum wires subjected to a magnetic field", Phys. Rev. B, **57**, 4687 (1998). See Also the erratum in Phys. Rev. B., **58**, 10065 (1998).
7. S. Datta, *Surface Acoustic Wave Devices* (Prentice Hall, Englewood Cliffs, 1986).
8. P. J. Price, "Transport in semiconductor layers", Ann. Phys., **133**, 217 (1981).
9. B. K. Ridley, "The electron-phonon interaction in quasi two-dimensional semiconductor quantum well structures", J. Phys. C: Solid State Phys., **15**, 5899 (1982).
10. S. Bandyopadhyay, A. Svizhenko and M. A. Stroscio, "Why would anyone want to build a narrow channel (quantum wire) transistor?", Superlat. Microstruct., **27**, 67 (2000).
11. See, for example, R. Mickevicius and V. Mitin, "Acoustic phonon scattering in a rectangular quantum wire", Phys. Rev. B., **48**, 17194 (1993).
12. M. A. Stroscio, "Interaction between longitudinal optical phonon modes of a rectangular quantum wire and charge carriers of a one-dimensional electron gas", Phys. Rev. B., **40**, 6428 (1989).
13. M. A. Stroscio, K. W. Kim, M. A. Littlejohn and H. Chuang, "Polarization eigenvectors of surface optical phonon modes in a rectangular quantum wire", Phys. Rev. B., **42**, 1488 (1990).
14. K. W. Kim, M. A. Stroscio, A. Bhatt, R. Mickevicius and V. V. Mitin, "Electron optical phonon scattering rates in a rectangular semiconductor quantum wire", J. Appl. Phys., **70**, 319 (1991).
15. N. Telang and S. Bandyopadhyay, "Quenching of electron acoustic phonon scattering in a quantum wire by a magnetic field", Appl. Phys. Lett., **62**, 3161 (1993).
16. N. Telang and S. Bandyopadhyay, "Effects of a magnetic field on electron-phonon scattering in a quantum wire", Phys. Rev. B., **48**, 18002 (1993).
17. N. Bannov, V. Aristov, V. V. Mitin and M. A. Stroscio, "Electron relaxation times due to the deformation potential interaction of electrons with confined acosutic phonons in a free-standing quantum well", Phys. Rev. B., **51**, 9930 (1995).
18. A. Varfolomeev, D. Zaretsky, V. Pokalyakin, S. Tereshin, S. Pramanik and S. Bandyopadhyay, "Admittance of CdS nanowires embedded in porous alumina template", Appl. Phys. Lett., **88**, 113114 (2006).
19. H. Benisty, C. M. Sotomayor-Torres and C. Weisbuch, "Intrinsic mechanism for the poor luminescence properties of quantum box systems", Phys. Rev. B., **44**, 10945 (1991).
20. B. Kanchibotla, S. Pramanik, S. Bandyopadhyay and M. Cahay, "Transverse spin relaxation time in organic molecules", Phys. Rev. B., **78**, 193306 (2008).
21. A. Goussev and S. Bandyopadhyay, "Universal negative differential resistance in one-dimensional resistors", Phys. Lett. A, **309**, 240 (2003).

Chapter 6
Optical Properties of Solids

6.1 Interaction of Light with Matter: Absorption and Emission Processes

When light is incident on a solid, it is partially transmitted, partially reflected, and the rest is absorbed. We will focus on the absorption phenomenon since it is caused by electron–photon interaction within the solid and is a fundamental property of light interacting with matter. When absorption takes place, a photon in the incident light is absorbed to excite an electron from a lower energy state in the solid to a higher energy state in accordance with the energy conservation relation

$$E_{\mathrm{final}} - E_{\mathrm{initial}} = \hbar\omega_{\mathrm{l}}, \tag{6.1}$$

where E_{initial} and E_{final} are the initial and final energies of the electron and ω_{l} is the angular frequency of light or the photon that was absorbed.

In most semiconductors, the conduction band and higher bands are more or less devoid of electrons at any reasonable temperature, while the valence band and lower bands are mostly full. Therefore, a photon absorbed by a semiconductor will usually excite electrons from the valence band (or lower bands) to the conduction band (or higher bands) since Pauli Exclusion Principle will demand that the final state be initially unoccupied. In order to cause a transition from a state below the valence band to a state above the conduction band, it will take a very energetic photon (deep ultraviolet or x-ray); therefore, the origination point will usually be the valence band and the destination will be the conduction band. Of course, intraband processes, whereby a photon excites an electron from a low-lying energy state in either the valence or the conduction band to a higher energy state in the same band are also possible, but those are comparatively rare.

S. Bandyopadhyay, *Physics of Nanostructured Solid State Devices*,
DOI 10.1007/978-1-4614-1141-3_6,

In a semiconductor, semimetal or insulator,[1] there is clearly a minimum frequency ω_l of photons that can be absorbed because of (6.1), as long as we are restricting ourselves to transitions between the valence and the conduction band. Since the minimum energy difference between states in the conduction band and in the valence band is the bandgap energy E_g, the minimum value of $E_{final} - E_{initial}$ is E_g. Hence, the minimum frequency of photons that can be absorbed is $\omega_l^{min} = \omega_g = E_g/\hbar$. Intraband processes are not constrained by this requirement and photons of energy less than $\hbar\omega_g$ can be absorbed to excite an electron from a low-lying state to a higher lying state in the conduction band. Such processes are termed *free carrier absorption*, but they are rare in comparison with the valence-to-conduction band processes for two reasons: First, there are many more electrons in the mostly filled valence band than in the mostly empty conduction band so that valence-to-conduction band processes are much more probable than free carrier processes. Second, momentum must be conserved in the absorption process, which leads to the so-called *k-selection rule* that we will discuss next. Because of this rule, the electron's wavevectors in the initial and final states must be approximately equal. Two such states are not usually found in the same band. Hence, free carrier absorption may require violation of the k-selection rule, which makes it improbable. Only in quantum-confined systems (quantum wells, quantum wires, and quantum dots), intraband processes are somewhat strong, but in bulk three-dimensional structures, they are relatively weak. The only exception of course is metals where there is a very high concentration of electrons in the conduction band. That makes free carrier absorption quite strong in metals.

As far as a semiconductor, semimetal, or insulator is concerned, the value of ω_g is very important since it determines whether incident light will be absorbed. Frequencies lower than ω_g will be mostly transmitted and those above ω_g will be mostly absorbed. Therefore, ω_g is called the *absorption edge frequency* and it determines whether a semiconductor, semimetal, or insulator will appear transparent (mostly transmittive) or opaque (mostly absorbing) under illumination with light of a given frequency. A solid may appear transparent to yellow light, but opaque to blue light, if the frequency ω_g is in the green.

Let us now derive an expression for ω_g. In a semiconductor, semimetal, or insulator, (6.1) will yield

$$\begin{aligned}\hbar\omega_g = \hbar\omega_l^{min} &= E_g \texttt{ for bulk semiconductor structures}\\ &= E_g + \Delta_e^{2d} + \Delta_{hh}^{2d} \texttt{ for quasi two-dimensional structures}\\ &= E_g + \Delta_e^{1d} + \Delta_{hh}^{1d} \texttt{ for quasi one-dimensional structures}\\ &= E_g + \Delta_e^{0d} + \Delta_{hh}^{0d} \texttt{ for quasi zero-dimensional structures,}\end{aligned} \tag{6.2}$$

[1]For our purpose here, a semimetal is a semiconductor with a narrow bandgap and an insulator is a semiconductor with a wide bandgap.

where E_g is the band gap in bulk (energy difference between conduction and valence band edges), Δ_e is the energy of the lowest electron subband, and Δ_{hh} is the energy of the highest heavy hole subband in either a quantum well, or quantum wire, or quantum dot. We consider heavy holes rather than light holes since the heavy hole subband in a quantum-confined structure will be higher in energy than the light hole subband owing to the heavier effective mass of heavy holes.

The right-hand-side of the above equation is sometimes called the *effective bandgap energy* E_g^{eff}. It is smaller between the lowest electron subband and highest heavy hole subband than between the lowest electron subband and the highest light hole subband. However, for some polarizations of light, the heavy hole may not couple to the photon, which means that a photon of that polarization will not excite an electron from the heavy hole to the conduction band. In that case, the effective bandgap E_g^{eff} will be the energy difference between the lowest electron subband and the highest light hole subband, as long as the coupling between the photon and the light hole is nonzero.

The effective bandgap, and hence the absorption edge frequency ω_g, is smallest in bulk and progressively increases in quantum wells, quantum wires, and quantum dots. This happens because the confinement energies of the lowest electron subband and the highest heavy-hole (or light-hole) subband obey the following hierarchy:

$$\begin{aligned} \Delta_e^{2d} &< \Delta_e^{1d} < \Delta_e^{0d} \\ \Delta_{hh}^{2d} &< \Delta_{hh}^{1d} < \Delta_{hh}^{0d} \\ \Delta_{lh}^{2d} &< \Delta_{lh}^{1d} < \Delta_{lh}^{0d}. \end{aligned} \tag{6.3}$$

Hence, $\omega_g^{0d} > \omega_g^{1d} > \omega_g^{2d} > \omega_g^{3d}$. This is schematically depicted in Fig. 6.1.

The difference between the absorption edge frequency in quantum-confined structures (quantum well, quantum wire, or quantum dot) and in bulk (three-dimensional structures) is the *blue shift* in the absorption edge arising from quantum confinement of electrons and holes. Normally, this will be a measure of the quantum confinement. However, the situation is immensely complicated by the fact that blue shifts can arise from a host of other effects such as *Moss–Burstein shift* and strain. Moss–Burstein shift occurs in degenerate n-type semiconductors where the Fermi level is above the conduction band edge. At low temperatures, all states in the conduction band below the Fermi level are filled with electrons and a photon cannot excite an electron into these states because of the Pauli Exclusion Principle. Hence, the effective bandgap is now approximately the energy difference between the Fermi level and the valence band edge which is larger than the energy difference between the conduction and valence band edges, or the true bandgap. This will result in a blue shift of the absorption edge in even bulk structures. Strain, which is always present in heterostructures because of the lattice mismatch between the constituent materials, can introduce either a positive or a negative shift in the bandgap depending on material properties and the sign of the strain (compressive or tensile). If the shift is positive, then it results in a blue shift, while a negative shift

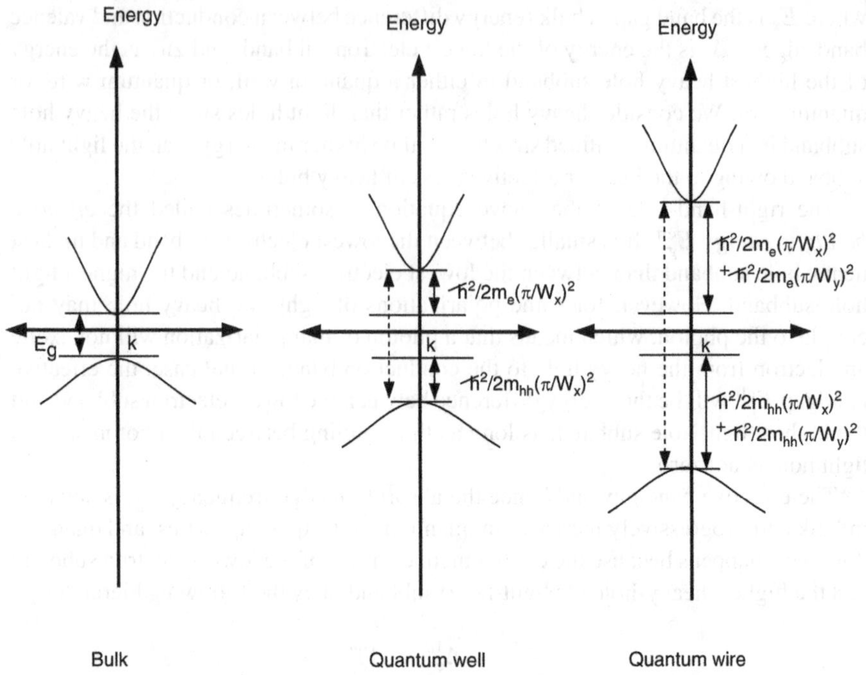

Fig. 6.1 The conduction and valence band dispersion relations in bulk, quantum wells, and quantum wires of a direct-gap semiconductor. The values of ω_g are shown by the *broken vertical arrows*. The quantum well is assumed to have a *thickness* of W_x and a rectangular potential profile with infinite barriers. The quantum wire is assumed to have a rectangular cross-section of dimensions $W_x \times W_y$ and a rectangular potential profile with infinite barriers in both transverse directions. Note that ω_g is largest in the wire and smallest in bulk

will cause a red shift in the absorption edge. Thus, it is difficult, if not impossible, to deduce unambiguously the degree of quantum confinement from experimental measurement of the blue shift in the absorption edge.

6.1.1 The k-selection Rule

In addition to energy conservation, momentum must also be conserved in the absorption process.[2] Therefore, we will require that

$$\vec{k}_{\text{final}} = \vec{k}_{\text{initial}} + \vec{q}_1, \tag{6.4}$$

where $\vec{k}_{\text{initial}}$ is the electron's initial wavevector, $\vec{k}_{\text{final}}$ is the electron's final wavevector, and $\vec{q}_1$ is the photon's wavevector.

[2]Except in a quantum dot, where the expected value of momentum in any direction is exactly zero.

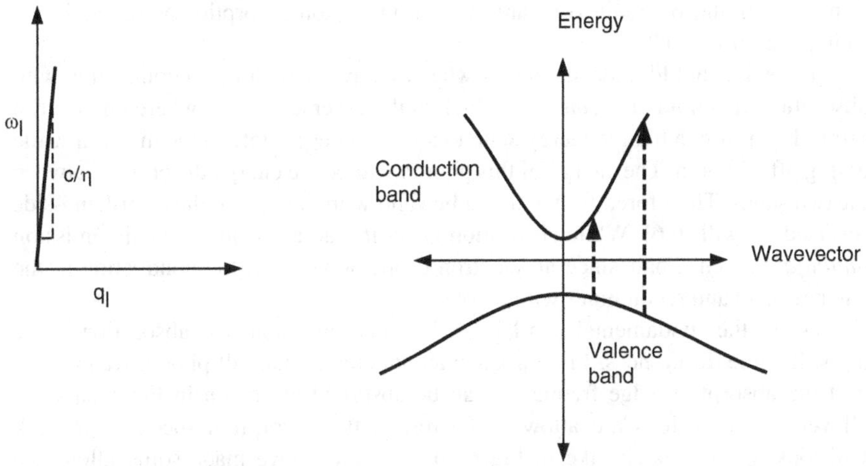

Fig. 6.2 (*Left*) Frequency versus wavevector (dispersion relation) of a photon in a TEM wave, where we have ignored the fact that the refractive index of any medium has a slight dependence on photon frequency. (*Right*) Absorption processes in a direct-gap semiconductor showing that k-selection rule mandates only vertical transitions in the Brillouin zone

If the photon is a free propagating wave, also known as a transverse electromagnetic wave (TEM mode), then the photon's dispersion relation is linear and given by

$$\omega_l = \frac{c}{\eta} q_l, \tag{6.5}$$

where c is the speed of light in vacuum and η is the refractive index of the solid. Normally η will depend on ω_l, which will cause photons of different frequencies to travel with different velocities in a medium. This will cause some deviation from linearity of the dispersion relation ω_l versus q_l, but we will ignore that effect here. The important point to note is that since c is a large quantity (3×10^8 m/s), the slope of the ω_l versus q_l is very large, as shown in the left panel of Fig. 6.2. Therefore, at reasonable values of ω_l, $q_l \ll k$, where k is the magnitude of the electron wavevector in the initial or final state. As an example, consider ultraviolet light which has a wavelength $\lambda_{UV} \sim 100$ nm, so that the corresponding value of $q_l = 2\pi/\lambda_{UV} = 6.28 \times 10^7\,\text{m}^{-1}$. On the other hand, the thermal DeBroglie wavevector of electrons (i.e., the DeBroglie wavevector of an electron with the thermal energy $k_B T$) at room temperature in the conduction band of a semiconductor like GaAs is $\sim 2 \times 10^8\,\text{m}^{-1}$. In this case, even at ultraviolet light frequencies, the photon wavevector is less than one-third the electron wavevector. Accordingly, the inequality $q_l \ll k$ generally holds, so that (6.4) becomes

$$\vec{k}_{\text{final}} \approx \vec{k}_{\text{initial}}. \tag{6.6}$$

This is known as the *k-selection rule* and it stipulates that initial and final state wavevectors of the electron must be approximately equal. That means only *vertical* transitions in the Brillouin zone are allowed in photon absorption as shown in the right panel of Fig. 6.2.

The reader should understand that what we have said so far in connection with absorption also holds for *emission* which is the reverse process where an electron drops down from a higher energy state to a lower energy state in the material while giving off a photon. The energy of this photon will be the energy difference between the two states. Therefore, (6.1) will still be valid with the sign of the right-hand-side reversed, as will (6.6). While absorption *generates* an electron–hole pair, emission *annihilates* such a pair since an electron drops down from the conduction to the valence band and recombines with a hole.

Despite the fundamental similarities between emission and absorption, there are still some differences. In a bulk semiconductor crystal, all photon frequencies past the absorption edge frequency can be absorbed as shown in Fig. 6.3a since all vertical transitions are allowed. Therefore, the absorption spectrum in bulk will look approximately like in Fig. 6.3b, where we have made some allowance for free carrier absorption below the absorption edge frequency. The absorption strength increases with frequency past the absorption edge because more and more conduction and valence band states become available with increasing photon frequency (remember that the density of states of electrons or holes is proportional to the square root of energy in bulk crystals). Hence, the absorption strength should increase proportionately with $\sqrt{\hbar(\omega_l - \omega_g)}$ if $\hbar\omega_l > E_g$. Consequently, the absorption spectrum will not have a "peak," but instead increase monotonically past the absorption edge.

This will change completely in a quantum well or quantum wire since in a two- or one-dimensional system, the density of states in either the conduction or the valence band does not increase with increasing energy. In a two-dimensional system, the density of states is independent of energy, while in a one-dimensional system, it decreases with increasing energy. Hence, the absorption spectrum in a strictly two-dimensional system should approximately saturate at photon energies above the effective bandgap, while in a strictly one-dimensional system it should peak at a photon energy equal to the effective bandgap, and then decrease with increasing photon energy. Thus, the shape of the absorption spectrum, more than anything else, is a good indicator of whether a structure is behaving as bulk (3-d), quantum well (2-d), or quantum wire (1-d).

Emission, on the other hand, behaves differently. Once an electron is excited to any arbitrary state in the conduction band, it almost immediately starts emitting phonons and drops down to the bottom of the conduction band *before* it can fall down into the valence band and recombine with a hole there. This happens because the phonon emission processes are much faster than the photon emission processes (the mean time between phonon emissions is of the order of 0.1 ps at room temperature in GaAs, whereas the mean time between photon emission is ~1 ns). Therefore, when the electron does ultimately make an interband transition and recombines with a hole to give off a photon, it had already fallen down to the bottom of the conduction band and the k-selection rule guarantees that it will recombine with a hole at the top of the valence band, as long as the bottom of the

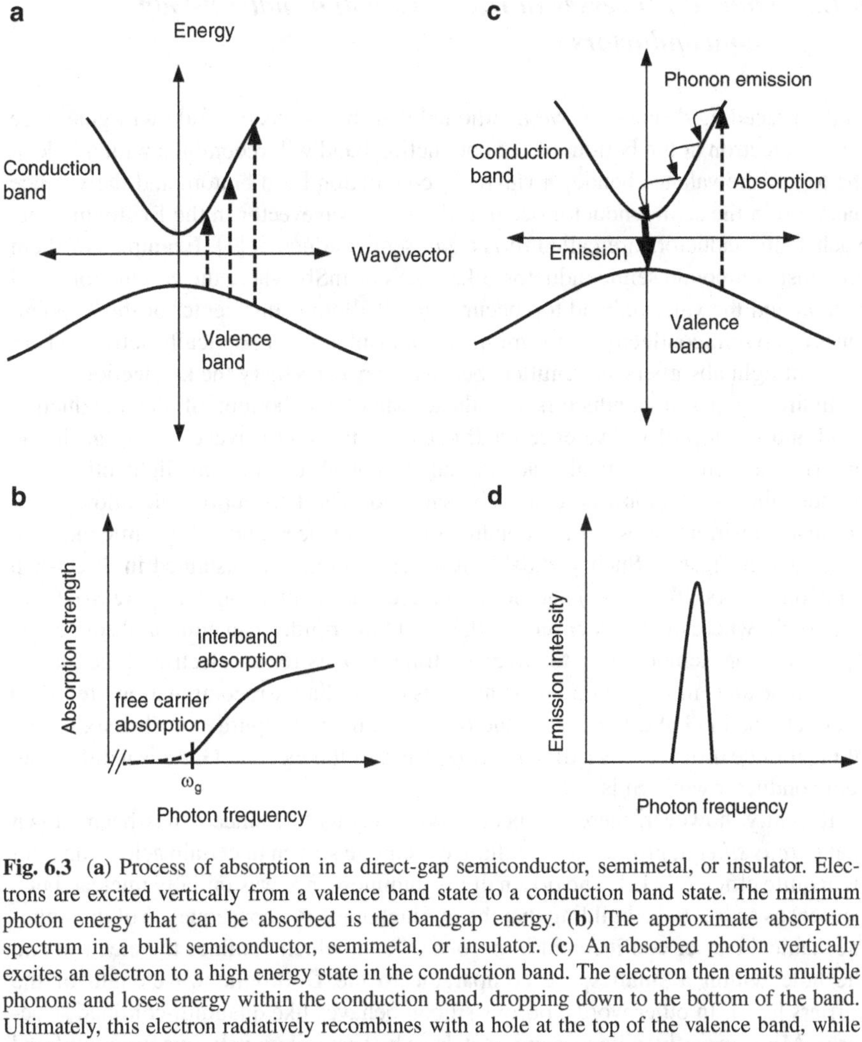

Fig. 6.3 (**a**) Process of absorption in a direct-gap semiconductor, semimetal, or insulator. Electrons are excited vertically from a valence band state to a conduction band state. The minimum photon energy that can be absorbed is the bandgap energy. (**b**) The approximate absorption spectrum in a bulk semiconductor, semimetal, or insulator. (**c**) An absorbed photon vertically excites an electron to a high energy state in the conduction band. The electron then emits multiple phonons and loses energy within the conduction band, dropping down to the bottom of the band. Ultimately, this electron radiatively recombines with a hole at the top of the valence band, while emitting a photon. The emitted photon therefore always has the bandgap energy. The point to note is that the phonon emission process is much faster than the photon emission process, which is why the emitted photon always has energy close to the bandgap energy. (**d**) The emission spectrum is always peaked around the absorption edge frequency ω_g

conduction band and the top of the valence band occur at the same wavevector in the Brillouin zone. This is shown in Fig. 6.3c. Consequently, the energy of the photon emitted is *always* around the bandgap energy. In other words, the emission spectrum, unlike the absorption spectrum, is *peaked* around ω_g, regardless of whether the system is three-, two-, one-, or zero-dimensional. This is shown in Fig. 6.3d.

6.1.2 Optical Processes in Direct-Gap and Indirect-Gap Semiconductors

In the preceding discussion, we mentioned that the k-selection rule will guarantee that an electron at the bottom of the conduction band will recombine with a hole at the top of the valence band, *provided* the conduction band bottom and the valence band top in the semiconductor occur at the same wavevector in the Brillouin zone. Such semiconductors are called *direct-gap* semiconductors [1]. Examples of them are most compound semiconductors like GaAs or InSb, where the conduction band bottom and the valence band top occur at the Brillouin zone center or the Γ point. Direct-gap semiconductors (or semimetals or insulators) are optically active and are efficient light absorbers and emitters because they can satisfy the k-selection rule.

Indirect-gap semiconductors are those where the bottom of the conduction band and the top of the valence band occur at different wavevectors (e.g., in Ge or Si). They are not optically active and do not absorb or emit light efficiently because the k-selection rule cannot be satisfied. The k-selection rule allows only *vertical* transitions between the conduction and valence bands of a semiconductor as shown in Fig. 6.4. Such vertical transitions can only be sustained in direct-gap semiconductors (like GaAs) and not in indirect-gap semiconductors (like Si). Only impure Si, where the k-selection rule does not hold rigidly, can emit or absorb some light. At room temperature, the average time it takes for an electron–hole pair to recombine and emit a photon (also known as the "radiative recombination lifetime") is about 1 ns in GaAs, but it is of the order of 1 ms in (impure) Si. This six orders of magnitude difference is a direct consequence of the fact that GaAs is a direct-gap semiconductor while Si is not.

Recently, however, there has been a twist to this tale since it has been shown that *porous silicon*, produced by etching crystalline silicon in certain acids, can emit light quite efficiently [2]. The origin of this luminescence was initially unclear, but a consensus seems to be building that the luminescence is a consequence of quantum-mechanical carrier confinement in porous silicon, which confines the electron and the hole within a small space comparable to the DeBroglie wavelength of the carriers [3, 4]. In other words, porous silicon behaves like quantum-confined structures. More recently, silicon nanocrystals, which are obviously quantum confined structures, have exhibited optical gain whereby the amount of light emitted by the material exceeds the amount of light absorbed [5]. This is a prerequisite for laser action, as we shall see later in this chapter. A silicon laser was finally demonstrated in 2004 [6]. Not only silicon, but germanium microcrystals emit light as well, even though germanium is an indirect-gap semiconductor [7]. In the case of germanium, it is possible that micro- or nanocrystals undergo an indirect-to-direct transition since the energy difference between the direct bandgap and indirect bandgap is rather small. This is not likely to happen in silicon where the difference is much larger. Thus, the origin of the luminescence in porous silicon is quite intriguing.

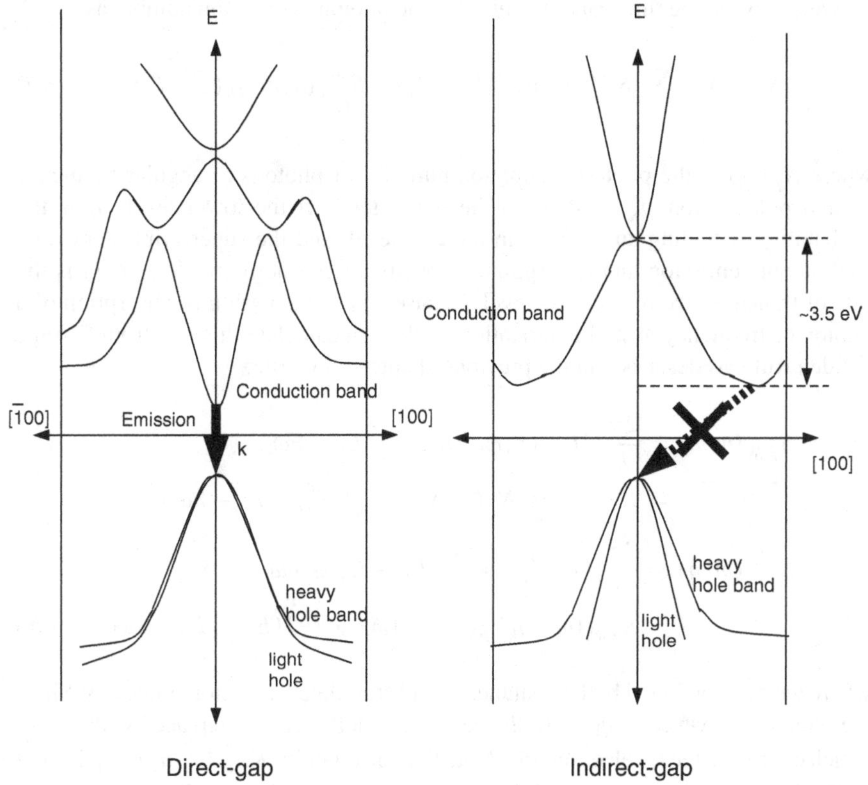

Fig. 6.4 Bandstructures of two hypothetical semiconductors plotted along the [100] crystallographic direction. The one on the *left* is a direct-gap semiconductor while the one on the *right* is indirect-gap. Transitions between the *bottom* of the conduction band and the *top* of the valence band are "vertical" in the former case, but not in the latter

6.2 Rate Equation in Two-Level Systems

Let us consider a solid where light is interacting with matter, i.e., photons are being absorbed and emitted as electrons are transitioning between two energy states. Such a system is designated as a *two-level system*. We will derive a rate equation in a *two-level* system, where absorption involves excitation of electrons from the lower energy level to the upper, and emission is the reverse process of decay from the upper to the lower. This equation will tell us the rate at which the photon occupation number (of a given frequency and polarization) is changing with time because of both absorption and emission taking place.

We can write the time-rate of change of the photon occupation number as:

$$\frac{\mathrm{d}}{\mathrm{d}t}N_{\hat{p}}(\omega_1) = \sum_{m,n} S^{\mathrm{em}}_{\hat{p},\omega_1}(m,n) f_m(1-f_n) - S^{\mathrm{ab}}_{\hat{p},\omega_1}(n,m) f_n(1-f_m), \tag{6.7}$$

where $N_{\hat{p}}(\omega_1)$ is the photon occupation number for photons of angular frequency ω_1 and polarization $\hat{p}$,[3] m denotes the upper level, n the lower level, f_i is the probability of finding an electron in the i-th level, and the superscripts "em" and "ab" denote emission and absorption, respectively. Furthermore, $S_{\hat{p},\omega_1}(i,j)$ is the rate of transition from level i to level j caused by the emission or absorption of a photon of frequency ω_1 and polarization $\hat{p}$. We can calculate this rate from Fermi's Golden Rule, as described in the previous chapter, and write:

$$\begin{aligned}
S^{\mathrm{ab}}_{\hat{p},\omega_1}(n,m) &= \frac{2\pi}{\hbar}|M_{n,m}(\hat{p},\omega_1)|^2 \delta(E_m - E_n - \hbar\omega_1) \\
&= K_{\hat{p},\omega_1}(n,m) N_{\hat{p}}(\omega_1) \delta_{\vec{k}_m,\vec{k}_n+\vec{q}_1} \delta(E_m - E_n - \hbar\omega_1) \\
S^{\mathrm{em}}_{\hat{p},\omega_1}(m,n) &= \frac{2\pi}{\hbar}|M_{m,n}(\hat{p},\omega_1)|^2 \delta(E_n - E_m + \hbar\omega_1) \\
&= K_{\hat{p},\omega_1}(m,n)[N_{\hat{p}}(\omega_1)+1] \delta_{\vec{k}_n,\vec{k}_m-\vec{q}_1} \delta(E_m - E_n - \hbar\omega_1),
\end{aligned} \tag{6.8}$$

where we have included both spontaneous and stimulated emission processes. Since the photon's wavevector $\vec{q}_1 \to 0$, the Kronecker deltas can be replaced with $\delta_{\vec{k}_n,\vec{k}_m}$, which expresses the k-selection rule. Note that the quantity $K_{\hat{p},\omega_1}(m,n)$ is equivalent to $(2\pi/\hbar)|V^{+}_{\omega,\vec{q},\hat{p}}|^2$ in Chap. 5, while $K_{\hat{p},\omega_1}(n,m)$ is equivalent to $(2\pi/\hbar)|V^{-}_{\omega,\vec{q},\hat{p}}|^2$.

It is obvious from the material in the previous chapter that the constant $K_{\hat{p},\omega_1}(i,j)$ will depend on the electron–photon interaction Hamiltonian (or the nature of the coupling between electrons and photons) as all matrix elements do. This quantity depends on the energy levels involved, but does not depend on which energy level is higher and which is lower, so that it is obvious that $K_{\hat{p},\omega_1}(m,n) = K_{\hat{p},\omega_1}(n,m)$.[4] Note that in Chap. 5 too, it always turned out that $|V^{+}_{\omega,\vec{q},\hat{p}}|^2 = |V^{-}_{\omega,\vec{q},\hat{p}}|^2$. We will calculate the quantity $K_{\hat{p},\omega_1}(m,n)$ from the interaction Hamiltonian in due time, but before we do that, we can derive certain important results from what we have already inferred so far.

[3] Since we are considering a single photon mode, the frequency and wavevector of the photon are uniquely related. Hence, there is no need to label the occupation number with *both* frequency and wavevector, unlike what we did in the case of phonons in the last chapter. If we consider multiple modes, then frequency alone is not enough to designate the mode since different modes can have different wavevectors for the same frequency. Since we considered multiple modes for phonons, it was necessary to label the phonon occupation number with both frequency and wavevector. That is not the case here with photons.

[4] Note that this makes $\frac{S^{\mathrm{em}}_{\hat{p},\omega_1}(m,n)}{S^{\mathrm{ab}}_{\hat{p},\omega_1}(n,m)} = \frac{N_{\hat{p}}(\omega_1)+1}{N_{\hat{p}}(\omega_1)}$.

Using (6.8) in (6.7), we immediately get that

$$\frac{\mathrm{d}}{\mathrm{d}t}N_{\hat{p}}(\omega_1) = \sum_{m,n} K_{\hat{p},\omega_1}(m,n)\delta_{\vec{k}_n,\vec{k}_m}\delta(E_m - E_n - \hbar\omega_1) \times \left\{\left[N_{\hat{p}}(\omega_1) + 1\right] f_m(1 - f_n) - N_{\hat{p}}(\omega_1) f_n(1 - f_m)\right\}. \tag{6.9}$$

At equilibrium, $N_{\hat{p}}(\omega_1)$ will be given by the Bose–Einstein factor and the electron occupation probability f_i will be given by the Fermi–Dirac factor. Therefore, (see (5.36)), the right-hand side of the preceding equation (or, more specifically, the term within the curly brackets) will vanish at equilibrium and the photon occupation number will reach steady state and not change with time. Recall Einstein's thermodynamic argument that we discussed in the previous chapter. The same argument (which we invoked for phonons in the previous chapter) applies here for photons as well. At equilibrium, there is no *net* emission or absorption to change the photon occupation number with time. If the system deviates from equilibrium, then photons will be emitted or absorbed to restore equilibrium distribution of both photons and electrons.

Equation (6.9) can be rewritten as [8]

$$\frac{\mathrm{d}}{\mathrm{d}t}N_{\hat{p}}(\omega_1) = -\frac{N_{\hat{p}}(\omega_1)}{\tau_{\hat{p}}(\omega_1)} + l_{\hat{p}}(\omega_1), \tag{6.10}$$

where

$$\frac{1}{\tau_{\hat{p}}(\omega_1)} = \sum_{m,n} K_{\hat{p},\omega_1}(m,n)\delta_{\vec{k}_n,\vec{k}_m}\delta(E_m - E_n - \hbar\omega_1)(f_n - f_m)$$

$$l_{\hat{p}}(\omega_1) = \sum_{m,n} K_{\hat{p},\omega_1}(m,n)\delta_{\vec{k}_n,\vec{k}_m}\delta(E_m - E_n - \hbar\omega_1) f_m(1 - f_n). \tag{6.11}$$

Here, the first term is related to the difference between absorption and stimulated emission, while the second term is clearly related to spontaneous emission since it is independent of the photon occupation number.

Consider now a solid on which light is incident. The energy density $U_{\hat{p}}(\omega_1)$ due to light inside the solid is the number of photons per unit volume multiplied by the energy of each photon. The light intensity $I_{\hat{p}}(\omega_1)$ is the light flux or the light energy crossing a unit area per unit time. Hence, we get

$$U_{\hat{p}}(\omega_1) = \frac{N_{\hat{p}}(\omega_1)\hbar\omega_1}{\Omega}$$

$$I_{\hat{p}}(\omega_1) = \frac{c}{\eta_{\hat{p}}(\omega_1)}U_{\hat{p}}(\omega_1)$$

$$L_{\hat{p}}(\omega_1) = \frac{l_{\hat{p}}(\omega_1)\hbar\omega_1}{\Omega}, \tag{6.12}$$

where Ω is the normalizing volume, and $L_{\hat{p}}(\omega_1)$ is called the luminescence. The luminescence is due to spontaneous emission.

The total derivative in (6.10) can be expanded in partial derivatives, so that we get

$$\frac{d}{dt}N_{\hat{p}}(\omega_1) = \frac{\partial}{\partial z}N_{\hat{p}}(\omega_1)\frac{dz}{dt} + \frac{\partial}{\partial t}N_{\hat{p}}(\omega_1) = \frac{\partial N_{\hat{p}}(\omega_1)}{\partial z}(c/\eta_{\hat{p}}(\omega_1)) + \frac{\partial N_{\hat{p}}(\omega_1)}{\partial t}$$
$$= -\frac{N_{\hat{p}}(\omega_1)}{\tau_{\hat{p}}(\omega_1)} + l_{\hat{p}}(\omega_1), \qquad (6.13)$$

where we have assumed that the direction of light propagation is the z-direction, so that the speed of propagation $\frac{dz}{dt} = (c/\eta_{\hat{p}}(\omega_1))$.

Multiplying the last equation throughout with $\hbar\omega_1/\Omega$, we get

$$\frac{\partial I_{\hat{p}}(\omega_1)}{\partial z} + \frac{\partial U_{\hat{p}}(\omega_1)}{\partial t} = L_{\hat{p}}(\omega_1) - \frac{U_{\hat{p}}(\omega_1)}{\tau_{\hat{p}}(\omega_1)}. \qquad (6.14)$$

The reader should compare the last equation with the general charge continuity equation (2.33) derived in Chap. 2, which can be rewritten in one-dimension as

$$\frac{\partial J}{\partial z} + \frac{\partial \rho}{\partial t} = G - R, \qquad (6.15)$$

with ρ being the charge density, J the current density, G the generation rate, and R the recombination rate. Obviously, there is a one-to-one correspondence between current density and light intensity, as well as light energy density and charge density. Similarly, there is a correspondence between generation and luminescence due to spontaneous emission, as well as between recombination and the difference between absorption and stimulated emission.

Under steady-state situation, partial derivatives with respect to time will vanish, leaving us with an equation of the form

$$\frac{\partial I_{\hat{p}}(\omega_1)}{\partial z} + \alpha_{\hat{p}}(\omega_1)I_{\hat{p}}(\omega_1) = L_{\hat{p}}(\omega_1), \qquad (6.16)$$

where

$$\alpha_{\hat{p}}(\omega_1) = \frac{\eta_{\hat{p}}(\omega_1)}{c\tau_{\hat{p}}(\omega_1)} = \frac{\eta_{\hat{p}}(\omega_1)}{c}\sum_{m,n} K_{\hat{p},\omega_1}(m,n)\delta_{\vec{k}_n,\vec{k}_m}\delta(E_m - E_n - \hbar\omega_1)(f_n - f_m)$$

$$L_{\hat{p}}(\omega_1) = \sum_{m,n} \frac{K_{\hat{p},\omega_1}(m,n)\hbar\omega_1}{\Omega}\delta_{\vec{k}_n,\vec{k}_m}\delta(E_m - E_n - \hbar\omega_1)f_m(1 - f_n). \qquad (6.17)$$

We used (6.11) and (6.12) to write down the above equation.

The last equation is the defining equation for the *absorption coefficient* $\alpha_{\hat{p}}(\omega_1)$ and *luminescence* $L_{\hat{p}}(\omega_1)$. In order to evaluate the absorption coefficient, all that remains to be done is to find the quantity $K_{\hat{p},\omega_1}(m,n)$ from the interaction Hamiltonian for electron–photon interaction. We will do that later in this chapter.

If we assume that $\alpha_{\hat{p}}(\omega_1)$ and $L_{\hat{p}}(\omega_1)$ are spatially invariant, then we can solve (6.16) to obtain

$$I_{\hat{p}}(\omega_1, z) = \frac{L_{\hat{p}}(\omega_1)}{\alpha_{\hat{p}}(\omega_1)} + \left(I_{\hat{p}}(\omega_1, 0) - \frac{L_{\hat{p}}(\omega_1)}{\alpha_{\hat{p}}(\omega_1)} \right) \mathrm{e}^{-\alpha_{\hat{p}}(\omega_1)z} \approx I_{\hat{p}}(\omega_1, 0)\mathrm{e}^{-\alpha_{\hat{p}}(\omega_1)z}, \tag{6.18}$$

if $\frac{L_{\hat{p}}(\omega_1)}{\alpha_{\hat{p}}(\omega_1)} \ll I_{\hat{p}}(\omega_1, 0)$. This last inequality is satisfied if either the incident intensity or the absorption coefficient is high and the luminescence is low.

The last equation shows how one can experimentally measure the absorption coefficient in a material. Assume that the material is fashioned into a thin film that has a thickness d. If light of sufficient intensity is incident on the film (so as to satisfy the previous inequality), then we can measure the transmitted intensity and evaluate the absorption coefficient as

$$\alpha_{\hat{p}}(\omega_1) = -\frac{1}{d} \ln \left(\frac{I_{\hat{p}}(\omega_1, d)}{I_{\hat{p}}(\omega_1, 0)} \right) = -\frac{1}{\mathrm{d}} \ln \left(\frac{I_{\text{transmitted}}}{I_{\text{incident}}} \right). \tag{6.19}$$

6.2.1 *Gain Medium and Lasing*

An interesting observation to make from (6.17) is that if we can somehow manage to "invert" the electron population and put more electrons up at the higher level than at the lower level, then $f_m > f_n$. This is known as "population inversion". In that case, the absorption coefficient will be *negative*, which means that there will be more light coming out of the medium than what went in, or was incident. In other words, the intensity will continue to grow, rather than diminish, with increasing distance of propagation into the medium, in accordance with (6.18). Such a medium exhibits optical *gain*, or negative absorption.[5] It therefore acts as a light amplifier instead of an attenuator, and this is at the heart of *light amplification by stimulated emission of radiation* (LASER). If the system has an intrinsic optical gain or amplification $A(\omega_1)$ at frequency ω_1, and we introduce a frequency selective feedback with feedback factor $\beta(\omega_1)$, then the closed loop gain at frequency ω_1, which is the light output O divided by the light input I is $A(\omega_1)/\left[1 + A(\omega_1)\beta(\omega_1)\right]$. This is shown in Fig. 6.5. If we now make $A(\omega_1)\beta(\omega_1) = -1$, then the closed loop gain diverges to infinity at the frequency ω_1. This means that the system can output light of frequency ω_1 without any input. This is precisely what an injection laser does; it produces laser light output without any light input.

Students of electrical engineering are very familiar with this concept and know that such a construct is called an "oscillator," which produces an output signal at the

[5]Population inversion is, however, not a necessary condition for optical gain. Gain can happen even without population inversion, but then one normally needs an additional energy source to absorb energy from. Typically this would involve a signal source and a pump source of light, with the latter providing the energy for the signal source to experience optical gain. See, for example, B. R. Mollow, Phys. Rev. A, **5**, 2217 (1972).

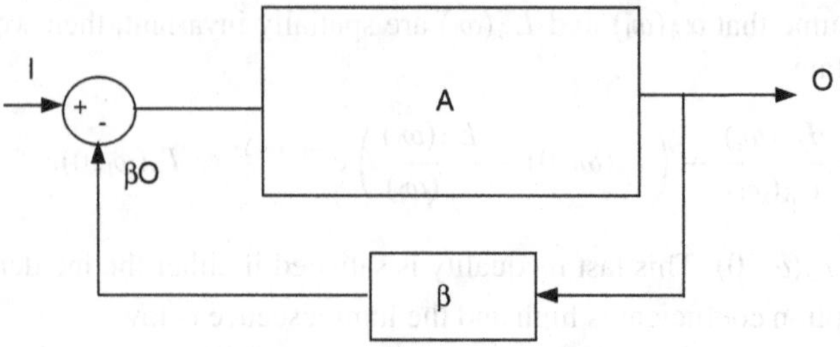

Fig. 6.5 An amplifier with feedback. The amplifier's intrinsic gain (or open-loop gain) is A and the closed loop gain will be $A/(1 + A\beta)$

frequency ω_l without any input. An injection laser is really such an oscillator since it produces light output without light input. Of course, there is energy input in the form of injection current so that energy conservation is not violated. On the other hand, an optically pumped laser will be more like an amplifier with feedback. Perhaps, an injection laser should have been called a LOSER as opposed to a LASER since it is technically an oscillator rather than an amplifier, but since the term "loser" has a negative connotation, it could not stick. We will have occasion to revisit lasers later.

6.3 The Three Important Optical Processes

When it comes to light interaction with matter, there are obviously three optical processes of interest that we need to focus on: absorption, stimulated emission, and spontaneous emission. Starting from (6.7), the rate of change of photon occupation number due to these three processes can be written as

$$\begin{aligned}
\left.\frac{\mathrm{d}}{\mathrm{d}t}N_{\hat{p}}(\omega_l)\right|_{\text{absorption}} &= \sum_{m,n} K_{\hat{p},\omega_l}(m,n)N_{\hat{p}}(\omega_l)\delta(E_m - E_n - \hbar\omega_l) \\
&\quad \times \delta_{\vec{k}_m,\vec{k}_n} f_n(1 - f_m) \\
\left.\frac{\mathrm{d}}{\mathrm{d}t}N_{\hat{p}}(\omega_l)\right|_{\text{stimulated emission}} &= \sum_{m,n} K_{\hat{p},\omega_l}(m,n)N_{\hat{p}}(\omega_l)\delta(E_m - E_n - \hbar\omega_l) \\
&\quad \times \delta_{\vec{k}_m,\vec{k}_n} f_m(1 - f_n) \\
\left.\frac{\mathrm{d}}{\mathrm{d}t}N_{\hat{p}}(\omega_l)\right|_{\text{spontaneous emission}} &= \sum_{m,n} K_{\hat{p},\omega_l}(m,n)\delta(E_m - E_n - \hbar\omega_l) \\
&\quad \times \delta_{\vec{k}_m,\vec{k}_n} f_m(1 - f_n).
\end{aligned} \tag{6.20}$$

Consider now a strictly 2-level system, for which we can remove the summation over m and n. We have seen that when equilibrium prevails, i.e., the electrons obey Fermi–Dirac statistics and the photons obey Bose–Einstein statistics, the rate of change in the photon occupation number due to absorption is exactly equal to the sum of those due to stimulated and spontaneous emission. In other words, the absorption and emission processes balance exactly and there is no net absorption or emission. This is the basis of Einstein's thermodynamic argument which holds at equilibrium.

Let us now consider deviations from equilibrium.

Case I. Assume that the electrons are still in equilibrium, but strong light input from the outside has put the photon system out of equilibrium and made $N_{\hat{p}}(\omega_l) \gg 1$. In this case, obviously stimulated emission will dominate over spontaneous emission since the ratio of the two rates is $N_{\hat{p}}(\omega_l)$, which is, by assumption, much larger than unity. But will stimulated emission also dominate over absorption? The answer is "no" since the ratio of the two rates is

$$\frac{\text{stimulated emission rate}}{\text{absorption rate}} = \frac{f_m(1-f_n)}{f_n(1-f_m)} = \frac{1/f_n - 1}{1/f_m - 1} = \exp\left[(E_n - E_m)/k_B T\right], \tag{6.21}$$

as long as the electrons are in *equilibrium* and obey Fermi–Dirac statistics. Since $E_m > E_n$, the above ratio is *less* than unity. Therefore, the following inequality will hold when the photons are far out of equilibrium but the electrons are at equilibrium:

$$\text{absorption rate} > \text{stimulated emission rate} > \text{spontaneous emission rate},$$

and the dominant process will be absorption. This actually makes perfect sense. If electrons are in equilibrium and the photons have been driven out of equilibrium by strong light input, absorption must rise above emission to remove the extra photons in an attempt to restore equilibrium.

Case II. Consider now the opposite situation where the electrons are driven out of equilibrium, but the photons are near equilibrium and $\hbar\omega_l \gg k_B T$ so that $N_{\hat{p}}(\omega_l) \ll 1$. In this case, spontaneous emission will obviously dominate over stimulated emission since the ratio of the two is $1/N_{\hat{p}}(\omega_l)$. But will the latter dominate over absorption? Since the electrons are out of equilibrium, we now have

$$\frac{\text{stimulated emission rate}}{\text{absorption rate}} = \exp\left[(E_n - F_n - E_m + F_m)/k_B T\right] \tag{6.22}$$

where F_m and F_n are the quasi-Fermi-levels associated with the energy states m and n.[6] The electron occupation probability for the i-th state is $1/[\exp(E_i - F_i)/$

[6] We are assuming that the concept of quasi Fermi level is still valid when the electron system is far out of equilibrium.

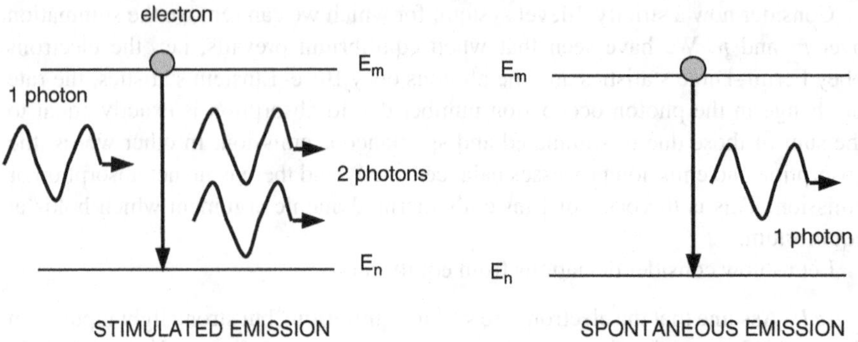

Fig. 6.6 The processes of stimulated and spontaneous emissions in a 2-level system

$k_B T + 1]$. Clearly, if $F_m - F_n > E_m - E_n$, so that $f_m > f_n$, i.e., if population inversion has occurred, then the following hierarchy will hold:

spontaneous emission rate > stimulated emission rate > absorption rate.

Here, the dominant process is spontaneous emission. This, too, makes perfect sense. Since the electron system has been driven out of equilibrium and a population inversion has been created, the extra electrons in the higher level will spontaneously emit photons and drop down to the lower level in an attempt to restore equilibrium. Thus, whichever process acts to restore equilibrium will invariably dominate.

Unfortunately, light that is produced as a result of spontaneous emission is *incoherent*, whereas laser light is expected to be *coherent*. Only stimulated emission produces coherent light. In that process, a photon stimulates an electron in a higher energy state and induces it to drop down to a lower state whose energy is exactly the photon's energy below the excited state. When that happens, a second photon is emitted which follows the first photon exactly in *phase* [9] (see Fig. 6.6). Thus, the two photons are coherent with each other and all subsequent photons that are emitted upon stimulation by these two photons are also coherent with them. In this fashion, a coherent population of photons builds up, which is responsible for the coherence of the laser light. In other words, we need the following hierarchy for a laser:

stimulated emission rate > spontaneous emission rate

stimulated emission rate > absorption rate.

We do not care if the spontaneous emission rate exceeds the absorption rate or not. The first condition requires that the photon population be far out of equilibrium so that $N_{\hat{p}}(\omega_l) > 1$ (recall that the ratio of the stimulated emission rate to spontaneous emission rate is $N_{\hat{p}}(\omega_l)$). The second condition requires that the electron population is also far out of equilibrium, resulting in population inversion, i.e., $f_m > f_n$ so that $F_m - F_n > E_m - E_n$. This last condition also ensures that the absorption coefficient is negative (see (6.17)), or that there is optical *gain*. Thus, a laser is a very nonequilibrium device where *both* electron and photon populations are placed far out of equilibrium to produce coherent light emission. In the end, the two conditions that we need for lasing are:

1. $N_{\hat{p}}(\omega_l) > 1$, so that the photons are far out of equilibrium.
2. $F_m - F_n > E_m - E_n$, so that the electrons are far out of equilibrium, resulting in population inversion and optical gain.

Suppose now that we do not need a laser but instead want a simple luminescent device such as a LED which emits *incoherent* light. In this case, we can allow spontaneous emission to dominate over stimulated emission since the light produced does not have to be coherent. In other words, we can allow $N_{\hat{p}}(\omega_l)$ to be less than unity. That means the photon system can be in equilibrium with $\hbar\omega_l \gg k_B T$. We, however, need the spontaneous emission to dominate over absorption (although stimulated emission need not) since otherwise, no net light will be emitted. The ratio of the two rates is

$$\begin{aligned}\frac{\text{spontaneous emission rate}}{\text{absorption rate}} &= \frac{1}{e^{\frac{\hbar\omega_l}{k_B T}} - 1} e^{(E_n - F_n - E_m + F_m)/k_B T} \\ &\approx \frac{\exp\left[(E_n - F_n - E_m + F_m)/k_B T\right]}{\exp\left[(E_m - E_n)/k_B T\right]} \\ &= \exp\left[(F_m - F_n)/k_B T\right], \end{aligned} \tag{6.23}$$

where we have made use of the fact (see (6.1)) that $E_m - E_n = \hbar\omega_l \gg k_B T$.

For incoherent light to be emitted, we need this ratio to exceed unity. Therefore, all we need for an incoherent emitter (like an LED) is that $F_m - F_n > 0$, whereas for a coherent emitter, we have a stronger condition, namely, $F_m - F_n > E_m - E_n$. In a semiconductor, E_m is the conduction band edge and E_n is the valence band edge, so that $E_m - E_n = E_g^{\text{eff}}$. If we want to make a semiconductor diode laser (which is used ubiquitously, such as in laser pointers), we will need to place the quasi Fermi level for electrons above the conduction band edge and the quasi Fermi level for holes below the valence band edge. This can only happen if both the p-side and the n-side of the p-n junction diode are *degenerately* doped. No laser diode can be made without degenerate doping. Therefore, to summarize, we need:

$$F_e - F_h > E_g^{\text{eff}} \ (\texttt{LASER})$$

$$F_e - F_h > 0 \ (\texttt{LED}),$$

where F_e is the quasi Fermi-level for electrons and F_h is the quasi Fermi-level for holes.

6.4 Van Roosbroeck–Shockley Relation

Consider (6.17) for a 2-level system where we can remove the summation over indices m and n since we are dealing with only two levels. The absorption coefficient can be related to the luminescence according to

$$
\begin{aligned}
\frac{c}{\eta_{\hat{p}}(\omega_l)}\alpha_{\hat{p}}(\omega_l) &= \frac{L_{\hat{p}}(\omega_l)}{\hbar\omega_l/\Omega}\frac{1/f_m - 1/f_n}{1/f_n - 1} \\
&= \frac{L_{\hat{p}}(\omega_l)}{\hbar\omega_l/\Omega}\left[\mathrm{e}^{\frac{(E_m-E_n)-(F_m-F_n)}{k_B T}} - 1\right] \\
&= \frac{L_{\hat{p}}(\omega_l)}{\hbar\omega_l/\Omega}\left[\mathrm{e}^{\frac{\hbar\omega_l-(F_m-F_n)}{k_B T}} - 1\right]. \qquad (6.24)
\end{aligned}
$$

Near equilibrium, $F_m \approx F_n = \mu_F$, so that if we denote quasi-equilibrium quantities with a subscript "0," then

$$
\frac{c}{\eta_{\hat{p}}(\omega_l)}\alpha_{\vec{p},0}(\omega_l) = \frac{L_{\vec{p},0}(\omega_l)}{\hbar\omega_l/\Omega}\left[\mathrm{e}^{\frac{\hbar\omega_l}{k_B T}} - 1\right], \qquad (6.25)
$$

or, conversely,

$$
L_{\vec{p},0}(\omega_l) = \frac{c}{\eta_{\hat{p}}(\omega_l)}\alpha_{\vec{p},0}(\omega_l)\,(\hbar\omega_l/\Omega)\,\frac{1}{\mathrm{e}^{\frac{\hbar\omega_l}{k_B T}} - 1}, \qquad (6.26)
$$

Using (6.12) in (6.24), we get that the rate of spontaneous emission $l_{\hat{p}}(\omega_l)$ of photons of frequency ω_l is given by

$$
l_{\hat{p}}(\omega_l) = \frac{c}{\eta_{\hat{p}}(\omega_l)}\alpha_{\hat{p}}(\omega_l)\frac{1}{\mathrm{e}^{\frac{\hbar\omega_l-(F_m-F_n)}{k_B T}} - 1}. \qquad (6.27)
$$

In order to find the total spontaneous emission rate, we integrate $l_{\hat{p}}(\omega_l)$ over all ω_l after multiplying by the photon density of states per unit volume given by (4.144). This yields

$$
\begin{aligned}
\int \mathrm{d}\omega_l [D_{ph}(\omega_l)/\Omega] l(\omega_l) &= \int \mathrm{d}\omega_l R^{sp}(\omega_l) \\
&= \int \mathrm{d}\omega_l \frac{\eta^2(\omega_l)}{c^2}\frac{\omega_l^2}{\pi^2}\alpha(\omega_l)\frac{1}{\mathrm{e}^{\frac{\hbar\omega_l-(F_m-F_n)}{k_B T}} - 1}, \qquad (6.28)
\end{aligned}
$$

where we have removed the polarization index since we are summing over polarization when we integrate over the photon density of states. In this process, we have defined a new quantity $R^{sp}(\omega_l)$ which is the spontaneous emission rate $l_{\hat{p}}(\omega_l)$ multiplied by the photon density of states per unit volume. This quantity is the rate of spontaneous emission per unit volume per unit photon frequency. It is clearly given by

$$
R^{sp}(\omega_l) = \frac{\eta^2(\omega_l)}{c^2}\frac{\omega_l^2}{\pi^2}\alpha(\omega_l)\frac{1}{\mathrm{e}^{\frac{\hbar\omega_l-(F_m-F_n)}{k_B T}} - 1}. \qquad (6.29)
$$

The last equation is referred to as the *Van Roosbroeck–Shockley relation* [10] and relates the spectrum of the spontaneous emission per unit volume per unit photon frequency per unit time $R^{\rm sp}(\omega_l)$ with the absorption spectrum $\alpha(\omega_l)$. The reader should note that the quantity $R^{\rm sp}(\omega_l)$ has the dimension of inverse volume.

The difference between off-equilibrium and equilibrium rates is

$$\Delta R^{\rm sp}(\omega_l) = R^{\rm sp}(\omega_l) - R_0^{\rm sp}(\omega_l) = R_0^{\rm sp}(\omega_l)\left[\frac{{\rm e}^{\frac{\hbar\omega_l}{k_{\rm B}T}} - 1}{{\rm e}^{\frac{\hbar\omega_l - \Delta F}{k_{\rm B}T}} - 1} - 1\right], \tag{6.30}$$

where $\Delta F = F_m - F_n$.

If the photon energy $\hbar\omega_l$ is much larger than the thermal energy $k_{\rm B}T$ and we are still not too far from equilibrium so that $\Delta F \ll \hbar\omega_l$,[7] then both ${\rm e}^{\frac{\hbar\omega_l - \Delta F}{k_{\rm B}T}}$ and ${\rm e}^{\frac{\hbar\omega_l}{k_{\rm B}T}}$ are far larger than unity. In that case, we get

$$\Delta R^{\rm sp}(\omega_l) \approx R_0^{\rm sp}(\omega_l)\left[{\rm e}^{\frac{\Delta F}{k_{\rm B}T}} - 1\right]. \tag{6.31}$$

In a nondegenerate semiconductor, we can relate ΔF to the electron (n) and hole (p) concentrations using the relations:

$$n = n_i {\rm e}^{(F_{\rm e} - E_i)/k_{\rm B}T}$$
$$p = n_i {\rm e}^{(E_i - F_{\rm h})/k_{\rm B}T}, \tag{6.32}$$

where $F_{\rm e} = F_m$ is the quasi Fermi level for electrons and $F_{\rm h} = F_n$ is the quasi Fermi level for holes. Therefore,

$$np = n_i^2 {\rm e}^{\frac{\Delta F}{k_{\rm B}T}}. \tag{6.33}$$

We can write $n = n_0 + \Delta n$ and $p = p_0 + \Delta p$, where n_0 and p_0 are the equilibrium electron and hole concentrations ($n_0 p_0 = n_i^2$). In any reasonably conducting material, $\Delta n = \Delta p$ for the sake of charge neutrality. Therefore,

$$np = (n_0 + \Delta n)(p_0 + \Delta p) = n_i^2 + \Delta n(n_0 + p_0) + (\Delta n)^2. \tag{6.34}$$

We can neglect the last term since it is much smaller than the others as long as we do not deviate too far from equilibrium. Consequently,

$$\frac{np - n_i^2}{n_i^2} = {\rm e}^{\frac{\Delta F}{k_{\rm B}T}} - 1 = \frac{\Delta n(n_0 + p_0)}{n_i^2}. \tag{6.35}$$

[7] Note that this condition implies an LED and not a LASER.

Substituting this result in (6.31), we get that as long as we are not too far from equilibrium,

$$\Delta R^{sp}(\omega_l) \approx R_0^{sp}(\omega_l)\frac{\Delta n(n_0 + p_0)}{n_i^2}. \tag{6.36}$$

The last equation clearly shows that unless equilibrium is violated and $\Delta n \neq 0$, there is no net spontaneous emission. We have already seen this before.

6.4.1 *The B-Coefficient*

When we are not too far from equilibrium, the integrated luminescence (integrated over all photon frequencies) is given by the expression

$$\int_0^{\infty} \Delta R^{sp}(\omega_l)\mathrm{d}\omega_l = \frac{\int_0^{\infty} R_0^{sp}(\omega_l)\mathrm{d}\omega_l}{n_i^2}\Delta n(n_0 + p_0) = B\Delta n(n_0 + p_0), \tag{6.37}$$

where

$$B = \frac{\int_0^{\infty} R_0^{sp}(\omega_l)d(\hbar\omega_l)}{n_i^2}. \tag{6.38}$$

The quantity B is called the *B-coefficient.* Note that the integrated luminescence $\int_0^{\infty} R_0^{sp}(\omega_l)\mathrm{d}\omega_l$ has the dimensions of inverse volume time. Physically, this quantity is the rate of radiative recombination per unit volume since radiative recombination produce the luminescence. Of course, the luminescence is due to spontaneous emission only, but spontaneous emission events far outnumber the stimulated emission events when the photon system is close to equilibrium. Therefore, we can write the integrated luminescence as $\Delta n/\tau_r$, where the quantity τ_r will be the radiative recombination lifetime, which is the average time it takes for an excited electron in the conduction band to decay spontaneously to the valence band and give off a photon. Therefore, from (6.37), we get

$$\frac{1}{\tau_r} = B(n_0 + p_0). \tag{6.39}$$

6.5 Semiconductor Lasers

We have seen earlier that in order to devise a semiconductor laser, which emits coherent light, we need to ensure that both photon and electron populations are far out of equilibrium, so that $N_{\hat{p}}(\omega_l) \gg 1$ and $F_e - F_h > E_g^{eff}$. The last condition implies population inversion, which results in optical gain. Gain alone, however, produces a light amplifier, but not a light oscillator, which is really what a laser is.

Recall Fig. 6.5 and recognize that in order to obtain an oscillator, we need to have a feedback loop and ensure that $A\beta = -1$. In the case of a semiconductor laser, the feedback loop is implemented with an optical "cavity." The cavity has partially reflecting walls surrounding the region with optical gain where stimulated emission (and the production of coherent photons) takes place. The coherent photons produced by stimulated emission are partially reflected back into the gain region by the cavity walls and stimulate further emission of coherent photons. This has two effects: first, it will eliminate the need for external stimulators, just as in an oscillator where there is no need for an external input to produce an output. Second, this will quickly build up the population of coherent photons within the cavity, so that sooner or later, we will have $N_{\hat{p}}(\omega_l) \gg 1$. Thus, in the end, we need *two* conditions to be fulfilled in a semiconductor laser[8]:

- The existence of optical gain caused by population inversion, which requires that $F_e - F_h > E_g^{eff}$
- The existence of a cavity for feedback, which also ensures that $N_{\hat{p}}(\omega_l) \gg 1$

Students of electrical engineering, with some knowledge of electronic circuitry, will immediately make a connection between this and circuit theory. They will recognize that these two conditions are the analog of Barkenhausen's criteria for making an electronic oscillator, i.e., $A > 1$ and $A\beta = -1$. This reinforces the notion that the laser is an oscillator and not a mere amplifier, contrary to what its name implies.

6.5.1 Threshold Condition for Lasing

In Fig. 6.7, we depict a semiconductor cavity. Let us assume that the walls of the cavity have reflection coefficients of $R_1(\omega_l)$ and $R_2(\omega_l)$ at the lasing frequency ω_l, which is the frequency of monochromatic light emanating from the laser. Let the length of the cavity be L. In a trip around the cavity, the intensity of the light will grow by a factor $R_1(\omega_l)R_2(\omega_l)\mathrm{e}^{[\alpha_g(\omega_l)-\alpha_i(\omega_l)]2L}$, where $\alpha_g(\omega_l)$ is the optical gain or negative absorption due to population inversion and $\alpha_i(\omega_l)$ is the intrinsic loss (positive absorption) in the cavity due to defects, etc. Note that $-\alpha_g(\omega_l)$ will be given by (6.17) with $f_m > f_n$ (population inversion).

In order for the photon population to increase with every round trip, we need

$$R_1(\omega_l)R_2(\omega_l)\mathrm{e}^{[\alpha_g(\omega_l)-\alpha_i(\omega_l)]2L} \geq 1. \tag{6.40}$$

The left-hand-side is called the "cavity gain" and it must exceed unity. This requires that $\alpha_g(\omega_l)$ not just merely exceed 0, but actually obey the inequality

[8] We will understand later that these conditions are sufficient but not necessary.

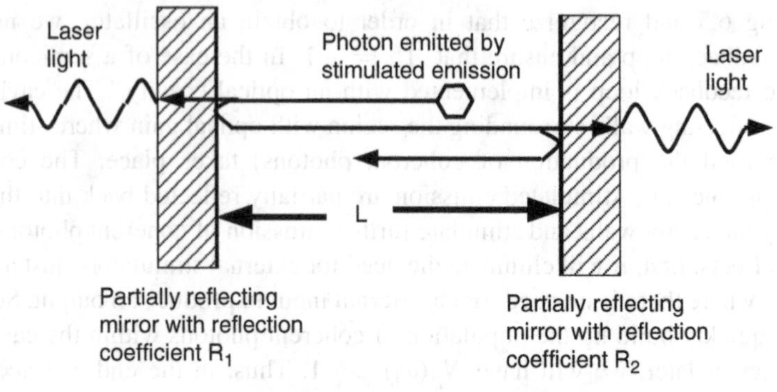

Fig. 6.7 A laser cavity. A photon produced by stimulated emission suffers multiple reflections back and forth between the cavity walls and finally escapes through one of the partially reflecting walls to produce laser light. A coherent beam of laser light will emerge from both walls if they are both partially reflecting. If one is completely reflecting and the other is partially reflecting, then the beam will of course emerge from only the latter

$$\alpha_g(\omega_l) \geq \alpha_i(\omega_l) + \frac{1}{2L}\ln\left(\frac{1}{R_1(\omega_l)R_2(\omega_l)}\right). \tag{6.41}$$

6.5.1.1 Threshold Current Density in a Semiconductor Diode Injection Laser

Semiconductor lasers are mostly of two types: optically pumped and electrically pumped.[9] The latter are also called "injection lasers." In optically pumped lasers, an external light source supplies photons with energy larger than the bandgap energy. These photons are at first absorbed to excite a large number of electrons from the nearly filled valence band to the nearly empty conduction band to cause population inversion. The electrons excited high up in the conduction band emit phonons and decay down to the bottom of the conduction band. From there, they drop down into the valence band to recombine radiatively with a hole when stimulated to do so by the photons in the cavity. In the process, another photon is emitted that follows in phase the stimulating photon. This way a coherent population of photons builds up, resulting in laser light being emitted.

Injection lasers do not have any external photon source exciting electrons from the valence to the conduction band. Instead, the excitation is achieved electrically. Most semiconductor injection lasers are made of p^+-n^+ diodes where both the

[9] There are many nonsemiconductor lasers, such as gas lasers where transitions occur between two atomic energy levels, instead of between the conduction and valence bands in a semiconductor, but we will restrict ourselves to semiconductors in this book.

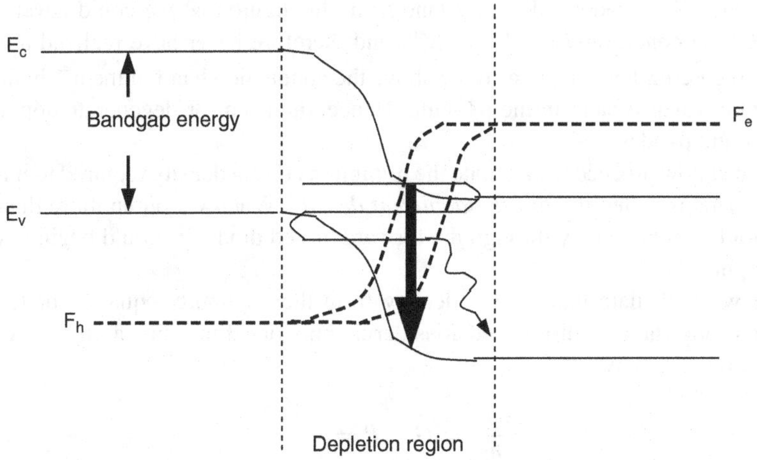

Fig. 6.8 A semiconductor diode injection laser. The band diagram of a p$^+$-n$^+$ semiconductor diode under strong forward bias is shown. The quasi Fermi levels are split by more than the bandgap energy in the depletion region which ensures population inversion and optical gain in this region. This region also has a slightly larger refractive index than the surrounding regions so that it acts as an optical cavity as well. A photon produced by stimulated emission within the cavity suffers multiple reflections between the edges of the depletion region and finally escapes through either the n$^+$-region or p$^+$-region. The latter regions should not reabsorb the photon to produce an electron–hole pair. The quasi Fermi levels have negligible slope in the n$^+$-region or p$^+$-region because these regions have very high carrier concentrations

p- and n-sides are degenerately doped. The diode is forward biased to drive a current through it which directly injects majority carrier electrons from the conduction band in the n-side and majority carrier holes from the valence band in the p-side into the depletion region. There, they recombine radiatively under stimulation by a photon and emit another photon that is in phase with the stimulating photon. This builds up a coherent population of photons within the depletion region. Because the depletion region has very few carriers and the surrounding regions (the quasi-neutral n- and p-bulk regions) have a high density of charge carriers and are much more conductive, the refractive index in the depletion region is slightly larger than in the surrounding regions so that the depletion region confines the photons and acts as an optical cavity as well. The heavily doped p$^+$-side and n$^+$-side are metallic and act as partially reflecting mirrors. Thus, the second condition for lasing is met. The only question then is whether the first condition has been met, i.e., whether there is population inversion and optical gain within the depletion region.

In Fig. 6.8, we show the bandstructure of a p$^+$-n$^+$ diode that is forward biased and carries a large current. Within the depletion region, the Fermi level is spatially varying and splits into quasi Fermi levels for electrons and holes. As long as the splitting exceeds the bandgap energy, the condition $F_e - F_h > E_g^{eff}$ is fulfilled and we have both population inversion and optical gain. That, coupled with the fact that the depletion region also acts as a (leaky) optical cavity, is sufficient to make this device lase. The reader will understand from this figure that we could never have satisfied the condition $F_e - F_h > E_g^{eff}$, and therefore never have realized a laser, unless the Fermi level happens to be above the conduction band in the n^+-bulk and below the valence band in the p^+-bulk. Hence, the need for degenerate doping of both n- and p-sides.

We can now proceed to calculate the minimum current density required to initiate lasing. This is called the *threshold current density*. When we pump more than the threshold current density through the forward biased diode, it should begin to emit laser light.

We will calculate this current density from the continuity equation or (2.33). Under steady state condition, and considering only one dimension along the length of the cavity, we have

$$\frac{\partial J_n}{\partial z} = G - R = \frac{e\Delta n}{\tau}$$
$$\frac{\partial J_p}{\partial z} = G - R = \frac{e\Delta p}{\tau} \tag{6.42}$$

where e is the electronic charge and $1/\tau$ is the total electron–hole recombination rate due to both radiative processes (that produce photons) and nonradiative processes (such as Auger recombination that do not produce photons). Calling $1/\tau_r$ the radiative recombination rate and $1/\tau_{nr}$ the nonradiative recombination rate, we get

$$\frac{1}{\tau} = \frac{1}{\tau_r} + \frac{1}{\tau_{nr}} \tag{6.43}$$

Next, we define a "quantum efficiency" Q as

$$Q = \frac{1/\tau_r}{1/\tau_r + 1/\tau_{nr}} = \frac{\tau_{nr}}{\tau_r + \tau_{nr}} = \frac{1/\tau_r}{1/\tau} = \frac{\tau}{\tau_r}, \tag{6.44}$$

so that

$$\frac{\Delta n}{\tau} = \frac{1}{Q}\frac{\Delta n}{\tau_r}$$
$$\frac{\Delta p}{\tau} = \frac{1}{Q}\frac{\Delta p}{\tau_r} \tag{6.45}$$

Integrating equation (6.42) over the z-coordinate, we get that

$$J = J_n + J_p = \frac{e\Delta n}{\tau}L + \frac{e\Delta p}{\tau}L = \frac{1}{Q}\frac{e\Delta n}{\tau_r}L + \frac{1}{Q}\frac{e\Delta p}{\tau_r}L = \frac{2}{Q}\frac{e\Delta n}{\tau_r}L, \tag{6.46}$$

since in any reasonably conducting material $\Delta n = \Delta p$. We have also assumed that the recombination is taking place only within the cavity which has a length L, and outside the cavity, the recombination rate is zero.

Next, using (6.39), we get

$$J = \frac{2}{Q} e \Delta n B L (n_0 + p_0). \tag{6.47}$$

Then, using (6.37), we obtain

$$J = \frac{2}{Q} e L \int_0^\infty \Delta R^{\mathrm{sp}}(\omega_1) \mathrm{d}\omega_1. \tag{6.48}$$

Since at the onset of lasing, we are reasonably far from equilibrium (although not too far), $\Delta R_{\hat{p}}^{\mathrm{sp}}(\omega_1) = R^{\mathrm{sp}}(\omega_1) - R_0^{\mathrm{sp}}(\omega_1) \approx R^{\mathrm{sp}}(\omega_1)$. Substituting this result in the previous equation, we get

$$J = \frac{2}{Q} e L \int_0^\infty R^{\mathrm{sp}}(\omega_1) d\omega_1. \tag{6.49}$$

It is customary to replace the integral $\int_0^\infty R^{\mathrm{sp}}(\omega_1) d\omega_1$ with the quantity $R_{\mathrm{peak}}^{\mathrm{sp}} \Delta\omega$, where $\Delta\omega$ is the full-width-at-half-maximum (FWHM) of the $R^{\mathrm{sp}}(\omega_1)$ spectrum and $R_{\mathrm{peak}}^{\mathrm{sp}}$ is the peak of this spectrum which occurs at photon energy close to the bandgap.

Thereafter, using (6.29) and assuming that $R^{\mathrm{sp}}(\omega_{\mathrm{g}}) = R_{\mathrm{peak}}^{\mathrm{sp}}$,[10] we get

$$\begin{aligned} J &\approx \frac{2}{Q} e L \frac{\eta_{\hat{p}}^2(\omega_{\mathrm{g}})}{c^2} \alpha_{\hat{p}}(\omega_{\mathrm{g}}) \omega_{\mathrm{g}}^2 \frac{1}{\mathrm{e}^{\frac{\hbar\omega_{\mathrm{g}} - \Delta F}{k_{\mathrm{B}} T}} - 1} \Delta\omega \\ &= \frac{2}{Q} e L \frac{\eta_{\hat{p}}^2(\omega_{\mathrm{g}})}{c^2} \alpha_{\mathrm{g}}(\omega_{\mathrm{g}}) \omega^2 \frac{1}{1 - \mathrm{e}^{\frac{\hbar\omega_{\mathrm{g}} - \Delta F}{k_{\mathrm{B}} T}}} \Delta\omega. \end{aligned} \tag{6.50}$$

where we have made use of the fact that optical gain is the negative of optical absorption so that $\alpha_{\mathrm{g}}(\omega_1) = -\alpha_{\hat{p}}(\omega_1)$.

Finally, using the threshold condition in (6.41), we get the expression for the threshold current density:

$$\begin{aligned} J_{\mathrm{th}} &= \frac{2}{Q} e L \frac{\eta_{\hat{p}}^2(\omega_{\mathrm{g}})}{c^2} \left[\alpha_i(\omega_{\mathrm{g}}) + \frac{1}{2L} \ln \left(\frac{1}{R_1(\omega_{\mathrm{g}}) R_2(\omega_{\mathrm{g}})} \right) \right] \omega_{\mathrm{g}}^2 \frac{\Delta\omega}{1 - \mathrm{e}^{\frac{\hbar\omega_{\mathrm{g}} - \Delta F}{k_{\mathrm{B}} T}}} \\ &= \frac{2}{Q} e L \frac{\eta_{\hat{p}}^2(\omega_1)}{c^2} \left[\alpha_i(\omega_1) + \frac{1}{2L} \ln \left(\frac{1}{R_1(\omega_1) R_2(\omega_1)} \right) \right] \omega_1^2 \frac{\Delta\omega}{1 - \mathrm{e}^{\frac{\hbar\omega_1 - \Delta F}{k_{\mathrm{B}} T}}}, \end{aligned} \tag{6.51}$$

where $\omega_{\mathrm{g}} = \omega_1$ is the lasing frequency.

Note that the threshold current density is guaranteed to be a positive quantity since $\Delta F > E_{\mathrm{g}}^{\mathrm{eff}} = \hbar\omega_{\mathrm{g}}$.

---

[10] This assumption is correct since the spontaneous emission spectrum does peak at ω_{g}.

The last equation is very instructive. First, it tells us that the threshold current density increases with decreasing quantum efficiency Q, so that it is important to ensure that the nonradiative transition rate is much smaller than the radiative transition rate. Otherwise, the quantum efficiency will be small and the threshold current density will be too high. This immediately tells us that it will be extremely hard to make a silicon laser since silicon is an indirect-gap semiconductor and the quantum efficiency will be very poor (the radiative recombination rate is much smaller that the nonradiative recombination rate).[11] Even in direct-gap semiconductors, Q may not be sufficiently large and sophisticated designs may be called for to increase this quantity. Second, the threshold current density increases with cavity length L so that it is beneficial to have a small cavity. However, the cavity length must be at least one-half of the wavelength; otherwise, efficient feedback is not possible. Third, the threshold current density increases with the so-called "linewidth" or FWHM $\Delta\omega$. This quantity increases with temperature, which makes it much more difficult to make a laser at higher temperatures. Before the advent of heterostructure lasers, which we will discuss next, room temperature lasers were rare and semiconductor diodes had to be cooled with cryogens to make them lase with reasonable current densities. Finally, the threshold current increases with the laser frequency ω_g or ω_l. Hence, it is much harder to make a high frequency laser than a low frequency one. Everything else being equal, it will be much harder to make a wide gap semiconductor like GaN lase (in the ultraviolet) than a narrow gap semiconductor like InSb lase (in the infrared).

Equation (6.51) is derived based on the premise that the semiconductor diode needs to produce enough optical gain to meet the condition in (6.41). However, optical gain, or equivalently, population inversion is only one condition for lasing. The other condition is that the photon system remains sufficiently out of equilibrium to make the photon occupation number exceed unity. That condition was never used in deriving the expression for the threshold current, and it was tacitly assumed that the cavity will ensure that $N_{\hat{p}}(\omega_l) \gg 1$. However, the cavity alone cannot ensure this. The rate of photon production *must be high enough* to guarantee that $N_{\hat{p}}(\omega_l) \gg 1$, and this requires that the radiative recombination rate is sufficiently high. Equation (6.51) gives the impression that we can reduce the threshold current arbitrarily by reducing the radiative recombination rate $1/\tau_r$. This is very deceptive and of course not true. If we reduce $1/\tau_r$, we may never meet the condition $N_{\hat{p}}(\omega_l) \gg 1$, and therefore never lase. This is not evident from (6.51), but has important consequences as we shall see later.

6.6 Different Types of Semiconductor Lasers

In this section, we discuss a few popular types of semiconductor lasers that have shown major improvements over the traditional diode laser.

---

[11] Raman lasers have been made out of silicon, but their principle of operation is different.

6.6.1 Double Heterostructure Lasers

In the depletion region of a homostructure semiconductor diode (see Fig. 6.8), there is a strong electric field that drives the negatively charged electron in one direction and the positively charged hole in the opposite direction. Since the two particles run away from each other, the radiative recombination rate decreases. As discussed, this may make it impossible to ensure that $N_{\hat{p}}(\omega_l) \gg 1$. Therefore, in order to make the recombination rate high enough to meet the lasing condition, excess current needs to be pumped (see (6.46) to appreciate how the recombination rate depends on the current and vice versa). This actually increases the threshold current required to initiate lasing.

The solution to this problem is to make a heterostructure diode, where the depletion region (or cavity) occurs in a narrower gap semiconductor (like GaAs) placed between two wider gap semiconductors (like AlGaAs). The energy band diagram of such a heterostructure diode is shown in Fig. 6.9. Note that unless the electric field in the narrow gap semiconductor is extremely high, the electron and the hole remain in close proximity with each other because of the confining potential caused by the conduction and valence band offsets between the two materials. This increases the radiative recombination rate and hence increases the rate of photon production, thereby decreasing the threshold current density.

The double heterostructure laser emits the laser light from the sides as shown in the bottom panel of Fig. 6.9. The wider gap materials do not reabsorb the emitted light since the photon energy is less than the bandgap energy in the wider gap layers. The wider gap layers are known as "cladding layers."

6.6.2 Quantum Well Lasers

A quantum well laser is a double heterostructure laser where the narrow gap material layer is very thin and forms a quantum well. Only the wide gap material on both sides of the quantum well are doped p- and n-type, respectively, while the quantum well layer itself (the narrow gap material) need not be doped at all. This reduces the nonradiative recombination rate in the active region (quantum well) since dopants introduce traps that could increase the nonradiative rate. The result is an increase in the quantum efficiency Q over that of the double heterostructure laser.

6.6.3 Quantum Cascade Lasers

In a quantum cascade laser [11], the electronic transitions responsible for photon emission are not between valence and conduction band states, but instead between quantized subband states within the conduction band. This structure consists of

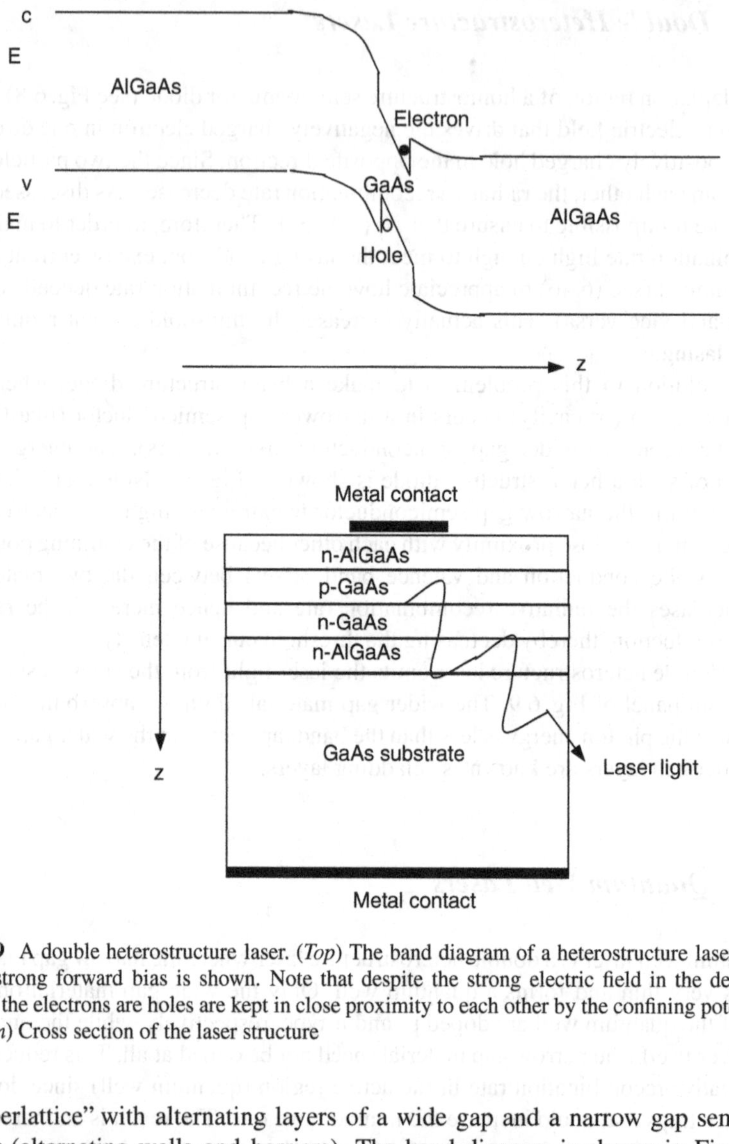

Fig. 6.9 A double heterostructure laser. (*Top*) The band diagram of a heterostructure laser diode under strong forward bias is shown. Note that despite the strong electric field in the depletion region, the electrons are holes are kept in close proximity to each other by the confining potentials. (*Bottom*) Cross section of the laser structure

a "superlattice" with alternating layers of a wide gap and a narrow gap semiconductor (alternating wells and barriers). The band diagram is shown in Fig. 6.10. An electron tunnels into an upper subband within a quantum well and then emits a photon and falls down into a lower subband. This electron then tunnels to the next well and undergoes the same process in a cascading fashion. In this way, one electron can generate *many* photons resulting is extraordinary quantum yield of photons, whereas in all previous constructs, one electron could have at best generated a single photon. The maximum number of photons generated per electron is the number of quantum wells in the superlattice. The essential requirement is that the lower subband bottom energy of one well must be precisely aligned with the

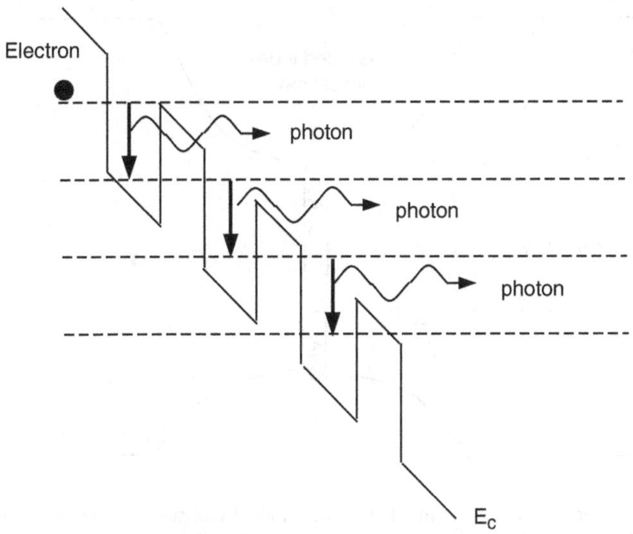

Fig. 6.10 Quantum cascade laser. The conduction band diagram for a superlattice under bias. An electron tunnels into the first well, emits a photon, then resonantly tunnels into the next and emits yet another photon and so one in a cascading fashion. The upper energy level in any well must be precisely aligned with the lower energy level in the preceding well for resonant tunneling through the barrier to occur

higher subband bottom energy in the next one so that the electron can resonantly tunnel from the first well to the second. Such precise alignments of energy levels allow for very little fabrication tolerance and each layer must be built with nearly atomic precision. Another issue of some concern is that the dwell time in a well must exceed the radiative recombination lifetime in that well; otherwise, the electron will exit that well without emitting a photon. If the excited state in the first well is aligned in energy with an excited state in the second well, then the electron can resonantly tunnel out of the first well before it can emit a photon. This will decrease the quantum yield of photons and is best avoided.

6.6.4 Graded Index Separate Confinement Heterostructure Lasers

In quantum well lasers, the quantum well not only confines electron–hole pairs, but serves as the cavity for photons as well. The well typically has higher refractive index than the barrier[12] so that emitted light will concentrate inside the well, making it act as the cavity. The well is also expected to act as a waveguide for the laser light and transport the light to the sides whence it is emitted. Unfortunately, because the well's transverse dimension is much smaller than the wavelength of the light emitted

[12]Narrower gap materials tend to have higher dielectric constant and hence higher refractive index.

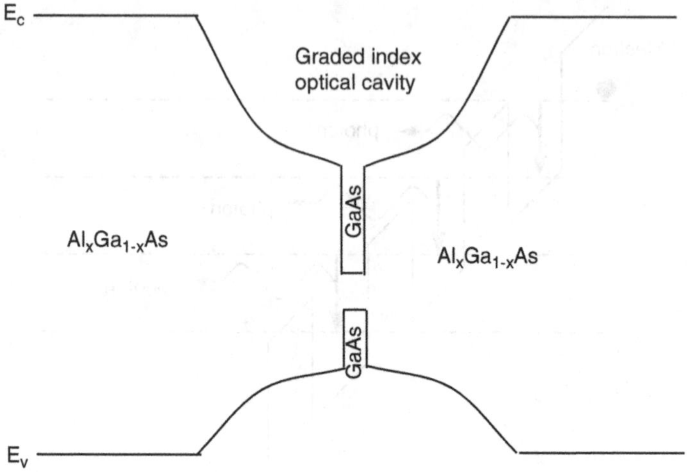

Fig. 6.11 The energy band diagram of a graded index separate confinement heterostructure (GRINSCH) laser. The electrons and holes are confined in the narrower well (GaAs) whereas the photons are confined in the parabolic well ($Al_xGa_{1-x}As$ with varying mole fraction x)

(the well width may be ~10 nm, while the wavelength of light emitted is slightly shorter than 1 μm in GaAs), it acts as a poor waveguide. The cut-off wavelength in the waveguide is shorter than the wavelength of light emitted, which makes the wave evanescent and unable to propagate far enough. Therefore, it is often better to delineate separate confinement regions for electrons/holes and photons, with the latter being much larger (larger than one-half the wavelength of light). One way to achieve this is to employ a heterostructure where the bandgap of the barrier layer is gradually increased away from the interface with the well layer. The band diagram is shown in Fig. 6.11. The well material is GaAs and the cladding layer is $Al_xGa_{1-x}As$, where the mole fraction of Al (the value of x) is gradually increased from about 0.2 to 0.45 away from the interface. The parabolic well confines light while the much narrower rectangular well confines electrons and holes. Since the bandgap of the wider gap material is "graded," the refractive index within the optical cavity is graded as well. Such laser is called a "graded index separate confinement heterostructure" laser, with the acronym GRINSCH.

6.6.5 Distributed Feedback Lasers

In distributed feedback lasers, a diffraction grating is etched closed to the junction of a p-n junction diode. The diffraction grating acts as a frequency selective filter and reflects light of a particular frequency back into the cavity, thereby stimulating the emission of more photons of the same frequency. Thus, such lasers are extremely monochromatic and have a narrow linewidth $\Delta\omega$, which reduces the threshold current.

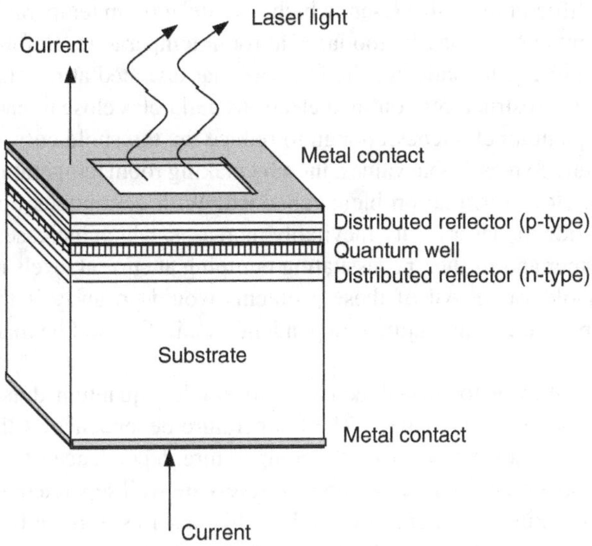

Fig. 6.12 A vertical cavity surface emitting laser structure. The current and light both propagate in the same direction perpendicular to the quantum well plane

6.6.6 *Vertical Cavity Surface Emitting Lasers*

All of the previous lasers that we have discussed so far emit light from the sides. Therefore, the direction of propagation of light is in the plane of the p-n junction and hence perpendicular to the direction of propagation of the current. Sometimes, this is inconvenient and there is a need for a laser which emits light from the top surface as opposed to the sides. Here, both light and current will propagate in the same direction, i.e., perpendicular to the plane of the p-n junction. The structure for such a laser is shown in Fig. 6.12, where the light and current both flow in the vertical direction and the laser emits from the surface.

6.6.7 *Quantum Dot Lasers*

A vexing problem with semiconductor lasers is that the threshold current density increases with temperature, roughly according to the empirical relation

$$\frac{J_{\text{th}}(T)}{J_{\text{th}}(0)} = e^{T/T_0}. \tag{6.52}$$

This makes it difficult to realize lasers which operate at room temperature since the threshold current density may be too large at room temperature. Before the advent of heterostructures, semiconductor diode lasers that operated at room temperature were a rarity. Heterostructures confined electrons and holes close to each other and increased the quantum efficiency enough to reduce the threshold current density at room temperature to reasonable values, thereby making room temperature operation commonplace. However, one problem remained. With continuous operation, the temperature inside the laser cavity inevitably increases, which increases the threshold current further and further, necessitating pumping at current levels much higher than the threshold value. All of these problems would go away if the threshold current could be made temperature independent, i.e., if T_0 could be made infinitely large.

A possible pathway to this objective is to employ quantum dots, instead of quantum wells, for the laser cavity. The temperature dependence of the threshold current density accrues mostly from the temperature dependence of the linewidth $\Delta\omega$. In a quantum dot, as long as the subband levels are well separated in energy, the absorption and luminescence spectra will look like a series of delta functions with ideally zero width, since both the electron and hole states are completely discretized in energy. Hence, emission can take place only at discrete photon energies in accordance with (6.1). Because of the complete discretization of electron and hole states into well-defined energy levels, there are no available energy states *around* the allowed levels that electrons and holes can occupy. Hence, there is no uncertainty in the energy of photon emission (other than the fundamental uncertainty due to the Heisenberg Principle, which we can ignore) to cause any significant broadening of the emission spectrum in a quantum dot. Therefore, $\Delta\omega \to 0$. Moreover, because there are no available energy states around the allowed levels, temperature plays no role in the broadening since the occupation of states around the allowed levels (which would have been temperature dependent) is a nonissue. Therefore, the linewidth is both small and temperature independent, so that the threshold current density will also be both small and temperature independent. This was shown rigorously in [12] and it evoked a lot of interest in quantum dot lasers, where the active region confining electrons and holes are fashioned into quantum dots with three-dimensional carrier confinement. These lasers indeed have low threshold [13] and show temperature independence of the threshold current [14].

Quantum dot lasers have other advantages as well. We will show later that the absorption coefficient in any semiconductor structure is proportional to the electron-hole joint density of states. In quantum dots, the latter quantity diverges when the photon energy equals the energy separation between an electron subband and a hole subband. Therefore, for such photon energies, the gain $\alpha_g(\omega_l)$ will be very large, which will make it easier to fulfill the condition in (6.41). Thus, one expects stronger lasing in quantum dots.

The only possible drawback of quantum dot lasers is that they may be vulnerable to the phononbottleneck effect discussed in the previous chapter [15]. In *ideal* quantum dots in which the confining potentials for electrons and holes are the same, radiative recombination between electrons and holes can occur only if the subband indices (i, j, k) for electrons and (i', j', k') for holes are such that $i = i'$, $j = j'$, and $k = k'$. Suppose now that the electron is excited (by current or anything else) to a subband that does not satisfy the above relation. Therefore, in order to recombine radiatively, the electron must first decay to a different subband in the quantum dot that does satisfy the above relation. This decay must be nonradiative, or phonon assisted, since radiative recombination is not possible between different subbands in the same band (in this case, the conduction band) of the quantum dot. But if the energy separation between the two subbands in the conduction band do not match any available phonon energy, this intraband decay becomes impossible, or very slow, owing to the requirement of energy conservation. This reduces the effective electron-hole radiative recombination rate drastically, thereby impeding laser action.

Quantum dot lasers continue to inspire significant research and the interested reader is referred to many excellent treatise on this topic, such as [16, 17].

6.7 Calculation of $K_{\hat{p},\omega_1}(m,n)$ in Bulk Semiconductor

So far, we have avoided calculating the quantity $K_{\hat{p},\omega_1}(m,n)$ which governs both absorption and emission (see (6.17)). In this section, we will complete this task using quantum mechanics.

From (6.8), we immediately see that

$$\begin{aligned} K_{\hat{p},\omega_1}(m,n)\left[N_{\hat{p}}(\omega_1)+\frac{1}{2}\pm\frac{1}{2}\right]\delta_{\vec{k}_m,\vec{k}_n+\vec{q}_1} \\ = K_{\hat{p},\omega_1}(n,m)\left[N_{\hat{p}}(\omega_1)+\frac{1}{2}\pm\frac{1}{2}\right]\delta_{\vec{k}_m,\vec{k}_n+\vec{q}_1} = \frac{2\pi}{\hbar}\left|M_{\pm}(\hat{p},\omega_1)\right|^2, \end{aligned} \tag{6.53}$$

where $M_{+}(\hat{p},\omega_1) = M_{m,n}(\hat{p},\omega_1)$ and $M_{-}(\hat{p},\omega_1) = M_{n,m}(\hat{p},\omega_1)$. Obviously, these two quantities are equal since we have shown earlier that $K_{\hat{p},\omega_1}(m,n) = K_{\hat{p},\omega_1}(n,m)$.

As in the case of phonons (recall Chap. 5 material), we will view the photon's interaction with the electron to result in a wave-like perturbation in the electron's Hamiltonian that varies periodically in time and space. This perturbation $H_{e-p}(\vec{r},t)$ can be expanded in a Fourier series:

$$\begin{aligned} H_{e-p}(\vec{r},t) &= V^{+}_{e-p}(\vec{r},t) + V^{-}_{e-p}(\vec{r},t) \\ &= \sum_{\omega_1,\vec{q}_1,\hat{p}} V^{+}_{\hat{p},\omega_1,\vec{q}_1} e^{i(\omega_1 t - \vec{q}_1\cdot\vec{r})} + \sum_{\omega_1,\vec{q}_1,\hat{p}} V^{-}_{\hat{p},\omega_1,\vec{q}_1} e^{-i(\omega_1 t - \vec{q}_1\cdot\vec{r})}. \end{aligned} \tag{6.54}$$

Since the frequency and wavevector of the photon are uniquely related as $\omega_1 = [c/\eta(\omega_1)]q_1$ for any polarization, we need not sum over both frequency and wavevector in the preceding equation. Summation over either one (e.g., the frequency) is sufficient.[13] Hence, we can write

$$H_{e-p}(\vec{r},t) = V^+_{e-p}(\vec{r},t) + V^-_{e-p}(\vec{r},t) = \sum_{\omega_1,\hat{p}} V^+_{\hat{p},\omega_1} e^{i(\omega_1 t - \vec{q}_1\cdot\vec{r})} + \sum_{\omega_1,\hat{p}} V^-_{\hat{p},\omega_1} e^{-i(\omega_1 t - \vec{q}_1\cdot\vec{r})}. \tag{6.55}$$

Focusing on the Fourier component with frequency ω_1 and polarization $\hat{p}$, we get that the time-dependent matrix element due to this component is (see (5.6))

$$H^{\omega_1,\hat{p}}_{e-p}(t) = M_+(\hat{p},\omega_1)e^{i\omega_1 t} + M_-(\hat{p},\omega_1)e^{-i\omega_1 t}, \tag{6.56}$$

where, using (5.31), we obtain

$$M_+(\hat{p},\omega_1) = \sqrt{N_{\hat{p}}(\omega_1)+1}\frac{1}{\Omega}\int_0^\infty d^3\vec{r} e^{-i\vec{k}_n\cdot\vec{r}} u^*_{\vec{k}_n,n}(\vec{r}) V^+_{\hat{p},\omega_1} e^{-i\vec{q}_1\cdot\vec{r}} e^{i\vec{k}_m\cdot\vec{r}} u_{\vec{k}_m,m}(\vec{r})$$
$$M_-(\hat{p},\omega_1) = \sqrt{N_{\hat{p}}(\omega_1)}\frac{1}{\Omega}\int_0^\infty d^3\vec{r} e^{-i\vec{k}_m\cdot\vec{r}} u^*_{\vec{k}_m,m}(\vec{r}) V^-_{\hat{p},\omega_1} e^{i\vec{q}_1\cdot\vec{r}} e^{i\vec{k}_n\cdot\vec{r}} u_{\vec{k}_n,n}(\vec{r}), \tag{6.57}$$

where k_m is the electron's wavevector in the final state (in the m-th band of the crystal) and k_n is the electron's wavevector in the initial state (in the n-th band of the crystal).

In deriving the above, we made use of the fact that in a crystal, the wavefunctions of electrons and holes will be Bloch states of the form $e^{i\vec{k}\cdot\vec{r}} u_{\vec{k},j}(\vec{r})$, where $\vec{k}$ is the particle's wavevector and $u_{\vec{k},j}(\vec{r})$ is the Bloch function in the j-th band. We have to remember, however, that in the case of emission, the initial state's wavefunction is an electron wavefunction and the final state's wavefunction is a hole's wavefunction if the transition involves the conduction and the valence band. For absorption, the opposite will be true.

In order to find $K_{\hat{p},\omega_1}(m,n)$, which is our ultimate objective, we will proceed as follows. First, we will find the electron–photon interaction Hamiltonian $H_{e-p}(\vec{r},t)$ from first principles. Then we will use (6.55) to find $V^{\pm}_{\hat{p},\omega_1}$. Next, we will utilize (6.57) to find $M_\pm(\hat{p},\omega_1)$. Finally, we will use (6.53) to find $K_{\hat{p},\omega_1}(m,n)$.

Before we can derive expressions for $H_{e-p}(\vec{r},t)$, we have to understand how an electron will respond to light or an electromagnetic wave that is made up of photons. Any electromagnetic wave or photon has an electric field and a magnetic field with which the electron will interact. However, in quantum mechanics, we never deal with fields, but instead deal with potentials that the fields generate. These potentials will determine the interaction Hamiltonian.

[13] We did not do this for phonons because there are multiple phonon modes—acoustic, optical, etc.

The fields in an electromagnetic wave generate both vector and scalar potentials $\vec{A}(\vec{r},t)$ and $V(\vec{r},t)$, such that the electric field and the magnetic flux density in the wave can be related to these potentials as [18]

$$\vec{\mathcal{E}}(\vec{r},t) = -\vec{\nabla}_r V(\vec{r},t) - \frac{\partial \vec{A}(\vec{r},t)}{\partial t}$$
$$\vec{B}(\vec{r},t) = \vec{\nabla}_r \times \vec{A}(\vec{r},t). \tag{6.58}$$

Note that the vector "potential" does not have the unit or dimension of potential energy unlike the scalar potential. However, the quantity $c\vec{A}(\vec{r},t)$ has the unit of potential energy, while the quantity $e\vec{A}(\vec{r},t)$ has the unit of momentum.

Since an electromagnetic field certainly has a nonzero magnetic field (otherwise it will not be an electro"magnetic" field), $\vec{B}(\vec{r},t) \neq 0$ and hence $\vec{A}(\vec{r},t) \neq 0$. While we know how to include the scalar potential $V(\vec{r},t)$ in the Hamiltonian as a simple additive term, we do not yet know how to include the magnetic vector potential. That is trickier business. The magnetic vector potential does not simply add another term to the Hamiltonian; instead, it modifies the electron's momentum.

In Chap. 3, we showed that the operator for the momentum $\vec{p}_{op}$ is $-i\hbar\vec{\nabla}_r$. This momentum is called the *canonical momentum.* A particle, however, always responds to *kinematic* (or mechanical) momentum $m\left(d\vec{r}/dt\right)$ which is *different* from the canonical momentum in the presence of a vector potential. The kinematic momentum operator for a charged particle, which is sometimes represented as $\vec{\Pi}_{op}$, is given by [19,20]

$$\vec{\Pi}_{op} = \vec{p}_{op} + q\vec{A}(\vec{r},t) = -i\hbar\vec{\nabla}_r + q\vec{A}(\vec{r},t), \tag{6.59}$$

where q is the particle's charge (for an electron, $q = -e$). One immediately sees that while $\vec{p}_{op}$ is time and space invariant, $\vec{\Pi}_{op}$ is not since the magnetic vector potential can change in space and time.

We have not derived (6.59), nor can we derive it rigorously from first principles, but we can offer arguments as to why it makes sense. We cannot derive the Schrödinger equation (or conservation of energy, for that matter) rigorously from first principles either, but we can offer arguments as to why they make sense. We will now offer an argument as to why (6.59) makes sense.

Special Topic 1: Gauge transformations and gauge invariance

Consider the following transformations to the scalar and vector potentials:

$$V(\vec{r},t) \rightarrow V(\vec{r},t) - \frac{\partial \Lambda(\vec{r},t)}{\partial t}$$
$$\vec{A}(\vec{r},t) \rightarrow \vec{A}(\vec{r},t) + \vec{\nabla}_r \Lambda(\vec{r},t), \tag{6.60}$$

where $\Lambda(\vec{r}, t)$ is any arbitrary space- and time-dependent scalar. From (6.58), we can immediately see that these potential transformations—which are called *gauge transformations*—do *not* affect the electric field and magnetic flux density at all. Therefore, the electric field and magnetic flux density are *gauge invariant*. Any physically measurable quantity must remain gauge invariant. That is because potentials are always undefined to the extent of arbitrary quantities and that arbitrariness should not impact anything that we measure.

It can be shown mathematically that in the presence of an electromagnetic field (i.e., nonzero magnetic vector potential), the canonical momentum $\vec{p}$ is *not* gauge invariant, but the kinematic momentum $\vec{\Pi}$ is, *as long as* we define the latter according to (6.59) [20]. This justifies (6.59) and therefore the definition of the kinematic momentum given by this equation. We emphasize that we have not rigorously derived (6.59), but we have shown that it indeed makes sense since it allows the kinematic momentum to be gauge invariant.

Equation (6.59) also implies that an electromagnetic field has kinematic momentum since it has a nonzero vector potential. This is not something new since we have already postulated that a photon has a momentum $\hbar q_l$ along the direction of light propagation. Not only does it have linear momentum but it has angular momentum as well. Each photon behaves like a spinning bullet and carries an angular momentum of $\hbar$. Richard Feynman thought of a delightful experiment to underscore this fact [21]. Consider a circular disk, made of an insulating material, that is centered on a frictionless concentric spindle around which the disk can rotate freely. This gadget is shown in Fig. 6.13. Around the periphery, there are equally spaced grooves, each of which contains a charge Q. These charges do not leak out quickly since the surrounding material is insulating and will not support a current.

A superconducting solenoid is wound around the spindle and carriers a constant (steady-state) supercurrent without a battery. This current gives rise to a magnetic field directed upward with an associated flux density $\vec{B}$ given by Maxwell's first equation (Ampere's law):

$$\vec{\nabla}_r \times \vec{B} = \mu \vec{J}, \tag{6.61}$$

where $\vec{J}$ is the supercurrent density and μ is the magnetic permeability.

Suddenly, the ambient temperature is raised above the superconducting transition temperature of the solenoid (made of a type I superconductor), which drives the coil normal and ultimately cuts off the current since there is no battery. That, in turn, will change the magnetic flux density and bring it down to zero sooner or later. Since the flux density will change with time, there will be a resultant time derivative $\partial \vec{B}/\partial t$, which according to Maxwell's second equation (or Faraday's law of induction) will give rise to an electromagnetic field with a circulating electric field $\vec{\mathcal{E}}$ given by

$$\vec{\nabla}_r \times \vec{\mathcal{E}} = -\frac{\partial \vec{B}}{\partial t}. \tag{6.62}$$

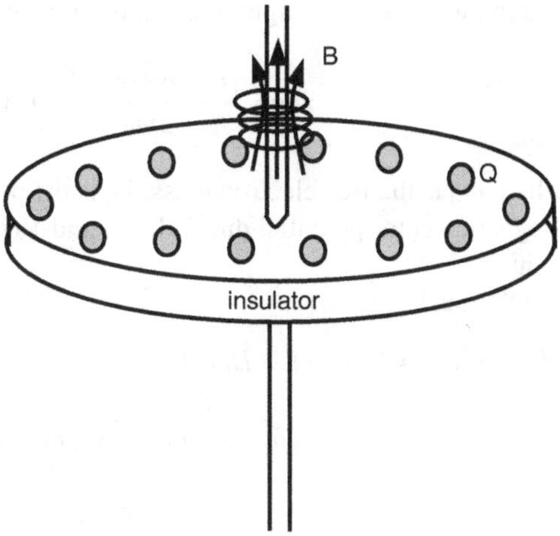

Fig. 6.13 A circular disk supported around a concentric frictionless spindle. The disk material is insulating which allows one to maintain fixed charges of Q in each groove cut around the periphery. A superconducting coil is wound around the spindle which initially carriers a supercurrent and causes a magnetic field directed vertically up. Adapted from [21] with permission from Perseus Books Group

The circulating electric field will exert a force on each charge Q in a direction tangential to the circumference of the disk. Since this force is in the same sense in every groove, it will exert a torque on the disk and attempt to make it rotate around the spindle. The all-important question is whether the disk will, in fact, rotate. What might prevent it from rotating is the principle of conservation of angular momentum. Since the disk was initially stationary and had no angular momentum, it would appear that after the current stops, the angular momentum should remain zero and the disk should not rotate. So, will the disk rotate or not?

Feynman did not answer this question in his book, but the answer is obviously "yes." The disk will rotate. It will not rotate because the electrons in the solenoid were going about a circular path and since electrons have mass, the current had some angular momentum that was mystically transferred to the disk. That is not the correct reasoning. The correct reasoning is that the electromagnetic field generated after the supercurrent cuts off has nonzero angular momentum and supplies the angular momentum for the disk to rotate. This is a beautiful gedanken experiment to underscore the fact that photons have angular momentum, in addition to linear momentum.

6.7.1 The Electron–Photon Interaction Hamiltonian

Let us now return to the subject of $H_{e-p}(\vec{r},t)$. In the absence of light, the electron's Hamiltonian in a crystal will be

$$H_0 = \frac{|\vec{p}|^2}{2m_0} + V_{\rm L}(\vec{r}), \tag{6.63}$$

and in the presence of light, the Hamiltonian will be

$$H = \frac{|\vec{p} - e\vec{A}(\vec{r},t)|^2}{2m_0} + V_L(\vec{r}) + V(\vec{r},t), \tag{6.64}$$

where m_0 is the free electron mass, V_L is the periodic lattice potential, $\vec{A}(\vec{r},t)$ is the magnetic vector potential due to light, and $V(\vec{r},t)$ is the scalar potential caused by light.

Obviously,

$$\begin{aligned} H_{e-p}(\vec{r},t) &= H(\vec{r},t) - H_0(\vec{r}) \\ &= -\frac{e}{2m_0}[\vec{p}\cdot\vec{A}(\vec{r},t) + \vec{A}(\vec{r},t)\cdot\vec{p}] + \frac{e^2}{2m_0}\vec{A}(\vec{r},t)\cdot\vec{A}(\vec{r},t) + V(\vec{r},t). \end{aligned} \tag{6.65}$$

Remember that the operator $\vec{p} = -i\hbar\vec{\nabla}_r$. Therefore, the commutator

$$\vec{p}\cdot\vec{A}(\vec{r},t) - \vec{A}(\vec{r},t)\cdot\vec{p} = -i\hbar\vec{\nabla}_r\cdot\vec{A}(\vec{r},t) \neq 0 \text{ (generally)}. \tag{6.66}$$

However, for a TEM or TE wave, the divergence of the magnetic vector potential $\vec{\nabla}_r\cdot\vec{A}(\vec{r},t)$ is always zero (see Problem 6.5). Therefore, $\vec{p}$ and $\vec{A}(\vec{r},t)$ will commute and we can write the electron–photon interaction Hamiltonian as

$$H_{e-p}(\vec{r},t) = -\frac{e}{m_0}\vec{A}(\vec{r},t)\cdot\vec{p} + \frac{e^2}{2m_0}|\vec{A}(\vec{r},t)|^2 + V(\vec{r},t). \tag{6.67}$$

From (6.58), we see that the electric field has two components due to the gradient of the scalar potential and the time-derivative of the vector potential. The first component is a conservative field since its curl is zero, while the second component is a nonconservative field. Since potentials are always undefined to the extent of an arbitrary constant, we can apportion the electric field in any way we like between the two components. The simplest approach will be to have only one component and not the other. Accordingly, we will make the first component zero and write the electric field entirely as the time derivative of the vector potential. That is

$$\vec{\mathcal{E}}(\vec{r},t) = -\frac{\partial\vec{A}(\vec{r},t)}{\partial t}. \tag{6.68}$$

This choice makes $\vec{\nabla}_r V(\vec{r},t) \equiv 0$, so that $V(\vec{r},t)$ becomes spatially invariant. Therefore, we can summarily set $V(\vec{r},t) = 0$ in (6.67) since scalar potentials are undefined to the extent of an arbitrary constant.

Now, if the electromagnetic field is relatively weak, then we can neglect the square term in the magnetic vector potential in comparison with the linear term in (6.67) and write the electron–photon interaction Hamiltonian as

$$H_{e-p}(\vec{r},t) \approx -\frac{e}{m_0}\vec{A}(\vec{r},t)\cdot\vec{p} = \frac{ie\hbar}{m_0}\vec{A}(\vec{r},t)\cdot\vec{\nabla}_r. \tag{6.69}$$

This leaves us with only more step to go. We simply have to derive an expression for the vector potential and that will give us the electron–photon interaction Hamiltonian $H_{e-p}(\vec{r},t)$.

In order to find the expression for the vector potential, we proceed as follows. The *instantaneous* electric field in a monochromatic electromagnetic wave[14] of angular frequency ω_1 and polarization $\hat{p}$ is [18]

$$\vec{\mathcal{E}}(\vec{r},t) = \hat{p}\mathcal{E}_0 \sin(\omega_1 t - \vec{q}_1 \cdot \vec{r}), \tag{6.70}$$

where $\hat{p}$ is the unit vector along the polarization direction. For a TE or TEM wave, this direction is perpendicular to the propagation direction (or direction of the photon's wavevector), so that

$$\hat{p} \cdot \vec{q}_1 = 0. \tag{6.71}$$

The amplitude $\mathcal{E}_0$ must be found from conservation of energy. We know that the energy in the wave alternates between electric and magnetic energy and the energy density in the electric field is $(1/2)\epsilon\mathcal{E}_0^2$, where ϵ is the dielectric constant of the medium in which the wave is propagating. Therefore, we can write from conservation of energy

$$\frac{1}{2}\epsilon\mathcal{E}_0^2 = \frac{\hbar\omega_1}{\Omega}, \tag{6.72}$$

where Ω is the normalizing volume.

Using this result in (6.70), we obtain

$$\vec{\mathcal{E}}(\vec{r},t) = -\mathrm{i}\hat{p}\sqrt{\frac{\hbar\omega_1}{2\epsilon\Omega}}\left[\mathrm{e}^{\mathrm{i}(\omega_1 t - \vec{q}_1\cdot\vec{r})} - \mathrm{e}^{-\mathrm{i}(\omega_1 t - \vec{q}_1\cdot\vec{r})}\right]. \tag{6.73}$$

Next, using (6.68), we get that

$$\vec{A}(\vec{r},t) = \hat{p}\sqrt{\frac{\hbar}{2\epsilon\omega_1\Omega}}\left[\mathrm{e}^{\mathrm{i}(\omega_1 t - \vec{q}_1\cdot\vec{r})} + \mathrm{e}^{-\mathrm{i}(\omega_1 t - \vec{q}_1\cdot\vec{r})}\right] + \vec{\Upsilon}(\vec{r}), \tag{6.74}$$

where $\vec{\Upsilon}(\vec{r})$ is the constant of integration and is a time-independent quantity.

Once again, the vector potential $\vec{\Upsilon}(\vec{r})$ does not give rise to any electric field since it is time independent and hence can be set equal to zero. This is also the right thing to do since otherwise it will give rise to a spurious magnetic field in accordance with (6.58). This choice is sometimes called the Weyl gauge. Thereafter, substitution of (6.74) in (6.69) yields

$$H_{e-p}(\vec{r},t) = \frac{\mathrm{i}e\hbar}{m_0}\sqrt{\frac{\hbar}{2\epsilon\omega_1\Omega}}\left[\mathrm{e}^{\mathrm{i}(\omega_1 t - \vec{q}_1\cdot\vec{r})} + \mathrm{e}^{-\mathrm{i}(\omega_1 t - \vec{q}_1\cdot\vec{r})}\right]\hat{p}\cdot\vec{\nabla}_{\mathrm{r}}$$

[14]The instantaneous field is always a real quantity and cannot be complex.

$$= \frac{ie\hbar}{m_0}\sqrt{\frac{\hbar}{2\epsilon\omega_1\Omega}}\left[e^{i(\omega_1 t-\vec{q}_1\cdot\vec{r})} + e^{-i(\omega_1 t-\vec{q}_1\cdot\vec{r})}\right]\partial_{\hat{p}}. \tag{6.75}$$

where $\partial_{\hat{p}}$ is the derivative operator in the direction of the photon's polarization.

Comparing the last equation with (6.55), it is easy to see that

$$V^{+}_{\hat{p},\omega_1} = V^{-}_{\hat{p},\omega_1} = \frac{ie\hbar}{m_0}\sqrt{\frac{\hbar}{2\epsilon\omega_1\Omega}}\partial_{\hat{p}}. \tag{6.76}$$

Next, using (6.57), we find

$$\begin{aligned} M_+(\hat{p},\omega_1) &= \sqrt{N_{\hat{p}}(\omega_1)+1} \\ &\times \frac{1}{\Omega}\int_0^\infty d^3\vec{r} e^{-i\vec{k}_n\cdot\vec{r}} u^*_{\vec{k}_n,n}(\vec{r}) \frac{ie\hbar}{m_0}\sqrt{\frac{\hbar}{2\epsilon\omega_1\Omega}}\partial_{\hat{p}} e^{-i\vec{q}_1\cdot\vec{r}} e^{i\vec{k}_m\cdot\vec{r}} u_{\vec{k}_m,m}(\vec{r}) \\ M_-(\hat{p},\omega_1) &= \sqrt{N_{\hat{p}}(\omega_1)} \\ &\times \frac{1}{\Omega}\int_0^\infty d^3\vec{r} e^{-i\vec{k}_m\cdot\vec{r}} u^*_{\vec{k}_m,m}(\vec{r}) \frac{ie\hbar}{m_0}\sqrt{\frac{\hbar}{2\epsilon\omega_1\Omega}}\partial_{\hat{p}} e^{i\vec{q}_1\cdot\vec{r}} e^{i\vec{k}_n\cdot\vec{r}} u_{\vec{k}_n,n}(\vec{r}). \end{aligned} \tag{6.77}$$

Finally, using the last equation in (6.53), we get

$$\begin{aligned} &K_{\hat{p},\omega_1}(n,m)\delta_{\vec{k}_m,\vec{k}_n+\vec{q}_1} \\ &= \frac{2\pi}{\hbar}\left|\int_0^\infty d^3\vec{r}\frac{1}{\sqrt{\Omega}} e^{-i\vec{k}_n\cdot\vec{r}} u^*_{\vec{k}_n,n}(\vec{r}) \frac{ie\hbar}{m_0}\sqrt{\frac{\hbar}{2\epsilon\omega_1\Omega}} e^{-i\vec{q}_1\cdot\vec{r}} \frac{1}{\sqrt{\Omega}}\partial_{\hat{p}}\left[e^{i\vec{k}_m\cdot\vec{r}} u_{\vec{k}_m,m}(\vec{r})\right]\right|^2 \end{aligned} \tag{6.78}$$

$$\begin{aligned} &K_{\hat{p},\omega_1}(m,n)\delta_{\vec{k}_m,\vec{k}_n+\vec{q}_1} \\ &= \frac{2\pi}{\hbar}\left|\int_0^\infty d^3\vec{r}\frac{1}{\sqrt{\Omega}} e^{-i\vec{k}_m\cdot\vec{r}} u^*_{\vec{k}_m,m}(\vec{r}) \frac{ie\hbar}{m_0}\sqrt{\frac{\hbar}{2\epsilon\omega_1\Omega}} e^{i\vec{q}_1\cdot\vec{r}} \frac{1}{\sqrt{\Omega}}\partial_{\hat{p}}\left[e^{i\vec{k}_n\cdot\vec{r}} u_{\vec{k}_n,n}(\vec{r})\right]\right|^2, \end{aligned} \tag{6.79}$$

where, as usual, the asterisk denotes complex conjugate. The above equation immediately reaffirms that $K_{\hat{p},\omega_1}(n,m) = K_{\hat{p},\omega_1}(m,n)$

Now,

$$\partial_{\hat{p}}\left[\frac{1}{\sqrt{\Omega}} e^{i\vec{k}\cdot\vec{r}} u_{\vec{k},j}(\vec{r})\right] = e^{i\vec{k}\cdot\vec{r}}\left(ik_p + \partial_{\hat{p}}\right)\frac{1}{\sqrt{\Omega}} u_{\vec{k},j}(\vec{r}). \tag{6.80}$$

Using this result in (6.79), we get

$$K_{\hat{p},\omega_1}(m,n)\delta_{\vec{k}_m,\vec{k}_n+\vec{q}_1}$$
$$= \frac{\pi e^2\hbar^2}{m_0^2\epsilon\omega_1\Omega}\left|\int_0^\infty d^3\vec{r}\frac{1}{\Omega}e^{i(\vec{k}_n-\vec{k}_m+\vec{q}_1)\cdot\vec{r}}u^*_{\vec{k}_m,m}(\vec{r})(ik_p+\partial_{\hat{p}})u_{\vec{k}_n,n}(\vec{r})\right|^2. \quad (6.81)$$

We can simplify the integral in the last equation a lot more by noting that the Bloch functions $u_{\vec{k},j}(\vec{r})$ repeat periodically with period equal to the lattice constant or the dimension of the unit cell (recall the Bloch Theorem in Chap. 4). Therefore, the last integral is significant only if the phase $(\vec{k}_n - \vec{k}_m + \vec{q}_1)\cdot\vec{r} = 0$. Otherwise, the integrand will have different phases in different unit cells and the integral will tend to average out to zero. Hence, we can rewrite the last equation as

$$K_{\hat{p},\omega_1}(m,n)\delta_{\vec{k}_m,\vec{k}_n+\vec{q}_1} = \frac{\pi e^2\hbar^2}{m_0^2\epsilon\omega_1\Omega}\left|\int_0^\infty d^3\vec{r}\frac{1}{\Omega}u^*_{\vec{k}_m,m}(\vec{r})(ik_p+\partial_{\hat{p}})u_{\vec{k}_n,n}(\vec{r})\right|^2\delta_{\vec{k}_m,\vec{k}_n+\vec{q}_1}$$
$$\Rightarrow K_{\hat{p},\omega_1}(m,n) = \frac{\pi e^2\hbar^2}{m_0^2\epsilon\omega_1\Omega}\left|\int_0^\infty d^3\vec{r}\frac{1}{\Omega}u^*_{\vec{k}_m,m}(\vec{r})(ik_p+\partial_{\hat{p}})u_{\vec{k}_n,n}(\vec{r})\right|^2. \quad (6.82)$$

The last equation can be expanded as

$$K_{\hat{p},\omega_1}(m,n) = \frac{\pi e^2\hbar^2}{m_0^2\epsilon\omega_1\Omega}\left|ik_p\int_0^\infty d^3\vec{r}\frac{1}{\Omega}u^*_{\vec{k}_m,m}(\vec{r})u_{\vec{k}_n,n}(\vec{r}) + \int_0^\infty d^3\vec{r}\frac{1}{\Omega}u^*_{\vec{k}_m,m}(\vec{r})\partial_{\hat{p}}u_{\vec{k}_n,n}(\vec{r})\right|^2$$
$$= \frac{\pi e^2}{m_0^2\epsilon\omega_1\Omega}\left|\int_0^\infty d^3\vec{r}\frac{1}{\Omega}u^*_{\vec{k}_m,m}(\vec{r})(-i\hbar\partial_{\hat{p}})u_{\vec{k}_n,n}(\vec{r})\right|^2, \quad (6.83)$$

where we have made use of the orthogonality of the Bloch functions in different bands at the same wavevector value (see (4.59)), i.e., $\int_0^\infty d^3\vec{r}u^*_{\vec{k},m}(\vec{r})u_{\vec{k},n}(\vec{r}) = 0$, if $m \neq n$. Note that the k-selection rule mandated that $\vec{k}_m = \vec{k}_n$, but $m \neq n$.

Since $(-i\hbar\partial_{\hat{p}})$ is a Hermitian operator, we get that

$$K_{\hat{p},\omega_1}(m,n) = K_{\hat{p},\omega_1}(n,m) = \frac{\pi e^2}{m_0^2\epsilon\omega_1\Omega}|\hat{p}\cdot\vec{\Upsilon}_{m,n}|^2, \quad (6.84)$$

where the "momentum matrix element" $\vec{\Upsilon}_{m,n}$ is defined by the relation

$$\vec{\Upsilon}_{m,n} = \int_0^\infty d^3\vec{r}\frac{1}{\Omega}u^*_{\vec{k},m}(\vec{r})(-i\hbar\vec{\nabla}_r)u_{\vec{k},n}(\vec{r}). \quad (6.85)$$

Using (6.17), we can now write

$$
\begin{aligned}
\alpha_{\hat{p}}(\omega_l) &= \frac{\eta_{\hat{p}}(\omega_l)}{c} \sum_{m,n} \frac{\pi e^2}{m_0^2 \epsilon \omega_l \Omega} |\hat{p} \cdot \vec{\Upsilon}_{m,n}|^2 \delta(E_m - E_n - \hbar\omega_l)(f_n - f_m) \\
&= \frac{\eta_{\hat{p}}(\omega_l)}{c} \sum_{\text{cond, val}} \frac{\pi e^2}{m_0^2 \epsilon \omega_l \Omega} \left| \hat{p} \cdot \vec{\Upsilon}_{\text{cond,val}} \right|^2 \\
&\quad \times \sum_{\vec{k}} \delta \left[E_{\text{cond}}(\vec{k}) - E_{\text{val}}(\vec{k}) - \hbar\omega_l \right] \left[f_{\text{val}}(\vec{k}) - f_{\text{cond}}(\vec{k}) \right] \\
L_{\hat{p}}(\omega_l) &= \sum_{m,n} \frac{\pi e^2}{m_0^2 \epsilon \omega_l \Omega^2} \left| \hat{p} \cdot \vec{\Upsilon}_{m,n} \right|^2 \delta\left[E_m - E_n - \hbar\omega_l\right] f_m(1 - f_n) \\
&= \sum_{\text{cond,val}} \frac{\pi e^2}{m_0^2 \epsilon \omega_l \Omega^2} \left| \hat{p} \cdot \vec{\Upsilon}_{\text{cond,val}} \right|^2 \\
&\quad \times \sum_{\vec{k}} \delta \left[E_{\text{cond}}(\vec{k}) - E_{\text{val}}(\vec{k}) - \hbar\omega_l \right] f_{\text{cond}}(\vec{k}) \left[1 - f_{\text{val}}(\vec{k}) \right],
\end{aligned}
\tag{6.86}
$$

where "cond" refers to conduction band states and "val" refers to valence band states. In the case of absorption, it is customary to assume that $f_{\text{val}}(\vec{k}) \approx 1$ (all states in the valence band are filled) and $f_{\text{cond}}(\vec{k}) \approx 0$ (all states in the conduction band are empty), so that the factor $\left[f_{\text{val}}(\vec{k}) - f_{\text{cond}}(\vec{k}) \right]$ can be omitted in the expression for the absorption coefficient.

In the literature on optical properties, one often comes across the term "oscillator strength." This quantity is related to the matrix element and is defined as

$$
O_{\hat{p}}(\omega_l) = \frac{2m_0\epsilon}{e^2\pi\hbar} K_{\hat{p},\omega_l}(m,n), \tag{6.87}
$$

so that it is given by

$$
O_{\hat{p}}(\omega_l) = \frac{2m_0}{\hbar\omega_l} \left| \hat{p} \cdot \vec{\Upsilon}_{\text{cond,val}} \right|^2 . \tag{6.88}
$$

6.7.2 *Polarization Dependence of Absorption and Emission*

From (4.91), we can deduce the momentum matrix element between different pairs of conduction and valence band states at or near $k = 0$. They are given in Table 6.1, where Υ is a constant.

From the above table, we see that absorption between different bands is *polarization sensitive*. For example, light in which the electric field is polarized

Table 6.1 Momentum matrix element $\vec{\Upsilon}_{\text{cond,val}}$ between different pairs of conduction and valence band states

	Conduction (upspin)	Conduction (downspin)
Heavy-hole (upspin)	$\frac{\Upsilon}{\sqrt{2}}[\hat{x}+\mathrm{i}\hat{y}]$	0
Heavy-hole (downspin)	0	$\frac{\mathrm{i}\Upsilon}{\sqrt{2}}[\hat{x}-\mathrm{i}\hat{y}]$
Light-hole (upspin)	$-\mathrm{i}\Upsilon\sqrt{\frac{2}{3}}\hat{z}$	$\frac{\mathrm{i}\Upsilon}{\sqrt{6}}[\hat{x}+\mathrm{i}\hat{y}]$
Light-hole (downspin)	$\frac{\Upsilon}{\sqrt{6}}[\hat{x}-\mathrm{i}\hat{y}]$	$\Upsilon\sqrt{\frac{2}{3}}\hat{z}$
Split-off (upspin)	$\frac{\Upsilon}{\sqrt{3}}\hat{z}$	$\frac{\Upsilon}{\sqrt{3}}[\hat{x}+\mathrm{i}\hat{y}]$
Split-off (downspin)	$\frac{-\mathrm{i}\Upsilon}{\sqrt{3}}[\hat{x}-\mathrm{i}\hat{y}]$	$\frac{\mathrm{i}\Upsilon}{\sqrt{3}}\hat{z}$

in the z-direction will induce transitions between conduction and light hole bands of parallel spins, or between conduction and split-off bands of parallel spins. On the other hand, if the electric field is polarized in the x-y plane, then it can induce transitions between conduction and heavy hole bands of parallel spins, or between conduction and light hole bands of anti-parallel spins, or between conduction and split-off bands of anti-parallel spins.

It may trouble the reader that transitions between parallel and anti-parallel states are possible, and that for such states, $\Upsilon \neq 0$. We discuss this in the special topic section next.

Special Topic 2: Photon and phonon transitions between anti-parallel spin states

Consider two spin states that are mutually anti-parallel. We will write their wavefunctions as $[\Psi_\uparrow(x,y,z)]$ and $[\Psi_\downarrow(x,y,z)]$, where these wavefunctions are actually "spinors" or 2×1 column vectors. We will very briefly discuss spinors in Chap. 7, but that is not important for the ensuing discussion. Since the spin states are anti-parallel, the following orthogonality condition holds:

$$\iiint \mathrm{d}x\mathrm{d}y\mathrm{d}z[\Psi_\uparrow(x,y,z)]^\dagger[\Psi_\downarrow(x,y,z)] = 0, \tag{6.89}$$

where the "dagger" represents Hermitian conjugate.

In the case of phonons (see (5.12)), the matrix element for transition (emission or absorption) between anti-parallel spin states will be (recall Chap. 5 material)

$$\begin{aligned} M^{\pm}_{\omega,\vec{q},\hat{p},\vec{k}_2,\vec{k}_1}\Big|_{\text{phonons}} &= \sqrt{N_{\omega_n,\vec{q}_n,\vec{p}_n} + (1/2) \pm (1/2)}\frac{V^{\pm}_{\omega,\vec{q},\hat{p}}}{\Omega} \\ &\quad \times \iiint \mathrm{d}x\mathrm{d}y\mathrm{d}z[\Psi_\uparrow(x,y,z)]^\dagger[\Psi_\downarrow(x,y,z)] \\ &= 0. \end{aligned} \tag{6.90}$$

whereas for photons, we have just seen that it is

$$M^{\pm}_{\omega,\vec{q},\hat{p},\vec{k}_2,\vec{k}_1}\Big|_{\text{photons}} = \sqrt{N_{\omega_n,\vec{q}_n,\hat{p}_n} + (1/2) \pm (1/2)}\frac{\mathcal{V}^{\pm}_{\omega,\vec{q},\hat{p}}}{\Omega}$$
$$\times \iiint \mathrm{d}x\mathrm{d}y\mathrm{d}z[\Psi_{\uparrow}(x,y,z)]^{\dagger}\partial_{\hat{p}}[\Psi_{\downarrow}(x,y,z)]$$
$$\neq 0. \tag{6.91}$$

All we have done here is to replace the scalar wavefunctions with spinor wavefunctions and use the orthogonality of spinors describing anti-parallel spins given in (6.89). Thus, barring unusual circumstances, phonon transitions between anti-parallel spin states will be forbidden, but photon transitions will be allowed. This happens only because of the difference between the matrix elements for phonon and photon transitions.

Consider now a quantum well with carrier confinement along the z-direction. Here, light polarized in the plane of the well (x-y plane) will induce transitions from the heavy-hole band to the conduction band (between parallel spins) as well as from the light-hole band to the conduction band (anti-parallel spins). These two absorptions can be distinguished since quantum confinement lifts the degeneracy between the light and heavy-hole bands owing to the difference in the effective masses. As a result, the absorption edges will be at different photon frequencies. The ratio of the two absorption strengths at the absorption edge will be 3:1, based on the matrix elements in Table 6.1. On the other hand, light polarized perpendicular to the plane of the quantum well will couple only to the light holes and induce transitions between light-hole bands and conduction bands of parallel spins. It will not couple to the heavy-hole bands at all. Polarization dependence of absorption is frequently used to distinguish between heavy- and light-hole transitions.

6.8 Absorption and Emission in Semiconductor Nanostructures (Quantum Wells and Wires)

In quantum wells and wires, there can be "band mixing" effects which mixes up the wavefunctions of two or more bands, particularly in the valence band where the light- and heavy-hole wavefunctions are no longer completely distinct and the actual wavefunction of either particle is a mixture (or superposition) of both wavefunctions. We will neglect this effect here since it makes matters immensely complicated, and assume instead that the wavefunction in any given band is characterized by a well-defined Bloch function specific to that band.

We now proceed to calculate the effective matrix element $K_{\hat{p},\omega_l}(m,n)$ in quantum wells and wires, which will allow us to determine the absorption coefficient and luminescence in these systems as a function of photon frequency. In other words, we will determine the absorption and luminescence *spectrum*.

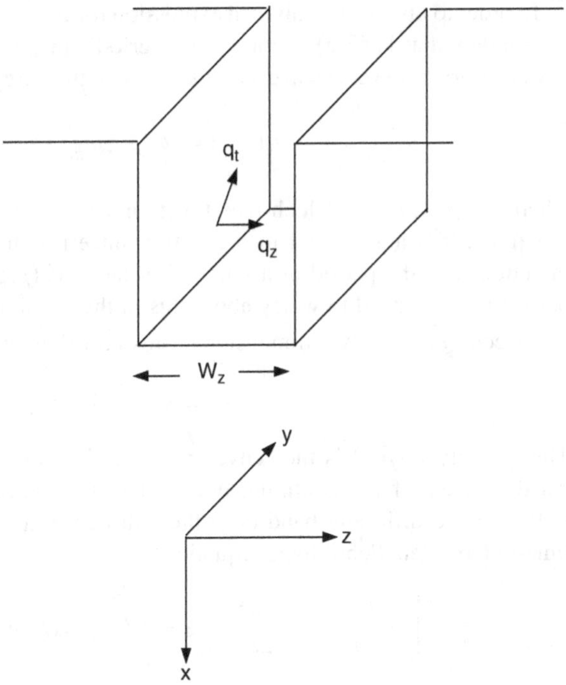

Fig. 6.14 A quantum well in the x-y plane. The photon's wavevector component in the x-y plane is $\vec{q}_t$ and along the z-direction is q_z

6.8.1 *Quantum Wells*

Consider a quantum well (2-DEG) in the x-y plane as shown in Fig. 6.14. The electron motion is confined in the z-direction by potential walls. The radial vector in the x-y plane is given by

$$\vec{\rho} = x\hat{x} + y\hat{y}. \tag{6.92}$$

The time-independent Schrödinger equation governing an electron in the quantum well is

$$\left[-\frac{\hbar^2}{2m_0}\vec{\nabla}^2_{\vec{\rho}} - \frac{\hbar^2}{2m_0}\frac{\partial^2}{\partial z^2} + V(z) + V_{\mathrm{L}}(\vec{\rho}, z) \right] \psi(\vec{\rho}, z) = E\psi(\vec{\rho}, z), \tag{6.93}$$

where m_0 is the free electron mass, $V(z)$ is the confining potential in the z-direction, and $V_{\mathrm{L}}(\vec{\rho}, z)$ is the lattice potential due to the atoms in the crystal. This latter potential is *not* periodic in the z-direction. That is because there are material discontinuities at the boundaries between the well region and barrier regions, where both the amplitude and the period of $V_{\mathrm{L}}(\vec{\rho}, z)$ can change abruptly, disrupting the periodicity. Moreover, note that since $V_{\mathrm{L}}(\vec{\rho}, z)$ is a "mixed term" that depends on both $\vec{\rho}$ and z, we cannot write the wavefunction $\psi(\vec{\rho}, z)$ as the *product* of a $\vec{\rho}$-dependent term and a z-dependent term.

In order to obtain an analytical expression for $K_{\vec{p},\omega_1}(m,n)$, we will have to ignore the subtlety that $V_L(\vec{\rho},z)$ is not exactly periodic in z, and write the wavefunction at any arbitrary transverse wavevector k_t in the i-th energy band as:

$$\psi_i(\vec{\rho},z) = \varphi_i(\vec{\rho},z)u_{\vec{k}_t,i}(\vec{\rho},z), \tag{6.94}$$

where $u_{\vec{k}_t,i}(\vec{\rho},z)$ is a Bloch-like function which we will call a "pseudo Bloch function". It is not a true Bloch function since it is not exactly periodic in the z-direction with the period of a unit cell because $V_L(\vec{\rho},z)$ is not quite periodic in z, but we will not need to worry about this in the ensuing discussion. The transverse wavevector $\vec{k}_t$ is the two-dimensional wavevector in the x-y plane given by

$$\vec{k}_t = k_x\hat{x} + k_y\hat{y}. \tag{6.95}$$

The quantity $\varphi_i(\vec{\rho},z)$ is the "envelope" wavefunction, which is no longer a plane wave because of the confining potential in the z-direction [compare (4.35) and (6.94)]. Instead, in any band (say, the i-th band), it will obey the effective mass time-independent Schrödinger equation

$$\left[-\frac{\hbar^2}{2m_i^*}\vec{\nabla}_{\vec{\rho}}^2 - \frac{\hbar^2}{2m_i^*}\frac{\partial^2}{\partial z^2} + V(z)\right]\varphi_i(\vec{\rho},z) = E_i\varphi_i(\vec{\rho},z), \tag{6.96}$$

where m_i^* is the effective mass in the i-th band. Equation (6.94) is the more general form of the Bloch Theorem given in (4.35).

In the last equation, there are fortunately no mixed terms in the Hamiltonian that depend on both $\vec{\rho}$ and z. Hence, we can write the envelope wavefunction as a product:

$$\varphi_i(\vec{\rho},z) = \xi_p(\vec{\rho})\phi_i(z). \tag{6.97}$$

Substitution of this product solution in the Schrödinger equation for the envelope wavefunction immediately leads to two decoupled equations (decoupling $\vec{\rho}$- and z-dependent terms):

$$\begin{gathered} -\frac{\hbar^2}{2m_i^*}\vec{\nabla}_{\vec{\rho}}^2\xi_i(\vec{\rho}) = \epsilon_1\xi_i(\vec{\rho}) \\ \left[-\frac{\hbar^2}{2m_i^*}\frac{\partial^2}{\partial z^2} + V(z)\right]\phi_i(z) = \epsilon_2\phi_i(z), \end{gathered} \tag{6.98}$$

where $E_i = \epsilon_1 + \epsilon_2$. Here ϵ_1 is the energy of motion in the x-y plane and ϵ_2 is the energy of motion in the z-direction.

The solution of the first equation is

$$\xi_i(\vec{\rho}) = \frac{1}{\sqrt{A}}e^{i\vec{k}_t\cdot\vec{\rho}}, \tag{6.99}$$

where $k_t = \sqrt{2m_i\epsilon_1}/\hbar$ and A is a normalizing area in the x-y plane, while the solution of the second equation depends on the shape of the confining potential $V(z)$ (for rectangular confining potential with infinite barrier height, the solution will be particle-in-a-box states $\sqrt{\frac{2}{W_z}}sin\left(\frac{m\pi}{W_z}\right)$, with W_z being the well width and m an integer denoting the subband index in the quantum well). We will write the solution of the second equation as $\phi_{i,m}(z)$. Note that the confining potential $V(z)$, regardless of its shape, will break up every energy band into denumerably infinite subbands. Therefore, the total wavefunction in the m-th subband of the i-th energy band can be written as

$$\psi_{i,m}(\vec{\rho},z) = \frac{1}{\sqrt{A}}\mathrm{e}^{\mathrm{i}\vec{k}_t\cdot\vec{\rho}}u_{\vec{k}_t,i}(\vec{\rho},z)\phi_{i,m}(z). \tag{6.100}$$

We are now prepared to find the effective matrix element $K_{\hat{p},\omega_1}(i,m : j,n)$ in a quantum well or 2-DEG associated with transition between the m-th subband of the i-th energy band and the n-th subband of the j-th energy band. Following the derivation of the bulk case,

$$K_{\hat{p},\omega_1}(i,m : j,n) = \left|\int_0^\infty\int_{-\infty}^\infty d^2\vec{\rho}\mathrm{d}z\left\{\frac{1}{\sqrt{A}}\mathrm{e}^{-\mathrm{i}\vec{k}_t'\cdot\vec{\rho}}u^*_{\vec{k}_t',i}(\vec{\rho},z)\phi^*_{i,m}(z)\mathrm{e}^{\mathrm{i}\vec{q}_t\cdot\vec{\rho}}\mathrm{e}^{\mathrm{i}q_z z}\right.\right.$$
$$\left.\left.\partial_{\hat{p}}\left(\frac{1}{\sqrt{A}}\mathrm{e}^{\mathrm{i}\vec{k}_t\cdot\vec{\rho}}u_{\vec{k}_t,j}(\vec{\rho},z)\phi_{j,n}(z)\right)\right\}\right|^2, \tag{6.101}$$

where $\vec{q}_t$ is the photon's wavevector component in the x-y plane and q_z is the component along the z-axis. The quantities $\vec{k}_t$ and $\vec{k}_t'$ are the initial and final wavevectors in the x-y plane. Since the photon's momentum is small, we will assume that $\vec{q}_t \to 0$ and $q_z \to 0$, which immediately yields

$$K_{\hat{p},\omega_1}(i,m : j,n) \approx \left|\int_0^\infty\int_{-\infty}^\infty d^2\vec{\rho}\mathrm{d}z\left\{\frac{1}{\sqrt{A}}\mathrm{e}^{-\mathrm{i}\vec{k}_t'\cdot\vec{\rho}}u^*_{\vec{k}_t',i}(\vec{\rho},z)\phi^*_{i,m}(z)\right.\right.$$
$$\left.\left.\partial_{\hat{p}}\left(\frac{1}{\sqrt{A}}\mathrm{e}^{\mathrm{i}\vec{k}_t\cdot\vec{\rho}}u_{\vec{k}_t,j}(\vec{\rho},z)\phi_{j,n}(z)\right)\right\}\right|^2, \tag{6.102}$$

where $\phi_{i,m}(z)$ will be the subband wavefunction of a hole and $\phi_{j,n}(z)$ the subband wavefunction of an electron in the case of emission involving the conduction and valence band. The reverse will be true in the case of absorption. For intersubband transitions within, say, the conduction band, both wavefunctions will be electron wavefunctions.

Next, we will carry out the double integration on the right-hand side and find analytical expressions for the effective matrix element $K_{\hat{p},\omega_1}(i,m : j,n)$ for two different polarizations of light.

6.8.1.1 For Light Polarized in the Plane of the Quantum Well

When light is polarized in the plane of the quantum well, the polarization vector lies in the x-y plane, so that (6.102) yields

$$
\begin{aligned}
K_{\hat{p},\omega_1}(i,m:j,n) &\approx \left| \int_0^\infty \int_{-\infty}^\infty d^2\vec{\rho}\mathrm{d}z \left\{ \frac{1}{A} \mathrm{e}^{-\mathrm{i}\left(\vec{k}_t'-\vec{k}_t\right)\cdot\vec{\rho}} \phi_{i,m}^*(z)\phi_{j,n}(z) u_{\vec{k}_t',i}^*(\vec{\rho},z) \right.\right. \\
&\qquad \left.\left. \left(\mathrm{i}\vec{k}_p + \partial_{\hat{p}}\right) u_{\vec{k}_t,j}(\vec{\rho},z) \right\} \right|^2 \\
&= \left| \int_{-\infty}^\infty \mathrm{d}z \phi_{i,m}^*(z)\phi_{j,n}(z) \int_0^\infty d^2\vec{\rho} \left\{ \frac{1}{A} \mathrm{e}^{-\mathrm{i}\left(\vec{k}_t'-\vec{k}_t\right)\cdot\vec{\rho}} u_{\vec{k}_t',i}^*(\vec{\rho},z) \right.\right. \\
&\qquad \left.\left. \left(\mathrm{i}\vec{k}_p + \partial_{\hat{p}}\right) u_{\vec{k}_t,j}(\vec{\rho},z) \right\} \right|^2 \\
&\approx \left| \int_{-\infty}^\infty \mathrm{d}z \phi_{i,m}^*(z)\phi_{j,n}(z) \int_0^\infty d^2\vec{\rho} \left\{ u_{\vec{k}_t',i}^*(\vec{\rho},z) \right.\right. \\
&\qquad \left.\left. \left(\mathrm{i}\vec{k}_p + \partial_{\hat{p}}\right) u_{\vec{k}_t,j}(\vec{\rho},z) \frac{1}{A} \delta_{\vec{k}_t',\vec{k}_t} \right\} \right|^2 \\
&= \left| \int_{-\infty}^\infty \mathrm{d}z \phi_{i,m}^*(z)\phi_{j,n}(z) \int_0^\infty d^2\vec{\rho} \left\{ u_{\vec{k}_t,i}^*(\vec{\rho},z) \right.\right. \\
&\qquad \left.\left. \left(\mathrm{i}\vec{k}_p + \partial_{\hat{p}}\right) \frac{1}{A} u_{\vec{k}_t,j}(\vec{\rho},z) \right\} \right|^2 .
\end{aligned} \tag{6.103}
$$

The first thing we notice from the preceding equation is that the k-selection rule is upheld (because of the Kronecker delta $\delta_{\vec{k}_t',\vec{k}_t}$) so that $\vec{k}_t' = \vec{k}_t$ or $\vec{k}_t^{\text{initial}} = \vec{k}_t^{\text{final}}$. As before, we obtained the Kronecker delta by noting that the Bloch function repeats in every unit cell and therefore the integral involving them would have averaged out to nearly zero if the phase $\left(\vec{k}_t' - \vec{k}_t\right) \cdot \vec{\rho}$ were nonzero. The only difference between the result in the last equation and the result for the bulk case is that in the last equation, the wavevector is not a three-dimensional vector, but rather a two-dimensional vector in the plane of the quantum well.

We will now make a crucial approximation which we will call the *envelope approximation*. The pseudo Bloch function $u_{\vec{k}_t,i}(\vec{\rho},z)$ varies much more rapidly in the z-direction than the envelope wavefunction $\phi_{i,m}(z)$ since the former varies over a unit cell of the crystal and the latter varies over the width of the quantum well which extends over several unit cells. Therefore, in the integral over the z-coordinate, we can replace the product of the envelope wavefunctions with its average value (which is a constant independent of z) and then pull this average value outside the integral over z. This is a common approximation in nanostructures (see, for example, [22]). Accordingly,

$$K_{\hat{p},\omega_1}(i,m:j,n) = \left| \langle \phi^*_{i,m}(z)\phi_{j,n}(z) \rangle \left\{ \mathrm{i}\vec{k}_p \int_{-\infty}^{\infty} \int_0^{\infty} \mathrm{d}z d^2\vec{\rho} \frac{1}{A} u^*_{\vec{k}_t,i}(\vec{\rho},z) u_{\vec{k}_t,j}(\vec{\rho},z) + \int_{-\infty}^{\infty} \int_0^{\infty} \mathrm{d}z d^2\vec{\rho} \frac{1}{A} u^*_{\vec{k}_t,i}(\vec{\rho},z) \left(\hat{p}\cdot\vec{\nabla}_{\vec{\rho}}\right) u_{\vec{k}_t,j}(\vec{\rho},z) \right\} \right|^2, \tag{6.104}$$

where $\langle \phi^*_{i,m}(z)\phi_{j,n}(z) \rangle$ is the average of the product of the envelope wavefunctions (averaged over the z-coordinate).

Because of the confining potential $V(z)$, the envelope wavefunctions decay rapidly in the barrier regions. As a result, let us say that their magnitudes are negligible outside the domain $[-L_z, L_z']$ for any i, m, j, n. Then,

$$\langle \phi^*_{i,m}(z)\phi_{j,n}(z) \rangle \approx \frac{\int_{-L_z}^{L_z'} \mathrm{d}z \phi^*_{i,m}(z)\phi_{j,n}(z)}{L_z + L_z'} \approx \frac{\int_{-\infty}^{\infty} \mathrm{d}z \phi^*_{i,m}(z)\phi_{j,n}(z)}{L}, \tag{6.105}$$

where $L = L_z + L_z'$. Note that it is entirely possible that $\langle \phi^*_{i,m}(z)\phi_{j,n}(z) \rangle \neq 0$ when $m \neq n$, as long as $i \neq j$. This is because the confining potentials in two different bands i and j may be different, so that $\phi_{i,m}$ and $\phi_{j,n}$ may not be orthogonal. This is clearly the case when there is an electric field transverse to the plane of the quantum well, as in the quantum-confined Stark effect discussed in Problem 3.8.

Using this result in (6.104), we obtain

$$\begin{aligned} K_{\hat{p},\omega_1}(i,m:j,n) &= \left| \int_{-\infty}^{\infty} \mathrm{d}z \phi^*_{i,m}(z)\phi_{j,n}(z) \right|^2 \\ &\times \left| \mathrm{i}\vec{k}_p \int_{-\infty}^{\infty} \int_0^{\infty} \mathrm{d}z d^2\vec{\rho} \frac{1}{AL} u^*_{\vec{k}_t,i}(\vec{\rho},z) u_{\vec{k}_t,j}(\vec{\rho},z) \right. \\ &\left. + \int_{-\infty}^{\infty} \int_0^{\infty} \mathrm{d}z d^2\vec{\rho} \frac{1}{AL} u^*_{\vec{k}_t,i}(\vec{\rho},z) \left(\hat{p}\cdot\vec{\nabla}_{\vec{\rho}}\right) u_{\vec{k}_t,j}(\vec{\rho},z) \right|^2 \\ &= \left| \int_{-\infty}^{\infty} \mathrm{d}z \phi^*_{i,m}(z)\phi_{j,n}(z) \right|^2 \\ &\times \left| i\vec{k}_{\hat{p}} \int_{-\infty}^{\infty} \int_0^{\infty} d^3\vec{r} \frac{1}{\Omega} \overline{u}^*_{\vec{k}_t,i}(\vec{r}) \overline{u}_{\vec{k}_t,j}(\vec{r}) \right. \\ &\left. + \int_0^{\infty} d^3\vec{r} \frac{1}{\Omega} \overline{u}^*_{\vec{k}_t,i}(\vec{r}) \left(\hat{p}\cdot\vec{\nabla}_{\vec{r}}\right) \overline{u}_{\vec{k}_t,j}(\vec{r}) \right|^2, \end{aligned} \tag{6.106}$$

where $\Omega = AL$ and $\overline{u}_{\vec{k}_t,q}(\vec{r}) = u_{\vec{k}_t,q}(\vec{\rho}, z)$. We have put a "bar" over $\overline{u}_{\vec{k}_t,q}(\vec{r})$ just to remind the reader that it is a pseudo-Bloch function, as opposed to a true Bloch function, since it is not really periodic in the z-direction.

We will assume (tacitly) that the pseudo-Bloch functions share the same orthogonality property with true Bloch functions that are given in (4.59), so that $\int_0^\infty d^3\vec{r}\,\overline{u}^*_{\vec{k}_t,i}(\vec{r})\overline{u}_{\vec{k}_t,j}(\vec{r}) = \delta_{i,j}$, where the δ is a Kronecker delta. This will yield:

$$K_{\hat{p},\omega_1}(i,m:j,n) = \left|\int_{-\infty}^{\infty} \mathrm{d}z\phi^*_{i,m}(z)\phi_{j,n}(z)\right|^2$$

$$\times\left|\mathrm{i}\vec{k}_p\delta_{i,j} + \int_0^\infty d^3\vec{r}\frac{1}{\Omega}\overline{u}^*_{\vec{k}_t,i}(\vec{r})\left(\hat{p}\cdot\vec{\nabla}_{\vec{r}}\right)\overline{u}_{\vec{k}_t,j}(\vec{r})\right|^2. \quad (6.107)$$

Next, we will derive expressions for the effective matrix element $K_{\hat{p},\omega_1}(i,m:j,n)$ in the case of interband and intraband transitions.

Interband transitions ($i \neq j$)

For interband transitions, the first term in the right-hand side of (6.107) vanishes because of the Kronecker delta and we are left with

$$K_{\hat{p},\omega_1}(i,m:j,n) = \left|\int_{-\infty}^{\infty} \mathrm{d}z\phi^*_{i,m}(z)\phi_{j,n}(z)\right|^2 \left|\int_0^\infty d^3\vec{r}\frac{1}{\Omega}\overline{u}^*_{\vec{k}_t,i}(\vec{r})(\hat{p}\cdot\vec{\nabla}_{\vec{r}})\overline{u}_{\vec{k}_t,j}(\vec{r})\right|^2$$

$$= \frac{1}{\hbar^2}\left|\int_{-\infty}^{\infty} \mathrm{d}z\phi^*_{i,m}(z)\phi_{j,n}(z)\right|^2 \left|\hat{p}\cdot\vec{\Upsilon}'_{i,j}\right|^2, \quad (6.108)$$

where

$$\vec{\Upsilon}'_{i,j} = \int_0^\infty d^3\vec{r}\frac{1}{\Omega}\overline{u}^*_{\vec{k}_t,i}(\vec{r})\left(-\mathrm{i}\hbar\vec{\nabla}_{\vec{r}}\right)\overline{u}_{\vec{k}_t,j}(\vec{r}). \quad (6.109)$$

Equation (6.108) tells us that absorption and emission strengths in a quantum well are proportional to the squared magnitude of the overlap between the subband wavefunctions of the initial and final states. This is the basis for Problem 3.8 in Chap. 3.

For light polarized in the x-y plane, the unit polarization vector $\hat{p} = a\hat{x} + b\hat{y}$, where a and b are real quantities and $a^2 + b^2 = 1$. Therefore, using Table 6.1 and assuming that the pseudo-Bloch functions are close enough to true Bloch functions that $\vec{\Upsilon}'_{i,j} \approx \vec{\Upsilon}_{i,j}$, we get that absorption and luminescence involving electron and heavy-hole transition in a symmetric quantum well are proportional to

$$\left|\hat{p}\cdot\vec{\Upsilon}'_{i,j}\right|^2 \propto \left|(a\hat{x} + b\hat{y})\cdot\frac{1}{\sqrt{2}}(\hat{x} \pm \mathrm{i}\hat{y})\right|^2 = \frac{1}{2}, \quad (6.110)$$

whereas absorption and luminescence involving electron and light-hole transition in a symmetric quantum well are proportional to

$$\left|\hat{p}\cdot\vec{\Upsilon}'_{i,j}\right|^2 \propto \left|(a\hat{x}+b\hat{y})\cdot\frac{1}{\sqrt{6}}(\hat{x}\pm \mathrm{i}\hat{y})\right|^2 = \frac{1}{6}. \tag{6.111}$$

Therefore, the heavy hole transition strength is three times that of the light hole transition strength in a symmetric quantum well with light polarized in the plane of the well.

Intraband, intersubband transitions ($i = j,\ m \neq n$)

The subband wavefunctions $\phi_{i,m}(z)$ and $\phi_{i,n}(z)$ corresponding to different subbands within the same band ($i = j,\ m \neq n$) are orthogonal, i.e.,

$$\int_{-\infty}^{\infty} \mathrm{d}z\phi^*_{i,m}(z)\phi_{i,n}(z) = \delta_{m,n}; \tag{6.112}$$

hence, we immediately see from (6.108) that $K_{\hat{p},\omega_1}(i,m : i,n) = 0$, i.e, intraband, intersubband transitions are not allowed if light is polarized in the plane of the quantum well. This is true even in an asymmetric quantum well where an electric field breaks inversion symmetry in the z-direction (recall the discussion of quantum-confined Stark effect in Chap. 3). It is true because $\phi_{i,m}(z)$ and $\phi_{i,n}(z)$ are two nondegenerate solutions of the same Hamiltonian which is a hermitian operator. Therefore, they remain orthogonal if $m \neq n$, even in an asymmetric quantum well. Consequently, light polarized in the plane of the quantum well cannot induce intraband intersubband transitions (within all the approximations we made).

6.8.1.2 Light Polarized Perpendicular to the Plane of the Quantum Well

In this case, the polarization vector is in the z-direction and (6.102) yields the effective matrix element as

$$\begin{aligned}
K_{\hat{p},\omega_1}(i,m : j,n) \approx &\left|\int_{-\infty}^{\infty}\mathrm{d}z\phi^*_{i,m}(z)\int_0^{\infty} d^2\vec{\rho}\left\{\frac{1}{\sqrt{A}}\mathrm{e}^{-\mathrm{i}\vec{k}'_t\cdot\vec{\rho}}\right.\right.\\
&\left.\left. u^*_{\vec{k}'_t,i}(\vec{\rho},z)\partial_{\hat{p}}\left(\frac{1}{\sqrt{A}}\mathrm{e}^{\mathrm{i}\vec{k}_t\cdot\vec{\rho}}u_{\vec{k}_t,j}(\vec{\rho},z)\phi_{j,n}(z)\right)\right\}\right|^2\\
= &\left|\int_{-\infty}^{\infty}\mathrm{d}z\phi^*_{i,m}(z)\int_0^{\infty} d^2\vec{\rho}\left\{\frac{1}{A}\mathrm{e}^{-\mathrm{i}\left(\vec{k}'_t-\vec{k}_t\right)\cdot\vec{\rho}}u^*_{\vec{k}'_t,i}(\vec{\rho},z)\right.\right.\\
&\left.\left.\frac{\partial}{\partial z}\left(u_{\vec{k}_t,j}(\vec{\rho},z)\phi_{j,n}(z)\right)\right\}\right|^2
\end{aligned}$$

$$= \left| \int_{-\infty}^{\infty} \mathrm{d}z\phi_{i,m}^*(z) \int_0^{\infty} d^2\vec{\rho} \left\{ \phi_{j,n}(z) u_{\vec{k}_t',i}^*(\vec{\rho},z) \frac{\partial}{\partial z} \left(u_{\vec{k}_t,j}(\vec{\rho},z) \right) \right.\right.$$
$$\left.\left. + u_{\vec{k}_t',i}^*(\vec{\rho},z) u_{\vec{k}_t,j}(\vec{\rho},z) \phi_{i,m}^*(z) \frac{\partial}{\partial z} \left(\phi_{j,n}(z) \right) \right\} \frac{1}{A} \delta_{\vec{k}_t',\vec{k}_t} \right|^2, \quad (6.113)$$

where we see that once again the k-selection rule is validated because $\vec{k}_t' = \vec{k}_t$.

We will now invoke the "envelope approximation" since the envelope wavefunctions vary in z much more slowly than the pseudo-Bloch functions. This will lead us to:

$$K_{\hat{p},\omega_1}(i,m:j,n) \approx \left| \langle \phi_{i,m}^*(z) \phi_{j,n}(z) \rangle \int_{-\infty}^{\infty} \int_0^{\infty} \mathrm{d}z d^2\vec{\rho} \frac{1}{A} u_{\vec{k}_t,i}^*(\vec{\rho},z) \frac{\partial}{\partial z} \left(u_{\vec{k}_t,j}(\vec{\rho},z) \right) \right.$$
$$\left. + \left\langle \phi_{i,m}^*(z) \frac{\partial \phi_{j,n}(z)}{\partial z} \right\rangle \int_{-\infty}^{\infty} \int_0^{\infty} \mathrm{d}z d^2\vec{\rho} \frac{1}{A} u_{\vec{k}_t,i}^*(\vec{\rho},z) u_{\vec{k}_t,j}(\vec{\rho},z) \right|^2$$
$$= \frac{1}{\hbar^2} \left| \int_{-\infty}^{\infty} \mathrm{d}z\phi_{i,m}^*(z) \phi_{j,n}(z) \right.$$
$$\times \int_0^{\infty} d^3\vec{r} \frac{1}{\Omega} u_{\vec{k}_t,i}^*(\vec{r}) \left[\hat{z} \cdot \left(-\mathrm{i}\hbar \vec{\nabla}_r \right) \right] u_{\vec{k}_t,j}(\vec{r})$$
$$\left. + \int_{-\infty}^{\infty} \mathrm{d}z\phi_{i,m}^*(z) \left[-\mathrm{i}\hbar \frac{\partial}{\partial z} \right] \phi_{j,n}(z) \times \int_0^{\infty} d^3\vec{r} \frac{1}{\Omega} \overline{u}_{\vec{k}_t,i}^*(\vec{r}) \overline{u}_{\vec{k}_t,j}(\vec{r}) \right|^2$$
$$= \frac{1}{\hbar^2} \left| \left\{ \int_{-\infty}^{\infty} \mathrm{d}z\phi_{i,m}^*(z) \phi_{j,n}(z) \right\} \times \left[\hat{z} \cdot \vec{\Upsilon}_{i,j}' \right] \right.$$
$$\left. + \left\{ \int_{-\infty}^{\infty} \mathrm{d}z\phi_{i,m}^*(z) \left[-\mathrm{i}\hbar \frac{\partial}{\partial z} \right] \phi_{j,n}(z) \right\} \times \delta_{i,j} \right|^2, \quad (6.114)$$

where $\vec{\Upsilon}_{i,j}'$ has been defined in (6.109) and we have assumed, as before, that the pseudo-Bloch functions at the same wavevector k_t in different bands are orthonormal.

Interband transitions ($i \neq j$)

For interband transitions, the Kronecker delta makes the second term within the modulus vanish, leaving us with

$$K_{\hat{p},\omega_1}(i,m:j,n) = \frac{1}{\hbar^2} \left| \int_{-\infty}^{\infty} \mathrm{d}z\phi_{i,m}^*(z) \phi_{j,n}(z) \right|^2 \left| \hat{z} \cdot \vec{\Upsilon}_{i,j}' \right|^2. \quad (6.115)$$

We can see immediately that light polarized perpendicular to the plane of the quantum well does *not* induce any transition involving the heavy-hole since from Table 6.1, $\hat{z} \cdot \vec{\Upsilon}_{i,j}' = 0$. However, it does induce light-hole transitions.

Intraband, intersubband transitions ($i = j$, $m \neq n$)

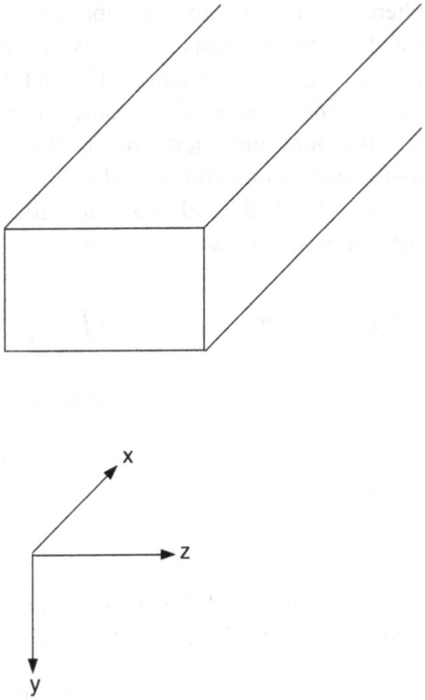

Fig. 6.15 A generic quantum wire with rectangular cross-section

Invoking the orthogonality of the wavefunctions of different subbands in the same band (see (6.112), we get

$$K_{\hat{p},\omega_1}(i,m:i,n) = \left| \int_{-\infty}^{\infty} \mathrm{d}z \phi^*_{i,m}(z) \frac{\partial \phi_{i,n}(z)}{\partial z} \right|^2. \tag{6.116}$$

The right-hand side is not zero. Thus, unlike light polarized in the plane of the quantum well, light polarized perpendicular to the plane can induce intraband, intersubband transitions.

6.8.2 *Quantum Wires*

Consider a quantum wire whose axis is along the x-direction, as shown in Fig. 6.15. The electron can move freely in the x-direction but its motion is confined in the y- and z-directions.

Following the 2DEG case, we can write the wavefunction in the i-th band of the quasi one-dimensional electron gas (1-DEG) as

$$\psi(x,y,z) = \frac{1}{\sqrt{L}} \mathrm{e}^{\mathrm{i}k_x x} \tilde{u}_{k_x,i}(x,y,z) \varphi_{i,m}(y) \phi_{i,m'}(z), \tag{6.117}$$

where m and m' are the subband indices associated with quantization in the y-and z-direction, respectively. We have put a tilde over the pseudo-Bloch function $\tilde{u}_{k_x,i}(x,y,z)$ just to highlight the fact that it is not globally periodic in the y- and z-directions because of the material discontinuities in those two directions. Note also that now there are two envelope wavefunctions $\varphi(y)$ and $\phi(z)$ because of confinement along both y- and z-directions.

As in the three- and two-dimensional cases, the effective matrix element in the one-dimensional case is given by

$$K_{\hat{p},\omega_1}(i,m,m':j,n,n') = \left| \int_{-\infty}^{\infty}\int_{-\infty}^{\infty}\int_{-\infty}^{\infty} \mathrm{d}x\mathrm{d}y\mathrm{d}z \left\{ \frac{1}{\sqrt{L}} \mathrm{e}^{-\mathrm{i}k'_x x} \tilde{u}^*_{k'_x,i}(x,y,z) \right.\right.$$
$$\times\, \varphi^*_{i,m}(y)\phi^*_{i,m'}(z)\mathrm{e}^{\mathrm{i}(q_x x+q_y y+q_z z)}$$
$$\left.\left. \times\, \partial_{\hat{p}}\left(\frac{1}{\sqrt{L}}\mathrm{e}^{\mathrm{i}k_x x}\tilde{u}_{k_x,j}(x,y,z)\varphi_{j,n}(y)\phi_{j,n'}(z)\right)\right\}\right|^2. \tag{6.118}$$

Once again, we will account for the fact that the photon's momentum (and hence wavevector) is negligible, so that the last equation can be approximated as

$$K_{\hat{p},\omega_1}(i,m,m':j,n,n') = \left| \int_{-\infty}^{\infty}\int_{-\infty}^{\infty}\int_{-\infty}^{\infty} \mathrm{d}x\mathrm{d}y\mathrm{d}z \left\{ \frac{1}{\sqrt{L}} \mathrm{e}^{-\mathrm{i}k'_x x} \tilde{u}^*_{k'_x,i}(x,y,z) \right.\right.$$
$$\times\, \varphi^*_{i,m}(y)\phi^*_{i,m'}(z)$$
$$\left.\left. \times\, \partial_{\hat{p}}\left(\frac{1}{\sqrt{L}}\mathrm{e}^{\mathrm{i}k_x x}\tilde{u}_{k_x,j}(x,y,z)\varphi_{j,n}(y)\phi_{j,n'}(z)\right)\right\}\right|^2. \tag{6.119}$$

We will now proceed to calculate the effective matrix element for two different light polarizations.

6.8.2.1 Light Polarized Along the Wire Axis

For light polarized along the x-axis, the effective matrix element is given by

$$K_{\hat{p},\omega_1}(i,m,m':j,n,n') = \left| \int_{-\infty}^{\infty}\int_{-\infty}^{\infty}\int_{-\infty}^{\infty} \mathrm{d}x\mathrm{d}y\mathrm{d}z \left\{ \frac{1}{L} \mathrm{e}^{\mathrm{i}(k'_x-k_x)x} \varphi^*_{i,m}(y)\varphi_{j,n}(y) \right.\right.$$
$$\left.\left. \times\, \phi^*_{i,m'}(z)\phi_{j,n'}(z)\tilde{u}^*_{k'_x,i}(x,y,z)\left(\mathrm{i}k_x + \frac{\partial}{\partial x}\right)\tilde{u}_{k_x,j}(x,y,z)\right\}\right|^2$$

$$
\begin{aligned}
&= \left| \int_{-\infty}^{\infty} \int_{-\infty}^{\infty} \int_{-\infty}^{\infty} \mathrm{d}x\mathrm{d}y\mathrm{d}z \left\{ \varphi_{i,m}^{*}(y)\varphi_{j,n}(y)\phi_{i,m'}^{*}(z)\phi_{j,n'}(z) \right.\right. \\
&\quad \left.\left. \times \tilde{u}_{k_x',i}^{*}(x,y,z)\left(\mathrm{i}k_x + \frac{\partial}{\partial x}\right)\tilde{u}_{k_x,j}(x,y,z)\frac{1}{L}\delta_{k_x',k_x} \right\} \right|^2 \\
&= \left| \int_{-\infty}^{\infty} \int_{-\infty}^{\infty} \int_{-\infty}^{\infty} \mathrm{d}x\mathrm{d}y\mathrm{d}z \left\{ \frac{1}{L}\varphi_{i,m}^{*}(y)\varphi_{j,n}(y)\phi_{i,m'}^{*}(z)\phi_{j,n'}(z) \right.\right. \\
&\quad \left.\left. \times \tilde{u}_{k_x,i}^{*}(x,y,z)\left(\mathrm{i}k_x + \frac{\partial}{\partial x}\right)\tilde{u}_{k_x,j}(x,y,z) \right\} \right|^2 \\
&\approx \left| \langle \varphi_{i,m}^{*}(y)\varphi_{j,n}(y)\rangle \left\langle \phi_{i,m'}^{*}(z)\phi_{j,n'}(z)\right\rangle \right. \\
&\quad \times \int_{-\infty}^{\infty} \int_{-\infty}^{\infty} \int_{-\infty}^{\infty} \mathrm{d}x\mathrm{d}y\mathrm{d}z \left\{ \frac{1}{L}\tilde{u}_{k_x,i}^{*}(x,y,z) \right. \\
&\quad \left.\left. \left(\mathrm{i}k_x + \frac{\partial}{\partial x}\right)\tilde{u}_{k_x,j}(x,y,z) \right\} \right|^2 \\
&= \left| \int_{-\infty}^{\infty} \mathrm{d}y\varphi_{i,m}^{*}(y)\varphi_{j,n}(y) \right|^2 \left| \int_{-\infty}^{\infty} \mathrm{d}z\phi_{i,m'}^{*}(z)\phi_{j,n'}(z) \right|^2 \\
&\quad \times \left| \mathrm{i}k_x \int_{-\infty}^{\infty} \int_{-\infty}^{\infty} \int_{-\infty}^{\infty} \mathrm{d}x\mathrm{d}y\mathrm{d}z \frac{1}{\Omega}\tilde{u}_{k_x,i}^{*}(x,y,z)\tilde{u}_{k_x,j}(x,y,z) \right. \\
&\quad \left. + \int_{-\infty}^{\infty} \int_{-\infty}^{\infty} \int_{-\infty}^{\infty} \mathrm{d}x\mathrm{d}y\mathrm{d}z \frac{1}{\Omega}\tilde{u}_{k_x,i}^{*}(x,y,z)\frac{\partial}{\partial x}\left[\tilde{u}_{k_x,j}(x,y,z)\right] \right|^2 \\
&= \left| \int_{-\infty}^{\infty} \mathrm{d}y\varphi_{i,m}^{*}(y)\varphi_{j,n}(y) \right|^2 \left| \int_{-\infty}^{\infty} \mathrm{d}z\phi_{i,m'}^{*}(z)\phi_{j,n'}(z) \right|^2 \\
&\quad \times \left| \mathrm{i}k_x\delta_{i,j} \right. \\
&\quad \left. + \frac{1}{\Omega} \int_{-\infty}^{\infty} \int_{-\infty}^{\infty} \int_{-\infty}^{\infty} \mathrm{d}x\mathrm{d}y\mathrm{d}z\tilde{u}_{k_x,i}^{*}(x,y,z)\frac{\partial}{\partial x}\left[\tilde{u}_{k_x,j}(x,y,z)\right] \right|^2 \\
&= \left| \int_{-\infty}^{\infty} \mathrm{d}y\varphi_{i,m}^{*}(y)\varphi_{j,n}(y) \right|^2 \left| \int_{-\infty}^{\infty} \mathrm{d}z\phi_{i,m'}^{*}(z)\phi_{j,n'}(z) \right|^2 \\
&\quad \times \frac{1}{\hbar^2} \left| \mathrm{i}\hbar k_x\delta_{i,j} + \hat{x}\cdot\vec{\Upsilon}_{ij}'' \right|^2,
\end{aligned} \tag{6.120}
$$

where

$$
\vec{\Upsilon}_{ij}'' = \int_0^{\infty} d^3\vec{r}\frac{1}{\Omega}\tilde{u}_{k_x,i}^{*}(\vec{r})\left(-\mathrm{i}\hbar\vec{\nabla}_r\right)\left[\tilde{u}_{k_x,j}(\vec{r})\right]. \tag{6.121}
$$

Note the Kronecker delta in the second line of the preceding equation which ensures that the k-selection rule is valid as far as the electron's wavevector along the wire axis is concerned, i.e., $k_x^{\text{initial}} = k_x^{\text{final}}$.

For interband transitions (i ≠ j)

$$K_{\hat{p},\omega_1}(i,m,m':j,n,n') = \frac{1}{\hbar^2}\left|\int_{-\infty}^{\infty} \mathrm{d}y\varphi_{i,m}^*(y)\varphi_{j,n}(y)\right|^2 \left|\int_{-\infty}^{\infty} \mathrm{d}z\phi_{i,m'}^*(z)\phi_{j,n'}(z)\right|^2 \times \left|\hat{x}\cdot\vec{\Upsilon}_{ij}''\right|^2, \tag{6.122}$$

For intraband, intersubband transitions (i = j, m ≠ n and/or m′ ≠ n′)

$$K_{\hat{p},\omega_1}(i,m,m':j,n,n') = 0, \tag{6.123}$$

since

$$\int_{-\infty}^{\infty} \mathrm{d}y\varphi_{i,m}^*(y)\varphi_{i,n}(y) = \delta_{m,n}$$
$$\int_{-\infty}^{\infty} \mathrm{d}z\phi_{i,m'}^*(z)\phi_{i,n'}(z) = \delta_{m',n'}. \tag{6.124}$$

Therefore, light polarized along the wire axis cannot induce intersubband transitions within the same band. Compare this with the quantum well case.

6.8.2.2 For Light Polarized Perpendicular to the Wire Axis

Say, light is polarized in the y-direction. In that case, ignoring the photon's momentum, the effective matrix element is given by

$$\begin{aligned}
K_{\hat{p},\omega_1}(i,m,m':j,n,n') &\approx \Bigg|\int_{-\infty}^{\infty}\int_{-\infty}^{\infty}\int_{-\infty}^{\infty} \mathrm{d}x\mathrm{d}y\mathrm{d}z\Bigg\{\frac{1}{\sqrt{L}}\mathrm{e}^{-\mathrm{i}k_x'x}\tilde{u}_{k_x',i}^*(x,y,z) \\
&\quad \times\varphi_{i,m}^*(y)\phi_{i,m'}^*(z) \\
&\quad \times\frac{\partial}{\partial y}\left(\frac{1}{\sqrt{L}}\mathrm{e}^{\mathrm{i}k_xx}\tilde{u}_{k_x,j}(x,y,z)\varphi_{j,n}(y)\phi_{j,n'}(z)\right)\Bigg\}\Bigg|^2 \\
&= \Bigg|\frac{1}{L}\int_{-\infty}^{\infty}\int_{-\infty}^{\infty}\int_{-\infty}^{\infty} \mathrm{d}x\mathrm{d}y\mathrm{d}z\Bigg\{\mathrm{e}^{-\mathrm{i}(k_x'-k_x)x}\tilde{u}_{k_x',i}^*(x,y,z) \\
&\quad \times \varphi_{i,m}^*(y)\phi_{i,m'}^*(z)\phi_{j,n'}(z)\frac{\partial}{\partial y}\left[\tilde{u}_{k_x,j}(x,y,z)\varphi_{j,n}(y)\right]\Bigg\}\Bigg|^2 \\
&= \Bigg|\frac{1}{L}\int_{-\infty}^{\infty}\int_{-\infty}^{\infty}\int_{-\infty}^{\infty} \mathrm{d}x\mathrm{d}y\mathrm{d}z\Bigg\{\delta_{k_x',k_x}\tilde{u}_{k_x',i}^*(x,y,z)\times \\
&\quad \varphi_{i,m}^*(y)\phi_{i,m'}^*(z)\phi_{j,n'}(z) \\
&\quad \times\left[\varphi_{j,n}(y)\frac{\partial\tilde{u}_{k_x,j}(x,y,z)}{\partial y}+\tilde{u}_{k_x,j}(x,y,z)\frac{\partial\varphi_{j,n}(y)}{\partial y}\right]\Bigg\}\Bigg|^2
\end{aligned}$$

$$
\begin{aligned}
&= \Bigg| \langle \varphi_{i,m}^*(y)\varphi_{j,n}(y)\rangle \left\langle \phi_{i,m'}^*(z)\phi_{j,n'}(z)\right\rangle \\
&\times \left\{ \frac{1}{L}\int_{-\infty}^{\infty}\int_{-\infty}^{\infty}\int_{-\infty}^{\infty} \mathrm{d}x\mathrm{d}y\mathrm{d}z\tilde{u}_{k_x,i}^*(x,y,z)\frac{\partial \tilde{u}_{k_x,j}(x,y,z)}{\partial y} \right. \\
&+ \left\langle \varphi_{i,m}^*(y)\frac{\partial \varphi_{j,n}(y)}{\partial y}\right\rangle \left\langle \phi_{i,m'}^*(z)\phi_{j,n'}(z)\right\rangle \\
&\times \left. \frac{1}{L}\int_{-\infty}^{\infty}\int_{-\infty}^{\infty}\int_{-\infty}^{\infty} \mathrm{d}x\mathrm{d}y\mathrm{d}z\tilde{u}_{k_x,i}^*(x,y,z)\tilde{u}_{k_x,j}(x,y,z) \right\} \Bigg|^2 \\
&= \Bigg| \int_{-\infty}^{\infty} \mathrm{d}y\varphi_{i,m}^*(y)\varphi_{j,n}(y)\int_{-\infty}^{\infty}\mathrm{d}z\phi_{i,m'}^*(z)\phi_{j,n'}(z) \\
&\times \frac{1}{\Omega}\int_{-\infty}^{\infty}\int_{-\infty}^{\infty}\int_{-\infty}^{\infty} \mathrm{d}x\mathrm{d}y\mathrm{d}z\tilde{u}_{k_x,i}^*(x,y,z)\frac{\partial \tilde{u}_{k_x,j}(x,y,z)}{\partial y} \\
&+ \int_{-\infty}^{\infty} \mathrm{d}y\varphi_{i,m}^*(y)\frac{\partial \varphi_{j,n}(y)}{\partial y}\int_{-\infty}^{\infty}\mathrm{d}z\phi_{i,m'}^*(z)\phi_{j,n'}(z) \\
&\times \frac{1}{\Omega}\int_{-\infty}^{\infty}\int_{-\infty}^{\infty}\int_{-\infty}^{\infty} \mathrm{d}x\mathrm{d}y\mathrm{d}z\tilde{u}_{k_x,i}^*(x,y,z)\tilde{u}_{k_x,j}(x,y,z) \Bigg|^2 \\
&= \frac{1}{\hbar^2}\left| \int_{-\infty}^{\infty}\mathrm{d}z\phi_{i,m'}^*(z)\phi_{j,n'}(z)\right|^2 \\
&\times \left\{ \left| \int_{-\infty}^{\infty}\mathrm{d}y\varphi_{i,m}^*(y)\varphi_{j,n}(y)\right|^2 \left|\hat{y}\cdot\vec{\Upsilon}_{ij}''\right|^2 \right. \\
&+ \left. \left| \int_{-\infty}^{\infty}\mathrm{d}y\varphi_{i,m}^*(y)(-\mathrm{i}\hbar)\frac{\partial \varphi_{j,n}(y)}{\partial y}\right|^2 \delta_{i,j} \right\}. \qquad (6.125)
\end{aligned}
$$

As always, the k-selection rule is valid, i.e., $k_x^{\text{initial}} = k_x^{\text{final}}$.

For interband transitions $(i \neq j)$

$$
\begin{aligned}
K_{\hat{p},\omega_1}(i,m,m':j,n,n') &= \frac{1}{\hbar^2}\left|\int_{-\infty}^{\infty}\mathrm{d}y\varphi_{i,m}^*(y)\varphi_{j,n}(y)\right|^2 \left|\int_{-\infty}^{\infty}\mathrm{d}z\phi_{i,m'}^*(z)\phi_{j,n'}(z)\right|^2 \\
&\times \left|\hat{y}\cdot\vec{\Upsilon}_{ij}''\right|^2. \qquad (6.126)
\end{aligned}
$$

For intraband, intersubband transitions $(i = j, m \neq n)$

$$
K_{\hat{p},\omega_1}(i,m,m':i,n,n') = \left|\int_{-\infty}^{\infty}\mathrm{d}y\varphi_{i,m}^*(y)\frac{\partial \varphi_{j,n}(y)}{\partial y}\right|^2 \delta_{m',n'}. \qquad (6.127)
$$

The right-hand side is not zero as long as $m' = n'$, and hence light polarized perpendicular to the wire axis can induce intersubband transitions within the same band. Note, however, that the indices of the initial and final subbands associated with confinement in the direction perpendicular to light polarization must be the same, i.e., $m' = n'$.

6.9 Excitons

So far, we have described the *interband* absorption process as the generation of an electron–hole pair by an absorbed photon. Similarly, the *interband* emission process has been described as the annihilation of a pair accompanied by the emission of a photon. In this simple picture, the electron energy will be (assuming a parabolic dispersion relation in the conduction band)

$$E_e = E_{c0} + \frac{\hbar^2 k_e^2}{2m_e^*}, \tag{6.128}$$

while the hole energy will be (again, assuming a parabolic dispersion relation in the valence band)

$$E_h = E_{v0} - \frac{\hbar^2 k_h^2}{2m_h^*}, \tag{6.129}$$

where E_{c0} and E_{v0} are the conduction and valence band edge energies, k_e and k_h are the electron and hole wavevectors, and m_e^* and m_h^* are the electron and hole effective masses, respectively.

This construct ignores a crucial effect, namely that the electron and hole, being oppositely charged, actually attract each other so that they do not behave as independent particles. They are Coulomb coupled and their motions are correlated. In the limit of strong attraction, they will behave as one composite particle called an *exciton*.

Because of the mutual Coulomb attraction, the energy difference between the electron state and the hole state will not quite be the bandgap energy E_g, but instead, it will be $E_g - E_b$, where E_b is the Coulomb attraction energy (called the exciton binding energy) between the coupled electron–hole pair. Therefore, light absorption in the solid will initiate not at photon energy equal to E_g, but instead at slightly lower photon energy equal to $E_g - E_b$. This "below-bandgap" absorption is called excitonic absorption. In fact, excitons introduce an energy level E_{exc} in the bandgap at energy E_b below the conduction band edge E_{c0} and an electron can be excited from the top of the valence band E_{v0} to this state while absorbing a photon of energy $E_g - E_b$ (see Fig. 6.16). Since the exciton level is a discrete level, there is a gap in the available states between the exciton level and the conduction band edge. Thus, if everything were ideal, we should see a discrete absorption line at photon energy $E_g - E_b$, no absorption for photon energies between $E_g - E_b$ and E_g, and then strong absorption for photon energies exceeding E_g. Thus, the excitons should introduce an "exciton peak" in the absorption spectrum. For this to be visible, the temperature must be low enough that $k_B T < E_b$. Otherwise, two things can happen both of which will wash out the exciton peak. First, thermal fluctuations can dissociate (or "ionize") the exciton and break apart the electron and hole, so that they behave as free and independent particles. This will eliminate the level E_{exc}. Second,

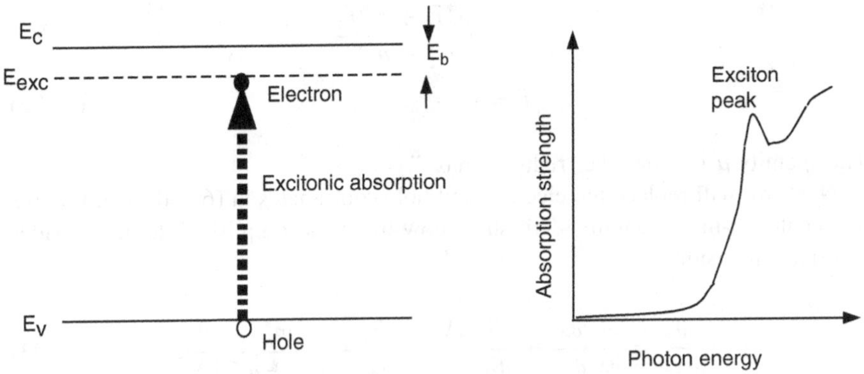

Fig. 6.16 (*Left*) The conduction band edge, valence band edge and the exciton level in a semiconductor (shown in real space). An absorption event may involve excitation of an electron from the valence band edge to the exciton level which is located below the conduction band edge by an amount equal to the exciton binding energy. The binding energy is the energy that binds the electron and hole making up the exciton. The photon that is absorbed in this process has energy less than the bandgap by an amount equal to the binding energy. (*Right*) The absorption spectrum showing the exciton absorption peak in a bulk semiconductor

thermal broadening of the exciton level may merge the exciton absorption peak with the bandedge absorption. That will make the exciton peak indistinguishable from normal band-to-band absorption.

In order to find the exciton absorption energy, we need to calculate E_b. To do that quantum-mechanically, we must solve the two-particle Schrödinger equation for the electron and hole, taking into account the Coulomb attraction between them. To do this, we will use the effective mass Schrödinger equation, so that we do not have to account for the lattice potential $V_L(\vec{r})$ explicitly.

This two-particle equation will be

$$\left[-\frac{\hbar^2}{2m_e^*}\nabla_e^2 - \frac{\hbar^2}{2m_h^*}\nabla_h^2 - \frac{e^2}{4\pi\epsilon\left|\vec{r}_e - \vec{r}_h\right|}\right]\psi(\vec{r}_e, \vec{r}_h) = E\psi(\vec{r}_e, \vec{r}_h), \tag{6.130}$$

where $\vec{r}_e$ and $\vec{r}_h$ are the electron and hole coordinates, ∇_e and ∇_h are the gradient operators for the electron and the hole, and ϵ is the dielectric constant.

Since the Coulomb coupling term involves *both* electron and hole coordinates, we cannot write the wavefunction as the product of two wavefunctions—one involving the electron coordinate and the other the hole coordinate—and attempt a variable separation solution. However, a different kind of variable separation solution does become possible if we employ relative and center-of-mass quantities:

$$M = m_e^* + m_h^*$$
$$\frac{1}{\mu} = \frac{1}{m_e^*} + \frac{1}{m_h^*}. \tag{6.131}$$

$$\vec{R} = \frac{m_e^* \vec{r}_e + m_h^* \vec{r}_h}{m_e^* + m_h^*}$$
$$\vec{r} = \vec{r}_e - \vec{r}_h. \tag{6.132}$$

The quantity μ is called the "reduced mass."

Next, we will replace the electron and hole coordinates in (6.130) with relative and center-of-mass quantities. To show how this is accomplished, let us consider just one dimension:

$$\frac{\partial}{\partial x_e} = \frac{\partial}{\partial x}\frac{\partial x}{\partial x_e} + \frac{\partial}{\partial X}\frac{\partial X}{\partial x_e} = \frac{\partial}{\partial x} + \frac{m_e^*}{m_e^* + m_h^*}\frac{\partial}{\partial X}. \tag{6.133}$$

Consequently,

$$\frac{\partial^2}{\partial x_e^2} = \frac{\partial^2}{\partial x^2} + \left(\frac{m_e^*}{m_e^* + m_h^*}\right)^2 \frac{\partial^2}{\partial X^2} + \frac{2m_e^*}{m_e^* + m_h^*}\frac{\partial^2}{\partial x \partial X}, \tag{6.134}$$

and

$$\frac{\partial^2}{\partial x_h^2} = \frac{\partial^2}{\partial x^2} + \left(\frac{m_h^*}{m_e^* + m_h^*}\right)^2 \frac{\partial^2}{\partial X^2} - \frac{2m_h^*}{m_e^* + m_h^*}\frac{\partial^2}{\partial x \partial X}, \tag{6.135}$$

Using the previous two equations, we immediately see that

$$-\frac{\hbar^2}{2m_e^*}\frac{\partial^2}{\partial x_e^2} - \frac{\hbar^2}{2m_h^*}\frac{\partial^2}{\partial x_h^2} = -\frac{\hbar^2}{2M}\frac{\partial^2}{\partial X^2} - \frac{\hbar^2}{2\mu}\frac{\partial^2}{\partial x^2}. \tag{6.136}$$

Similar relations hold for the y- and z-coordinates. Therefore, (6.130) becomes

$$\left[-\frac{\hbar^2}{2M}\nabla_{\vec{R}}^2 - \frac{\hbar^2}{2\mu}\nabla_{\vec{r}}^2 - \frac{e^2}{4\pi\epsilon|\vec{r}|}\right]\psi(\vec{R},\vec{r}) = E\psi(\vec{R},\vec{r}), \tag{6.137}$$

where, in spherical coordinates,

$$\nabla_\rho^2 = \frac{1}{\rho^2}\frac{\partial}{\partial\rho}\left(\rho^2\frac{\partial}{\partial\rho}\right) + \frac{1}{\rho^2 \sin\theta}\frac{\partial}{\partial\theta}\left(\sin\theta\frac{\partial}{\partial\theta}\right) + \frac{1}{\rho^2\sin^2\theta}\frac{\partial^2}{\partial\phi^2}, \tag{6.138}$$

with ρ being the radial vector, θ the polar angle, and ϕ the azimuthal angle.

In (6.137) there are no terms that involve both $\vec{R}$ and $\vec{r}$, so that the center-of-mass motion and the relative motion are decoupled.[15] Therefore, we can write a product

[15] This is a commonplace occurrence in physics. The analogous problem in classical mechanics is that of two masses connected by a spring. If we toss this object up against gravity and try to study the dynamics of the two masses individually using Newton's laws, it becomes an enormously

solution for the wavefunction:

$$\psi(\vec{R},\vec{r}) = \zeta(\vec{R})\xi(\vec{r}). \tag{6.139}$$

which then allows us to obtain a variable separation solution as follows:

If we substitute this solution in (6.137), we immediately get two decoupled equations:

$$-\frac{\hbar^2}{2M}\nabla^2_{\vec{R}}\zeta(\vec{R}) = E_{\rm K}\zeta(\vec{R})$$
$$-\frac{\hbar^2}{2\mu}\nabla^2_{\vec{r}} - \frac{e^2}{4\pi\epsilon|\vec{r}|}\xi(\vec{r}) = E_n\xi(\vec{r}). \tag{6.140}$$

The first equation has the solution

$$\zeta(\vec{R}) = \frac{1}{\sqrt{\Omega}}{\rm e}^{{\rm i}\vec{K}\cdot\vec{R}}, \tag{6.141}$$

where $K = \sqrt{2ME_{\rm K}}/\hbar$. Clearly, $E_{\rm K}$ is the kinetic energy due to the center-of-mass motion of the exciton.

The second equation is a familiar equation in atomic physics. It is the equation describing the lone electron in a hydrogen atom and the solution for the wavefunction $\xi(\vec{r}) = \xi(r,\theta,\phi)$ can be found in many atomic physics textbooks. In spherical coordinate system, the solution can be written as the product of a function in the radial vector r and a function in polar and azimuthal angles θ and ϕ. The first function is Laguerre polynomials and the second is spherical harmonics or associated Legendre polynomials. Each Laguerre polynomial is indexed by an integer n which is the *principal quantum number* that determines the size of the electron orbit around the hydrogen nucleus (proton). The wavefunction in the smallest orbit (known as the 1s state for $n = 1$) is independent of θ and ϕ since the orbit is spherical and is given by

$$\xi_{n=1s}(r) = \frac{1}{\sqrt{\pi a_{\rm B}^3}}{\rm e}^{-r/a_{\rm B}}, \tag{6.142}$$

where $a_{\rm B}$ is the so-called *effective Bohr radius* and is given by

$$a_{\rm B} = \frac{4\pi\epsilon\hbar^2}{\mu e^2}, \tag{6.143}$$

---

complicated affair since their motions are coupled and correlated. However, the center-of-mass motion and the relative motion are decoupled, so that it is common practice to study the motion of the center-of-mass and the relative motion. The exciton is an electrical analog of two masses connected by a spring, with the Coulomb interaction acting as the spring. Therefore, it is no surprise that the center-of-mass motion and the relative motion will turn out to be decoupled and uncorrelated.

and

$$E_n = -\frac{e^2}{4n^2\pi\epsilon(2a_B)}. \tag{6.144}$$

The quantity $|E_n|$ is the exciton binding energy. Obviously, since n can take different integer values, we will have different excitons such as 1s-exciton, 2s-exciton, etc. The 1s-exciton has the highest binding energy and therefore most readily observed in experiments. Excitons that behave like hydrogen atoms in this fashion are sometimes called Mott–Wannier excitons.

Finally, there is one little subtlety that needs to be mentioned. It is well known that in quantum mechanics, two *indistinguishable* particles experience *exchange interaction*. That comes about owing to the fact that the combined wavefunction of two indistinguishable particles must be anti-symmetric under exchange of the particles' coordinates. This is a fundamental principle of quantum mechanics like the Pauli Exclusion Principle. Thus, exchange interaction is purely of quantum-mechanical origin and has no classical analog.

An electron and a proton are distinguishable particles because they have different charge and mass; hence, there is no exchange interaction between them. By the same token, an electron and a hole might appear to be distinguishable particles since they have opposite charges and different effective masses. However, a hole is a fictitious particle that is simply the absence of an electron. Consequently, there can be exchange interaction between an electron and its own absence, meaning a hole. This interaction is typically weak compared to the Coulomb interaction and therefore, we neglected it, although it can be accounted for using numerical methods [23]. Since we were interested in analytical expressions only, we ignored the exchange interaction, but there may be some rare occasions when it may not be inconsequential.

6.9.1 Excitons in Quantum-Confined Nanostructures

The two particle Schrödinger equation (6.130) cannot be solved so easily in a quantum-confined structure such as a quantum well, wire, or dot. Reverting to Cartesian coordinates, this equation will become

$$\left[-\frac{\hbar^2}{2m_e^*}\left(\frac{\partial^2}{\partial x_e^2}+\frac{\partial^2}{\partial y_e^2}+\frac{\partial^2}{\partial z_e^2}\right)-\frac{\hbar^2}{2m_h^*}\left(\frac{\partial^2}{\partial x_h^2}+\frac{\partial^2}{\partial y_h^2}+\frac{\partial^2}{\partial z_h^2}\right)\right.$$
$$\left.-\frac{e^2}{4\pi\epsilon\sqrt{(x_e-x_h)^2+(y_e-y_h)^2+(z_e-z_h)^2}}+V_e(z_e)+V_h(z_h)\right]\psi(x_e,y_e,z_e,x_h,y_h,z_h)$$
$$= E\psi(x_e,y_e,z_e,x_h,y_h,z_h)\ \texttt{(quantum wells)}$$

$$\left[-\frac{\hbar^2}{2m_e^*}\left(\frac{\partial^2}{\partial x_e^2}+\frac{\partial^2}{\partial y_e^2}+\frac{\partial^2}{\partial z_e^2}\right)-\frac{\hbar^2}{2m_h^*}\left(\frac{\partial^2}{\partial x_h^2}+\frac{\partial^2}{\partial y_h^2}+\frac{\partial^2}{\partial z_h^2}\right)\right.$$

$$- \frac{e^2}{4\pi\epsilon\sqrt{(x_e - x_h)^2 + (y_e - y_h)^2 + (z_e - z_h)^2}}$$

$$\left. + V_e(z_e) + V_h(z_h) + V_e(y_e) + V_h(y_h) \right] \psi(x_e, y_e, z_e, x_h, y_h, z_h)$$

$$= E\psi(x_e, y_e, z_e, x_h, y_h, z_h) \text{ (quantum wires)}$$

$$\left[-\frac{\hbar^2}{2m_e^*}\left(\frac{\partial^2}{\partial x_e^2} + \frac{\partial^2}{\partial y_e^2} + \frac{\partial^2}{\partial z_e^2}\right) - \frac{\hbar^2}{2m_h^*}\left(\frac{\partial^2}{\partial x_h^2} + \frac{\partial^2}{\partial y_h^2} + \frac{\partial^2}{\partial z_h^2}\right) \right.$$

$$- \frac{e^2}{4\pi\epsilon\sqrt{(x_e - x_h)^2 + (y_e - y_h)^2 + (z_e - z_h)^2}} + V_e(z_e) + V_h(z_h)$$

$$\left. + V_e(y_e) + V_h(y_h) + V_e(x_e) + V_h(x_h) \right] \psi(x_e, y_e, z_e, x_h, y_h, z_h)$$

$$= E\psi(x_e, y_e, z_e, x_h, y_h, z_h) \text{ (quantum dots)}, \quad (6.145)$$

where $V_e(i_e)$ and $V_h(i_h)$ are the confining potentials for electrons and holes in the i-direction, respectively.

Because of the confining potential terms, the last equation cannot be solved analytically and only numerical solutions are possible. The effect of the confinement is to push the electron closer to the hole (since now there are two influences that bring the electron and hole together: the Coulomb attraction and the walls of the confining potential that squeeze the electron and hole wavefunctions into a small volume of space). The confinement therefore results in an increase in the binding energy. However, if the confining potentials present barriers of *finite* height, then making the confining space (width of the quantum well, for example) smaller and smaller may be counterproductive. If we decrease the width, we raise the subband bottom energies and ultimately the wavefunction leaks out of the well. At that point, the electron and hole wavefunctions are expanded by further confinement instead of being squeezed and the exciton binding energy begins to decrease with further decrease in well width. Therefore, the exciton binding energy has a nonmonotonic dependence on well width. Starting from a large width, it first increases when the well width decreases, peaks at a certain width, and then decreases with further decrease in the width. This dependence is shown qualitatively in Fig. 6.17.

6.9.1.1 Dielectric Confinement

In quantum-confined structures (quantum wells, wires, and dots), a narrow bandgap material with higher dielectric constant is surrounded by a wide bandgap material with lower dielectric constant. The exciton is confined in the narrow gap material which has the higher dielectric constant, so that the electric field lines emanating from the hole and sinking in the electron are further concentrated within a small volume by the difference in the dielectric constant. This is known as "dielectric confinement" and can work in concert with quantum confinement of electrons and

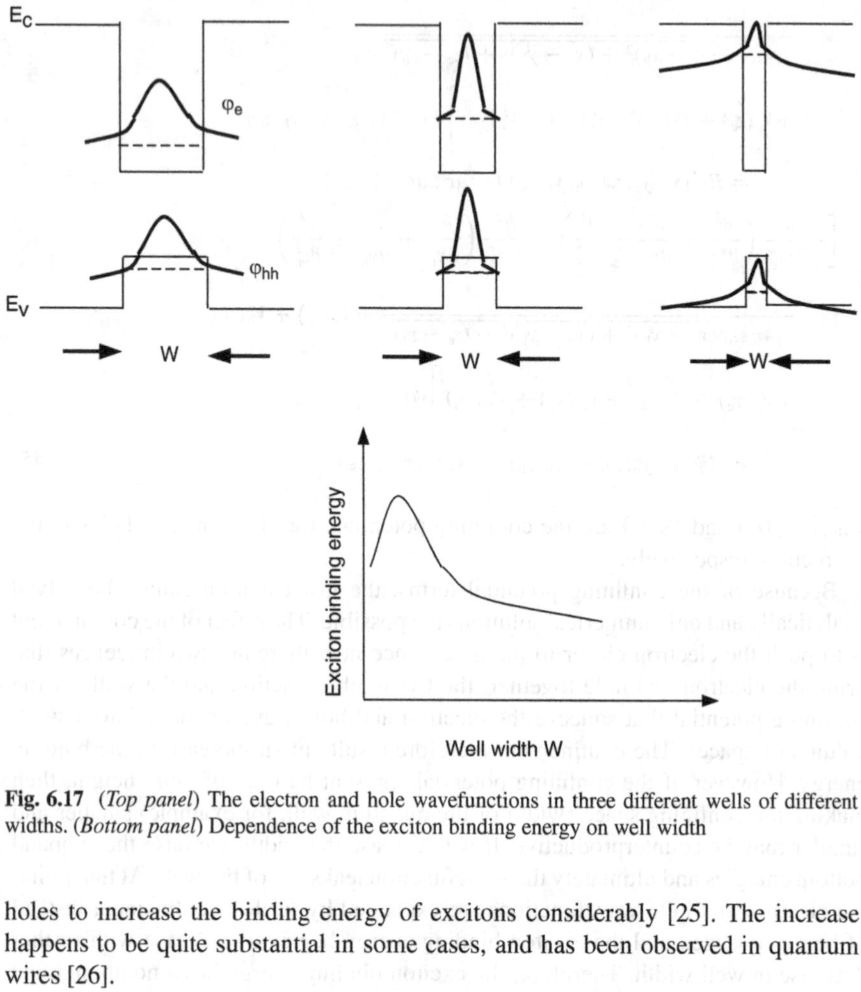

Fig. 6.17 (*Top panel*) The electron and hole wavefunctions in three different wells of different widths. (*Bottom panel*) Dependence of the exciton binding energy on well width

holes to increase the binding energy of excitons considerably [25]. The increase happens to be quite substantial in some cases, and has been observed in quantum wires [26].

6.9.2 Bleaching of Exciton Absorption: Nonlinear Optical Properties

In (6.18), we assumed that the absorption coefficient $\alpha_{\hat{p}}(\omega_l)$ is *independent* of the incident intensity. If that is not true, then the optical properties will become *nonlinear*.

The absorption coefficient due to exciton absorption may indeed depend on the incident intensity at high intensities. When the photon flux (intensity) is very high, a large density of excitons is generated as numerous photons are absorbed. This has two effects. First, some of the excitons ionize into free electron-hole pairs by collision or thermal fluctuation, and these pairs raise the conductivity of the medium.

The Coulomb attraction between the electron and hole is then increasingly *screened*. The Coulomb interaction terms becomes $e^2 \exp[-\vec{\lambda} \cdot \vec{r}]/(4\pi\epsilon|\vec{r}|)$, where λ is the screening length which becomes shorter and shorter with increasing conductivity. When the screening length becomes comparable to the effective Bohr radius of the exciton, the binding energy nearly vanishes and the exciton absorption peak is no longer distinguishable from band edge absorption. We call this phenomenon "bleaching" of the exciton absorption.

A second effect that may cause bleaching is *phase space filling*. Because an exciton consists of an electron and hole, both of which are spin-1/2 particles, the exciton itself should be a boson. That it is, but at very high densities, it loses its Bose character and starts to behave like a fermion and obeys Pauli Exclusion Principle! The exciton is one of very few particles that has this peculiar behavior (another one is magnon). But because of this characteristic, when a very high density of excitons is generated, the available phase space for excitons begins to fill up and further generation of excitons by photon absorption is blocked by Pauli Exclusion Principle. Therefore, at high incident intensities, the exciton absorption may be bleached out as a result of this effect.

Bleaching makes the absorption coefficient associated with exciton absorption intensity-dependent. Such nonlinearities have found many device applications in frequency mixers, limiters, second harmonic generators, and other types of optical devices. For a review of this field, see, for example, [24].

6.10 Polariton Laser

This chapter will be incomplete without describing a very different type of "laser" that operates in a way that is distinct from the way the standard laser operates. The standard laser discussed in the preceding sections requires optical gain and optical feedback to behave like a light "oscillator" and produce a steady stream of coherent light (photon) output. The optical gain is caused by population inversion and the optical feedback is due to reflections of the light between the mirrors of the cavity. However, there may be other ways of creating coherent light output that may not require population inversion at all. An example of that is the *polariton laser*.

There are many types of "waves" in nature: electromagnetic waves (like light), acoustic waves (like sound) and matter waves (like electrons). Waves whose quanta are bosons (such as photons or phonons) can form coherent states. Fermions, on the other hand, do not form coherent states by themselves, but two fermions can couple to form bosons. Examples of such entities are excitons and Cooper pairs. Such couples can form coherent states which are called Bose–Einstein condensate. There is, however, a fundamental difference between coherent states of photons and coherent states of excitons (in a Bose–Einstein condensate); one is the constituent of electromagnetic waves and the other is the constituent of matter waves. Let us now examine how coherent states of excitons can form.

When an electromagnetic wave is confined to a cavity, it becomes a guided wave and its dispersion relation [ω versus k] becomes *nonlinear*. In that case, the photon develops a slight effective mass (it is no longer zero). In a particular type of planar cavity, the mass of the photon becomes $m_{\text{ph}} \approx (\eta\lambda_c/2W)m_0$, where η is the refractive index of the medium, W is the cavity width, and λ_c is a constant [Compton wavelength]. By changing the width W of the cavity, one can control the photon's effective mass, or equivalently, its dispersion relation. Using this technique, one can "match" the dispersion relation of the photons with that of the excitons that are formed within the cavity. There will be strong coupling at energies (or wavevectors) where the dispersion curve of the exciton intersects the dispersion curve of the photon[16] and this can lead to the formation of a new particle called the "polariton" (which is a coupled photon–exciton system). The polariton is a light particle, much lighter than the electron or the hole in the exciton. A collection of polaritons can form a coherent state.

When the exciton density in the cavity is very high, the excitons behave like fermions rather than bosons and will not form a coherent state. In this condition, the polariton is very short-lived. It is destroyed quickly because of exciton screening, phase space filling, and dephasing. This regime is called the *weak coupling regime* where photons and excitons couple weakly and polariton formation is not favorable. Such a system is analogous to a conventional laser; the photons will stimulate the excitons to recombine radiatively and the emitted photon population will build up coherently to form a laser.

At the other extreme, when the exciton density is low and the polariton is long-lived, the latter can form a coherent state. This state is spawned by pumping the cavity with photons at a specific angle [27]. The polaritons scatter to the lowest energy mode and crowd there tending to form a Bose–Einstein condensate and a coherent state. When the excitonic character of the polaritons in this coherent state vanishes, the coherent state becomes the coherent state of photons, which is equivalent to a laser [28, 29]. This is known as a *polariton laser*, which does *not* require population inversion. The advantage of this is that no large current is required to produce population inversion and the threshold power will be low.

There have been many demonstrations of polariton lasing at low temperatures and ultimately at room temperature [30]. The room temperature laser consisted of a GaN quantum well sandwiched between two distributed Bragg reflectors made of multiple periods of alternating Si_3N_4/SiO_2 on one side and multiple periods of alternating AlInN/AlGaN on the other to form a microcavity. The threshold power for the onset of lasing was 1 mW which is one order of magnitude smaller than the threshold power for (In,Ga)N quantum well vertical cavity surface emitting lasers (VCSELs). The use of GaN as the cavity material made room temperature operation possible since the exciton binding energy is large in GaN, which prevents the excitons from thermally dissociating into free electrons and holes at room temperature. The device structure is schematically depicted in Fig. 6.18.

[16]Remember that the dispersion relation governing the center-of-mass motion of excitons is $E_K = \frac{\hbar^2 K^2}{2M}$.

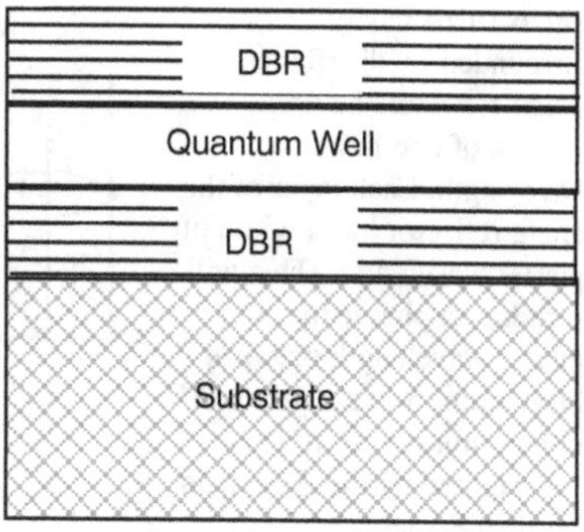

Fig. 6.18 Structure of a semiconductor microcavity for a polariton laser

6.11 Photonic Crystals

We mentioned in Chap. 3 that there is a fundamental mathematical similarity between Maxwell's equation and Schrödinger equation, which is why the physics of an electromagnetic wave propagating through a medium with spatially varying refractive index is analogous to the physics of an electron wave propagating through a solid with spatially varying potential. This analogy was the inspiration behind a system known as *photonic crystals.*

In Chap. 4, we found that the existence of a *periodically* varying potential in a semiconductor crystal causes bandgaps to open up at certain energies, where no electron state can exist, or equivalently, the density of states becomes zero. The bandgaps open up because periodically placed atoms partially reflect electron waves traveling through the crystal. At certain energies, the reflected waves will interfere destructively, so that there will be no propagating wave. For these energies, there will be no real-valued wavevectors, and this is what we call the bandgap. Clearly, the bandgap is a consequence of periodic distributed reflection of electron waves.

We can employ a similar strategy to create an equivalent "photonic crystal" that will have a "photonic bandgap." Using the analogy between spatially varying potential and spatially varying refractive index, the photonic crystal will be a periodic medium with periodically varying refractive index. The periodically varying refractive index will cause distributed reflection of electromagnetic waves propagating through it and open up bandgaps at wavelengths (or frequencies) where the interference between the reflected waves becomes destructive. In the bandgap, the density of states of photons will be zero, or at least very small. Now, if we make the photonic bandgap *overlap* with the crystal's bandgap, then the frequency of photons emitted by radiative recombination of electrons and holes in the crystal (i.e., by stimulated or spontaneous emission) will be in the common bandgap, where the density of states of photons is ideally zero! This has interesting consequences.

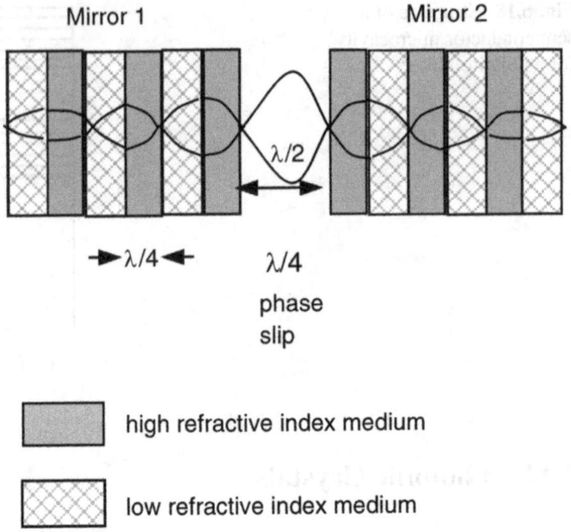

Fig. 6.19 A cavity surrounded by distributed Bragg reflectors having a spacing of one-half wavelength. Each layer of the Bragg reflector has a width of quarter wavelength. This will introduce a phase slip of a quarter wavelength between the standing waves. Adapted with permission from [31]. Copyrighted by the American Physical Society

Any photon created within the photonic bandgap of course could not propagate because the associated electromagnetic wave will be evanescent, but more importantly, from (6.28) we see that the rate of spontaneous emission per unit volume per unit photon frequency is

$$R^{\mathrm{sp}}(\omega_{\mathrm{l}}) = [D_{\mathrm{ph}}(\omega_{\mathrm{l}})/\Omega]l(\omega_{\mathrm{l}}). \tag{6.146}$$

Since the frequency ω_{l} is in the photonic bandgap, $D_{\mathrm{ph}}(\omega_{\mathrm{l}}) \approx 0$, and hence the rate of spontaneous emission falls to nearly zero. By this strategy, we can *suppress spontaneous emission*.

This problem was studied by Yablonovitch [31] who considered a cavity as shown in Fig. 6.19. The structure is shown in one-dimension, but can exist in all three dimensions. Such a structure can act as a photonic crystal and suppress spontaneous emission in the cavity mode.

Suppression of spontaneous emission has several applications [31]. For example, it can reduce electron–hole recombination rate (see (6.38) and (6.39)) and this is desirable in bipolar junction transistors where electron–hole recombination in the base degrades the transistor's short-circuit current gain. By fashioning the base out of a photonic crystal, one can increase the transistor's short-circuit current gain significantly.

There is also application in solar cells. By suppressing spontaneous emission and therefore electron–hole recombination, one can increase the photocollection efficiency in solar cells by preventing the photogenerated electron–hole pairs from recombining before they reach the current collecting contacts. This will increase the efficiency of the solar cell.

However, the most important application of photonic crystals is in reducing the threshold current of injection lasers. From (6.49), it is clear that the threshold current is proportional to $\int R^{\mathrm{sp}}(\omega_1)\mathrm{d}\omega_1$. Therefore, by reducing $R^{\mathrm{sp}}(\omega_1)$ with the use of photonic crystals, one can reduce the threshold current dramatically in injection lasers. Of course, one has to also maintain adequate stimulated emission by ensuring that $N_{\hat{p}}(\omega_1) \gg 1$, but that can be achieved through cavity feedback employing cavities with excellent quality factor.

Photonic crystals are now in an advanced state of art. Semiconductor laser using cavities formed within photonic crystals have been demonstrated [32] and continues to be a topic of significant interest.

6.12 Negative Refraction

Recently, there has been considerable interest in the phenomenon of negative refraction because of its many intriguing applications. Materials that can exhibit negative refraction have been made possible by advances in nanofabrication. Therefore, research on negative refraction has become an important area of nanophotonics.

To understand how the refractive index of a medium can become negative, consider a lossless medium of relative dielectric constant ϵ_{r} and relative magnetic permeability μ_{r}. The ratio of the speed of light in vacuum to that in the medium is the refractive index η which is given by

$$\eta^2 = \epsilon_{\mathrm{r}}\mu_{\mathrm{r}}. \tag{6.147}$$

One usually expresses the refractive index by taking the *positive* square-root of the right-hand-side so that

$$\eta = \sqrt{\epsilon_{\mathrm{r}}\mu_{\mathrm{r}}}. \tag{6.148}$$

This is the appropriate thing to do if both ϵ_{r} and μ_{r} are positive quantities. However, if both are negative, then one should take the negative square root, so that

$$\eta = -\sqrt{\epsilon_{\mathrm{r}}\mu_{\mathrm{r}}}. \tag{6.149}$$

In that case, the refractive index is both real and negative [33]. Media that have negative refractive index have not been found in nature (although they might exist), but have been artificially engineered. Such media are popularly referred to as *metamaterials*. Interest in metamaterials with negative refractive index received a boost recently since it was realized that they can have two intriguing applications: (1) superlensing, and (2) invisibility cloaking.

6.12.1 Superlens

Conventional wisdom dictates that no lens can focus an electromagnetic wave into an area smaller than a square wavelength. This happens because of the following reason.

Consider a monochromatic electromagnetic wave (of frequency ω_l) being focused by a lens whose axis is the z-axis. The electric field in the wave is expressed as a Fourier expansion

$$\vec{E}(\vec{r},t) = \sum_{\hat{p},k_x,k_y} \vec{E}_{\hat{p},k_x,k_y} \exp[i(k_x x + k_y y + k_z z - \omega_l t)], \tag{6.150}$$

where, as always, $\hat{p}$ is the polarization.

If we substitute this in Maxwell's equation, we will get

$$k_z = +\sqrt{\eta^2\omega_l^2/c^2 - k_x^2 - k_y^2} \quad \texttt{if } \omega_l^2/c^2 \geq k_x^2 + k_y^2, \tag{6.151}$$

where c is the speed of light in vacuum.

However, for large transverse wavevectors $\sqrt{k_x^2 + k_y^2}$, it is very possible that $\omega_l^2/c^2 < k_x^2 + k_y^2$, so that we will get instead

$$k_z = +i\sqrt{k_x^2 + k_y^2 - \eta^2\omega_l^2/c^2}. \tag{6.152}$$

The waves with imaginary wavevector k_z are evanescent and quickly decay in amplitude with distance. They are removed from the image formed by the lens at the focal length. Since only propagating states with transverse wavevector $\sqrt{k_x^2 + k_y^2} = \sqrt{\left[\frac{2\pi}{\lambda_x}\right]^2 + \left[\frac{2\pi}{\lambda_y}\right]^2}$ smaller than $\eta\omega_l/c$ can contribute to the image, the maximum resolution of the image will be $2\pi c/\omega_l\eta = \lambda_l/\eta$, which is the wavelength of light in the lens medium. This is a fundamental limitation set forth by the laws of electromagnetics and cannot be overcome by building a better lens.

There is, however, a nontraditional alternative to a lens. A material with negative refractive index can focus light rays even if it is in the form of a rectangular slab with no curvature. This was illustrated by Pendry [34] and is shown in Fig. 6.20.

Pendry [34] also showed that the transmission probability through the negative refractive index medium is

$$T = \exp\left(-i\sqrt{\eta^2\omega_l^2/c^2 - k_x^2 - k_y^2 d}\right), \tag{6.153}$$

where d is the thickness of the slab. For propagating waves $\left[\eta^2\omega_l^2/c^2 > k_x^2 + k_y^2\right]$, there is no amplification or decay as the wave traverses the slab, since the magnitude

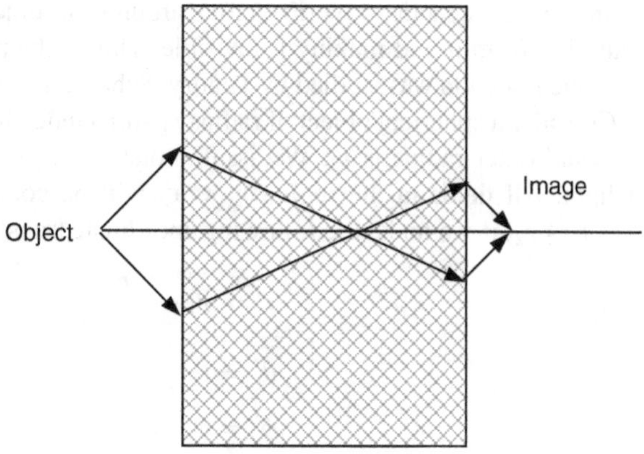

Fig. 6.20 A lens implemented with a material having a negative refractive index. The negative refractive index bends light to a negative angle with the surface normal in keeping with Snell's law. Light diverging from a point source converges to a point within the medium and upon leaving the medium will converge to a focus. Even though the slab has no curvature, it acts as a convex lens. Adapted with permission from [34]. Copyrighted by the American Physical Society

of the transmission probability remains unity. However, for evanescent waves $\left[\eta^2\omega_1^2/c^2 < k_x^2 + k_y^2\right]$, the transmission probability will be

$$T = \exp\left(\sqrt{k_x^2 + k_y^2 - \eta^2\omega_1^2/c^2}d\right), \tag{6.154}$$

which implies that an evanescent wave is *amplified* as it travels through the medium. Since now both propagating and evanescent waves can reach the focus and contribute to the resolution of the image, there is no physical hindrance to perfect reconstruction of any image and there is no reason why light cannot be focused to an area smaller than the square of the wavelength. Basically, the limit on resolution has been removed by avoiding decay of the evanescent waves and allowing them to be amplified instead.

The above theory assumes lossless material since we take both ϵ_r and μ_r to be real, albeit negative. Losses within the medium will make these quantities complex and will affect the transmission probability. This will have a deleterious effect on the superlens' resolution. This effect has been recently analyzed by Wee and Pendry [35].

6.12.2 *Invisibility Cloaking*

If metamaterials of arbitrary dielectric constant and magnetic permeability were available, then by shrouding an object with a metamaterial, we can force the electric displacement vector, the magnetic flux density vector, and the Poynting vector in an electromagnetic wave to detour around the object, thereby preventing any scattering from it and making it invisible. This is the principle behind "cloaking." This phenomenon was studied by Pendry, Schurig and Smith [36].

Consider a spherical object of radius r_1 surrounded by a shell whose inner radius is r_1 and outer radius is r_2. We assume that $r_2 \gg \lambda$, where λ is the wavelength of light. All fields in the region $r < r_2$ will be compressed within the annulus $r_1 \leq r \leq r_2$, if we make the following coordinate transformations:

$$\begin{aligned} r' &= r_1 + r\frac{r_2 - r_1}{r_2} \\ \theta' &= \theta \\ \phi' &= \phi. \end{aligned} \tag{6.155}$$

in spherical coordinate system.

Pendry, et al. showed that the electric displacement vector, the magnetic flux density vector and the Poynting vector in an electromagnetic wave will detour around the sphere enclosed within the shell provided the following conditions are met: the relative dielectric constant and the relative magnetic permeability can take any value within the sphere $r \leq r_1$, but in the shell $r_1 \leq r_2$, they must take the values

$$\begin{aligned} \epsilon_{r'} = \mu_{r'} &= \frac{r_2}{r_2 - r_1}\frac{(r' - r_1)^2}{r'} \\ \epsilon_{\theta'} = \mu_{\theta'} &= \frac{r_2}{r_2 - r_1} \\ \epsilon_{\phi'} = \mu_{\phi'} &= \frac{r_2}{r_2 - r_1}. \end{aligned} \tag{6.156}$$

Outside the shell, i.e., for $r > r_2$, we must have

$$\epsilon_{r'} = \mu_{r'} = \epsilon_{\theta'} = \mu_{\theta'} = \epsilon_{\phi'} = \mu_{\phi'} = 1. \tag{6.157}$$

Using a ray-tracing approach based on numerical integration of Hamilton's equations obtained by taking the geometric limit of Maxwell's equations for anisotropic and inhomogeneous media, Pendry et al. showed that rays will bend around the spherical object as shown in Fig. 6.21, thereby concealing the object. Unfortunately, this works only at one frequency and is not "broadband." Nonetheless, this is a remarkable phenomenon.

6.13 Summary

In this chapter, we learned that

1. Absorption of light in a semiconductor, semimetal, or insulator is caused by the excitation of an electron from the valence band to the conduction band while

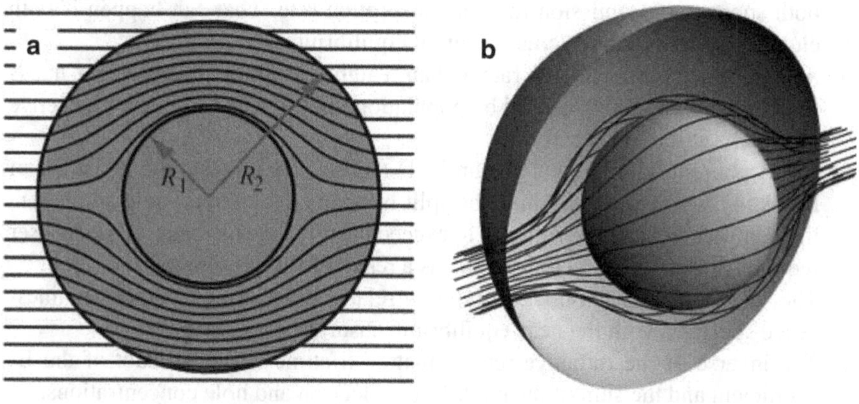

Fig. 6.21 Rays bending around a spherical object enclosed in a metamaterial shell, resulting in cloaking of the object. Reproduced from [36] with permission from the American Association for the Advancement of Science. Copyright American Association for the Advancement of Science, 2006

obeying the k-selection rule. The latter guarantees that the wavevector of the electron in the final state (in the conduction band) is approximately equal to the wavevector in the initial state (in the valence band), i.e., only vertical transitions in the Brillouin zone are allowed.

2. In a semiconductor, semimetal, or insulator, there is a minimum energy of the photon that can be absorbed. This energy is the effective bandgap energy.
3. The k-selection rule makes indirect-gap semiconductors inefficient light emitters. Direct-gap semiconductors are far more efficient light emitters.
4. The intensity and luminescence obey an equation that is similar to the charge continuity equation. Solution of this equation shows that if the incident intensity and the absorption coefficient are high, and the luminescence is low, then the intensity decays exponentially with distance into a material. The decay constant is the absorption coefficient which has the unit of inverse distance.
5. In a medium where there is "population inversion," meaning that a higher energy level is more occupied by electrons than a lower energy level, the absorption will be negative, i.e., there will be optical gain or amplification. This is at the heart of light amplification in a laser.
6. If photons have been driven out of equilibrium by external optical stimulus while the electrons remain in equilibrium, then absorption will dominate over stimulated and spontaneous emission since that would drive the system back towards global equilibrium.
7. If electrons are driven out of equilibrium while the photons remain in equilibrium, spontaneous emission will dominate over stimulated emission and absorption, since that would restore equilibrium.
8. Spontaneous emission produces incoherent light, while stimulated emission produces coherent light. For a laser, we need stimulated emission rate to exceed

both spontaneous emission rate and absorption rate. That can happen if both electron and photons systems are out of equilibrium.

9. A laser is a light oscillator rather than a light amplifier. Therefore, it needs both light amplification caused by population inversion and frequency selective feedback caused by a cavity.
10. An incoherent light emitter like an LED merely requires that the quasi Fermi levels in a semiconductor diode be split by nonzero amount, while a coherent light emitter requires the splitting to exceed the effective bandgap. A diode laser requires degenerate p- and n-doping as a result of this requirement.
11. The Van Roosbroeck–Shockley relation relates the near-equilibrium luminescence spectrum with the near-equilibrium absorption spectrum.
12. The inverse of the radiative recombination lifetime is the product of the B-coefficient and the sum of the equilibrium electron and hole concentrations.
13. There are different kinds of semiconductor lasers that improve over the traditional diode laser. Mostly, they reduce the threshold current for the onset of lasing.
14. In the presence of an electromagnetic field, the electron's canonical momentum is replaced by the kinematic momentum. The latter operator is space and time dependent.
15. Physical quantities must be gauge invariant.
16. An electromagnetic field (or photon) has both linear and angular momenta.
17. Absorption and emission are polarization dependent. Polarization selects the bands between which transitions are induced.
18. In a quantum well, light polarized parallel to the plane of the well cannot induce transitions within subbands of the same band (intraband, inter-subband transitions), but can induce transitions between different bands (interband transitions). The latter is three times stronger for heavy-hole to conduction band transition, than for light-hole to conduction band transition.
19. Light polarized perpendicular to a quantum well can induce both intraband, intersubband transitions, and interband transitions.
20. In a quantum wire, light polarized along the wire's axis cannot induce transitions within subbands of the same band (intraband, inter-subband transitions), but can induce transitions between different bands (interband transitions).
21. Light polarized perpendicular to a quantum wire's axis can induce both intraband, intersubband transitions and interband transitions.
22. Coulomb attraction between photogenerated electron–hole pairs leads to the formation of excitons, which introduce a level below the conduction band edge. The energy separation between the conduction band edge and the exciton level is the exciton binding energy. This new level gives rise to an exciton absorption peak below the bandedge frequency.
23. Excitons behave like hydrogen atoms in the simplest picture.
24. Quantum confinement and dielectric confinement increase the exciton binding energy.

25. When the intensity of incident light is very high, exciton ionization and phase space filling cause bleaching of the exciton absorption peak. That gives rise to associated nonlinear optical effects.
26. A photon in a guided wave has a slight mass.
27. A polariton is a coupled exciton–photon particle and is a boson. It has very light mass which is more than that of a photon, but much less than that of an electron or hole.
28. A polariton laser works via Bose–Einstein condensation of polaritons which produces a coherent state of polaritons. Operation of this laser does not require population inversion. Hence, its threshold is low. Polariton lasers have been demonstrated at room temperature using a GaN cavity. Polariton formation is favored in this material since the excitons are long-lived as a consequence of having large binding energy.
29. A photonic crystal is the electromagnetic analog of a material crystal. It is made using a three-dimensional periodic structure consisting of alternating layers of materials with high and low refractive index. Photons of certain frequencies cannot propagate in a photonic crystal, resulting in the formation of photonic bandgaps.
30. The density of states of photons in the photonic bandgap is ideally zero.
31. If the photonic bandgap overlaps with the crystal's bandgap, then spontaneous emission of light due to radiative recombination of electrons and holes is suppressed severely. This has applications in increasing the short-circuit current gain of bipolar junction transistors, collection efficiencies of solar cells, and reducing the threshold current densities in injection lasers.
32. Negative refraction, achieved using metamaterials, can produce a superlens able to focus an object within an area much smaller than the square of the wavelength of light. This allows subwavelength resolution. This ability accrues from the fact that evanescent waves are amplified, rather than attenuated, in traversing a material with negative refractive index.
33. Metamaterials can produce an invisibility cloak. They work by diverting the electric field, magnetic field, and Poynting vector of an electromagnetic wave around an object shielded by the metamaterial. This makes the object invisible since it does not scatter the electromagnetic wave.

Problems

Problem 6.1. Refer to Fig. 6.4 and note that the next highest band above the conduction band in the indirect-gap semiconductor has its minimum at the Brillouin zone center. It is possible to have vertical transitions from this valley to the valence band peak since the latter also occurs at the zone center. Therefore, we could have made the indirect-gap semiconductor emit light efficiently if we could have made this valley lower in energy than the valley near the X-point (zone edge). One way to accomplish this would have been to fashion a quantum dot, where the Γ-valley will have an effective energy increase of

$$\Delta_{\Gamma}^{0d} = (\hbar^2/2m_{\Gamma})\left[(\pi/W_x)^2 + (\pi/W_y)^2 + (\pi/W_z)^2\right],$$

and the X-valley will have an effective energy increase of

$$\Delta_{X}^{0d} = (\hbar^2/2m_{X})\left[(\pi/W_x)^2 + (\pi/W_y)^2 + (\pi/W_z)^2\right],$$

where m_{Γ} and m_X are the effective masses in the two valleys and W_x, W_y, W_z are the edges of the quantum dot.

Assume that in bulk, the energy difference between the two valleys in the indirect-gap semiconductor is about 3.5 eV. Therefore, if we could make: (1) $\Delta_X^{0d} > \Delta_{\Gamma}^{0d}$, and (2) $\Delta_X^{0d} - \Delta_{\Gamma}^{0d} > 3.5$ eV, then we might have succeeded in making the indirect-gap semiconductor optically active and an efficient light emitter.

Show that if m_{Γ} were larger than m_X, then in a cubic quantum dot, the indirect-gap material would have been an efficient light emitter if

$$\frac{1}{m_X} - \frac{1}{m_{\Gamma}} = \frac{E_{\text{crit}}}{3\hbar^2}\left(\frac{W}{\pi}\right)^2,$$

where W is the edge dimension of the dot and $E_{\text{crit}} = 7$ eV.

Problem 6.2. Consider a cubic GaAs quantum dot of edge 10 nm. Assume that the dot is placed in air so that the potential barriers at the boundaries have infinite heights. The bulk bandgap of this material is 1.42 eV and the effective masses of various particles are: electrons=0.067 m_0, heavy holes=0.45 m_0, and light holes=0.082m_0, where m_0 is the free electron mass. Sketch the ideal absorption spectrum in this dot for photon energy 0–2.5 eV.

Problem 6.3. Show that while the canonical momentum operators $\vec{p}_i$ and $\vec{p}_j$ commute, the kinematic momentum operators $\vec{\Pi}_i$ and $\vec{\Pi}_j$ do not if $\vec{A} \neq 0$ and is not spatially invariant.

Problem 6.4. Use (6.71) to show that for a TEM or TE wave, the magnetic vector potential given by (6.74) (with $\vec{B}(\vec{r}) = 0$) obeys the condition

$$\vec{\nabla}_r \cdot \vec{A}(\vec{r}, t) = 0.$$

Problem 6.5. For light polarized in the x-y plane, show that the strengths of absorption from the light hole band to the conduction band, and from the split-off band to the conduction band, bear the ratio 1:2. Show also that for light polarized in the z-direction, that ratio is reversed and becomes 2:1.

Problem 6.6. Consider one conduction band and one valence band, as well as one polarization of light. Complete the summations over wavevector states $\vec{k}$ in (6.86) in one-, two-, and three-dimensional systems to find expressions for the absorption coefficient as a function of photon energy in each of these systems due to transitions from the valence to the conduction band. Assume that all states in the conduction band are empty and all states in the valence band are filled. Find an expression for the electron–hole *joint* density of states.

Solution. We will solve this problem for three dimensions. The reader can easily extend this method to two- and one-dimension.

From (6.86), we get that

$$\begin{aligned}
\alpha_{\hat{p}}(\omega_l) &= \frac{\eta_{\hat{p}}(\omega_l)}{c}\frac{\pi e^2}{m_0^2\epsilon\omega_l\Omega}\left|\hat{p}\cdot\vec{\Upsilon}_{\text{cond,val}}\right|^2 \\
&\quad\times\sum_{\vec{k}}\delta\left(E_{\text{cond}}(\vec{k})-E_{\text{val}}(\vec{k})-\hbar\omega_l\right) \\
&= \frac{\eta_{\hat{p}}(\omega_l)}{c}\frac{\pi e^2}{m_0^2\epsilon\omega_l\Omega}\left|\hat{p}\cdot\vec{\Upsilon}_{\text{cond,val}}\right|^2\frac{\Omega}{8\pi^3}\int d^3\vec{k}\,\delta\left(E_g+\frac{\hbar^2k^2}{2\mu}-\hbar\omega_l\right) \\
&= \frac{\eta_{\hat{p}}(\omega_l)e^2}{4\pi c m_0^2\epsilon\omega_l}\left|\hat{p}\cdot\vec{\Upsilon}_{\text{cond,val}}\right|^2\int k\mathrm{d}k^2\delta\left(\frac{\hbar^2}{2\mu}\left[k^2-\frac{2\mu}{\hbar^2}\{\hbar\omega_l-E_g\}\right]\right) \\
&= \frac{\eta_{\hat{p}}(\omega_l)e^2}{4\pi c m_0^2\epsilon\omega_l}\left|\hat{p}\cdot\vec{\Upsilon}_{\text{cond,val}}\right|^2\left(\frac{2\mu}{\hbar^2}\right)^{3/2}\sqrt{\hbar\omega_l-E_g},
\end{aligned}\tag{6.158}$$

where μ is the reduced mass given in (6.131).

Consider an electron and a hole with the same wavevector k. We can write the energy of the electron *plus* the energy of the hole measured from their respective band edges as

$$E=\frac{\hbar^2k^2}{2m_e}+\frac{\hbar^2k^2}{2m_h}=\frac{\hbar^2k^2}{2\mu}.$$

Defining the joint density of states following the usual prescription:

$$D^j(E)dE=D^{3-d}(k)d^3k=\frac{\Omega}{4\pi^3}4\pi k^2\mathrm{d}k=\frac{\Omega}{2\pi^2}k\mathrm{d}k^2,$$

we get

$$\begin{aligned}
D^j(E)\frac{\mathrm{d}E}{\mathrm{d}k^2} &= \frac{\Omega}{2\pi^2}k \\
D^j(E)\frac{\hbar^2}{2\mu} &= \frac{\Omega}{2\pi^2}\frac{\sqrt{2\mu E}}{\hbar}.
\end{aligned}$$

This yields the electron–hole joint density of states in three dimensions as

$$D^j(E) = \frac{1}{2\pi^2}\left(\frac{2\mu}{\hbar^2}\right)^{3/2}\sqrt{E}.$$

The absorption is therefore proportional to the electron–hole joint density of states with E replaced by $\hbar\omega_l - E_g$.

Taking the derivative of the absorption coefficient with respect to $\hbar\omega_l$ and setting it equal to zero, we find that the absorption peaks at the photon energy equal to twice the bandgap energy. Thus, the absorption has a nonmonotonic dependence on photon energy. At first, it increases with photon energy since the density of available final states increases, but then it decreases with photon energy since the electron–photon coupling (or the vector potential) decreases. Near the bandedge, Fig. 6.3b is still pretty valid.

Problem 6.7. Using the van Roosbroeck–Shockley relation, and the result of the previous problem, determine the photon frequency where the near-equilibrium luminescence spectrum peaks at a temperature T in a bulk semiconductor.

Problem 6.8. In the ideal situation, do you expect the exciton absorption peak due to an exciton formed by a heavy-hole and that formed by a light hole to be resolvable in the absorption spectrum in a bulk semiconductor? Explain.

Problem 6.9. Calculate the Bohr radius of the heavy hole exciton in GaAs which has a relative dielectric constant of 12, electron effective mass=0.067m_0, heavy hole mass=0.45m_0.

Problem 6.10. Calculate the threshold current density in a GaAs-$Al_xGa_{1-x}As$ double heterostructure laser where the optical cavity length is $\sim 1\,\mu$m, quantum efficiency is 50 the intrinsic cavity loss is 10^3 cm^{-1}, the photon energy density is 0.1 Joules cm^{-3}, the linewidth is 10 meV in photon energy, the temperature is 500 K inside the cavity, and ΔF is twice the bandgap. Plot the threshold current as a function of ΔF from bandgap energy to three times the bandgap energy.

Problem 6.11. In Fig. 6.22, we show the energy band diagram of four diodes. Explain which one is acting as an LED, which one is acting as a laser and which one is not emitting any light at all. Which of these diodes that is not acting as a laser could be made to act as one if more current is injected into it? Explain exhaustively.

Problem 6.12. If we treat the electron as a free particle (or, equivalently, make the parabolic band approximation), then the electron's wavefunction would have been the plane wave $\frac{1}{\sqrt{\Omega}}e^{i\vec{k}\cdot\vec{r}}$. In that case, the Bloch function would be considered a constant equal to unity. Show that this will yield, $K^a_{\hat{p},\omega_l} = K^e_{\hat{p},\omega_l} = \frac{\pi e^2\hbar^2 k_p^2}{m_0^2\epsilon\omega_l\Omega}\delta_{\vec{k}',\vec{k}+\vec{q}_l}$. Note that although such a treatment still gives us the k-selection rule, it is not completely correct. Absorption and emission are band phenomena and bandstructure effects cannot be summarily neglected. In other words, we need to account for the Bloch function explicitly.

Fig. 6.22 Energy band diagram of four p-n junction diodes

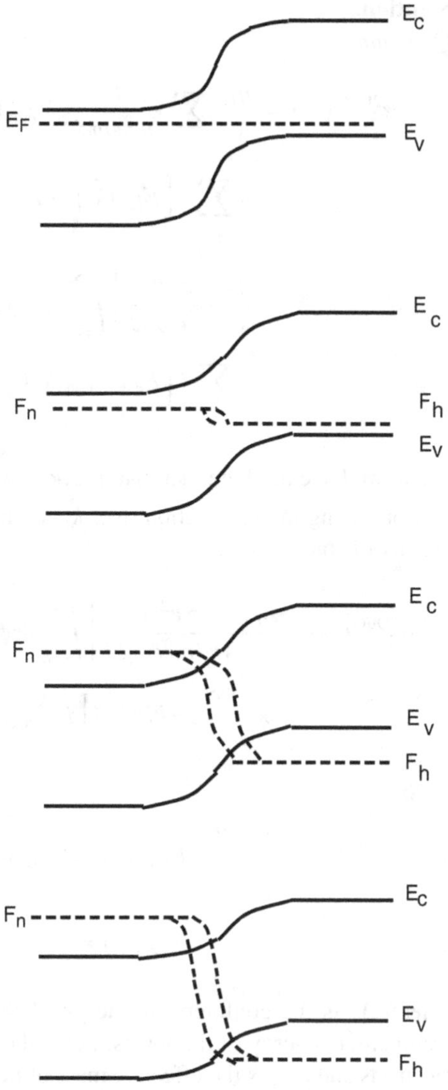

Problem 6.13. Consider a quantum well and light polarized in the plane of the well. Derive an expression for the absorption coefficient as a function of the photon frequency due to transitions from the first heavy-hole subband to the first electron subband. Assume that the hole subband is completely filled with electrons and the electron subband is completely empty.

Repeat the same problem for a quantum wire with light polarized along the wire axis.

Solution.
Quantum well

$$
\begin{aligned}
\alpha_{\hat{p}}^{\mathrm{2DEG}}(\omega_{\mathrm{l}}) &= \frac{\eta_{\hat{p}}}{c}\sum_{i,m,j,n}\frac{\pi e^2}{m_0^2\epsilon\omega_{\mathrm{l}}\Omega}K_{\hat{p},\omega_{\mathrm{l}}}(i,m:j,n) \\
&\quad\times\sum_{\vec{k}_t}\delta\left[E_{i,m}(\vec{k}_t)-E_{j,n}(\vec{k}_t)-\hbar\omega_{\mathrm{l}}\right]\left[f_{j,n}(\vec{k}_t)-f_{i,m}(\vec{k}_t)\right] \\
&= \frac{\eta_{\hat{p}}}{c}\frac{\pi e^2}{m_0^2\epsilon\omega_{\mathrm{l}}\Omega}\left|\int_{-\infty}^{\infty}\mathrm{d}z\phi_{\mathrm{electron}}*(z)\phi_{\text{heavy-hole}}(z)\right|^2\left|\hat{p}\cdot\vec{\Upsilon}'_{ij}\right|^2 \\
&\quad\times\sum_{\vec{k}_t}\delta\left[E_{\mathrm{electron},m}(\vec{k}_t)-E_{\text{heavy-hole},n}(\vec{k}_t)-\hbar\omega_{\mathrm{l}}\right],
\end{aligned}
$$

where we have used the fact that $f_{j,n}(\vec{k}_t) = 1$ and $f_{i,m}(\vec{k}_t) = 0$.

Converting the summation over $\vec{k}_t$ to an integration using the two-dimensional density of states, we get

$$
\begin{aligned}
\alpha_{\hat{p}}^{\mathrm{2DEG}}(\omega_{\mathrm{l}}) &= \frac{\eta_{\hat{p}}}{c}\frac{\pi e^2 A}{4\pi^2 m_0^2\epsilon\omega_{\mathrm{l}}\Omega}\left|\int_{-\infty}^{\infty}\mathrm{d}z\phi_{\mathrm{electron}}*(z)\phi_{\text{heavy-hole}}(z)\right|^2\left|\hat{p}\cdot\vec{\Upsilon}'_{ij}\right|^2 \\
&\quad\times\int_0^{\infty}2\pi k_t dk_t\delta\left[E_{\mathrm{electron},m}(\vec{k}_t)-E_{\text{heavy-hole},n}(\vec{k}_t)-\hbar\omega_{\mathrm{l}}\right].
\end{aligned}
$$

Since

$$
\begin{aligned}
E_{i,m}(\vec{k}_t) &= E_{\mathrm{c0}}+\Delta_{\mathrm{e}}+\frac{\hbar^2k_t^2}{2m_{\mathrm{e}}^*} \\
E_{j,n}(\vec{k}_t) &= E_{\mathrm{v0}}-\Delta_{\mathrm{hh}}-\frac{\hbar^2k_t^2}{2m_{\mathrm{hh}}^*},
\end{aligned}
$$

where Δ_{e} is the confinement energy of the lowest electron subband, Δ_{hh} is the confinement energy of the lowest heavy-hole subband, m_{e}^* is the effective mass of electrons and m_{hh}^* is the effective mass of heavy-holes, we get:

$$
\begin{aligned}
\alpha_{\hat{p}}^{\mathrm{2DEG}}(\omega_{\mathrm{l}}) &= \frac{\eta_{\hat{p}}}{c}\frac{e^2 A}{4m_0^2\epsilon\omega_{\mathrm{l}}\Omega}\left|\int_{-\infty}^{\infty}\mathrm{d}z\phi_{\mathrm{electron}}*(z)\phi_{\text{heavy-hole}}(z)\right|^2\left|\hat{p}\cdot\vec{\Upsilon}'_{ij}\right|^2 \\
&\quad\times\int_0^{\infty}dk_t^2\delta\left[E_{\mathrm{g}}+\Delta_{\mathrm{e}}+\Delta_{\mathrm{hh}}+\frac{\hbar^2k_t^2}{2\mu}-\hbar\omega_{\mathrm{l}}\right],
\end{aligned}
$$

where $E_{\mathrm{g}} = E_{\mathrm{c0}} - E_{\mathrm{v0}}$ is the bulk bandgap, and μ is the reduced mass given by

$$
\frac{1}{\mu} = \frac{1}{m_{\mathrm{e}}^*}+\frac{1}{m_{\mathrm{hh}}^*}.
$$

Completing the integration over the delta function, we get:

$$\alpha_{\hat{p}}^{\text{2DEG}}(\omega_l) = \frac{\eta_{\hat{p}}}{c}\frac{e^2 A}{4m_0^2\epsilon\omega_l\Omega}\left|\int_{-\infty}^{\infty} dz\phi_{\text{electron}} * (z)\phi_{\text{heavy-hole}}(z)\right|^2 \left|\hat{p}\cdot\vec{\Upsilon}'_{ij}\right|^2$$
$$\times\int_0^{\infty} dk_t^2 \frac{2\mu}{\hbar^2}\delta\left[k_t^2 - \frac{2\mu}{\hbar^2}\left\{\hbar\omega_l - \left(E_g + \Delta_e + \Delta_{hh}\right)\right\}\right].$$

The integration over the delta function will be $2\mu/\hbar^2$ only if the value of k_t^2 found by setting the argument of the delta function to zero lies within the range of integration from 0 to ∞. Otherwise, this integral will vanish. Since any negative k_t^2 is outside the range of integration, the integral will be $2\mu/\hbar^2$ as long as $\hbar\omega_l \geq E_g + \Delta_e + \Delta_{hh}$, and will be zero otherwise. Therefore, we will get that

$$\alpha_{\hat{p}}^{\text{2DEG}}(\omega_l) = \frac{\eta_{\hat{p}}}{c}\frac{e^2\mu A}{2\hbar^2 m_0^2\epsilon\omega_l\Omega}\left|\int_{-\infty}^{\infty} dz\phi_{\text{electron}} * (z)\phi_{\text{heavy-hole}}(z)\right|^2 \left|\hat{p}\cdot\vec{\Upsilon}'_{ij}\right|^2$$
$$\Theta\left(\hbar\omega_l - E_g + \Delta_e + \Delta_{hh}\right)$$
$$= \frac{\eta_{\hat{p}}}{c}\frac{e^2\mu}{2\hbar^2 m_0^2\epsilon\omega_l L}\left|\int_{-\infty}^{\infty} dz\phi_{\text{electron}} * (z)\phi_{\text{heavy-hole}}(z)\right|^2 \left|\hat{p}\cdot\vec{\Upsilon}'_{ij}\right|^2$$
$$\Theta\left(\hbar\omega_l - E_g + \Delta_e + \Delta_{hh}\right),$$

where Θ is the Heaviside function.

Note that the absorption coefficient is independent of the photon energy since the two-dimensional electron–hole joint density of states is independent of carrier energy. Once again, the absorption coefficient is proportional to the joint density of states.

Quantum wire

$$\alpha_{\hat{p}}^{\text{1DEG}}(\omega_l) = \frac{\eta_{\hat{p}}}{c}\sum_{i,m,m',j,n,n'}\frac{\pi e^2}{m_0^2\epsilon\omega_l\Omega}K_{\hat{p},\omega_l}(i,m,m':j,n,n')$$
$$\times\sum_{k_x}\delta\left[E_{i,m,m'}(k_x) - E_{j,n,n'}(k_x) - \hbar\omega_l\right]\left[f_{j,n,n'}(k_x) - f_{i,m,m'}(k_x)\right]$$
$$= \frac{\eta_{\hat{p}}}{c}\frac{\pi e^2}{m_0^2\epsilon\omega_l\Omega}\left|\int_{-\infty}^{\infty} dy\phi_{\text{electron},m} * (y)\phi_{\text{heavy-hole},n}(y)\right|^2$$
$$\times\left|\int_{-\infty}^{\infty} dz\phi_{\text{electron},m'} * (z)\phi_{\text{heavy-hole},n'}(z)\right|^2$$
$$\times\left|\hat{x}\cdot\vec{\Upsilon}''_{ij}\right|^2\sum_{k_x}\delta\left[E_{\text{electron},m,m'}(k_x) - E_{\text{heavy-hole},n,n'}(k_x) - \hbar\omega_l\right].$$

where we have assumed as usual that the occupation probability of the initial state is unity and that of the final state is zero.

Converting the summation over k_x to an integration using the one-dimensional density of states:

$$\begin{aligned}
\alpha_{\hat{p}}^{\mathrm{1DEG}}(\omega_1) &= \frac{\eta_{\hat{p}}}{c}\frac{\pi e^2}{m_0^2\epsilon\omega_1\Omega}\left|\int_{-\infty}^{\infty}\mathrm{d}y\phi_{\mathrm{electron},m}*(y)\phi_{\mathrm{heavy\text{-}hole},n}(y)\right|^2 \\
&\quad\times\left|\int_{-\infty}^{\infty}\mathrm{d}z\phi_{\mathrm{electron},m'}*(z)\phi_{\mathrm{heavy\text{-}hole},n'}(z)\right|^2\left|\hat{x}\cdot\vec{\Upsilon}_{ij}''\right|^2\frac{L}{2\pi} \\
&\quad\times\int_{-\infty}^{\infty}\mathrm{d}k_x\delta\left[E_{\mathrm{g}}+\Delta_{\mathrm{e}}'+\Delta_{\mathrm{hh}}'+\frac{\hbar^2k_x^2}{2\mu}-\hbar\omega_1\right] \\
&= \frac{\eta_{\hat{p}}}{c}\frac{\pi e^2}{m_0^2\epsilon\omega_1\Omega}\left|\int_{-\infty}^{\infty}\mathrm{d}y\phi_{\mathrm{electron},m}*(y)\phi_{\mathrm{heavy\text{-}hole},n}(y)\right|^2 \\
&\quad\times\left|\int_{-\infty}^{\infty}\mathrm{d}z\phi_{\mathrm{electron},m'}*(z)\phi_{\mathrm{heavy\text{-}hole},n'}(z)\right|^2\left|\hat{x}\cdot\vec{\Upsilon}_{ij}''\right|^2\frac{L}{2\pi} \\
&\quad\times\int_{-\infty}^{\infty}\mathrm{d}k_x\frac{2\mu}{\hbar^2}\delta\left[k_x^2-\frac{2\mu}{\hbar^2}\left\{\hbar\omega_1-\left(E_{\mathrm{g}}+\Delta_{\mathrm{e}}'+\Delta_{\mathrm{hh}}'\right)\right\}\right] \\
&= \frac{\eta_{\hat{p}}}{c}\frac{e^2L\mu}{m_0^2\hbar^2\epsilon\omega_1\Omega}\left|\int_{-\infty}^{\infty}\mathrm{d}y\phi_{\mathrm{electron},m}*(y)\phi_{\mathrm{heavy\text{-}hole},n}(y)\right|^2 \\
&\quad\times\left|\int_{-\infty}^{\infty}\mathrm{d}z\phi_{\mathrm{electron},m'}*(z)\phi_{\mathrm{heavy\text{-}hole},n'}(z)\right|^2\left|\hat{x}\cdot\vec{\Upsilon}_{ij}''\right|^2 \\
&\quad\times\int_{-\infty}^{\infty}\mathrm{d}k_x\left\{\frac{\delta\left[k_x-\sqrt{\frac{2\mu}{\hbar^2}\left\{\hbar\omega_1-\left(E_{\mathrm{g}}+\Delta_{\mathrm{e}}'+\Delta_{\mathrm{hh}}'\right)\right\}}\right]}{2\sqrt{\frac{2\mu}{\hbar^2}\left\{\hbar\omega_1-\left(E_{\mathrm{g}}+\Delta_{\mathrm{e}}'+\Delta_{\mathrm{hh}}'\right)\right\}}}\right. \\
&\qquad\left.+\frac{\delta\left[k_x+\sqrt{\frac{2\mu}{\hbar^2}\left\{\hbar\omega_1-\left(E_{\mathrm{g}}+\Delta_{\mathrm{e}}'+\Delta_{\mathrm{hh}}'\right)\right\}}\right]}{2\sqrt{\frac{2\mu}{\hbar^2}\left\{\hbar\omega_1-\left(E_{\mathrm{g}}+\Delta_{\mathrm{e}}'+\Delta_{\mathrm{hh}}'\right)\right\}}}\right\} \\
&= \frac{\eta_{\hat{p}}}{c}\frac{e^2L}{2m_0^2\epsilon\omega_1\Omega}\left(\frac{2\mu}{\hbar^2}\right)^{1/2}\left|\hat{x}\cdot\vec{\Upsilon}_{ij}''\right|^2 \\
&\quad\times\left|\int_{-\infty}^{\infty}\mathrm{d}y\phi_{\mathrm{electron},m}*(y)\phi_{\mathrm{heavy\text{-}hole},n}(y)\right|^2 \\
&\quad\times\left|\int_{-\infty}^{\infty}\mathrm{d}z\phi_{\mathrm{electron},m'}*(z)\phi_{\mathrm{heavy\text{-}hole},n'}(z)\right|^2 \\
&\quad\times\frac{1}{\sqrt{\hbar\omega_1-\left(E_{\mathrm{g}}+\Delta_{\mathrm{e}}'+\Delta_{\mathrm{hh}}'\right)}}.
\end{aligned}$$

Note that the absorption coefficient is once again proportional to the electron-hole joint density of states. Note also that it has a singularity and diverges when the photon energy is equal to the energy separation between the first electron subband and the first heavy-hole subband. This singularity is due to the van-Hove singularity in the electron-hole joint density of states in one-dimension.

References

1. S. M. Sze, *Physics of Semiconductor Devices*, 2nd. edition, (John Wiley & Sons, New York, 1981).
2. L. T. Canham, "Silicon quantum wire array fabrication by electrochemical and chemical dissolution of wafers", Appl. Phys. Lett., **57**, 1046 (1990); L. T. Canham, W. Y. Leong, M. I. J. Beale, T. I. Cox and L. Taylor, "Efficient visible electroluminescence from highly porous silicon under cathodic bias", Appl. Phys. Lett., **61**, 2563 (1992).
3. A. G. Cullis and L. T. Canham, "Visible light emission due to quantum size effects in highly porous crystalline silicon", Nature, **353**, 335 (1991).
4. T. K. Sham, D. T. Jiang, I. Coulthard, J. W. Lorimer, X. H. Feng, K. H. Tan, S. P. Frigo, R. A. Rosenberg, D. C. Houghton and B. Bryskiewicz, "Origin of luminescence from porous silicon deduced by synchrotron-light-induced optical luminescence", Nature, **363**, 331 (1993).
5. L. Pavesi, L. Dal Negro, C. Mazzoleni, G. Franzo and F. Priolo, "Optical gain in silicon nanocrystals", Nature, **408**, 440 (2000).
6. O. Boyraz and B. Jalali, "Demonstration of a silicon Raman laser", Optics Express, **12**, 5269 (2004).
7. Y. Maeda, N. Tsukamoto, Y. Yazawa, Y. Kanemitsu and Y. Masumoto, "Visible photoluminescence of Ge microcrystals embedded in SiO_2 glassy matrices", Appl. Phys. Lett., **59**, 3168 (1991).
8. S. Datta, *Quantum Phenomena*, Modular Series on Solid State Devices, Eds. R. F. Pierret and G. W. Neudeck, (Addison-Wesley, Reading, 1989).
9. G. Gamow and J. M. Cleveland, *Physics: Foundations and Frontiers*, (Prentice Hall, Englewood Cliffs, 1960).
10. W. van Roosbroeck and W. Shockley, "Photon radiative recombination of electrons and holes in germanium", Phys. Rev., **94**, 1558 (1954).
11. J. Faist, F. Capasso, D. L. Civco, C. Sirtori, A. L. Hutchinson and A. Y. Cho, "Quantum cascade laser", Science, **264**, 553 (1994).
12. Y. Arakawa and H. Sakaki, "Multidimensional quantum well laser and temperature dependence of its threshold current", Appl. Phys. Lett., **40**, 939 (1982).
13. P. G. Elisiv, H. Lee, T. Liu, R. C. Newell, L. F. Lester and K. J. Malloy, "Ground state emission and gain in ultralow-threshold InAs-InGaAs quantum dots lasers", IEEE J. Sel. Top. Quant. Electron., **7**, 135 (2001).
14. S. Fathpour, Z. Mi. P. Bhattacharya, A. R. Kovsh, S. S. Mikhrin, I. L. Krestnikov, A. V. Kozhukov and N. N. Ledentsev, "The role of Auger recombination in the temperature dependent output characteristics ($T_0 = \infty$) of p-doped 1.3 μm quantum dot lasers", Appl. Phys. Lett., **85**, 5164 (2004).
15. H. Benisty, C. M. Sotomayor-Torres and C. Weisbuch, "Intrinsic mechanism for the poor luminescence properties of quantum box systems", Phys. Rev. B., **44**, 10945 (1991).
16. D. Bimberg, M. Grundmann and N. N. Ledentsov, *Quantum Dot Heterostructures*, (Wiley, 1999).
17. P. Bhattacharya, Z. Mi, J. Yang, D. Basu and D. Saha, "Quantum dot lasers: From promise to high performance devices", J. Crystl. Growth, **311**, 1625 (2009).

18. J. D. Jackson, *Classical Electrodynamics*, (Academic Press, New York, 1975).
19. R. P. Feynman, R. B. Leighton and M. Sands, *The Feynman Lectures on Physics*, Vol. III, Ch. 21, (Addison-Wesley, Reading, 1965).
20. J. J. Sakurai, *Modern Quantum Mechanics*, (Addison-Wesley, Reading, 1985).
21. R. P. Feynman, R. B. Leighton and M. Sands, *The Feynman Lectures on Physics*, Vol. II, Ch. 17, (Addison-Wesley, Reading, 1965).
22. U. Bockelmann and G. Bastard, "Interband absorption in quantum wires. I. Zero magnetic field case", Phys. Rev. B., **45**, 1688 (1992).
23. P. G. Rohner, "Calculation of the exchange energy for excitons in the two-body model", Phys. Rev. B, **3**, 433 (1971).
24. D. S. Chemla, D. A. B. Miller and P. W. Smith, "Non-linear optical properties of multiple quantum well structures for optical signal processing", Chapter 5, in *Semiconductors and Semimetals*, Ed. Raymond Dingle (Academic Press, San Diego, 1987).
25. L. V. Keldysh, "Excitons in semiconductor-dielectric nanostructures", Phys. Stat. Sol. A, **164**, 3 (1997).
26. E. A. Muljarov, E. A. Zhukov, V. S. Dneprovskii and Y. Masumoto, "Dielectrically enhanced excitons in semiconductor-insulator quantum wires: Theory and experiment", Phys. Rev. B, **62**, 7420 (2000).
27. P. G. Savvidis, J. J. Baumberg, R. M. Stevenson, M. S. Skolnick, D. M. Whittaker and J. S. Roberts, "Angle resonant stimulated polariton amplifier", Phys. Rev. Lett., **84**, 1547 (2000); "Asymmetric angular emission in semiconductor microcavities", Phys. Rev. B, **62**, R13278 (2000).
28. A. Imamoḡlu and R. J. Ram, "Quantum dynamics of exciton lasers", Phys. Lett. A, **214**, 193 (1996).
29. A. Imamoḡlu, R. J. Ram, S. Pau and Y. Yamamoto, "Non-equilibrium condenstates and lasers without inversion: Exciton-polariton lasers", Phys. Rev. A, **53**, 4250 (1996).
30. S. Christopoulos, G. Baldassarri Höger von Högersthal, A. J. D. Grundy, P. G. Lagoudakis, A. V. Kavokin, J. J. Baumberg, G. Christmann, R. Butté, E. Feltin, J-F Carlin and N. Grandjean, "Room temperature polariton lasing in semiconductor microcavities", Phys. Rev. Lett., **98**, 126405 (2007).
31. E. Yablonovitch, "Inhibited spontaneous emission in solid state physics and electronics", Phys. Rev. Lett., **58**, 2059 (1987).
32. O. Painter, R. K. Lee, A. Scherer, A. Yarive, J. D. O'Brien, P. D. Dapkus and I. Kim, "Two-dimensional photonic bandgap defect mode laser", Science, **284**, 1819 (1999).
33. V. G. Veselago, "The electrodynamics of substances with simultaneously negative values of ϵ and μ", Sov. Phys. Usp. **10**, 509 (1968).
34. J. B. Pendry, "Negative refraction makes a perfect lens", Phys. Rev. Lett., **85**, 3966 (2000).
35. W. H. Wee and J. B. Pendry, "Universal evolution of perfect lenses", Phys. Rev. Lett., **106**, 165503 (2011).
36. J. B. Pendry, D. Schurig and D. R. Smith, "Controlling electromagnetic fields", Science, **312**, 1780 (2006).

Chapter 7
Magnetic Field Effects in a Nanostructured Device

7.1 Electrons in a Magnetic Field

The behavior of an electron in a magnetic field is rich in physics. Many interesting phenomena, such as the Hall effect, Shubnikov–deHaas oscillations (oscillatory variation in the conductance of a sample as a function of inverse magnetic flux density), etc. are manifested when a structure is placed in a static magnetic field. These phenomena often provide valuable information about material parameters such as carrier concentration and carrier mobility within the structure. Moreover, a magnetic field can be harnessed for novel device applications, an example being the QCLE that we will discuss in this chapter. Magnetic fields can also lead to exotic effects such as in the integer and fractional quantum Hall effects (FQHE)—both of which have earned the Nobel Prize in physics. Suffice it to say then that it is important to understand how an electron in a nanostructured solid interacts with a magnetic field. That is the subject of this chapter.

A magnetic field has two basic effects on an electron: (1) it alters the electron's motion and behavior since it applies a Lorentz force $e\vec{v}\times\vec{B}$ on a moving electron which changes its trajectory and affects its wavefunction and energy states, and (2) it brings to the foreground effects that have to do with the electron's quantum-mechanical "spin." The electron's spin is like a tiny magnetic moment arising from the electron spinning about its own axis. Although this picture is somewhat crude and cannot explain many features such as why the spin angular momentum has a fixed magnitude of $\hbar/2$, it is nonetheless a convenient mental tool to visualize the spin [1, 2]. Most importantly, an electron's spin has an associated tiny magnetic moment $\vec{\mu}$ which will interact with any magnetic flux density $\vec{B}$ through the interaction Hamiltonian $-\vec{\mu}\cdot\vec{B}$. Therefore, when a magnetic field is present, spin related effects can no longer be ignored since the interaction will lift the spin degeneracy and make the energy states of an electron *spin-dependent*. In other words, we cannot account for spin by just multiplying the density of states with a factor of 2; instead, we must treat it in its own right and account for it explicitly.

S. Bandyopadhyay, *Physics of Nanostructured Solid State Devices*, 341
DOI 10.1007/978-1-4614-1141-3_7, © Springer Science+Business Media, LLC 2012

If we do account for spin explicitly, and furthermore treat the electron within relativistic quantum mechanics as opposed to the usual nonrelativistic quantum mechanics that we have seen in all the preceding chapters, then the equation which yields the "spin-dependent" wavefunction of the electron is no longer the familiar Schrödinger equation, but the so-called *Dirac equation* which we discuss next.

7.2 Dirac's Equation and Pauli's Equation

The Dirac equation reads [2]

$$\left[\left(i\hbar\frac{\partial}{c\partial t}+eA_0\right)-\{\alpha_x\}\left(-i\hbar\frac{\partial}{\partial x}+eA_x\right)-\{\alpha_y\}\left(-i\hbar\frac{\partial}{\partial y}+eA_y\right)\right.$$
$$\left.-\{\alpha_z\}\left(-i\hbar\frac{\partial}{\partial z}+eA_z\right)-\alpha_0 m_0 c\right][\psi(x,y,z,t)]=0, \tag{7.1}$$

where $\vec{A} = (A_0, A_x, A_y, A_z)$ is the 4-component magnetic vector potential introduced by magnetic fields, electric fields, etc. (in relativity, space and time are treated on an equal footing so that there are four dimensions – three in space and one in time), m_0 is the free electron mass, c is the speed of light in vacuum, α-s are 4×4 matrices, and $[\psi(x, y, z, t)]$ is a 4×1 column vector, whose first two rows pertain to the electron's spin-dependent wavefunction and the last two actually pertain to that of the associated positron (relativistic quantum mechanics couples matter and anti-matter so that an electron's and positron's wavefunctions are inseparable). Thus, when we take spin into account, the coupled electron-positron wavefunction is no longer a scalar, but a 4-component "tensor" $[\psi(x, y, z, t)]$. The matrices α appearing in (7.1) are given by

$$\{\alpha_0\} = \begin{bmatrix} 1 & 0 & 0 & 0 \\ 0 & 1 & 0 & 0 \\ 0 & 0 & -1 & 0 \\ 0 & 0 & 0 & -1 \end{bmatrix},$$

$$\{\alpha_x\} = \begin{bmatrix} 0 & 0 & 0 & 1 \\ 0 & 0 & 1 & 0 \\ 0 & 1 & 0 & 0 \\ 1 & 0 & 0 & 0 \end{bmatrix},$$

$$\{\alpha_y\} = \begin{bmatrix} 0 & 0 & 0 & -i \\ 0 & 0 & i & 0 \\ 0 & -i & 0 & 0 \\ i & 0 & 0 & 0 \end{bmatrix},$$

$$\{\alpha_z\} = \begin{bmatrix} 0 & 0 & 1 & 0 \\ 0 & 0 & 0 & -1 \\ 1 & 0 & 0 & 0 \\ 0 & -1 & 0 & 0 \end{bmatrix}. \tag{7.2}$$

In the nonrelativistic limit (or within the framework of nonrelativistic quantum mechanics), the electron and positron can be decoupled, leaving us with a simpler equation for the electron alone. This equation is known as the *Pauli equation* and has the form [1]

$$i\hbar\frac{\partial}{\partial t}\begin{bmatrix}\phi_1(x,y,z,t)\\ \phi_2(x,y,z,t)\end{bmatrix} = \left\{\left[\frac{\left|-i\hbar\vec{\nabla}_r - e\vec{A}(x,y,z,t)\right|^2}{2m_0} + V(x,y,z,t)\right]\begin{bmatrix}1 & 0\\ 0 & 1\end{bmatrix} + H_{\text{Zeeman}} + H_{\text{spin-orbit}}\right\}\begin{bmatrix}\phi_1(x,y,z,t)\\ \phi_2(x,y,z,t)\end{bmatrix}, \tag{7.3}$$

where the spin-dependent wavefunction is now a 2×1 component "spinor" with components $\phi_1(x, y, z, t)$ and $\phi_2(x, y, z, t)$, $\vec{A}$ is the vector potential associated with the magnetic field, H_{Zeeman} and $H_{\text{spin-orbit}}$ are 2×2 matrices representing the Zeeman interaction (in a magnetic field) and spin–orbit interaction, and $V(x, y, z, t)$ is the potential energy term which includes the lattice periodic potential and every other scalar potential. The Zeeman interaction arises when an electron is placed in an external magnetic field and is due to the magnetic moment of the spin interacting with that field through the $-\vec{\mu} \cdot \vec{B}$ term. On the other hand, the spin–orbit interaction is due to the magnetic moment of the spin interacting with an effective magnetic field of flux density B_{eff} through the $-\vec{\mu} \cdot \vec{B}_{\text{eff}}$ term. The effective magnetic field arises because of Lorentz transformation. An electron in an electric field accelerates and in the rest frame of the electron, the electric field that the electron sees Lorentz transforms into an effective magnetic field which depends on the electron's velocity. This is the origin of $\vec{B}_{\text{eff}}$. When the spin interacts with this effective magnetic field $\vec{B}_{\text{eff}}$, it gives rise to spin–orbit interaction [1].

Note that even in the nonrelativistic case, the electron's wavefunction is not a scalar, but a 2×1 "spinor" that has two components ϕ_1 and ϕ_2. Such a wavefunction allows us to determine the spin angular momentum vector $\vec{S} = S_x\hat{x} + S_y\hat{y} + S_z\hat{z}$ as follows:

$$\begin{aligned} S_x(x,y,z,t) &= (\hbar/2)\left[\phi_1^*(x,y,z,t)\ \phi_2^*(x,y,z,t)\right][\sigma_x]\begin{bmatrix}\phi_1(x,y,z,t)\\ \phi_2(x,y,z,t)\end{bmatrix} \\ &= (\hbar/2)\left[\phi_1^*(x,y,z,t)\ \phi_2^*(x,y,z,t)\right]\begin{bmatrix}0 & 1\\ 1 & 0\end{bmatrix}\begin{bmatrix}\phi_1(x,y,z,t)\\ \phi_2(x,y,z,t)\end{bmatrix} \\ &= \hbar\text{Re}\left[\phi_1^*(x,y,z,t)\phi_2(x,y,z,t)\right], \end{aligned} \tag{7.4}$$

$$
\begin{aligned}
S_y(x,y,z,t) &= (\hbar/2)\left[\phi_1^*(x,y,z,t)\ \phi_2^*(x,y,z,t)\right]\left[\sigma_y\right]\begin{bmatrix}\phi_1(x,y,z,t)\\ \phi_2(x,y,z,t)\end{bmatrix}\\
&= (\hbar/2)\left[\phi_1^*(x,y,z,t)\ \phi_2^*(x,y,z,t)\right]\begin{bmatrix}0 & -\mathrm{i}\\ \mathrm{i} & 0\end{bmatrix}\begin{bmatrix}\phi_1(x,y,z,t)\\ \phi_2(x,y,z,t)\end{bmatrix}\\
&= \hbar \mathrm{Im}\left[\phi_1^*(x,y,z,t)\phi_2(x,y,z,t)\right],
\end{aligned} \tag{7.5}
$$

$$
\begin{aligned}
S_z(x,y,z,t) &= (\hbar/2)\left[\phi_1^*(x,y,z,t)\ \phi_2^*(x,y,z,t)\right]\left[\sigma_z\right]\begin{bmatrix}\phi_1(x,y,z,t)\\ \phi_2(x,y,z,t)\end{bmatrix}\\
&= (\hbar/2)\left[\phi_1^*(x,y,z,t)\ \phi_2^*(x,y,z,t)\right]\begin{bmatrix}1\,0\\ 0\,1\end{bmatrix}\begin{bmatrix}\phi_1(x,y,z,t)\\ \phi_2(x,y,z,t)\end{bmatrix}\\
&= (\hbar/2)\left[|\phi_1(x,y,z,t)|^2 - |\phi_2(x,y,z,t)|^2\right].
\end{aligned} \tag{7.6}
$$

where Re stands for the real part, Im stands for the imaginary part, and the superscript * (asterisk) represents complex conjugate.

The quantities $[\sigma_x]$, $[\sigma_y]$, and $[\sigma_z]$ are called *Pauli spin matrices* and are given by

$$
\begin{aligned}
\sigma_x &= \begin{bmatrix}0\,1\\ 1\,0\end{bmatrix},\\
\sigma_y &= \begin{bmatrix}0 & -\mathrm{i}\\ \mathrm{i} & 0\end{bmatrix}\\
\sigma_z &= \begin{bmatrix}1 & 0\\ 0 & -1\end{bmatrix}
\end{aligned} \tag{7.7}
$$

Just as in the Schrödinger picture, the expectation value of any operator O can be determined as

$$
\langle O\rangle(t) = \int\!\!\int\!\!\int \mathrm{d}x\mathrm{d}y\mathrm{d}z\left[\phi_1^*(x,y,z,t)\ \phi_2^*(x,y,z,t)\right] O \begin{bmatrix}\phi_1(x,y,z,t)\\ \phi_2(x,y,z,t)\end{bmatrix}. \tag{7.8}
$$

The reader should have understood by now that dealing with spin is a little more involved mathematically, but really no more complex conceptually. However, in this book, we will not be dealing with spin explicitly using Dirac's or Pauli's equation; instead, we will treat the electron as a spinless particle and only focus on how a magnetic field affects its dynamics. This will allow us to restrict ourselves to the Schrödinger picture without the need to venture into the Dirac equation or the Pauli equation. This will also allow us to deal with a scalar wavefunction instead of a spinor wavefunction.

There is, however, a vast body of literature dealing with spin-related phenomena. This has recently received significant boost with the advent of the field of "spintronics" which deals with how to exploit the spin degree of freedom of an electron to fashion novel devices and find new ways of processing information, including quantum information that quantum computers deal with. That entire field has burgeoned into a major enterprise and the interested reader is referred to many excellent reviews [3] and a textbook [1] on this topic. In this book, we will not be discussing spin-related issues, but instead treat the electron as a "spinless" entity.

7.3 The "Spinless" Electron in a Magnetic Field

If the electron is viewed as a spinless particle, then we can describe its dynamics in a magnetic field using the familiar Schrödinger equation with the Hamiltonian given by (6.64) but with two differences. We will replace the free electron mass with the effective mass, which will allow us to ignore the lattice potential, and we will assume that the magnetic field is time independent. That will make the magnetic vector potential time invariant as well, so that the time-dependent Schrödinger equation describing a (spinless) electron in a static (but spatially varying) magnetic field will be

$$i\hbar \frac{\partial \psi(\vec{r},t)}{\partial t} = \left[\frac{\left| -i\hbar\vec{\nabla}_r - e\vec{A}(\vec{r}) \right|^2}{2m^*} + V(\vec{r},t) \right] \psi(\vec{r},t). \tag{7.9}$$

The time-invariant magnetic vector potential due to a static magnetic field of flux density $\vec{B}(\vec{r})$ is given by (see (6.58))

$$\vec{B}(\vec{r}) = \vec{\nabla}_r \times \vec{A}(\vec{r}). \tag{7.10}$$

Obviously, there are many choices of the magnetic vector potential that can satisfy the last equation. All of them are legitimate choices and each of them constitutes a so-called "gauge." We will now solve (7.9) in various scenarios to find the electron's wavefunction and energy states in a magnetic field.

7.4 An Electron in a Two-Dimensional Electron Gas (Quantum Well) Placed in a Static and Uniform Magnetic Field

Consider a 2-DEG with a static and spatially uniform magnetic field of flux density B directed perpendicular to its plane as shown in Fig. 7.1. We will first find the wavefunction and allowed energy states of a nonrelativistic spinless electron in this system and then use them to derive several interesting properties.

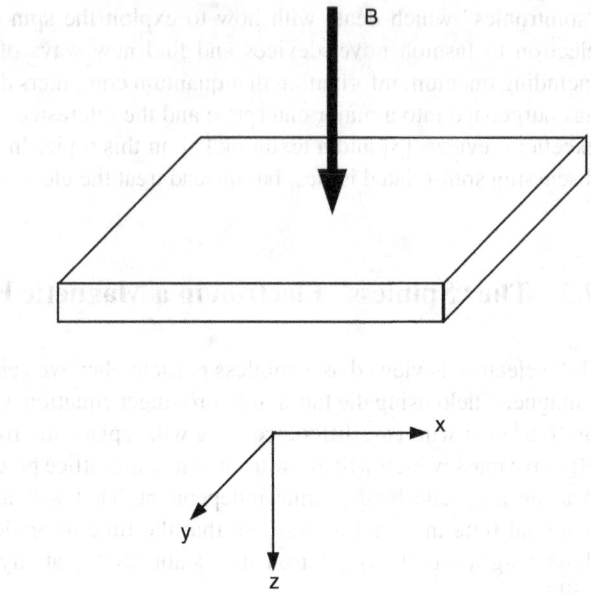

Fig. 7.1 A two dimensional electron gas with a static and uniform magnetic field directed perpendicular to its plane

Clearly, all the potentials that the electron sees are time independent. Therefore, we can use the time-independent Schrödinger Equation to find the wavefunction and energy eigenstates of the electron. This equation is

$$\left[\frac{\left|-i\hbar\vec{\nabla}_r - e\vec{A}(\vec{r})\right|^2}{2m^*} + V(\vec{r})\right]\psi(\vec{r}) = E\psi(\vec{r}). \tag{7.11}$$

The magnetic vector potential $\vec{A}(\vec{r})$ must satisfy (7.10) and we have to pick a suitable gauge. For this purpose, assume that the plane of the electron gas is in the x-y plane and the magnetic field is directed along the z-axis as shown in Fig. 7.1. Then, the scalar components of the magnetic vector potential A_x, A_y, and A_z must satisfy

$$\det\begin{vmatrix} \hat{x} & \hat{y} & \hat{z} \\ \frac{\partial}{\partial x} & \frac{\partial}{\partial y} & \frac{\partial}{\partial z} \\ A_x & A_y & A_z \end{vmatrix} = B\hat{z}. \tag{7.12}$$

One possible choice that satisfies the previous equation is $A_x = -By$ and $A_y = A_z = 0$. Such a choice is called a *Landau gauge*. Another possible choice is $A_x = -A_y = -By/2$ and $A_z = 0$. This is called a *symmetric gauge* for obvious reasons. We can make either of these choices for the magnetic vector potential, but the former is mathematically more convenient. Therefore, making that choice, the Schrödinger equation for the spinless electron becomes

$$\left\{\frac{\left(-i\hbar\frac{\partial}{\partial x}+eBy\right)^2}{2m^*}-\frac{\hbar^2}{2m^*}\frac{\partial^2}{\partial y^2}+\left[-\frac{\hbar^2}{2m^*}\frac{\partial^2}{\partial z^2}+V(z)\right]\right\}\psi(x,y,z)=E\psi(x,y,z), \tag{7.13}$$

where $V(z)$ is the confining potential in the z-direction that confines the 2-DEG, $\psi(x, y, z)$ is the electron wavefunction, and E is the energy eigenstate. The preceding equation can be expanded as

$$\left\{-\frac{\hbar^2}{2m^*}\frac{\partial^2}{\partial x^2}-i\hbar\frac{eB}{m^*}y\frac{\partial}{\partial x}+\frac{(eBy)^2}{2m^*}-\frac{\hbar^2}{2m^*}\frac{\partial^2}{\partial y^2}+\left[-\frac{\hbar^2}{2m^*}\frac{\partial^2}{\partial z^2}+V(z)\right]\right\}\psi(x,y,z)$$
$$= E\psi(x,y,z). \tag{7.14}$$

Since in the last equation there are no "mixed terms" involving two or more coordinates, the wavefunction can be written as a product solution:

$$\psi(x,y,z)=\phi(x)\zeta(y)\xi(z). \tag{7.15}$$

We should note that the Hamiltonian in (7.14) is invariant in x, meaning that there are no terms that depend on the x-coordinate. Therefore, the x-component of the wavefunction will be a plane wave propagating in the x-direction, i.e.,

$$\phi(x)=\frac{1}{\sqrt{L_x}}e^{ik_x x}, \tag{7.16}$$

where L_x is a normalizing length.

Substituting the last result in (7.14), we get

$$\left\{\frac{\hbar^2k_x^2}{2m^*}+\hbar k_x\frac{eB}{m^*}y+\frac{(eBy)^2}{2m^*}-\frac{\hbar^2}{2m^*}\frac{\partial^2}{\partial y^2}+\left[-\frac{\hbar^2}{2m^*}\frac{\partial^2}{\partial z^2}+V(z)\right]\right\}\zeta(y)\xi(z)=E\zeta(y)\xi(z). \tag{7.17}$$

We can break the last equation up into two equations in such a way that it will decouple motion in the x-y plane from motion along the z-direction:

$$\left[\frac{\hbar^2k_x^2}{2m^*}+\hbar k_x\frac{eB}{m^*}y+\frac{(eBy)^2}{2m^*}\right]\zeta(y)\xi(z)-\frac{\hbar^2}{2m^*}\frac{\partial^2\zeta(y)}{\partial y^2}\xi(z)=\Lambda\zeta(y)\xi(z)$$
$$-\frac{\hbar^2}{2m^*}\frac{\partial^2\xi(z)}{\partial z^2}\zeta(y)+V(z)\zeta(y)\xi(z)=\Xi\zeta(y)\xi(z), \tag{7.18}$$

where

$$E=\Lambda+\Xi. \tag{7.19}$$

Dividing the first line of (7.18) throughout by $\xi(z)$ and the second line throughout by $\zeta(y)$, we obtain

$$\left[\frac{\hbar^2 k_x^2}{2m^*} + \hbar k_x \frac{eB}{m^*} y + \frac{(eBy)^2}{2m^*} - \frac{\hbar^2}{2m^*}\frac{\partial^2}{\partial y^2}\right]\zeta(y) = \Lambda\zeta(y)$$
$$\left[-\frac{\hbar^2}{2m^*}\frac{\partial^2}{\partial z^2} + V(z)\right]\xi(z) = \Xi\xi(z), \quad (7.20)$$

It is clear now that the first line in the preceding equation describes motion in the x-y plane while the second line describes (confined) motion along the z-direction. Thus, these two motions have been decoupled.

After a little algebra, we can write the first line of (7.20) as

$$\left[-\frac{\hbar^2}{2m^*}\frac{\partial^2}{\partial y^2} + \frac{1}{2}m^*\omega_c^2(y + y_0)^2\right]\zeta(y) = \Lambda\zeta(y), \quad (7.21)$$

where

$$\omega_c = \frac{eB}{m^*}$$
$$y_0 = \frac{\hbar k_x}{eB} = k_x r_c^2$$
$$r_c = \sqrt{\frac{\hbar}{eB}}. \quad (7.22)$$

The quantity ω_c is called the *cyclotron frequency*, y_0 is the y-coordinate of *cyclotron orbit center*, and r_c is the *cyclotron radius*. The origin of these names will become clear shortly.

The solution of the second line in (7.20) depends on the shape of the confining potential $V(z)$ along the z-direction. If the confining potential is rectangular with infinite barrier heights, then the solution will be particle-in-a-box states, i.e.,

$$\xi(z) = \xi_n(z) = \sqrt{\frac{2}{W_z}}\sin\left(\frac{n\pi z}{W_z}\right)$$
$$\Xi = \Xi_n = \frac{\hbar^2}{2m^*}\left(\frac{n\pi}{W_z}\right)^2, \quad (7.23)$$

where n is an integer denoting the subband index in the z-direction, and W_z is the width of the potential well in the z-direction, or the thickness of the two-dimensional electron gas.

Equation (7.21) should be recognized as the Schrödinger equation describing simple harmonic motion along the y-axis centered at $y = -y_0$. This equation has well-known solutions for the eigenstates (wavefunctions) and eigenenergies (allowed energy levels), which are [4]:

$$\zeta(y) = \zeta_m(y) = \left(\frac{\alpha}{\pi^{1/2}2^m m!}\right)^{1/2} \mathcal{H}_m\left[\alpha(y + y_0)\right] \mathrm{e}^{-(1/2)\alpha^2(y+y_0)^2}$$

$$\Lambda = \Lambda_m = \left(m + \frac{1}{2}\right)\hbar\omega_c, \tag{7.24}$$

where m is an integer denoting different modes of simple harmonic motion, $\mathcal{H}_m$ is the Hermite polynomial of the m-th order and $\alpha = \sqrt{m^*\omega_c/\hbar} = 1/r_c$. Note that the wavefunction of the lowest mode ($m = 1$)decays exponentially with the square of the displacement from the center of the simple harmonic motion located at $y = -y_0$. The decay constant is $1/\alpha = r_c$, which means that at a distance of r_c from the center, the wavefunction has decayed by a factor of $e^{1/2}$ or by a factor of~ 1.65 from the (maximum) value at the center.

Summing up, if we assume that the confining potential (in the direction of the magnetic field) that confines the electron to the two-dimensional layer is rectangular with infinite barriers, then the total energy and the wavefunction of the electron in a perpendicular magnetic field are

$$E = E_{m,n} = \Lambda_m + \Xi_n = \left(m + \frac{1}{2}\right)\hbar\omega_c + \frac{\hbar^2}{2m^*}\left(\frac{n\pi}{W_z}\right)^2$$

$$\psi(x, y, z) = \phi(x)\zeta_m(y)\xi_n(z) = \frac{1}{\sqrt{L_x}}\mathrm{e}^{ik_x x}$$

$$\times \left(\frac{\alpha}{\pi^{1/2}2^m m!}\right)^{1/2} \mathcal{H}_m[\alpha(y + y_0)]\mathrm{e}^{-(1/2)\alpha^2(y+y_0)^2} \times \sqrt{\frac{2}{W_z}}\sin\left(\frac{n\pi z}{W_z}\right). \tag{7.25}$$

Note from (7.22) that with increasing magnetic field, $y_0 \to 0$, so that a stronger magnetic field pushes the center of the simple harmonic motion more towards the center of the y-axis. Note also that the expectation value of the y-position is $\langle y \rangle = \langle \zeta_m(y)|y|\zeta_m(y)\rangle = \int \mathrm{d}y \zeta_m^*(y) y \zeta_m(y) = -y_0$ for every m.

7.4.1 Landau Levels and Landau Orbits

It is interesting to observe from (7.24) that the energy associated with motion in the x-y plane (Λ_m) is independent of the electron's wavevector. In fact, the allowed energy states (for different values of m) form discrete levels that are uniformly

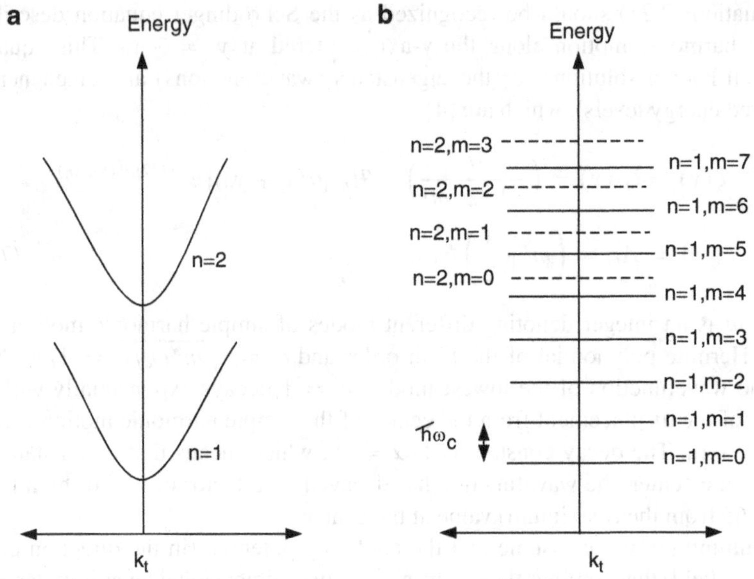

Fig. 7.2 Energy dispersion relations of a spinless electron in a two-dimensional electron gas where $k_t^2 = k_x^2 + k_y^2$. (**a**) Without a magnetic field, and (**b**) with a magnetic field directed perpendicular to the plane

spaced in energy with a spacing of $\hbar\omega_c$. These levels are called *Landau levels*. In Fig. 7.2, we show the energy dispersion relations of electrons in the 2-DEG with and without the magnetic field. The latter is found from (7.25).

We can now infer a few facts about the nature of an electron's trajectory in a 2-DEG with a magnetic field directed perpendicular to the plane. First, note that if the electron has a nonzero k_x, then it will have nonzero displacement along the x-direction. The only way that can happen, i.e., the electron can execute simple harmonic motion along the y-axis while having nonzero displacement along the x-axis, is if it goes around in a circular trajectory in the x-y plane. These trajectories are called *Landau orbits* or *cyclotron orbits*. The orbit is obviously centered at $y = -y_0$ along the y-axis. For any given magnetic field strength, electrons with different wavevectors k_x will execute circular orbital motion centered at different y-coordinates since y_0 depends on k_x. Electrons with higher wavevectors will have the centers of their cyclotron motion located farther away from the origin of the y-axis.

Another way to show that the motion in the x-y plane results in at least closed orbits is to show that the expectation values of x- and y-component of the velocity are both identically zero, which means that the electron has no *translational* motion

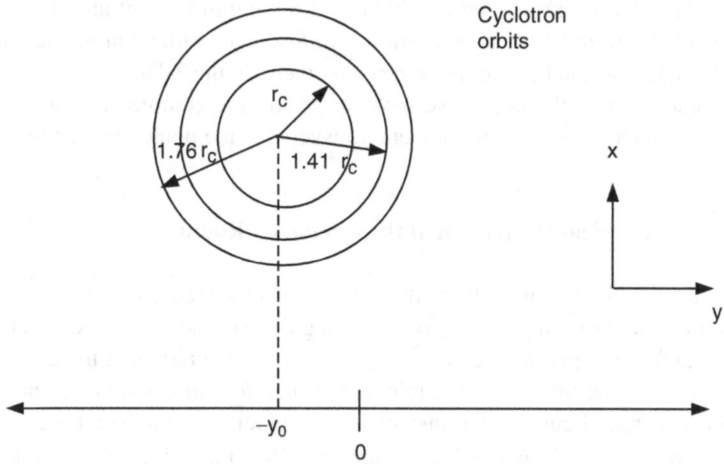

Fig. 7.3 Cyclotron motion of an electron in a two-dimensional electron gas with a magnetic field directed perpendicular to its plane

along either the x- or the y-direction. This is consistent with circular motion. We show this as follows:

$$\begin{aligned}\langle v_x \rangle &= \frac{\langle p_x - eA_x \rangle}{m^*} = \frac{\langle p_x \rangle - e\langle A_x \rangle}{m^*} = \frac{\langle p_x \rangle + eB\langle y \rangle}{m^*} = \frac{\hbar k_x - eBy_0}{m^*} \\ &= \frac{\hbar k_x - \hbar k_x}{m^*} = 0 \\ \langle v_y \rangle &= \frac{\langle p_y - eA_y \rangle}{m^*} = \frac{\langle p_y \rangle}{m^*} = \frac{\left\langle \zeta_m(y) \left| -i\hbar \frac{\partial}{\partial y} \right| \zeta_m(y) \right\rangle}{m^*} = 0. \end{aligned} \tag{7.26}$$

Note that the electron's velocity depends on the kinematic momentum and not the canonical momentum. Since $\langle v_x \rangle = \langle v_y \rangle = 0$, the motion of the electron is definitely a closed orbit in the x-y plane, which turns out to be circular. The radius of the circle will be the classical displacement from the center along the y-direction, which is $1/\alpha = r_c = \sqrt{\frac{\hbar}{eB}}$. With increasing magnetic field strength, the radius becomes smaller. The cyclotron orbits are shown in Fig. 7.3. Circles of different radii $\sqrt{m} r_c$ [$m =$ integer] correspond to different Landau levels for different integral values of m. Electrons in a circle with radius $\sqrt{m} r_c$ has kinetic energies of $(m + 1/2)\hbar\omega_c$ (see Problem 7.1). The time to complete one period of rotation in any given circle is always the same and is equal to $1/\omega_c = m^*/(eB)$.

It is intriguing to find that despite the fact that there are no potential barriers in the x-y plane, the electron's motion in this plane is *not free*, but is constrained to closed

Landau orbits. Thus, the magnetic field acts as a confining agent and this kind of confinement is referred to as *magnetostatic confinement*, while confinement due to potential barriers would be *electrostatic confinement*. In the 2-DEG with a magnetic field perpendicular to the plane, we have magnetostatic confinement in the plane (x-y plane) and electrostatic confinement transverse to the plane (z-direction).

7.4.1.1 Semiclassical Derivation of the Cyclotron Radius

We can show that the radius of the m-th cyclotron orbit will be $\sqrt{m}r_c = \sqrt{m\hbar/(eB)}$ by invoking semiclassical physics. An electron going around in a circle experiences a centripetal force of magnitude m^*v^2/R_c where R_c is the radius of the circle. This force must lie in the plane of the circle and is therefore directed along the radius because it is perpendicular to the instantaneous velocity vector. The Lorentz force on the electron is $e\vec{v} \times \vec{B}$, which is also directed along the radius of the circle since $\vec{v}$ lies in the x-y plane and $\vec{B}$ is perpendicular to this plane. Setting this two forces equal to each other, we get

$$evB = \frac{m^*v^2}{R_c}, \tag{7.27}$$

which yields

$$R_c = \frac{m^*v}{eB} = \frac{\hbar k}{eB} = \frac{h}{\lambda eB}, \tag{7.28}$$

where we have utilized the DeBroglie relation that $m^*v = h/\lambda$, with λ being the DeBroglie wavelength of the electron.

We will now "quantize" the orbit, i.e., enforce the condition that the circumference should be an integral multiple of the DeBroglie wavelength. In other words, $2\pi R_c = m\lambda$. Substitution of this result in the previous equation immediately yields that

$$R_c = \sqrt{\frac{m\hbar}{eB}} = \sqrt{m}r_c. \tag{7.29}$$

This analysis also reveals that it is the classical Lorentz force due to the perpendicular magnetic field that makes an electron go around in circular Landau orbits or cyclotron orbits. We then impose the bit of quantum mechanics, i.e., enforce the condition that the circumference is quantized in integral multiples of the DeBroglie wavelength. That yields the expression for the cyclotron radius R_c. Since the area of any orbit is πR_c^2, the area of cyclotron orbits is quantized to integral multiples of $\pi r_c^2 = \pi \frac{\hbar}{eB} = \frac{h}{2eB}$. Therefore, the magnetic flux enclosed by any orbit is an integral multiple of $B \times \frac{h}{2eB} = h/(2e)$, which depends only on universal constants (Planck's constant and the charge of an electron). The quantity h/e is called the *fundamental flux quantum*.

7.4.1.2 Degeneracy of Landau Levels

Each allowed value of the radius $R_c = \sqrt{m} r_c$ corresponds to a Landau orbit. Each orbit has a specific energy $(m + 1/2)\hbar\omega_c$ and constitutes a *Landau level*. Any Landau level can contain a number of orbiting electrons, all having the same radius $R_c = \sqrt{m} r_c$, but their centers are located at different locations $-y_0$ along the y-coordinate. Since $y_0 = \hbar k_x/(eB)$, these electrons will all have different k_x and hence belong to different wavevector states. We will now find the number of allowed states in a Landau level, which will then permit us to determine the maximum number of electrons in a Landau level, since each state can accommodate a maximum of 2 electrons (of opposite spins) according to the Pauli Exclusion Principle.[1] The number of allowed states in a Landau level is that state's *degeneracy*.

Consider a *finite* 2-DEG of length L (along x-direction) and width W (along y-direction) placed a magnetic field of flux density B directed perpendicular to its plane. We will assume that the 2DEG is ideal, i.e., its thickness is so small so that only the lowest transverse subband ($n = 1$) is occupied with electrons. Higher transverse subbands ($n > 1$) are so much higher in energy, that they are never occupied by electrons.

Different states in the 2DEG are characterized by different y_0 values. But since y_0 must remain within the boundaries of the 2DEG, we must have

$$y_0^{\max} - y_0^{\min} = W. \tag{7.30}$$

so that

$$k_x^{\max} - k_x^{\min} = \frac{eBW}{\hbar}. \tag{7.31}$$

Remember now that in a crystal with periodic boundary conditions, allowed k_x-states are separated by the interval $2\pi/L$, where L is the dimension of the crystal along the x-direction, i.e., its length (see Chap. 4). Therefore, the number of k-states in any Landau level is

$$N = \frac{k_x^{\max} - k_x^{\min}}{2\pi/L} = \frac{eBWL}{h} = \frac{\Phi}{h/e}, \tag{7.32}$$

where $\Phi = BWL$ is the flux enclosed by the entire 2DEG and h/e is the fundamental flux quantum. This number is the *degeneracy* of each Landau level and it is the number of fundamental flux quanta enclosed by the entire 2DEG. Note that this number increases linearly with magnetic field strength since Φ is proportional to the field strength.

[1]This assumes that the electrons are not spin polarized. However, in a very strong magnetic field, the electrons could be completely spin polarized, i.e., every electron could point either along the field or opposite to it, depending on whether the so-called gyromagnetic ratio, or Landé g-factor, in the material is positive or negative. If the electrons are completely spin-polarized, then Pauli Exclusion Principle will allow only one electron to be accommodated in each state.

Although we had been discussing the spinless electron, this should not stop us from interjecting spin-related considerations into the discussion from now on. For the time being, consider the situation when the electrons are not spin polarized, i.e., the magnetic field is too weak to polarize spins and both spin orientations (parallel and anti-parallel to the magnetic field) are allowed. According to Pauli Exclusion Principle, we can put at most two electrons in each k-state, where these two electrons have opposite spins. Consequently, in a *fully filled* Landau level, there are $2N = 2\Phi/(h/e)$ electrons. Thus, each Landau level is N-fold degenerate and can accommodate up to $2N = 2e\Phi/h$ electrons. All of them have the same energy $(m+1/2)\hbar\omega_c$! Since there can be so many electrons with the same energy in strong magnetic fields, they interact and can form complex many-body ground states that can lead to exotic phenomena such as FQHE that we will briefly discuss later.

If the magnetic field is so strong as to completely spin polarize the electrons, then we can put only one electron in each k-state. In that case, in a fully filled Landau level, there will be $N = \Phi/(h/e)$ electrons.

7.4.1.3 Landau Level Filling

Consider an ideal 2-DEG where the electrons are not spin polarized. Suppose that there are p fully filled Landau levels. Since each Landau level contains $2N$ electrons, the total number of electrons in the 2DEG is

$$N_s = p \times 2N = \frac{2p\Phi}{h/e}. \tag{7.33}$$

Consequently, the two-dimensional electron density is

$$n_s = \frac{N_s}{WL} = 2p\frac{B}{h/e}. \tag{7.34}$$

Therefore, the number of fully filled Landau levels at low temperatures is

$$p = \frac{hn_s}{2eB} \quad [\texttt{electrons are not spin-polarized}]. \tag{7.35}$$

This number is sometimes called the "filling factor." It is easy to show that if the electrons are completely spin polarized, then the filling factor will be

$$p' = \frac{hn_s}{eB} \quad [\texttt{electrons are completely spin-polarized}]. \tag{7.36}$$

Note that the number of fully filled Landau levels *decreases* with increasing magnetic field strength. This should not be surprising since the degeneracy of each Landau level ($N = e\Phi/h$) increases with field strength. As a result, each level can accommodate more and more electrons as the field strength increases. Since

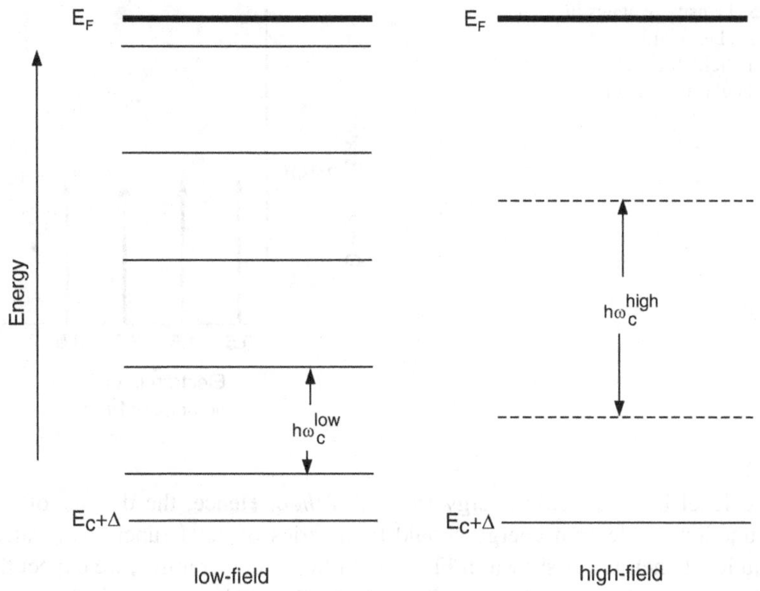

Fig. 7.4 Landau levels below the fixed Fermi level at two different magnetic field strengths in a 2-DEG where only one transverse subband ($n = 1$) is filled. Here E_c is the bulk conduction band edge energy and Δ is the confinement energy of the lowest transverse subband (which is the energy of the lowest subband without any magnetic field present)

the number of electrons is fixed (it does not depend on magnetic field), therefore it stands to reason that the number of fully filled Landau levels should decrease with increasing field strength.

This is also consistent with the fact that the Fermi level in the 2-DEG must be fixed in energy independent of the magnetic field. The field cannot change n_s or N_s and the Fermi level is determined by these two quantities. Now, in accordance with Fermi Dirac statistics, the number of fully filled Landau levels at low temperatures is the number below the Fermi level (see Fig. 7.4). With increasing magnetic field strength, the spacing $\hbar\omega_c$ between consecutive levels increases. Therefore, if the Fermi level is fixed in energy, then obviously the number of Landau levels below the Fermi level should decrease with increasing field strength. In other words, (7.35) and (7.36) make perfect sense.

7.4.2 *Density of States in a Two-Dimensional Electron Gas in a Perpendicular Magnetic Field*

Since every electron in the ideal 2DEG in a perpendicular magnetic field is in a Landau level, there are ideally no allowed electron states outside these levels. Each

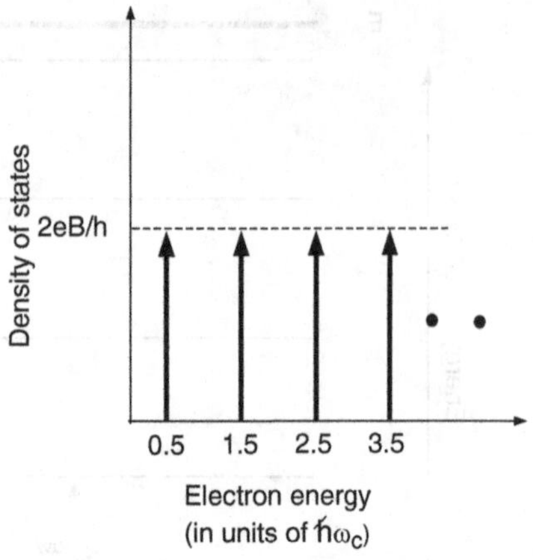

Fig. 7.5 Density of states in an ideal 2-DEG with a magnetic field directed perpendicular to its plane

Landau level has a specific energy $(m + 1/2)\hbar\omega_c$. Hence, the density of states as a function of electron energy should be a series of delta functions located at Landau level energies as shown in Fig. 7.5. At finite temperatures, we expect these delta functions to be somewhat broadened due to fluctuations in the electron energy (electrons will absorb and emit phonons which will alter their energies), but we will ignore that effect here and consider the temperature to be 0 K, where such effects are nonexistent.[2]

Next, we must find the heights of these delta functions. Let us call the height of the m-th peak A_m. By definition, the two-dimensional density of states $D^{2-d}(E)$ is related to the two-dimensional electron density at 0 K temperature as

$$n_s = \int_0^{E_F} D^{2-d}(E)\mathrm{d}E = \int_0^{E_F} A_m \delta[E - (m + 1/2)\hbar\omega_c]\mathrm{d}E = p \times A, \quad (7.37)$$

where we have assumed that $A_m = A$, independent of m. Here, p is the number of Landau levels below the Fermi level μ_F, i.e., p is the number of fully filled Landau levels (when the spins are unpolarized). Comparing the last equation with (7.34), we immediately get

$$A = \frac{2eB}{h}, \quad (7.38)$$

which indeed is independent of the Landau level index m, justifying our assumption. Therefore, all the delta functions have equal heights of $2eB/h$. It is easy to show that if the spins are completely polarized, then the height of the delta functions will be eB/h (we will simply have to replace p with p' in (7.37)).

---

[2]Even at 0 K, disorder and impurities in the 2-DEG could broaden the delta functions, but we are considering an ideal 2-DEG with no disorder or impurities.

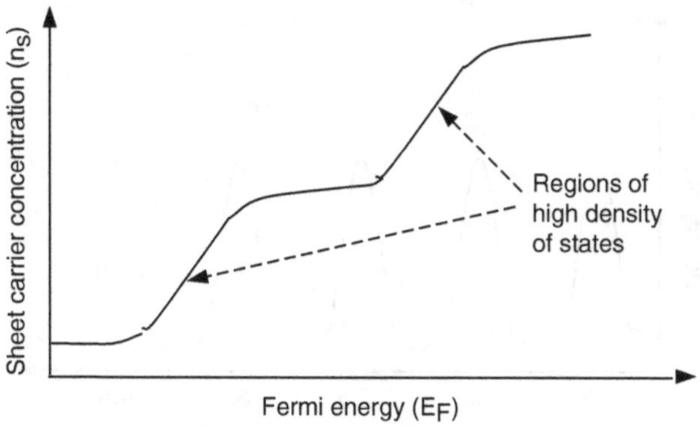

Fig. 7.6 Sheet carrier concentration versus Fermi energy

7.4.2.1 Fermi Level Location

Since there are no electron states outside Landau levels, it might appear that the Fermi level should always coincide with the highest occupied Landau level, i.e., it must be pinned at a Landau level. This actually does not happen and the Fermi level is often between two Landau levels. The reason for this was discussed in [5].

From (7.37), we get that a change ΔE_{F} in the Fermi energy will cause a corresponding change Δn_s in the carrier concentration given by

$$\Delta n_s = D^{2-d}(E)\Delta E_{\mathrm{F}} = \frac{2eB}{h}\delta[E - (m + 1/2)\hbar\omega_{\mathrm{c}}]\Delta E_{\mathrm{F}}. \qquad (7.39)$$

In Fig. 7.6, we show n_s versus E_{F} based on integrating the previous equation (where we have allowed some broadening of the delta function). We find that a slight change in n_s can cause a very large change in E_{F} if the Fermi energy is not located at an energy where the density of states is high. Therefore, in order to stabilize the Fermi level, it should be located where the density of states peaks, i.e., at a Landau level energy. This is our basis for thinking that the Fermi level should be pinned at a Landau level.

The reason why that does not always happen is because impurities and disorder can give rise to a decent density of states in between any two Landau levels [6]. This is because cyclotron orbits can get stuck around impurities forming localized states. Also current vortices form around impurities in a magnetic field [7] and result in "magnetic bound states" that cause additional states to appear between two consecutive Landau levels. As a result, the actual density of states plot may not look like Fig. 7.5 but rather like Fig. 7.7, where the density of states is nonzero between two Landau levels. The spurious states caused by disorder, which appear in the gaps between Landau levels, can stabilize the Fermi level between two Landau levels, so that the Fermi level need not be pinned at a Landau level.

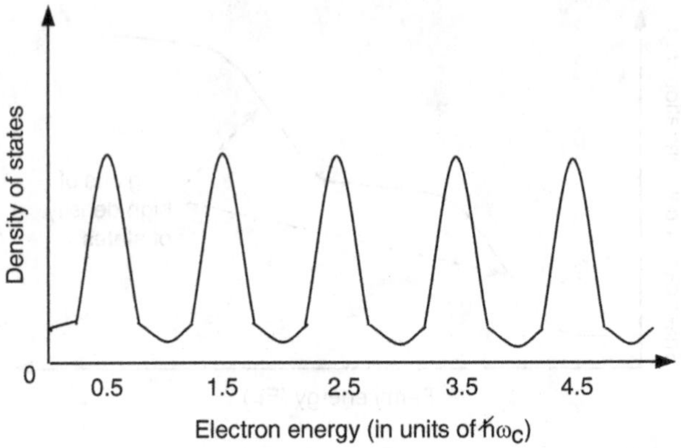

Fig. 7.7 Actual density of states in a disordered 2DEG placed in a perpendicular magnetic field

7.4.3 Shubnikov–deHaas Oscillations in a Two-Dimensional Electron Gas

We now describe a phenomenon that will cause the conductance or resistance of a 2DEG placed in a perpendicular magnetic field to oscillate as a function of the inverse flux density. These oscillations are called Shubnikov–deHaas oscillations. They are frequently used to extract the sheet carrier concentration in the 2DEG since the period of the oscillation depends on the sheet carrier concentration.

Consider an ideal 2DEG in a perpendicular magnetic field of flux density B. Assume that the sheet carrier concentration is n_s and also that the electrons are not spin polarized. Then the number of fully filled Landau levels is p given by (7.35). If this number turns out to be a noninteger, say 11.2, then 11 Landau levels are fully filled and the 12th one is partially filled.

As we increase B, the number of filled Landau levels decreases as more and more of them rise above the Fermi level μ_F and get depopulated. Every time a Landau level crosses the Fermi level, the density of states at the Fermi level peaks. We will see in the next chapter that conductance measured under low biases and at low temperature is proportional to the density of states at the Fermi level since carriers with Fermi energy carry most of the current. Therefore, every time a Landau level crosses the Fermi level, the conductance peaks. As a result, the resistance of the 2DEG is minimum when p is an integer (e.g., $p = 11$) and maximum when p is a half-integer (e.g., $p = 11.5$). If two successive conductance peaks occur at magnetic flux densities B_n and B_{n+1} corresponding to $p = n$ and $p = n + 1$, respectively, then

$$n + 1 - n = \frac{hn_s}{2e}\left(\frac{1}{B_{n+1}} - \frac{1}{B_n}\right). \tag{7.40}$$

Consequently, we will expect to see the conductance (and equivalently resistance) oscillate as a function of inverse magnetic flux density $1/B$. These oscillations are called *Shubnikov–deHaas oscillations.* The period of this oscillation $\Delta(1/B) = \frac{1}{B_{n+1}} - \frac{1}{B_n}$ allows us to find the sheet carrier concentration from the relation

$$n_s = \frac{2e}{h} \frac{1}{\frac{1}{B_{n+1}} - \frac{1}{B_n}} = \frac{2e}{h} \frac{1}{\Delta(1/B)}. \tag{7.41}$$

These oscillations are not observed in weak magnetic fields and their amplitudes increase with increasing magnetic field strength. That is because the time taken to complete a cyclotron orbit is $1/\omega_c = m^*/(eB)$ and this time decreases with increasing field strength. The more the number of times an orbit is completed before scattering disrupts the cyclotron motion, the stronger will be the amplitude of Shubnikov–deHaas oscillations. A stronger magnetic field allows more orbits to be circled before scattering occurs, and results in a larger amplitude of the oscillation.

We can roughly state that the condition for observing these oscillations is that at least one orbit should be circled before scattering takes place, which corresponds to the requirement that

$$\frac{1}{\omega_c} < \tau, \tag{7.42}$$

where τ is the mean time between collisions. Therefore, we will need the magnetic field to be strong enough to ensure that $\omega_c \tau > 1$.

The quantity τ is called the *quantum lifetime* or *single particle lifetime* and could be considerably different from the ensemble averaged momentum relaxation time $\langle \tau_m \rangle$. Recall from (2.40) that for a nondegenerate carrier population and for elastic scattering, the momentum relaxation time $\tau_m(k)$ is $1/\sum_{\vec{k}'} S(\vec{k}, \vec{k}')(1 - \cos\alpha)$, where α is the angle between $\vec{k}$ and $\vec{k}'$. On the other hand, the quantum lifetime $\tau(k)$ will be $1/\sum_{\vec{k}'} S(\vec{k}, \vec{k}')$. Thus, the two would be equal only if the scattering mechanisms in the sample are both elastic and isotropic. Otherwise, they could be vastly different. For example, if mobility is primarily determined by impurity scattering, which prefers small angle scattering, then $\langle \tau_m \rangle \gg \tau$ [8]. Das Sarma and Stern have shown that the ratio $\langle \tau_m \rangle/\tau$ could exceed 100 in high-mobility 2-DEGs [9].

We can rewrite (7.42) as

$$\omega_c \tau = \frac{eB}{m^*} \frac{\tau}{\langle \tau_m \rangle} \langle \tau_m \rangle = \frac{e\langle \tau_m \rangle}{m^*} \frac{\tau}{\langle \tau_m \rangle} B = \mu \frac{\tau}{\langle \tau_m \rangle} B > 1, \tag{7.43}$$

where μ is the carrier mobility. Therefore, oscillations will start to show up when the product of the mobility and magnetic flux density exceeds the ratio $\langle \tau_m \rangle/\tau$. Normally, one would expect higher mobility samples to start exhibiting Shubnikov–deHaas oscillations at lower magnetic fields, but this may not be always true since higher mobility samples also tend to have a higher $\langle \tau_m \rangle/\tau$ ratios.

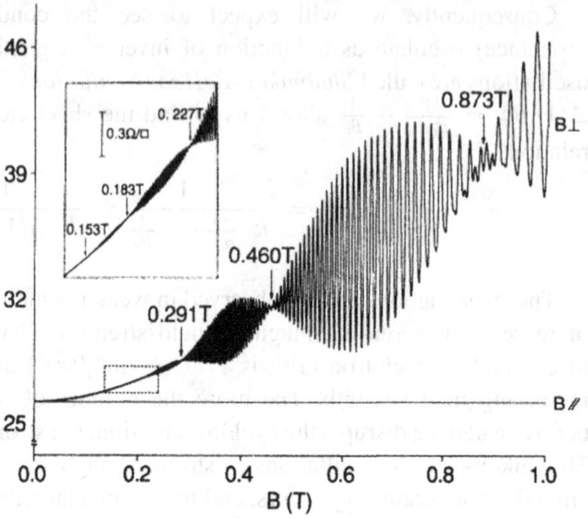

Fig. 7.8 Beating pattern in Shubnikov–deHaas oscillations due to spin splitting of subbands in a 2-DEG as a result of spin–orbit interaction. The magnetoresistance in the boxed region is amplified for clarity in the inset. Reproduced from [11] with permission from the American Physical Society. Copyright American Physical Society 1989. See http://prb.aps.org/abstract/PRB/v39/i2/p1411_1

7.4.3.1 Determining the Quantum Lifetime from the Amplitude Decay of Shubnikov–deHaas Oscillations

The amplitude of the Shubnikov–deHaas oscillations increases with increasing magnetic field because an electron can complete more cyclotron orbits in a higher magnetic field before it scatters. Thus, the oscillations have a decaying envelope going from higher to lower magnetic fields. From this envelope, one can extract the quantum lifetime.

Coleridge has derived an approximate analytical expression for the amplitude of resistance oscillations as a function of magnetic field at a temperature T [10]. It is:

$$\Delta R(B,T) = 4R_0 \frac{2\pi^2 k_B T}{\hbar\omega_c} \frac{1}{\sinh\left(2\pi^2 k_B T/\hbar\omega_c\right)} \exp\left[-\frac{\pi}{\omega_c \tau}\right], \tag{7.44}$$

where R_0 is the zero-field resistance. This provides a method to determine the quantum lifetime τ from the decaying envelope of the oscillations. One plots the amplitude as a function of inverse magnetic flux density in a log-linear plot. Ideally, this plot should be linear (otherwise, one fits a straight line to the plot) and from the slope one can extract the quantum lifetime based on (7.44). Such a plot is called a "Dingle plot" and has been successfully applied to extract the quantum lifetimes in both two- and quasi-one-dimensional structures.

7.4.3.2 Beating Patterns in Shubnikov–deHaas Oscillations

Sometimes the Shubnikov–deHaas oscillations exhibit beating as shown in Fig. 7.8, indicating that there is a mixture of two or more frequencies. There are many possible causes of this, but the two primary ones are (1) parallel layer formation, and (2) spin splitting. Often, a 2DEG really consists of two or more parallel layers with slightly different sheet carrier concentrations. Thus, in accordance with (7.41), there will be two or more distinct periods of Shubnikov–deHaas oscillations and their mixture causes the beating pattern. Beating patterns are often observed in high electron mobility transistors (HEMT) where a 2DEG forms in the gate insulator as well as at the interface between the gate insulator and the underlying semiconductor. The dopants are always placed in the gate insulator causing space charge band bending and a resulting potential barrier at the interface. The barrier causes incomplete electron transfer from the dopants into the semiconductor. Therefore, a 2DEG with low carrier concentration forms in the insulator and another 2DEG with higher carrier concentration forms in the semiconductor (along the interface) as shown in Fig. 7.9. These two 2DEG-s cause the beating pattern because they have different sheet carrier concentrations.

Sometimes, the magnetic field and spin–orbit interaction causes spin splitting of each Landau level into two sublevels as shown in Fig. 7.10. We ignored that effect in our analysis since we decided to ignore spin-related phenomena at the very outset. However, they exist, and can show up in Shubnikov–deHaas oscillations. Because of Fermi–Dirac statistics, two spin–split sublevels will have slightly different sheet carrier concentration and hence there will be two frequencies that will cause beating of the Shubnikov–deHaas oscillations [11]. An example of this is shown in Fig. 7.8. The Shubnikov–deHaas oscillations can yield the carrier concentrations in the two different layers in a HEMT, or of two different spin polarizations in a 2DEG, *individually*, whereas Hall effect would have given us the total carrier concentration only. Thus, Shubnikov–deHaas oscillations are a more powerful measurement tool than the Hall effect as far as determining sheet carrier concentration is concerned.

7.5 An Electron in a One-Dimensional Electron Gas (Quantum Wire) Placed in a Static and Uniform Magnetic Field

In this section, we will study the physics of electrons in a quasi 1-DEG (quantum wire) placed in a static and uniform magnetic field. We will find that there are some significant differences between quasi one- and two-dimensional systems.

7.5.1 Quantum Wire with Parabolic Confinement Potential Along the Width

Consider a quasi one-dimensional structure (or "quantum wire") where electron motion is free along the x-direction. Confining potentials restrict the motion in the y- and z- directions. We will assume that the confining potential in the z-direction (along the thickness of the wire) is rectangular with infinite barriers, while the

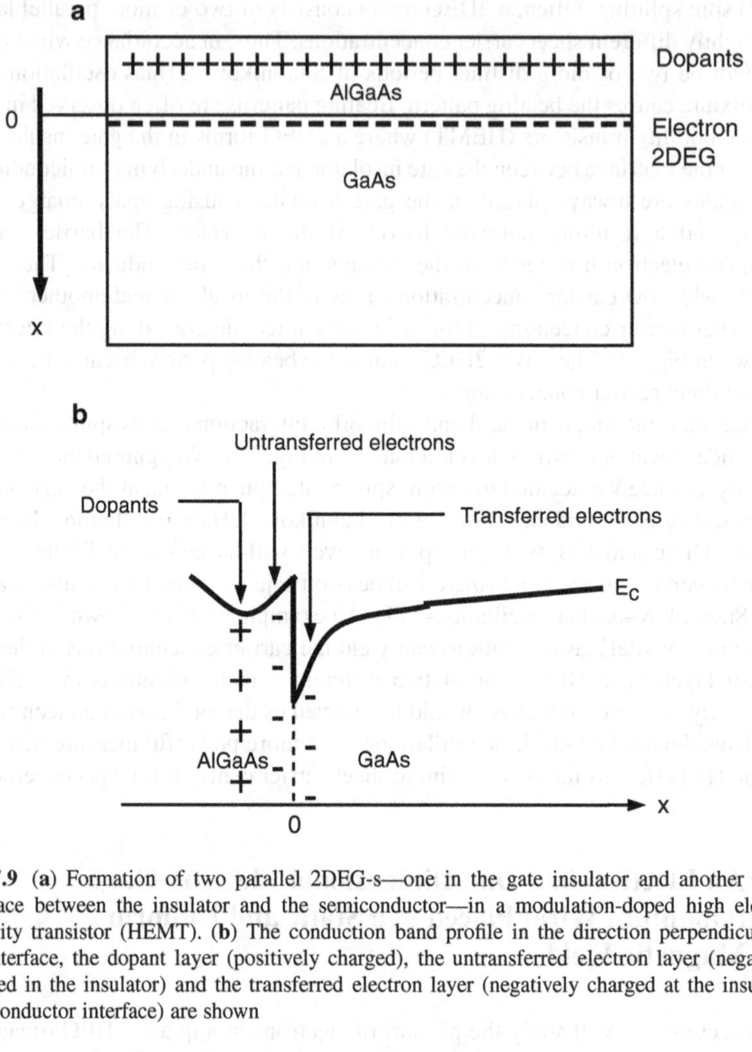

Fig. 7.9 (**a**) Formation of two parallel 2DEG-s—one in the gate insulator and another at the interface between the insulator and the semiconductor—in a modulation-doped high electron mobility transistor (HEMT). (**b**) The conduction band profile in the direction perpendicular to the interface, the dopant layer (positively charged), the untransferred electron layer (negatively charged in the insulator) and the transferred electron layer (negatively charged at the insulator-semiconductor interface) are shown

confining potential in the y-direction (along the wire's width) is parabolic and varies quadratically with distance from the axis of the wire in both $+y$ and $-y$ directions. Such a quantum wire can be realized in practice by using split gates (metallic gate with a slit in the middle) delineated on a 2DEG or quantum well. This system is shown in Fig. 7.11. A negative potential applied to the metal depletes electrons in the 2DEG from underneath the gate, leaving a narrow sliver of electrons under the slit, which forms a quantum wire. The confining potential in the y-direction in such a system is close to being parabolic and can be approximated as

$$V_y(y) \approx \frac{1}{2} m^* \omega_0^2 y^2, \tag{7.45}$$

where ω_0 is called the "curvature" of the confining potential in space.

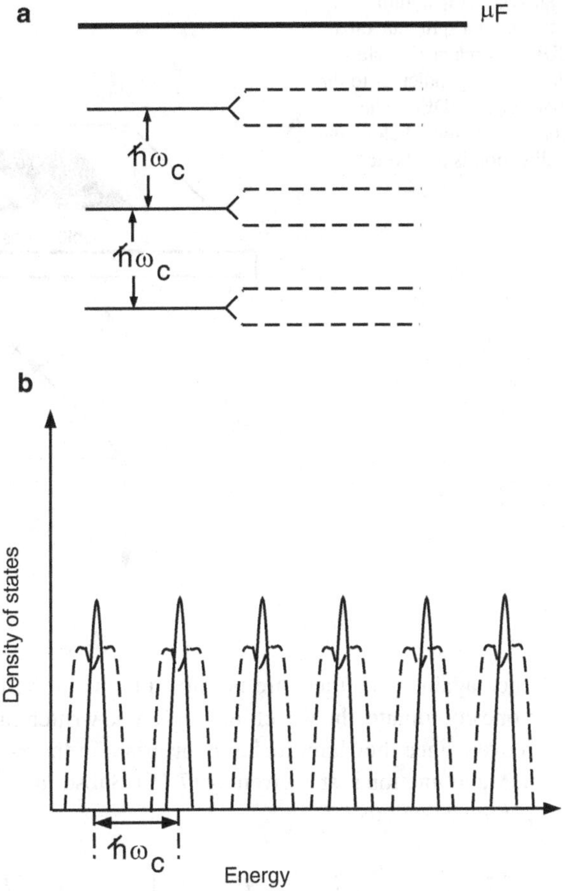

Fig. 7.10 (**a**) The Zeeman effect due to the magnetic field, or other spin–orbit interaction, can split every Landau level into two sublevels containing electrons of anti-parallel spins. The Landau levels are shown in *solid lines* while the sublevels are shown in *broken lines*. The Fermi level is denoted by μ_F. If the spin splitting is due to spin–orbit interaction, then it can be wavevector dependent. (**b**) The density of states versus energy in a two dimensional electron gas with a transverse magnetic field present. The *solid* (*broken*) *lines* show the density of states without (with) spin splitting. Each density of states peak corresponding to a Landau level splits into two in the presence of spin splitting

We will assume that a magnetic field is directed in the z-direction. In that case, the Schrödinger equation describing the spinless electron in the quantum wire (assuming the same Landau gauge for the magnetic vector potential as in the case of the 2DEG in the previous section) is

$$\left\{-\frac{\hbar^2}{2m^*}\frac{\partial^2}{\partial x^2} - i\hbar\frac{eB}{m^*}y\frac{\partial}{\partial x} + \frac{(eBy)^2}{2m^*} - \frac{\hbar^2}{2m^*}\frac{\partial^2}{\partial y^2} + \left(\frac{1}{2}m^*\omega_0^2 y^2\right) + \left[-\frac{\hbar^2}{2m^*}\frac{\partial^2}{\partial z^2} + V(z)\right]\right\}\psi(x,y,z) = E\psi(x,y,z). \tag{7.46}$$

which is the same as (7.14) except for the y^2-dependent term (enclosed in parentheses) which represents the parabolic confining potential in the y-direction.

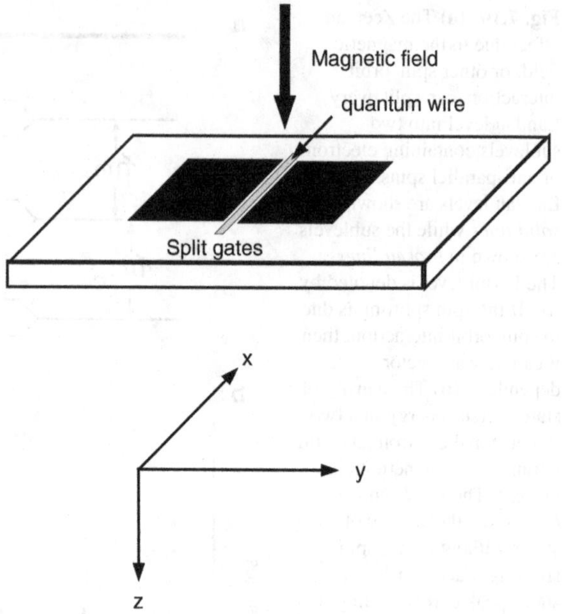

Fig. 7.11 A quantum wire defined by a split gate on a 2DEG. A magnetic field is directed perpendicular to the plane of the 2DEG. The confining potential along the y-direction is parabolic

Once again, since there are no mixed terms in the Hamiltonian involving more than one coordinate, the wavefunction can be written in a product form as in (7.15). Moreover, since the Hamiltonian is invariant in the x-coordinate, the x-component of the wavefunction is still given by (7.16). Substituting this wavefunction in (7.46), we get the equations

$$\left[\frac{\hbar^2 k_x^2}{2m^*} + \hbar k_x \frac{eB}{m^*} y + \frac{(eBy)^2}{2m^*} + \frac{1}{2} m^* \omega_0^2 y^2 - \frac{\hbar^2}{2m^*}\frac{\partial^2}{\partial y^2}\right]\zeta(y) = \Lambda \zeta(y)$$

$$-\left[\frac{\hbar^2}{2m^*}\frac{\partial^2}{\partial z^2} + V(z)\right]\xi(z) = \Xi \xi(z), \qquad (7.47)$$

where $E = \Lambda + \Xi$. Here, we have once again decoupled the x-y motion from the z-motion, as we did before in the 2DEG case.

The first line of the preceding equation can be written as

$$\left[-\frac{\hbar^2}{2m^*}\frac{\partial^2}{\partial y^2} + \frac{1}{2} m^* \omega_c^2 (y + y_0)^2 \frac{1}{2} m^* \omega_0^2 y^2\right]\zeta(y) = \Lambda \zeta(y), \qquad (7.48)$$

where the various quantities are defined by (7.22).

Simplifying, we get

$$\left[-\frac{\hbar^2}{2m^*}\frac{\partial^2}{\partial y^2}+\frac{1}{2}m^*\frac{\omega_c^2\omega_0^2}{\omega^2}y_0^2+\frac{1}{2}m^*\omega^2\left(y+\frac{\omega_c^2}{\omega^2}y_0\right)^2\right]\zeta(y)=\Lambda\zeta(y), \quad (7.49)$$

where $\omega^2=\omega_0^2+\omega_c^2$.

Transposing the constant term from the left-hand-side to the right-hand-side, we obtain

$$\left[-\frac{\hbar^2}{2m^*}\frac{\partial^2}{\partial y^2}+\frac{1}{2}m^*\omega^2\left(y+\frac{\omega_c^2}{\omega^2}y_0\right)^2\right]\zeta(y)=\left[\Lambda-\frac{1}{2}m^*\frac{\omega_c^2\omega_0^2}{\omega^2}y_0^2\right]\zeta(y). \quad (7.50)$$

This equation is the familiar Schrödinger equation describing simple harmonic motion centered at coordinate $y=-\frac{\omega_c^2}{\omega^2}y_0$. The solutions of this equation yield the wavefunction and energy eigenstate associated with y-component of the motion:

$$\zeta(y)=\zeta_m(y)=\left(\frac{\beta}{\pi^{1/2}2^m m!}\right)^{1/2}\mathcal{H}_m\left[\beta\left(y+\frac{\omega_c^2}{\omega^2}y_0\right)\right]\mathrm{e}^{-(1/2)\beta^2\left(y+\frac{\omega_c^2}{\omega^2}y_0\right)^2}$$

$$\Lambda=\Lambda_m=\left(m+\frac{1}{2}\right)\hbar\omega+\frac{1}{2}m^*\frac{\omega_c^2\omega_0^2}{\omega^2}y_0^2, \quad (7.51)$$

where $\beta=\sqrt{m^*\omega/\hbar}\neq\sqrt{m^*\omega_c/\hbar}$. Therefore, $\beta\neq\alpha=1/r_c$. This then yields the total energy and eigenstate as

$$E=E_{m,n}=\Lambda_m+\Xi_n=\left(m+\frac{1}{2}\right)\hbar\omega+\frac{1}{2}m^*\frac{\omega_c^2\omega_0^2}{\omega^2}y_0^2+\frac{\hbar^2}{2m^*}\left(\frac{n\pi}{W_z}\right)^2$$

$$=\left(m+\frac{1}{2}\right)\hbar\omega+\frac{\hbar^2k_x^2}{2m^*\left(\frac{\omega}{\omega_0}\right)}+\frac{\hbar^2}{2m^*}\left(\frac{n\pi}{W_z}\right)^2$$

$$\psi(x,y,z)=\phi(x)\zeta_m(y)\xi_n(z)=\frac{1}{\sqrt{L_x}}\mathrm{e}^{ik_xx}$$

$$\times\left(\frac{\beta}{\pi^{1/2}2^m m!}\right)^{1/2}\mathcal{H}_m\left[\beta\left(y+\frac{\omega_c^2}{\omega^2}y_0\right)\right]\mathrm{e}^{-(1/2)\beta^2\left(y+\frac{\omega_c^2}{\omega^2}y_0\right)^2}$$

$$\times\sqrt{\frac{2}{W_z}}sin\left(\frac{n\pi z}{W_z}\right), \quad (7.52)$$

where obviously L_x is the normalizing length in the x-direction and W_z is the width of the quantum well in the z-direction.

It is easy to see that these equations become identical to the corresponding equations for the two-dimensional case when ω_0 becomes zero, because then the confinement in the y-direction vanishes and the quantum wire becomes a quantum well.

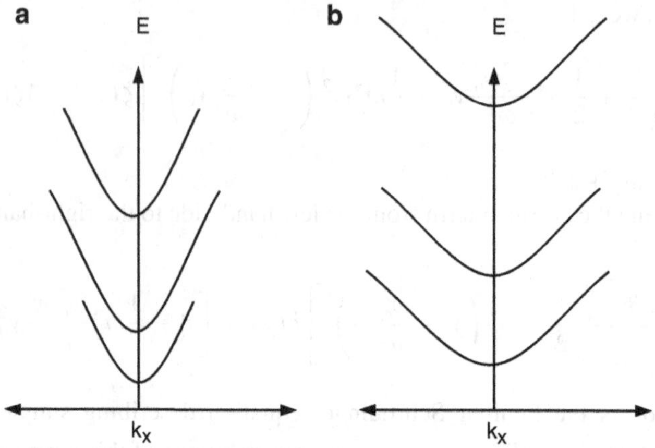

Fig. 7.12 Dispersion relations in different hybrid magneto-electric subbands at (**a**) low magnetic fields, and (**b**) high magnetic fields in a quantum wire with parabolic confinement potential in the direction transverse to the magnetic field and the wire axis

In Fig. 7.12, we plot the energy dispersion relation E versus k_x. Note that with increasing magnetic field, the energy spacing between the subbands increases and the curvatures of the parabolas decrease, as if the effective mass defined as $\hbar^2/\left(\frac{\partial^2 E}{\partial k_x^2}\right)$ increases. The effective mass in this case is $\left[1 + (\omega_c/\omega_0)^2\right] m^*$, which obviously increases with increasing magnetic field strength. Ultimately, in the limit of infinitely strong magnetic field, the curvatures reduce to zero, the effective mass becomes infinity, and the dispersion curves become totally flat and resemble the dispersion curves of Landau levels in the 2-DEG shown in Fig. 7.2b. The reader should understand that this happens because in the presence of a very strong magnetic field, the cyclotron orbit diameters $2\sqrt{m}r_c = 2\sqrt{m\hbar/(m^*\omega_c)}$ ($m =$ integer) become much smaller than the effective width of the quantum wire which is $2\sqrt{\hbar/(m^*\omega_0)}$. At that point, the electron goes around in cyclotron orbits without ever becoming quite aware of the confinement in the y-direction due to the electrostatic potential $V(y)$. Since an electron in a cyclotron orbit has no translational velocity, its effective mass is infinity. The magnetic field causes *magnetostatic confinement* by binding the electron into cyclotron orbits, while the parabolic confining potential causes *electrostatic confinement*. When the magnetostatic confinement overwhelms the electrostatic confinement (this happens when $\omega_c \gg \omega_0$, or when $B \gg m^*\omega_0/e$), the electron no longer feels the electrostatic confinement in the y-direction and the difference between the one-dimensional case and the two-dimensional case gets blurred.

The states given by (7.51) are called *hybrid magnetoelectric states* since they are a consequence of both magnetostatic confinement and electrostatic confinement in the y-direction. Note that in the 2-DEG , we had magnetostatic confinement in the x-y plane (because the magnetic field is in the z-direction) and electrostatic confinement in the z-direction. Here, we have magnetostatic confinement in the x-y plane (once again because the magnetic field is in the z-direction) and electrostatic confinements along both y- and z-directions. The y-component of motion therefore experiences both magnetostatic and electrostatic confinements.

Note also that the electrostatic confinement represented by the curvature ω_0 and the magnetostatic confinement represented by the curvature ω_c add to create an effective magneto-electric confinement represented by the curvature ω, where $\omega^2 = \omega_0^2 + \omega_c^2$.

Finally, note that the simple harmonic motion is now centered at $-\frac{\omega_c^2}{\omega^2} y_0$. If the electrostatic confinement is much stronger than the magnetostatic confinement, i.e., $\omega_0 \gg \omega_c$ (which happens at weak magnetic field strengths), then the center is at

$$\begin{aligned} y_{\text{center}} &= \frac{\omega_c^2}{\omega^2} y_0 \approx \frac{\omega_c^2}{\omega_0^2} y_0 \\ &= \frac{1}{\omega_0^2} \left(\frac{eB}{m^*}\right)^2 \frac{\hbar k_x}{eB} = \frac{1}{\omega_0^2} \frac{eB}{m^*} \frac{\hbar k_x}{m^*} = \frac{e\hbar k_x}{(m^* \omega_0)^2} B. \end{aligned} \tag{7.53}$$

Therefore, with increasing magnetic field, the center of simple harmonic motion, y_{center}, gets pushed farther and farther away from the origin of the y-coordinate. That is, it gets pushed away from the axis of the wire toward the edges and the electrons tend to localize at the edges of the quantum wire. This behavior is *opposite* to that in a 2DEG, where increasing magnetic field moved the center closer to the origin of the y-axis. Unlike a quantum wire, a quantum well has infinite extent along the y-axis and hence no "edges." Note two important features: (1) electrons with opposite signs of the wavevectors (and hence possibly opposite velocities) get pushed toward opposite edges. We expect that from the classical picture because the Lorentz force $e\vec{v} \times \vec{B}$ acts in opposite directions on oppositely traveling electrons; (2) electrons and holes with the same k_x get pushed toward opposite edges in a quantum wire since their charges have opposite signs e and $-e$. This is the basis of the QCLE that we will discuss later. The traditional Hall effect has a different basis. There, an electric field is applied along the length of the wire so that electrons and holes drift in opposite directions and have opposite signs of k_x. Therefore, their opposite charges actually push them toward the *same* edge.[3] This will tend to prevent building up of the Hall voltage and indeed the Hall effect is weak in bipolar materials that have nearly equal concentrations of electrons and holes. Only if the concentrations of electrons and holes are vastly different (as in a unipolar material that is either n-type or p-type), will a significant Hall voltage develop between the two edges and give rise to the Hall effect.

[3] Classically, the Lorentz force $q\vec{v} \times \vec{B}$ [q = particle charge] on the electron and hole have the same direction since both $\vec{v}$ and q are opposite for the two particles. Therefore, both particles are pushed in the same direction.

7.5.2 Quantum Wire with Rectangular Confinement Potential Along the Width

So far, we have considered a quantum wire with parabolic confining potential in the y-direction (direction mutually perpendicular to both the wire axis and the magnetic field). Let us now consider a quantum wire with *rectangular* cross-section where the confining potentials along both width (y-direction) and thickness (z-direction) are rectangular with infinite barriers. A magnetic field is applied in the z-direction. We are interested in finding the wavefunctions and energy dispersion relations of the hybrid magneto-electric states in this case, for which we need to solve the time-independent Schrödinger equation. Unfortunately, the Schrödinger equation does not have an analytical solution in this case and must be solved numerically, as described below.

The effective mass time-independent Schrödinger equation will be

$$\left\{\frac{\left[-\mathrm{i}\hbar\vec{\nabla}_r - e\vec{A}(x,y,z)\right]^2}{2m^*} + V(y) + V(z)\right\}\psi(x,y,z) = E\psi(x,y,z), \quad (7.54)$$

where the confining potentials are:

$$\begin{aligned} V(y) &= 0 \; (|y| \le W_y/2) \\ &= \infty \; (|y| > W_y/2) \\ V(z) &= 0 \; (|z| \le W_z/2) \\ &= \infty \; (|z| > W_z/2), \end{aligned} \quad (7.55)$$

with W_z being the thickness and W_y the width.

The magnetic field is directed along the z-direction and the associated flux density is $B\hat{z}$. Choosing the Landau gauge, we write the magnetic vector potential as

$$\vec{A}(x,y,z) = -By\hat{x}. \quad (7.56)$$

Since there are no mixed terms in the Hamiltonian, we can write the wavefunction in a product form:

$$\psi(x,y,z) = \varphi(x)\phi(y)\zeta(z). \quad (7.57)$$

Furthermore, because the Hamiltonian is invariant in x, the x-component of the wavefunction is a plane wave given by

$$\varphi(x) = \frac{1}{\sqrt{L_x}}\mathrm{e}^{\mathrm{i}k_x x}. \quad (7.58)$$

The reader can easily verify that we can now write the Schrödinger equation as two equations after decoupling the z-component of motion:

$$\frac{\partial^2 \phi(y)}{\partial y^2} + \frac{2m^*}{\hbar^2}\epsilon_1 \phi(y) - \left(\frac{y}{r_c^2}\right)\phi(y) + 2\frac{y}{r_c^2}k_x \phi(y) - k_x^2 \phi(y) = 0$$

$$\frac{\partial^2 \zeta(z)}{\partial z^2} + \frac{2m^*}{\hbar^2}\epsilon_2 \zeta(z) = 0. \quad (7.59)$$

where r_c is the cyclotron radius given by $r_c = \sqrt{\hbar/(eB)}$, and $E = \epsilon_1 + \epsilon_2$.

The solution of the second subequation in the preceding equation (subject to the boundary conditions in (7.55)) are particle-in-a-box states $\zeta(z) = (2/W_z)^{1/2}\sin(n\pi z/W_z)$. The solution of the first subequation is more challenging. We would like to solve it as an eigenvalue problem (subject to the boundary conditions in (7.55)) where we find the values of the wavevector k_x for a given value of ϵ_1 and repeat this procedure for different ϵ_1 to find the energy dispersion relation (ϵ_1 versus k_x) of the hybrid magneto-electric states. At the same time, the eigenvectors will give us the wavefunctions $\phi(y)$ of the hybrid magneto-electric states. Unfortunately, this is not straightforward since the first subequation is *not* an eigenequation in k_x for a given ϵ_1 because it is not linear in k_x. We therefore use the following transformation:

Let

$$\xi(y) = k_x \phi(y). \quad (7.60)$$

We can then write the nonlinear equation as two coupled linear equations:

$$\begin{bmatrix} 0 & 1 \\ \frac{\partial^2}{\partial y^2} + \frac{2m^*}{\hbar^2}\epsilon_1 - \left(\frac{y}{r_c^2}\right)^2 & \frac{2y}{r_c^2} \end{bmatrix} \begin{pmatrix} \phi(y) \\ \xi(y) \end{pmatrix} = k_x \begin{pmatrix} \phi(y) \\ \xi(y) \end{pmatrix}. \quad (7.61)$$

The last equation was solved numerically in [12], subject to the boundary condition in (7.55), using a finite difference scheme. The y-domain was discretized into N grid points, resulting in a $2N \times 2N$ matrix which has $2N$ different k_x eigenvalues ($k_x^{[1]}, \pm k_x^{[2]}, \ldots \pm k_x^{[2N]}$) for any given energy ϵ_1, with corresponding eigenvectors that give the wavefunction $\phi_n(y)$. A pair of $k_x^{[n]}$ values will correspond to a magneto-electric subband (the two members of the pair yield the positive and negative wavevectors at energy ϵ_1 in the corresponding subband). In this process, one obtains as many subbands as the number of grid points N. Some of the $k_x^{[n]}$ values will turn out to be imaginary or complex, and the others will be real. The subbands with real k_x have their minimum energy (subband bottom) below ϵ_1 and constitute propagating states, while subbands with nonreal k_x have their minimum energies (subband bottom) above ϵ_1 and are evanescent. The dispersion relations for different (propagating) magneto-electric subbands are shown *schematically* in Fig. 7.13. These dispersion relations are plotted as ϵ_1 versus $k_x + (eB/\hbar)y_0$, where

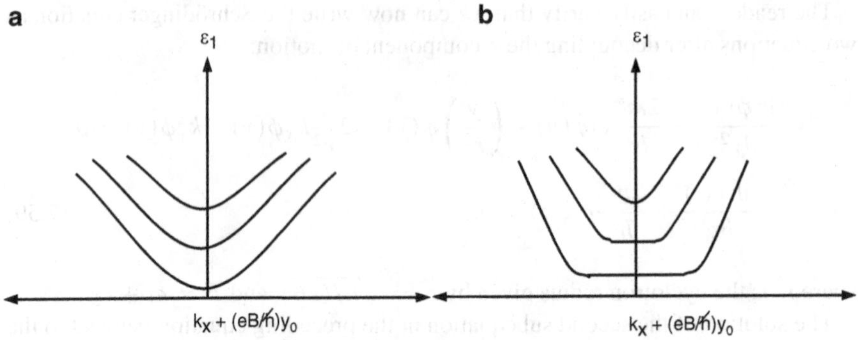

Fig. 7.13 Dispersion relations in different magneto-electric subbands in a quantum wire with rectangular confining potential at (**a**) low magnetic fields, and (**b**) high magnetic fields in a quantum wire with rectangular confining potential in the direction transverse to the magnetic field and the wire axis. Only the dispersion relations of propagating states are shown

$y_0 = \langle \phi_{n,k_0}(y)|y|\phi_{n,k_0}(y)\rangle$. The quantity k_0 is the wavevector at which the bottom of the magneto-electric subband occurs, i.e., where $\partial \epsilon_1/\partial k_x = 0$. Note that y_0 is different in different magneto-electric subbands because $\phi_{n,k_0}(y)$ is different. The reason we plot the dispersion relations as ϵ_1 versus $k_x + (eB/\hbar)y_0$, instead of as ϵ_1 versus k_x, is because then the curves are symmetric about the energy axis. Otherwise, if we plotted them as ϵ_1 versus k_x, then each curve will be displaced horizontally from the energy axis by an amount $-(eB/\hbar)y_0$, which would be different in different subbands since y_0 is subband dependent. It is important to understand that in a magnetic field, the electron's velocity $v_x \neq \hbar k_x/m^*$, even if the bandstructure is parabolic, but $v_x = (1/\hbar)\partial\epsilon_1/\partial k_x$.

7.5.2.1 Edge States

The dispersion relation in Fig. 7.13 can also be viewed as a plot of *energy versus the y-coordinate*, where $k_x = 0$ corresponds to the center of the quantum wire ($y = 0$). We state this here without explicit proof because it gives us some insight into why the dispersion relations are so different at high and low magnetic fields. At high fields, the lowest subband dispersion relations have "flat" bottoms and the velocity of an electron $v_x = (1/\hbar)\partial E/\partial k_x$ is zero there. Since the bottoms straggle $k_x = 0$ (or, equivalently, $y = 0$), this means that an electron's velocity is zero at or around the wire's axis. We expect that to happen at high fields when the cyclotron radius $\sqrt{m}r_c = \sqrt{m\hbar/(eB)} \ll W_y/2$ ($m =$ the subband index or the Landau level index) for the lowest subbands. In that case, an electron at or near the wire's axis, in the lowest subbands (which have the smallest cyclotron radii), can complete a cyclotron orbit without ever hitting the walls of the wire located at $y = \pm W_y/2$. Since the expectation value of the velocity operator of an electron in a cyclotron orbit is zero (see (7.26)), the dispersion relations of the lowest subbands should have flat bottoms (i.e., $(1/\hbar)\partial\epsilon_1/\partial k_x = 0$) around $k_x = 0$, which indeed they do. On the other hand, at low magnetic fields, $r_c > W_y/2$ and an electron cannot complete a cyclotron orbit without hitting the walls, even if it is in the lowest subband with $m = 1$. Since the electron cannot complete a closed cyclotron orbit, it will never have zero translational velocity in the x-direction. Consequently, the dispersion relations at low magnetic fields will *not* have flat bottoms. Actually, even at high magnetic fields, the *higher* subbands do not have flat bottoms because they have larger cyclotron radii equal to $\sqrt{m}r_c$ ($m \gg 1$) which do not fulfill the requirement $\sqrt{m}r_c < W_y/2$. An electron in the higher subbands therefore cannot execute cyclotron motion and have zero translational velocity even at relatively high magnetic fields. As a result, the higher subbands do not have flat bottoms.

Note also that at high magnetic fields, only states with large k_x values in the lowest subbands have nonzero slopes in the dispersion relations and therefore have nonzero velocities. *That means only electrons that are far from the center of the wire, or are close to either edge of the wire, have nonzero translational velocity and can carry current.* These states are justifiably called *edge states* because they hug the edges of the wire, and are the only states that can carry current in a high magnetic field. Note also that the slope of the dispersion relation is positive at one edge and negative at the other edge. Hence, edge states localized at opposite edges carry current in *opposite* directions. This is not unlike the parabolic confinement case where electrons localized at opposite edges carry current in opposite directions.

In Fig. 7.14, we show schematically the wavefunctions $\phi^+(y)$ and $\phi^-(y)$ in the lowest magneto-electric subband for two states at the same energy but with oppositely directed wavevectors $\pm k_x$. The value of k_x was chosen such that the magnitude of the slope of the dispersion curve at that wavevector is large. Note that at low magnetic fields, the two wavefunctions are only slightly displaced from each other in the y-direction and still have a significant overlap (the mutual displacement would have been zero at zero magnetic field), but at high magnetic fields, they are very much displaced from each other and have virtually no overlap. In fact, the wavefunction for positive velocity is localized along the left edge of the wire, while that for negative velocity is localized at the right edge. These are edge state wavefunctions. Therefore, at high magnetic fields, all current carrying states, that have large translational velocities (i.e., large slopes in the dispersion curves), are localized at one edge or the other.

In Fig. 7.15b, we show the electron trajectories in a quantum wire at high magnetic field. The trajectories at the center are closed cyclotron orbits, while those at the edges are incomplete orbits that hit the walls, rebound, and loop around to form what are known as "skipping orbits." The latter obviously travel along the wire, while hugging the walls, and have nonzero effective velocities. The skipping orbits make up the edge states. Note that skipping orbits at opposite edges are skipping along in opposite directions and hence carry current in opposite directions.

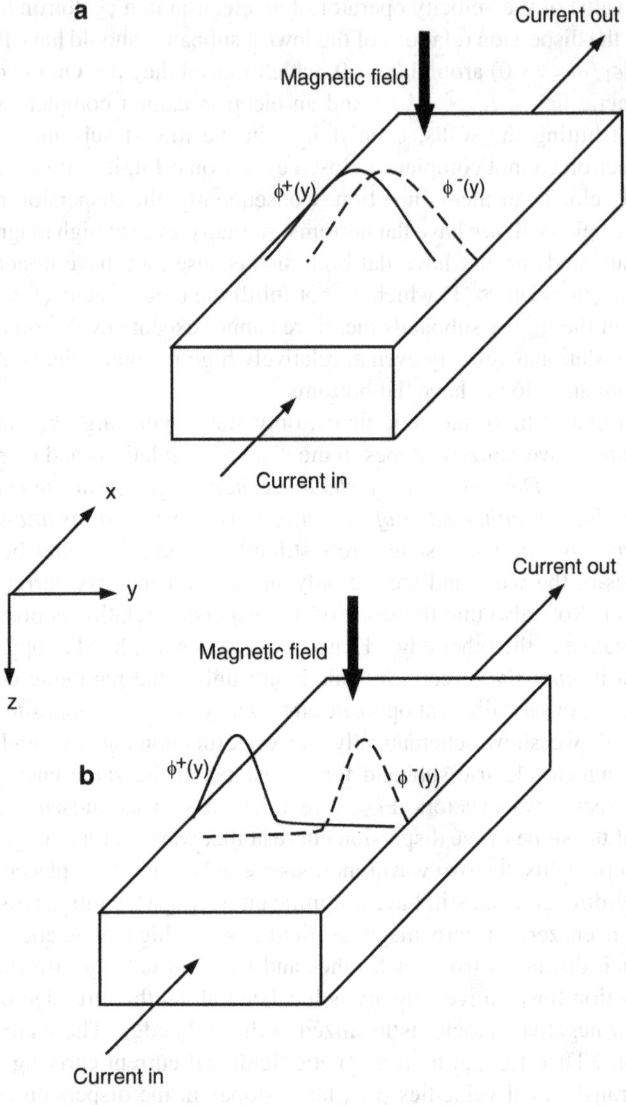

Fig. 7.14 The wavefunctions $\phi^+(y)$ and $\phi^-(y)$ in the lowest magneto-electric subband for two states at the same energy but with oppositely directed wavevectors $\pm k_x$: (**a**) low magnetic fields, and (**b**) high magnetic fields. The wavevector k_x is chosen such that the slope of the dispersion relation of the lowest magneto-electric subband at that wavevector is relatively large. At high magnetic fields, the wavefunctions are those of edge states

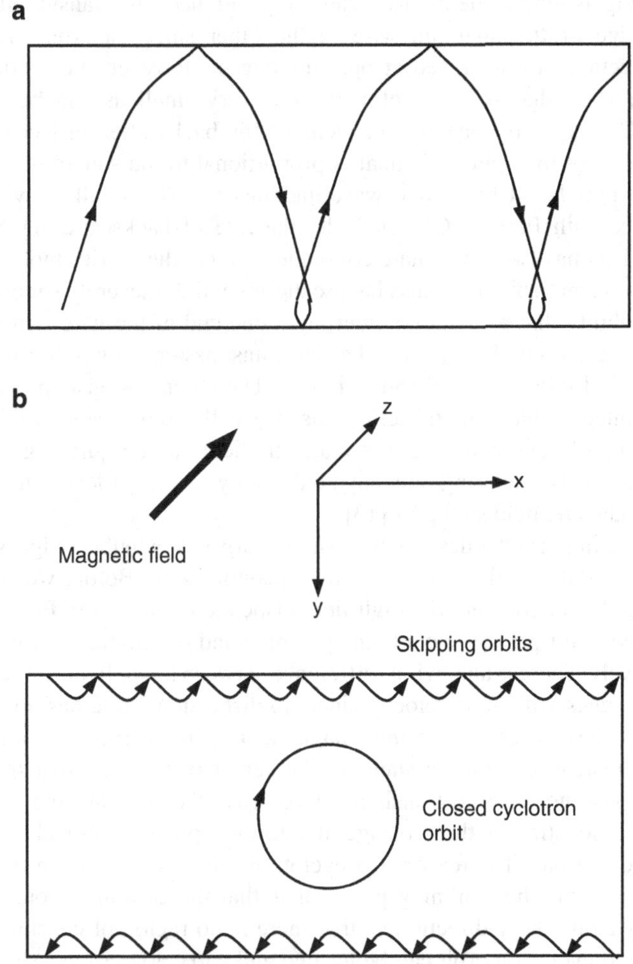

Fig. 7.15 Electron trajectories in real space in a quantum wire: (**a**) low magnetic field, and (**b**)high magnetic field where cyclotron orbits form at the center and skipping orbits result at the edges

7.5.3 Edge State Phenomena

Edge states give rise to many intriguing phenomena, but one that we can predict almost immediately is suppression of backscattering in a quantum wire. In a quantum wire, an electron can either travel forward along the wire, or backward since sideways motion is restricted by the walls of the wire. A forward traveling electron can be scattered forward with little change in momentum, or scattered backward with a large change in momentum. The backscattering event will cause large momentum relaxation and degrade mobility rapidly, while forward scattering will have little effect on mobility since it does not cause much momentum relaxation.

In a strong magnetic field, backscattering will need to transfer an electron from one edge of the quantum wire to the other since electrons traveling in opposite directions are localized at opposite edges. However, the spatial overlap between the two edge state wavefunctions is very small as can be seen from Fig. 7.14b. Consequently, the matrix element for backscattering (recall that for phonon scattering, the matrix element is proportional to the square of the overlap between the initial and final state wavefunctions) is very small as well. Hence, in accordance with Fermi's Golden Rule, the rate of backscattering $S(k, k')$ is very small. This has two immediate consequences: (1) the carrier mobility should increase in a magnetic field because backscattering will be severely suppressed, and (2) the probability that an electron entering at one end of the wire will exit at the other end is very high. That means that the transmission probability through the wire is high and reflection probability is low. The latter has an important role to play in the integer quantum Hall effect that we will discuss later in this chapter. The quenching of backscattering in a magnetic field can be quite significant and the energy-averaged scattering rate may go down by several orders of magnitude at reasonable magnetic field strengths [13].

Finally, an important question that should arise is whether edge states and cyclotron orbit states will *always* form in a quantum wire. Before we answer this question, we should compare the high-field dispersion relations in Figs. 7.12b and 7.13b, which are for parabolic confining potential and rectangular confining potential, respectively. There is an obvious difference between them. In the parabolic case, there are no states with zero velocity since no dispersion curve has a flat bottom. Every state carries current at any finite magnetic field. In the rectangular case, there is a clear demarcation between states in the center of the wire that form closed cyclotron orbits and have no translational velocity (thereby carrying no current), and edge states localized at the two edges that form skipping orbits and carry current in opposite directions. The reason that cyclotron orbits isolated from the walls do not form in a parabolic confining potential is that the confining potential varies continuously along the y-direction so that there is no region of constant potential where a closed cyclotron orbit can be located and carry no current. Also note that a parabolic confining potential has no "edge" and therefore it makes no sense to talk about edge states in a parabolic potential. That is why the demarcation between edge states and cyclotron orbit states is lost in the parabolic case.

7.5.4 Shubnikov–deHaas Oscillations in a Quasi One-Dimensional Electron Gas

We saw earlier that the conductance of a 2DEG oscillates in a magnetic field and is periodic in $1/B$ where B is the flux density. These are Shubnikov–deHaas oscillations. An interesting question is whether the same oscillations can occur in a 1DEG. We know that in order for these oscillations to be manifested, the electrons must be able to complete cyclotron orbits and that can happen only if the diameter of the smallest orbit $2r_c$ is smaller than the effective width of the quantum wire, which is $2\sqrt{\hbar/(m^*\omega_0)}$ in the case of parabolic confining potential. Therefore, Shubnikov–deHaas oscillations will appear when the magnetic flux density $B > 1/\mu$ *and* $B > m^*\omega_0/e$, or $\omega_c > \omega_0$. In the case of rectangular confining potential, these oscillations will appear when $B > 1/\mu$ *and* $r_c = \sqrt{\hbar/(eB)} < W_y/2$, i.e., when $B > 2(h/e)(1/\pi W_y^2)$. However, the interplay between the electrostatic confinement and magnetostatic confinement can cause complicated features in the magnetoresistance oscillations that are not seen in normal Shubnikov–deHaas oscillations [14].

7.6 The Quantum-Confined Lorentz Effect

Consider a quantum wire with parabolic confining potential along its width (y-direction in Fig. 7.14). A weak magnetic field of flux density B is applied in the z-direction. The centers of the hybrid magneto-electric wavefunctions for electrons and holes are obtained from (7.53):

$$y_{\text{center}}^{\text{electron}} = \frac{|e|\hbar k_x^{\text{e}} B}{\left(m_{\text{e}}\omega_0^{\text{e}}\right)^2}$$

$$y_{\text{center}}^{\text{hole}} = -\frac{|e|\hbar k_x^{\text{h}} B}{\left(m_{\text{h}}\omega_0^{\text{h}}\right)^2}, \qquad (7.62)$$

where k_x^{e} is the wavevector of the electron along the wire's length, k_x^{h} is the wavevector of the hole, m_{e} is the electron's effective mass, m_{h} is the hole's effective mass, ω_0^{e} is the curvature of the confining potential for electrons, and ω_0^{h} is that for holes. Note that we have accounted for the fact that electrons and holes have opposite charges and hence there is a minus sign preceding the right-hand side of the second line in the last equation.

It is now obvious that the magnetic field pushes the center of the hole's wavefunction and that of the electron's wavefunction in *opposite* directions from the wire's axis as long as k_x^{e} and k_x^{h} have the same sign (because then $y_{\text{center}}^{\text{electron}}$ and $y_{\text{center}}^{\text{hole}}$ will have opposite signs). In the case of optical absorption, $k_x^{\text{e}} = k_x^{\text{h}}$ by virtue of the celebrated k-selection rule that we discussed in Chap. 6. Therefore, the electron and hole coupled by a photon in the optical absorption process will have their wavefunctions skewed in opposite directions by the magnetic field as shown in Fig. 7.16. Insofar as the strength of absorption (for light polarized along the wire axis) is proportional to the square of the overlap between the electron and hole wavefunctions (recall Chap. 6 material), this strength will decrease in a magnetic field because the overlap will be reduced. Therefore, the magnetic field will quench optical absorption, much like the electric field did in Quantum-Confined Stark Effect that was discussed in Chap. 3.

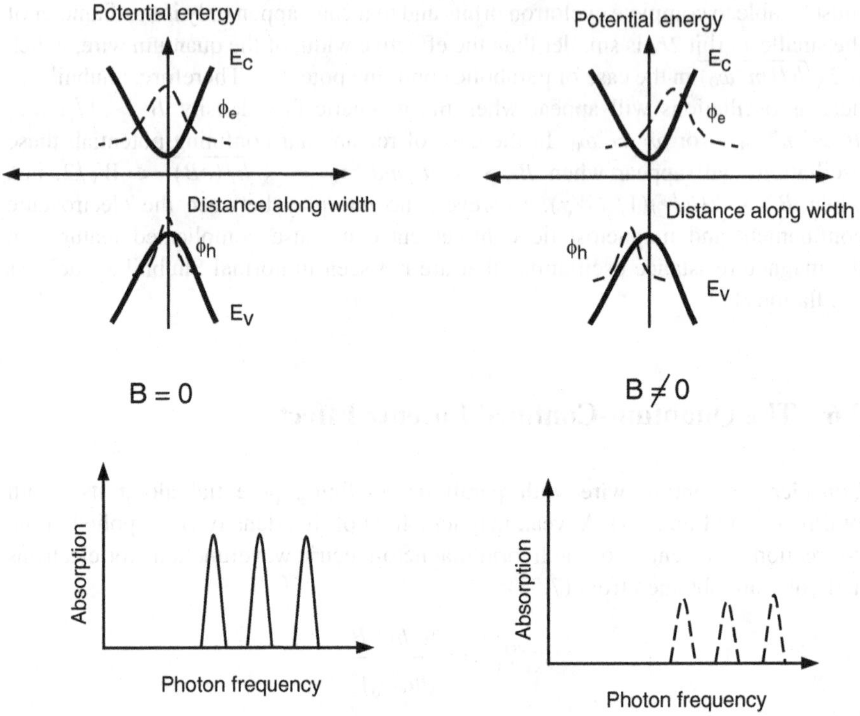

Fig. 7.16 Quantum-Confined Lorentz Effect. In a quantum wire (shown with parabolic confining potential along the width), a magnetic field will skew the electron and hole wavefunctions of two states with the same wavevector in opposite directions. This will reduce the overlap between the wavefunctions and quench absorption. The electron and hole wavefunctions are shown with *broken lines*. From the spread of the wavefunctions shown and assuming that the curvatures of the conduction and the valence bands due to parabolic confinement are the same, can you tell whether the electron or the hole has the lighter effective mass in this case? In the bottom panel, the absorption spectra are shown in the absence (*left*) and presence (*right*) of a magnetic field. Note that the magnetic field quenches absorption and induces a *blue shift* in the absorption peak. The different absorption peaks are due to transitions between different electron and hole subbands. The emission spectrum (or photoluminescence spectrum) will exhibit similar behavior

We call this magnetic field induced effect *QCLE* [15] because its classical analog is related to the Lorentz force. The Lorentz force $q\vec{v} \times \vec{B}$ on an electron and a hole *moving in the same direction* will be oppositely directed because the particles have opposite charges. Hence, a magnetic field will force them apart. Now, two particles that participate in absorption are always moving in the same direction by virtue of the k-selection rule. Hence, a magnetic field will push them away from each other, which reduces the overlap between their wavefunctions and consequently the absorption strength. Note that this is very different from the Hall effect where electrons and holes are drifting under an applied electric field and therefore moving in opposite directions. Consequently, the Lorentz forces on them act in the same direction and pushes them in the same direction. However, in the case of optical absorption, the particles must be moving in the same direction by virtue of the k-selection rule, so that the Lorentz force on them acts in opposite directions and keeps them apart.

Although both QCLE and Quantum-Confined Stark Effect quench optical absorption, there are some obvious differences between them. First, the former is induced by a magnetic field which always increases the energy separation between electron and hole subbands since $\omega_0 \to \sqrt{\omega_0^2 + \omega_c^2}$. Therefore, the quenching of absorption is always accompanied by a *blue shift* in the absorption spectrum. In the case of the Quantum-Confined Stark Effect, the quenching is typically accompanied by a *red shift*. Second, the electric field used in Quantum-Confined Stark Effect tends to make the confining potential shallower since it tilts the bands, which will ultimately allow the electron and hole to escape from the confining potential. In contrast, the magnetic field in QCLE makes the confining potential stronger since $\omega_0 \to \sqrt{\omega_0^2 + \omega_c^2}$, which means that the curvature of the confining potential increases. These differences determine which would be preferred in optical modulation applications under a given set of conditions.

7.7 The Hall Effect in a Two-Dimensional Electron Gas

Consider an ideal two-dimensional electron gas (2-DEG) in the x-y plane as shown in Fig. 7.1. A magnetic field is applied perpendicular to the plane in the z-direction. Now, if a current flows in the x-direction, then a Lorentz force will appear on the electrons which will be directed in the y-direction. This will cause the electrons to be deflected in the y-direction and a charge imbalance will result which will give rise to a Hall electric field (see Fig. 1.3). The force due to this field will eventually balance the Lorentz force and steady state will be reached. If we call the steady-state Hall field $\mathcal{E}_y^{\text{Hall}}$, then

$$\mathcal{E}_y^{\text{Hall}} = v_x B_z, \tag{7.63}$$

where v_x is the x-directed velocity of the electron and B_z is the magnetic flux density (in the z-direction).

Integrating the last equation over the y-coordinate, we obtain:

$$V_{\text{Hall}} = V_y = \int_0^W \mathcal{E}_y^{\text{Hall}} dy = v_x B_z \int_0^W dy = v_x B_z W, \tag{7.64}$$

where W is the width of the 2-DEG and V_{Hall} is the Hall voltage that is generated.

The current density in the 2-DEG is given by

$$J_x = nev_x, \tag{7.65}$$

where n is the electron concentration. The resulting current flowing in the x-direction is

$$I_x = WtJ_x = ntWev_x = n_sWev_x, \tag{7.66}$$

where $n_s = nt$ is the sheet electron concentration. This yields that $v_x = I_x/(Wen_s)$.

Substituting the last result in (7.64), we get

$$\frac{V_{\text{Hall}}}{I_x} = \frac{V_y}{I_x} = \frac{B_z}{en_s}. \tag{7.67}$$

Normally, because of the Hall electric field in the y-direction, a current could flow in that direction as well. Therefore, we can write:

$$\begin{bmatrix} V_x \\ V_y \end{bmatrix} = \begin{bmatrix} R_{xx} & R_{xy} \\ R_{yx} & R_{yy} \end{bmatrix} \begin{bmatrix} I_x \\ I_y \end{bmatrix}, \tag{7.68}$$

where the square matrix is the resistance tensor. The diagonal term R_{xx} is called the "longitudinal resistance" and the off-diagonal term R_{yx} is called the "Hall resistance."

If we were carrying out a Hall experiment to measure the Hall resistance, we would have connected a voltmeter across the sample's width to measure the Hall voltage $V_{\text{Hall}} = V_y$. Since an ideal voltmeter would have had infinite impedance, no current would have flowed in the y-direction and consequently, $I_y = 0$. Therefore, (7.68) would have yielded the Hall resistance as

$$R_{\text{Hall}} = R_{yx} = \frac{V_y}{I_x} = \frac{V_{\text{Hall}}}{I_x} = \frac{B_z}{en_s}. \tag{7.69}$$

Consider the situation when the electrons are not spin polarized. Since (see (7.35)) $n_s = p/(2eB_z/h)$, where p is the number of fully filled Landau levels (sometimes called the "filling factor"), we immediately get that ideally the Hall resistance should be *quantized* and given by

$$R_{\text{Hall}} = R_{yx} = \frac{1}{p}\frac{h}{2e^2} = \frac{1}{p}\frac{1}{4\alpha_f c\epsilon_0}, \tag{7.70}$$

where c is the speed of light in vacuum, ϵ_0 is the dielectric constant of vacuum, and α_f is a dimensionless constant known as the "fine-structure constant," with a value of $1/137 = 0.00729720$. This is a remarkable result since it predicts that the Hall resistance will not depend on the electron concentration in the sample and, in fact, will be independent of all sample parameters! It will depend only on universal constants h and e.

At very high magnetic fields, the electrons may be fully spin polarized and this will remove the spin degeneracy, which means that each spin-degenerate level splits into two distinct levels at different energies. Consequently, for the same carrier concentration n_s, the number of fully filled Landau levels will be $p' = 2p$ (see (7.35) and (7.36)). In that case, the last equation will become

$$R_{\text{Hall}} = R_{yx} = \frac{1}{p'}\frac{h}{e^2} = \frac{1}{p}\frac{h}{2e^2}, \tag{7.71}$$

which is the same result as before, since n_s has not changed. Therefore, the Hall resistance is *quantized* in units of h/e^2, which is 25.56 k-ohms.

The last equation predicts that if we measure the Hall resistance of an ideal spin-polarized two-dimensional electron gas (2-DEG) while progressively decreasing p' by successive depopulation of spin-nondegenerate Landau levels with an increasing magnetic field, then we should observe the Hall resistance increase in *steps*, with the last step having a height of h/e^2, or 25.56 kΩ corresponding to $p' = 1$. This will happen as long as the magnetic field is strong enough to cause complete spin polarization of all the electrons. The steps occur because the quantity p' is an *integer*, so that every time a spin-nondegenerate Landau level is depopulated, the Hall resistance jumps by precisely $\frac{1}{p'(p'+1)}\frac{h}{e^2}$. The quantity p' is referred to as the "filling factor" since it is the number of Landau levels filled with electrons. From the height of any resistance step, one can infer the corresponding filling factor.

In an actual experiment, we could progressively decrease p' in two ways: (1) We could hold the Fermi level in the 2-DEG invariant by holding the sheet carrier concentration n_s constant and vary the magnetic field strength which increases the separation between successive Landau levels (as well as the spin splitting within each level) since the separation is $\hbar\omega_c = \hbar(eB/m^*)$. This would decrease p' with increasing magnetic field strength, as is obvious from (7.36). Alternately, (2) we could hold the magnetic field constant so that the energy separation between successive Landau levels and the spin splitting do not change, but we vary the position of the Fermi level (with respect to the Landau levels) by varying the sheet carrier concentration n_s. This too would change p' in accordance with (7.36) (Fig. 7.17).

In the year 1980, a group of experimentalists reported measuring the Hall resistance of a 2-DEG formed in the inversion layer of a metal-oxide-semiconductor-field-effect-transistor as a function of the sheet carrier concentration n_s [16]. The latter quantity was varied using the voltage applied to the gate terminal of the transistor. These measurements were carried out at low temperature (liquid helium) and very high magnetic field strengths (~15 Tesla) so that the electrons were nearly fully spin polarized. It was found that the Hall resistance indeed decreased in steps as p' was increased by increasing n_s with the gate voltage. The height of each step was precisely $(1/n)h/e^2$ [$n = p'(p'+1) =$ an integer] as predicted by (7.71) and it did not depend on any sample parameter. In fact the step height agreed with the value $(1/n)h/e^2$ with surprising precision (1 part in 10^5). The experimental plot of the Hall resistance versus the gate voltage is shown in Fig. 7.18. Another intriguing observation was that whenever the Hall resistance R_{yx} was in a plateau, the longitudinal resistance R_{xx} reached zero. This phenomenon came to be

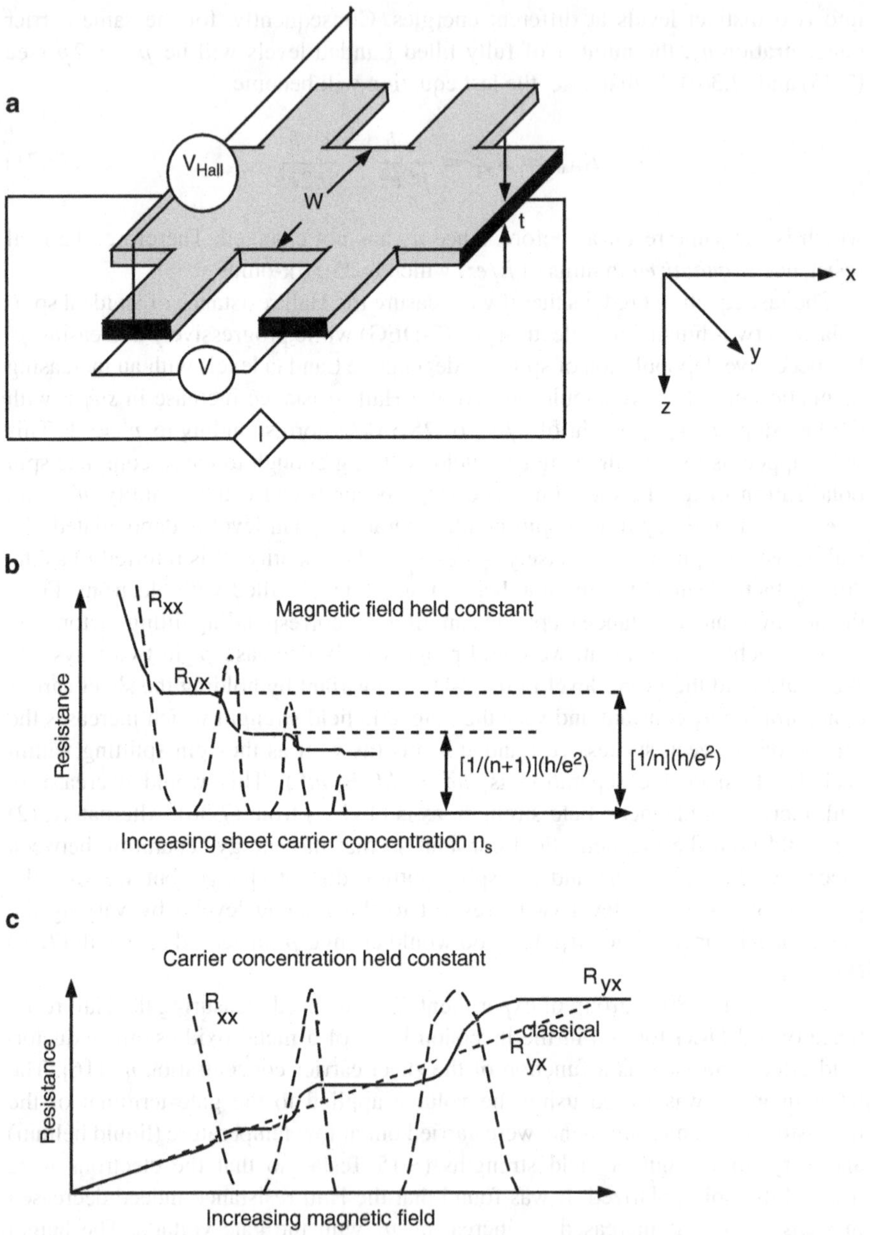

Fig. 7.17 (**a**) Schematic of a Hall bar sample for measuring Hall resistance and longitudinal resistance. (**b**) The quantized Hall resistance $R_{yx} = V_{\text{Hall}}/I$ (*solid line*) and the longitudinal resistance $R_{xx} = V/I$ (*broken line*) as a function of the sheet carrier concentration in a 2-DEG when the magnetic flux density B is held constant at a high value. (**c**) The same quantities as a function of the magnetic flux density B when the sheet carrier concentration n_s is held constant. The dotted line shows the classical Hall resistance which increases linearly with the magnetic flux density. This line often intersects the quantum Hall plateaus at the mid-points. In order to observe such quantized Hall resistance, the temperature must be kept low and the voltage V must be low as well. These plots are highly idealized for illustrative purposes

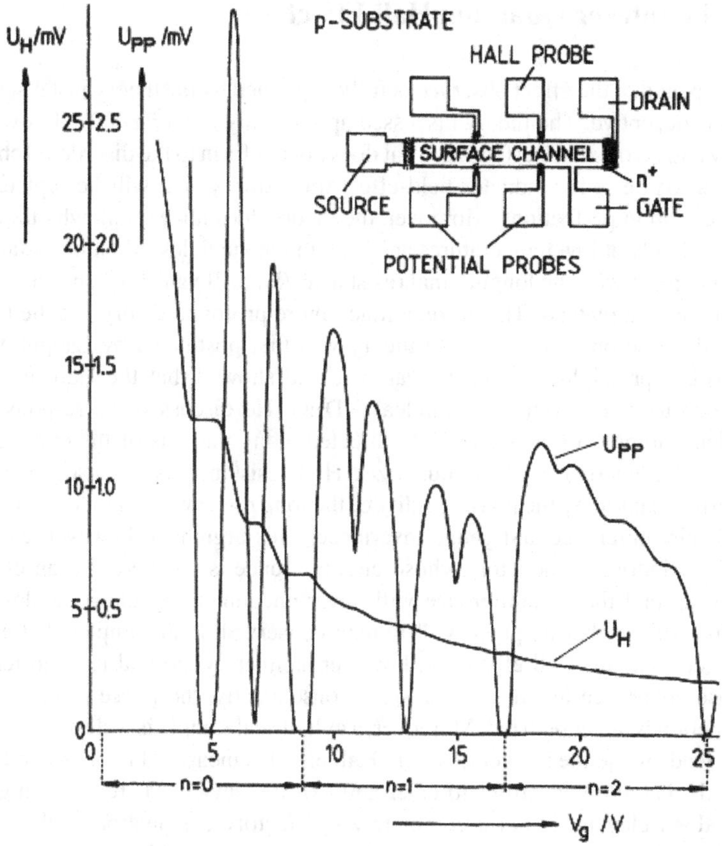

Fig. 7.18 The Hall voltage U_{H} (or, equivalently, the Hall resistance R_{yx}) and the voltage between voltage probes U_{pp} (or, equivalently, the longitudinal resistance R_{xx}) as a function of the gate voltage in the inversion layer of a MOSFET at a temperature of 1.5 K with a constant magnetic field of 18 Tesla directed perpendicular to the layer. The inset shows the schematic of the experimental structure. The distance between current probes is 400 μm, the distance between voltage probes is 130 μm and the width is 50 μm. Reproduced from [16] with permission from the American Physical Society. Copyright American Physical Society 1980. See *http://prl.aps.org/abstract/PRL/v44/i6/p494_1*

known as the celebrated *integer quantum Hall effect.* Since then, the quantum Hall resistance—which always agrees so precisely with the value $(1/n)h/e^2$—has been proposed as a metrological standard of resistance. To date, this resistance has been measured with an accuracy of few parts per billion and always agrees with the value $(1/n)(h/e^2)$ [17].

7.8 The Integer Quantum Hall Effect

It may appear that the effect observed is fully explained by the theory just discussed, but that is deceptive. The theory discussed applies only to an *ideal* 2-DEG with no impurity and no other imperfection. That does not conform to the disordered channel of a metal-oxide-semiconductor-field-effect-transistor, which will be replete with impurities and imperfections. Moreover, the theory does not explain why the effect is observed only at low temperatures and high magnetic fields. Most importantly, it does not explain why the longitudinal resistance R_{xx} will vanish whenever the Hall resistance is in a plateau. Therefore, a much more profound theory is called for to explain all these observations. That theory was first postulated by Laughlin [18], who invoked principles of gauge invariance and showed that the Hall resistance should be quantized even in a nonideal 2-DEG. He discussed the response of a cylindrical sample to a magnetic field directed along the axis of the cylinder and explained the emergence of the quantized Hall resistance as a consequence of a supercurrent caused by the phase rigidity of the long range wavefunction around the sample's circumference and gauge invariance. This argument, however, can only apply to cylindrical conductors whose circumference is so short that an electron traveling around the circumference will never encounter any phase randomizing collision event so that its phase will remain conserved. Real samples that exhibit the integer quantum Hall effect are, however, neither cylindrical nor shorter than the phase coherence length of electrons. Consequently, the phase rigidity of the wavefunction is never realized. Moreover, a cylindrical sample is a closed conductor with no leads while the integer quantum Hall effect is measured in open conductors with leads that allow electrons to enter and exit the device. Therefore, Laughlin's theory, albeit elegant, is not a completely satisfactory explanation of the integer quantum Hall effect. Later, an alternate theory was advanced by Büttiker [19] that did not require phase rigidity of the wavefunction, or the other assumptions. We will discuss it in Chap. 9.

7.9 The Fractional Quantum Hall Effect

In 1982, another group of experimentalists reported observing quantized Hall plateaus, but with a different twist. They found that at very high magnetic fields and low temperatures, the Hall resistance in their 2-DEG samples were quantized to the value

$$R_{\text{Hall}} = \frac{1}{l}\frac{h}{e^2} = \frac{1}{l}\frac{1}{2\alpha_f c\epsilon_0}, \tag{7.72}$$

where the quantity l was a *fraction* ($l = 1/3$) [20]. This was called the *fractional quantum Hall effect*. Since then, many other "fractions" have been discovered corresponding to

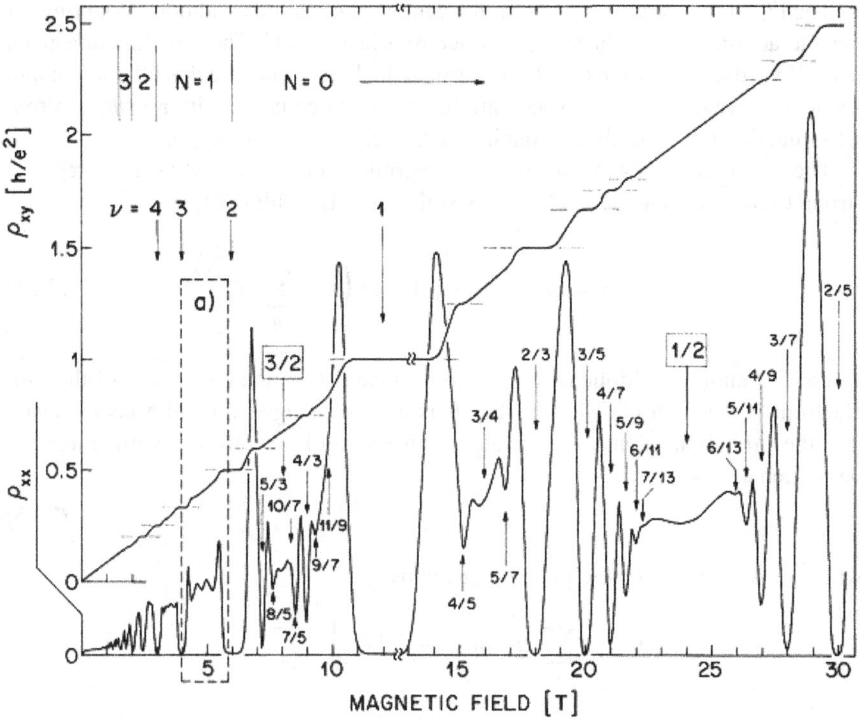

Fig. 7.19 The longitudinal (ρ_{xx}) and Hall (ρ_{xy}) resistances of a two dimensional electron gas as a function of magnetic field at a temperature of 150 mK. Reproduced from [22] with permission of the American Physical Society. Copyright American Physical Society 1987. See *http://prl.aps.org/abstract/PRL/v59/i15/p1776_1*

$$l = \frac{m}{2m+1} = \frac{1}{3}, \frac{2}{5}, \frac{3}{7}, \frac{4}{9}, \ldots\ldots$$
$$l = \frac{m}{4m+1} = \frac{1}{5}, \frac{2}{9}, \frac{3}{13}, \ldots\ldots$$
$$l = 1 - \frac{m}{2m+1} = \frac{2}{3}, \frac{3}{5}, \frac{4}{7}, \frac{5}{9}, \ldots\ldots$$
$$l = 1 - \frac{m}{4m+1} = \frac{4}{5}, \frac{7}{9}, \frac{10}{13}, \ldots\ldots \quad (7.73)$$

Additionally, the fractions 2/7, 3/11, and 4/15 have also been observed. All observed fractions have an odd denominator [21], with the sole exception of a fraction 5/2 [22] where it was later found that the 2-DEG was not completely spin polarized [23]. Figure 7.19 shows the longitudinal and Hall resistances as a function of magnetic field in a 2-DEG, reported in [22].

The FQHE is a *many-body effect* that cannot be explained without accounting for the interaction between the numerous electrons in a 2-DEG. The wavefunction given in (7.25) is that of a *single* isolated electron, which does not take this interaction into account. If we do take it into account then we must come up with a *multi-electron* wavefunction that describes all the interacting electrons in a sample.

The multi-electron wavefunction in the ground state of a 2-DEG subjected to a strong transverse magnetic field was postulated by Laughlin [24] as

$$\psi_n(x, y) = \Pi_{j<k}(\zeta_j - \zeta_k)^n \exp\left(-\frac{1}{4}\sum_q |\zeta_q|^2\right), \tag{7.74}$$

where Π denotes multiplication, $\zeta = x + \mathrm{i}y$, and ζ_q is the coordinate of the q-th electron in the 2-DEG. The quantity n is an odd integer. Laughlin also showed that the probability density associated with this multi-electron wavefunction can be written as [24]

$$|\psi_n(x, y)|^2 = \mathrm{e}^{-\beta\Delta}, \tag{7.75}$$

where Δ is a dimensionless potential given by

$$\Delta = -\sum_{j<k} 2n^2 \ln|\zeta_j - \zeta_k| + \frac{1}{2}n\sum_q |z_q|^2. \tag{7.76}$$

and $\beta = 1/n$.

The all-important question now is what will be the quantum Hall resistance of a 2-DEG described by the above multi-electron wavefunction? Before we proceed with this question, we need to understand first the concept of a "quasi-particle," since it will be vital in answering this question. In order to comprehend what a quasi-particle is, consider the fact that in a real solid there are numerous mobile electrons moving around against a background of fixed positive charges due to the immobile ions. The ionic charges balance those of the electrons to make the entire solid charge-neutral. But now, because the positive background charges mildly attract the mobile electrons, the electrons are really not all that free and mobile, but instead behave collectively like a viscous fluid (or "jelly"). Hence, the electron sea is sometimes called a "jellium."

Next, we need to understand that the electrons in a jellium are not oblivious of each other. They mutually *interact* via Coulomb interaction so that their motions are not independent, but correlated. Therefore, when an extra electron is introduced into jellium, all other electrons will tend to move away from the intruder because of Coulomb repulsion. This will expose a region around the intruder that is positively charged because of the background ions (see Fig. 7.20). The positively charged annulus gets permanently attached to the intruding electron and moves with it wherever it goes [25]. Thus, the electron is "dressed" with a positively charged cloud which screens its negative charge partially. This composite entity (the electron with its surrounding positively charged cloud) is a quasi-particle, or quasi-electron, and

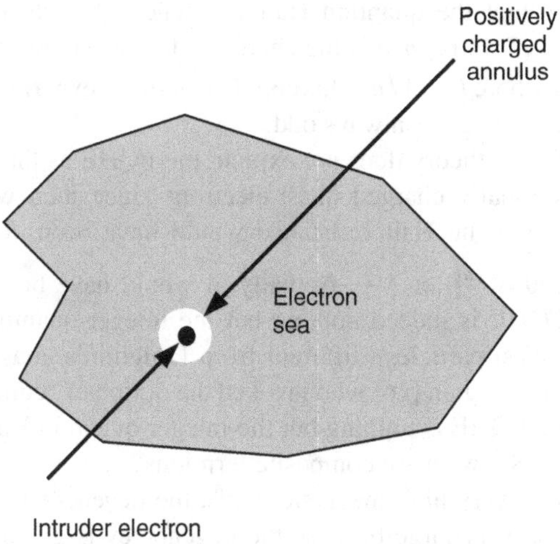

Fig. 7.20 An electron introduced into jellium pushes away the neighboring electrons owing to Coulomb repulsion, exposing an annulus of positive charges surrounding it. The positive charges are due to the background positive ions in jellium. This positively charged annulus gets permanently attached to the electron and moves with it, partially screening its charge. The electron and its surrounding annulus constitute a "quasi-particle" whose effective charge is obviously smaller (less negative) than the bare electron's charge. Adapted with permission from [25]

obviously its charge is less than that of the bare electron because of the screening effect of the annulus (the electron's "dress"). It should be obvious now that whenever we are dealing with many interacting electrons, we really should not think in terms of bare electrons (which ignores the interaction with all other electrons), but instead think in terms of quasi-electrons.

Not only does the charged annulus dress a bare electron, but there are other effects that can contribute to the dressing as well. For example, the background ions are not completely immobile. At a finite temperature, they are vibrating in unison and a quantum of this collective excitation is a phonon. When an electron passes through the vibrating lattice, the ions in its path are slightly attracted to the passing electron because of Coulomb attraction. This responsive motion of the ions can be thought of as a phonon mode and we can think of this phonon dressing the electron as well. We had alluded to phonon dressing in Chap. 5. This will obviously make the effective mass of the quasi-electron heavier because of the drag the ions exert. Thus, it is not only the other electrons, but the background ions could also dress an electron. Such dressing is the cornerstone of Cooper pairing of electrons that leads to superconductivity, but unfortunately we cannot delve further into this subject here since it is beyond the scope of this book.

Laughlin showed that in a high magnetic field, the many-body ground state of a two-dimensional sea of interacting electrons, described by the wavefunction in (7.74), behaves like an incompressible quantum fluid. The quasi-electrons in this fluid carry fractional charge of $1/n$ each, where n is an odd integer. He then proceeded to show that the quantum Hall resistance of such a fluid should be $R_{\text{Hall}} = \frac{h}{e^* e}$, where $e^* = (1/n)e$ is the charge of the quasi-electron. Accordingly, $R_{\text{Hall}} = \frac{nh}{e^2} = \frac{1}{l}\frac{h}{e^2}$, where $l = 1/n$. This could explain the experimental observation of the $l = 1/3$ effect since n is always odd.

Note that Laughlin's theory does *not* explain the FQHE as the integer quantum Hall effect of fractionally charged quasi-electrons since then we would simply replace e with e^* and the Hall resistance would have been $R_{\text{Hall}} = \frac{h}{(e^*)^2} = \frac{n^2 h}{e^2} = \frac{1}{l^2}\frac{h}{e^2}$, instead of $\frac{nh}{e^2}$ or $\frac{1}{l}\frac{h}{e^2}$. Actually, it would have been very satisfying to show that the FQHE is indeed nothing but the integer quantum Hall effect of some appropriate quasi-particles (different from Laughlin's quasi-particles). That objective was attained by Jain [26] who invoked the notion of "composite fermions" and showed that the FQHE *is* nothing but the integer quantum Hall effect of these composite fermions. So, what are composite fermions?

Jain argued that at very high magnetic fields, the degeneracy of each occupied Landau level will be very large because the degeneracy is given by (7.32) and it increases with magnetic field strength. Since the degeneracy is the number of k-states in each Landau level and each k-state is occupied by an electron, there will be several electrons in the *same* Landau level—all sharing the same energy. These electrons will repel each other (because of Coulomb interaction) and that repulsion will lift the degeneracy and open up gaps in the energy spectrum (recall the material in Chap. 3 where we discussed degenerate perturbation theory).

The Coulomb repulsions in a magnetic field generate "composite fermions" which are quasi-particles. Basically, they are electrons coupled to (or dressed with) an even number of current vortices (or cyclotron orbits). The vortex dressing screens the Coulomb interactions between bare electrons and reduces the total energy of electron ensemble, thus stabilizing the ground state. Jain then showed that the fractional quantum Hall effect is essentially the integer quantum Hall effect of composite fermions. The composite fermion state at filling factor i is equivalent to an electron state at filling factor

$$p' = \frac{i}{2ij + 1}, \tag{7.77}$$

where i and j are integers. This is able to explain the fractions in (7.73). Thus, this picture of the FQHE is not only elegant, but it is also comprehensive.

7.10 Summary

In this Chapter, we learned that

1. A magnetic field has two major effects on an electron: (1) it changes its trajectory in space, affecting its wavefunction and energy states, and (2) it resolves spin degeneracy and mandates that we treat spin effects explicitly.

2. If we account for spin explicitly, then a relativistic electron is described by the Dirac equation, which yields the spinorial wavefunction of the coupled electron-positron. This wavefunction is a 4×1 component spinor.
3. In the nonrelativistic limit, the electron and positron can be decoupled and the spinorial wavefunction of each is described by a 2×1 component spinor. For an electron, this spinor is found by solving the Pauli equation.
4. The spinorial wavefunction allows us to determine the x-, y-, and z-components of the spin angular momentum. Thus, we can treat the electron's spin explicitly if we deal with the spinorial wavefunction instead of the normal scalar wavefunction obtained by solving the Schrödinger equation.
5. In the Schrödinger equation describing a "spinless" electron placed in a magnetic field, the magnetic field is taken into account through the magnetic vector potential.
6. The magnetic vector potential is arbitrary and depends on the chosen gauge. However, the expectation value of any physical observable calculated from the wavefunction must be gauge-independent.
7. An electron in a 2-DEG subjected to a transverse magnetic field executes cyclotron motion. In other words, its trajectory is closed cyclotron orbits or Landau orbits.
8. The expectation value of the drift velocity of the electron in any direction along the 2-DEG is exactly zero since the Landau orbits are closed circular orbits.
9. In the m-th Landau orbit, the electron energy is given by $(m + 1/2)\hbar\omega_c$, where ω_c is the angular frequency with which the electron goes around the landau orbit. The angular frequency is determined by the flux density of the transverse magnetic field B. It is given by $\omega_c = eB/m^*$, where m^* is the electron's effective mass.
10. The radius of the m-th Landau orbit is $\sqrt{m\hbar/(eB)}$.
11. The magnetic flux enclosed by the m-th Landau orbit is $m(h/2e)$.
12. Landau levels are equally spaced in energy with a spacing of $\hbar\omega_c$.
13. The degeneracy of a Landau level (number of k-states in a Landau level) is $\Phi/(h/e)$, where Φ is the total flux enclosed by the 2-DEG. The degeneracy is also the area of the 2-DEG divided by the area enclosed by the smallest Landau orbit. The degeneracy increases linearly with the magnetic field strength.
14. In a 2-DEG with sheet electron concentration n_s, the number of filled Landau level (also called the "filling factor") is $hn_s/(2eB)$ if the electrons are not spin polarized. Here, B is the magnetic flux density. If the electrons are completely spin polarized, then the filling factor is $hn_s/(eB)$. The filling factor decreases with increasing magnetic field strength.
15. The density of states in a 2-DEG placed in a transverse magnetic field of flux density B is a series of equally spaced delta functions in energy. The energy spacing between neighboring delta functions is $\hbar\omega_c$ and the height of the delta functions is $2eB/h$ if the electrons are not spin polarized and eB/h if they are fully spin polarized.
16. The conductance of a 2-DEG placed in a transverse magnetic field is oscillatory in the inverse of the magnetic flux density. The period is related to the

sheet carrier concentration as $n_s = (2e/h)(1/\text{period})$ if the electrons are not spin-polarized. These oscillations can be observed at magnetic fields strong enough where the conditions $\omega_c\tau > 1$ is maintained, where τ is the quantum lifetime of the carriers.

17. The electron states in a quantum wire subjected to a magnetic field perpendicular to the axis are hybrid magneto-electric states which result from simultaneous magnetostatic confinement (due to cyclotron orbit formation) and electrostatic confinement (due to the walls of the quantum wire).
18. At high magnetic fields, electrons in a quantum wire tend to localize along the edges forming "edge states." Electrons traveling in one direction along the axis localize at one edge and those traveling in the opposite direction localize along the other edge.
19. The edge states correspond to skipping orbits. At high magnetic field strengths, electrons at the edges of the quantum wire move along skipping orbits, while electrons in the center of the wire execute closed cyclotron orbit motion and do not possess translational velocity along the axis of the wire.
20. A magnetic field applied transverse to the axis of a quantum wire can quench optical absorption and emission from the wire and result in a blue shift of the emission spectrum. The effect that causes this is called the quantum confined Lorentz effect which is the magnetic analog of the quantum-confined Stark effect.
21. In an ideal 2-DEG, the Hall resistance should be quantized to $(1/n)(h/2e^2)$ if the electrons are not spin polarized and $(1/n)(h/e^2)$ if the electrons are fully spin polarized, where n is an integer (it is the filling factor). Thus, the Hall resistance should not depend on any sample parameter, but instead depend solely on universal constants.
22. Even in a nonideal 2-DEG, the Hall resistance is quantized to the above values at low temperatures and high magnetic fields. The Hall resistance as a function of magnetic field exhibit plateaus that are quantized to the above values with remarkable accuracy. When the Hall resistance is in a plateau, the longitudinal resistance vanishes. This is called the integer quantum Hall effect.
23. At very high magnetic fields and at very low temperatures, the Hall resistance of a nonideal 2-DEG is quantized to $(1/l)(h/e^2)$, where l is a fraction with odd denominator. This is called the FQHE and is the result of an exotic many body ground state forming under high magnetic fields. This ground state has fractional charged excitations.
24. The FQHE can be explained by invoking "composite fermions" which are electrons dressed with current vortices (Landau orbits). The dressing reduces the Coulomb repulsion between bare electrons and stabilizes the many-body ground state.

Problems

Problem 7.1. Consider a 2-DEG with a magnetic field directed perpendicular to its plane. Find the expectation values $\langle v_x^2\rangle$ and $\langle v_y^2\rangle$ in Landau orbits.

Solution

$$\langle v_x^2\rangle = \left\langle\left(\frac{p_x - eA_x}{m^*}\right)^2\right\rangle = \frac{\langle p_x^2\rangle - e\langle p_x A_x + A_x p_x\rangle + e^2\langle A_x^2\rangle}{(m^*)^2}$$

$$= \frac{\hbar^2 k_x^2 + 2e\hbar k_x B\langle y\rangle + e^2B^2\langle y^2\rangle}{(m^*)^2}$$

$$= \frac{\hbar^2 k_x^2 - 2e\hbar k_x B y_0 + e^2B^2\langle y^2\rangle}{(m^*)^2}$$

$$= \frac{e^2B^2(\langle y^2\rangle - y_0^2)}{(m^*)^2}.$$

Note that

$$\langle y^2\rangle = \langle \zeta_m(y)|y^2|\zeta_m(y)\rangle = \left(m+\frac{1}{2}\right)\frac{\hbar}{m^*\omega_c} + y_0^2.$$

Substituting the last result in the preceding equation, we get

$$\langle v_x^2\rangle = \left(m+\frac{1}{2}\right)\frac{\hbar\omega_c}{m^*}$$

Note also that $\frac{1}{2}m^*\langle v_x^2\rangle = \frac{1}{2}\left(m+\frac{1}{2}\right)\hbar\omega_c$. But the total energy of motion in the x-y plane is $\left(m+\frac{1}{2}\right)\hbar\omega_c$. Therefore,

$$\frac{1}{2}m^*\langle v_x^2\rangle = \frac{1}{2}\left(m+\frac{1}{2}\right)\hbar\omega_c$$

$$\frac{1}{2}m^*\langle v_y^2\rangle = \frac{1}{2}\left(m+\frac{1}{2}\right)\hbar\omega_c.$$

Problem 7.2. Consider a quantum wire with a magnetic field directed along the axis of the wire, which is the x-direction. The confining potentials along the y- and z-directions are parabolic and given by

$$V_y(y) = \frac{1}{2}m^*\omega_y^2 y^2$$

$$V_z(z) = \frac{1}{2}m^*\omega_z^2 z^2$$

Find the solutions of the wavefunctions and the energy eigenstates when $\omega_y = \omega_z$. These states are known as *Fock Darwin states* (see C. G. Darwin, Proc. Camb. Phil. Soc., **27**, 86 (1930)). An analytical solution exists for $\omega_y \neq \omega_z$, as well, but it is far more complicated (see, B. Schuh, J. Phys. A: Math. Gen., **18**, 803 (1985)).

Problem 7.3. Someone carried out Shubnikov–deHaas oscillation measurements in a quantum well where only the lowest subband was occupied by electrons. A beating pattern was observed and a Fourier transform of the oscillation produced two frequencies $\Delta(1/B)_1$ and $\Delta(1/B)_2$. It was determined that the beating was due to spin splitting of the subband. Show that the spin splitting energy E_s is

$$E_s = \frac{e\hbar}{m^*}[\Delta(1/B)_1 - \Delta(1/B)_2],$$

where m^* is the effective mass.

Solution. Calling $n_s^\uparrow$ and $n_s^\uparrow$ the carrier concentrations in the two spin levels, we get

$$n_s^\uparrow = \frac{e}{h}\Delta(1/B)_1$$
$$n_s^\downarrow = \frac{e}{h}\Delta(1/B)_1$$

Note that these expressions are different from (7.41) by a factor of 2 since the spin degeneracy of the subband has been resolved.

At low temperatures, the Fermi–Dirac function can be approximation as $1 - \Theta(E - \mu_F)$, where Θ is the Heaviside (or unit step) function. Therefore, using (4.130) and (4.131), we get

$$n_s^\uparrow = \frac{m^*}{2\pi\hbar^2}\int_0^\infty \mathrm{d}E\,\Theta(E - E_\uparrow)[1 - \Theta(E - \mu_F)] = \frac{m^*}{2\pi\hbar^2}(\mu_F - E_\uparrow)$$
$$n_s^\downarrow = \frac{m^*}{2\pi\hbar^2}\int_0^\infty \mathrm{d}E\,\Theta(E - E_\downarrow)[1 - \Theta(E - \mu_F)] = \frac{m^*}{2\pi\hbar^2}(\mu_F - E_\downarrow).$$

Consequently,

$$E_s = E_\downarrow - E_\uparrow = \frac{2\pi\hbar^2}{m^*}\left(n_s^\uparrow - n_s^\downarrow\right) = \frac{e\hbar}{m^*}[\Delta(1/B)_1 - \Delta(1/B)_2].$$

Problem 7.4. Show that the kinematic momentum operator $\vec{\Pi}_{op} = \vec{p}_{op} + e\vec{A}$ satisfies the relation

$$\vec{\Pi}_{op} \times \vec{\Pi}_{op} = -\mathrm{i}e\vec{B}\hbar,$$

where $\vec{B}$ is the magnetic flux density.

Problem 7.5. Consider a quantum wire with a parabolic confining potential along its width. Since the potential is parabolic, the wire does not really have a well-defined "width" unlike in the case of rectangular confining potential. Hence, we define an "effective width" as $1/\beta$, since the wavefunction of the lowest mode ($m = 1$) decays by a factor of $\sqrt{e}$ over this distance from the center of the wire (see (7.51)). In the absence of any magnetic field, the value of β is $\beta_0 = \sqrt{m^*\omega_0/\hbar}$. This quantity depends on both the effective mass of the particle and the curvature of the confining potential ω_0.

Let us assume that both electrons and holes have the same effective width $1/\beta$. In this case, show that had it not been for the fact that electrons and holes have opposite charges, there would be no QCLE, since $y^{\text{electron}}_{\text{center}}$ and $y^{\text{hole}}_{\text{center}}$ would have been equal. In other words, the effective mass difference alone can cause the Quantum-Confined Stark Effect since the heavier mass particle is always skewed more by an electric field, but the effective mass difference alone is not sufficient to cause the QCLE.

Solution. Since the effective widths are equal, we have

$$m_e\omega_e = m_h\omega_h,$$

where ω_e and ω_h are the curvatures of the parabolic confining potentials in the y-direction for the two particles, respectively and m_e and m_h are the electron and hole effective masses.

Using (7.62) and noting that for optical absorption $k_x^e = k_x^h$, we get that

$$y^{\text{electron}}_{\text{center}} = \frac{q_{\text{electron}}}{q_{\text{hole}}} y^{\text{hole}}_{\text{center}}$$

Therefore, if the electron and hole had the same charge, both wavefunctions will be affected equally and the overlap between them will not change in a magnetic field. Consequently, there would be no QCLE despite the fact that the particles have different effective masses.

In a quantum wire with rectangular confining potential along the width, there is a well-defined "width" which is of course the same for both electrons and holes. In this case, the opposite sign of the charges is primarily responsible for the QCLE.

Problem 7.6. Consider a 2-DEG in the x-y plane with a uniform electric field $\mathcal{E}$ in the -y-direction. A magnetic field of flux density B is directed in the z-direction. The effective mass Hamiltonian for this system is given by

$$H = \frac{(\vec{p} - e\vec{A})^2}{2m^*} + e\mathcal{E}y.$$

Show that the wavefunction of an electron in this 2DEG is given by

$$\psi_{k_x,m} = e^{ik_x x}\zeta_m(y),$$

where ζ_m are the simple harmonic oscillator wavefunctions given by (7.24) and

$$y_0 = \frac{1}{\omega_c}\left[\frac{\hbar k_x}{m^*} + \frac{\mathcal{E}}{B}\right].$$

Additionally, show that the eigenenergies are given by

$$\Lambda_m = \left(m + \frac{1}{2}\right)\hbar\omega_c - e\mathcal{E}y_0 + \frac{1}{2}m^*\left(\frac{\mathcal{E}}{B}\right)^2,$$

where ω_c is the cyclotron frequency eB/m^*.

Problem 7.7. Starting with (1.11) and (1.13), show that the *classical* Hall resistance of an ideal 2-DEG can be expressed as $R_{yx}^{\text{classical}} = (n_\phi/n_s)(h/e^2)$, where n_ϕ is the number of fundamental flux quanta per unit area in the 2-DEG, and n_s is the sheet carrier concentration.

Solution. Refer to the Hall bar sample shown in Fig. 7.17a. The Hall voltage is measured with an ideal voltmeter of infinite impedance which makes the current density flowing in the y-direction $J_{yy} \equiv 0$. Using (1.13), we can then write

$$\rho_{yx}J_x = E_y.$$

Let the width of the Hall bar be W and the thickness t. Multiplying the last equation with the product Wt, we get

$$\rho_{yx}J_xWt = E_yWt, \text{ or}$$
$$\rho_{yx}I = V_{\text{Hall}}t.$$

Using (1.11), we obtain $\rho_{yx} = B/(en)$. Since the classical Hall resistance is $R_{yx}^{\text{classical}} = V_{\text{Hall}}/I = \rho_{yx}/t$, we get

$$R_{yx}^{\text{classical}} = \frac{B}{ent} = \frac{B}{en_s}.$$

By definition,

$$n_\phi = \frac{B}{h/e}.$$

Using this result in the last expression for $R_{yx}^{\text{classical}}$, we obtain the desired result

$$R_{yx}^{\text{classical}} = \frac{n_\phi}{n_s}\frac{h}{e^2}.$$

Problem 7.8. Show that the number of filled Landau levels in a spin-unpolarized 2-DEG and spin-polarized 2-DEG will be given respectively by

$$p = \frac{n_s}{2n_\phi}$$

$$p' = \frac{n_s}{n_\phi}.$$

References

1. S. Bandyopadhyay and M. Cahay, *Introduction to Spintronics*, (CRC Press, Boca Raton, 2008).
2. Sin-itoro Tomonoga, *The Story of Spin*, (The University of Chicago Press, 1997).
3. I Zutic, J. Fabian and S. Das Sarma, "Spintronics: Fundamentals and Applications", Rev. Mod. Phys., **76**, 323 (2004).
4. L. I. Schiff, *Quantum Mechanics*. 3rd. edition, (McGraw Hill, New York, 1955).
5. S. Datta, *Electronic Transport in Mesoscopic Systems*, (Cambridge University Press, Cambridge, 1995).
6. R. E. Prange and S. M. Girvin eds., *The Quantum Hall Effect*, (Spinger, New York, 1987).
7. S. Chaudhuri, S. Bandyopadhyay and M. Cahay, "Current, potential, electric field and Fermi carrier distributions around localized elastic scatterers in phase coherent quantum magnetotransport", Phys. Rev. B., **47**, 12649 (1993).
8. P. T. Coleridge, "Small angle scattering in two-dimensional electron gases", Phys. Rev. B., **44**, 3793 (1991)
9. S. Das Sarma and F. Stern, "Single particle relaxation time versus scattering time in an impure electron gas", Phys. Rev. B., **32**, 8442 (1985).
10. P. T. Coleridge, R. Stoner and R. Fletcher, "Low field transport coefficients in GaAs/$Al_xGa_{1-x}As$ heterostructures", Phys. Rev. B., **39**, 1120 (1989).
11. B. Das, D. C. Miller, S. Datta, R. Reifenberger, W. P. Hong, P. K. Bhattacharya, J. Singh and M. Jaffe, "Evidence for spin-splitting in $In_xGa_{1-x}As/In_{0.52}Al_{0.48}As$ heterostructures as $B \to 0$", Phys. Rev. B., **39**, 1411 (1989).
12. S. Chaudhuri and S. Bandyopadhyay, "Numerical calculation of hybrid magneto- electric states in an electron waveguide", J. Appl. Phys., **71**, 3027 (1992).
13. N. Telang and S. Bandyopadhyay, "Quenching of electron-acoustic-phonon scattering in a quantum wire by a magnetic field", Appl. Phys. Lett., **62**, 3161 (1993).
14. Y. Nakamura, T. Inoshita and H. Sakaki, "Novel magneto-resistance oscillations in laterally modulated two dimensional electrons with 20 nm periodicity formed on vicinal GaAs (111)B substrates", Physica E, **2**, 944 (1998).
15. A. Balandin and S. Bandyopadhyay, "Quantum confined Lorentz effect in a quantum wire", J. Appl. Phys., **77**, 5924 (1995).
16. K. v. Klitzing, G. Dorda and M. Pepper, "New method for the high-accuracy determination of the fine-structure constant based on quantized Hall resistance", Phys. Rev. Lett., **45**, 494 (1980).
17. B. Jeckelmann and B. Jeanneret, "The quantum Hall effect as an electrical resistance standard", Rep. Prog. Phys., **64**, 1603 (2001).
18. R. B. Laughlin, "Quantized Hall conductivity in two dimensions", Phys. Rev. B., **23**, 5632 (1981).
19. M. Büttiker, "Absence of backscattering in the quantum Hall effect in multiprobe conductors", Phys. Rev. B., **38**, 9375 (1988).

20. D. C. Tsui, H. L. Störmer and A. C. Gossard, "Two-dimensional magnetotransport in the extreme quantum limit", Phys. Rev. Lett., **48**, 1559 (1982).
21. *The Quantum Hall Effect*, eds. R. E. Prange and S. M. Girvin, (Springer-Verlag, New York, 1987).
22. R. Willet, J. P. Eisenstein, H. L. Störmer, D. C. Tsui, A. C. Gossard and J. H. English, "Observation of an even denominator quantum number in the fractional quantum Hall effect", Phys. Rev. Lett., **59**, 1776 (1987).
23. J. P. Eisenstein, R. Willet, H. L. Störmer, D. C. Tsui, A. C. Gossard and J. H. English, "Collapse of the even denominator fractional quantum Hall effect in tilted fields", Phys. Rev. Lett., **61**, 997 (1988).
24. R. B. Laughlin, "Anomalous quantum Hall effect: An incompressible quantum fluid with fractionally charged excitations", Phys. Rev. Lett., **50**, 1395 (1983).
25. R. D. Mattuck, *A Guide to Feynman Diagrams in the Many Body Problem*, 2nd. edition (Dover, 1992).
26. J. K. Jain, "Microscopic theory of the fractional quantum Hall effect", Adv. Phys., **41**, 105 (1992).

Chapter 8
Quantum Transport Formalisms

8.1 When Is a Quantum-Mechanical Model for Transport Necessary?

In Chap. 1 we had mentioned that as long as electrons behave as classical particles, traveling through a solid obeying Newton's laws of motion, charge transport can be described adequately by the classical drift–diffusion model unless nonlocal effects exist. If they do exist, then the drift–diffusion model should be replaced with the (still classical) Boltzmann transport model.[1] However, there are some solid-state phenomena that cannot be described within a classical framework at all. Examples of them are electron interference and tunneling where electrons behave as waves propagating through a solid, instead of as particles obeying Newtonian mechanics. The wave nature mandates a proper quantum mechanical treatment. Therefore, the important question is: when is the wave nature manifested?

Events that mask the wave nature will always promote the particle nature of electrons, and vice versa. This is a consequence of the famous Bohr's Complementarity Principle [1]—later extended by Englert [2]—which posits that the wave nature and the particle nature tend to be mutually exclusive (or complementary). When one is prominent, the other is muted, although there are some rare, remarkable situations when both can coexist [3–6]. Events that suppress the wave nature are those that scramble or randomize a wave's phase so that interference effects, which require phase coherence, can no longer be observed. If such events occur frequently, then an electron behaves more like a classical particle than a quantum-mechanical wave and a classical description will be adequate. However, if such events are rare, then the quantum-mechanical wave nature can be manifested and a quantum-mechanical treatment of electron transport will be necessary.

[1]Sometimes the Boltzmann transport model is called "semi-classical" instead of "classical". This is only because the scattering rate $S(\vec{k}, \vec{k}')$ is calculated using Fermi's Golden Rule or an equivalent quantum-mechanical prescription. Other than that, the Boltzmann model is entirely classical.

S. Bandyopadhyay, *Physics of Nanostructured Solid State Devices*,

DOI 10.1007/978-1-4614-1141-3_8,

The most common cause of phase randomization is so-called *phase-breaking scattering* where the internal state of the scatterer changes as a result of the scattering event [7]. An obvious example of such an event is electron–phonon scattering where a phonon is either absorbed or emitted, so that the internal state of the scatterer (the phonon) changes drastically (it is either annihilated or created). Other examples of phase-breaking scattering are electron–electron, electron–hole, or hole–hole collisions, where the scatterer (an electron or a hole in this case) changes its momentum and energy as a result of the collision. Any time the scatterer changes its energy, the carrier must also change its energy since the total energy of the scatterer plus the carrier must be conserved. Therefore, inelastic collisions, which cause a carrier to change its energy, are always phase breaking since they will inevitably also cause the scatterer to change its energy and therefore its internal state. Some elastic collisions, such as scattering by magnetic impurities, can also be phase-breaking since there the internal magnetic moment of the impurity changes when collision occurs (the electron can flip its spin due to such scattering). Thus, there is a wide assortment of scattering events that can break the phase.

If a device is so small that a carrier can traverse it without encountering any (or too many) phase-breaking scattering events, then we expect to see the wave nature of the carrier more or less preserved. Such a device should exhibit quantum-mechanical effects (such as tunneling through a potential barrier, or electron interference) which cannot be captured within any classical formalism, and will need a proper quantum-mechanical transport model. Such devices have been termed *mesoscopic devices* since their dimensions are in the realm between the truly microscopic (single atoms and molecules) and macroscopic (exceeding a few micrometers).

In order for a solid structure to qualify as a phase-coherent mesoscopic device,[2] the transit time τ_t of a carrier through that structure should be smaller than the mean time between phase-breaking collisions, which we will call τ_ϕ. Therefore, a necessary (but not a sufficient) condition is

$$\tau_t < \tau_\phi. \tag{8.1}$$

We can recast the above condition into a condition on the physical dimension of the structure. Obviously, the physical dimension of a mesoscopic structure should be smaller than the distance a charge carrier travels in the time τ_ϕ. This distance depends on the circumstances. If transport is ballistic and there is no collision at all, then this distance is

$$L_\phi = v_i \tau_\phi + \frac{1}{2}\frac{e\mathcal{E}}{m^*}\tau_\phi^2, \tag{8.2}$$

where v_i is the velocity with which the carrier entered the device and $\mathcal{E}$ is the electric field accelerating the carrier (of effective mass m^*).

[2]It is very important to understand that "mesoscopic" does not mean that the device is smaller than a particular dimension (like 10 or 100 nm), but rather it is small enough to exhibit quantum-mechanical effects.

If the carrier population is nondegenerate (e.g., in a semiconductor where the Fermi level is well below the conduction band edge), then v_i is the injection velocity. On the other hand, if the carrier population is degenerate (e.g., in a metal or a semiconductor in which the Fermi level is well above the conduction band edge), then v_i will be essentially the Fermi velocity v_F, which is related to the Fermi energy E_F as $E_F = (1/2)m^* v_F^2$, if the bandstructure is parabolic. Moreover, the acceleration due to the electric field can be neglected because v_F is typically large, so that we can write $L_\phi \approx v_F \tau_\phi$.

Next, consider the situation when transport is diffusive, i.e., there are elastic collisions (which do not randomize phase). In this case, a carrier will be traveling under two distinct influences—drift and diffusion. The distance traveled under both influences in time τ_ϕ is approximately

$$L_\phi = \left[-\frac{\mu(\mathcal{E})\mathcal{E}}{2D(\mathcal{E})} + \sqrt{\frac{\mu(\mathcal{E})\mathcal{E}}{2D(\mathcal{E})} + \frac{1}{D(\mathcal{E})\tau_\phi}} \right]^{-1}, \tag{8.3}$$

where $\mathcal{E}$ is the electric field causing drift, $\mu(\mathcal{E})$ is the field-dependent carrier mobility (see Fig. 1.9), and $D(\mathcal{E})$ is the field-dependent diffusion coefficient.

It is easy to see that under weak electric field, when diffusion dominates over drift, the above expression reduces to

$$L_\phi \approx \sqrt{D(\mathcal{E})\tau_\phi}, \tag{8.4}$$

whereas under strong electric fields, when drift dominates over diffusion,

$$L_\phi \approx \mu(\mathcal{E})\mathcal{E}\tau_\phi. \tag{8.5}$$

If the carrier population is nondegenerate, then the mobility and diffusion coefficient are related by the Einstein relation $D(\mathcal{E}) = (k_B T/e)\mu(\mathcal{E})$. On the other hand, if the carrier population is degenerate, then $D(\mathcal{E}) \approx (1/\eta)v_F^2 \tau_m$, where $\eta = 3$ for bulk structures (3-d), 2 for quantum wells (2-d), and 1 for quantum wires (1-d), and τ_m is the momentum relaxation time. Thus, we have the following relations:

$$\begin{aligned}
L_\phi &= v_i \tau_\phi + \frac{1}{2}\frac{e\mathcal{E}}{m^*}\tau_\phi^2 \quad \texttt{(ballistic, nondegenerate)} \\
&= v_F \tau_\phi \quad \texttt{(ballistic, degenerate)} \\
&= \left[-\frac{e\mathcal{E}}{2k_B T} + \sqrt{\frac{e\mathcal{E}}{2k_B T} + \frac{e}{k_B T \mu(\mathcal{E})\tau_\phi}} \right]^{-1} \quad \texttt{(diffusive, nondegenerate)} \\
&= \left[-\frac{e\eta\mathcal{E}}{2m^* v_F^2} + \sqrt{\frac{e\eta\mathcal{E}}{2m^* v_F^2} + \frac{\eta}{v_F^2 \tau_m(\mathcal{E})\tau_\phi}} \right]^{-1} \quad \texttt{(diffusive, degenerate)}
\end{aligned}$$

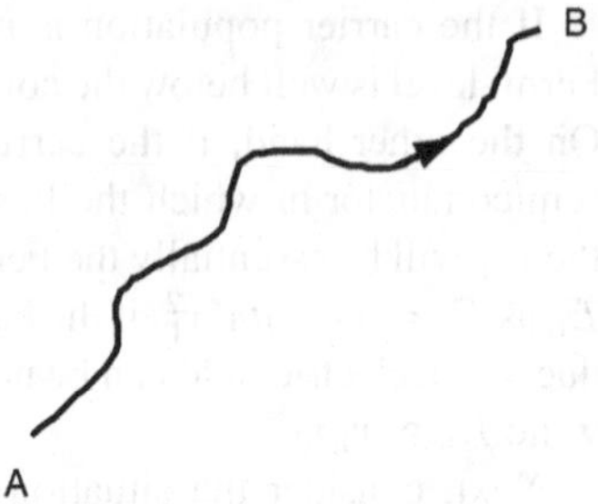

Fig. 8.1 A charge carrier's phase-coherent trajectory of length S from one point in space to another

The quantity L_ϕ is called the *phase-breaking length* and the quantity τ_ϕ is called the *phase-breaking time* or "phase-memory time" since an electron will retain memory of its phase for this duration. As long as a device's dimension along the direction of carrier transport (or current flow) is smaller than L_ϕ, the device will qualify as "mesoscopic." It still may or may not exhibit quantum-mechanical effects like interference of electron waves. Being mesoscopic is a necessary, but not a sufficient condition for manifesting quantum-mechanical wave-like attributes. There are other conditions that must be fulfilled to observe quantum mechanical wave properties, which we address next.

Consider an electron traveling from point A to point B along a path of length S as shown in Fig. 8.1. Assume that at any finite temperature T, there is a spread ΔE in the energy of the electron, which is approximately equal to $k_B T$, where k_B is the Boltzmann constant.

The spread in the electron's energy will result in a spread in the phase shift $\Delta\phi$ accumulated by the electron in traversing the path. If the latter spread exceeds $\sim \pi$, then the electron's phase is effectively uncertain and we cannot expect to see clear-cut interference effects. Here, the phase has been scrambled not by inelastic scattering, but by the finite temperature. Consequently, there is a maximum spread in the energy ΔE (and a maximum temperature T) that we can tolerate. This energy is called the "correlation energy" E_c—a term adopted from electromagnetics [8]. The corresponding temperature is called the "correlation temperature" T_c. We can determine E_c as follows:

$$\Delta\phi = \frac{S}{\lambda(E + E_c)} - \frac{S}{\lambda(E)} \approx SE_c \frac{d(1/\lambda)}{dE} = S\frac{E_c}{2\pi}\frac{dk}{dE} = S\frac{E_c}{hv} \approx \pi, \qquad (8.6)$$

where $\lambda(E)$ is the DeBroglie wavelength of an electron of energy E, $k = 2\pi/\lambda$, and v is the electron velocity.

Now, consider a device of length L (dimension along the direction of current flow). If transport is ballistic, then an electron entering the device goes straight through since it does not encounter any scattering that can cause it to deviate from its course. Hence, $S = L$ and we get that the correlation energy is

$$E_c = k_B T_c \approx \pi hv/L \quad \texttt{(ballistic transport)} \qquad (8.7)$$

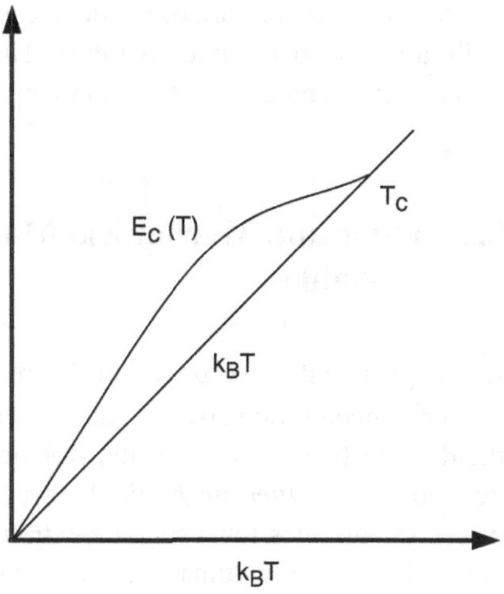

Fig. 8.2 Determining the correlation temperature (or Thouless temperature) in a sample whose dimensions exceed the phase-breaking length

whereas in a diffusive device at low electric field strengths (diffusion dominates over drift), the electron will execute a diffusive walk before exiting the device so that $S = vL^2/D$, where D is the diffusion constant. In that case,

$$E_c = k_B T_c \approx \pi h D/L^2 \quad \texttt{(diffusive transport)} \tag{8.8}$$

The temperature T_c is also called the "Thouless temperature" after Thouless who first pointed out its importance [9]. Note that the Thouless temperature is inversely proportional to L (ballistic transport) or L^2 (diffusive transport), so that shorter devices can exhibit quantum-mechanical interference effects at higher temperatures. Thus, *small device dimension and low temperature are conducive to the observation of quantum-mechanical effects arising from phase coherence of electron waves in a solid.* In the end, at least two conditions must be fulfilled in order to observe quantum-mechanical effects such as wave interference:

$$\begin{aligned} L &< L_\phi \\ T &< T_c. \end{aligned} \tag{8.9}$$

If a device's length exceeds the phase breaking length L_ϕ, then transport is certainly not ballistic. Nor is it phase conserving. In this case, the correlation energy is equal to the Thouless energy [10, 11]

$$E_c = hD/L_\phi^2 = h/\tau_\phi. \tag{8.10}$$

Typically, the phase breaking time decreases with increasing temperature T and varies as $\tau_\phi \sim 1/T^p$, where the exponent p is usually between 0.33 and 1 [12]. Therefore, to find the maximum allowed temperature T_c for the observation of quantum-mechanical wave interference effects, we must solve the equation $k_B T = \hbar/\tau_\phi(T)$ as shown in Fig. 8.2. The solution will yield the Thouless temperature T_c.

Once the temperature of the device exceeds T_c, interference effects decay *slowly* with further increase in temperature. This is helpful since it may make it possible to observe interference effects at relatively elevated temperatures in some cases.

8.2 Quantum Mechanical Model to Calculate Transport Variables

In Chaps. 1 and 2 we provided the recipe to calculate two transport variables—carrier concentration $n(\vec{r},t)$ and current density $\vec{J}(\vec{r},t)$—*classically* from either the drift–diffusion model or the Boltzmann transport model. Later, in Chap. 3, we derived the *quantum-mechanical* expressions for the expectation values of these two transport variables for a *single electron* (see (3.22) and (3.29)). If we now wish to derive the same quantum mechanical expressions for an entire ensemble of electrons in an actual device, then we simply have to sum up the single-electron expressions over all available electron states labeled by the wavevector $\vec{k}$, keeping in mind that each $\vec{k}$-state can be filled up with one electron of a given spin with probability $f(\vec{r},\vec{k},t)$. We can then account for spin degeneracy (i.e., each $\vec{k}$-state can actually accommodate up to two electrons of opposite spins) by multiplying the result with a factor of 2.

Based on the above discussion, (3.22) and (3.29) yield that the quantum-mechanical expressions for the expectation values of carrier concentration and current density in an ensemble of electrons will be

$$n^{\mathrm{QM}}(\vec{r},t) = \sum_{\vec{k}} |\psi_{\vec{k}}(\vec{r},t)|^2 f(\vec{r},\vec{k},t)$$

$$\vec{J}^{\mathrm{QM}}(\vec{r},t) = -\sum_{\vec{k}} \left\{ \frac{ie\hbar}{2m^*} \left[\psi_{\vec{k}}(\vec{r},t) \left(\vec{\nabla}_{\mathrm{r}} \psi_{\vec{k}}(\vec{r},t) \right)^* - \psi^*_{\vec{k}}(\vec{r},t) \left(\vec{\nabla}_{\mathrm{r}} \psi_{\vec{k}}(\vec{r},t) \right) \right] \right.$$
$$\left. + \frac{1}{m^*} \left(e\vec{A}(\vec{r},t) \left| \psi_{\vec{k}}(\vec{r},t) \right|^2 \right) \right\} f(\vec{r},\vec{k},t), \qquad (8.11)$$

where the subscript "QM" obviously stands for "quantum-mechanical". Here, the wavefunctions are assumed to be normalized properly so that $\int_0^\infty d^3\vec{r}\,|\psi_{\vec{k}}(\vec{r},t)|^2 = 1$. Note that the current expression has an additional term involving the magnetic vector potential $\vec{A}(\vec{r},t)$, which did not appear in (3.29) since there we had not considered such things as a magnetic field that will transform the momentum operator from $\vec{p}_{\mathrm{op}}$ to $\vec{p}_{\mathrm{op}} - e\vec{A}(\vec{r},t)$. If we account for this transformation, then the additional term shown here will appear in a magnetic field. Finally, note that the summation in the first line (carrier concentration) is a *scalar* addition while in the second line (current density), it is a *vector* addition.

In a bulk (three-dimensional) structure, the summation over the wavevector $\vec{k}$ can be converted into integrals in wavevector space via the usual density-of-states. This will yield

$$n^{\mathrm{QM}}(\vec{r}_3,t) = \frac{\Omega}{4\pi^3}\int_0^\infty d^3\vec{k}\,\left|\psi_{\vec{k}}(\vec{r}_3,t)\right|^2 f(\vec{r}_3,\vec{k},t)$$

$$\vec{J}^{\mathrm{QM}}(\vec{r}_3,t) = -\frac{\Omega}{4\pi^3}\int_0^\infty d^3\vec{k}\, f\left(\vec{r}_3,\vec{k},t\right) \times\left\{\frac{ie\hbar}{2m^*}\left[\psi_{\vec{k}}(\vec{r}_3,t)\left(\vec{\nabla}_{r3}\psi_{\vec{k}}(\vec{r}_3,t)\right)^* - \psi^*_{\vec{k}}(\vec{r}_3,t)\left(\vec{\nabla}_{r3}\psi_{\vec{k}}(\vec{r}_3,t)\right)\right] + \frac{1}{m^*}\left(e\vec{A}(\vec{r}_3,t)\left|\psi_{\vec{k}}(\vec{r}_3,t)\right|^2\right)\right\}, \tag{8.12}$$

where, as always, $\vec{r}_3$ is the spatial coordinate in three dimensions ($\vec{r}_3 = x\hat{x} + y\hat{y} + z\hat{z}$). The quantity Ω is the normalizing volume, but since the wavefunctions are assumed to be normalized, they will be proportional to $1/\sqrt{\Omega}$, so that Ω will cancel out and the above expressions will be independent of Ω.

By the same token, in a two-dimensional structure, we will have

$$n_s^{\mathrm{QM}}(\vec{r}_2,t) = \frac{A}{2\pi^2}\sum_{m=1}^{M}\int_0^\infty d^2\vec{k}\,\left|\psi_{\vec{k},m}(\vec{r}_2,t)\right|^2 f(\vec{r}_2,\vec{k},t)$$

$$\vec{J}_s^{\mathrm{QM}}(\vec{r}_2,t) = -\frac{A}{2\pi^2}\sum_{m=1}^{M}\int_0^\infty d^2\vec{k}\, f(\vec{r}_2,\vec{k},t) \times\left\{\frac{ie\hbar}{2m^*}\left[\psi_{\vec{k},m}(\vec{r}_2,t)\left(\vec{\nabla}_{r2}\psi_{\vec{k},m}(\vec{r}_2,t)\right)^* - \psi^*_{\vec{k},m}(\vec{r}_2,t)\left(\vec{\nabla}_{r2}\psi_{\vec{k},m}(\vec{r}_2,t)\right)\right] + \frac{1}{m^*}\left(e\vec{A}(\vec{r}_2,t)\left|\psi_{\vec{k},m}(\vec{r}_2,t)\right|^2\right)\right\}, \tag{8.13}$$

where m is the subband index in a two dimensional structure and $\vec{r}_2$ is the spatial coordinate in the two-dimensional plane ($\vec{r}_2 = x\hat{x} + y\hat{y}$). Similarly, in a one-dimensional structure, we will have

$$n_l^{\mathrm{QM}}(\vec{r}_1,t) = \frac{2L}{\pi}\sum_{m=1}^{M}\sum_{p=1}^{P}\int_0^\infty \mathrm{d}k\,\left|\psi_{k,m,p}(\vec{r}_1,t)\right|^2 f\left(\vec{r}_1,k,t\right)$$

$$\vec{J}_l^{\mathrm{QM}}(\vec{r}_1,t) = -\frac{2L}{\pi}\sum_{m=1}^{M}\sum_{p=1}^{P}\int_0^\infty \mathrm{d}k\, f\left(\vec{r}_1,\vec{k},t\right) \times\left\{\frac{ie\hbar}{2m^*}\left[\psi_{k,m,p}(\vec{r}_1,t)\left(\vec{\nabla}_{r1}\psi_{k,m,p}(\vec{r}_1,t)\right)^* - \psi^*_{k,m,p}(\vec{r}_1,t)\left(\vec{\nabla}_{r1}\psi_{k,m,p}(\vec{r}_1,t)\right)\right] + \frac{1}{m^*}\left(e\vec{A}(\vec{r}_1,t)\left|\psi_{k,m,p}(\vec{r}_1,t)\right|^2\right)\right\}, \tag{8.14}$$

where m and p are the two transverse subband indices characterizing a subband in a one-dimensional structure and $\vec{r}_1$ is the spatial coordinate along the length. In two dimensions, the normalized wavefunction will be proportional to $1/\sqrt{A}$ and in one dimension, it will be proportional to $1/\sqrt{L}$, so that the transport variables are always independent of normalizing area or length.

A very important question that arises now is how many terms should one include in the summations that we have in the preceding equations, i.e., what should be the values of M and P? The correct answer is that M and P should be large enough that if we increase them any further, the carrier concentration or the current density should not change appreciably. At very low temperatures, we need to only include the subbands that are below the Fermi level μ_F because electrons occupy these subbands while all other (higher energy) subbands are relatively unoccupied.

8.2.1 *How to Calculate the Wavefunction*

At this point it is clear that we have to find the wavefunction $\psi_{\vec{k}}(\vec{r}_3, t)$ (in 3-d), or $\psi_{\vec{k},m}(\vec{r}_2, t)$ (in 2-d), or $\psi_{k,m,p}(\vec{r}_1, t)$ (in 1-d), before we can evaluate the carrier concentration or the current density. To do this, we must solve the effective mass Schrödinger equation in a mesoscopic structure subject to the appropriate boundary conditions. This equation is:

$$\left[\frac{(\vec{p}_{op} - e\vec{A})^2}{2m^*} + V(\vec{r}, t)\right]\psi_{\vec{k}}(\vec{r}, t) = E_{\vec{k}}\psi_{\vec{k}}(\vec{r}, t) \tag{8.15}$$

Solving it may not be a trivial job.

Before we proceed to solve the Schrödinger equation, we should recall that inelastic collisions destroy phase coherence and hence the potential term $V(\vec{r}, t)$ should not contain any term associated with electron-inelastic scatterer interaction. Actually, any such term would have been nonhermitian, which will immediately violate the continuity equation and hence the principle of conservation of charge (see Problem 3.8). A nonhermitian Hamiltonian will also violate time-reversal symmetry (see the second part of Problem 3.8) which means that dissipation violates time-reversal symmetry of the Schrödinger equation. To make a long story short, Schrödinger equation cannot handle dissipation and more complicated formalisms will be required to treat dissipation properly within a quantum-mechanical framework. We will visit this issue later in this chapter, but for the time being, suffice it to say that we will exclude all dissipative interactions such as inelastic collisions.

Schrödinger equation can handle *nondissipative* time-dependent interactions (such as interaction of an electron with a *coherent* radiation source like a laser or with a coherent time-varying magnetic field), but to make matters simple, we will exclude all time-dependent interactions and restrict ourselves only to *time-independent* potentials $V(\vec{r})$. In other words, we will be limiting ourselves to steady state situations exclusively. This would allow us to consider transport through any arbitrary time-invariant potential profile and thus permit treatment of *elastic* scattering due to imperfections in a solid-state device since this type of scattering potential is time-independent. Consequently, the type of Schrödinger equation that we will be dealing with is

$$\left[\frac{(\vec{p}_{\text{op}} - e\vec{A})^2}{2m^*} + V(\vec{r})\right]\psi_{\vec{k}}(\vec{r}) = E_{\vec{k}}\psi_{\vec{k}}(\vec{r}). \tag{8.16}$$

We have solved such an equation in Chap. 7 considering only confining potential terms $V(z)$ and $V(y)$ in the y- and z-directions, and periodic boundary conditions in the x-direction. Examples are (7.14) for a 2-d system, and (7.46) and (7.59) for 1-d systems. However, a device in general will have two additional complications. First, the device will be of finite length and there will be contacts at the two ends of the device causing a discontinuity in the potential profile at the interfaces with the two contacts. Thus, periodic boundary condition in the x-direction is clearly inappropriate because that works only if the Hamiltonian is invariant in x, i.e., it does not depend on the x-coordinate. The discontinuities in the potential profile in the x-direction owing to contacts will make the Hamiltonian of any finite structure depend on the x-coordinate, so that periodic boundary conditions are no longer valid. In fact, these discontinuities will cause reflections of the electron waves at the contact interfaces, which will have to be taken into account. Second, there can be defects and disorder within the device causing impurity scattering potentials $V_s(\vec{r})$. The way to handle these two additional complications is to adopt a method such as the one described below.

8.2.1.1 The Wavefunction in a Device and the Transport Variables

Consider a *quasi* one-dimensional device as shown in Fig. 8.3. Any device of finite width and thickness can be thought of as a quasi one-dimensional device with multiple transverse subbands occupied by electrons. Hence, our treatment is perfectly general.

Electrons are incident on this quasi one-dimensional device from both the left and the right contacts with a spectrum of wavevectors. A wave incident from one particular subband in the left contact can be partially reflected back into all other subbands in the left contact and partially transmitted into all subbands inside the device. The same is true of an electron incident from the right contact. We will assume that current flows along the length of the device in the x-direction. The subbands within the contact will be labeled with unprimed indices m and p (representing quantization of motion along the y- and z-direction, respectively), while the subbands within the device will be labeled with primed indices m' and p'.

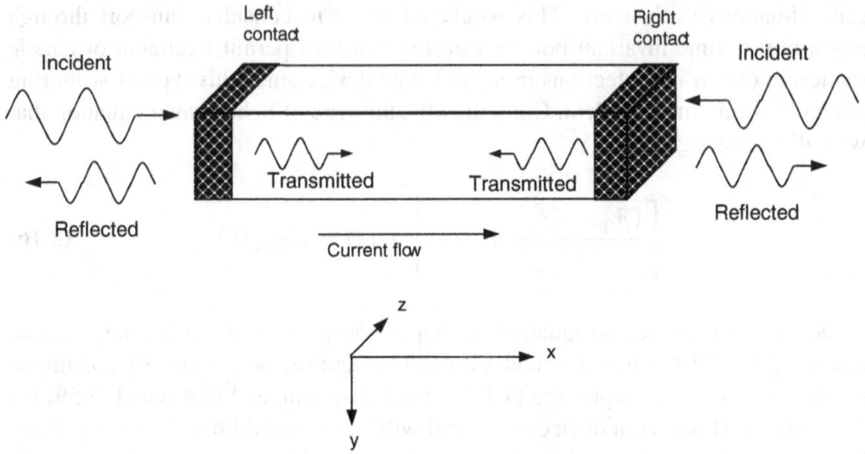

Fig. 8.3 The wavefunction of an electron inside a quasi one-dimensional device is composed of electron waves entering from the *left* and *right* contacts. The waves impinging on the device from either contact is partially transmitted and partially reflected. The components that are transmitted into the device contribute to the carrier concentration and current density

The wavefunction at any location x within the device, due to an electron that impinged on the device with total energy E from the subband (m, p) in the left contact, can be written as a coherent superposition of left and right traveling waves [13]:

$$\psi^{\mathrm{L}}_{\mathrm{E},m,p}(x,y,z) = \sum_{m'=1}^{M'}\sum_{p'=1}^{P'}\{A_{m,p;m',p'}(x)\exp(\mathrm{i}k_{m',p'}x)\phi_{m',p'}(y,z)$$
$$+B_{m,p;m',p'}(x)\exp(\mathrm{i}k_{-m',-p'}x)\phi_{-m',-p'}(y,z)\}, \tag{8.17}$$

where the double summation extends over both propagating and evanescent states inside the device. Here, $\phi_{m',p'}(y,z)$ and $\phi_{-m',-p'}(y,z)$ are the transverse components of the wavefunction within the device in the subband with indices m' and p'. The subscripts $-m', -p'$ indicate that the quantity in question corresponds to a wavefunction with oppositely directed velocity compared to the one with subscript m', p' (see Problem 8.1). Likewise, the wavevectors $k_{m',p'}$ and $k_{-m',-p'}$ are the wavevectors of oppositely traveling states with total energy E in the (m', p')-th subband. These wavevectors will have to be found from the $E - k$ relation in the device, such as those shown in Fig. 7.12 or Fig. 7.13. For propagating states, these wavevectors are real, whereas for evanescent states, these wavevectors will be complex or imaginary. An elastic scatterer inside the device can cause intersubband coupling, i.e., an electron in one subband can be reflected by the scatterer into a different subband without change in the total energy. This effect is fully incorporated within the coefficients $A_{m,p;m',p'}(x)$ and $B_{m,p;m',p'}(x)$.

In the foregoing, we assumed that the potential within the device is *spatially invariant*, except perhaps for localized *point* scatterers that cause elastic scattering. If there is a spatially varying potential $V(x, y, z)$ within the device, then we will have to slightly modify our approach as we briefly discuss in Sect. 8.4. However, this does not add anything fundamental to the discussion; therefore, we ignore this subtlety for the time being.

By analogy with (8.17), the wavefunction of an electron that is injected into the device with total energy E from the subband (m, p) in the right contact can be written as

$$\psi^R_{E,m,p}(x, y, z) = \sum_{m'=1}^{M'} \sum_{p'=1}^{P'} \{C_{m,p;m',p'}(x) \exp(ik_{-m',-p'}(x - L))\phi_{-m',-p'}(y, z) + D_{m,p;m',p'}(x) \exp(ik_{m',p'}(x - L))\phi_{m',p'}(y, z)\}, \tag{8.18}$$

where L is the length of the device.

The problem of determining the wavefunctions $\psi^{\mathrm{L}}_{\mathrm{E},m,p}(x, y, z)$ and $\psi^R_{E,m,p}(\mathrm{x, y, z})$ now boils down to finding the coefficients $A_{m,p;m',p'}(x)$, $B_{m,p;m',p'}(x)$, $C_{m,p;m',p'}(x)$ and $D_{m,p;m',p'}(x)$. Once we have found those coefficients and therefore the wavefunctions, we can determine the linear carrier concentration and the current density due to electrons entering the quasi one-dimensional device from either the left or the right contact. The former is determined as follows:

$$\left[n^{\mathrm{QM}}_l(x, y, z)\right]_{\mathrm{left}} = \sum_{m=1}^{M} \sum_{p=1}^{P} \int_0^\infty \mathrm{d}E D^{1-d}(E) \left|\psi^{\mathrm{L}}_{\mathrm{E},m,p}(x, y, z)\right|^2 f\left(E + \epsilon_{m,p} - \mu^{\mathrm{L}}_{\mathrm{F}}\right)$$

$$\left[\vec{J}^{\mathrm{QM}}(x, y, z)\right]_{\mathrm{left}} = -\sum_{m=1}^{M} \sum_{p=1}^{P} \frac{ie\hbar}{2m^*} \int_0^\infty \mathrm{d}E D^{1-d}(E) f\left(E + \epsilon_{m,p} - \mu^{\mathrm{L}}_{\mathrm{F}}\right) \times \left\{ \left[\psi^{\mathrm{L}}_{\mathrm{E},m,p}(x, y, z) \left(\vec{\nabla}_{\mathrm{r}}\psi^{\mathrm{L}}_{\mathrm{E},m,p}(x, y, z)\right)^* - \left[\psi^{\mathrm{L}}_{\mathrm{E},m,p}(x, y, z)\right]^* \left(\vec{\nabla}_{\mathrm{r}}\psi^{\mathrm{L}}_{\mathrm{E},m,p}(x, y, z)\right)\right] + \frac{1}{m^*}\left(e\vec{A}(x, y, z) \left|\psi^{\mathrm{L}}_{\mathrm{E},m,p}(x, y, z)\right|^2\right)\right\}, \tag{8.19}$$

where $\epsilon_{m,p}$ is the energy at the bottom of the (m, n)-th subband and $\mu^{\mathrm{L}}_{\mathrm{F}}$ is the Fermi level in the left contact. Since the contacts are viewed as nearly infinite reservoirs of electrons, the carrier population in either the right or the left contact is assumed to be in local thermodynamic equilibrium described by a local Fermi level. Additionally, we have converted the density-of-states in wavevector space to that in energy space in order to have a single integration variable, namely energy, which is conserved in phase-coherent transport. Note that we are using the one-dimensional density-of-states $D^{1-d}(E)$ since the device is quasi one-dimensional.

We can use (8.19) to find the contributions to the linear carrier concentration and current density due to electrons impinging on the device from the right contact if we replace the Fermi level μ_F^L with the Fermi level μ_F^R in the right contact, the quantity $\epsilon_{m,p}$ with $\epsilon'_{m,p}$ where the latter is the energy at the bottom of the (m, n)-th subband in the right contact, and the wavefunction $\psi^L_{E,m,p}(x, y, z)$ with $\psi^R_{E,m,p}(x, y, z)$. Finally, the *total* linear carrier concentration and the *total* current density is found by adding the contributions from left and right:

$$n_l^{QM}(x, y, z) = \left[n^{QM}(x, y, z)\right]_{\text{left}} + \left[n^{QM}(x, y, z)\right]_{\text{right}}$$
$$\vec{J}^{QM}(x, y, z) = \left[\vec{J}^{QM}(x, y, z)\right]_{\text{left}} + \left[\vec{J}^{QM}(x, y, z)\right]_{\text{right}}. \quad (8.20)$$

Once again, note that the carrier concentration is found by scalar addition of the two contributions, while the current density is found from the vector addition of the two contributions.

Note that if there is an electrical "bias" across the device, resulting in a potential drop V between the left and right contacts, then $\mu_F^L - \mu_F^R = qV$. Sometimes, the local Fermi levels μ_F^L and μ_F^R are referred to as local "chemical potentials." Engineers favor the term "Fermi level," while physicists and chemists seem to prefer the term "chemical potential."

8.2.1.2 Should We Have Added the Wavefunctions or the Transport Variables?

It might be troubling to some that we did not add the wavefunctions due to electrons impinging from the left and right, and then calculate the carrier concentration and current density from the total wavefunction. Instead, what we did was to calculate the carrier concentrations and current densities due to the left-incident and right-incident electrons individually, and then add them up to find the total quantities. Obviously, the two approaches would *not* have yielded the same result. So, which one is correct?

If we had added the wavefunctions first, then we would have tacitly assumed that the wavefunction of an electron coming from the left contact and that of an electron coming from the right contact are mutually phase coherent. This is impossible since we view these electrons as two *different* electrons. Even though electrons are indistinguishable particles, only the wavefunctions of the *same* electron (e.g., wavefunctions associated with traversal of two different paths in either real space or wavevector space) can be mutually coherent. That is why we say that an electron can interfere only with itself, and not with another electron. This is a fundamental property of all fermions. The famous Young's double slit experiment with electrons [14] involves interference between the wavefunctions of the *same* electron going through two different slits as shown in Fig. 8.4.

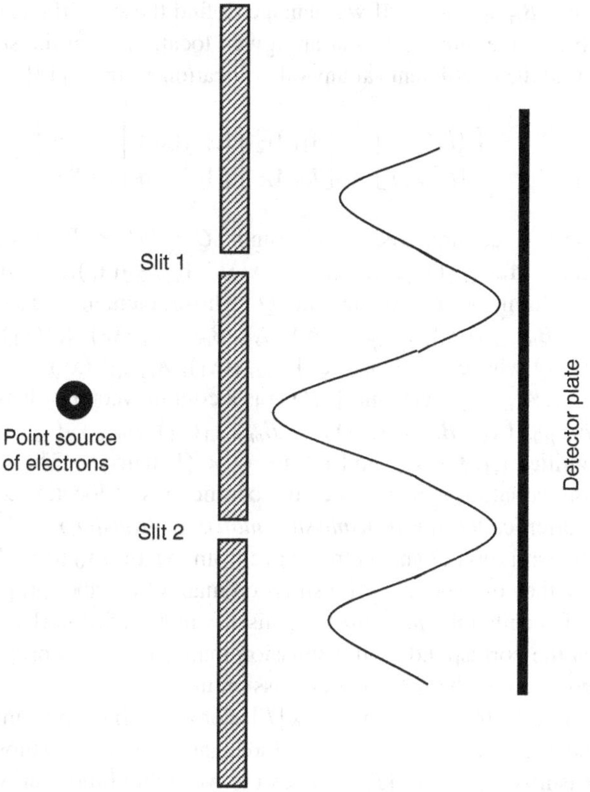

Fig. 8.4 Young's double slit experiment where an electron interferes with itself to produce interference fringes on a detector plate

Since the wavefunctions of an electron entering from the left and from the right are mutually *incoherent*, we should *not* add their wavefunctions and then calculate the transport variables from the composite wavefunction. Instead, we should calculate the contributions of the left-incident and right-incident electrons to the transport variables individually and then add them up, which is exactly what we did.

A more sophisticated treatment of how to find the carrier concentration and the current density (involving the concept of a "density matrix") is available.[3]

8.2.2 *The Transmission Matrix Method*

Let us now return to the problem of finding the wavefunction $\psi^{L}_{E,m,p}(x, y, z)$ in a disordered structure, which is equivalent to finding the unknown coefficients

[3]See, for example, Supriyo Datta, "Nanoscale device modeling: The Green's function method", Superlat. Microstruct., Vol. 28, 253 (2000).

$A_{m,p;m',p'}(x)$ and $B_{m,p;m',p'}(x)$. If we manage to find these coefficients (which are actually traveling wave amplitudes) at any given location x_0 in the structure, then we can easily find the coefficients at any other location x_1 from [13]

$$\begin{bmatrix} \{l\}(x_1) \\ \{r\}(x_1) \end{bmatrix} = \begin{bmatrix} \mathbf{t}_{11} & \mathbf{t}_{12} \\ \mathbf{t}_{21} & \mathbf{t}_{22} \end{bmatrix} \begin{bmatrix} \{l\}(x_0) \\ \{r\}(x_0) \end{bmatrix}, \tag{8.21}$$

where $\{l\}(x_0)$ is a column vector of length $Q = M' \times P'$ whose elements are $A_{m,p;1,1}(x_0)$, $A_{m,p;1,2}(x_0)\ldots\ldots A_{m,p;2,1}(x_0)$, $A_{m,p;2,2}(x_0)\ldots\ldots A_{m,p;M,P}(x_0)$, $\{r\}(x_0)$ is a column vector of length Q whose elements are $B_{m,p;1,1}(x_0)$, $B_{m,p;1,2}(x_0)\ldots\ldots B_{m,p;2,1}(x_0)$, $B_{m,p;2,2}(x_0)\ldots\ldots B_{m,p;M,P}(x_0)$, $\{l\}(x_1)$ is a column vector of length Q whose elements are $A_{m,p;1,1}(x_1)$, $A_{m,p;1,2}(x_1)\ldots\ldots A_{m,p;2,1}(x_1)$, $A_{m,p;2,2}(x_1)\ldots\ldots A_{m,p;M,P}(x_1)$, and $\{r\}(x_1)$ is a column vector of length Q whose elements are $B_{m,p;1,1}(x_1)$, $B_{m,p;1,2}(x_1)\ldots\ldots B_{m,p;2,1}(x_1)$, $B_{m,p;2,2}(x_1)\ldots\ldots B_{m,p;M,P}(x_1)$. The quantities $\mathbf{t}_{11}$, $\mathbf{t}_{12}$, $\mathbf{t}_{21}$, and $\mathbf{t}_{22}$ are $Q \times Q$ matrices. The square matrix in the preceding equation which relates the coefficients at location x_1 to those at location x_0 is often called the *transmission matrix* or *transfer matrix* $[T]$ since it describes the transmission of an electron wave from location x_0 to x_1. If x_1 is to the right of x_0, then the corresponding transmission matrix describes propagation from left to right and is called the *forward* transmission matrix. Similarly, if x_1 is to the left of x_0, then the corresponding transmission matrix describes propagation from right to left and is called the *reverse* transmission matrix.

One property of the transmission matrix $[T]$ is obvious. If the transmission matrix $[T_{01}]$ relates the coefficients (or wave amplitudes) at location x_1 to those at location x_0 and the transmission matrix $[T_{12}]$ relates the wave amplitudes at location x_2 to those at location x_1, then the transmission matrix $[T_{02}]$ relating the wave amplitudes at location x_2 to those at location x_0 is found by multiplying the other two matrices *in the proper order* as follows:

$$[T_{02}] = [T_{12}] \times [T_{01}]. \tag{8.22}$$

Thus, cascading transmission matrices is a simple task as long as we remember to maintain the proper order. The order is important since matrices do not necessarily commute, i.e., $[T_{12}] \times [T_{01}] \neq [T_{01}] \times [T_{12}]$.

We still have two tasks to complete. First, we need to know the coefficients $A_{m,p;m',p'}(x)$ and $B_{m,p;m',p'}(x)$ at *some location* in the structure so that we can find them everywhere else using forward and reverse transmission matrices, and second, we need to know the elements of the $Q \times Q$ forward and reverse transmission matrix. Only then can we find the wavefunction $\psi^{\mathrm{L}}_{\mathrm{E},m,p}(x, y, z)$ due to left-impinging electrons at arbitrary locations, which will yield the carrier concentration $[n^{\mathrm{QM}}(x, y, z)]_{\mathrm{left}}$ and the current density $[\vec{J}^{\mathrm{QM}}(x, y, z)]_{\mathrm{left}}$ anywhere in the device. Similarly, if we know the coefficients $C_{m,p;m',p'}(x)$ and $D_{m,p;m',p'}(x)$ at *any location* in the structure, then using forward and reverse transmission matrices, we can determine the carrier concentration $[n^{\mathrm{QM}}(x, y, z)]_{\mathrm{right}}$ and the current density $[\vec{J}^{\mathrm{QM}}(x, y, z)]_{\mathrm{right}}$ everywhere in the device.

We will address the task of finding $A_{m,p;m',p'}(x)$, $B_{m,p;m',p'}(x)$, $C_{m,p;m',p'}(x)$ and $D_{m,p;m',p'}(x)$ a little later and instead tackle the task of finding the forward and reverse transmission matrices first. Although there are easy recipes for finding the elements of these transmission matrices (see, for example, the Special Topic 4 of Chap. 3 which tells us how to calculate reflection and transmission amplitudes), they are seldom used in practice. The reason is that it is actually very inconvenient to deal with transmission matrices because of the problem posed by evanescent states. These states have imaginary or complex wavevectors and therefore elements of the transmission matrices representing free propagation of an evanescent wave within any region will have the form $e^{\pm \kappa x}$. Consequently, when transmission matrices are cascaded by multiplication, these terms may build up like $e^{\pm \kappa x_1} \times e^{\pm \kappa x_2} \times e^{\pm \kappa x_3} \ldots$, which will quickly blow up for positive exponents and cause numerical instability. In order to mitigate this problem, one usually deals with *scattering matrices* instead of *transmission matrices*.The transmission and scattering matrices are completely equivalent. There is a one-to-one correspondence between them and they contain the same information, so that it does not matter which one we deal with. The elements of a scattering matrix can be related to the elements of the corresponding transmission matrix and vice versa, so that if we can find one these matrices, we can immediately deduce the other.

The advantage of scattering matrices over transmission matrices is that if we use a particular type of scattering matrix known as current scattering matrix, then the subblock of this matrix that deals only with propagating states will always be unitary. Because of this property, which we prove later, these matrices are numerically better behaved than transmission matrices. In other words, they show far less tendency to blow up when we cascade them. This is one reason why one prefers scattering matrices over transmission matrices. Another reason is that the transmission matrix makes perfect sense for a 2-port network where we relate the amplitudes at one port to those at the other. It makes less sense if we have an N-port network ($N > 2$), where we have to inter-relate amplitudes at more than two ports. A scattering matrix, on the other hand, makes perfect sense in this scenario since it relates outgoing amplitudes at every port, regardless of how many we have, with incoming waves at these ports.

8.2.3 The Scattering Matrix

The relation between the transmission and the scattering matrices for a 2-port network can be understood by referring to Fig. 8.5 which shows a device with incident, reflected and transmitted waves at both edges. At the left edge, the amplitudes of the incident and reflected waves are **a** and **b**, whereas at the right edge, the amplitudes of the incident and reflected waves are **c** and **d**. The *forward* transmission matrix relates the amplitudes at the right edge to those at the left edge, while the *reverse* transmission matrix will relate the amplitudes at the left edge to those at the right edge. The scattering matrix, on the other hand, relates the reflected

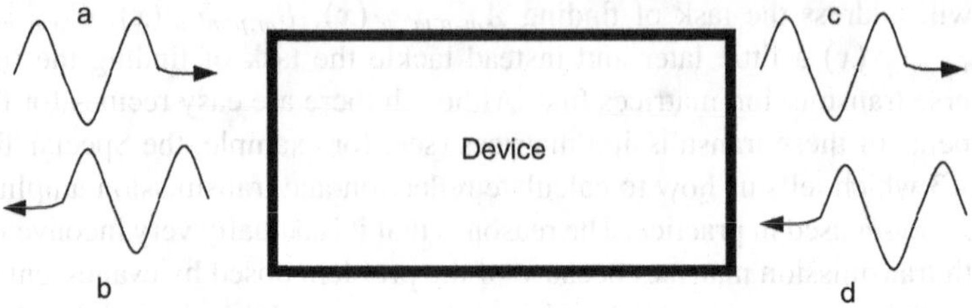

Fig. 8.5 The incoming and outgoing waves at the *left* and *right* edges of a device

amplitudes (traveling away from the device at either edge) to the incident amplitudes (traveling towards the device). In other words, the forward transmission matrix is the square matrix in

$$\begin{bmatrix} \{c\} \\ \{d\} \end{bmatrix} = \begin{bmatrix} \mathbf{t}_{11} & \mathbf{t}_{12} \\ \mathbf{t}_{21} & \mathbf{t}_{22} \end{bmatrix} \begin{bmatrix} \{a\} \\ \{b\} \end{bmatrix}, \tag{8.23}$$

the reverse transmission matrix is the square matrix in

$$\begin{bmatrix} \{b\} \\ \{a\} \end{bmatrix} = \begin{bmatrix} \mathbf{t'}_{11} & \mathbf{t'}_{12} \\ \mathbf{t'}_{21} & \mathbf{t'}_{22} \end{bmatrix} \begin{bmatrix} \{d\} \\ \{c\} \end{bmatrix}, \tag{8.24}$$

while the scattering matrix is the square matrix in

$$\begin{bmatrix} \{b\} \\ \{c\} \end{bmatrix} = \begin{bmatrix} \mathbf{s}_{11} & \mathbf{s}_{12} \\ \mathbf{s}_{21} & \mathbf{s}_{22} \end{bmatrix} \begin{bmatrix} \{a\} \\ \{d\} \end{bmatrix}. \tag{8.25}$$

It is obvious that there is a one-to-one correspondence between either the forward or the reverse transmission matrix and the scattering matrix. The elements of either transmission matrix can be related to the elements of the scattering matrix via the relations given in (8.26) and (8.28) in Special Topic 1 below.

Special Topic 1: Relation between the elements of the transmission matrices and those of the scattering matrix

The elements of the scattering matrix and the forward transmission matrix are related by the following relations:

$$\begin{aligned} \mathbf{t}_{11} &= \mathbf{s}_{21} - \mathbf{s}_{22}\mathbf{s}_{12}^{-1}\mathbf{s}_{11} \\ \mathbf{t}_{12} &= \mathbf{s}_{22}\mathbf{s}_{12}^{-1} \\ \mathbf{t}_{21} &= -\mathbf{s}_{12}^{-1}\mathbf{s}_{11} \\ \mathbf{t}_{22} &= \mathbf{s}_{12}^{-1}. \end{aligned} \tag{8.26}$$

$$
\begin{aligned}
\mathbf{s}_{11} &= -\mathbf{t}_{22}^{-1}\mathbf{t}_{21} \\
\mathbf{s}_{12} &= \mathbf{t}_{22}^{-1} \\
\mathbf{s}_{21} &= \mathbf{t}_{11} - \mathbf{t}_{12}\mathbf{t}_{22}^{-1}\mathbf{t}_{21} \\
\mathbf{s}_{22} &= \mathbf{t}_{12}\mathbf{t}_{22}^{-1}.
\end{aligned} \tag{8.27}
$$

The elements of the scattering matrix and the reverse transmission matrix are related by the following relations:

$$
\begin{aligned}
\mathbf{t}'_{11} &= \mathbf{s}_{12} - \mathbf{s}_{11}\mathbf{s}_{21}^{-1}\mathbf{s}_{22} \\
\mathbf{t}'_{12} &= \mathbf{s}_{11}\mathbf{s}_{21}^{-1} \\
\mathbf{t}'_{21} &= -\mathbf{s}_{21}^{-1}\mathbf{s}_{22} \\
\mathbf{t}'_{22} &= \mathbf{s}_{21}^{-1}.
\end{aligned} \tag{8.28}
$$

These relations are actually quite easy to prove (see Problem 8.2).

Special Topic 2: Unitarity and reciprocity of the current-scattering-matrix

8.2.3.1 Unitarity

We have defined the scattering matrix $[S]$ in terms of the incoming and outgoing wave amplitudes in different modes. We could adopt a slightly different convention and define a so-called "current-scattering-matrix" $[S']$ in terms of the current amplitudes in different modes. For this purpose, we have to assume that every mode carries current so that each one is a propagating mode and none is an evanescent mode.

Since the magnitude of the current associated with a mode is proportional to the velocity v and the squared amplitude of the wavefunction, it is clear that

$$
s'_{mn} = \sqrt{v_m/v_n} s_{mn}, \tag{8.29}
$$

where v_i is the velocity in the i-th mode.

Let us label the current amplitudes of the outgoing waves in Fig. 8.5 as b_l ($l = 1, 2, \ldots L$; and L is the total number of modes impinging from the right and left added together). Similarly, let us label the current amplitudes of all the incoming waves as a_l. The current in any mode is proportional to the squared current amplitude of that mode, so that the total current entering a device is proportional to $\sum_{l=1}^{L} |a_l|^2$, and the total current exiting the device is proportional to $\sum_{l=1}^{L} |b_l|^2$. In unipolar transport, there is no recombination or generation within the device; hence, the continuity equation of Chap. 1 (or simply current conservation) will dictate that the entering current must equal the exiting current, or

$$\sum_{l=1}^{L} |a_l|^2 = \sum_{l=1}^{L} |b_l|^2. \tag{8.30}$$

We can write the above equality in a matrix form as

$$\{a\}^\dagger\{a\} = \{b\}^\dagger\{b\}, \tag{8.31}$$

where $\{c\}$ is an arbitrary column vector containing the corresponding current mode amplitudes, and $\{c\}^\dagger$ is the hermitian adjoint of $\{c\}$, which is a row vector containing the complex conjugates of the current mode amplitudes.

Since $\{b\} = [S']\{a\}$,

$$\{a\}^\dagger\{a\} = ([S']\{a\})^\dagger[S']\{a\} = \{a\}^\dagger[S']^\dagger[S']\{a\}. \tag{8.32}$$

Therefore,

$$[S']^\dagger[S'] = [I], \tag{8.33}$$

where $[I]$ is the identity matrix. This proves that the current-scattering-matrix is unitary. The normal scattering matrix $[S]$, defined in terms of wave amplitudes rather than current amplitudes, however, is *not* unitary.

There is one last word on this. In proving the unitarity of the current scattering matrix, we assumed that all modes carry current, which is clearly untrue if one or more of them are evanescent modes. What happens if evanescent modes are included in the current scattering matrix? Clearly, the whole matrix will no longer be unitary, but the subblock that deals with the propagating modes will still remain unitary.

8.2.3.2 Reciprocity

The time-independent Schrödinger equation in a magnetic field is

$$\left[\frac{\left(-i\hbar\vec{\nabla} - e\vec{A}(\vec{r})\right)^2}{2m^*} + V(\vec{r})\right]\Psi(\vec{r}) = E\Psi(\vec{r}). \tag{8.34}$$

If we take the complex conjugate of the above equation, we obtain

$$\left[\frac{\left(i\hbar\vec{\nabla} - e\vec{A}(\vec{r})\right)^2}{2m^*} + V(\vec{r})\right]\Psi^*(\vec{r}) = E\Psi^*(\vec{r}), \tag{8.35}$$

where $\vec{A}(\vec{r})$ and $V(\vec{r})$ are assumed to be real, which they must be since otherwise the Hamiltonian would not have been hermitian.

Next, if we reverse the magnetic field $\vec{B}(\vec{r})$ so that $\vec{A}(\vec{r})$ changes sign, then the last equation can be written as

$$\left[\frac{\left(-i\hbar\vec{\nabla} - e\vec{A}(\vec{r})\right)^2}{2m^*} + V(\vec{r})\right]\Psi^*(\vec{r}) = E\Psi^*(\vec{r}), \tag{8.36}$$

Comparing (8.34) and (8.36), we see that $\Psi(\vec{r})\big|_{+B}$ and $\Psi^*(\vec{r})\big|_{-B}$ satisfy the same exact equation. Therefore, these two solutions must be equal, so that

$$\Psi(\vec{r})|_{+B} = \Psi^*(\vec{r})|_{-B} \tag{8.37}$$

In other words, the wavefunctions for oppositely directed magnetic fields are merely complex conjugates of each other. These wavefunctions are time-reversed pairs. If we know the wavefunction in a magnetic flux density B, then we can find the wavefunction in a magnetic flux density—B by simply taking its complex conjugate.

Now, since $\{b\} = [S']_{+B}\{a\}$, we can take its complex conjugate and write

$$\{b^*\} = [S']^*_{+B}\{a^*\}. \tag{8.38}$$

Whenever we take complex conjugate of a wavefunction, we turn an incoming plane wave $e^{i\vec{k}\cdot\vec{r}}$ to $e^{-i\vec{k}\cdot\vec{r}}$ which is an outgoing wave since it is traveling in the opposite direction. Since reversing a magnetic field changes a wavefunction to its complex conjugate, or equivalently changes an incoming wave to an outgoing one and vice versa, the preceding equation tells us that

$$\{a^*\} = [S']_{-B}\{b^*\}, \tag{8.39}$$

which, upon inverting, yields

$$\{b^*\} = [S']^{-1}_{-B}\{a^*\}. \tag{8.40}$$

Comparing (8.38) and (8.40), we get

$$[S']^*_{+B} = [S']^{-1}_{-B}. \tag{8.41}$$

But, the matrix $[S']_{-B}$ is unitary, which means that its inverse is its hermitian adjoint. Therefore, $[S']^{-1}_{-B} = [S']^{\dagger}_{-B}$, which means that we can write the previous equation as

$$[S']^*_{+B} = [S']^{\dagger}_{-B}. \tag{8.42}$$

Next, we take the complex conjugate of both sides and get

$$[S']_{+B} = [S']^T_{-B}, \tag{8.43}$$

where the superscript T stands for "transpose."

The last equation tells us that if we reverse the magnetic field, then we must interchange the rows with columns of the current-scattering-matrix to get the new current-scattering-matrix. If there is no magnetic field present, then

$$[S'] = [S']^T, \tag{8.44}$$

which means that the current-scattering-matrix is symmetric. Once again, note that the normal scattering matrix is not necessarily symmetric, since in deriving symmetricity, we had to assume unitarity, which the normal scattering matrix does not possess. Moreover, evanescent modes will make the current scattering matrix nonunitary, so that if we include evanescent modes, then we can no longer guarantee that the current scattering matrix will remain completely symmetric.

The "reciprocity" implied by the last two equations works only at low bias conditions since in deriving it, we have tacitly assumed that reversing the magnetic field does not affect the potential $V(\vec{r})$. Obviously, that will not be true if we take Hall effect into account, since the Hall voltage reverses when we reverse magnetic field. But the Hall voltage is proportional to the current flowing through the sample. Therefore, it will be negligible if the current is very small, meaning that we have a tiny bias across the sample. In that case, we can ignore the Hall voltage and assume that the potential $V(\vec{r})$ is invariant under reversal of the magnetic field.

8.2.3.3 Cascading Scattering Matrices

The rule for cascading normal scattering matrices $[S]$ and current scattering matrices $[S']$ is the same, but it is not so simple as in the case of cascading transmission matrices $[T]$. The latter are cascaded by simply multiplying the transmission matrices of individual sections in the proper order, but the rule for cascading scattering matrices is very different.

If $[S_{01}]$ is the scattering matrix describing the region between x_0 and x_1, and $[S_{12}]$ is the scattering matrix describing the region between x_1 and x_2, then the composite scattering matrix describing the region between x_0 and x_2 is given by [16]:

$$[S_{02}] = [S_{01}] \bigodot [S_{12}]. \tag{8.45}$$

where $\bigodot$ does not represent simple product. Instead, the elements of $[S_{02}]$, which are themselves square matrices, will be given by [16]

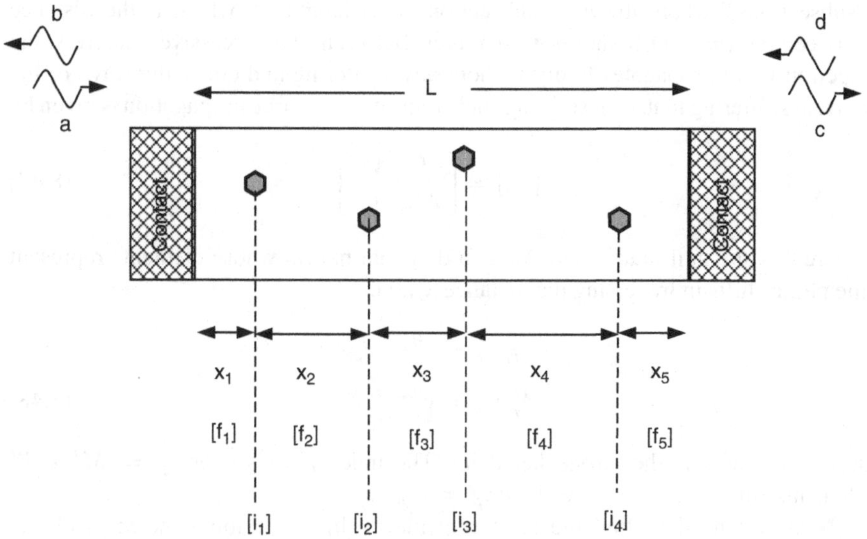

Fig. 8.6 A quasi one-dimensional resistor viewed from the *top*. The hexagons represent point elastic scatterers with delta function scattering potentials. An electron wave propagates freely between any two consecutive scatterers and the scattering matrix representing such a free-propagation section is designated as $[f_m]$. The scattering matrix for the n-th point scatterer is designated as $[i_n]$. The scattering matrix for the entire device is found by cascading the scattering matrices of all sections—free propagation sections and scattering sections—successively. Adapted with permission from [16]. Copyrighted by the American Physical Society

$$\begin{aligned}
[\mathbf{s}_{11}]_{02} &= [\mathbf{s}_{11}]_{01} + [\mathbf{s}_{12}]_{01}[\mathbf{s}_{11}]_{12}[\mathbf{I} - [\mathbf{s}_{22}]_{01}[\mathbf{s}_{11}]_{12}]^{-1}[\mathbf{s}_{21}]_{01} \\
[\mathbf{s}_{12}]_{02} &= [\mathbf{s}_{12}]_{01}\{\mathbf{I} + [\mathbf{s}_{11}]_{12}[\mathbf{I} - [\mathbf{s}_{22}]_{01}[\mathbf{s}_{11}]_{12}]^{-1}[\mathbf{s}_{22}]_{01}\}[\mathbf{s}_{12}]_{12} \\
[\mathbf{s}_{21}]_{02} &= [\mathbf{s}_{21}]_{12}[\mathbf{I} - [\mathbf{s}_{22}]_{01}[\mathbf{s}_{11}]_{12}]^{-1}[\mathbf{s}_{21}]_{01} \\
[\mathbf{s}_{22}]_{02} &= [\mathbf{s}_{22}]_{01} + [\mathbf{s}_{21}]_{12}[\mathbf{I} - [\mathbf{s}_{22}]_{01}[\mathbf{s}_{11}]_{12}]^{-1}[\mathbf{s}_{22}]_{01}[\mathbf{s}_{12}]_{12},
\end{aligned} \tag{8.46}$$

where **I** is the identity matrix. Clearly, cascading scattering matrices is a more difficult task than cascading transmission matrices. Nevertheless, scattering matrices are preferred over transmission matrices since they show less tendency to diverge upon cascading when evanescent modes are included in the calculation.

8.2.4 Finding the Composite Current Scattering Matrix of a Disordered Region

Consider now a device as shown in Fig. 8.6. It is disordered and has elastic scatterers randomly distributed within the structure. In order to find the scattering matrix describing any arbitrary section of this device, we will break up that section into "subsections," where the m-th subsection has a length x_m which is the distance between the $(m-1)$-th and m-th scatterer. Between two successive scatterers, the electron wave propagates freely without any scattering and either the wave or the current scattering matrix describing such a subsection of free propagation is given by

$$[f_m] = \begin{bmatrix} \mathbf{0} & \mathbf{\Lambda}_m \\ \mathbf{\Lambda'}_m & \mathbf{0} \end{bmatrix} \tag{8.47}$$

where $\mathbf{0}$ is the null matrix and $\mathbf{\Lambda}_m$ is a diagonal matrix whose elements represent the phase shifts in traversing the distance x_m, i.e.,

$$\begin{aligned} (\Lambda_m)_{\alpha\beta} &= \mathrm{e}^{\mathrm{i}k_\beta x_m}\delta_{\alpha\beta} \\ (\Lambda'_m)_{\alpha\beta} &= \mathrm{e}^{\mathrm{i}k_{-\beta} x_m}\delta_{\alpha\beta} \end{aligned} \tag{8.48}$$

where the delta is the Kronecker delta. The index β runs over $Q = M' \times P'$ elements since $k_\beta = k_{m',p'}$, while $k_{-\beta} = k_{-m',-p'}$.

Next, we need to find the scattering matrix $[i_m]$ describing the m-th elastic scatterer. Once we have done that, we can find the scattering matrix describing an entire section by cascading the scattering matrices of the successive subsections:

$$[S'] = [f_1] \otimes [i_1] \otimes [f_2] \otimes [i_2] \cdots \otimes [f_L], \tag{8.49}$$

if there are L free propagating sections and $L-1$ scatterers along the section's length. The entire device can constitute one section, so that we can easily find the scattering matrix of the entire device using this approach. Note that the scattering matrix has a size $2Q \times 2Q$.

8.2.5 The Scattering Matrix for an Elastic Scatterer

Assume that an elastic scatterer is located at $(0, y_0, z_0)$. Then the wavefunction of an electron of energy E in subband (i, j) impinging from the left side on the scatterer can be written as

$$\psi^l_{E,i,j}(x,y,z) = \mathrm{e}^{\mathrm{i}k_{i',j'}x}\phi_{i',j'}(y,z) + \sum_{i'=1}^{I'}\sum_{j'=1}^{J'} r_{i,j;i',j'}\mathrm{e}^{\mathrm{i}k_{-i',-j'}x}\phi_{-i',-j'}(y,z), \quad x \le 0$$

$$\psi^r_{E,i,i}(x,y,z) = \sum_{i'=1}^{I'}\sum_{j'=1}^{J'} t_{i,j;i',j'}\mathrm{e}^{\mathrm{i}k_{i',j'}x}\phi_{i',j'}(y,z), \quad x \ge 0. \tag{8.50}$$

Here, $r_{i,j;i',j'}$ and $t_{i,j;i',j'}$ denote the reflection and transmission amplitudes of an electron being reflected or transmitted from the (i, j) subband to the (i', j') subband within the device by the scatterer. Similarly, the wavefunction of an electron with energy E in subband (i, j) impinging from the right side of the scatterer can be written as

$$\chi^r_{E,i,j}(x,y,z) = e^{ik_{-i',-j'}x}\phi_{-i',-j'}(y,z) + \sum_{i'=1}^{I'}\sum_{j'=1}^{J'} r'_{i,j;i',j'} e^{ik_{i',j'}x}\phi_{i',j'}(y,z), \quad x \geq 0$$

$$\chi^l_{E,i,j}(x,y,z) = \sum_{i'=1}^{I'}\sum_{j'=1}^{J'} t'_{i,j;i',j'} e^{ik_{-i',-j'}x}\phi_{-i',-j'}(y,z), \quad x \leq 0. \tag{8.51}$$

The scattering matrix across the N-th scatterer is defined as

$$[i_N] = \begin{bmatrix} \mathbf{r}_N & \mathbf{t}'_N \\ \mathbf{t}_N & \mathbf{r}'_N \end{bmatrix} \tag{8.52}$$

where the individual elements of the matrix are themselves $Q \times Q$ matrices ($Q = I' \times J' = M' \times P'$), where Q is also the number of modes (propagating + evanescent) that we consider.

We will view the scatterers as *delta scatterers* or point scatterers of strength γ, i.e., the scattering potential due to the N-th scatterer located at $(0, y_0, z_0)$ is $\gamma_N\delta(x)\delta(y - y_0)\delta(z - z_0)$. In that case, if we enforce continuity of the wavefunction and its first derivative across a scatterer for waves impinging from the left, we will get

$$\psi^r_{E,i,j}(0,y,z) = \psi^l_{E,i,j}(0,y,z)$$

$$\frac{\partial}{\partial x}\psi^r_{E,i,j}(0,y,z) - \frac{\partial}{\partial x}\psi^l_{E,i,j}(0,y,z) = \frac{2m^*}{\hbar^2}\gamma_N\delta(y - y_0)\delta(z - z_0)\psi_{E,i,j}(0,y,z). \tag{8.53}$$

We can repeat the procedure for waves impinging from the right and we will get

$$\chi^r_{E,i,j}(0,y,z) = \chi^l_{E,i,j}(0,y,z)$$

$$\frac{\partial}{\partial x}\chi^r_{E,i,j}(0,y,z) - \frac{\partial}{\partial x}\chi^l_{E,i,j}(0,y,z) = \frac{2m^*}{\hbar^2}\gamma\delta(y - y_0)\delta(z - z_0)\chi_{E,i,j}(0,y,z). \tag{8.54}$$

The last two equations allow one to compute the elements of the scattering matrix across an elastic delta scatterer. This was done in [16] which showed that the elements of the current scattering matrix $[i_N]$ for the N-th scatterer located at $(0, y_0, z_0)$ will be given by

$$\mathbf{t}_N = \left[I + ia^N_{++}\right]^{-1}$$

$$\mathbf{r}'_N = -\left[I + ia^N_{+-}\right]^{-1}\left(ia^N_{+-}\right)$$

$$\mathbf{r}_N = -\left[I + ia_{-+}^N\right]^{-1}\left(ia_{-+}^N\right)$$
$$\mathbf{t}'_N = \left[I + ia_{--}^N\right]^{-1} \tag{8.55}$$

where

$$\left[a_{++}^N\right]_{\alpha\beta} = \frac{m^*\gamma_N\phi_{i,j}^*(y_0,z_0)\phi_{i',j'}^*(y_0,z_0)}{\hbar^2\sqrt{k_{i,j}k_{i',j'}}} = \frac{m^*\gamma_N\phi_\alpha^*(y_0,z_0)\phi_\beta^*(y_0,z_0)}{\hbar^2\sqrt{k_\alpha k_\beta}}$$

$$\left[a_{+-}^N\right]_{\alpha\beta} = \frac{m^*\gamma_N\phi_{-i,-j}^*(y_0,z_0)\phi_{i',j'}^*(y_0,z_0)}{\hbar^2\sqrt{k_{i,j}k_{-i',-j'}}} = \frac{m^*\gamma_N\phi_{-\alpha}^*(y_0,z_0)\phi_\beta^*(y_0,z_0)}{\hbar^2\sqrt{k_\alpha k_{-\beta}}}$$

$$\left[a_{-+}^N\right]_{\alpha\beta} = \frac{m^*\gamma_N\phi_{i,j}^*(y_0,z_0)\phi_{-i',-j'}^*(y_0,z_0)}{\hbar^2\sqrt{k_{-i,-j}k_{i',j'}}} = \frac{m^*\gamma_N\phi_\alpha^*(y_0,z_0)\phi_{-\beta}^*(y_0,z_0)}{\hbar^2\sqrt{k_{-\alpha}k_\beta}}$$

$$\left[a_{--}^N\right]_{\alpha\beta} = \frac{m^*\gamma_N\phi_{-i,-j}^*(y_0,z_0)\phi_{-i',-j'}^*(y_0,z_0)}{\hbar^2\sqrt{k_{-i,-j}k_{-i',-j'}}} = \frac{m^*\gamma_N\phi_{-\alpha}^*(y_0,z_0)\phi_{-\beta}^*(y_0,z_0)}{\hbar^2\sqrt{k_{-\alpha}k_{-\beta}}}, \tag{8.56}$$

where α is a composite index representing two indices i and j, i.e., $\alpha = 1$ corresponds to $i = 1$ and $j = 1$, $\alpha = 2$ corresponds to $i = 1$ and $j = 2$, $\alpha = 3$ corresponds to $i = 1$ and $j = 3$, $\alpha = J$ corresponds to $i = 1$ and $j = J$, $\alpha = J + 1$ corresponds to $i = 2$ and $j = 1$, and so on. Similarly, β is a composite index representing two indices i' and j'. Obviously, $\alpha = 1, 2, 3, \ldots .Q$ and $\beta = 1, 2, 3, \ldots .Q$, where $Q = I \times J = M' \times P'$.

The current scattering matrix described here is valid for point scatterers only. Scattering matrices for extended scatterers in a magnetic field has been studied by Uryu and Ando [17].

Once we have found the current scattering matrix across an elastic point-scatterer using the technique outlined, we can use (8.49) along with (8.47) and (8.48) to determine the composite current scattering matrix for any arbitrary section of the device. We can then convert the current scattering matrix to wave scattering matrix using (8.29) and subsequently deduce that section's forward and reverse transmission matrices using the relations given in (8.26) and (8.28). This is one of the two tasks that we have to complete in order to find the wavefunction due to either a left-incident or a right-incident electron impinging from a particular subband in either contact with a given energy. The other task was to determine the coefficients $A_{m,p;m',p'}(x)$, $B_{m,p;m',p'}(x)$, $C_{m,p;m',p'}(x)$, and $D_{m,p;m',p'}(x)$ at some arbitrary location within the device. We address this matter next.

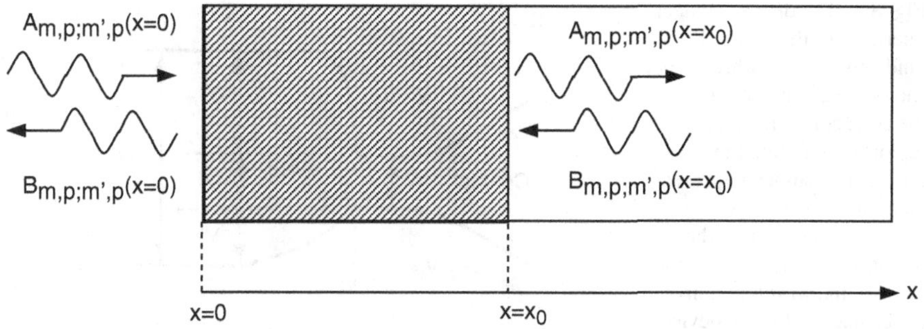

Fig. 8.7 Determining the wave amplitudes at any arbitrary location $x = x_0$ due to an electron incident from the left contact located at $x = 0$

8.2.6 Finding the Coefficients $A_{m,p;m',p'}(x)$, $B_{m,p;m',p'}(x)$, $C_{m,p;m',p'}(x)$, and $D_{m,p;m',p'}(x)$ at Some Location Within the Device

In order to determine $A_{m,p;m',p'}(x)$ and $B_{m,p;m',p'}(x)$ at an arbitrary location x_0 in the device, we adopt the following approach: Choose a section of length x_0 whose left edge coincides with the left edge of the device itself at $x = 0$. Since we can determine the forward transmission matrix for the section, we can determine $A_{m,p;m',p'}(x = x_0)$ and $B_{m,p;m',p'}(x = x_0)$ as soon as we know the incident and reflected wave amplitudes at the left edge of the device. These amplitudes are nothing but the quantities $A_{m,p;m',p'}(x = 0)$ and $B_{m,p;m',p'}(x = 0)$ evaluated at the left edge. Therefore, all we have to do is find $A_{m,p;m',p'}(x = 0)$ and $B_{m,p;m',p'}(x = 0)$. This concept is illustrated in Fig. 8.7.

We can repeat this process for an electron impinging from the right contact to find $C_{m,p;m',p'}(x = x_0)$ and $D_{m,p;m',p'}(x = x_0)$. In this case, we choose a section of length $L - x_0$ and make its right edge coincide with the right edge of the device. Since we can find the reverse transmission matrix of this section, we can find the wave amplitudes at the left edge of the section, which are the quantities $C_{m,p;m',p'}(x = x_0)$ and $D_{m,p;m',p'}(x = x_0)$ once we know the incident and reflected wave amplitudes at the right edge at $x = L$. The latter are $C_{m,p;m',p'}(x = L)$ and $D_{m,p;m',p'}(x = L)$. Therefore, what we really need in the end are just four quantities: $A_{m,p;m',p'}(0)$, $B_{m,p;m',p'}(0)$, $C_{m,p;m',p'}(L)$, and $D_{m,p;m',p'}(L)$.

In order to determine $A_{m,p;m',p'}(x = 0)$ and $B_{m,p;m',p'}(x = 0)$, consider the interface of the device with the left lead as shown in Fig. 8.8.

In the domain [Y', Z'], the wavefunction in the lead is zero since the boundaries of the lead are assumed to present infinite barriers for the electron. Only in the domain [Y, Z], the lead wavefunction is nonzero.

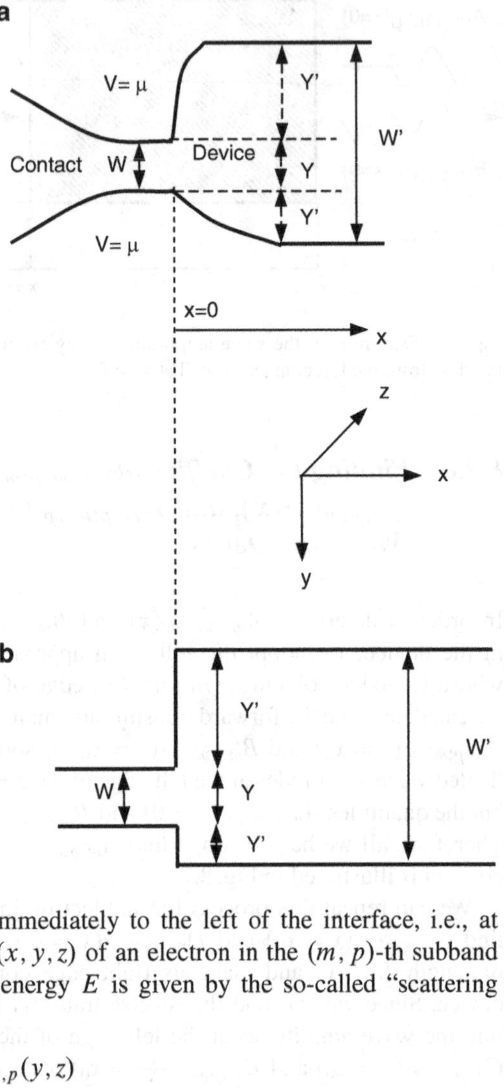

Fig. 8.8 The device–contact interface. In the region Y' (and similarly Z' which is not shown), the wavefunction in the contact is zero since the region surrounding the contact is assumed to present an infinite barrier to the electron ($V = \infty$). In the region Y and similarly Z (not shown), the wavefunctions in the contact and in the device, as well as their first derivatives, are continuous at the device–contact interface located at $x = 0$. (**a**) An interface with arbitrary shape, and (**b**) an abrupt interface

Consider the domain [Y, Z]. Immediately to the left of the interface, i.e., at $x = 0-$, the wavefunction $\varphi^{\mathrm{L}}_{\mathrm{E},m,p}(x, y, z)$ of an electron in the (m, p)-th subband in the left lead (or contact) with energy E is given by the so-called "scattering state" [15]:

$$\varphi^{\mathrm{L}}_{\mathrm{E},m,p}(x = 0-, y, z) = \mathrm{e}^{\mathrm{i}\overline{k}_{m,p}x}\xi_{m,p}(y, z) + \sum_{\mu=1}^{M}\sum_{\nu=1}^{P} R_{m,p;\mu,\nu}\mathrm{e}^{\mathrm{i}\overline{k}_{-\mu,-\nu}x}\xi_{-\mu,-\nu}(y, z)[y \in Y, z \in Z], \tag{8.57}$$

where $\overline{k}_{\mu,\nu}$ is the wavevector and $\xi_{\mu,\nu}(y, z)$ is the transverse component of the wavefunction of a state in the (μ, ν)-th subband in the left lead traveling from left to right. Obviously, $\xi_{-\mu,-\nu}(y, z)$ is the wavefunction, $\overline{k}_{-\mu,-\nu}x$ is the wavevector of the oppositely traveling state in the same subband of the lead, and $R_{m,p;\mu,\nu}$ is the reflection coefficient (amplitude) for an electron incident from the (m, p)-th subband in the left lead to be reflected back into the (μ, ν)-th subband in the same lead.

Immediately to the right of the interface, i.e., at $x = 0+$, the wavefunction of the electron transmitted into the device is given by the scattering state [15]:

$$\psi^{\mathrm{L}}_{\mathrm{E},m,p}(x = 0+, y, z) = \sum_{m'=1}^{M'} \sum_{p'=1}^{P'} T_{m,p;m',p'} \mathrm{e}^{\mathrm{i}k_{m',p'}x} \phi_{m',p'}(y, z), \tag{8.58}$$

where $T_{m,p;m',p'}$ is the transmission amplitude for an electron incident from the (m, p)-th subband in the left lead to be transmitted into the (m', p')-th subband in the device.

Since the wavefunction must be continuous at $x = 0$, we can deduce that

$$\psi^{\mathrm{L}}_{\mathrm{E},m,p}(x = 0, y, z) = \sum_{m'=1}^{M'} \sum_{p'=1}^{P'} T_{m,p;m',p'} \phi_{m',p'}(y, z). \tag{8.59}$$

By comparing the last equation with (8.17), we get

$$\sum_{m'=1}^{M'} \sum_{p'=1}^{P'} A_{m,p;m',p'}(x = 0)\phi_{m',p'}(y, z) + B_{m,p;m',p'}(x = 0)\phi_{-m',-p'}(y, z)$$
$$= \sum_{m'=1}^{M'} \sum_{p'=1}^{P'} T_{m,p;m',p'} \phi_{m',p'}(y, z). \tag{8.60}$$

Additionally, the first derivative of the wavefunction must be continuous at $x = 0$. This yields

$$\sum_{m'=1}^{M'} \sum_{p'=1}^{P'} k_{m',p'} A_{m,p;m',p'}(x = 0)\phi_{m',p'}(y, z) + k_{-m',-p'} B_{m,p;m',p'}(x = 0)\phi_{-m',-p'}(y, z)$$
$$= \sum_{m'=1}^{M'} \sum_{p'=1}^{P'} k_{m',p'} T_{m,p;m',p'} \phi_{m',p'}(y, z). \tag{8.61}$$

Taking the projection of (8.60) and (8.61) on the wavefunction $\phi_{k',l'}(y, z)$ in the device, we get

$$\sum_{m'=1}^{M'} \sum_{p'=1}^{P'} A_{m,p;m',p'}(x = 0)\langle \phi_{k',l'}(y, z)|\phi_{m',p'}(y, z)\rangle$$
$$+ B_{m,p;m',p'}(x = 0)\langle \phi_{k',l'}(y, z)|\phi_{-m',-p'}(y, z)\rangle$$
$$= \sum_{m'=1}^{M'} \sum_{p'=1}^{P'} T_{m,p;m',p'} \langle \phi_{k',l'}(y, z)|\phi_{m',p'}(y, z)\rangle$$

$$\sum_{m'=1}^{M'}\sum_{p'=1}^{P'}\{k_{m',p'}A_{m,p;m',p'}(x=0)\langle\phi_{k',l'}(y,z)|\phi_{m',p'}(y,z)\rangle$$

$$+\,k_{-m',-p'}B_{m,p;m',p'}(x=0)\langle\phi_{k',l'}(y,z)|\phi_{-m',-p'}(y,z)\rangle\}$$

$$=\sum_{m'=1}^{M'}\sum_{p'=1}^{P'}k_{m',p'}T_{m,p;m',p'}\langle\phi_{k',l'}(y,z)|\phi_{m',p'}(y,z)\rangle. \tag{8.62}$$

where, as always, the overlap integral

$$\langle\xi_{a,b}(y,z)|\zeta_{c,d}(y,z)\rangle=\int_{-\infty}^{\infty}\int_{-\infty}^{\infty}\mathrm{d}y\mathrm{d}z\xi^*_{a,b}(y,z)\zeta_{c,d}(y,z). \tag{8.63}$$

The ensuing algebra will appear less daunting if we switch over to composite indices $\pm\alpha\equiv(\pm m,\pm p)$, $\pm\beta\equiv(\pm m',\pm p')$ and $\pm\gamma\equiv(\pm k',\pm l')$. Each of these indices (α, β, γ) runs from 1 to Q and labels a subband; α labels a subband in the left lead, while β and γ label subbands in the device. We are considering a total of Q subbands in both the lead and the device. It does not matter that we are considering equal number of subbands in both the device and the lead since these subbands span both propagating and evanescent modes or states. Thus, if there are fewer propagating subbands in the lead than there are in the device, then we will be considering more evanescent states in the lead, and vice versa.

A little reflection should convince the reader that in any magnetic field, $\langle\phi_\beta(y,z)|\phi_\gamma(y,z)\rangle=\delta_{\beta\gamma}$ as long as β and γ have the same sign. If they have opposite signs, then in zero magnetic field, this relation still holds, but in a nonzero magnetic field, it no longer holds. In fact, in a very strong magnetic field, when oppositely traveling edge states are localized at opposite edges of the quasi one-dimensional device and lead, with no overlap between their wavefunctions, $\langle\phi_\beta(y,z)|\phi_\gamma(y,z)\rangle=0$ if β and γ have opposite signs.

Thus, in *any arbitrary magnetic field*, we get

$$\sum_{\beta=1}^{Q}\{A_{\alpha\beta}(x=0)\delta_{\beta\gamma}+B_{\alpha\beta}(x=0)\langle\phi_\gamma(y,z)|\phi_{-\beta}(y,z)\rangle\}=\sum_{\beta=1}^{Q}T_{\alpha\beta}\delta_{\beta\gamma}$$

$$\sum_{\beta=1}^{Q}\{k_\beta A_{\alpha\beta}(x=0)\delta_{\beta\gamma}+k_{-\beta}B_{\alpha\beta}(x=0)\langle\phi_\gamma(y,z)|\phi_{-\beta}(y,z)\rangle\}=\sum_{\beta=1}^{Q}k_\beta T_{\alpha\beta}\delta_{\beta\gamma}, \tag{8.64}$$

which can be recast as

$$A_{\alpha\gamma}(x=0)+\sum_{\beta=1}^{Q}B_{\alpha\beta}(x=0)\langle\phi_\gamma(y,z)|\phi_{-\beta}(y,z)\rangle=T_{\alpha\gamma}$$

$$k_\gamma A_{\alpha\gamma}(x=0) + \sum_{\beta=1}^{Q} k_{-\beta} B_{\alpha\beta}(x=0)\langle\phi_\gamma(y,z)|\phi_{-\beta}(y,z)\rangle = k_\gamma T_{\alpha\gamma}. \tag{8.65}$$

Multiplying the first line of the preceding equation throughout with k_γ and subtracting the second line from it, we get

$$\sum_{\beta=1}^{Q}(k_\gamma - k_{-\beta}) B_{\alpha\beta}(x=0)\langle\phi_\gamma(y,z)|\phi_{-\beta}(y,z)\rangle = 0. \tag{8.66}$$

Let us now define the quantity $U_{\beta\gamma} = \left(k_\gamma - k_{-\beta}\right)\langle\phi_\gamma(y,z)|\phi_{-\beta}(y,z)\rangle$. Then, we can write the previous equation as

$$\sum_{\beta=1}^{Q} B_{\alpha\beta}(x=0) U_{\beta\gamma} = 0. \tag{8.67}$$

This equation can be recast in a matrix form:

$$\begin{bmatrix} B_{11} & B_{12} & \dots\dots\dots & B_{1Q} \\ B_{12} & B_{22} & \dots\dots\dots & B_{2Q} \\ \dots & \dots & \dots\dots\dots & \dots \\ \dots & \dots & \dots\dots\dots & \dots \\ \dots & \dots & \dots\dots\dots & \dots \\ B_{Q1} & B_{Q2} & \dots\dots\dots & B_{QQ} \end{bmatrix} \begin{bmatrix} U_{11} & U_{12} & \dots\dots\dots & U_{1Q} \\ U_{12} & U_{22} & \dots\dots\dots & U_{2Q} \\ \dots & \dots & \dots\dots\dots & \dots \\ \dots & \dots & \dots\dots\dots & \dots \\ \dots & \dots & \dots\dots\dots & \dots \\ U_{Q1} & U_{Q2} & \dots\dots\dots & U_{QQ} \end{bmatrix}$$
$$= \begin{bmatrix} 0 & 0 & \dots\dots\dots & 0 \\ 0 & 0 & \dots\dots\dots & 0 \\ \dots & \dots & \dots\dots\dots & \dots \\ \dots & \dots & \dots\dots\dots & \dots \\ \dots & \dots & \dots\dots\dots & \dots \\ 0 & 0 & \dots\dots\dots & 0 \end{bmatrix}, \tag{8.68}$$

or

$$[\mathbf{B}][\mathbf{U}] = [\mathbf{0}], \tag{8.69}$$

where the elements of the $Q \times Q$ matrix [**B**] are $B_{\alpha\beta}(x=0)$, the elements of the $Q \times Q$ matrix [**U**] are $U_{\beta\gamma}$ and [**0**] is the $Q \times Q$ null matrix.

Since [**U**] is not a null matrix, we conclude that [**B**] must be null, so $B_{\alpha\beta}(x=0) = 0$ for every α, β. Using these results in (8.65), we immediately arrive at

$$\begin{aligned} A_{m,p;m',p'}(x=0) &= T_{m,p;m',p'} \\ B_{m,p;m',p'}(x=0) &= 0. \end{aligned} \tag{8.70}$$

Our job now is to find $T_{m,p;m',p'}$. For this purpose, consider the interface between the left lead and the device as shown in Fig. 8.8. We will enforce continuity of the wavefunction and its first derivative with respect to x at $x = 0$ to get

$$\begin{aligned} \varphi^{\mathrm{L}}_{\mathrm{E},m,p}(x=0,y,z) &= \psi^{\mathrm{L}}_{\mathrm{E},m,p}(x=0,y,z) \\ \left.\frac{\mathrm{d}\left[\varphi^{\mathrm{L}}_{\mathrm{E},m,p}(x,y,z)\right]}{\mathrm{d}x}\right|_{x=0} &= \left.\frac{\mathrm{d}\left[\psi^{\mathrm{L}}_{\mathrm{E},m,p}(x,y,z)\right]}{\mathrm{d}x}\right|_{x=0}, \end{aligned} \tag{8.71}$$

which yields

$$\begin{aligned} \xi_{m,p}(y,z) + \sum_{\mu=1}^{M}\sum_{\nu=1}^{P} R_{m,p,\mu,\nu}\xi_{-\mu,-\nu}(y,z) &= \sum_{m'=1}^{M'}\sum_{p'=1}^{P'} T_{m,p;m',p'}\phi_{m',p'}(y,z) \\ \overline{k}_{m,p}\xi_{m,p}(y,z) + \sum_{\mu=1}^{M}\sum_{\nu=1}^{P} \overline{k}_{-\mu,-\nu}R_{m,p,\mu,\nu}\xi_{-\mu,-\nu}(y,z) &= \sum_{m'=1}^{M'}\sum_{p'=1}^{P'} k_{m',p'}T_{m,p;m',p'}\phi_{m',p'}(y,z). \end{aligned} \tag{8.72}$$

for $y \in Y$ and $z \in Z$ as shown in Fig. 8.8. Since the wavefunctions in the contact are zero in the regions $y \in Y'$ and $z \in Z'$, the wavefunctions in the boundary of the device must vanish outside those regions which have no interface with the contact.

Solution of (8.72) is cumbersome if the junction has arbitrary geometry as shown in Fig. 8.8a, but becomes simpler if the junction is abrupt, as shown in Fig. 8.8b. In the latter case, if an abrupt interface is located at $x = 0$, we will have

$$\sum_{m'=1}^{M'}\sum_{p'=1}^{P'} T_{m,p;m',p'}\phi_{m',p'}(y,z) = 0 \tag{8.73}$$

for $y \in Y'$ and $z \in Z'$.

Equations (8.72) and (8.73) can be solved with two different approaches known as (1) the *mode-matching* method, and (2) the *real-space matching method.* We elucidate them below.

8.2.6.1 The Mode-Matching Method

We will switch to the composite indices $\pm\alpha \equiv (\pm m, \pm p)$, $\pm\beta \equiv (\pm m', \pm p')$ and $\pm\gamma \equiv (\pm\mu, \pm\nu)$, where the indices α and γ label subbands or modes in the left lead, and the index β labels subbands or modes in the device. We will use these indices to rewrite (8.72) which matches wavefunctions and their first derivatives at the device-lead junction. Because of this reason, the technique has been termed the "mode-matching method."

Using the composite indices, we recast (8.72) as

$$\xi_\alpha(y,z) + \sum_{\gamma=1}^{Q} R_{\alpha\gamma}\xi_{-\gamma}(y,z) = \sum_{\beta=1}^{Q} T_{\alpha\beta}\phi_\beta(y,z) \quad [y \in Y, z \in Z]$$

$$\overline{k}_\alpha\xi_\alpha(y,z) + \sum_{\gamma=1}^{Q} \overline{k}_{-\gamma}R_{\alpha\gamma}\xi_{-\gamma}(y,z) = \sum_{\beta=1}^{Q} k_\beta T_{\alpha\beta}\phi_\beta(y,z) \quad [y \in Y, z \in Z]$$

$$0 = \sum_{\beta=1}^{Q} T_{\alpha\beta}\phi_\beta(y,z) \quad [y \in Y', z \in Z']. \tag{8.74}$$

Next, if we take projections on the wavefunction $\xi_{-\delta}(y,z)$, then we get that

$$\langle\xi_{-\delta}(y,z)|\xi_\alpha(y,z)\rangle + \sum_{\gamma=1}^{Q} R_{\alpha\gamma}\langle\xi_{-\delta}(y,z)|\xi_{-\gamma}(y,z)\rangle = \sum_{\beta=1}^{Q} T_{\alpha\beta}\langle\xi_{-\delta}(y,z)|\phi_\beta(y,z)\rangle \quad [y \in Y, z \in Z]. \tag{8.75}$$

$$\overline{k}_\alpha\langle\xi_{-\delta}(y,z)|\xi_\alpha(y,z)\rangle + \sum_{\gamma=1}^{Q} R_{\alpha\gamma}\overline{k}_{-\gamma}\langle\xi_{-\delta}(y,z)|\xi_{-\gamma}(y,z)\rangle = \sum_{\beta=1}^{Q} T_{\alpha\beta}k_\beta\langle\xi_{-\delta}(y,z)|\phi_\beta(y,z)\rangle \quad [y \in Y, z \in Z]. \tag{8.76}$$

Note that in any arbitrary magnetic field, $\langle\xi_{-\delta}(y,z)|\xi_{-\gamma}(y,z)\rangle = \delta_{\gamma\delta}$ if, *and only if*, the range of y and z extended from $-\infty$ to $+\infty$ However, we are unfortunately constrained to the domain [Y, Z] and hence this relation does not hold.

The last two equations can be cast in a matrix form

$$\begin{bmatrix} R\ T \end{bmatrix}\begin{bmatrix} \mathbf{A}\ \mathbf{B} \\ \mathbf{C}\ \mathbf{D} \end{bmatrix} = \begin{bmatrix} \mathbf{E}\ \mathbf{F} \end{bmatrix}, \tag{8.77}$$

where **A**, **B**, **C**, **D**, **E** and **F** are $Q \times Q$ matrices given by (in the interval $[y \in Y, z \in Z]$)

$$\begin{aligned} A_{mn} &= \langle\xi_{-n}(y,z)|\xi_{-m}(y,z)\rangle \\ B_{mn} &= \langle\xi_{-n}(y,z)|\phi_m(y,z)\rangle \\ C_{mn} &= \overline{k}_{-m}\langle\xi_{-n}(y,z)|\xi_{-m}(y,z)\rangle \\ D_{mn} &= k_m\langle\xi_{-n}(y,z)|\phi_m(y,z)\rangle \end{aligned}$$

$$
\begin{aligned}
E_{mn} &= \langle \xi_{-n}(y,z)|\xi_m(y,z)\rangle \\
F_{m,n} &= \overline{k}_m \langle \xi_{-n}(y,z)|\xi_m(y,z)\rangle \\
T_{mn} &= T_{\alpha\beta} \\
R_{mn} &= R_{\alpha\gamma}.
\end{aligned}
\tag{8.78}
$$

Note that in the absence of any magnetic field, $\mathbf{A} = \mathbf{E}$ and $\mathbf{C} = -\mathbf{F}$.

For an electron impinging from the right with energy E in the (m, p)-th subband in the right contact, we can write an equation similar to (8.77). Thereafter, we can combine them to yield a composite equation of the form

$$
\begin{bmatrix} R & T \\ T' & R' \end{bmatrix} \begin{bmatrix} \mathbf{A} & \mathbf{B} \\ \mathbf{C} & \mathbf{D} \end{bmatrix} = \begin{bmatrix} \mathbf{E} & \mathbf{F} \\ \mathbf{G} & \mathbf{H} \end{bmatrix},
\tag{8.79}
$$

where

$$
\begin{aligned}
G_{mn} &= \langle \kappa_{-n}(y,z)|\kappa_m(y,z)\rangle \\
H_{m,n} &= \overline{k}_m \langle \kappa_{-n}(y,z)|\kappa_m(y,z)\rangle \\
T'_{mn} &= T'_{ij} \\
R'_{mn} &= R'_{ik},
\end{aligned}
\tag{8.80}
$$

given that $\kappa_l(y,z)$ is the transverse component of the wavefunction in the l-th mode (l is a composite index) in the right lead, T'_{ij} is the transmission amplitude from the i-th mode in the right lead into the j-th mode in the device, and $R'_{i,k}$ is the reflection amplitude from the i-th mode in the right lead back into the k-th mode in the right lead.

Solution of (8.79) will yield the transmission coefficient $T_{\alpha\beta} = T_{m,p;m',p'}$ for every α and β, as well as T'_{ij} for every i and j.

Once $T_{m,p;m',p'}$ has been found, we can find the wavefunction $\psi^{\mathrm{L}}_{\mathrm{E},m,p}(x,y,z)$ anywhere in the device due to an electron impinging with energy E from the (m, p)-th subband in the left contact. Similarly, T'_{ij} will yield the wavefunction $\psi^{R}_{E,m,p}(x,y,z)$ anywhere in the device due to an electron impinging with energy E from the (m, p)-th subband in the right contact. This will then allow us to find the carrier concentration and current density everywhere within a device using (8.19) and (8.20).

8.2.6.2 The Real Space Matching Method

The "real space-matching method" is similar in spirit to the "mode-matching method". The difference is that in the mode matching method, we write the wavefunction as a linear combination of modal or subband wavefunctions (in both the contact and the device). Then, we match the wavefunction and its first derivative at the device–contact interface mode by mode, taking advantage of the orthogonality between modes that propagate in the same direction in the same region. By contrast, in the real space-matching method, we match the wavefunction and its first derivative at discrete regions in space over the interface between the device and the contact. Sometimes, this is numerically advantageous, especially if the interface has an odd geometry.

Following reference [18, 19], we first discretize the boundary between the lead and the device into a finite number of grid points $y_1, y_2, \ldots.. y_M$ and $z_1, z_2, \ldots\ldots.. z_N$, and then define an averaging operator

$$\langle\lambda| = \langle m, n| = \frac{1}{\sqrt{y_{m+1} - y_m}} \frac{1}{\sqrt{z_{n+1} - z_n}} \int_{y_m}^{y_{m+1}} \int_{z_n}^{z_{n+1}} dy dz, \qquad (8.81)$$

where, once again, λ is a composite index spanning the two indices m and n. The values of M and N determine the number of modes that we are effectively including in the analysis.

Operating on a function of y and z with this operator yields a quantity that is proportional to the average of the function over the interval $[(y_m, z_n), (y_{m+1}, z_{n+1})]$. We operate on (8.74) with this operator to get

$$\begin{aligned} &\langle\lambda|\xi_\alpha(y,z)\rangle + \sum_{\gamma=1}^{Q} R_{\alpha\gamma}\langle\lambda|\xi_{-\gamma}(y,z)\rangle \\ &= \sum_{\beta=1}^{Q} T_{\alpha\beta}\langle\lambda|\phi_\beta(y,z)\rangle \quad [y \in Y, z \in Z]. \end{aligned} \qquad (8.82)$$

$$\begin{aligned} &\overline{k}_\alpha\langle\lambda|\xi_\alpha(y,z)\rangle + \sum_{\gamma=1}^{Q} R_{\alpha\gamma}\overline{k}_{-\gamma}\langle\lambda|\xi_{-\gamma}(y,z)\rangle \\ &= \sum_{\beta=1}^{Q} T_{\alpha\beta}k_\beta\langle\lambda|\phi_\beta(y,z)\rangle \quad [y \in Y, z \in Z] \end{aligned} \qquad (8.83)$$

$$0 = \sum_{\beta=1}^{Q} T_{\alpha\beta}\langle\lambda|\phi_\beta(y,z)\rangle \quad [y \in Y', z \in Z'] \qquad (8.84)$$

Clearly, all we have done is replace $\langle\xi_{-\delta}(y,z)|$ in (8.75) and (8.76) with $\langle\lambda|$. Therefore, in the end, we will get the same equation as (8.79), except now

$$\begin{aligned} A_{m\lambda} &= \langle\lambda|\xi_{-m}(y,z)\rangle \\ B_{m\lambda} &= \langle\lambda|\phi_m(y,z)\rangle \\ C_{m\lambda} &= \overline{k}_{-m}\langle\lambda|\xi_{-m}(y,z)\rangle \end{aligned}$$

$$
\begin{aligned}
D_{m\lambda} &= k_m \langle \lambda | \phi_m(y,z) \rangle \\
E_{m\lambda} &= \langle \lambda | \xi_m(y,z) \rangle \\
F_{m,\lambda} &= \overline{k}_m \langle \lambda | \xi_m(y,z) \rangle \\
G_{m\lambda} &= \langle \lambda | \kappa_m(y,z) \rangle \\
H_{m,\lambda} &= \overline{k}_m \langle \lambda | \kappa_m(y,z) \rangle .
\end{aligned} \tag{8.85}
$$

Solution of the modified (8.79) will yield the transmission coefficients and hence the wavefunctions due to electrons impinging from the left and right with a given energy. In turn, this will yield the carrier concentration and current density everywhere within the device.

8.3 The Importance of Evanescent Modes

In the previous section, we included *both* propagating and evanescent modes in the calculation of the wavefunctions and ultimately the carrier concentration and current density. At first sight, it may appear odd that we are including evanescent modes in the calculation of current density since such modes do not carry any current. Only the propagating modes carry current. However, the current carrying capabilities of the propagating modes, or, in other words, the transmission coefficients of the propagating modes which determine the amount of current they conduct, are *affected by the evanescent modes*. Hence, the evanescent modes have an indirect but strong influence on the current density.

This feature was beautifully expounded by Bagwell [20] who showed that the transmission coefficients of propagating modes are *renormalized* by the evanescent modes. That analysis is outlined below.

Consider a quasi one-dimensional current-carrying structure with a scatterer located somewhere in the structure. For simplicity, we will assume that there is no magnetic field.

From (8.17) and (8.18), it is clear that in the absence of any magnetic field, the wavefunction anywhere in the device can be written generally as

$$
\psi(x,y,z) = \sum_{m'=1}^{M'} \sum_{p'=1}^{P'} c_{m',p'}(x) \phi_{m',p'}(y,z). \tag{8.86}
$$

Since the wavefunction must satisfy the effective-mass Schrödinger equation

$$
-\frac{\hbar^2}{2m^*} \nabla^2 \psi(x,y,z) + V_{\text{scat}}(x,y,z) \psi(x,y,z) = E \psi(x,y,z), \tag{8.87}
$$

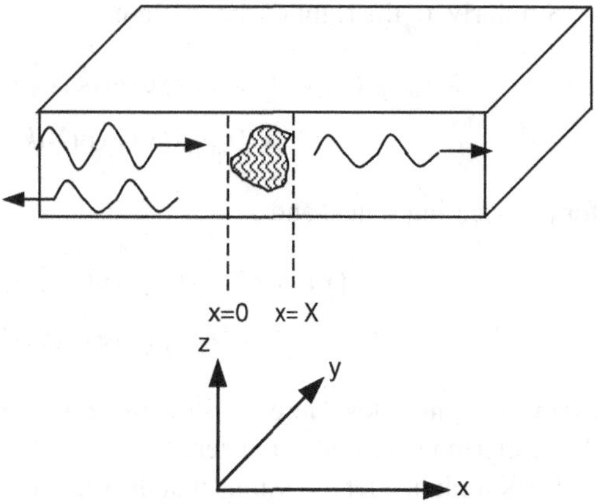

Fig. 8.9 An extended scatterer in a quasi-one-dimensional structure. Waves that are incident from the *left* are partially reflected and partially transmitted

where $V_{\text{scat}}(x, y, z)$ is the scattering potential due to the scatterer, it is easy to see that the Fourier coefficients $c_{m',p'}(x)$ will satisfy the equation

$$\frac{d^2 c_{m',p'}(x)}{dx^2} + k^2_{m',p'} c_{m',p'}(x) = \sum_{i,j} \Gamma_{m',p';i,j}(x) c_{i,j}(x), \quad (8.88)$$

where

$$E = \frac{\hbar^2 k^2_{m',p'}}{2m^*} + \epsilon_{m',p'}$$

$$\Gamma_{m',p';i,j}(x) = \frac{2m^*}{\hbar^2} \int_{-\infty}^{\infty} \int_{-\infty}^{\infty} \mathrm{d}y \mathrm{d}z \phi^*_{m',p'}(y,z) V_{\text{scat}}(x,y,z) \phi_{i,j}(y,z), \quad (8.89)$$

$\epsilon_{m',p'}$ is the energy at the bottom of the (m', p')-th subband, and E is the energy of the electron (remember that E is a good quantum number since we are considering phase-coherent, dissipationless transport).

To the left of the scatterer (see Fig. 8.9), the wavefunction will be a coherent superposition of left and right traveling waves, so that

$$c_{m',p'}(x) = A_{m',p'}(x) \exp(\mathrm{i} k_{m',p'} x) + B_{m',p'}(x) \exp(-\mathrm{i} k_{m',p'} x), \quad x < 0 \quad (8.90)$$

if the (m', p')-th mode is a propagating mode (i.e., $E \geq \epsilon_{m',p'}$, and $k_{m',p'}$ is real), while

$$c_{m',p'}(x) = A_{m',p'}(x) \exp(-\kappa_{m',p'} x) + B_{m',p'}(x) \exp(\kappa_{m',p'} x), \quad x < 0 \quad (8.91)$$

if the (m', p')-th mode is an evanescent mode (i.e., $E < \epsilon_{m',p'}$, and $k_{m',p'}$ is imaginary). We set $k_{m',p'} = \mathrm{i}\kappa_{m',p'}$ to get the last equation. Note that in the absence of any magnetic field, $k_{-m',-p'} = -k_{m',p'}$.

Similarly, to the right of the scatterer

$$c_{m',p'}(x) = C_{m',p'}(x)\exp(\mathrm{i}k_{m',p'}(x-X)) + D_{m',p'}(x)\exp(-\mathrm{i}k_{m',p'}(x-X)), \quad x > 0 \tag{8.92}$$

for propagating states, and

$$c_{m',p'}(x) = C_{m',p'}(x)\exp(-\kappa_{m',p'}(x-X)) + D_{m',p'}(x)\exp(\kappa_{m',p'}(x-X)), \quad x > 0 \tag{8.93}$$

for evanescent states. Here, X is the finite length of the scattering region (i.e., where the scattering potential is nonzero).

For simplicity, let us assume that the scatterer is a delta scatterer located at $x = 0$, $y = y_0$, and $z = z_0$. Its potential will be written as

$$V_{\mathrm{scat}}(x, y, z) = \gamma\delta(x)\delta(y - y_0)\delta(z - z_0), \tag{8.94}$$

where γ is the scattering strength and it can be either negative (attractive scatterer) or positive (repulsive scatterer).

Integrating (8.88) over the coordinate x across the delta function, we get

$$\left.\frac{\mathrm{d}c_{m',p'}(x)}{\mathrm{d}x}\right|_{x=0+} - \left.\frac{\mathrm{d}c_{m',p'}(x)}{\mathrm{d}x}\right|_{x=0-} = \sum_{i,j} \overline{\Gamma}_{m',p';i,j}c_{i,j}(0), \tag{8.95}$$

where

$$\overline{\Gamma}_{m',p';i,j} = \frac{2m^*\gamma}{\hbar^2}\phi^*_{m',p'}(y_0, z_0)\phi_{i,j}(y_0, z_0). \tag{8.96}$$

Using (8.90)–(8.93), and noting that for a delta-scatterer, $X = 0$, we get

$$\mathrm{i}k_{m',p'}[C_{m',p'}(0) - D_{m',p'}(0)] - \mathrm{i}k_{m',p'}[A_{m',p'}(0) - B_{m',p'}(0)] = \sum_{i,j} \overline{\Gamma}_{m',p';i,j}[A_{i,j}(0) - B_{i,j}(0)], \tag{8.97}$$

for a propagating mode, and

$$-\kappa_{m',p'}[C_{m',p'}(0) - D_{m',p'}(0)] + \kappa_{m',p'}[A_{m',p'}(0) - B_{m',p'}(0)] = \sum_{i,j} \overline{\Gamma}_{m',p';i,j}[A_{i,j}(0) - B_{i,j}(0)], \tag{8.98}$$

for an evanescent mode.

Continuity of the wavefunction across the delta scatterer guarantees that

$$c_{m',p'}(x = 0+) = c_{m',p'}(x = 0-), \tag{8.99}$$

or

$$A_{m',p'}(0) + B_{m',p'}(0) = C_{m',p'}(0) + D_{m',p'}(0). \tag{8.100}$$

Next, let us consider that electrons are incident from the left in the propagating mode (m', p'). In that case, the transmission amplitude into the mode (i, j) (regardless of whether it is propagating or evanescent) will, by definition, be given by

$$T_{m',p';i,j} = \frac{C_{i,j}(0)}{A_{m',p'}(0)}. \tag{8.101}$$

We will only consider the situation when electrons are incident from the left. Then $D_{m',p'}(0) = 0$ for all m', p' in the case of propagating modes. In the case of evanescent modes, we cannot let any mode to grow with distance, although we can certainly allow them to attenuate with distance. This means $A_{m',p'}(0) = 0$ and $D_{m',p'}(0) = 0$ for all m', p' in the case of evanescent modes.

Before proceeding further, let us switch back to the composite index convention so that $\alpha \equiv (m', p')$ and $\beta \equiv (i, j)$. In that case, (8.97), (8.98), (8.100), and (8.101) can be rewritten as

$$\begin{aligned} ik_\alpha[C_\alpha(0) - D_\alpha(0)] - ik_\alpha[A_\alpha(0) - B_\alpha(0)] &= \sum_\beta \overline{\Gamma}_{\alpha\beta}[A_\beta(0) - B_\beta(0)] \\ -\kappa_\alpha[C_\alpha(0) - D_\alpha(0)] + \kappa_\alpha[A_\alpha(0) - B_\alpha(0)] &= \sum_\beta \overline{\Gamma}_{\alpha\beta}[A_\beta(0) - B_\beta(0)] \\ A_\alpha(0) + B_\alpha(0) &= C_\alpha(0) + D_\alpha(0) \\ T_{\alpha\beta} &= \frac{C_\beta(0)}{A_\alpha(0)}. \end{aligned} \tag{8.102}$$

Now consider the scenario when there are two propagating modes (modes 1 and 2) and two evanescent modes (modes 3 and 4) in a structure. An electron is incident in mode 2 and gets coupled to all other modes (1, 3, and 4) owing to the scatterer. Using the last equation, we can write

$$\begin{bmatrix} 0 \\ -2ik_2 \\ 0 \\ 0 \end{bmatrix} = \begin{bmatrix} \Gamma_{11} - 2ik_1 & \Gamma_{12} & \Gamma_{13} & \Gamma_{14} \\ \Gamma_{21} & \Gamma_{22} - 2ik_2 & \Gamma_{23} & \Gamma_{24} \\ \Gamma_{31} & \Gamma_{32} & \Gamma_{33} + 2\kappa_3 & \Gamma_{34} \\ \Gamma_{41} & \Gamma_{42} & \Gamma_{43} & \Gamma_{44} + 2\kappa_4 \end{bmatrix} \begin{bmatrix} T_{21} \\ T_{22} \\ T_{23} \\ T_{24} \end{bmatrix}. \tag{8.103}$$

If we eliminate mode 4 and retain only one evanescent mode (mode 3) instead of two evanescent modes, then we will get the 3×3 matrix equation

$$\begin{bmatrix} 0 \\ -2ik_2 \\ 0 \end{bmatrix} = \begin{bmatrix} \tilde{\Gamma}_{11} - 2ik_1 & \tilde{\Gamma}_{12} & \tilde{\Gamma}_{13} \\ \tilde{\Gamma}_{21} & \tilde{\Gamma}_{22} - 2ik_2 & \tilde{\Gamma}_{23} \\ \tilde{\Gamma}_{31} & \tilde{\Gamma}_{32} & \tilde{\Gamma}_{33} + 2\kappa_3 \end{bmatrix} \begin{bmatrix} T_{21} \\ T_{22} \\ T_{23} \end{bmatrix}, \tag{8.104}$$

where

$$\tilde{\Gamma}_{ij} = \Gamma_{ij}\frac{2\kappa_4}{\Gamma_{44} + 2\kappa_4}. \tag{8.105}$$

Now consider the situation where we arbitrarily decide to truncate the highest evanescent mode (mode 4) from (8.103). This would have yielded

$$\begin{bmatrix} 0 \\ -2ik_2 \\ 0 \end{bmatrix} = \begin{bmatrix} \Gamma_{11} - 2ik_1 & \Gamma_{12} & \Gamma_{13} \\ \Gamma_{21} & \Gamma_{22} - 2ik_2 & \Gamma_{23} \\ \Gamma_{31} & \Gamma_{32} & \Gamma_{33} + 2\kappa_3 \end{bmatrix} \begin{bmatrix} T_{21} \\ T_{22} \\ T_{23} \end{bmatrix}, \tag{8.106}$$

Equation (8.106) is clearly *not* the same as (8.104) since $\Gamma \neq \tilde{\Gamma}$. Hence, the former equation will yield very *different* solutions for the transmission coefficients T_{21} and T_{22} for the *propagating* modes than (8.104) would. This analysis, reproduced from [20] clearly shows that transmission coefficients computed for the propagating modes depend on the evanescent modes that we consider. Thus, even though evanescent modes do not themselves carry current, they affect the transmission probability of the current-carrying propagating modes and hence indirectly determine the current. Consequently, we should always include evanescent modes in our calculation.

The next question that arises is how many evanescent modes do we need to include in our calculation since there are denumerably infinite number of them. We had mentioned earlier that the subblock of the current scattering matrix involving only propagating modes remains unitary, even though the entire matrix does not, if we include evanscent modes. It should now be clear that this will happen only *if we have included enough number of evanescent modes to calculate the elements of this subblock since these elements are renormalized by the evanescent modes.* Therefore, test of unitarity of this subblock will tell us if we have included enough evanescent modes. If it is unitary or close to unitary, then we have included enough evanescent modes; otherwise not. Frequently, the number of evanescent modes that we have to include in our calculation so as to make the subblock for propagating states unitary is far greater than the number of propagating modes in the structure! This is particularly true if the geometry of the structure is such that the dimensions do not vary slowly in space, but vary abruptly.

8.4 Spatially Varying Potential and Self-Consistent Solutions

In Sect. 8.2, we described how to find the wavefunction within a device in which the potential is spatially *invariant* except for localized elastic scatterers whose potentials are delta functions in space. From the wavefunction, we can determine transport variables such as carrier concentration and current density anywhere within the device. However, in any realistic situation, the potential within a device will not be spatially invariant between two successive scatterers. Instead, it will vary in space due to such things as doping variations, space charges, and external electric fields.

Handling spatially varying potentials requires only a slight modification. In (8.47), we provided the scattering matrix for a region of free propagation (between two successive scatterers) where the potential was assumed to be constant in space. If, instead, the potential was varying in space, we would have represented that potential as a series of steps as in Fig. 3.7. The wave scattering matrix for such a section (say, the m-th section of free propagation), for any given wavevector k, can be written as

$$[S_m] = \begin{bmatrix} r_{k,m} & t'_{k,m} \\ t_{k,m} & r'_{k,m} \end{bmatrix}. \tag{8.107}$$

where $t_{k,m}$ is the transmission amplitude of an electron propagating from left to right through the section, $r_{k,m}$ is the corresponding reflection amplitude, $t'_{k,m}$ is the reverse transmission amplitude, which is the transmission amplitude of an electron propagating from right to left, and $r'_{k,m}$ is the corresponding reflection amplitude. The last equation follows from the definition of either the wave or the current scattering matrix.

In order to find the elements of the above wave scattering matrix, we adopt the technique described in Special Topic 4 of Chap. 3, which provides the prescription for finding the transmission amplitude $t_{k,m}$ and reflection amplitude $r_{k,m}$ for any arbitrary section m of free propagation, within which the potential is varying in space. We break up such a section into a number of subsections, within each of which we assume the potential to be constant. We then cascade the transfer matrices of these subsections to find the total transfer matrix $[W]_m$ for the m-th section and use (3.93) and (3.94) to find $t_{k,m}$ and $r_{k,m}$. By proceeding in a similar manner, we can solve the problem of reverse propagation from right to left and deduce the coefficients $t'_{k,m}$ and $r'_{k,m}$. This will yield the wave scattering matrix $[S_m]$ for the m-th free propagation region when the potential within that region is spatially varying.

If the potential is varying somewhat gradually in space, then we can ignore "mode mixing" whereby an electron entering the device in one particular mode or subband gets reflected or transmitted into another mode by the spatially varying potential. In that case, we will replace the matrix in (8.47) with

$$[f'_m] = \begin{bmatrix} \mathbf{R_m} & \mathbf{T'_m} \\ \mathbf{T_m} & \mathbf{R'_m} \end{bmatrix}. \tag{8.108}$$

where

$$\begin{aligned} (R_m)_{\alpha\beta} &= r_{k_\beta,m}\delta_{\alpha\beta} \\ (R'_m)_{\alpha\beta} &= r'_{k_\beta,m}\delta_{\alpha\beta} \\ (T_m)_{\alpha\beta} &= t_{k_\beta,m}\delta_{\alpha\beta} \\ (T'_m)_{\alpha\beta} &= t'_{k_\beta,m}\delta_{\alpha\beta}. \end{aligned} \tag{8.109}$$

The rest of the analysis is the same as before. Finally, we end up with the wavefunction everywhere within the device even when the potential within it is varying in space between two successive scatterers. From this wavefunction, we can deduce any transport variable of interest such as carrier concentration and current density.

We can also incorporate self-consistency into our solution if we use the computed carrier concentrations $n(x,y,z)$ to renormalize the spatially varying potential $V(x,y,z)$ within the device using the Poisson equation

$$-\nabla^2 V(x,y,z) = e\left[N_D^+(x,y,z) - n(x,y,z)\right] \Big/ (\epsilon_r\epsilon_0), \tag{8.110}$$

where $N_D^+(x,y,z)$ is the ionized donor concentration and $\epsilon_r\epsilon_0$ is the dielectric constant. We will then have to recompute the wavefunctions and carrier concentrations from the new potential and continue this cycle iteratively until convergence is achieved and a self-consistent solution to the carrier concentration and current density emerges. Needless to say, such a method is computationally burdensome.

8.5 Terminal Currents

In dealing with electronic devices, we are frequently not concerned about the spatial distribution of the current density or carrier concentration within the device. More often than not, we are only interested in finding out what will be the currents flowing in or out of certain terminals if we apply specific voltages to these terminals. In other words, we seek only terminal currents.

There are elegant and simple quantum-mechanical formalisms that will allow us to determine terminal currents if transport is *phase coherent*, or dissipationless, i.e., there are no phase-breaking inelastic collisions occurring within the device. We will examine them in this section.

8.5.1 A Two-Terminal Device and Tsu–Esaki Equation

Consider first a two-terminal device shown in Fig. 8.10. The device is connected to two contacts (each assumed to be an infinite electron reservoir) by two quasi one-dimensional leads. When an electron enters any contact, it immediately loses any phase information and thermalizes within the contacts. Each contact is assumed to be in local thermodynamic equilibrium so that the electron occupation probabilities in these two contacts are given by the Fermi–Dirac factors:

$$f_1(\vec{r},\vec{k},t) = f(E-\mu_1) = \frac{1}{e^{\frac{E-\mu_1}{k_B T}}+1}$$

$$f_2(\vec{r},\vec{k},t) = f(E-\mu_2) = \frac{1}{e^{\frac{E-\mu_2}{k_B T}}+1} \tag{8.111}$$

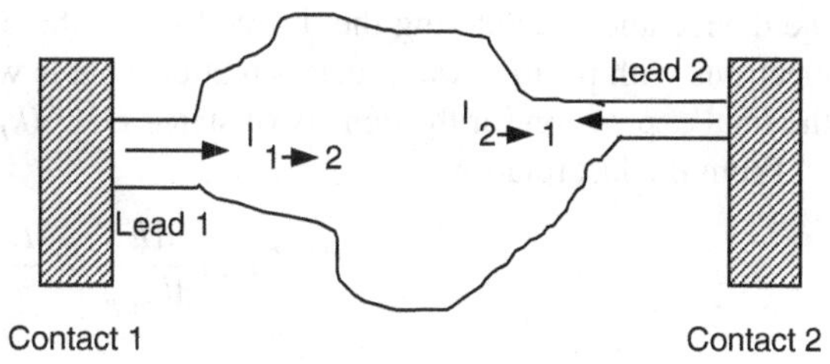

Fig. 8.10 A two-terminal device of arbitrary geometry, with quasi one-dimensional leads carrying current in one particular direction

where μ_1 is the Fermi level (or chemical potential) in lead 1 and μ_2 is that in lead 2. If a potential V is applied between the two contacts, then

$$\mu_1 - \mu_2 = e\mathrm{V}, \tag{8.112}$$

where e is the electron charge.

From Chap. 2 material, we surmise that the current density injected in lead 1 is

$$J_1 = \frac{e}{\Omega}\sum_{\vec{k}} v_x(k_x) f_1(\vec{r}, \vec{k}, t) = \frac{e}{\Omega}\sum_{k_{m,p}, m, p} v_{m,p}(k_{m,p}) f(E - \mu_1), \tag{8.113}$$

where Ω is the nomalizing volume. Note that for a parabolic band, the energy E of an electron in the quasi one-dimensional lead will be

$$E = \frac{\hbar^2 k_{m,p}^2}{2m^*} + \epsilon_{m,p}$$

$$v_{m,p} = \frac{\hbar k_{m,p}}{m^*}, \tag{8.114}$$

where $\epsilon_{m,p}$ is the energy at the bottom of the (m, p)-th subband and $k_{m,p}$ is the wavevector along the lead of an electron injected with energy E from the contact into the (m, p)-th subband in the lead.

Since the wavevectors $k_{m,p}$ form a continuum of states, we can rewrite (8.113) as

$$J_1 = \frac{e}{\Omega}\sum_{m,p}\int_0^\infty \mathrm{d}k_{m,p} D^{1-d}(k_{m,p}) v_{m,p}(k_{m,p}) f(E - \mu_1)$$

$$= \frac{e}{\Omega}\sum_{m,p}\int_{\epsilon_{m,p}}^\infty \mathrm{d}E D^{1-d}(E) v_{m,p}(E) f(E - \mu_1), \tag{8.115}$$

where $D^{1-d}(E)\mathrm{d}E = D^{1-d}(k_{m,p})\mathrm{d}k_{m,p} = (L/\pi)\mathrm{d}k_{m,p}$, where L is the normalizing length. Note that the limits of the integration over $k_{m,p}$ are from 0 to ∞ and not from $-\infty$ to ∞, since electrons with negative wavevectors are going away from the device and not entering the device to contribute to current J_1. Therefore, only electrons with positive wavevectors matter. That is why the factor of 2 is missing in the final expression for the density of states $D^{1-d}(k_{m,p})$ as well.

From the last relation,

$$D^{1-d}(E)\frac{\mathrm{d}E}{\mathrm{d}k_{m,p}} = \frac{L}{\pi}, \tag{8.116}$$

and using (8.114) to evaluate the derivative, we get that

$$D^{1-d}(E)v_{m,p}(E) = \frac{2L}{h}. \tag{8.117}$$

Note that the last relation will hold even for a nonparabolic band since $v_{m,p}(E) = (1/\hbar)(\partial E/\partial k_{m,p})$.

Substituting this result in the last expression for J_1, we get

$$J_1 = \frac{e}{\Omega}\sum_{m,p}\frac{2L}{h}\int_{\epsilon_{m,p}}^{\infty}\mathrm{d}Ef(E-\mu_1) = \frac{2e}{Ah}\sum_{m,p}\int_{\epsilon_{m,p}}^{\infty}\mathrm{d}Ef(E-\mu_1), \tag{8.118}$$

where A is the normalizing area ($\omega = AL$).

The net current injected into lead 1 is then

$$I_1 = J_1 A = \frac{2e}{h}\sum_{m,p}\int_{\epsilon_{m,p}}^{\infty}\mathrm{d}Ef(E-\mu_1). \tag{8.119}$$

Let $\hat{\tau}_{1\to 2}^{m,p,m',p'}(E)$ be the transmission probability for an electron injected from the (m,p)-th subband in lead 1 with energy E to transmit into the (m',p')-th subband in lead 2.[4] Therefore, if we call $I_{1\to 2}$ the current flowing from lead 1 into lead 2, then

$$I_{1\to 2} = \frac{2e}{h}\sum_{m,p}\int_{\epsilon_{m,p}}^{\infty}\mathrm{d}E\,\tau_{1\to 2}^{m,p}(E)f(E-\mu_1), \tag{8.120}$$

where

$$\tau_{1\to 2}^{m,p}(E) = \sum_{m',p'}\hat{\tau}_{1\to 2}^{m,p,m',p'}(E). \tag{8.121}$$

Next, we will define

$$T_{1\to 2}(E) = \sum_{m,p}\tau_{1\to 2}^{m,p}(E)\Theta(E-\epsilon_{m,p}), \tag{8.122}$$

[4]We deal with transmission probabilities rather than complex transmission amplitudes since electrons are phase-incoherent in the leads. Since different modes are mutually orthogonal in the leads, we should add the transmission probabilities and understand that there is no interference between the transmission amplitudes of different modes within the leads.

where Θ is the unit step function (or Heaviside function). This allows us to write

$$I_{1\to 2} = \frac{2e}{h}\int_0^\infty \mathrm{d}E T_{1\to 2}(E) f(E-\mu_1). \tag{8.123}$$

Similarly, the current flowing from lead 2 into lead 1 is

$$I_{2\to 1} = \frac{2e}{h}\int_0^\infty \mathrm{d}E T_{2\to 1}(E) f(E-\mu_2), \tag{8.124}$$

where

$$T_{2\to 1}(E) = \sum_{m',p'}\left[\sum_{m,p}\hat{\tau}_{2\to 1}^{m',p',m,p}(E)\right]\Theta(E-\epsilon_{m',p'}), \tag{8.125}$$

and $\hat{\tau}_{2\to 1}^{m',p',m,p}(E)$ is the transmission probability for an electron injected from the (m',p')-th subband in lead 2 to transmit into the (m,p)-th subband in lead 1 with energy E (E is the electron's final energy in lead 1).

Therefore, the *net* current flowing into lead 1 or flowing out of lead 2 in steady state is

$$I = I_{1\to 2} - I_{2\to 1} = \tfrac{2e}{h}\int_0^\infty \mathrm{d}E\left[T_{1\to 2}(E) f(E-\mu_1) - T_{2\to 1}(E) f(E-\mu_2)\right], \tag{8.126}$$

while the net current flowing into lead 2 or flowing out of lead 1 is simply the negative of that.

In equilibrium, assuming that there are no thermal gradients and the effective mass is spatially invariant, the spatial gradient of the Fermi level is zero everywhere (recall the Special Topic 3 in Chap. 2), implying that $\mu_1^{\text{equil}} = \mu_2^{\text{equil}} = \mu_{\text{F}}$. Also, no net current flows in equilibrium. Therefore, we can conclude

$$0 = \frac{2e}{h}\int_0^\infty \mathrm{d}E\left[T_{1\to 2}^{\text{equil}}(E) - T_{2\to 1}^{\text{equil}}(E)\right] f(E-\mu_{\text{F}}), \tag{8.127}$$

which immediately tells us that

$$T_{1\to 2}^{\text{equil}}(E) = T_{2\to 1}^{\text{equil}}(E). \tag{8.128}$$

We have just proved the *reciprocity* or *symmetricity* of the transmission probability in equilibrium, i.e., the probability of transmitting from lead 1 to lead 2 with a given energy in lead 1 at equilibrium is the same as the probability of transmitting from lead 2 to lead 1 with the same energy in lead 1. But what happens when there is a bias across the structure and a current flows so that we are out of equilibrium? In that case, the reciprocity is still valid as long as we define the transmission coefficient properly. Note that when there is a bias across a structure, the wavevector (or velocity) with which an electron enters the structure is not the wavevector (or velocity) with which it exits, as shown in Fig. 8.11.

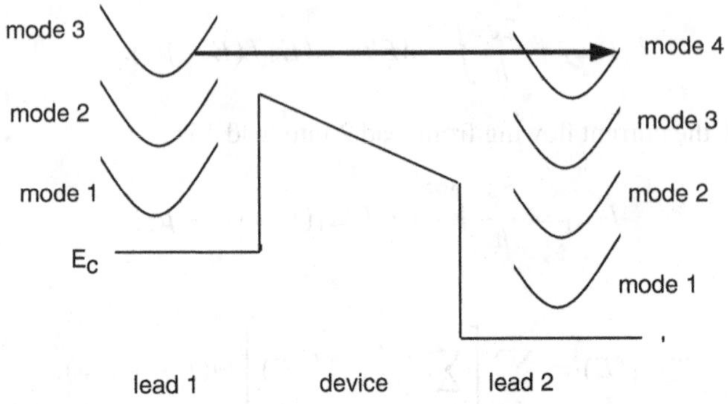

Fig. 8.11 The wavevector with which an electron enters a structure under bias is not necessarily the wavevector with which it exits. The entering and exiting mode numbers may not be the same either. The parabolas in the leads depict the energy-wavevector dispersion relations of different modes

If the electron enters the device with wavevector $k_{m,p}$ and exits the device with wavevector $k'_{m',p'}$, we should redefine the quantity $\hat{\tau}_{1\to 2}^{m,p,m',p'}(E)$ in (8.121) as

$$\hat{\tau}_{1\to 2}^{m,p,m',p'}(E) \to \hat{\tau}_{1\to 2}^{m,p,m',p'}(E)\frac{v(k_{m',p'})}{v(k_{m,p})}, \tag{8.129}$$

where $v(k)$ is the velocity of an electron with wavevector k. As long as we use the redefined $\hat{\tau}_{1\to 2}^{m,p,m',p'}(E)$ in calculating the quantity $T_{1\to 2}(E)$, we can always write (even under bias and out of equilibrium)

$$T_{1\to 2}(E) = T_{2\to 1}(E), \tag{8.130}$$

as long as there is no magnetic field.

The last equality follows from the general principle of time-reversal symmetry and the symmetric nature of the current scattering matrix discussed in Special Topic 2 of this chapter. Inelastic collision, however, will violate time reversal symmetry and invalidate (8.130). To see how this might happen, consider the toy example of electron transport through a device containing a potential barrier as shown in Fig. 8.12 (S. Datta, private communication). If there is an inelastic collision at the location indicated, resulting in energy loss (e.g., phonon emission), then the probability of going from lead 1 to lead 2 is virtually zero if the electron enters with the energy shown, but the probability of going from lead 2 to lead 1 is virtually 1. Therefore, inelastic collisions appear to violate reciprocity. We deduced the reciprocity property in Special Topic 2 of this chapter from the Schrödinger equation which does *not* capture inelastic collisions or other dissipative processes.

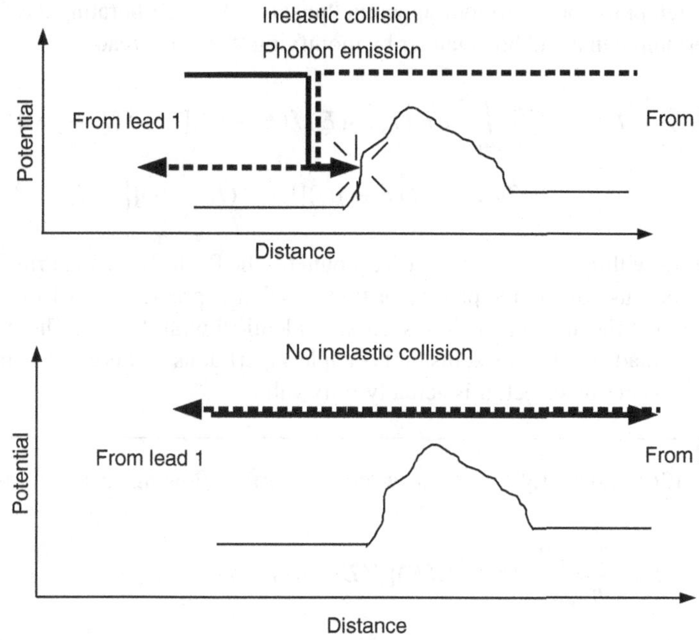

Fig. 8.12 Figure illustrating how an inelastic collision can invalidate reciprocity of transmission and (8.130). An electron enters the device whose potential profile is shown. The *arrows* indicate the direction of electron trajectory. If the electron enters from lead 1 with the energy shown and an inelastic (energy loss) collision occurs before encountering the barrier, the transmission probability will plummet to nearly zero. If the electron enters with the same energy from lead 2, and an inelastic collision occurs at the same location, the transmission probability will still be very high (close to 1). If there is no inelastic event, as shown in the *lower panel*, then it is plausible that the transmission probability at a fixed energy will be the same regardless of whether the electron entered from lead 1 or lead 2

Consequently, there is no reason to believe that reciprocity will hold in the presence of inelastic collisions, and indeed it does not. However, in phase coherent transport, (8.130) is always valid.

Special Topic 3: What about the Pauli Exclusion Principle?

In deriving (8.126), we never made any mention of the fact that an electron incident with energy E from the left contact can never make it into the right contact unless there is an empty state at that energy in the right contact. The electron obviously cannot go into a filled state because of Pauli Exclusion Principle. We tacitly assumed that there is always an empty state at energy E in the right contact for the electron to go into. This is, of course, not true in general, because the probability of finding an empty state at energy E in the right contact is $1 - f(E - \mu_2)$, which is less than unity.

We can attempt to correct for our apparent "mistake" by incorporating the Pauli Exclusion Principle in an ad hoc manner by modifying (8.126) to read

$$I = I_{1\to 2} - I_{2\to 1} = \frac{2e}{h}\int_0^\infty \mathrm{d}E\Big\{T_{1\to 2}(E)f(E-\mu_1)[1-f(E-\mu_2)] - T_{2\to 1}(E)f(E-\mu_2)[1-f(E-\mu_1)]\Big\}, \quad (8.131)$$

where the terms within the square brackets account for the Pauli Exclusion Principle.

However, because of the reciprocity of transmission expressed in (8.130), it is easy to verify that the above equation is actually identical with (8.126). Therefore, reciprocity has made the Pauli Exclusion Principle superfluous in this case! In other words, (8.126) is not incorrect; it is actually fully valid.

Using (8.126) and (8.130), we can write the current flowing in a 2-terminal device as

$$I = \frac{2e}{h}\int_0^\infty \mathrm{d}E\, T_{1\to 2}(E)[f(E-\mu_1) - f(E-\mu_2)]. \quad (8.132)$$

For historical reasons, this equation is sometimes referred to as the *Tsu–Esaki equation* since Tsu and Esaki used it to analyze the current versus voltage characteristics of resonant tunneling diodes which we discuss in the next chapter.

8.5.1.1 Tsu–Esaki Formula for Three-Dimensional Leads

The Tsu–Esaki formula derived above is valid for leads of any dimensionality—one, two, or three—since any polydimensional lead can be viewed as a quasi one-dimensional lead with numerous energy subbands (or transverse modes) occupied by electrons. However, if we prefer not to do that and view leads as three-dimensional, then it should be obvious from the preceding discussion that the current density J, flowing in a 2-terminal device in response to a bias voltage V, can be written as:

$$\begin{aligned} J_{3-d} &= \frac{e}{\Omega}\sum_{k_x}\sum_{k_y}\sum_{k_z}\Big\{v_x(k_x,k_y,k_z)\tau_{1\to 2}(k_x,k_y,k_z) \\ &\quad \times\big[f(E(k_x,k_y,k_z)-\mu_1) - f(E(k_x,k_y,k_z)-\mu_2)\big]\Big\} \\ &= \frac{e}{\Omega}\int_0^\infty d^3\vec{k}\, D_q^{3-d}(\vec{k})v_x(\vec{k})\tau_{1\to 2}(\vec{k})\Big[f(E(\vec{k})-\mu_1) - f(E(\vec{k})-\mu_2)\Big] \\ &= \frac{e}{\Omega}\int_0^\infty \mathrm{d}E D_q^{3-d}(E)v_x(E)\tau_{1\to 2}(E)\,[f(E-\mu_1) - f(E-\mu_2)], \end{aligned} \quad (8.133)$$

if we view the leads as "three-dimensional." Here, D_q^{3-d} is the three dimensional density of states, Ω is the normalizing volume, $\vec{k} = k_x\hat{x}+k_y\hat{y}+k_z\hat{z}$, and $\tau_{1\to2}(E) = t_{1\to2}(E)v'(E)/v(E)$, where $t_{1\to2}(E)$ is the transmission probability of an electron with energy E to transmit from the injecting to the extracting lead, $v(E)$ is the velocity with which the electron enters from the injecting lead and $v'(E)$ is the velocity with which it exits into the extracting lead. Reciprocity holds to guarantee that $\tau_{1\to2}(E) = \tau_{2\to1}(E)$.

Before we conclude this topic, we wish to stress two important points. First, it is clear that the maximum value of the quantity $\tau_{1\to2}^{m,p}(E)$ is unity since this quantity is the fraction of the current entering the device from the left lead with energy E that exits into the right lead. Therefore, from (8.122), we can see that the maximum value of $T_{1\to2}(E)$ is not 1, but Q, where Q is the total number of modes (or subbands) in the leads that can contribute to current (propagating states). Second, using (8.112), we can write (8.132) and (8.133) as

$$I = \frac{2e}{h}\int_0^\infty \mathrm{d}E T_{1\to2}(E,\mathrm{V})\Big[f(E-\mu_1) - f(E+q\mathrm{V}-\mu_1)\Big]. \tag{8.134}$$

$$J_{3-d} = \frac{e}{\Omega}\int_0^\infty \mathrm{d}E D_q^{3-d}(E)v_x(E)\tau_{1\to2}(E,\mathrm{V})\Big[f(E-\mu_1) - f(E+q\mathrm{V}-\mu_1)\Big], \tag{8.135}$$

where we have highlighted the fact that the transmission probability associated with transmitting through a structure at an energy E may depend on the bias across the structure since the bias may change the potential profile inside the device and affect the transmission through it. The bias will also affect $v'(E)$, thereby affecting $\tau_{1\to2}(E,\mathrm{V})$. The dependence of $\tau_{1\to2}(E,\mathrm{V})$ on the bias voltage V gives rise to the negative differential resistance in the double barrier resonant tunneling device, which we will see in the next chapter.

We will now derive a formula for the conductance of a two-terminal nondissipative device which is valid at low temperatures and low bias voltages. For this purpose, we proceed as follows:

Let us make Taylor series expansions of the occupation probability terms in the previous equation:

$$\begin{aligned} f(E-\mu_1) &= f(E-\mu_\mathrm{F}) + (\mu_\mathrm{F}-\mu_1)\frac{\partial f(E-\mu_\mathrm{F})}{\partial E} + \texttt{higher order terms} \\ f(E-\mu_2) &= f(E-\mu_\mathrm{F}) + (\mu_\mathrm{F}-\mu_2)\frac{\partial f(E-\mu_\mathrm{F})}{\partial E} + \texttt{higher order terms} \end{aligned} \tag{8.136}$$

Therefore, the term within the square brackets in (8.132) can be written as

$$\begin{aligned} &f(E-\mu_1)-f(E-\mu_2) \\ &\quad=(\mu_2-\mu_1)\frac{\partial f(E-\mu_F)}{\partial E}+\texttt{higher order terms} \\ &\quad=-e\mathrm{V}\frac{\partial f(E-\mu_F)}{\partial E}+\texttt{terms involving higher powers of V}. \end{aligned} \tag{8.137}$$

If the voltage across the device is sufficiently small, we can ignore the higher order terms and write

$$f(E-\mu_1)-f(E-\mu_2)\approx -e\mathrm{V}\frac{\partial f(E-\mu_F)}{\partial E}. \tag{8.138}$$

Substituting the last result in the Tsu–Esaki equation (8.132), we get

$$I=\frac{2e^2\mathrm{V}}{h}\int_0^\infty \mathrm{d}E\,T_{1\to 2}(E)\left[-\frac{\partial f(E-\mu_F)}{\partial E}\right]. \tag{8.139}$$

This equation shows that at small voltages, the current is *linearly proportional* to the voltage, so that we can write

$$I=G\mathrm{V}, \tag{8.140}$$

where $G=I/\mathrm{V}$ is, by definition, the conductance of the two-terminal device. Clearly,

$$G=\frac{2e^2}{h}\int_0^\infty \mathrm{d}E\,T_{1\to 2}(E)\left[-\frac{\partial f(E-\mu_F)}{\partial E}\right]. \tag{8.141}$$

We can use (8.111) to evaluate the partial derivative in the preceding equation. This yields

$$G=\frac{e^2}{2k_BTh}\int_0^\infty \mathrm{d}E\,T_{1\to 2}(E)\mathrm{sech}^2\left[\frac{E-\mu_F}{2k_BT}\right]. \tag{8.142}$$

Since the quantity $\mathrm{sech}^2(x)$ is sharply peaked around $x=0$, it is obvious that only electrons within a narrow range of the Fermi energy μ_F will contribute to current if the bias across the device is small.

Equation (8.140) is an expression of Ohm's law where current is linearly proportional to voltage. Because of this linearity, this regime of transport is called *linear response transport*.

The next question is how small does the bias V have to be in order to experience linear response transport? We visit this a little later, but first let us impose an additional condition that the temperature is also low. In that case

$$\left[-\frac{\partial f(E-\mu_{\mathrm{F}})}{\partial E}\right] = \frac{1}{4k_{\mathrm{B}}T}\mathrm{sech}^2\left[\frac{E-\mu_{\mathrm{F}}}{2k_{\mathrm{B}}T}\right] = \delta(E-\mu_{\mathrm{F}}), \tag{8.143}$$

where the delta is the Dirac delta function.

Using this result in either (8.141) or (8.142), we get

$$G = \frac{2e^2}{h}\int_0^{\infty} \mathrm{d}E\, T_{1\to 2}(E)\delta(E-\mu_{\mathrm{F}}) = \frac{2e^2}{h}T_{1\to 2}(\mu_{\mathrm{F}}), \tag{8.144}$$

which clearly shows that what determines the linear response conductance of a two-terminal nondissipative device at small voltages and low temperatures is the transmission probability of electrons at the Fermi level μ_{F}.

The last relation is the celebrated *Landauer 2-terminal conductance formula*. It provides the very important insight that the conductance of a nondissipative device is determined solely by the transmission of electrons that carry current. The more "transparent" a device is to these electrons, the more conductive it will be. This simple picture, which should be intuitive, makes it a remarkable result.

8.5.2 When Is Response Linear?

One final question we need to ponder is how low does the bias voltage V have to be in order for linear response conditions to prevail? It appears that it should be low enough to allow us to neglect the higher order terms in the Taylor series expansion, i.e.,

$$e\mathrm{V}\left[-\frac{\partial f(E-\mu_{\mathrm{F}})}{\partial E}\right] = \frac{e\mathrm{V}}{4k_{\mathrm{B}}T}\mathrm{sech}^2\left[\frac{E-\mu_{\mathrm{F}}}{2k_{\mathrm{B}}T}\right] \ll 1. \tag{8.145}$$

Since the maximum value of $\mathrm{sech}(x)$ is unity, it seems that we will have to ensure that

$$|\mathrm{V}| \ll \frac{4k_{\mathrm{B}}T}{e}. \tag{8.146}$$

The above condition is actually wrong and misleading. It is clearly wrong since it tells us (erroneously) that we can never have linear response transport at very low temperatures when $T \to 0$. Moreover, since both linear response and low temperature are needed for the Landauer formula to hold, it appears that there can be no situation when the Landauer formula will be valid!

A little reflection will allow us to deduce the correct condition for linear response. We start with the Tsu–Esaki formula (8.132) and note that *if the transmission probability* is independent of energy over the range of electron energies that contribute to current, then we can write the Tsu–Esaki equation as

$$I = \frac{2e}{h} T_{1\to 2} \int_0^\infty \mathrm{d}E[f(E-\mu_1) - f(E-\mu_2)]. \tag{8.147}$$

At low temperatures, when $T \to 0$, we can write

$$\begin{aligned} f(E-\mu_1) &\approx 1 - \Theta(E-\mu_1) \\ f(E-\mu_2) &\approx 1 - \Theta(E-\mu_2) \\ f(E-\mu_1) - f(E-\mu_2) &\approx \Theta(E-\mu_2) - \Theta(E-\mu_1), \end{aligned} \tag{8.148}$$

where Θ is the Heaviside function.

Therefore,

$$\begin{aligned} \lim_{T\to 0} \int_0^\infty \mathrm{d}E[f(E-\mu_1) - f(E-\mu_2)] &= \int_0^\infty \mathrm{d}E[\Theta(E-\mu_2) - \Theta(E-\mu_1)] \\ &= \int_{\mu_2}^{\mu_1} \mathrm{d}E = \mu_1 - \mu_2 = e\mathrm{V}. \end{aligned} \tag{8.149}$$

Substitution of this result in (8.147) shows that

$$I = \frac{2e^2}{h} T_{1\to 2}\mathrm{V}, \tag{8.150}$$

which means that linear response (and the Landauer formula) can hold at *arbitrarily large* voltages as long as the temperature is low and the transmission coefficient of electrons contributing to current *at that voltage* is energy independent. Of course, the latter condition will be violated at too large a voltage, which is why the voltage must be reasonably small, but we definitely do not need to ensure that $|\mathrm{V}| \ll \frac{4kT}{e}$.

The question now is what happens if the voltage V is so large that the transmission coefficient of electrons contributing to the current is *no longer* independent of energy? In that case, Landauer's formula can still hold, but we have to redefine the transmission coefficient. The following analysis that extends the Landauer formula to such large voltages is adapted from Bagwell [21].

We can write

$$f(E-\mu_F) = [1 - \Theta(E-\mu_F)] \otimes \left[-\frac{\partial f(E)}{\partial E}\right], \tag{8.151}$$

where $\otimes$ denotes the convolution operation, Θ is the usual Heaviside function, and $\partial f(E)/\partial E = (1/4k_B T)\mathrm{sech}^2(E/2k_B T)$.

The convolution of two functions $A(E)$ and $B(E)$ has the usual meaning:

$$A(E) \otimes B(E) = \int_{-\infty}^{\infty} A(E-E')B(E')\mathrm{d}E' = \int_{-\infty}^{\infty} A(E')B(E-E')\mathrm{d}E'. \tag{8.152}$$

The difference of the Fermi–Dirac factors can therefore be written as

$$f(E-\mu_1)-f(E-\mu_2)=[\Theta(E-\mu_2)-\Theta(E-\mu_1)]\otimes\left[-\frac{\partial f(E)}{\partial E}\right]$$
$$=[\Theta(E-\mu_2)-\Theta(E-\mu_2-e\text{V})]\otimes\left[-\frac{\partial f(E)}{\partial E}\right]. \tag{8.153}$$

The last equation shows that the difference of the Fermi–Dirac factors is the convolution of two functions, one of which depends only on the voltage V, while the other depends only on temperature. This equation therefore separates out the effects of finite voltage and finite temperature.

Using this insight, the Tsu–Esaki equation (8.132) is recast as

$$I=\frac{2e}{h}T_{1\to2}(\mu_2)\otimes W(\mu_2)\otimes\left[-\left.\frac{\partial f(\eta)}{\partial\eta}\right|_{\eta=\mu_2}\right], \tag{8.154}$$

where $W(x)=\Theta(x+e\text{V})-\Theta(x)$.

Using the associative and commutative properties of the convolution operation, we can write the previous equation as

$$\begin{aligned}
I&=\frac{2e}{h}T_{1\to2}(\mu_2)\otimes W(\mu_2)\otimes\left[-\left.\frac{\partial f(\eta)}{\partial\eta}\right|_{\eta=\mu_2}\right]\\
&=\frac{2e}{h}W(\mu_2)\otimes T_{1\to2}(\mu_2)\otimes\left[-\left.\frac{\partial f(\eta)}{\partial\eta}\right|_{\eta=\mu_2}\right]\\
&=\frac{2e}{h}W(\mu_2)\otimes\overline{T}(\mu_2)\\
&=\frac{2e}{h}\int_{-\infty}^{\infty}[\Theta(E'+e\text{V})-\Theta(E')]\overline{T}(\mu_2-E')\text{d}E'\\
&=\frac{2e}{h}\int_{-e\text{V}}^{0}\overline{T}(\mu_2-E')\text{d}E'\\
&=\frac{2e^2\text{V}}{h}\hat{T}.
\end{aligned} \tag{8.155}$$

where $\overline{T}(\mu_2)=-\int_{-\infty}^{\infty}\text{d}x\,T_{1\to2}(\mu_2-x)\,\partial f(\eta)/\partial\eta|_{\eta=x}$ and $\hat{T}$ is the energy-averaged value of $\overline{T}(E)$ with E being between μ_2 and μ_1. The last result shows that the Landauer formula can be extended to finite voltages and temperatures if we replace the actual transmission probability with the redefined transmission probability $\hat{T}$.

8.5.3 Landauer Formula and Quantized Conductance of Point Contacts

A dramatic experimental validation of the Landauer formula was reported by two groups in the 1980s [22, 23]. They measured the linear response conductance of a structure commonly referred to as a "quantum point contact." It consists of two very closely spaced metallic gates delineated by lithography on a 2-DEG (recall the split gates of Fig. 7.11). The geometry of the gates is shown in Fig. 8.13a. Note that the gates are tapered and come very close together at one end. When a negative potential is applied simultaneously to both gates, the electron gas underneath them is depleted, leaving a narrow sliver of a conducting channel (a quasi one-dimensional electron gas) in the constriction between the gates. The width of this channel progressively decreases along the length because of the tapered shape of the gates. Since the resistance of the channel is inversely proportional to its width, the most resistive region is the one pinched between the gates where they come closest to each other. This region therefore dominates the resistance of the entire quasi one-dimensional channel. Since this region is very short, the probability of finding a scatterer within this region is quite small. In other words, we expect that transmission probability of every propagating mode through this region will be unity.[5]

The number of propagating modes in the constriction is the number of subbands occupied by electrons. Since the thickness of the electron gas is vanishingly small and the potential profile along the width of the constriction is approximately parabolic with curvature ω (see Fig. 8.13b), the energy separation between the subbands is $\hbar\omega$. Hence, at low temperatures, the number N_{occ} of occupied subbands is the largest integer that satisfies the condition

$$(N_{\text{occ}} + 1/2)\,\hbar\omega \leq E_{\text{F}}, \tag{8.156}$$

where E_{F} is the Fermi energy.

Fig. 8.13 (**a**) A quantum point contact consists of a constricted channel of electrons pinched between two negatively biased split gates delineated on a two-dimensional electron gas. The constriction is a quasi one-dimensional channel (quantum wire) whose width can be varied by varying the split-gate voltage (more negative bias on the split gates makes the channel narrower). Current flows in the x-direction when a small potential is applied between the source and drain contacts. (**b**) the potential profile along the y-direction is approximately parabolic. The curvature of the parabola, and hence the energy separation between the subbands in the conducting channel, can be increased by making the split-gate potential more negative. The subband energy levels are depicted by the *broken lines* and the energy spacing between them increases as the split-gate bias is made increasingly negative. (**c**) The conductance of the quantum point contact decreases in well-defined steps of height $2e^2/h$ as the split-gate potential is made more and more negative and subbands are successively depopulated. Observation of these steps is a dramatic experimental validation of the Landauer linear response two-terminal conductance formula

[5]The region is, however, long enough that the probability of an evanescent mode tunneling through it is virtually zero.

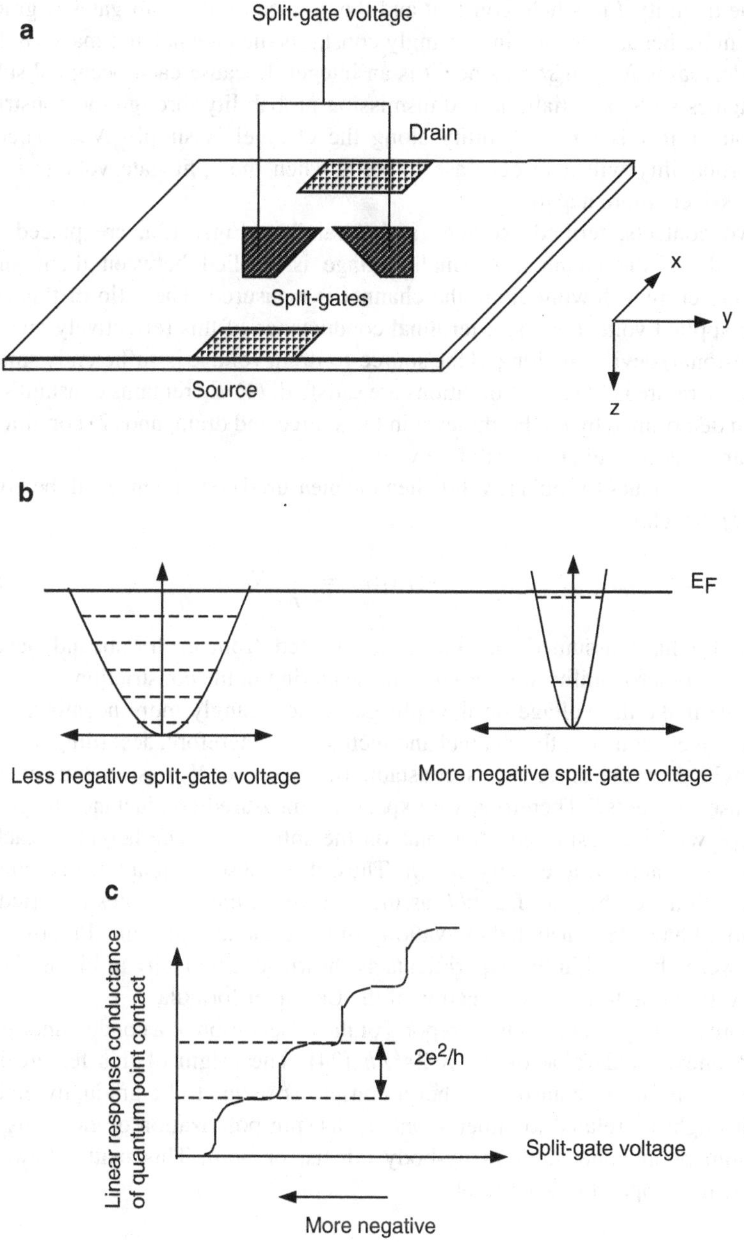
a
Split-gate voltage
Drain
Split-gates
Source
x
y
z
b
E_F
Less negative split-gate voltage
More negative split-gate voltage
c
Linear response conductance of quantum point contact
2e²/h
Split-gate voltage
More negative

The quantity E_F is held constant and the voltage on the split gates is gradually made more negative which increasingly constricts the channel and makes ω larger. This decreases N_{occ} in *steps* since it is an integer. Because each occupied subband propagates with essentially unit transmission probability through the constriction, the total transmission probability along the channel is simply N_{occ}. Therefore, this probability will also decrease in steps when the split gate voltage is made progressively more negative.

Two contacts, termed "source" and "drain" in Fig. 8.13a, are placed at the two ends of the channel. A small voltage is applied between them and the resulting current flowing along the channel is measured. The ratio of this current to the applied voltage is the 2-terminal conductance of this (effectively quasi one-dimensional) device. As long as the source-to-drain voltage is sufficiently small and the temperature is low, two conditions are satisfied: (1) E_F remains constant since it is then determined by the Fermi level in the source and drain, and (2) conditions for the Landauer formula to be valid prevail.

If the Landauer formula is valid, then the measured conductance will be given by (8.144), which is

$$G = \frac{2e^2}{h} T(\mu_F) = \frac{2e^2}{h} N_{occ}, \tag{8.157}$$

where we have assumed that electrons injected from each subband have unit transmission probability since there is no scattering in the constriction.

If we make the voltage on the split gates increasingly more negative, we will progressively constrict the channel and increase ω. Therefore, according to (8.156), N_{occ} will decrease since E_F is constant. But because N_{occ} is an integer, it will decrease in "steps." Therefore, we expect the measured conductance to go down in steps with increasing negative bias on the split gates. The height of each step should be quantized to exactly $2e^2/h$. Thus, if we observe quantized conductance steps with a step height of $2e^2/h$ as the voltage on the split gates is varied, then we would have demonstrated the validity of the Landauer formula. Precisely these steps were observed in the experiments as shown schematically in Fig. 8.13c. This observation is a dramatic vindication of the Landauer formula.

More recently, there has been reports of the observation of a conductance plateau at the unexpected value of $0.7 \times 2e^2/h$ [24]. The origin of this feature is still somewhat mired in controversy, but a consensus seems to be gradually emerging that it might be related to either spontaneous spin polarization of electrons in the quantum point contact, or many-body effects, or both. This matter however is outside the scope of this textbook.

8.5.4 The Büttiker Multiprobe Formula

The Landauer conductance formula is valid for a 2-terminal device. In order to treat devices with multiple (more than two) terminals, let us focus on (8.126). When the bias across the sample is vanishingly small and we are in the linear response regime, we can make a Taylor series expansion of the Fermi–Dirac factors as in (8.136) and write (neglecting higher order terms)

$$
\begin{aligned}
I = I_{1\to 2} - I_{2\to 1} &= \frac{2e}{h}\int_0^\infty \mathrm{d}E\left[T_{1\to 2}(E)\left\{f(E-\mu_F) + (\mu_F - \mu_1)\frac{\partial f(E-\mu_F)}{\partial E}\right\}\right. \\
&\quad \left. - T_{2\to 1}(E)\left\{f(E-\mu_F) + (\mu_F - \mu_2)\frac{\partial f(E-\mu_F)}{\partial E}\right\}\right] \\
&= \frac{2e}{h}\int_0^\infty \mathrm{d}E\,[T_{1\to 2}(E)\mu_1 - T_{2\to 1}(E)\mu_2]\left[-\frac{\partial f(E-\mu_F)}{\partial E}\right] \\
&= \frac{2e}{h}\left[\hat{T}_{1\to 2}\mu_1 - \hat{T}_{2\to 1}\mu_2\right],
\end{aligned} \tag{8.158}
$$

where

$$
\hat{T}_{p\to q} = \int_0^\infty \mathrm{d}E\,T_{p\to q}(E)\left[-\frac{\partial f(E-\mu_F)}{\partial E}\right]. \tag{8.159}
$$

Equation (8.158) is the same as the Landauer formula since $\mu_1 - \mu_2 = e\mathrm{V}$ and $\hat{T}_{1\to 2} = \hat{T}_{2\to 1}$ because of reciprocity.

Note that since we made no assumption about low temperature, (8.158) is valid at any arbitrary temperature. However, if the temperature happens to be low, then

$$
\hat{T}_{p\to q} = \int_0^\infty \mathrm{d}E\,T_{p\to q}(E)\delta(E-\mu_F) = T_{p\to q}(\mu_F). \tag{8.160}
$$

Equation (8.158) is valid for a two-terminal device and gives the *net* current flowing into either terminal of such a device as

$$
I = I_1^{\text{net}} = -I_2^{\text{net}} = \frac{2e}{h}\left[\hat{T}_{1\to 2}\mu_1 - \hat{T}_{2\to 1}\mu_2\right]. \tag{8.161}
$$

We can generalize (8.158) to the case of a device with more than two terminals. Such a device is shown in Fig. 8.14. Clearly, the net current flowing into the i-th terminal of an N-terminal device will be given by

$$
I_i^{\text{net}} = \frac{2e}{h}\sum_{j,j\neq i}^{N}\left[\hat{T}_{i\to j}\mu_i - \hat{T}_{j\to i}\mu_j\right]. \tag{8.162}
$$

Since the term for $i = j$ will vanish within the summation, we might as well have written

$$
I_i^{\text{net}} = \frac{2e}{h}\sum_{j}^{N}\left[\hat{T}_{i\to j}\mu_i - \hat{T}_{j\to i}\mu_j\right]. \tag{8.163}
$$

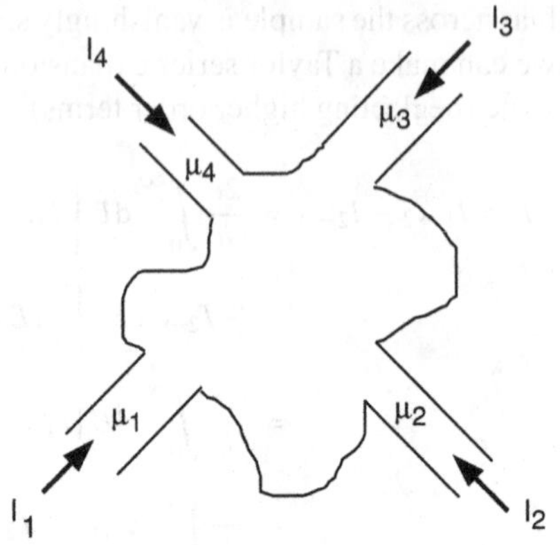

Fig. 8.14 A multiterminal device

At low temperatures,

$$I_i^{\text{net}} = \frac{2e}{h} \sum_{j,j \neq i}^{N} \left[T_{i \to j}(\mu_{\text{F}})\mu_i - T_{j \to i}(\mu_{\text{F}})\mu_j \right]. \tag{8.164}$$

If no magnetic field is present, then we can use the reciprocity relation $T_{i \to j}(\mu_{\text{F}}) = T_{j \to i}(\mu_{\text{F}})$ and simplify the last equation to

$$I_i^{\text{net}} = \frac{2e}{h} \sum_{j,j \neq i}^{N} T_{i \to j}(\mu_{\text{F}})[\mu_i - \mu_j]. \tag{8.165}$$

For any given propagating mode or channel in the lead, the transmission from one lead into all other leads, plus the reflection into the same lead must add up to unity. Thus, if there are M propagating channels or modes in the i-th lead, then

$$R_{i \to i}(\mu_{\text{F}}) + \sum_{j,j \neq i}^{NT_{i \to j}} (\mu_{\text{F}}) = M, \tag{8.166}$$

where $R_{i \to i}$ is the reflection probability of an electron incident from the i-the terminal back into the i-th terminal.

Substituting this result into the preceding equation, we get that at low temperatures and voltages, the net current carried into the i-th terminal of a multi-terminal device is

$$I_i^{\text{net}} = \frac{2e}{h} \left[M - R_{i \to i}(\mu_{\text{F}})\mu_i - \sum_{j,j \neq i}^{N} T_{j \to i}(\mu_{\text{F}})\mu_j \right]. \tag{8.167}$$

The last equation is due to Büttiker [25] and is known as *Büttiker's multiprobe formula*. It expresses the current in any terminal of a phase coherent device as a function of the potentials at all other terminals and the transmission and reflection probabilities of electrons at the local Fermi levels in these terminals. Note that it shows immediately that any phase coherent device is intrinsically *nonlocal* in nature, since if we change the potential at any one terminal without changing anything else (e.g., carrier concentration in the device), then the currents at *all other terminals* will change, which is the hallmark of nonlocality. This formula is very helpful in understanding a number of phase-coherent phenomena, including an alternate explanation for the integer quantum Hall effect, as we will see in the next chapter.

8.6 The 2-probe and 4-probe Landauer Formulas

Consider a 2-terminal device shown in Fig. 8.15a whose conductance is measured by injecting a known current I from a current source into one of the terminals, extracting it from the other, and measuring the resulting potential difference V between the same two terminals with a voltmeter. The conductance that we will measure in this 2-probe set up is $G_{2\text{-probe}} = I/\mathrm{V}$. We can also write this conductance as $G_{12,12}$ where $G_{ij,kl}$ represents the conductance measured with i, j as the current probes and k, l as the voltage probes. Let us see what the Büttiker multiprobe formula will predict for the 2-probe conductance of the device, assuming that transport is phase coherent and we are in the linear response, low-temperature regime.

We can write (8.164) in the matrix form

$$\begin{bmatrix} I_1 \\ I_2 \end{bmatrix} = \frac{2e}{h} \begin{bmatrix} T_{1\to 2} & -T_{2\to 1} \\ -T_{1\to 2} & T_{2\to 1} \end{bmatrix} \begin{bmatrix} \mu_1 \\ \mu_2 \end{bmatrix}. \tag{8.168}$$

Clearly, $I_1 = -I_2 = I$, and reciprocity dictates that $T_{1\to 2} = T_{2\to 1} = \mathcal{T}$. Therefore, the last equation reduces to

$$\begin{bmatrix} I \\ -I \end{bmatrix} = \frac{2e}{h} \begin{bmatrix} \mathcal{T} & -\mathcal{T} \\ -\mathcal{T} & \mathcal{T} \end{bmatrix} \begin{bmatrix} \mu_1 \\ \mu_2 \end{bmatrix}, \tag{8.169}$$

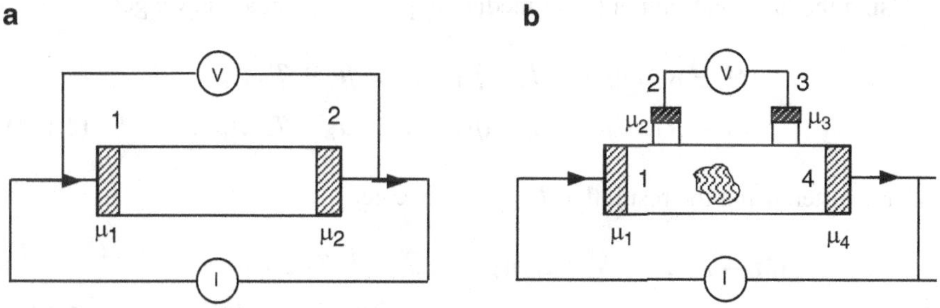

Fig. 8.15 (**a**) A 2-probe setup, and (**b**) a 4-probe setup, for measuring conductance of a device

which immediately yields

$$\frac{I}{\mu_1 - \mu_2} = \frac{2e}{h}\mathcal{T}. \tag{8.170}$$

As always, we assume that the contacts are in local thermodynamic equilibrium characterized by chemical potentials μ_1 and μ_2 and that an electron equilibrates (by dissipation) immediately upon entering a contact. In that case, since the difference between the Fermi energies (or chemical potentials) in the two probes is the electron charge times the applied voltage, i.e., $\mu_1 - \mu_2 = e\mathrm{V}$, we get

$$G_{2\text{-probe}} = G_{12,12} = \frac{I}{\mathrm{V}} = \frac{2e^2}{h}\mathcal{T}, \tag{8.171}$$

which is immediately recognized as the familiar Landauer formula in (8.144). Therefore, everything is consistent and appears to make sense.

Let us now investigate a slightly more complicated situation shown in Fig. 8.15b where we have a 4-probe set up for measuring conductance. Here, current is injected into one probe (probe 1), extracted from another (probe 4), while the resulting voltage drop over the device is measured between two *other* probes (probes 2 and 3). The conductance of the device (between the latter two probes) is now given by $G_{4\text{-probe}} = G_{14,23} = I/\mathrm{V}$.

If we apply Büttiker's multiprobe formula (8.167) to the latter scenario, we will get the following matrix equation:

$$\begin{bmatrix} I_1 \\ I_2 \\ I_3 \\ I_4 \end{bmatrix} = \frac{2e}{h} \begin{bmatrix} M - R_{1\to1} & -T_{2\to1} & -T_{3\to1} & -T_{4\to1} \\ -T_{1\to2} & M - R_{2\to2} & -T_{3\to2} & -T_{4\to2} \\ -T_{1\to3} & -T_{2\to3} & M - R_{3\to3} & -T_{4\to3} \\ -T_{1\to4} & -T_{2\to4} & -T_{3\to4} & M - R_{4\to4} \end{bmatrix} \begin{bmatrix} \mu_1 \\ \mu_2 \\ \mu_3 \\ \mu_4 \end{bmatrix}. \tag{8.172}$$

Next, we will assume that the voltmeter connected between the two voltage probes is an ideal voltmeter, which will draw no current and hence result in

$$I_2 = -I_3 = 0. \tag{8.173}$$

Using the last condition in the immediately preceding equation, we get

$$\begin{aligned} [M - R_{2\to2}]\mu_2 &= T_{1\to2}\mu_1 + T_{3\to2}\mu_3 + T_{4\to2}\mu_4 \\ [M - R_{3\to3}]\mu_3 &= T_{1\to3}\mu_1 + T_{2\to3}\mu_2 + T_{4\to3}\mu_4. \end{aligned} \tag{8.174}$$

Thereafter, using the result that $I_1 = -I_4$, we get

$$\begin{aligned} &[M - R_{1\to1}]\mu_1 - T_{2\to1}\mu_2 - T_{3\to1}\mu_3 - T_{4\to1}\mu_4 \\ &\quad = -[M - R_{4\to4}]\mu_4 + T_{1\to4}\mu_1 + T_{2\to4}\mu_2 + T_{3\to4}\mu_3. \end{aligned} \tag{8.175}$$

Using the reciprocity relation $T_{ij} = T_{ji}$, we can use (8.174) to express μ_1 and μ_4 in terms of μ_2 and μ_3 as follows:

$$\mu_1 = \frac{[(M - R_{2\to2})T_{3\to4} + T_{2\to3}T_{2\to4}]\,\mu_2 - [(M - R_{3\to3})T_{2\to4} + T_{2\to3}T_{3\to4}]\,\mu_3}{T_{1\to2}T_{3\to4} - T_{1\to3}T_{2\to4}}$$
$$\mu_4 = \frac{[(M - R_{2\to2})T_{1\to3} + T_{1\to2}T_{2\to3}]\,\mu_2 - [(M - R_{3\to3})T_{1\to2} + T_{1\to3}T_{2\to3}]\,\mu_3}{T_{1\to3}T_{2\to4} - T_{1\to2}T_{3\to4}}. \tag{8.176}$$

Now, from (8.172), we obtain

$$\begin{aligned} I_1 &= \frac{2e}{h}\{(M - R_{1\to1})\mu_1 - T_{2\to1}\mu_2 - T_{3\to1}\mu_3 - T_{4\to1}\mu_4\} \\ &= \frac{2e}{h}\left\{\left[(M - R_{1\to1})\frac{(M - R_{2\to2})T_{3\to4} + T_{2\to3}T_{2\to4}}{T_{1\to2}T_{3\to4} - T_{1\to3}T_{2\to4}} - \right.\right. \\ &\qquad \left. T_{1\to4}\frac{(M - R_{2\to2})T_{1\to3} + T_{1\to2}T_{2\to3}}{T_{1\to3}T_{2\to4} - T_{1\to2}T_{3\to4}} - T_{1\to2}\right]\mu_2 \\ &\qquad - \left[(M - R_{1\to1})\frac{(M - R_{3\to3})T_{2\to4} + T_{2\to3}T_{3\to4}}{T_{1\to2}T_{3\to4} - T_{1\to3}T_{2\to4}} - \right. \\ &\qquad \left.\left. T_{1\to4}\frac{(M - R_{3\to3})T_{1\to2} + T_{1\to3}T_{2\to3}}{T_{1\to3}T_{2\to4} - T_{1\to2}T_{3\to4}} - T_{1\to3}\right]\mu_3\right\}. \end{aligned} \tag{8.177}$$

The conductance measured in the 4-probe setup will be $G_{4\text{-probe}} = I/V = eI_1/(\mu_2 - \mu_3)$. Clearly, we can find this result from the last equation, and (8.175) and (8.176) that relate μ_2 to μ_3, but then the resulting expression for $G_{4\text{-probe}}$ will be quite cumbersome (we leave it to the reader to work it out). A much simpler expression can be obtained if we make the following assumptions:

1. First, we will assume that the voltage probes are *weakly coupled* to the device. It is as if there are tunnel barriers in leads 2 and 3 that make the transmission probability of electrons passing through these leads very small. Let us call these transmission probabilities ϵ, where $\epsilon \to 0$, and assume that they are the same in leads 2 and 3 since these leads are nominally identical.
2. Next, we will assume that the current probes are strongly coupled to the device, i.e., there are no tunnel barriers in these probes and the probability of transmitting from these probes into the device is unity.
3. Finally, we will also assume that the bulk of the scattering encountered by electrons going from one current probe (say, probe 1) to the other current probe (probe 4) is encountered in the region between the voltage probes, so that we can lump all scattering into one scatterer located between the voltage probes. This is shown by the shaded scatterer in Fig. 8.15b. Let the transmission probability associated with traversing this scatterer be $\mathcal{T}$.

In that case, we can see that

$$\begin{aligned}
T_{1\to 4} &= \mathcal{T} \\
T_{1\to 2} &= T_{3\to 4} = \epsilon(M + \mathcal{R}) \\
T_{2\to 4} &= T_{3\to 1} = \epsilon\mathcal{T} \\
T_{2\to 3} &= \epsilon^2\mathcal{T},
\end{aligned} \tag{8.178}$$

where $\mathcal{R}$ is the reflection probability associated with reflection from the lumped scatterer, and $\mathcal{T} + \mathcal{R} = M$. To understand where the preceding relations come from, consider the fact that an electron can reach probe 2 from probe 1 by either heading straight from 1 to 2 or by starting out from 1 toward probes 3 and 4, but then reflecting off the lumped scatterer with a probability $\mathcal{R}$ to journey toward probe 2. Once the electron enters lead 2, it will ultimately emerge out of that lead by transmitting through the barrier in lead 2 with probability ϵ. Therefore, the total transmission probability $T_{1\to 2}$ will be $\epsilon(M + \mathcal{R})$. Similarly, an electron can reach probe 4 from probe 2 in only one way: transmit through the barrier in lead 2 with probability ϵ and then traverse the lumped scatterer with a probability $\mathcal{T}$ so that $T_{2\to 4} = \epsilon\mathcal{T}$. Finally, an electron trying to reach probe 3 from probe 2 has an arduous task. It has to transmit through lead 2 with the small probability ϵ, then transmit through the lumped scatterer with probability $\mathcal{T}$ and finally transmit through lead 3 with small probability ϵ. Therefore, $T_{2\to 3} = \epsilon^2\mathcal{T}$, which is vanishingly small.

Using (8.178) in (8.177), and making use of the fact $\epsilon \ll 1$, we get

$$\begin{aligned}
I_1 = I &\approx \frac{2e}{h} M\mathcal{T} \frac{2(M + \mathcal{R}) + 2\mathcal{T}}{(M + \mathcal{R})^2 - \mathcal{T}^2} (\mu_2 - \mu_3) \\
&= \frac{2e^2}{h}\frac{\mathcal{T}}{\mathcal{R}} \mathrm{V}.
\end{aligned} \tag{8.179}$$

Therefore, with all the three assumptions that we made, we end up with

$$G_{\text{4-probe}} = G_{14,23} = I/\mathrm{V} \approx \frac{2e^2}{h}\frac{M\mathcal{T}}{\mathcal{R}} = \frac{2e^2}{h}\frac{M\mathcal{T}}{M - \mathcal{T}}. \tag{8.180}$$

Curiously, Landauer [26, 27] too had derived the last equation for the conductance of a device in phase coherent linear response transport, but without introducing the 4-probe concept and without making all the assumptions that we made. His derivation was very different from the one outlined here (it took into account spatially localized electric fields caused by the scatterers), but nevertheless yielded the same final result as (8.180). The 2-probe result (8.144) or (8.171), on the other hand, was originally derived by Economou and Soukoulis [28], as well as Fisher and Lee [29], who did not use the type of derivation that we have used, but instead used the so-called Kubo formula which yields the linear response conductance based on an entirely different formalism. For some time, there was immense disagreement as to which formula—(8.171) or (8.180)—yields the correct linear response conductance of a device when transport is phase coherent [30].

The first individuals to shed some slight on this matter were Engquist and Anderson [31] who introduced the notion of using a 4-probe setup—with separate current and voltage leads—to justify (8.180). With this setup, they derived (8.180) while tacitly assuming weak coupling of the device to the voltage probes. Their derivation was slightly different from the one outlined here which is reproduced from Büttiker [32], but the essential ingredients were similar. In the end, it is clear that (8.180) refers to a 4-probe measurement with weakly coupled voltage probes and (8.171) refers to a 2-probe measurement. Hence, we will call the expressions in these equations the 4- and 2-probe Landauer conductances, respectively, and their reciprocals will be the 4- and 2-probe Landauer resistances.

8.6.1 The Contact Resistance

Transport experimentalists are very familiar with the fact that a two-terminal measurement of resistance never yields the true resistance of a device. Instead, it yields a resistance which is the sum of the true resistance and the so-called "contact resistance" R_{contact}. The latter comes about because the leads and the device are dissimilar entities and therefore a junction resistance will inevitably materialize where the two meet. However, a four-terminal measurement will eliminate the contact resistance and yield the true resistance. In other words, we can roughly write

$$R_{\text{2-probe}} = R_{\text{4-probe}} + R_{\text{contact}}. \tag{8.181}$$

Consider a perfect conductor that has no scatterer and transport through it is completely ballistic. The transmission probability through such a device is unity for every channel and hence, if there are M propagating channels, then $\mathcal{T} = M$. The 4-probe resistance of this device is $R_{\text{4-probe}} = 1/G_{\text{4-probe}} = (h/2e^2)[(M - \mathcal{T})/\mathcal{T}] = (h/2e^2)[(M - M)/\mathcal{T}] = 0$. This makes perfect sense. If there is no scatterer, there is nothing to "resist" the flow of current and hence the resistance should vanish. However, the 2-probe resistance is $R_{\text{2-probe}} = (h/2e^2)[1/\mathcal{T}] = (h/2e^2)[1/M]$, which does not vanish.

Substituting this result in (8.181), we immediately get that

$$R_{\text{contact}} = \frac{h}{2e^2}\frac{1}{M}. \tag{8.182}$$

The identification of the excess resistance (difference between the 2-probe and 4-probe resistance formulas in the case of ballistic transport) as contact resistance is due to Imry [33]. Therefore, the last equation gives an expression for the contact resistance in the linear response, phase coherent regime.

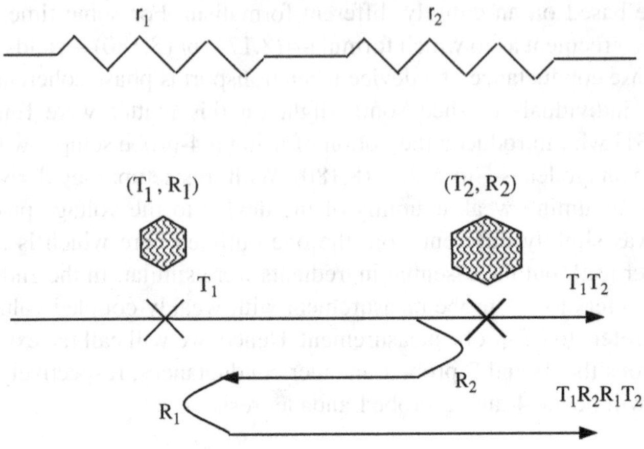

Fig. 8.16 Two resistors r_1 and r_2 in series. Each resistor is represented as a lumped scatterer characterized by a transmission and reflection probability (T, R). The ray diagram for finding the overall transmission through the series combination is shown

8.6.2 The Addition Law of Series Resistance

All students of engineering and physics are very familiar with the addition law of series resistance. Consider two resistors r_1 and r_2 in series. Normally, we will expect that the series resistance of the two will be $r_1 + r_2$. Assume now that each resistor is an ideal quantum wire with only one occupied subband so that there is a single channel for transport, or, in other words, a single propagating mode in each resistor. In the phase coherent regime, when the resistance of either resistor is given by the Landauer formula, we will show that indeed the series resistance of the two will be $r_1 + r_2$ if we use the 4-probe formula, but *not* if we use the 2-probe formula [34]. This is a surprising result, but makes perfect sense.

In Fig. 8.16, we show two of the denumerably infinite number of so-called Feynman paths for finding the transmission probability through the two resistors in terms of the transmission and reflection probabilities associated with each resistor. The Feynman paths depict the possible real space trajectories of an electron determined by multiple reflections. To find the total transmission through the series combination, we will merely sum the transmission probabilities associated with each Feynman path. The reason why we will add the transmission *probabilities* and not the transmission *amplitudes* associated with these paths is that in the contacts, the electrons thermalize (equilibrate with the contacts) and lose all phase information. Therefore, we must add probabilities and not complex amplitudes that contain phase information. Clearly, the total transmission probability through the series combination is

$$\begin{aligned} T_{\text{series}} &= T_1 \left(1 + R_1 R_2 + R_1^2 R_2^2 + \ldots\ldots\ldots\right) T_2 \\ &= \frac{T_1 T_2}{1 - R_1 R_2}. \end{aligned} \tag{8.183}$$

Therefore,

$$\frac{1}{T_{\text{series}}} = \frac{1 - R_1 R_2}{T_1 T_2} = \frac{1 - (1 - T_1)(1 - T_2)}{T_1 T_2}$$
$$= \frac{1}{T_1} + \frac{1}{T_2} - 1, \tag{8.184}$$

where we have used the fact that $T_n + R_n = 1$ for single-channeled transport.

Now, if we use the 2-probe formula, then the series resistance will be

$$r_{\text{series}} = \frac{h}{2e^2}\frac{1}{T_{\text{series}}} = \frac{h}{2e^2}\frac{1}{T_1} + \frac{h}{2e^2}\frac{1}{T_2} - \frac{h}{2e^2}$$
$$= r_1 + r_2 - \frac{h}{2e^2} \neq r_1 + r_2. \tag{8.185}$$

We are off from the expected formula by the contact resistance of either resistor! In the 2-probe formula, we have implicitly included the contact resistance of each resistor, so that we have counted it twice. Hence, we must subtract one contact resistance to get the correct result.

However, if we use the 4-probe formula, then the series resistance will be

$$r_{\text{series}} = \frac{h}{2e^2}\frac{1 - T_{\text{series}}}{T_{\text{series}}} = \frac{h}{2e^2}\frac{1 - T_1}{T_1} + \frac{h}{2e^2}\frac{1 - T_2}{T_2}$$
$$= r_1 + r_2, \tag{8.186}$$

which is the expected result. The 4-probe formula always yields the expected series resistance since, in this case, there is no issue of the contact resistance.

In (8.184), the quantities T_1 and T_2 can be viewed as the transmission probabilities through two sections of a resistor of lengths L_1 and L_2, respectively, while the total length of the resistor is $L = L_1 + L_2$. Therefore, we can rewrite (8.184) as

$$\frac{1}{T_L} = \frac{1}{T_{L_1}} + \frac{1}{T_{L_2}} - 1, \tag{8.187}$$

If we are seeking an expression for the transmission probability through a section as a function of its length, the following function satisfies the last equation:

$$T(L) = \frac{L_0}{L + L_0}, \tag{8.188}$$

where L_0 is the length at which the transmission probability becomes 0.5. What this expression shows is that at small length $L \ll L_0$, the transmission probability is independent of L and nearly unity, while for large length $L \gg L_0$, the transmission probability falls off as the inverse of the length L.

8.6.3 Computing the Landauer Resistances by Finding the Transmission Probability Through a Device

A truly wonderful feature of the Landauer formulas is that in order to find either the 2- or the 4-probe Landauer resistance[6] of a device in linear response and at low temperatures, all that we need to know is the total transmission probability of electrons that are at the Fermi level in the contacts. Therefore, we need a handy prescription to find this quantity.

The transmission probability through the device can be generally found by solving the Schrödinger equation in a one-dimensional spatially varying potential which includes elastic scattering potentials and all other potentials due to external electric fields and space charges. The potential varies spatially in the direction of transmission, or current flow. Equation (3.93) provides the transmission amplitude t_k for an electron incident on the device from the injecting contact with wavevector k . Since the matrix elements W_{pq} are always purely real [35], we can write

$$\mathcal{T}(k) = |t_k|^2 = \frac{4k^2}{k^2(W_{11} + W_{22})^2 + (W_{12}k^2 - W_{21})^2}. \tag{8.189}$$

The last relation gives us the transmission probability of any electron incident with wavevector k on a device containing a spatially varying potential in the direction of current flow. In order to find the Landauer resistances of such a device, we need to find the transmission probabilities of an electron at the Fermi energy E_F for every propagating mode, i.e., for every occupied subband in the leads. We will call these probabilities $T_{p,q}$, where p and q are the subband indices in the leads.

We proceed by first finding the wavevector $k_{p,q}$ for an electron at the Fermi level in the (p, q)-th subband by using the relation

$$E_F = \frac{\hbar^2 k_{p,q}^2}{2m_l} + \epsilon_{p,q}, \tag{8.190}$$

where $\epsilon_{p,q}$ is the energy at the bottom of the (p, q)-th subband and m_l is the effective mass of an electron in the leads.[7] We then set $k = k_{p,q}$ in (8.189) to find T_{pq}, and from that we can find the total transmission probability $\mathcal{T}$ of all propagating modes at the Fermi level since $\mathcal{T} = \sum_{p,q}^{\text{propagating modes}} T_{pq}$. This then allows us to find the Landauer resistances of any device containing a spatially varying potential in the direction of current flow.

---

[6]Whenever we mention 4-probe Landauer resistance, we tacitly imply that the probes are weakly coupled to the device so that the 4-probe formula is valid.

[7]The quantity $\epsilon_{p,q}$ is of course found from the geometry and transverse dimensions of the leads. For example, if the cross-section of the leads is a rectangle of dimensions W_y and W_z, then $\epsilon_{p,q} = (\hbar^2/2m_l)[(p\pi/W_y)^2 + (q\pi/W_z)^2]$.

8.7 The Landau–Vlasov Equation or the Collisionless Quantum Boltzmann Transport Equation

So far, we have described quantum transport in the framework of the Schrödinger equation, which is very different from the Boltzmann formalism that we used to describe classical transport in Chap. 2. In Boltzmann formalism, we solve the BTE to find the classical Boltzmann distribution function whose various moments are transport variables, such as carrier concentration, current density, etc. It is therefore natural to ask if there is a quantum-mechanical equivalent of the BTE which will yield perhaps a "quantum-mechanical distribution function" whose moments might yield transport variables. The answer to that question is "yes," but before we examine this further, let us revisit the reasons why the Boltzmann distribution function cannot describe phase coherent quantum phenomena.

The Boltzmann distribution function is fundamentally a classical construct. It is a "probability" and hence always positive indefinite, i.e., its value can be positive or zero, but never negative. Such a quantity will be incapable of describing quantum-mechanical phenomena like destructive interference where two nonzero distribution functions will have to superpose to produce a null since at the point of destructive interference, the probability of finding an electron should vanish. That can only happen if one of the distribution functions has a positive value and the other a negative value. But negative values are not permissible for the Boltzmann distribution function, which is why it can never describe quantum reality. Of course the Boltzmann distribution function has other shortcomings too. It is simultaneously a function of spatial and wavevector coordinates $\vec{r}$ and $\vec{k}$, which is not permissible in the quantum world since $\vec{r}$ and $\vec{k}$ are Fourier transform pairs and therefore must obey the Heisenberg Uncertainty relation $\Delta\vec{r} \cdot \Delta\vec{k} \geq 1/2$.

It is, however, possible to define an alternate quantum-mechanical function that obeys an equation which looks like the BTE mathematically. Such a function will not be positive indefinite and will not even necessarily be a real quantity, which is perfectly fine since it will not have anything to do with "probability." The equation that describes the evolution of this function in time t, real space $\vec{r}$ and wavevector space $\vec{k}$ has exactly the same mathematical appearance as the BTE, except that the right-hand side is zero. Since the right-hand side of the BTE gives us the time rate of change of the distribution function due to collisions, this equation will have the appearance of the collisionless BTE describing ballistic transport. Therefore, it is sometimes referred to as the "collisionless quantum BTE." Landau and Vlasov had used this equation extensively to study ballistic transport in plasmas, which is why it is also sometimes referred to as the Landau–Vlasov equation. We will derive it next.

Let us start with the Schrödinger equation governing a single electron in a time- and space-dependent potential. It is

$$i\hbar \frac{\partial \psi(\vec{r},t)}{\partial t} = \left[-\frac{\hbar^2}{2m^*}\nabla^2 + V(\vec{r},t) \right] \psi(\vec{r},t). \tag{8.191}$$

We can treat the potential $V(\vec{r},t)$ as a perturbation, so that the unperturbed wavefunctions are the free electron wavefunctions

$$\psi_0(\vec{r},t) = \frac{1}{\sqrt{\Omega}} \mathrm{e}^{\mathrm{i}\left(\vec{k}'\cdot\vec{r} - E_{k'}t/\hbar\right)}, \tag{8.192}$$

where

$$E_{k'} = \frac{\hbar^2 k'^2}{2m^*}. \tag{8.193}$$

The perturbed wavefunctions can be written as usual:

$$\psi(\vec{r},t) = \sum_{\vec{k}'} \overline{C_{\vec{k}'}}(t) \frac{1}{\sqrt{\Omega}} \mathrm{e}^{\mathrm{i}\left(\vec{k}'\cdot\vec{r} - E_{k'}t/\hbar\right)} = \sum_{\vec{k}'} C_{\vec{k}'}(t) \frac{1}{\sqrt{\Omega}} \mathrm{e}^{\mathrm{i}\vec{k}'\cdot\vec{r}}, \tag{8.194}$$

where

$$C_{\vec{k}'}(t) = \overline{C_{\vec{k}'}}(t) e^{-\mathrm{i}E_{k'}t/\hbar}. \tag{8.195}$$

Substituting (8.194) in (8.192), we get

$$\begin{aligned} \mathrm{i}\hbar \sum_{\vec{k}'} \frac{\partial C_{\vec{k}'}(t)}{\partial t} \frac{1}{\sqrt{\Omega}} \mathrm{e}^{\mathrm{i}\vec{k}'\cdot\vec{r}} &= \sum_{\vec{k}'} C_{\vec{k}'}(t) \left[-\frac{\hbar^2}{2m^*} \nabla^2 \right] \frac{1}{\sqrt{\Omega}} \mathrm{e}^{\mathrm{i}\vec{k}'\cdot\vec{r}} + \sum_{\vec{k}'} V(\vec{r},t) C_{\vec{k}'}(t) \frac{1}{\sqrt{\Omega}} \mathrm{e}^{\mathrm{i}\vec{k}'\cdot\vec{r}} \\ &= \sum_{\vec{k}'} C_{\vec{k}'}(t) E_{k'} \frac{1}{\sqrt{\Omega}} \mathrm{e}^{\mathrm{i}\vec{k}'\cdot\vec{r}} + \sum_{\vec{k}'} V(\vec{r},t) C_{\vec{k}'}(t) \frac{1}{\sqrt{\Omega}} \mathrm{e}^{\mathrm{i}\vec{k}'\cdot\vec{r}}. \end{aligned} \tag{8.196}$$

If we multiply the last equation throughout by $\frac{1}{\sqrt{\Omega}} \mathrm{e}^{-\mathrm{i}\vec{k}\cdot\vec{r}}$, integrate over all space and use the orthonormality condition

$$\int_0^\infty d^3\vec{r} \frac{1}{\sqrt{\Omega}} \mathrm{e}^{-\mathrm{i}\vec{k}\cdot\vec{r}} \frac{1}{\sqrt{\Omega}} \mathrm{e}^{\mathrm{i}\vec{k}'\cdot\vec{r}} = \int_0^\infty d^3\vec{r} \frac{1}{\Omega} \mathrm{e}^{-\mathrm{i}(\vec{k}-\vec{k}')\cdot\vec{r}} = \frac{1}{\Omega} \delta(\vec{k}-\vec{k}') = \delta_{\vec{k},\vec{k}'}, \tag{8.197}$$

where the first delta is a Dirac delta function and the last one is a Kronecker delta, then we get

$$\begin{aligned} \mathrm{i}\hbar \sum_{\vec{k}'} \frac{\partial C_{\vec{k}'}(t)}{\partial t} \delta_{\vec{k},\vec{k}'} &= \sum_{\vec{k}'} C_{\vec{k}'}(t) E_{k'} \delta_{\vec{k},\vec{k}'} + \sum_{\vec{k}'} C_{\vec{k}'}(t) \frac{1}{\Omega} \int_0^\infty d^3\vec{r} V(\vec{r},t) \mathrm{e}^{-\mathrm{i}(\vec{k}-\vec{k}')\cdot\vec{r}} \\ &= \sum_{\vec{k}'} C_{\vec{k}'}(t) E_{k'} \delta_{\vec{k},\vec{k}'} + \sum_{\vec{k}'} C_{\vec{k}'}(t) V_{\vec{k}-\vec{k}'}(t), \end{aligned} \tag{8.198}$$

where

$$V_{\vec{k}-\vec{k}'}(t) = \frac{1}{\Omega}\int_0^\infty d^3\vec{r}\,V(\vec{r},t)\mathrm{e}^{-\mathrm{i}(\vec{k}-\vec{k}')\cdot\vec{r}}. \tag{8.199}$$

This quantity is the Fourier component of the potential $V(\vec{r},t)$ at the wavevector $\vec{k}-\vec{k}'$

In (8.200), the Kronecker delta will remove the summations, so that this equation reduces to

$$\frac{\partial C_{\vec{k}}(t)}{\partial t} + \frac{i}{\hbar}C_{\vec{k}}(t)E_k = \sum_{\vec{k}'} C_{\vec{k}'}(t)\frac{V_{\vec{k}-\vec{k}'}(t)}{\mathrm{i}\hbar}. \tag{8.200}$$

Taking the complex conjugate of both sides, we obtain

$$\frac{\partial C^*_{\vec{k}}(t)}{\partial t} - \frac{i}{\hbar}C^*_{\vec{k}}(t)E_k = -\sum_{\vec{k}'} C^*_{\vec{k}'}(t)\frac{V^*_{\vec{k}-\vec{k}'}(t)}{\mathrm{i}\hbar}. \tag{8.201}$$

Now, the potential $V(\vec{r},t)$ had to have been a real quantity since the Hamiltonian must always be Hermitian. Therefore, $V^*_{\vec{k}-\vec{k}'}(t) = V^*_{\vec{k}'-\vec{k}}(t)$, which yields

$$\frac{\partial C^*_{\vec{k}}(t)}{\partial t} - \frac{i}{\hbar}C^*_{\vec{k}}(t)E_k = -\sum_{\vec{k}'} C^*_{\vec{k}'}(t)\frac{V_{\vec{k}'-\vec{k}}(t)}{\mathrm{i}\hbar}. \tag{8.202}$$

Multiplying the last equation by $C_{\vec{k}+\vec{q}}(t)$, where $\vec{q}$ is an arbitrary wavevector, we obtain

$$\frac{\partial C^*_{\vec{k}}(t)}{\partial t}C_{\vec{k}+\vec{q}}(t) - \frac{i}{\hbar}C^*_{\vec{k}}(t)C_{\vec{k}+\vec{q}}(t)E_k = -\sum_{\vec{k}"} C^*_{\vec{k}"}(t)C_{\vec{k}+\vec{q}}(t)\frac{V_{\vec{k}"-\vec{k}}(t)}{\mathrm{i}\hbar}. \tag{8.203}$$

where we have replaced $\vec{k}'$ with $\vec{k}$" since both $\vec{k}'$ and $\vec{k}$" are dummy variables for summation.

Since (8.200) is valid for any arbitrary wavevector $\vec{k}$, we can replace $\vec{k}$ with $\vec{k}+\vec{q}$ in this equation to get

$$\frac{\partial C_{\vec{k}+\vec{q}}(t)}{\partial t} + \frac{i}{\hbar}C_{\vec{k}+\vec{q}}(t)E_{|\vec{k}+\vec{q}|} = \sum_{\vec{k}'} C_{\vec{k}'}(t)\frac{V_{\vec{k}+\vec{q}-\vec{k}'}(t)}{\mathrm{i}\hbar}. \tag{8.204}$$

Multiplying the last equation throughout by $C^*_{\vec{k}}(t)$, we obtain

$$C^*_{\vec{k}}(t)\frac{\partial C_{\vec{k}+\vec{q}}(t)}{\partial t} + \frac{i}{\hbar}C^*_{\vec{k}}(t)C_{\vec{k}+\vec{q}}(t)E_{|\vec{k}+\vec{q}|} = \sum_{\vec{k}'} C^*_{\vec{k}}(t)C_{\vec{k}'}(t)\frac{V_{\vec{k}+\vec{q}-\vec{k}'}(t)}{\mathrm{i}\hbar}. \tag{8.205}$$

Adding (8.203) and (8.205), we find

$$\frac{\partial\left[C^*_{\vec{k}}(t)C_{\vec{k}+\vec{q}}(t)\right]}{\partial t}+\frac{i}{\hbar}C^*_{\vec{k}}(t)C_{\vec{k}+\vec{q}}(t)\left[E_{|\vec{k}+\vec{q}|}-E_{\mathrm{k}}\right]$$
$$=\sum_{\vec{k}'}C^*_{\vec{k}}(t)C_{\vec{k}'}(t)\frac{V_{\vec{k}+\vec{q}-\vec{k}'}(t)}{i\hbar}-\sum_{\vec{k}"}C^*_{\vec{k}"}(t)C_{\vec{k}+\vec{q}}(t)\frac{V_{\vec{k}"-\vec{k}}(t)}{i\hbar}. \tag{8.206}$$

Since both $\vec{k}'$ and $\vec{k}$" are dummy variables for summation, let

$$\vec{k}'=\vec{k}+\vec{q}-\vec{q}'$$
$$\vec{k}"=\vec{k}+\vec{q}'. \tag{8.207}$$

This will make $\vec{q}'$ the new dummy variable for summation.
With the last variable transformations, (8.206) becomes

$$\frac{\partial\left[C^*_{\vec{k}}(t)C_{\vec{k}+\vec{q}}(t)\right]}{\partial t}+\frac{i}{\hbar}C^*_{\vec{k}}(t)C_{\vec{k}+\vec{q}}(t)\left[E_{|\vec{k}+\vec{q}|}-E_{\mathrm{k}}\right]$$
$$=\sum_{\vec{q}'}\frac{V_{\vec{q}'}(t)}{i\hbar}\left[C^*_{\vec{k}}(t)C_{\vec{k}+\vec{q}-\vec{q}'}(t)-C_{\vec{k}+\vec{q}}(t)C^*_{\vec{k}+\vec{q}'}(t)\right]. \tag{8.208}$$

Now, define a quantum *correlation function* $\hat{f}\left(\vec{k}_1,\vec{k}_2,t\right)$ as

$$\hat{f}(\vec{k}_1,\vec{k}_2,t)=C^*_{\vec{k}_1}(t)C_{\vec{k}_2+\vec{k}_1}(t). \tag{8.209}$$

With this definition, (8.208) becomes

$$\frac{\partial\hat{f}(\vec{k},\vec{q},t)}{\partial t}+\frac{i}{\hbar}\hat{f}(\vec{k},\vec{q},t)\left[E_{|\vec{k}+\vec{q}|}-E_{\mathrm{k}}\right]$$
$$=\sum_{\vec{q}'}\frac{V_{\vec{q}'}(t)}{i\hbar}[\hat{f}(\vec{k},\vec{q}-\vec{q}',t)-\hat{f}(\vec{k}+\vec{q},\vec{q}-\vec{q}',t)]. \tag{8.210}$$

We can now make a Taylor series expansion of the energy $E_{|\vec{k}+\vec{q}|}$ and write it as

$$E_{|\vec{k}+\vec{q}|}=E_{\mathrm{k}}+\vec{q}\cdot\vec{\nabla}_{\vec{k}}E_{\mathrm{k}}+\frac{|\vec{q}|^2}{2!}\nabla^2_{\vec{k}}E_{\mathrm{k}}+\ldots. \tag{8.211}$$

If $|\vec{q}|\ll|\vec{k}|$, then we can neglect the second and higher order terms involving $|\vec{q}|$ in the Taylor expansion and write

$$E_{|\vec{k}+\vec{q}|}-E_{\mathrm{k}}\approx\vec{q}\cdot\vec{\nabla}_{\vec{k}}E_{\mathrm{k}}. \tag{8.212}$$

Similarly, using Taylor series expansion, we find that for small $\vec{q}$,

$$\hat{f}(\vec{k},\vec{q}-\vec{q}',t) - \hat{f}(\vec{k}+\vec{q},\vec{q}-\vec{q}',t) \approx -\vec{q}\cdot\vec{\nabla}_{\mathrm{k}}\hat{f}(\vec{k},\vec{q}-\vec{q}',t). \tag{8.213}$$

Equation (8.210) can then be written as

$$\frac{\partial \hat{f}(\vec{k},\vec{q},t)}{\partial t} + \left[\mathrm{i}\vec{q}\cdot\frac{1}{\hbar}\vec{\nabla}_{\mathrm{k}}E_{\mathrm{k}}\right]\hat{f}(\vec{k},\vec{q},t) = \sum_{\vec{q}'}\mathrm{i}\vec{q}\frac{V_{\vec{q}'}(t)}{\hbar}\cdot\vec{\nabla}_{\mathrm{k}}\hat{f}(\vec{k},\vec{q}-\vec{q}',t). \tag{8.214}$$

or

$$\frac{\partial \hat{f}(\vec{k},\vec{q},t)}{\partial t} + \mathrm{i}\vec{q}\cdot\vec{v}_k\hat{f}(\vec{k},\vec{q},t) = \mathrm{i}\vec{q}\frac{V_{\vec{q}'}(t)}{\hbar}\otimes\vec{\nabla}_{\mathrm{k}}\hat{f}(\vec{k},\vec{q},t). \tag{8.215}$$

where $\otimes$ denotes the convolution operation and $\vec{v}_k = \frac{1}{\hbar}\vec{\nabla}_{\mathrm{k}}E_{\mathrm{k}}$.

We now take the inverse Fourier transform of both sides of the preceding equation, i.e., we multiply both sides by $\mathrm{e}^{\mathrm{i}\vec{q}\cdot\vec{r}}$ and integrate over all $\vec{q}$, to obtain

$$\frac{\partial \hat{f}(\vec{k},\vec{r},t)}{\partial t} + \vec{v}_k\cdot\vec{\nabla}_{\mathrm{r}}\hat{f}(\vec{k},\vec{r},t) = \frac{\vec{\nabla}_{\mathrm{r}}V(\vec{r},t)}{\hbar}\cdot\vec{\nabla}_{\mathrm{k}}\hat{f}(\vec{k},\vec{r},t), \tag{8.216}$$

where

$$\hat{f}(\vec{k},\vec{r},t) = \int d^3\vec{q}\,\mathrm{e}^{\mathrm{i}\vec{q}\cdot\vec{r}}\hat{f}(\vec{k},\vec{q},t). \tag{8.217}$$

Since force is the negative gradient of potential energy, we can write

$$\vec{F}(\vec{r},t) = -\vec{\nabla}_{\mathrm{r}}V(\vec{r},t), \tag{8.218}$$

which will reduce the preceding equation to

$$\frac{\partial \hat{f}(\vec{r},\vec{k},t)}{\partial t} + \vec{v}_k\cdot\vec{\nabla}_{\mathrm{r}}\hat{f}(\vec{r},\vec{k},t) + \frac{\vec{F}(\vec{r},t)}{\hbar}\cdot\vec{\nabla}_{\mathrm{k}}\hat{f}(\vec{r},\vec{k},t) = 0. \tag{8.219}$$

The last equation has the same appearance as the BTE. Instead of the "distribution function," we have the quantum correlation function $\hat{f}(\vec{r},\vec{k},t)$. The latter is *not* the quantum-mechanical probability of finding at electron at position $\vec{r}$ with wavevector $\vec{k}$ at time t. It is not positive indefinite and clearly can be complex (not real). It is also a quantum-mechanically legitimate function since $\vec{k}$ and $\vec{r}$ are not Fourier transform pairs and therefore do not obey the uncertainty relation ($\vec{q}$ and $\vec{r}$ are Fourier transform pairs, but $\vec{q}$ is not $\vec{k}$).

It may seem that since the right-hand side is zero, this equation can only apply to the collisionless regime. That is actually not entirely correct since elastic (nondissipative) scattering caused by impurities can be incorporated in the potential $V(\vec{r},t)$ as a scattering potential, as long as the potential is varying slowly in space.

The slow variation is required since we assumed in the derivation that the Fourier transform of the position variable $\vec{r}$, i.e. the quantity $\vec{q}$, is small. That assumption is valid as long as all potentials are varying slowly in space so that they have no large wavevector (Fourier) components.

The correlation function $\hat{f}(\vec{r}, \vec{k}, t)$ does obey the collisionless BTE, but it is not immediately obvious what its use might be. We can actually formulate a quantum correlation function in a different way so that it will not only obey the collisionless BTE but its zeroth moment will give the expectation value of the particle density and the first moment will give the expectation value of the current density at position $\vec{r}$ at time t. Such a function will be an exact quantum-mechanical analog of the Boltzmann distribution function. We proceed to derive this quantum correlation function and show that it obeys the collisionless BTE.

8.8 A More Useful Collisionless Boltzmann Transport Equation

The Schrödinger equation for a particle with wavefunction $\psi(\vec{r}_1, t)$ is

$$i\hbar \frac{\partial \psi(\vec{r}_1, t)}{\partial t} = H(\vec{r}_1, t)\psi(\vec{r}_1, t). \tag{8.220}$$

Multiplying the above equation throughout by the wavefunction $\psi^*(\vec{r}_2, t)$, we obtain

$$i\hbar \psi^*(\vec{r}_2, t) \frac{\partial \psi(\vec{r}_1, t)}{\partial t} = H(\vec{r}_1, t)[\psi(\vec{r}_1, t)\psi^*(\vec{r}_2, t)]. \tag{8.221}$$

where the Hamiltonian $H(\vec{r}_1, t)$ does not act on $\psi^*(\vec{r}_2, t)$ since the former is a function of $\vec{r}_1$ and the latter is a function of $\vec{r}_2$.

Similarly, the Schrödinger equation for a particle with wavefunction $\psi(\vec{r}_2, t)$ is

$$i\hbar \frac{\partial \psi(\vec{r}_2, t)}{\partial t} = H(\vec{r}_2, t)\psi(\vec{r}_2, t). \tag{8.222}$$

which, upon complex conjugation, becomes

$$-i\hbar \frac{\partial \psi^*(\vec{r}_2, t)}{\partial t} = H(\vec{r}_2, t)\psi^*(\vec{r}_2, t). \tag{8.223}$$

where we have used the fact that $H^*(\vec{r}_2, t) = H(\vec{r}_2, t)$ since the Hamiltonian is hermitian as long as there is no dissipation.

Multiplying the last equation throughout by $-\psi(\vec{r}_1, t)$ yields

$$i\hbar \psi(\vec{r}_1, t) \frac{\partial \psi^*(\vec{r}_2, t)}{\partial t} = -H(\vec{r}_2, t)[\psi(\vec{r}_1, t)\psi^*(\vec{r}_2, t)]. \tag{8.224}$$

Adding (8.221) and (8.224), we get

$$i\hbar\frac{\partial g(\vec{r}_1,\vec{r}_2,t)}{\partial t} = [H(\vec{r}_1,t) - H(\vec{r}_2,t)]g(\vec{r}_1,\vec{r}_2,t)$$
$$= \left[-\frac{\hbar^2}{2m^*}\left(\nabla_{r_1}^2 - \nabla_{r_2}^2\right) + V(\vec{r}_1,t) - V(\vec{r}_2,t)\right] g(\vec{r}_1,\vec{r}_2,t), \quad (8.225)$$

where

$$g(\vec{r}_1,\vec{r}_2,t) = \psi(\vec{r}_1,t)\psi^*(\vec{r}_2,t). \quad (8.226)$$

We can rewrite (8.225) as

$$i\hbar\frac{\partial g(\vec{r}_1,\vec{r}_2,t)}{\partial t} = \left[-\frac{\hbar^2}{2m^*}(\vec{\nabla}_{r_1} + \vec{\nabla}_{r_2})(\vec{\nabla}_{r_1} - \vec{\nabla}_{r_2}) + V(\vec{r}_1,t) - V(\vec{r}_2,t)\right] g\left(\vec{r}_1,\vec{r}_2,t\right). \quad (8.227)$$

We will next define center-of-mass and relative coordinates as

$$\vec{R} = \frac{\vec{r}_1 + \vec{r}_2}{2}$$
$$\vec{r} = \vec{r}_1 - \vec{r}_2 \quad (8.228)$$

which will yield (recall the variable transformation carried out for excitons in Chap. 6)

$$i\hbar\frac{\partial g(\vec{R},\vec{r},t)}{\partial t} = \left[-\frac{\hbar^2}{m^*}\vec{\nabla}_{\vec{R}}\cdot\vec{\nabla}_{\vec{r}} + V(\vec{R}+\vec{r}/2,t) - V(\vec{R}-\vec{r}/2,t)\right] g(\vec{R},\vec{r},t) \quad (8.229)$$

We can expand the potentials in a Taylor series and retain only up to the first order terms if the potential is slowly varying in space. This transforms the last equation to

$$i\hbar\frac{\partial g(\vec{R},\vec{r},t)}{\partial t} = \left[-\frac{\hbar^2}{m^*}\nabla_{\vec{R}}\cdot\nabla_{\vec{r}} + \vec{r}\cdot\vec{\nabla}_R V(\vec{R},t)\right] g(\vec{R},\vec{r},t)$$
$$= \left[-\frac{\hbar^2}{2m^*}\vec{\nabla}_{\vec{R}}\cdot\vec{\nabla}_{\vec{r}} - \vec{r}\cdot\vec{F}(\vec{R},t)\right] g(\vec{R},\vec{r},t), \quad (8.230)$$

where $\vec{F}(\vec{R},t)$ is the force (negative gradient of potential).

If we now take the Fourier transform of the preceding equation with respect to the relative coordinate $\vec{r}$, i.e., if we multiply throughout by $e^{i\vec{k}\cdot\vec{r}}$ and integrate over all $\vec{r}$, then we get

$$\frac{\partial f(\vec{R},\vec{k},t)}{\partial t} + \frac{\hbar\vec{k}}{m^*}\cdot\vec{\nabla}_{\vec{R}} f(\vec{R},\vec{k},t) + \frac{\vec{F}(\vec{R},t)}{\hbar}\cdot\vec{\nabla}_{\vec{k}} f(\vec{R},\vec{k},t), \quad (8.231)$$

where

$$\tilde{f}(\vec{R},\vec{k},t) = \int d^3\vec{r}e^{-i\vec{k}\cdot\vec{r}}g(\vec{R},\vec{r},t). \tag{8.232}$$

In other words, the correlation function $\tilde{f}(\vec{R},\vec{k},t)$ is the Fourier transform of $g(\vec{R},\vec{r},t)$ with respect to the relative coordinate $\vec{r}$.

Note that the new correlation function $\tilde{f}(\vec{R},\vec{k},t)$ not only obeys an equation that looks identical with the collisionless BTE but it is a very meaningful function. If we integrate this function over all wavevector, then from (8.232), we get

$$\begin{aligned}\int d^3\vec{k}\tilde{f}(\vec{R},\vec{k},t) &= \int d^3\vec{r}\left[\int d^3\vec{k}e^{-i\vec{k}\cdot\vec{r}}\right]g(\vec{R},\vec{r},t) \\ &= \int d^3\vec{r}\delta(\vec{r})g(\vec{R},\vec{r},t) \\ &= g(\vec{R},0,t) \\ &= \psi(\vec{R},t)\psi^*(\vec{R},t) \\ &\propto n(\vec{R},t).\end{aligned} \tag{8.233}$$

Therefore, the zeroth moment of the new correlation function gives the particle density. Similarly, it can be shown that the first moment gives the particle current density [36]. In other words, this new correlation function is an exact quantum-mechanical analog of the Boltzmann distribution function. Its shortcoming, however, is that it can only describe ballistic transport or transport in the presence of slowly varying elastic scattering potentials which do not cause dissipation. If we wish to incorporate dissipation into our quantum transport model, then that becomes a seriously difficult task, which we address next.

8.9 Quantum Correlations

Consider an electron interacting with a spatially and temporally varying potential, as in a real device. Its wavefunction can be expanded in a complete orthonormal set (recall Chap. 3 material):

$$\Psi(\vec{r},t) = \frac{1}{\sqrt{\Omega}}\sum_{\vec{k}} C_{\vec{k}}(t)e^{i\vec{k}\cdot\vec{r}}. \tag{8.234}$$

We can define a density matrix as

$$\rho_{\vec{k},\vec{k}'}(t) = C^*_{\vec{k}}(t)C_{\vec{k}'}(t), \tag{8.235}$$

where, as usual, the asterisk denotes complex conjugate. In the full matrix form, the density matrix will be written as

$$[\rho(t)] = \begin{bmatrix} C^*_{\vec{k}_1}(t)C_{\vec{k}_1}(t) & C^*_{\vec{k}_1}(t)C_{\vec{k}_2}(t) & \cdots & \cdots & \cdots & C^*_{\vec{k}_1}(t)C_{\vec{k}_n}(t) \\ C^*_{\vec{k}_2}(t)C_{\vec{k}_1}(t) & C^*_{\vec{k}_2}(t)C_{\vec{k}_2}(t) & \cdots & \cdots & \cdots & C^*_{\vec{k}_2}(t)C_{\vec{k}_n}(t) \\ \cdots & \cdots & \cdots & \cdots & \cdots & \cdots \\ \cdots & \cdots & \cdots & \cdots & \cdots & \cdots \\ \cdots & \cdots & \cdots & \cdots & \cdots & \cdots \\ C^*_{\vec{k}_n}(t)C_{\vec{k}_1}(t) & C^*_{\vec{k}_n}(t)C_{\vec{k}_2}(t) & \cdots & \cdots & \cdots & C^*_{\vec{k}_n}(t)C_{\vec{k}_n}(t) \end{bmatrix}. \tag{8.236}$$

Clearly, a diagonal element $C^*_{\vec{k}_m}(t)C_{\vec{k}_m}(t) = |C_{\vec{k}_m}|^2(t)$ is the probability of finding the electron in the wavevector state $\vec{k}_m$ with wavefunction $(1/\sqrt{\Omega})\mathrm{e}^{\mathrm{i}\vec{k}_m\cdot\vec{r}}$ at time t, and is the Boltzmann distribution function $f(\vec{k}_m, t)$. The diagonal elements are all positive indefinite and real. The off-diagonal elements, on the other hand, give us the phase correlations between the different amplitudes. They can be complex quantities and are not positive indefinite. Therefore, the density matrix contains *more information* than the Boltzmann distribution function since the off-diagonal terms contain additional information which the Boltzmann distribution function does not have. Moreover, the density matrix is also inherently quantum-mechanical in nature since the phase relations between the different wavevector states is a quantum-mechanical attribute. Note that while the diagonal terms of the density matrix provide information about auto-correlations between the same wavevector states, the off-diagonal terms provide information about cross-correlations between different wavevector states.

Instead of working with amplitudes in wavevector space, we could have also used amplitudes in real space and written them as $C_{\vec{r}}(t)$ instead of $C_{\vec{k}}(t)$, so that (8.234) would have transformed to

$$\Psi(\vec{r}, t) = \frac{1}{\sqrt{\Omega}} \sum_{\vec{r}} C_{\vec{r}}(t)\mathrm{e}^{\mathrm{i}\vec{k}\cdot\vec{r}}. \tag{8.237}$$

We can then define a correlation function

$$-\mathrm{i}G^<(\vec{r}_1, t_1; \vec{r}_2, t_2) = \left\langle C^*_{\vec{r}_1}(t_1)C_{\vec{r}_2}(t_2)\right\rangle, \tag{8.238}$$

where i is the usual imaginary square-root of -1 and the brackets denote ensemble averaging over all the available states of the system, regardless of whether the system is in equilibrium or nonequilibrium. Note that this function tells us the correlation between the amplitude at position $\vec{r}_1$ at time t_1 and the amplitude at position $\vec{r}_2$ at time t_2.

Next, we can define center-of-mass and relative coordinates in the usual fashion:

$$\vec{R} = \frac{\vec{r}_1 + \vec{r}_2}{2}$$
$$\vec{r} = \vec{r}_1 - \vec{r}_2$$
$$T = \frac{t_1 + t_2}{2}$$
$$t = = t_1 - t_2, \tag{8.239}$$

and write the correlation function as

$$-\mathrm{i}G^{<}(\vec{r}_1, t_1; \vec{r}_2, t_2) = -\,\mathrm{i}G^{<}(\vec{R}, T; \vec{r}, t) = \left\langle C^*_{\vec{R}+\vec{r}/2}(T + t/2) C_{\vec{R}-\vec{r}/2}(T - t/2) \right\rangle. \tag{8.240}$$

Then we Fourier transform the relative variables $\vec{r}$ and t into $\vec{k}$ and ω, so that we obtain a new correlation function $W(\vec{k}, \vec{R}, \omega, T)$ given by:

$$\begin{aligned} W(\vec{k}, \vec{R}, \omega, T) &= -\mathrm{i} \int d^3\vec{r} \mathrm{e}^{-\mathrm{i}\vec{k}\cdot\vec{r}} \int \mathrm{d}t \mathrm{e}^{-\mathrm{i}\omega t} G^{<}(\vec{R}, T; \vec{r}, t) \\ &= \int d^3\vec{r} \mathrm{e}^{-\mathrm{i}\vec{k}\cdot\vec{r}} \int \mathrm{d}t \mathrm{e}^{-\mathrm{i}\omega t} \left\langle C^*_{\vec{R}+\vec{r}/2}(T + t/2) C_{\vec{R}-\vec{r}/2}(T - t/2) \right\rangle. \end{aligned} \tag{8.241}$$

This transformation is called the Wigner–Weyl transformation and the function $W(\vec{k}, \vec{R}, \omega, T)$ is called the *Wigner distribution function.*

If we integrate the Wigner distribution function over all wavevector and frequency, then we get

$$\begin{aligned} \int \frac{d^3\vec{k}}{(2\pi)^3} \int \frac{\mathrm{d}\omega}{2\pi} W(\vec{k}, \vec{R}, \omega, T) &= \int d^3\vec{r} \int \mathrm{d}t \left[\int \frac{d^3\vec{k}}{(2\pi)^3} \mathrm{e}^{-\mathrm{i}\vec{k}\cdot\vec{r}} \right] \left[\int \frac{\mathrm{d}\omega}{2\pi} \mathrm{e}^{-\mathrm{i}\omega t} \right] \\ &\quad \times \left\langle C^*_{\vec{R}+\vec{r}/2}(T + t/2) C_{\vec{R}-\vec{r}/2}(T - t/2) \right\rangle \\ &= \int d^3\vec{r} \int \mathrm{d}t \delta(\vec{r}) \delta(t) C^*_{\vec{R}+\vec{r}/2}(T + t/2) C_{\vec{R}-\vec{r}/2}(T - t/2) \\ &= \left\langle C^*_{\vec{R}}(T) C_{\vec{R}}(T) \right\rangle = \left\langle \left| C_{\vec{R}}(T) \right|^2 \right\rangle \\ &= -\mathrm{i}G^{<}(\vec{R}, T; \vec{R}, T), \end{aligned} \tag{8.242}$$

The quantity $\langle |C_{\vec{R}}(T)|^2 \rangle$ is the probability of finding an electron at position $\vec{R}$ at time T averaged over all available states of the system. Therefore, it is clearly proportional to the carrier concentration at position $\vec{R}$ at time T. Consequently, we can write

$$n(\vec{R},T) \propto -\mathrm{i}G^<(\vec{R},T;\vec{R},T) = \int \frac{d^3\vec{k}}{(2\pi)^3} \int \frac{\mathrm{d}\omega}{2\pi} W(\vec{k},\vec{R},\omega,T). \tag{8.243}$$

The Wigner distribution function $W(\vec{k},\vec{R},\omega,T)$ is therefore also a quantum mechanical analog of the Boltzmann distribution function, since integrating it over all wavevector and all frequencies yields the carrier concentration. In fact, it can also be shown that the current density at position $\vec{R}$ at time T is given by

$$\vec{J}(\vec{R},T) \propto -e \int \frac{d^3\vec{k}}{(2\pi)^3} \int \frac{\mathrm{d}\omega}{2\pi} \left[\frac{\hbar\vec{k} - e\vec{A}(\vec{R},T)}{m^*} \right] W(\vec{k},\vec{R},\omega,T), \tag{8.244}$$

where $\vec{A}(\vec{R},T)$ is a vector potential, due to such things as a magnetic field, at location $\vec{R}$ and time T. Therefore, the carrier concentration is like the zeroth moment and the current density is like the first moment of the Wigner distribution function. This is as far as the analogy between the Wigner function and the Boltzmann distribution function goes. The Wigner function does not satisfy an equation that looks like the BTE (unlike the correlation function $\tilde{f}(\vec{r},\vec{k},t)$), nor is it positive indefinite, nor does it violate the Heisenberg uncertainty principle since neither the pair $(\vec{R},\vec{k})$ nor the pair (ω, t) are Fourier transform pairs.

The reader can easily show that if we are interested in finding the carrier concentration at a particular energy E, or correspondingly at a particular frequency ω ($E = \hbar\omega$), then we will not be carrying out the integration over ω and the result will be

$$\begin{aligned} n(\omega,R,T) &\propto \int \mathrm{d}t e^{-\mathrm{i}\omega t} \left\langle C^*_{\vec{R}}(T+t/2) C_{\vec{R}}(T-t/2) \right\rangle \\ &= -\mathrm{i} \int \mathrm{d}t e^{-\mathrm{i}\omega t} G^<(\vec{R},T+t/2;\vec{R},T-t/2) = -\mathrm{i} \int \mathrm{d}t e^{-\mathrm{i}\omega t} G^<(\vec{R},t_1;R,t_2) \\ &= -\mathrm{i}G^<(\vec{R},T,\vec{R},\omega). \end{aligned} \tag{8.245}$$

Similarly, the current density at a particular frequency ω will be

$$\vec{J}(\omega,\vec{R},T) \propto \frac{e\hbar}{m^*} \left[\vec{\nabla}_R - \vec{\nabla}_{R'} \right] G^<(\vec{R},T,\vec{R}',\omega) \Big|_{\vec{R}=\vec{R}'} - \mathrm{i}\frac{e^2}{m^*} \vec{A}(\vec{R},T) G^<(\vec{R},T,\vec{R},\omega). \tag{8.246}$$

Once we have defined the correlation functions and the Wigner function, we are ready to introduce dissipation into quantum formalism. In order to do this, however, we need to use Green's functions which we discuss next.

Special Topic 4: A primer on Green's function for the Schrödinger equation

The time-independent Schrödinger equation for a free electron is $[E-H]\Psi(\vec{r})=0$. Consider now what happens if the right-hand side of this equation is replaced with a nonzero quantity I so that the equation is transformed to

$$[E - H]\overline{\Psi}(\vec{r}) = I. \tag{8.247}$$

We can call $\overline{\Psi}$ a "pseudo-wavefunction" since it is not really a solution of the actual Schrödinger equation. In fact, we can think of it as the "response" of the Schrödinger equation to an "input" I and write

$$\overline{\Psi}(\vec{r}) = [E - H]^{-1} I. \tag{8.248}$$

For convenience and simplicity, let us from now on just think in terms of one dimension, i.e., replace $\vec{r}$ with x. If we then try to find the pseudo-wavefunction at location x, resulting from a unit pulse input applied at location x', then this pseudo-wavefunction—which we will label as $G(x,x')$—will have to satisfy the equation

$$[E - H]G(x,x') = \left[E + \frac{\hbar^2}{2m^*}\frac{\partial^2}{\partial x^2} - V\right]G(x,x') = \delta(x - x'), \tag{8.249}$$

where the delta is a Dirac delta and represents a pulse of zero width.

This pseudo-wavefunction will be called a *Green's function* for the Schrödinger equation. Note that the Green's function has a correlative role; it *correlates* a quantity (pseudo-wavefunction) at location x, with a quantity (impulse or excitation) applied at location x'.

Physically, a unit excitation applied at x' should result in two plane waves propagating to the left and right of x' as shown in Fig. 8.17a. These would be the pseudo-wavefunction, or Green's function, so that we will have [34]

$$\begin{aligned} G(x,x') &= \Gamma^+ \exp[ik(x - x')] \quad (\texttt{for}\ x > x') \\ &= \Gamma^- \exp[-ik(x - x')] \quad (\texttt{for}\ x < x'), \end{aligned} \tag{8.250}$$

where $k = \sqrt{2m^*(E-V)}/\hbar$.

Clearly, in order to satisfy (8.249), the Green's function $G(x,x')$ must be continuous. Enforcing this continuity at $x = x'$ results in

$$G(x,x')|_{x=x'+} = G(x,x')|_{x=x'-}, \tag{8.251}$$

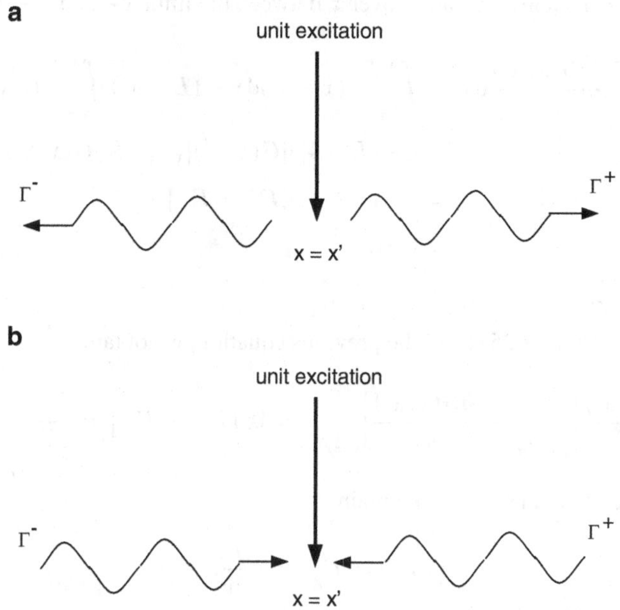

Fig. 8.17 Unit excitation applied at location x' causes two waves to travel (**a**) outwards, or (**b**) inwards

or

$$\Gamma^+ = \Gamma^-. \tag{8.252}$$

The first derivatives of the Green's function are

$$\left.\frac{\partial G(x,x')}{\partial x}\right|_{x=x'-} = -\lim_{x\to x'} \Gamma^- ik \exp\left[-ik(x-x')\right] = -ik\Gamma^-$$
$$\left.\frac{\partial G(x,x')}{\partial x}\right|_{x=x'+} = \lim_{x\to x'} \Gamma^+ ik \exp\left[ik(x-x')\right] = ik\Gamma^+. \tag{8.253}$$

Therefore,

$$\left.\frac{\partial G(x,x')}{\partial x}\right|_{x=x'+} - \left.\frac{\partial G(x,x')}{\partial x}\right|_{x=x'-} = ik\left[\Gamma^+ + \Gamma^-\right]. \tag{8.254}$$

From (8.249), we get

$$\frac{\hbar^2}{2m^*}\frac{\partial^2 G(x,x')}{\partial x^2} = \delta(x-x') - (E-V)G(x-x'). \tag{8.255}$$

Integrating the above equation over x between the limit $x-$ and $x+$, we obtain

$$\begin{aligned}\frac{\hbar^2}{2m^*}\int_{x'-}^{x'+}\frac{\partial^2 G(x,x')}{\partial x^2}\mathrm{d}x &= \int_{x'-}^{x'+}\delta(x-x')\mathrm{d}x - (E-V)\int_{x'-}^{x'+}G(x-x')\mathrm{d}x \\ &= 1-(E-V)[G(x,x')|_{x=x'+} - G(x,x')|_{x=x'-}] \\ &= 1-(E-V)[\Gamma^+ - \Gamma^-] \\ &= 1, \end{aligned} \tag{8.256}$$

since $\Gamma^+ = \Gamma^-$.

Therefore, from (8.254) and the previous equation, we obtain

$$\left.\frac{\partial G(x,x')}{\partial x}\right|_{x=x'+} - \left.\frac{\partial G(x,x')}{\partial x}\right|_{x=x'-} = \mathrm{i}k\left[\Gamma^+ + \Gamma^-\right] = \frac{2m^*}{\hbar^2}. \tag{8.257}$$

From (8.252) and (8.257), we obtain

$$\Gamma^+ = \Gamma^- = -\frac{i}{\hbar v}, \tag{8.258}$$

where $v = \hbar k/m^*$.

Using the last result in (8.250), we get

$$G(x,x') = -\frac{i}{\hbar v}\exp[\mathrm{i}k|x-x'|]. \tag{8.259}$$

Now, the unit excitation introduced at x' can also cause two plane waves to travel *toward* the location x', rather than away from it [34]. This is shown in Fig. 8.17b. This situation would give rise to an alternate solution for the Green's function:

$$G(x,x') = \frac{i}{\hbar v}\exp[-\mathrm{i}k|x-x'|]. \tag{8.260}$$

The first solution is called the *retarded* Green's function $G^{\mathrm{R}}(x,x')$ and the second is the *advanced* Green's function $G^{\mathrm{A}}(x,x')$. The difference between the two is that the former corresponds to outgoing waves and the latter to incoming waves. Note that one is *complex conjugate* or Hermitian conjugate of the other:

$$G^{\mathrm{R}}(x,x') = G^{\mathrm{A}}(x,x')^{\dagger}. \tag{8.261}$$

where the "dagger" represents Hermitian conjugate.

Which solution—the retarded or the advanced—is the acceptable solution for the Green's function can be decided by adding an arbitrary imaginary component to the energy of the electron [34]. This will transform (8.249) to

$$[E-H+\mathrm{i}\eta]G(x,x') = \left[E + \frac{\hbar^2}{2m^*}\frac{\partial^2}{\partial x^2} - V + \mathrm{i}\eta\right]G(x,x') = \delta(x-x'). \tag{8.262}$$

The wavevector k now becomes

$$k' = \frac{\sqrt{2m^*(E - V + i\eta)}}{\hbar} = \frac{\sqrt{2m^*(E - V)}}{\hbar}\sqrt{1 + \frac{i\eta}{E - V}}$$
$$\approx k\left[1 + \frac{i\eta}{2(E - V)}\right] = k(1 + i\delta), \qquad (8.263)$$

where $\delta = \frac{\eta}{2(E-V)}$. In that case,

$$G^{R}(x, x') = -\frac{i}{\hbar v}\exp\left[-\delta|x - x'| + ik|x - x'|\right]$$
$$G^{A}(x, x') = \frac{i}{\hbar v}\exp\left[\delta|x - x'| - ik|x - x'|\right]. \qquad (8.264)$$

In this case, the retarded Green's function is the only acceptable solution if η, $\delta > 0$, because then the retarded function is a decaying wave and the advanced function is a growing (and hence unphysical) wave. On the other hand, if $\eta, \delta < 0$, then the advanced Green's function is the only acceptable solution.

The preceding primer on Green's function is adapted from [34].

8.10 Quantum Transport in an Open System with Device-Contact Coupling and the Self-Energy Potential

In the Tsu–Esaki formalism, or the Landauer–Büttiker formalism, we have always viewed the device as a phase-coherent or dissipationless entity. All dissipation (loss of phase coherence) is assumed to take place in the contacts. Once an electron enters a contact, it immediately thermalizes and reaches equilibrium with its surroundings since the contact is an infinite reservoir whose equilibrium cannot be disturbed. This viewpoint ensures that Fermi–Dirac statistics is maintained in the contacts with a particular Fermi energy or chemical potential μ. The question that inevitably arises is how can one merge the dissipative contacts and the nondissipative device into one comprehensive description where there is *two-way* communication between the device and the contacts. It is not just that the contacts inject and extract carriers from the device while the device does nothing to the contacts; instead both affect one another.

If we view the device as a *closed system* that does not exchange electrons with the contacts, then the device would be the proverbial "box" in the particle-in-a-box description that we see in elementary quantum mechanics. The boundaries of the device will act as impenetrable infinite potentials or "hard walls" and the energy of an electron inside the device can assume only discrete values. However, if we couple the device to the contact in a manner that the device can exchange electrons with the contacts, then the discrete energy levels will broaden. Phenomenologically, the amount of broadening will be given by the Heisenberg uncertainty relation and will be of the order of $\hbar/\tau$, where $1/\tau$ is the rate at which an electron is entering or exiting into a contact. Such a system is an *open system* where particles are exchanged between contacts and the device.

8.10.1 Coupling of a Device to a Single Contact

To start with, let us focus on the simplest open system consisting of the device and one (not two) contact. In this case, there will be no current flow since we do not have two contacts to inject and extract electrons, and the device and contact will always be in equilibrium with each other. Quantum mechanically, such a coupled system will be described by an equation of the form

$$E \begin{bmatrix} \{\phi_\mathrm{d}\} \\ \{\phi_\mathrm{c}\} \end{bmatrix} = \begin{bmatrix} \mathbf{H}_\mathrm{d} & \mathbf{H}_\mathrm{d-c} \\ \mathbf{H}^\dagger_\mathrm{d-c} & \mathbf{H}_\mathrm{c} \end{bmatrix} \begin{bmatrix} \{\phi_\mathrm{d}\} \\ \{\phi_\mathrm{c}\} \end{bmatrix}, \tag{8.265}$$

where $\mathbf{H}_\mathrm{d}$ is the Hamiltonian describing the *isolated* device (if it were closed and not communicating with the contact), $\mathbf{H}_\mathrm{c}$ is likewise the Hamiltonian describing the isolated contact, $\mathbf{H}_\mathrm{d-c}$ is the device-contact coupling Hamiltonian that describes coupling between the contact and the device, and the superscript dagger represents Hermitian conjugate. Remember that all the Hamiltonians are matrices while $\{\phi_\mathrm{d}\}$ and $\{\phi_\mathrm{c}\}$ are column vectors.

The above equation couples the contact and the device through the coupling Hamiltonian $H_\mathrm{d-c}$. If we want to write an equation for the device alone, taking into account its coupling with the contact, that equation will have a form very different from the equation

$$[E\mathbf{I}_\mathrm{d} - \mathbf{H}_\mathrm{d}]\{\phi_\mathrm{d}\} = \{0\}, \tag{8.266}$$

which describes the isolated dissipationless device.[8] In this subsection, we will attempt to find this equation.

For the isolated contact not communicating with the device or anything else, the governing equation will be

$$[E\mathbf{I}_\mathrm{c} - \mathbf{H}_\mathrm{c}]\{\phi_\mathrm{c}\} = \{0\}, \tag{8.267}$$

where $\mathbf{I}_\mathrm{c}$ is the $N_\mathrm{c} \times N_\mathrm{c}$ identity matrix with N_c being the number of modes in the contact.

---

[8] The matrix $\mathbf{I}$ is the $N_d \times N_d$ identity matrix, where N_d is the number of modes in the device, and $\{0\}$ is a null column vector of size N_d.

If we now couple the contact to an external source such as a battery that injects and extracts electrons from the contact while maintaining it at a constant chemical potential μ, then the above equation will be modified to

$$[E\mathbf{I}_c - \mathbf{H}_c + i\eta\mathbf{I}_c]\{\phi_c\} = \{i_c\}. \tag{8.268}$$

where η is a small quantity that represents extraction of electrons from the contact by the external source and $\{i_c\}$ represents reinjection of electrons into the contact by the external source.

When the device-contact coupling is turned on, the contact infuses electrons in to the device so that the wavefunction $\{\phi_d\}$ within the device grows and some of it is scattered back into the contact. The net wavefunction in the contact then becomes a superposition of the original wavefunction $\{\phi_c\}$ and the scattered component $\{\phi_s\}$. Thereafter, we can modify (8.265) to write

$$\begin{bmatrix} E\mathbf{I}_d - \mathbf{H}_d & -\mathbf{H}_{d-c} \\ -\mathbf{H}_{d-c}^{\dagger} & E\mathbf{I}_c - \mathbf{H}_c + i\eta\mathbf{I}_c \end{bmatrix} \begin{bmatrix} \{\phi_d\} \\ \{\phi_c\} + \{\phi_s\} \end{bmatrix} = \begin{bmatrix} \{i_c\} \\ \{0\} \end{bmatrix}, \tag{8.269}$$

where we have assumed that turning on the device-contact coupling does not affect $\{i_c\}$.

We will assume that the number of modes in the contact is N_c and that in the device is N_d. In that case, $\mathbf{H}_d$ and $\mathbf{I}_d$ will be $N_d \times N_d$ matrices, $\mathbf{H}_c$ and $\mathbf{I}_c$ will be $N_c \times N_c$ matrices, $\mathbf{H}_{d-c}$ will be an $N_d \times N_c$ matrix, $\mathbf{H}_{d-c}^{\dagger}$ will be an $N_c \times N_d$ matrix, $\{\phi_c\}$ and $\{i_c\}$ will be $N_c \times 1$ column vectors, and $\{\phi_d\}$ will be an $N_d \times 1$ column vector.

Using (8.268) and (8.269), we can eliminate $\{i_c\}$ and write

$$[E\mathbf{I}_d - \mathbf{H}_d]\{\phi_d\} - \mathbf{H}_{d-c}\{\phi_s\} = \mathbf{H}_{d-c}\{\phi_c\} \tag{8.270}$$

and

$$[E\mathbf{I}_c - \mathbf{H}_c + i\eta\mathbf{I}_c]\{\phi_s\} - \mathbf{H}_{d-c}^{\dagger}\{\phi_d\} = \{0\}. \tag{8.271}$$

From (8.271), we can write

$$\{\phi_s\} = [E\mathbf{I}_c - \mathbf{H}_c + i\eta\mathbf{I}_c]^{-1}\mathbf{H}_{d-c}^{\dagger}\{\phi_d\}. \tag{8.272}$$

Now, integrating equation (8.262) over all x', we get $G^R(x) = [E - H + i\eta]^{-1}$. Therefore, we can write the above equation as

$$\{\phi_s\} = [G^R]\mathbf{H}_{d-c}^{\dagger}\{\phi_d\}, \tag{8.273}$$

where $[G^R] = [E\mathbf{I}_c - \mathbf{H}_c + i\eta\mathbf{I}_c]^{-1}$ is the retarded Green's function of the *isolated* contact since it involves only the Hamiltonian of the contact.

We can then substitute this result in (8.270) to get

$$[E\mathbf{I}_d - \mathbf{H}_d - \Sigma^R]\{\phi_d\} = \{\Phi\}, \tag{8.274}$$

where $\Sigma^R = \mathbf{H}_{d-c}[G^R]\mathbf{H}_{d-c}^{\dagger}$ and $\{\Phi\} = \mathbf{H}_{d-c}\{\phi_c\}$.

The last equation is an equation that describes the device, taking into account its coupling with the contact. Therefore, if we wish to embellish our analysis of the device, taking into account its coupling with the contact, we have to find three new quantities: $[G^{\mathrm{R}}]$, $\mathbf{H}_{\mathrm{d-c}}$ and $\{\phi_{\mathrm{c}}\}$. It is also clear from the above that the retarded Green's function for the device coupled to the contact—which we have termed the "open system"—is

$$G^{\mathrm{R}}_{\text{open-system}} \equiv G^{\mathrm{R}}_{\text{open-system}}(\vec{r}_1, \vec{r}_2, E) = [E\mathbf{I}_{\mathrm{d}} - \mathbf{H}_{\mathrm{d}}(\vec{r}) - \Sigma^R(\vec{r}_1, \vec{r}_2)]^{-1}. \quad (8.275)$$

The coupling of the device with the lead is accounted for in the term Σ^R. It is called the (retarded) *self-energy potential*. By analogy, we can also define an advanced self-energy potential as $\Sigma^A = [\Sigma^R]^\dagger$.

We can rewrite the last equation as

$$[E\mathbf{I}_{\mathrm{d}} - \mathbf{H}_{\mathrm{d}} - \Sigma^R]G^{\mathrm{R}}_{\text{open-system}} = [\mathbf{I}], \quad (8.276)$$

which is an equation in matrix representation since all the quantities in boldface are matrices. By analogy with (8.262), we can write it in the position representation as

$$[E - H^{\mathrm{op}}_{\mathrm{d}}]G^{\mathrm{R}}_{\text{open-system}}(\vec{r}, \vec{r}\,') - \int d^3\vec{\rho}\,\Sigma^R\vec{r}, \vec{\rho})G^{\mathrm{R}}_{\text{open-system}}(\vec{\rho}, \vec{r}\,')\mathrm{d}\vec{\rho} = \delta(\vec{r} - \vec{r}\,'). \quad (8.277)$$

We can think of the retarded Green's function $G^{\mathrm{R}}_{\text{open-system}}(\vec{r}, \vec{r}\,')$ as the wavefunction $\Psi(\vec{r})$ at position $\vec{r}$ somewhere within the device (which is coupled to the contact) due to an excitation at position $\vec{r}\,'$. With this notion, we can rewrite the preceding equation as (assuming $\vec{r} \neq \vec{r}\,'$)

$$E\Psi(\vec{r}) = H^{\mathrm{op}}_{\mathrm{d}}\Psi(\vec{r}) + \int d^3\vec{\rho}\,\Sigma^R(\vec{r}, \vec{\rho})\Psi(\vec{\rho})\mathrm{d}\vec{\rho}, \quad (8.278)$$

which is a Schrödinger-like equation where the self-energy potential (last term) clearly arises only because of the coupling to the contact. We say this because if the system were a closed system with no coupling with the contact, then it would be described by the Schrödinger equation without the last term on the right-hand side.

8.10.1.1 More on the Self-Energy Potential

The self-energy potential has two peculiarities: First, it is a *nonlocal* potential since it is a function of two coordinates $\vec{r}$ and $\vec{\rho}$. That means it does not depend solely on one location in space, which makes it inherently "nonlocal." Second, this potential operator is *nonhermitian* since

$$(\Sigma^R)^\dagger = \Sigma^A \neq \Sigma^R. \quad (8.279)$$

We would expect the self-energy potential to be nonhermitian since it represents charge and particle exchange between the contact and device. Since charge is no longer conserved *within* the device alone—although it must be conserved in the complete open system consisting of both device and contact —the self-energy potential could very well be nonhermitian. Hermiticity is associated with charge conservation and time reversal symmetry (see Problem 3.8). Since the self-energy potential represents charge nonconservation within the device alone, it is understandably nonhermitian. But what about the time-reversal symmetry issue? Violation of time-reversal symmetry is associated with dissipation. We can intuitively understand why the self-energy potential will be connected with violation of time reversal symmetry. In the Landauer–Büttiker picture, or Tsu–Esaki picture, there is no dissipation within the device and all dissipation takes place in the contacts. Therefore, any entity that connects the device to the contacts ensures dissipation and therefore is expectedly nonhermitian.

It is easy to show from the solution of Problem (3.8) that

$$i\hbar\vec{\nabla}\cdot\vec{J}(\vec{r}) = [H_{\rm d}^{\rm op}(\vec{r})\Psi(\vec{r})]^*\Psi(\vec{r}) - \Psi^*(\vec{r})[H_{\rm d}^{\rm op}(\vec{r})\Psi(\vec{r})]. \tag{8.280}$$

Using (8.278) to replace $H_{\rm d}^{\rm op}$ in the last equation, we get

$$\vec{\nabla}\cdot\vec{J}(\vec{r}) = \frac{i}{\hbar}\int d^3\vec{\rho}[\Sigma^R(\vec{r},\vec{\rho})\Psi^*(\vec{r})\Psi(\vec{\rho}) - \Sigma^A(\vec{\rho},\vec{r})\Psi^*(\vec{\rho})\Psi(\vec{r})], \tag{8.281}$$

where we have used the fact that Σ^A is the hermitian conjugate of Σ^R.

Integrating this equation over the entire device from the left edge of the device to the right, we obtain

$$\int_{\text{left-edge}}^{\text{right-edge}} \vec{\nabla}\cdot\vec{J}(\vec{r}) = \frac{i}{\hbar}\int_{\text{left-edge}}^{\text{right-edge}}\int d^3\vec{r}d^3\vec{\rho}\Psi^*(\vec{r})\Psi(\vec{\rho})[\Sigma^R(\vec{r},\vec{\rho}) - \Sigma^A(\vec{r},\vec{\rho})], \tag{8.282}$$

where we have interchanged the dummy variables $\vec{r}$ and $\vec{\rho}$ in the second term.

Using the divergence theorem, we can write the preceding equation as

$$\begin{aligned} I_{\text{exiting}} - I_{\text{entering}} &= \frac{i}{\hbar}\int_{\text{left-edge}}^{\text{right-edge}}\int d^3\vec{r}d^3\vec{\rho}\Psi^*(\vec{r})\Psi(\vec{\rho})[\Sigma^R(\vec{r},\vec{\rho}) - \Sigma^A(\vec{r},\vec{\rho})] \\ &= \frac{1}{\hbar}\int_{\text{left-edge}}^{\text{right-edge}}\int d^3\vec{r}d^3\vec{\rho}\Psi^*(\vec{r})\Psi(\vec{\rho})\Gamma(\vec{r},\vec{\rho}), \end{aligned} \tag{8.283}$$

where I_{entering} is the current entering the device at the left edge and I_{exiting} is the current exiting the device at the right edge, and $i\Gamma(\vec{r},\vec{\rho}) = \Sigma^R(\vec{r},\vec{\rho}) - \Sigma^A(\vec{r},\vec{\rho})$.

The immensely surprising result is that the current entering and exiting are not equal since $\Sigma^R(\vec{r},\vec{\rho}) \neq \Sigma^A(\vec{r},\vec{\rho})$ (or, equivalently, $\Gamma(\vec{r},\vec{\rho}) \neq 0$) owing to the fact that the self-energy potential is not hermitian. This seems to violate Kirchoff's current law!

To resolve this conundrum, one has to understand that the Green's function describes *coherent* propagation of an electron or hole. A nonzero $\Gamma(\vec{r},\vec{\rho})$ results in apparent loss of electrons (equivalent to gain of holes) or loss of holes (equivalent to gain of electrons). Physically, this represents end of the coherent evolution of an electron's (or hole's) trajectory in the device due to escape into the leads where thermalization (dissipation) takes place, or due to phase breaking scattering within the device.

Much of the above discussion is adapted from [34].

8.11 The Nonequilibrium Green's Function Formalism

So far, we have seen that the correlation function $-\mathrm{i}G^{<}(\vec{r}_1,t_1;\vec{r}_2,t_2)$ defined by Equation (8.238) and the Wigner distribution function $W(\vec{k},\vec{r},\omega,T)$ defined by (8.241) are effectively quantum-mechanical equivalents of the classical Boltzmann distribution function $f(\vec{r},\vec{k},t)$ since their zero-th and first moments are the transport variables carrier concentration and current density, respectively. We can define another correlation function $+iG^{>}(\vec{r}_1,t_1;\vec{r}_2,t_2)$ in the spirit of (8.238) and write it as

$$+\mathrm{i}G^{>}(\vec{r}_1,t_1;\vec{r}_2,t_2)=\langle C_{\vec{r}_1}(t_1)C^{*}_{\vec{r}_2}(t_2)\rangle, \tag{8.284}$$

Sometimes the function $-\mathrm{i}G^{<}$ is called the *electron correlation function* and the function $+iG^{>}$ is called the *hole correlation function*. Why these names came about has to do with the notion of the creation and annihilation operators. We discussed creation and annihilation operators for bosons in Chap. 5, but fermions too have their own creation and annihilation operators that obey different commutation rules from the Bose operators. In the language of second quantization, the coefficients $C_{\vec{r}}(t)$ are not numbers, but operators: $C^{*}_{\vec{r}_2}(t_2)$ is the creation operator that creates an electron at position $\vec{r}_2$ at time t_2 and $C_{\vec{r}_1}(t_1)$ is the annihilation operator that annihilates an electron at position $\vec{r}_1$ at time t_1, leaving behind a hole. Creation and annihilation operators for Fermions always occur in pairs since, in the end, we can neither create nor destroy a Fermion. The operator $C_{\vec{r}_1}(t_1)C^{*}_{\vec{r}_2}(t_2)$ will first create an electron at position $\vec{r}_2$ at time t_2 and then annihilate one at position $\vec{r}_1$ at time t_1, leaving behind a hole. Similarly, the operator $C^{*}_{\vec{r}_1}(t_1)C_{\vec{r}_2}(t_2)$ will annihilate an electron at position $\vec{r}_2$ at time t_2 and create one at position $\vec{r}_1$ at time t_1. Therefore, it seems logical to name the function $-\mathrm{i}G^{<}$ "electron" correlation function and the function $+\mathrm{i}G^{>}$ the "hole" correlation function.

Just as $-\mathrm{i}G^{<}(\vec{r}_1,t_1;\vec{r}_2,t_2)$ is the quantum-mechanical equivalent of the Boltzmann distribution function $f(\vec{r},\vec{k},t)$, $+iG^{>}(\vec{r}_1,t_1;\vec{r}_2,t_2)$ will be the quantum-mechanical equivalent of $1-f(\vec{r},\vec{k},t)$, simply because the hole is nothing but the absence of an electron.

Next, we need to find equivalents of the in-scattering rate $S_{\mathrm{in}}(\vec{k},\vec{k}')$ and out-scattering rate $S_{\mathrm{out}}(\vec{k},\vec{k}')$ that we encountered in Chap. 2 in connection with the BTE. The appropriate equivalents of these rates are the scattering potentials $\Sigma^{\text{in}}(\vec{r}_1, \vec{r}_2, E) = -\mathrm{i}\Sigma^{<}(\vec{r}_1, \vec{r}_2, E)$ and $\Sigma^{\text{out}}(\vec{r}_1, \vec{r}_2, E) = +\mathrm{i}\Sigma^{>}(\vec{r}_1, \vec{r}_2, E)$ which are computed by taking into account the self-energy potentials representing coupling of the device with contacts *and* scattering potentials associated with phase-breaking electron interactions with other electrons, phonons, etc. In general, we can write

$$\begin{aligned}\Sigma^{\text{in}}(\vec{r}_1, \vec{r}_2, E) &= \Sigma^{\text{in}}_{\text{leads}}(\vec{r}_1, \vec{r}_2, E) + \Sigma^{\text{in}}_{\text{interactions}}(\vec{r}_1, \vec{r}_2, E)\\ \Sigma^{\text{out}}(\vec{r}_1, \vec{r}_2, E) &= \Sigma^{\text{out}}_{\text{leads}}(\vec{r}_1, \vec{r}_2, E) + \Sigma^{\text{out}}_{\text{interactions}}(\vec{r}_1, \vec{r}_2, E).\end{aligned} \tag{8.285}$$

We will still assume that the leads are approximately in equilibrium since they are vast electron reservoirs, and therefore the electron occupation probability in the j-th lead is given by the Fermi–Dirac function $f_j(E)$. In that case,

$$\begin{aligned}\Sigma^{\text{in}}_{\text{leads}}(\vec{r}_1, \vec{r}_2, E) &= \mathrm{i}\sum_j f_j(E)\left[\Sigma^R_j(\vec{r}_1, \vec{r}_2) - \Sigma^A_j(\vec{r}_1, \vec{r}_2)\right]\\ &= \sum_j f_j(E)\Gamma_j(\vec{r}_1, \vec{r}_2)\\ \Sigma^{\text{out}}_{\text{leads}}(\vec{r}_1, \vec{r}_2, E) &= \mathrm{i}\sum_j \{1 - f_j(E)\}\left[\Sigma^R_j(\vec{r}_1, \vec{r}_2) - \Sigma^A_j(\vec{r}_1, \vec{r}_2)\right]\\ &= \sum_j \{1 - f_j(E)\}\Gamma_j(\vec{r}_1, \vec{r}_2).\end{aligned} \tag{8.286}$$

The procedure for calculating $\Gamma_j(\vec{r}_1, \vec{r}_2)$ in the j-th lead and the procedure for calculating $\Sigma^{\text{in}}_{\text{interactions}}$ and $\Sigma^{\text{out}}_{\text{interactions}}$ for different types of phase breaking interactions (electron–phonon, electron–electron, etc.) are described in [34].

The last thing that remains to be done is to formulate an equivalent of the BTE whose solution will yield the correlation functions. In the steady-state, this equation—known as the *quantum kinetic equation* [36]—is

$$\begin{aligned}-\mathrm{i}G^{<}(\vec{r}_1, \vec{r}_2, E) &= \int\!\!\int d^3\vec{\rho}_1 d^3\vec{\rho}_2 G^{\text{R}}(\vec{r}_1, \vec{\rho}_1, E)\Sigma^{\text{in}}(\vec{\rho}_1, \vec{\rho}_2, E)G^{\text{A}}(\vec{\rho}_2, \vec{r}_2, E)\\ \mathrm{i}G^{>}(\vec{r}_1, \vec{r}_2, E) &= \int\!\!\int d^3\vec{\rho}_1 d^3\vec{\rho}_2 G^{\text{R}}(\vec{r}_1, \vec{\rho}_1, E)\Sigma^{\text{out}}(\vec{\rho}_1, \vec{\rho}_2, E)G^{\text{A}}(\vec{\rho}_2, \vec{r}_2, E),\end{aligned} \tag{8.287}$$

where $G^{\text{R}}(\vec{r}_1, \vec{r}_2, E)$ is given by (8.275) and $G^{\text{A}}(\vec{r}_1, \vec{r}_2, E) = [G^{\text{R}}(\vec{r}_1, \vec{r}_2, E)]^{\dagger}$.

The above equation is in position representation. In matrix representation, this equation can be written as

$$\begin{aligned}-\mathrm{i}[\mathbf{G}^{<}] &= [\mathbf{G}^{\text{R}}][\mathbf{\Sigma}^{\text{in}}][\mathbf{G}^{\text{A}}]\\ +\mathrm{i}[\mathbf{G}^{>}] &= [\mathbf{G}^{\text{R}}][\mathbf{\Sigma}^{\text{out}}][\mathbf{G}^{\text{A}}].\end{aligned} \tag{8.288}$$

The self-energy potential Σ^R that appears in (8.275) is the sum of that due to coupling with the leads and that due to phase breaking interactions within the device, i.e.,

$$\Sigma^R(\vec{r}_1, \vec{r}_2, E) = \sum_j \Sigma_j^R(\vec{r}_1, \vec{r}_2, E) + \Sigma^R_{\text{interactions}}(\vec{r}_1, \vec{r}_2, E). \tag{8.289}$$

The procedure for calculating Σ_j^R in any lead and the procedure for calculating $\Sigma^R_{\text{interactions}}$ are described in [34].

One complication is that $\Sigma^R_{\text{interactions}}(\vec{r}_1, \vec{r}_2, E)$, $\Sigma^{\text{in}}_{\text{interactions}}(\vec{r}_1, \vec{r}_2, E)$ and $\Sigma^{\text{out}}_{\text{interactions}}(\vec{r}_1, \vec{r}_2, E)$ depend on $G^<(\vec{r}_1, \vec{r}_2, E)$ and $G^>(\vec{r}_1, \vec{r}_2, E)$, respectively. Therefore, these equations will have to be solved self-consistently. A flow chart for this self-consistent solution procedure is shown in Fig. 8.18. Once the correlation functions $G^<(\vec{r}_1, \vec{r}_2, E)$, and $G^>(\vec{r}_1, \vec{r}_2, E)$ are found, any transport variable can be calculated from these quantities as described earlier. For example, we can use (8.245) to find the carrier concentration and (8.246) to find the current density at any frequency ω, and then integrate over all ω to find the total carrier concentration and current density.

The formalism just described is called the *nonequilibrium Green's function formalism*. Its advantage is that it can handle dissipation and incoherent dynamics (due to coupling of a device with leads and phase-breaking collisions) within a quantum-mechanical formalism, which the other quantum formalisms (Schrödinger equation, Tsu–Esaki or Landauer–Büttiker) could not do.

8.12 Summary

1. Quantum mechanical effects arising from the wave nature of electrons, such as electron interference or tunneling through a barrier, are manifested when an electron's phase is not sufficiently randomized by either phase-breaking collisions or thermal averaging over energy. Therefore, the two conditions that must be fulfilled to experience quantum effects are: (1) the device dimension along the direction of current flow should be smaller than the phase-breaking length, which is the distance an electron can travel before its phase is randomized by phase-breaking collisions, and (2) the temperature should be lower than the correlation temperature or Thouless temperature, which is the temperature at which thermal averaging over energy introduces an uncertainty of π radians in the electron's phase.
2. Phase-breaking scattering events are those that change the internal state of the scatterer. Normally, these are inelastic collisions, although some elastic collisions that flip an electron's spin are also phase breaking. Inelastic collisions occur due to time-varying perturbations (e.g., lattice vibrations giving rise to phonons) while elastic collisions are caused by time-invariant scatterers like impurities. Electron–electron, electron–hole, and hole–hole scatterings are also inelastic and phase breaking.

Calculate scattering functions Σ^{in}_{leads} and Σ^{out}_{leads} [from $\Gamma(r_1, r_2)$] as well as self energy potential Σ^{R}_{lead} in every lead

Assume some self energy potential $\Sigma^{R}_{interactions}$ in the device owing to interactions. Using that, Σ^{R}_{lead}, and the device Hamiltonian Hd, calculate the retarded and advanced Green's functions G^R and G^A

Assume some $\Sigma^{in}_{interactions}$ and $\Sigma^{out}_{interactions}$ to find Σ^{in} and Σ_{out} Calculate the correlation functions $G^<$ and $G^>$ from all of the above quantities

Recalculate $\Sigma^{R}_{interactions}$, $\Sigma^{in}_{interactions}$ and $\Sigma^{out}_{interactions}$ from the correlation functions

Converged?

NO

YES

Calculate transport variables from the correlation functions

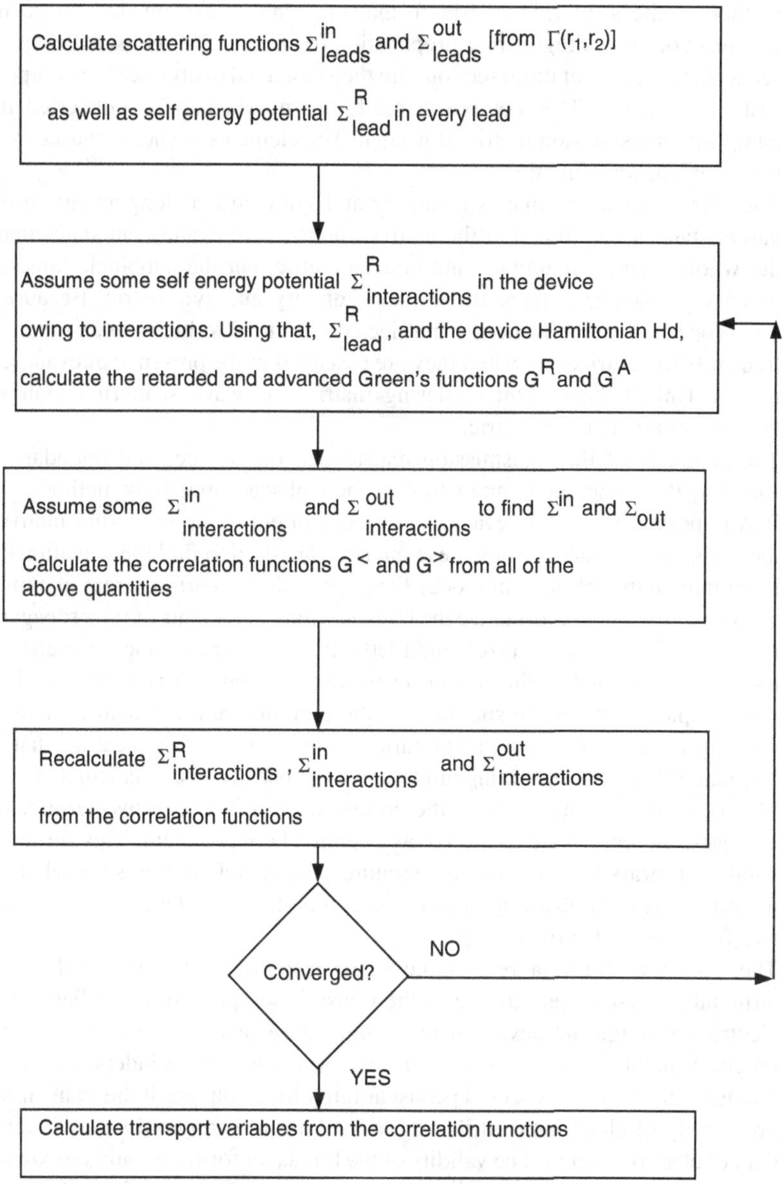

Fig. 8.18 Flow chart indicating the solution procedure for finding correlation functions and transport variables using the nonequilibrium Green's function formalism

3. In phase-coherent transport, the carrier concentration and current density at any location within a device can be found from the wavefunction. The wavefunction in a device containing elastic scatterers can be found by brute force solution of the Schrödinger equation, but a more elegant and numerically less demanding

method is the scattering matrix formalism. One breaks up the device into sections corresponding to free propagation (or free flight) and scattering. The scattering matrices of these sections are then cascaded to produce the composite scattering matrix. This composite scattering matrix can be converted to a composite transmission matrix, if needed. The elements of these matrices yield the complete wavefunction.

4. The current scattering matrix is unitary and symmetric as long as only propagating states are included in the matrix. Inclusion of evanescent states makes the whole matrix nonunitary and nonsymmetric, but the subblock consisting of only propagating states still remains unitary and symmetric. Because of this property, current scattering matrices do not tend to blow up as rapidly as transmission matrices do when they are cascaded in the presence of evanescent modes. Unlike the current scattering matrix, the wave scattering matrix is neither unitary, nor symmetric.
5. The elements of the transmission matrices at the device–lead boundary are found by the mode matching method or the real space matching method.
6. It is important to include evanescent modes in either the scattering matrix or the transmission matrix when these matrices are used to find the wavefunction. Even though the evanescent modes themselves do not carry current in a mesoscopic device, they renormalize the transmission coefficients of the propagating modes and hence affect current indirectly. If insufficient number of evanescent modes are included in the scattering or transmission matrix characterizing a mesoscopic structure, the subblock of the current scattering matrix involving only propagating states will not turn out to be unitary, indicating that the coefficients of the propagating modes have not been correctly calculated.
7. The Tsu–Esaki formula relates the current flowing between two terminals of a phase-coherent device to the voltage applied between them. This formula is valid at arbitrary voltages and temperatures, and merely requires knowledge of the transmission probability of electrons, going from one terminal to the other, as a function of electron energy.
8. The Landauer formula relates the linear-response conductance of a two-terminal phase-coherent device to the transmission probability of Fermi-level electrons through the device at low temperatures and low voltages, or more precisely, in the *linear response* regime when conductance is independent of the voltage. Linear response could persist at fairly high voltages if the transmission probability of electrons contributing to the current is approximately independent of electron energy. The validity of the Landauer formula could be extended to arbitrary temperatures and arbitrary voltages, if we replace the actual transmission probability with a renormalized energy-averaged transmission probability $\hat{T}$.
9. Büttiker's multiprobe formula yields the net current flowing in any terminal of a multiterminal phase coherent device at low voltages and low temperatures. It requires knowledge of the transmission probabilities associated with electron flow between different terminals, and the chemical potentials or Fermi energies in the those terminals.

10. The Landauer formula $G = (2e^2/h)\mathcal{T}$ can be interpreted as the "2-terminal conductance formula" (current and voltage leads are the same), while the modified Landauer formula $G = (2e^2/h)(M - \mathcal{T})/\mathcal{T}$ can be interpreted as the 4-terminal conductance formula (separate leads are used for current and voltage and the voltage leads are weakly coupled to the device), as long as the leads are weakly coupled to the device. The difference between the 2-terminal resistance and 4-terminal resistance is the contact resistance which can be thought of having a value $(h/2e^2)(1/M)$, where M is the total number of modes in the leads that contribute to current.
11. The Landau–Vlasov equation describes the evolution in time, real space, and wavevector space of a quantum correlation function. This equation has the same mathematical form as the Boltzmann Transport equation, except that the collision term is absent. A different quantum correlation function can be formulated which also obeys the collisionless Boltzmann Transport equation and whose moments are the transport variables. This correlation function is an quantum-mechanical exact analog of the classical Boltzmann distribution function.
12. The Wigner distribution function is also a quantum-mechanical equivalent of the Boltzmann distribution function. One can use this function to calculate transport variables. This function yields the energy-resolved transport variables, i.e., the value of a transport variable, such as current density, at a particular electron energy.
13. In order to include dissipative processes in a quantum-mechanical description, one must resort to retarded and advanced Green's function and self-energy potentials that describe incoherent processes due to coupling of a device with leads and phase-breaking collisions within the device. The self-energy potentials are nonhermitian since they represent dissipation. One solves the quantum kinetic equation (involving self-energy potentials and Green's function) self-consistently to find the correlation functions $G^<$ and $G^>$ and then calculate transport variables from various moments of $G^<$ and $G^>$.

Problems

Problem 8.1. In a magnetic field, is

$$\phi_{m',p'}(y,z) = \phi_{-m',-p'}(y,z) \;?$$

$$k_{m',p'} = k_{-m',-p'} \;?$$

Solution. The answer is no in both cases. In a magnetic field, the dispersion relations may not be symmetric about the wavevector (see Fig. 7.13; the horizontal axis is not k_x, but $k_x(eB/\hbar)y_0$), so that $k_{m',p'} \neq k_{-m',-p'}$. Additionally, the right traveling states and the left traveling states localize at opposite edges of a quantum wire (recall the discussion about "edge states" in the previous Chapter), so that $\phi_{m',p'}(y,z) \neq \phi_{-m',-p'}(y,z)$.

Problem 8.2. Derive the relations in (8.26) and (8.28).

Solution. From (8.25), we get

$$\mathbf{b} = \mathbf{s}_{11}\mathbf{a} + \mathbf{s}_{12}\mathbf{d}$$

$$\mathbf{c} = \mathbf{s}_{21}\mathbf{a} + \mathbf{s}_{22}\mathbf{d}$$

When $\mathbf{b} = 0$, $\mathbf{d} = -\mathbf{s}_{12}^{-1}\mathbf{s}_{11}\mathbf{a}$. Substituting this in the last relation, we get

$$\mathbf{c}|_{\mathbf{b}=0} = \mathbf{s}_{21}\mathbf{a} - \mathbf{s}_{22}\mathbf{s}_{12}^{-1}\mathbf{s}_{11}\mathbf{a}.$$

From (8.23), we get

$$\mathbf{c} = \mathbf{t}_{11}\mathbf{a} + \mathbf{t}_{12}\mathbf{b}$$

$$\mathbf{d} = \mathbf{t}_{21}\mathbf{a} + \mathbf{t}_{22}\mathbf{b}$$

Once again, when $\mathbf{b} = 0$, $\mathbf{t}_{11} = \mathbf{c}\mathbf{a}^{-1}$. Using the expression for $\mathbf{c}|_{\mathbf{b}=0}$, we immediately get that

$$\mathbf{t}_{11} = \mathbf{s}_{21} - \mathbf{s}_{22}\mathbf{s}_{12}^{-1}\mathbf{s}_{11}.$$

Use this approach to relate the other elements of the transmission matrix to elements of the wave scattering matrix so as to obtain the other relations in (8.26) and (8.28).

Problem 8.3. Using the result of Problem 3.11 and (8.189), show that the transmission probability of an electron with wavevector k through a finite periodic structure consisting of N periods is

$$\mathcal{T}_N = \frac{1}{\sin^2(N\theta)\left[\left(\frac{k^2W_{21}-W_{12}}{2k\sin\theta}\right)^2 - 1\right] + 1},$$

where θ was defined in Problem 3.11.

Problem 8.4. Show that the 4-probe Landauer resistance of N repeated (identical) sections $R_{4\text{-probe}}(N)$ is related to the 4-probe resistance $R_{4\text{-probe}}(1)$ of each section as

$$R_{4\text{-probe}}(N) = \left[\frac{\sin(N\theta)}{\sin\theta}\right]^2 R_{4\text{-probe}}(1).$$

In the limit $\theta \to \pm n\pi$, the above relation becomes

Fig. 8.19 A three terminal device. Incoming and outgoing waves at the three terminals are shown

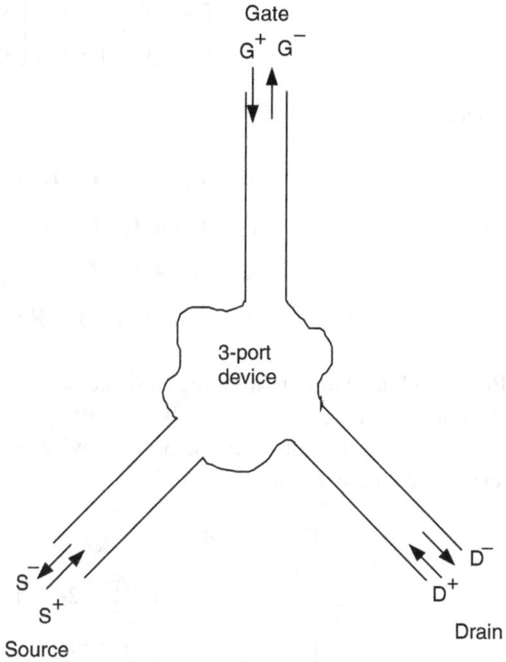

$$\lim_{\theta \to \pm n\pi} R_{4\text{-probe}}(N) = N^2 R_{4\text{-probe}}(1),$$

indicating that the resistance grows as the *square* of the length of the device, instead of linearly with the length as would be expected from the relation $R = \rho l/A$, where ρ is the resistivity of the material making up the resistor, l is the length, and A is the cross-sectional area.

Problem 8.5. Consider a three-terminal device as shown in Fig. 8.19 whose terminals will be labeled "source", "drain," and "gate" in analogy with a field effect transistor. The current scattering matrix for this device, relating the incoming and outgoing current amplitudes, will be defined according to

$$\begin{bmatrix} \mathbf{S}^- \\ \mathbf{D}^- \\ \mathbf{G}^- \end{bmatrix} = \begin{bmatrix} \mathbf{r}_{\mathrm{SS}} & \mathbf{t}_{\mathrm{SD}} & \mathbf{t}_{\mathrm{SG}} \\ \mathbf{t}_{\mathrm{DS}} & \mathbf{r}_{DD} & \mathbf{t}_{\mathrm{DG}} \\ \mathbf{t}_{\mathrm{GS}} & \mathbf{t}_{\mathrm{GD}} & \mathbf{r}_{\mathrm{GG}} \end{bmatrix} \begin{bmatrix} \mathbf{S}^+ \\ \mathbf{D}^+ \\ \mathbf{G}^+ \end{bmatrix},$$

where $\mathbf{S}^{\pm}$, etc. are $M \times 1$ column matrices if M is the number of modes in each of the terminals. Show that if

$$\mathbf{G}^+ = \mathbf{R}\mathbf{G}^-,$$

then we can relate the current amplitudes in the source and drain by the reduced current scattering matrix [37]

$$\begin{bmatrix} \mathbf{S}^- \\ \mathbf{D}^- \end{bmatrix} = \begin{bmatrix} \mathbf{r}\ \mathbf{t}' \\ \mathbf{t}\ \mathbf{r}' \end{bmatrix} \begin{bmatrix} \mathbf{S}^+ \\ \mathbf{D}^+ \end{bmatrix},$$

where

$$\mathbf{r} = \mathbf{r}_{\mathrm{SS}} + \mathbf{t}_{\mathrm{SG}}[\mathbf{I} - \mathbf{R}\mathbf{r}_{\mathrm{GG}}]^{-1}\mathbf{R}\mathbf{t}_{\mathrm{GS}}$$
$$\mathbf{t}' = \mathbf{t}_{\mathrm{SD}} + \mathbf{t}_{\mathrm{SG}}[\mathbf{I} - \mathbf{R}\mathbf{r}_{\mathrm{GG}}]^{-1}\mathbf{R}\mathbf{t}_{\mathrm{GD}}$$
$$\mathbf{t} = \mathbf{t}_{\mathrm{DS}} + \mathbf{t}_{\mathrm{DG}}[\mathbf{I} - \mathbf{R}\mathbf{r}_{\mathrm{GG}}]^{-1}\mathbf{R}\mathbf{t}_{\mathrm{GS}}$$
$$\mathbf{r}' = \mathbf{r}_{DD} + \mathbf{t}_{\mathrm{DG}}[\mathbf{I} - \mathbf{R}\mathbf{r}_{\mathrm{GG}}]^{-1}\mathbf{R}\mathbf{t}_{\mathrm{GD}}.$$

Problem 8.6. Show that taking into account the unitarity and symmetricity of the current scattering matrix (involving only propagating states), we can write the current scattering matrix for the 3-port device (whose last two ports are identical) in terms of a single parameter ϵ as [38]

$$[\mathbf{S}] = \begin{bmatrix} -\sqrt{1-2\epsilon} & \sqrt{\epsilon} & \sqrt{\epsilon} \\ \sqrt{\epsilon} & \frac{1}{2}\left(\sqrt{1-2\epsilon}-1\right) & \frac{1}{2}\left(\sqrt{1-2\epsilon}+1\right) \\ \sqrt{\epsilon} & \frac{1}{2}\left(\sqrt{1-2\epsilon}+1\right) & \frac{1}{2}\left(\sqrt{1-2\epsilon}-1\right) \end{bmatrix},$$

if we assume that all elements of the scattering matrix are real.

Problem 8.7. Consider a phase-coherent 2-terminal device. The transmission probability of electrons through this device at any temperature is given by the simple expression

$$T_{1\to 2}(E) = \Theta(E - E_0),$$

where E_0 is a constant energy and Θ is the unit step function or Heaviside function.

Show that in linear response regime, but at elevated temperatures, the conductance of this structure is given by

$$G_{\text{2-probe}} = \frac{e^2}{h}\left[1 + \tanh\left(\frac{E_{\mathrm{F}} - E_0}{2kT}\right)\right],$$

where E_{F} is the Fermi energy.

Using the above, show that at 0 K temperature,

$$\lim_{T\to 0} G_{\text{2-probe}} = \begin{matrix} \frac{2e^2}{h}\ \texttt{if}\ E_{\mathrm{F}} > E_0 \\ \frac{e^2}{h}\ \texttt{if}\ E_{\mathrm{F}} = E_0 \\ 0\ \texttt{if}\ E_{\mathrm{F}} < E_0 \end{matrix}$$

and then explain the meaning of this result in physical terms.

Problem 8.8. Consider a one-dimensional potential through which electron transmission probability is independent of bias voltage and depends only on the electron's wavevector component in the direction of current flow. Assume that the carrier population is nondegenerate so that Fermi–Dirac statistics can be approximated with Boltzmann statistics. For such a structure show that if transport is phase coherent, then the current–voltage relationship at finite temperature and voltage is

$$I = I_0[1 - \exp(-eV/k_BT)],$$

where

$$I_0 = eA\frac{m^*k_BT}{2\pi^2\hbar^3}\int_0^\infty dE_x\tau_{1\to2}(E_x)\exp[-E_x/(k_BT)],$$

and A is the cross-sectional area of the structure.

If transmission is via thermionic emission over a barrier, then

$$\tau_{1\to2}(E_x) = \Theta(E_x - \Phi),$$

where Θ is the Heaviside function and Φ is the barrier height. In this case, show that

$$\begin{aligned} I &= eA\frac{m^*(k_BT)^2}{2\pi^2\hbar^3}\exp\left[-\frac{\Phi}{k_BT}\right]\left(1 - \exp\left[-\frac{eV}{k_BT}\right]\right) \\ &= A^*T^2\exp\left[-\frac{\Phi}{k_BT}\right]\left(1 - \exp\left[-\frac{eV}{k_BT}\right]\right), \end{aligned}$$

where $A^* = eA\frac{m^*k_B^2}{2\pi^2\hbar^3}$ is the effective Richardson constant.

Solution. Using the Tsu–Esaki formula for three-dimensional leads in (8.133), we obtain

$$\begin{aligned} J_{3-d} = \frac{e}{\Omega}\sum_{k_x}\sum_{k_y}\sum_{k_z}&\{v_x(k_x)\tau_{1\to2}(k_x) \\ &\times\left[f(E(k_x,k_y,k_z) - \mu_1) - f(E(k_x,k_y,k_z) + eV - \mu_1)\right]\} \end{aligned}$$

Using the Boltzmann approximation, we can write this relation as

$$\begin{aligned} J = \frac{e}{\Omega}\int_0^\infty & d^2k_t D_q^{2-d}(k_t)\{\exp[-E_t(k_t)/(k_BT)] - \exp[-(E_t(k_t) + eV)/(k_BT)]\} \\ &\times\int_0^\infty dk_x D_q^{1-d}(k_x)\tau_{1\to2}(k_x)v_x(k_x)\exp[-E_x(k_x)/(k_BT)], \end{aligned}$$

where we have converted the summations to integrations using the density of states in wavevector space. Here, k_t is the transverse wavevector in the plane perpendicular to current flow, and k_x is the longitudinal wavevector (component in the direction of current flow). Similarly, E_t is the kinetic energy associated with transverse motion and E_x is that associated with longitudinal motion. If the bandstructure is parabolic, then

$$E_t(k_t) = \frac{\hbar^2 k_t^2}{2m^*}$$

$$E_x(k_x) = \frac{\hbar^2 k_x^2}{2m^*}.$$

Converting from wavevector space to energy space, we get

$$\begin{aligned}
J &= \frac{e}{\Omega}[1 - \exp(-eV/k_B T)] \int_0^\infty d^2 E_t D_q^{2-d}(E_t) \exp[-E_t/(k_B T)] \\
&\times \int_0^\infty \mathrm{d}E_x D_q^{1-d}(E_x)\tau_{1\to 2}(E_x) v_x(E_x) \exp[-E_x/(k_B T)] \\
&= -e\frac{m^* k_B T}{\pi\hbar^2}[1 - \exp(-eV/k_B T)][\exp(-E_t/(k_B T))]_0^\infty \\
&\times \frac{1}{2\pi\hbar}\int_0^\infty \mathrm{d}E_x \tau_{1\to 2}(E_x) \exp[-E_x/(k_B T)] \\
&= e\frac{m^* k_B T}{\pi^2\hbar^3}[1 - \exp(-eV/k_B T)] \int_0^\infty \mathrm{d}E_x \tau_{1\to 2}(E_x) \exp[-E_x/(k_B T)] \\
&= J_0[1 - \exp(-eV/k_B T)],
\end{aligned}$$

where

$$J_0 = I_0/A = e\frac{m^* k_B T}{\pi^2\hbar^3}\int_0^\infty \mathrm{d}E_x \tau_{1\to 2}(E_x) \exp[-E_x/(k_B T)].$$

Note that we have written the one-dimensional density of states D_q^{1-d} as $\frac{1}{2\pi\hbar v_x(E)}$ because: (1) only electrons with positive values of k_x enter the device and contribute to current, and (2) the spin degeneracy factor of 2 has been already absorbed in the two-dimensional density of states.

Solving the second part of this problem is trivial and is left for the reader.

References

1. N. Bohr, Naturwissenschaften, **16**, 245 (1928).
2. B-G Englert, "Fringe visibility and which-way information: An inequality", Phys. Rev. Lett., **77**, 2154 (1996).
3. J. C. Bose, in *Collected Physical Papers*, (Longmans and Green, London, 1927), pp. 44-49.

4. P. Ghose, D. Home and G. S. Agarwal, "An experiment to throw more light on light", Phys. Lett. A., **153**, 403 (1991).
5. Y. Mizobuchi and Y. Ohtake, "An experiment to throw more light on light", Phys. Lett. A., **168**, 1, (1992).
6. S. Rangwala and S. M. Roy, "Wave behavior and non-complementary particle behavior in the same experiment", Phys. Lett. A., **190**, 1 (1994).
7. A. J. Leggett, in *Nanostructure Physics and Fabrication*. Eds. M. A. Reed and W. P. Kirk, (Academic Press, Boston, 1989). p. 31.
8. A. D. Stone, "Magnetoresistance fluctuations in mesoscopic wires and rings", Phys. Rev. Lett., **54**, 2692 (1985).
9. J. T. Edwards and D. J. Thouless, "Numerical studies of localization in disordered systems", J. Phys. C, **5**, 807 (1972).
10. P. A. Lee and A. D. Stone, "Universal fluctuations in metals", Phys. Rev. Lett., **55**, 1622-1625 (1985).
11. R. A. Webb, in *Nanostructure Physics and Fabrication*. Eds. M. A. Reed and W. P. Kirk, (Academic Press, Boston, 1989). p. 43.
12. B. J. F. Lin, M. A. Paalanen, A. C. Gossard and D. C. Tsui, "Weak localization of two-dimensional electrons in GaAs-$Al_xGa_{1-x}As$ heterostructures", Phys. Rev. B., **29**, 927 (1983).
13. S. Chaudhuri, S. Bandyopadhyay and M. Cahay, "Current, potential, electric field, and Fermi carrier distributions around localized elastic scatterers in phase-coherent quantum magnetotransport", Phys. Rev. B., **47**, 12649 (1993).
14. R. P. Feynman, R. B. Leighton and M. Sands, *The Feynman Lectures on Physics*, Vol. III, (Addison-Wesley, Reading, 1965).
15. A. M. Kriman, N. C. Kluksdahl and D. K. Ferry, "Scattering states and distribution functions for microstructures", Phys. Rev. B., **36**, 5953 (1987).
16. M. Cahay, M. McLennan and S. Datta, "Conductance of an array of elastic scatterers: A scattering matrix approach", Phys. Rev. B., **37**, 10125 (1988).
17. S. Uryu and T. Ando, "Electronic states in anti-dot lattices: Scattering matrix formalism", Phys. Rev. B., **53**, 13613 (1996).
18. H. Rob Frohne, M. J. McLennan and S. Datta, "An efficient method for the analysis of electron waveguides", J. Appl. Phys., **66**, 2699 (1989).
19. R. Frohne and S. Datta, "Electron transfer between regions with different confining potentials", J. Appl. Phys., **64**, 4086 (1988).
20. P. F. Bagwell, "Evanescent modes and scattering in quasi-one-dimensional wires", Phys. Rev. B., **41**, 10354 (1990).
21. P. F. Bagwell and T. P. Orlando, "Landauer's conductance formula and its generalization to finite voltages", Phys. Rev. B., **40**, 1456 (1989).
22. B. J. Van Wees, H. van Houten, C. W. J. Beenakker, J. G. Williamson, L. P. Kouwenhoven, D. Van der Marel and C. T. Foxon, "Quantized conductance of point contacts in a two-dimensional electron gas", Phys. Rev. Lett., **60**, 848 (1988).
23. D. A. Wharam, T. J. Thornton, R. Newbury, M. Pepper, H. Ahmed, J. E. F. Frost, D. G. Hasko, D. C. Peacock, D. A. Ritchie and D. A. C. Jones, "One dimensional transport and the quantization of ballistic resistance", J. Phys. C: Solid State Phys., **21**, L209 (1988).
24. K. J. Thomas, J. T. Nicholls, M. Y. Simmons, M. Pepper, D. R. Mace and D. A. Ritchie, "Possible spin polarization in a one-dimensional electron gas", Phys. Rev. Lett., **77**, 135 (1996).
25. M. Büttiker, "Four-terminal phase-coherent conductance", Phys. Rev. Lett., **57**, 1761 (1986).
26. R. Landauer, "Spatial variation of currents and fields due to localized scatterers in metallic conduction", IBM J. Res. Develop., **1**, 223 (1957).
27. R. Landauer, "Electrical resistance of disordered one-dimensional lattices", Philos. Mag., **21**, 863 (1970).
28. E. M. Economou and C. M. Soukoulis, "Static conductance and scaling theory of localization in one dimension", Phys. Rev. Lett., **46**, 618 (1981).
29. D. S. Fisher and P. A. Lee, "Relation between conductivity and transmission matrix", Phys. Rev. B., **23**, 6851 (1981).

30. R. Landauer, "Can a length of perfect conductor have a resistance?" Phys. Lett. A, **85**, 91 (1981).
31. H. L. Engquist and P. W. Anderson, "Definition and measurement of the electrical and thermal resistances", Phys. Rev. B., **24**, 1151 (1981).
32. M. Büttiker, "Symmetry of electrical conduction", IBM J. Res. Develop., **32**, 317 (1988).
33. Y. Imry, in *Directions in Condensed Matter Physics*, Vol. 1, Eds. G. Grinstein and G. Mazenko, (World Scientific, Singapore, 1986).
34. S. Datta, *Electronic Transport in Mesoscopic Systems*, (Cambridge University Press, Cambridge, 1995).
35. M. Cahay and S. Bandyopadhyay, "Properties of the Landauer resistance of finite repeated structures", Phys. Rev. B., **42**, 5100 (1990).
36. L. P. Kadanoff and G. Baym, *Quantum Statistical Mechanics*, (Addison-Wesley, Redwood City, 1989).
37. S. Datta, "Quantum devices", Superlat. Microstruct., **6**, 83 (1989).
38. M. Büttiker, Y. Imry and M. Ya Azbel, "Quantum oscillations in one-dimensional normal metal rings", Phys. Rev. A., **30**, 1982 (1984).

Chapter 9
Quantum Devices and Mesoscopic Phenomena

9.1 Introduction

In this chapter, we will show how to apply some of the quantum transport formalisms that we discussed in the previous chapter to specific "quantum devices" and mesoscopic phenomena. Quantum devices are those in which quantum-mechanical effects that arise from phase coherence of electrons (e.g., tunneling, interference, etc.) are not only manifested, but undergird the basic device operation. In other words, they are central to the device's functioning. On the other hand, mesoscopic phenomena are those that are observed in mesoscopic structures whose dimensions along the direction of current flow are smaller than the phase-breaking length L_ϕ and the electron temperature is lower than the Thouless temperature. Mesoscopic phenomena arise from preservance of an electron's phase in the entire device. All true quantum devices use mesoscopic structures.

9.2 Double Barrier Resonant Tunneling Diodes

One of the earliest quantum devices that found widespread use as a prolific solid-state device is the Esaki tunnel diode, which exhibits negative differential resistance under forward bias. The dc current–voltage characteristic of this device has a region of negative slope in forward bias, which gives rise to the negative differential resistance. The negative differential resistance comes about because of Zener tunneling of electrons from the conduction to the valence band [1]. A more modern incarnation of this device is the double barrier resonant tunneling diode (DBRTD) which also exhibits negative differential resistance [2] and has the type of current–voltage characteristic shown in Fig. 9.1d. Theoretically, it can have many peaks and valleys with associated negative differential resistance. Experimentally, several peaks have been observed [3]. Devices with such characteristics (multiple peaks) have application in complex systems such as color image processors [4]. Moreover,

S. Bandyopadhyay, *Physics of Nanostructured Solid State Devices*,
DOI 10.1007/978-1-4614-1141-3_9,

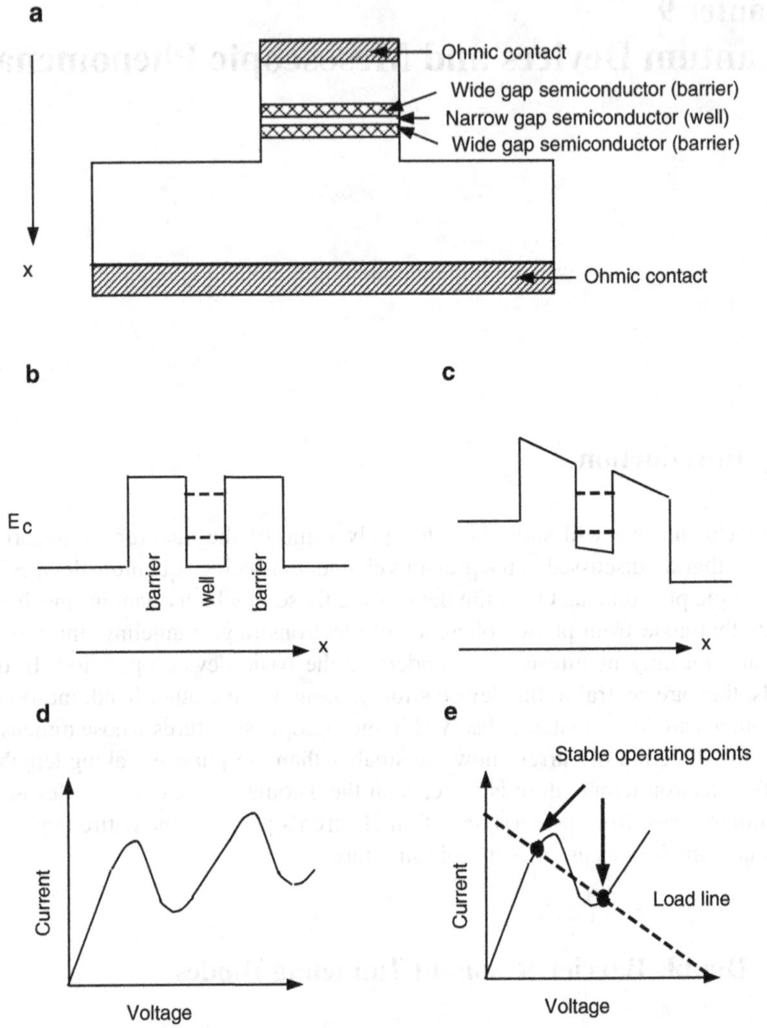

Fig. 9.1 (**a**) The basic structure of a double barrier resonant tunneling diode; (**b**) the conduction band diagram in the direction perpendicular to the heterointerfaces when no bias is applied between the two ohmic contacts; (**c**) the same conduction band diagram under bias (current flows perpendicular to heterointerfaces); (**d**) the ideal current versus voltage characteristic of the device, (**e**) connecting the device to a power supply through a load resistor yields two stable operating points on the load line, which can be used to encode binary bits 0 and 1 in digital logic

the negative differential resistance alone can be used to implement oscillators and the DBRTD is known to produce very high frequency oscillations approaching, or even exceeding, 1 THz [5]. Very few solid state devices can work at such high frequencies.

The DBRTD is also a versatile, multifunctional device. If it is connected to a power supply through a series resistor, then the load line intersects the current voltage characteristic of the DBRTD at two stable points which can encode binary logic bits 0 and 1. This is shown in Fig. 9.1e. There has been considerable work on logic applications of DBRTD.

The basic structure of a DBRTD is shown in Fig. 9.1a. It consists of multiple thin layers of different materials. Such a structure is usually fabricated with molecular beam epitaxy or metallo-organic-chemical-vapor-deposition techniques resulting in atomically sharp interfaces between the different layers. The essential feature of the device is that there is a narrow gap semiconductor layer sandwiched between two wide gap semiconductor layers, so that the ideal conduction band diagram in the direction perpendicular to heterointerfaces will look like Fig. 9.1b in the absence of any bias between the two ohmic contacts. Note that the wide gap semiconductors act as barriers and the narrow gap semiconductor acts like a quantum well.

When a bias is applied between the ohmic contacts, a current flows and the 2-terminal current can be calculated from the Tsu–Easki formula for three-dimensional leads. Clearly, what we have to do for this purpose is find the transmission probability $\tau_{1\to 2}(E)$ for any energy E. We can do this by the scattering matrix method.

Under bias, the conduction band diagram looks as in Fig. 9.1c. We will view this as a device composed of three sections—the first barrier, the well, and the second barrier. We will call their current scattering matrices $[S]_1$, $[S]_2$, and $[S]_3$.

Consider an electron traversing this device from one contact to the other, and let its wavevector in the well region be k_x. We have assumed that current flows in the x-direction. We will ignore the fact that any bias will cause some electric field in the well which will make the wavevector of an electron entering the well with a given energy vary with position, i.e., k_x will be a function of x. Furthermore, we will assume that there is no scattering—elastic or inelastic—within the device. All these assumptions will allow us to write

$$[S]_1 = \begin{bmatrix} r_1(k_x) & t_1'(k_x) \\ t_1(k_x) & r_1'(k_x) \end{bmatrix}, \tag{9.1}$$

$$[S]_2 = \begin{bmatrix} 0 & e^{ik_x L} \\ e^{ik_x L} & 0 \end{bmatrix}, \tag{9.2}$$

and

$$[S]_3 = \begin{bmatrix} r_3(k_x) & t_3'(k_x) \\ t_3(k_x) & r_3'(k_x) \end{bmatrix}, \tag{9.3}$$

where L is the width of the well and k_x is the (position-independent) wavevector in the well. The quantities T, r, t', and r' are the same as those defined in the previous chapter when we introduced the scattering matrix formulation.

The composite scattering matrix of this device is $[S] = [S]_1 \otimes [S]_2 \otimes [S]_3$ which is found from the rules of cascading scattering matrices in (8.46). From this recipe, we find that

$$[S] = \begin{bmatrix} r_{\mathrm{RTD}}(k_x) & t'_{\mathrm{RTD}}(k_x) \\ t_{\mathrm{RTD}}(k_x) & r'_{\mathrm{RTD}}(k_x) \end{bmatrix}, \tag{9.4}$$

where

$$\begin{aligned} r_{\mathrm{RTD}}(k_x) &= r_1(k_x) + \frac{t_1(k_x)t_1'(k_x)r_2(k_x)\mathrm{e}^{\mathrm{i}2k_xL}}{1 - r_1'(k_x)r_2(k_x)\mathrm{e}^{\mathrm{i}2k_xL}} \\ t'_{\mathrm{RTD}}(k_x) &= t_1'(k_x)t_2'(k_x)\mathrm{e}^{\mathrm{i}2k_xL}\left[1 + \frac{r_2(k_x)r_1'(k_x)\mathrm{e}^{\mathrm{i}2k_xL}}{1 - r_1'(k_x)r_2(k_x)\mathrm{e}^{\mathrm{i}2k_xL}}\right] \\ t_{\mathrm{RTD}}(k_x) &= \frac{t_1(k_x)t_2(k_x)\mathrm{e}^{\mathrm{i}k_xL}}{1 - r_1'(k_x)r_2(k_x)\mathrm{e}^{\mathrm{i}2k_xL}} \\ r'_{\mathrm{RTD}}(k_x) &= r_2'(k_x) + \frac{t_2(k_x)t_2'(k_x)r_1'(k_x)\mathrm{e}^{\mathrm{i}2k_xL}}{1 - r_1'(k_x)r_2(k_x)\mathrm{e}^{\mathrm{i}2k_xL}}. \end{aligned} \tag{9.5}$$

The last relation tells us that if an electron enters the well region with a wavevector k_x, then its transmission amplitude for transmitting through the entire DBRTD device is

$$t_{\mathrm{RTD}}(k_x) = \frac{t_1(k_x)t_2(k_x)\mathrm{e}^{\mathrm{i}k_xL}}{1 - r_1'(k_x)r_2(k_x)\mathrm{e}^{\mathrm{i}2k_xL}}. \tag{9.6}$$

We could have arrived at the above relation using the ray tracing method as shown in Fig. 9.2. We draw the Feynman paths associated with multiple reflection between the barriers and sum the amplitude of these paths to find the total transmission amplitude. This immediately yields

$$\begin{aligned} t_{\mathrm{RTD}}(k_x) &= t_1(k_x)\mathrm{e}^{\mathrm{i}k_xL}t_2(k_x) \\ &\quad + t_1(k_x)\mathrm{e}^{\mathrm{i}k_xL}r_2(k_x)\mathrm{e}^{\mathrm{i}k_xL}r_1'(k_x)\mathrm{e}^{\mathrm{i}k_xL}t_2(k_x) \\ &\quad + \ldots\ldots \\ &= \frac{t_1(k_x)t_2(k_x)\mathrm{e}^{\mathrm{i}k_xL}}{1 - r_1'(k_x)r_2(k_x)\mathrm{e}^{\mathrm{i}2k_xL}}. \end{aligned} \tag{9.7}$$

Let us now consider two idealizations: (1) the two barriers are identical so that $t_1(k_x) = t_2(k_x)$, and (2) the probability of reflecting off a barrier is the same whether the electron is incident from the left or right, i.e., $r_1'(k_x) = r_1(k_x) = r_2(k_x)$. These idealizations are clearly inappropriate under high bias, but we will make them in any case to illustrate the point we are about to make. Once we adopt these idealizations, we obtain

$$\tau_{1\to 2}(k_x) = t_{\mathrm{RTD}}(k_x) = \frac{|t_1(k_x)|^2\mathrm{e}^{\mathrm{i}[k_xL + 2\phi_t(k_x)]}}{1 - |r_1(k_x)|^2\mathrm{e}^{2i[k_xL+\phi_r(k_x)]}}, \tag{9.8}$$

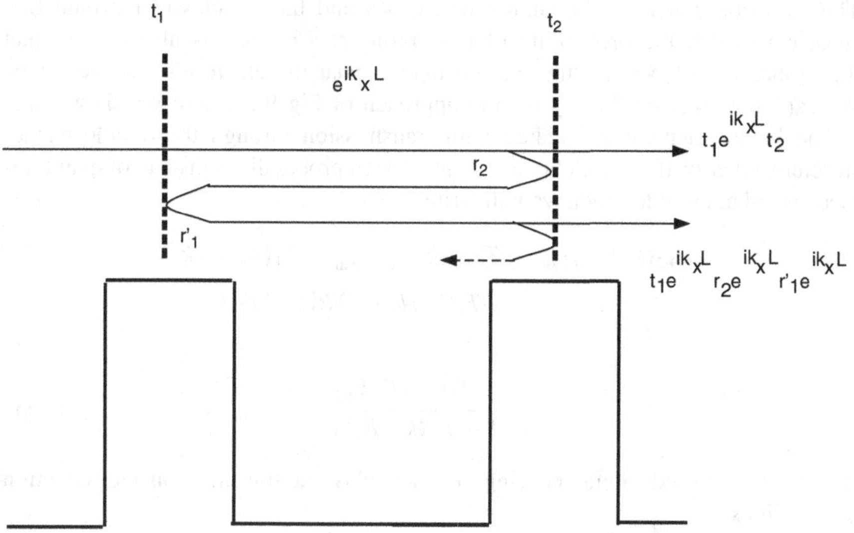

Fig. 9.2 Calculating the transmission through a double barrier resonant tunneling diode structure using the ray tracing method where the transmission amplitudes of all Feynman paths are summed up

where $\phi_t(k_x)$ is the phase of the transmission amplitude and $\phi_r(k_x)$ is the phase of the reflection amplitude.

Let us now examine what happens if we have the condition

$$k_x L + \phi_r(k_x) = n\pi, \tag{9.9}$$

where n is an integer.

In that case, using the unitarity of the scattering matrix $[S]_1$ which yields that $|t_1(k_x)|^2 + |r_1(k_x)|^2 = 1$ for every propagating mode, we find that[1]

$$t_{\mathrm{RTD}}(k_x) = \frac{|t_1(k_x)|^2}{|t_1(k_x)|^2} \mathrm{e}^{\mathrm{i}[k_x L + 2\phi_t(k_x)]} = \mathrm{e}^{\mathrm{i}[k_x L + 2\phi_t(k_x)]}. \tag{9.10}$$

The last relation shows that as long as the condition in (9.9)—called the "Fabry–Perot resonance condition"—is satisfied, $|\tau_{1\to 2}(k_x)| = |t_{\mathrm{RTD}}(k_x)|^2 = 1$, *even if* $|t_1(k_x)|^2 = |t_2(k_x)|^2 \ll 1$! This is a truly remarkable result. It shows that even though the electron has a very low probability of penetrating either barrier, the probability that it will penetrate both barriers is exactly 100% if the electron's wavevector is such that the resonance condition given by (9.9) is satisfied!

---

[1]Note that since we have neglected the effect of the bias, the current scattering matrix and the wave scattering matrix are the same, and each is unitary. Actually, only the sub-block relevant to propagating states will be unitary, but we have only considered propagating states here.

This is a consequence of quantum mechanics and has no classical explanation since classically, the probability of penetrating *two* barriers is always less than the probability of penetrating *one*, no matter what the electron's wavevector is. We can show this by the ray tracing approach of Fig. 9.2 where we show some of the Feynman paths associated with transmission through the double barrier structure. Clearly, if we use classical transmission probabilities instead of quantum-mechanical amplitudes, then we will write

$$\begin{aligned}\tau_{1\to 2}(k_x)|_{\text{classical}} &= T_{\text{RTD}}(k_x)|_{\text{classical}} = T_1(k_x)T_2(k_x)\\ &\quad + T_1(k_x)R_2(k_x)R_1'(k_x)T_2(k_x)\\ &\quad + \ldots\ldots\\ &= \frac{T_1(k_x)T_2(k_x)}{1 - R_1'(k_x)R_2(k_x)},\end{aligned} \tag{9.11}$$

where we have used capital lettering to denote classical transmission and reflection probabilities.

Now, if we assume identical barriers, then we get

$$\begin{aligned}\tau_{1\to 2}(k_x)|_{\text{classical}} &= T_{\text{RTD}}(k_x)|_{\text{classical}}\\ &= \frac{T_1^2(k_x)}{1 - R_1^2(k_x)} = \frac{T_1(k_x)}{1 + R_1(k_x)} = \frac{T_1(k_x)}{2 - T_1(k_x)} < T_1(k_x),\end{aligned} \tag{9.12}$$

where we have used the fact that $T_1(k_x) + R_1(k_x) = 1$, which accrues from current conservation. The last equation shows that classically, the transmission through two barriers ($\tau_{1\to 2}(k_x)|_{\text{classical}}$) is always less than the transmission through one ($T_1(k_x)$). This is in stark contrast with the quantum-mechanical result.

In Fig. 9.3, we plot the quantum-mechanically calculated transmission amplitude through each of two identical barriers spaced a certain distance L apart, as well as the transmission amplitude through both barriers, as a function of wavevector k_x. The transmission amplitude through both barriers reaches unity whenever the resonance condition of (9.9) is satisfied, but at those values of the wavevectors, the transmission through either barrier is much less than unity. Once again, this is a quantum-mechanical result that has no classical analog, since it shows that at resonance, the transmission through two consecutive barriers is actually *more* than the transmission through either one of them individually. In the 2-barrier case, the transmission amplitudes of the multiple reflection paths build up *coherently* when resonance occurs (i.e., all the Feynman paths interfere constructively), so that in the end, the transmission amplitude through the two interacting barriers reaches unity.

Let us revisit (9.9) for a moment. If the phase of the reflection coefficient is an even multiple of π, or is zero, then this condition reduces to

$$k_x^{\text{resonance}} = \frac{m\pi}{L}, \tag{9.13}$$

where m is an integer.

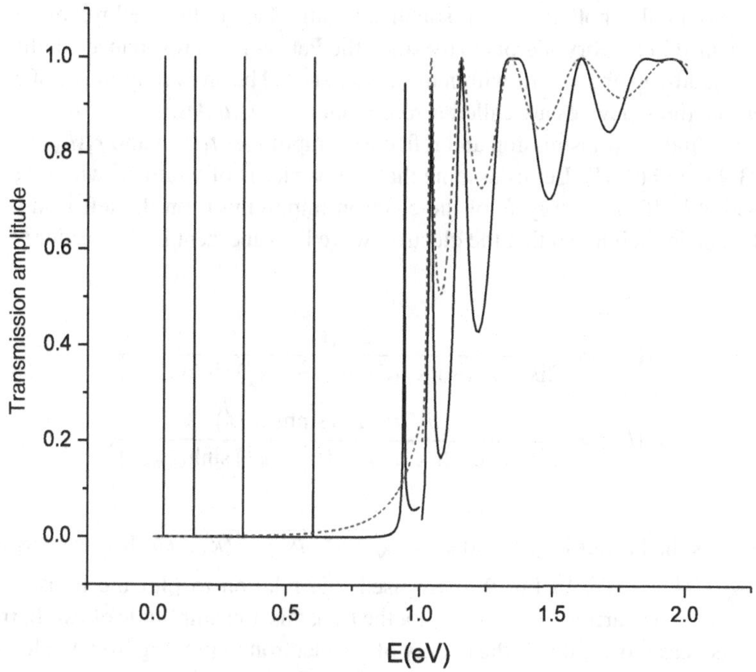

Fig. 9.3 The transmission amplitude associated with transmitting through one barrier (*broken curve*) and two identical barriers (*solid curve*) as a function of the incident electron's energy $E(k_x)$. Note that at low energies where the transmission amplitude through two barriers reaches unity (because of fulfilling the resonance condition), the transmission amplitude through a single barrier is much less than unity. While a single barrier is nearly opaque at these energies, two barriers happen to be completely transparent! This plot was generated by using (9.8) to find the transmission through two identical barriers and (9.15) and (9.16) to find the transmission through a single barrier. The parameters used in the calculation were: the barrier heights are 1 eV, the barriers widths are 5 nm, the well width is 10 nm and the electron's effective mass is assumed to be $0.1\,m_0$

The energies of the electron corresponding to these wavevectors are

$$E_m = \frac{\hbar^2}{2m^*}\left(k_x^{\text{resonance}}\right)^2 = \frac{\hbar^2}{2m^*}\left(\frac{m\pi}{L}\right)^2, \tag{9.14}$$

which are clearly the energies at the bottom of the subbands—or the so-called "subband levels"—in the rectangular quantum well if we assume that the barriers are infinitely high. Therefore, the transmission through the DBRTD device peaks when the incident electron's kinetic energy (corresponding to motion along the direction of current flow) equals the energy of a subband level in the well.

One should also note that the resonance taking place in the well region mirrors the resonance in a Fabry–Perot cavity since the Fabry–Perot resonance condition is that the round-trip phase shift within the cavity should be an even multiple of π. For this reason, these devices are called *resonant tunneling diodes.*[2]

We can find the transmission and reflection amplitudes $t_1(k_x)$ and $r_1(k_x)$ in (9.8) from (3.93) and (3.94). Let us assume that the barrier is of width L' while the well is of width L. If the energy E of the electron impinging from the left lead is less than the barrier height, so that the electron wave is evanescent in the barrier region, then

$$t_1(k_x) = \frac{2ik_x\kappa_x}{2ik_x\kappa_x \cosh(\kappa_x L') + (k_x^2 - \kappa_x^2)\sinh(\kappa_x L')}$$
$$r_1(k_x) = \frac{(2m^*\Phi_b/\hbar^2)\sinh(\kappa_x L')}{2ik_x\kappa_x \cosh(\kappa_x L') + (k_x^2 - \kappa_x^2)\sinh(\kappa_x L')}, \qquad (9.15)$$

where Φ_b is the barrier height and $\kappa_x = \sqrt{2m^*(\Phi_b - E(k_x))}/\hbar$. It is easy to show that $\frac{\hbar^2(k_x^2+\kappa_x^2)}{2m^*} = \Phi_b$. In Fig. 9.3, we used this relation to plot the transmission amplitude of one barrier and (9.8) to plot the transmission amplitude of two barriers.

It is also easy to see that if the energy of the electron impinging from the left lead is more than the barrier height so that the electron wave is propagating in the barrier region, then we will have

$$t_1(k_x) = \frac{2ik_x k'_x}{2ik_x k'_x \cos(k'_x L') + (k_x^2 + k'^2_x)\sin(k'_x L')}$$
$$r_1(k_x) = \frac{(2m^*\Phi_b/\hbar^2)\sin(k'_x L')}{2ik_x k'_x \cos(k'_x L') + (k_x^2 + k'^2_x)\sin(k'_x L')}, \qquad (9.16)$$

where $k'_x = \sqrt{2m^*(E(k_x) - \Phi_b)}/\hbar$. It is easy to see that $\frac{\hbar^2(k_x^2-k'^2_x)}{2m^*} = \Phi_b$. Thus, even if the electron is not tunneling through the barrier, but propagating above the barrier, the transmission characteristics will exhibit the same features as in the case of tunneling. We see this clearly in Fig. 9.3.

[2] An alternate explanation for the negative differential resistance observed in a DBRTD device has been proposed (S Luryi, "Frequency limit of double barrier resonant tunneling oscillators," Appl. Phys. Lett., **47**, 490 (1985)) based on energy and momentum conservation during tunneling from three-dimensional leads into a two-dimensional quantum well. However, it is not of coherent origin and does not require Fabry–Perot type resonance, which is why it is not discussed here.

9.2.1 Current–Voltage Characteristic of a Double Barrier Resonant Tunneling Diode

The current versus voltage relation of a DBRTD can be found from the expression derived in Problem 8.8 if the electron population is nondegenerate:

$$I = eA\frac{m^* k_B T}{\pi^2 \hbar^3}[1 - \exp(-e\mathrm{V}/k_B T)] \int_0^\infty \mathrm{d}E_x \tau_{1\to 2}(E_x, \mathrm{V}) \exp[-E_x/(k_B T)]. \tag{9.17}$$

If the electron population is degenerate, then we must use the full three-dimensional Tsu–Esaki formula given in (8.135). In either case, the characteristic is determined by the dependence of the transmission $\tau_{1\to 2}(E_x, \mathrm{V})$ on the voltage V applied across the structure.

The dependence of $\tau_{1\to 2}(E_x, \mathrm{V})$ on the bias voltage V can be understood from Fig. 9.4. As the voltage across the structure is increased, the conduction band bends down and brings the subband levels in the well into alignment with the energy of electrons impinging from the left contact. Every time a subband level comes into alignment, the resonance condition is met and the transmission through the device peaks, which gives rise to a peak in the current. In Fig. 9.4, we show the conductance band profile of the DBRTD at different bias voltages and indicate the corresponding operating point on the current–voltage characteristic.

9.2.2 Space Charge Effects and Self-Consistent Solution

In order to find the dependence of the transmission probability $\tau_{1\to 2}(E_x, \mathrm{V})$ on the bias potential V, one must find the conduction band profile inside the DBRTD device at different bias potentials *self-consistently* and then calculate the transmission probability using the method outlined in Special Topic 4 of Chap. 3. For self-consistent solutions, one solves the Schrödinger and Poisson equations iteratively or simultaneously. While the Schrödinger equation yields the wavefunction from which the carrier concentration everywhere within the DBRTD is inferred, the Poisson equation yields the potential profile inside the DBRTD based on knowledge of the carrier concentration. This potential profile is then fed back to the Schrödinger equation to find the new wavefunction and a new carrier concentration. The new carrier concentration is supplied to the Poisson solver to find the new potential profile. This process is repeated in a cyclic fashion until convergence is obtained. The final potential profile is used to calculate the transmission probability.

The current–voltage characteristic found from the self-consistent transmission probability $\tau_{1\to 2}(E_x, \mathrm{V})$ can be considerably different from that calculated without consideration of self-consistency [6] as shown in Fig. 9.5. Self-consistency is particularly important in calculating the peak current because electrons tunneling resonantly into the well can cause significant space charge effects.

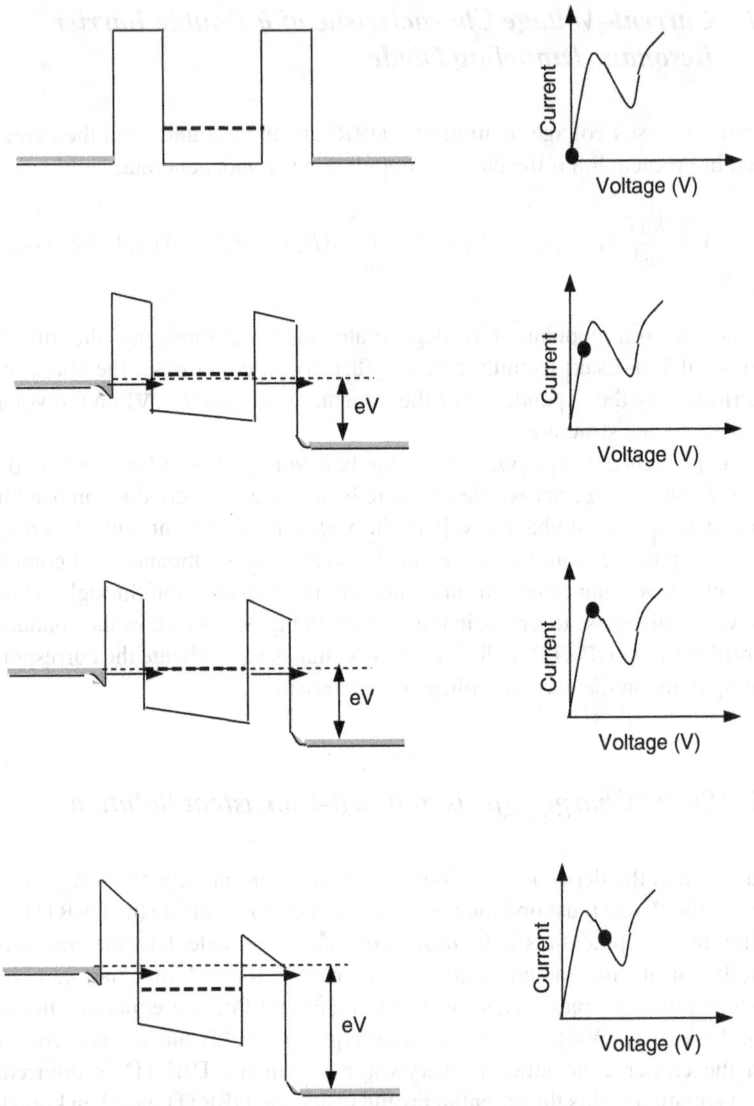

Fig. 9.4 The conduction band profile of an idealized double barrier resonant tunneling diode under different biases and the corresponding operating point on the current–voltage characteristic

9.2.3 *The Effect of Nonidealities on the Double-Barrier Resonant Tunneling Diode*

In the preceding discussion, we showed that the transmission amplitude through the DBRTD will reach unity when the resonance condition is fulfilled *provided* the transmission through each barrier is the same when the wavevector of the impinging

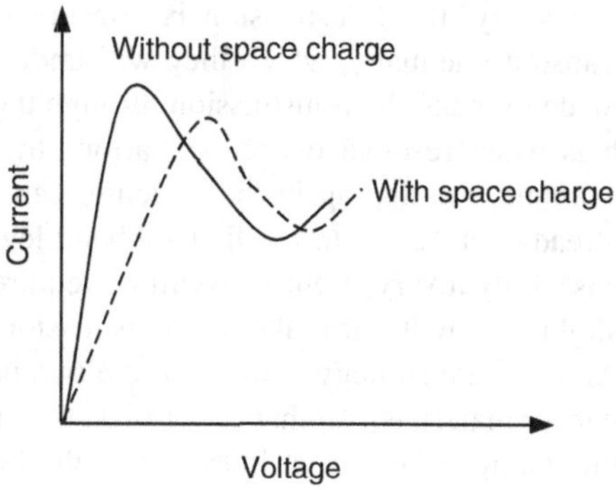

Fig. 9.5 The current–voltage characteristic of a double barrier resonant tunneling diode can be significantly modified by space charge effects. Space charge effects can alter the peak current and shift the voltage at which the current peaks

electron corresponds to the resonance condition. However, any bias will cause band bending in the conduction band so that the barrier heights seen by the resonantly transmitting electron will be different for the two barriers. This can be clearly seen in Fig. 9.4. The consequence of this is that the two barriers will transmit *unequally* and we can no longer assume that $t_1(k_x) = t_2(k_x)$. This has a disastrous effect on the magnitude of the transmission amplitude at resonance, as we show below.

As long as the transmission through each barrier is small and much less than unity, we can write the transmission amplitude through two barriers as (see Problem 9.2)

$$t_{\text{2-barriers}} = \frac{1}{C_1[|t_1||t_2|]^{-1} + C_2|t_1|/|t_2| + C_2|t_2|/|t_1| - (C_2/2)|t_1||t_2|}, \tag{9.18}$$

where

$$\begin{aligned} C_1 &= \left(1 - \mathrm{e}^{\mathrm{i}(2k_x L + \theta_r^{(1)} + \theta_r^{(2)})}\right) \mathrm{e}^{-\mathrm{i}\left(\theta_t^{(1)} + \theta_t^{(2)} + k_x L\right)} \\ C_2 &= \frac{1}{2} \mathrm{e}^{\mathrm{i}\left(k_x L + \theta_r^{(1)} + \theta_r^{(2)} - \theta_t^{(1)} - \theta_t^{(2)}\right)}. \end{aligned} \tag{9.19}$$

where $\theta_r^{(n)}$ and $\theta_t^{(n)}$ are the phases of the reflection and transmission amplitudes of the n-th barrier.

At resonance, when (9.9) is satisfied, $C_1 = 0$. Furthermore, since the transmission through each barrier is much less than unity, we can neglect second order terms in the transmission. Therefore, the transmission amplitude through two barriers will be

$$\begin{aligned} |t_{\text{resonance}}| &\approx \frac{2}{|t_1|/|t_2| + |t_2|/|t_1|} \\ &\approx 2\frac{|t_2|}{|t_1|} \ \texttt{if}\ |t_1| \gg |t_2| \\ &\approx 2\frac{|t_1|}{|t_2|} \ \texttt{if}\ |t_2| \gg |t_1|. \end{aligned} \tag{9.20}$$

Clearly, the transmission is considerably less than unity if the two barriers transmit unequally, which they will under bias. Therefore, every attempt must be made to make the transmission through the two barriers as equal as possible under bias when resonance occurs. Various tricks are employed for this purpose. For example, the doping in the structure can be tailored such that the Fermi level is already very close to the first subband level in the well under equilibrium. In this case, only a very slight bias will be required to reach resonance and hopefully that slight bias will not make the transmission through the two barriers very unequal. An alternate strategy is to make the two barriers unequally tall (by using different barrier materials), so that under bias, they present nearly equal potential barriers to the impinging electron. In that case, the transmission through the two barriers will be nearly equal when resonance takes place. Unfortunately, this trick does not work very well except at low temperatures. At higher temperatures, there is a significant spread in the energy of impinging electrons, which is much larger than the level broadening of the subband state in the well (which is the full width at half maximum of the transmission peak in Fig. 9.3). In that case, the transmission amplitude is fairly close to that of the more opaque barrier [7, 8]. Yet other tricks involve placing the double barrier resonant tunneling structure within the base of a heterojunction bipolar transistor and varying the energy of the impinging electron not by applying a voltage across the double barrier device, but by varying the base-emitter voltage of the transistor. In this process, there is no band bending in the double barrier device so that the barriers transmit almost equally and the transmission amplitude through the double barrier structure can approach unity [9].

9.2.4 Other Mechanisms for Negative Differential Resistance in a DBRTD

There are many mechanisms that can produce a negative differential resistance in a double barrier diode. The first is of course the normal resonant tunneling that we have discussed. The second mechanism that can also cause the appearance of negative differential resistance is tunneling from three dimensional leads into a two dimensional quantum well while conserving energy and momentum [13]. This mechanism does not require the Fabry–Perot type of resonance and is not discussed here.

The third mechanism is known as *incoherent tunneling* and is much more likely to occur at elevated temperatures when inelastic scattering will destroy the phase coherence of the electron wavefunction required for normal resonant tunneling. This process is a two-step process for a double barrier diode. Electrons first tunnel into the well through the first barrier and then tunnel out through the second barrier. It is the first step that produces negative differential resistance [10].

The fourth and final mechanism that can cause negative differential resistance in a periodic multiple-well/multiple-barrier structure (also known as a "superlattice") is *sequential resonant tunneling* [11, 12]. This is easy to understand by referring to

Fig. 9.6. It forms the basis of quantum cascade lasers that we discussed in Chap. 6. Under a bias or electric field, the subband levels in a quantum well are called a Stark ladder states (recall the quantum-confined Stark effect of Chap. 3). They are now quasi-bound states as opposed to bound states since an electron can always lower its energy by escaping from the well.

When a superlattice structure is biased with a voltage, electrons sequentially tunnel through the Stark ladder of states as shown in Fig. 9.6. The current peaks at voltages when a quasi bound state in the n-th well is degenerate in energy with a higher quasi bound state in the $(n + 1)$-th well. Electrons tunnel resonantly from the ground state in the $(n - 1)$-th well into an excited state in the n-th well, decay to the ground state in the n-th well by emitting phonons, and then tunnel resonantly into an excited state in the $(n + 1)$-th well to carry on the process. The tunneling amplitude (and hence the current) peaks at those voltages where the potential energy drop across a superlattice period eEd (E is the spatially averaged electric field and d is the superlattice period) equals the energy difference between an excited and a ground state in a well.

9.2.5 *Tunneling Time*

Resonant tunneling devices were the object of much attention over the last three decades primarily because they can operate at very high frequencies. It therefore behooves us to examine the possible upper limit on the operating frequency of a tunnel device. The maximum frequency probably can never exceed the inverse of the time it takes for an electron to tunnel through the structure, which would be the effective transit time through the device. This brings us to a very important topic of what is the time it takes for an electron to tunnel through a barrier.

This field has been somewhat controversial. It was first addressed in 1932 by MacColl [14] who studied the propagation and scattering of a wavepacket through a rectangular barrier and postulated that in such cases, the tunneling time will be given by

$$t_{\text{tun}} = \frac{m^*}{\hbar k} W, \tag{9.21}$$

where W is the barrier width and k is the wavevector of the incident (and transmitted) electron. This result makes intuitive sense since it says that the tunneling time is the distance tunneled through, divided by the velocity of the incident electron.

Hartman [15] found a similar result by representing the electron as a Gaussian wavepacket whose average energy was close to the barrier height (so that the barrier was relatively transparent). However, for wavepackets whose average energies were considerably less than the barrier height (so that the barrier was relatively opaque), he found that the tunneling time was independent of barrier width and is given by

$$t_{\text{tun}} = \frac{2m^*}{\hbar k \kappa}, \tag{9.22}$$

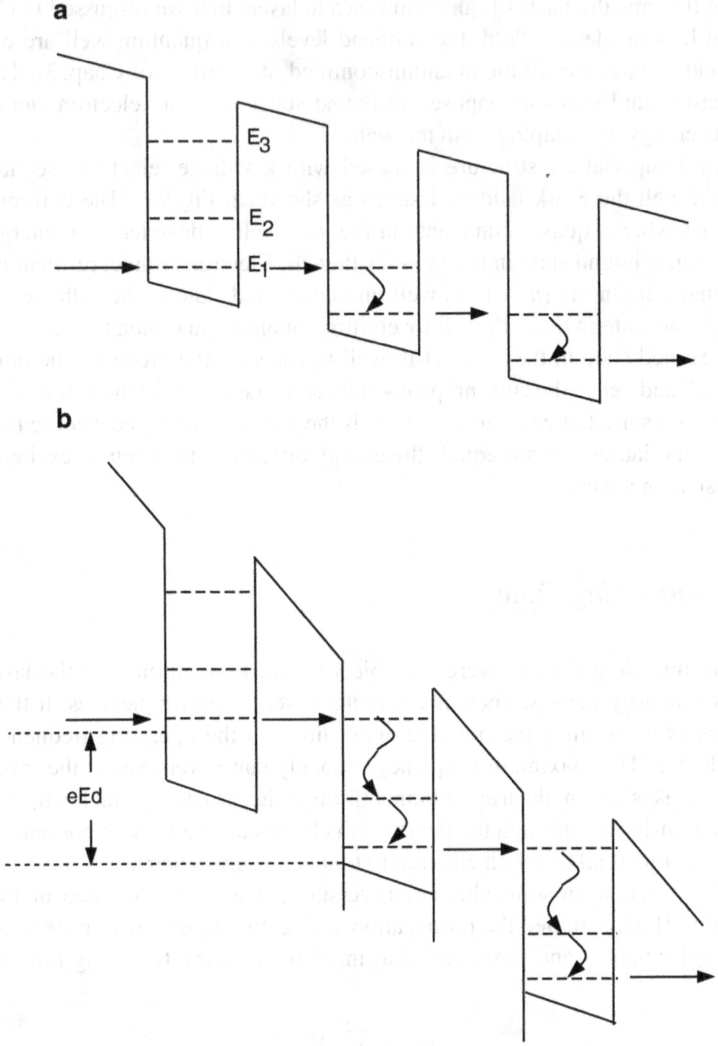

Fig. 9.6 Schematic illustration of sequential resonant tunneling through a superlattice consisting of multiple barriers and wells when the potential drop across a superlattice period (eEd, where E is the electric field due to the bias and d is the period of the superlattice) is equal to the energy difference between (**a**) the first excited state and the ground state of the wells, and (**b**) the second excited state and ground state of the wells. Adapted from F. Capasso, K. Mohammed and A. Y. Cho, "Resonant tunneling through double barriers, perpendicular quantum transport phenomena in superlattices, and their device applications", IEEE J. Quant. Elec. **QE-22**, 1853 (1986)

where $k = \sqrt{2m^* E/\hbar}$ and $\kappa = \sqrt{2m^*(\Phi_b - E)}/\hbar$, with E being the energy of the incident (and transmitted) electron and Φ_b being the barrier height.

Hagston [16] pointed out that the preceding approaches neglected the nonlocality of quantum mechanics in that the wavefunction must sense the barrier at some distance. He used the density matrix formalism to calculate a tunneling time and obtained the result:

$$t_{tun} = \frac{\hbar k}{2\kappa e \Phi_b}, \tag{9.23}$$

Barker [17] calculated the tunneling time using Wigner distribution functions and ended up with (9.21). A more careful analysis was carried out by Büttiker and Landauer [18] who modulated the potential barrier with a small ac voltage and then looked at the frequency response of the transmitted wave's modulated portion. From the roll-off point of the frequency response, they determined that the tunneling time for electrons whose energies are less than the barrier height is

$$t_{tun} = \int_0^W \frac{m^*}{\hbar\kappa(x)} dx, \tag{9.24}$$

where $\kappa(x) = \sqrt{2m^*[\Phi_b(x) - E]}/\hbar$. This result applies to barriers whose heights vary in space.

Let us take the last result and estimate what the tunneling time through a 1 eV high and 5 nm wide barrier will be if an electron is incident on it with an energy of 100 meV. We will assume that the electron's effective mass is $0.1\, m_0$. This yields a tunneling time of 28 fs, showing that an electron can tunnel through a barrier very quickly. The inverse of this time is 35 THz, showing that indeed tunneling devices should be capable of operating at very high frequencies unless burdened by extraneous parasitic elements. This is the reason why tunnel devices—and not just resonant tunneling diodes—have always captured the imagination of device engineers.

9.3 The Stub-Tuned Transistor

A microwave stub tuner is a device that typically looks like the object in Fig. 9.7a which consists of a T-shaped microwave waveguide. Port 2 has a movable "stub" which can be pushed in or pulled out mechanically. If we do that, then the microwave transmission from port 1 to port 3 changes.

Insofar as a phase coherent electron device behaves like an electron "waveguide" because of the essential similarity between the Maxwell's equation for the electric field in an electromagnetic wave and the Schrödinger equation for the wavefunction of an electron, we expect that one should be able to fashion an electron device that mirrors a microwave stub tuner. In such a device, we should be able to change

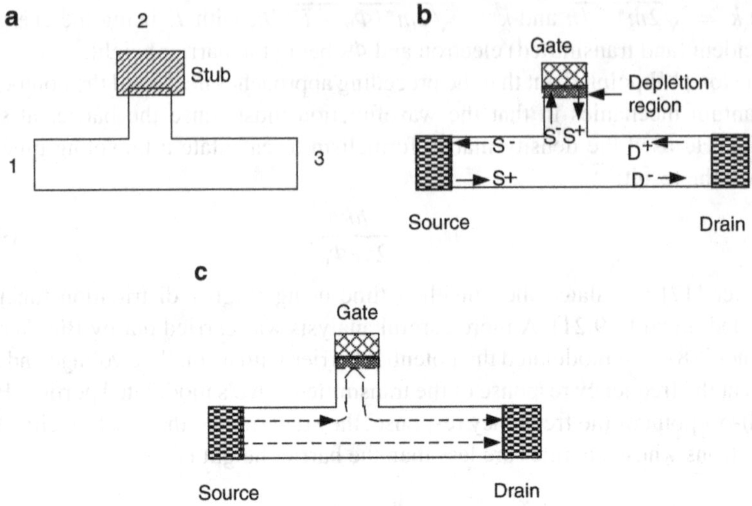

Fig. 9.7 (**a**) Schematic of a microwave stub tuner (a 3-port network); (**b**) a quantum interference transistor fashioned after the stub tuner where the incoming and outgoing current amplitudes are shown at the source, drain, and gate terminals; and (**c**) the two primary Feynman paths that interfere to produce the transistor action (modulation of the source-to-drain current by a negative gate potential that extends the depletion region under the gate and changes the length of the longer path to change the phase relationship between the two paths)

the transmission of charge carriers from one port to another by performing some stub-like action at the third terminal. Since transmission determines the current, this should change the current flowing between the first two terminals because of the stub action at the third terminal. If we can implement the stub with either a controlled voltage or a current, then we would have effectively realized a "transistor" since, by definition, a transistor is a 3-terminal device in which current between two of the terminals is modulated by a current or a potential applied at the third terminal.

Such a device is shown in Fig. 9.7b, where we will call ports 1 and 3 the "source" and "drain" terminals and port 2 the "gate" terminal in the spirit of the field effect transistor. A negative potential applied to the gate terminal will deplete the region under the gate of free electrons because of Coulomb repulsion. This is equivalent to pushing in a stub. We will show that this will indeed change the current flowing between the source and drain terminals, thereby realizing transistor action. Note that unlike the classical field effect transistor (FET), the source-to-drain current is modulated not by changing the carrier concentration in the electron waveguide with the gate voltage, but rather by controlling the quantum-mechanical interference of electron waves that start out from the source and reach the drain directly or after reflecting off various terminals. In Fig. 9.7c, we show two particular Feynman paths that are associated with electron waves transmitting directly from the source to the drain and transmitting after reflecting off the gate.[3] Clearly, interference between these two paths can be controlled by depleting the region underneath the gate which controls the length of the second path and hence the phase relationship between the two paths. When the phase difference between these two paths is an odd multiple of π, the two paths interfere *destructively* and the current exiting from the drain falls close to its minimum value since the transmission through the drain plummets. When the phase difference is an even multiple of π, the paths interfere *constructively* and the current approaches its maximum value. Thus, the source-to-drain current, or equivalently the source-to-drain conductance, can be modulated by the gate potential. We call this modulation "stub tuning." Such a device was proposed in the 1990s [19,20].

Note that the current scattering matrix of this device is given by that in Problem 8.5. Using that scattering matrix, we can analyze this device and find an expression for the source-to-drain conductance as a function of gate voltage, using Büttiker's multi probe formula. Calling the currents emerging from the source, the drain and the gate I_S, I_D, and I_G, respectively, we can write the multiprobe formula in (8.164) as

$$\begin{bmatrix} I_S \\ I_D \\ I_G \end{bmatrix} = \frac{2e}{h} \begin{bmatrix} T_{SD} + T_{SG} & -T_{DS} & -T_{GS} \\ -T_{SD} & T_{DS} + T_{DG} & -T_{GD} \\ -T_{SG} & -T_{DG} & T_{GS} + T_{GD} \end{bmatrix} \begin{bmatrix} \mu_S \\ \mu_D \\ \mu_G \end{bmatrix}, \tag{9.25}$$

where T_{ij} is the transmission probability from the i-th to the j-th terminal, and μ_i is the chemical potential or local Fermi level in the i-th terminal.

Since the gate is insulating and no current flows out of the gate, we can set the gate current to zero and get from the last equation that

$$I_G = \frac{2e}{h}[-T_{SG}\mu_S - T_{DG}\mu_D + (T_{GS} + T_{GD})\mu_G] = 0. \tag{9.26}$$

In the absence of any magnetic field, reciprocity of the transmission probability dictates that $T_{ij} = T_{ji}$, so that the last equation yields

$$\mu_G = -\frac{T_{GS}\mu_S + T_{GD}\mu_D}{T_{GS} + T_{GD}}. \tag{9.27}$$

From (9.25), we also get

$$I_D = \frac{2e}{h}[-T_{SD}\mu_S + (T_{SD} + T_{GD})\mu_D - T_{GD}\mu_G]. \tag{9.28}$$

---

[3] We loosely term them *Feynman paths*. Their transmission amplitudes are not equal and hence the terminology is not exact.

Combining the last two equations, we obtain the source-to-drain conductance as

$$G_{\mathrm{SD}} = e\frac{I_{\mathrm{D}}}{\mu_{\mathrm{D}} - \mu_{\mathrm{S}}} = \frac{2e^2}{h}\left[T_{\mathrm{SD}} + \frac{T_{\mathrm{GD}}T_{\mathrm{GS}}}{T_{\mathrm{GD}} + T_{\mathrm{GS}}}\right] = \frac{2e^2}{h}\left[T_{\mathrm{SD}} + \frac{1}{1/T_{\mathrm{GD}} + 1/T_{\mathrm{GS}}}\right]. \tag{9.29}$$

From Fig. 9.7, we can see that the geometry of the structure is such that clearly $T_{\mathrm{SD}} \gg T_{\mathrm{GS}}, T_{\mathrm{GD}}$ since in going from source to drain, the transmission path does not have to bend around a corner, whereas in going from the gate to the other terminals, the transmission path has to bend around a corner. Therefore, the last equation can be reduced to

$$G_{\mathrm{SD}} \approx \frac{2e^2}{h}T_{\mathrm{SD}}. \tag{9.30}$$

We can now use the result of Problem 8.5 to find T_{SD}. Clearly this quantity is $|\mathbf{t}|^2$, where

$$\mathbf{t} = \mathbf{t}_{\mathrm{DS}} + \mathbf{t}_{\mathrm{DG}}[\mathbf{I} - \mathbf{R}\mathbf{r}_{\mathrm{GG}}]^{-1}\mathbf{R}\mathbf{t}_{\mathrm{GS}}, \tag{9.31}$$

and $\mathbf{R}$ is the reflection coefficient at the gate.

The previous equation can be expanded in a geometric series:

$$\mathbf{t} = \mathbf{t}_{\mathrm{DS}} + \mathbf{t}_{\mathrm{DG}}[\mathbf{I} + \mathbf{R}\mathbf{r}_{\mathrm{GG}} + \mathbf{R}^2\mathbf{r}_{\mathrm{GG}}^2 + \ldots]\mathbf{R}\mathbf{t}_{\mathrm{GS}}. \tag{9.32}$$

Now, if $r_{\mathrm{GG}} \ll 1$, then we can approximate this to

$$\mathbf{t} \approx \mathbf{t}_{\mathrm{SD}} + \mathbf{t}_{\mathrm{GD}}\mathbf{R}\mathbf{t}_{\mathrm{SG}}, \tag{9.33}$$

so that

$$G_{\mathrm{SD}} \approx \frac{2e^2}{h}|\mathbf{t}|^2 \approx \frac{2e^2}{h}|\mathbf{t}_{\mathrm{SD}} + \mathbf{t}_{\mathrm{GD}}\mathbf{R}\mathbf{t}_{\mathrm{SG}}|^2, \tag{9.34}$$

The first term in the right-hand side is the transmission amplitude of the Feynman path going straight from the source to the drain while the second term is the transmission amplitude of the Feynman path going from the source to the drain after reflecting off the gate (recall the order in which transmission matrices are multiplied). These are the two paths shown by the broken lines in Fig. 9.7c. Clearly, the source-to-drain conductance is determined primarily by quantum-mechanical interference between these two paths. We can change the phase relationship between them by changing the phase of the coefficient $\mathbf{R}$ which we can write as $k_{\mathrm{F}}W$, where k_{F} is the Fermi wavevector in the device and W is the width of the depletion region underneath the gate. Using the gate potential, we can change W and hence the phase of $\mathbf{R}$. This will change the source-to-drain conductance and realize transistor action.

Devices such as the one just described are exotic and novel because they operate on the basis of quantum mechanics rather than classical physics, but it should be understood that they do not necessarily perform any better than conventional field effect transistors. In fact, in some respects they are worse since they have little fabrication tolerance [21] and will be clearly incapable of working at room temperature when the phase coherence length of electrons becomes too short. The electron's phase is a very delicate entity and it is extremely difficult to conserve it under most situations. As a result, these devices have remained theoretical curiosities rather than become practical transistors.

9.4 Aharonov–Bohm Quantum Interferometric Devices

In Chap. 7, we were exposed to the concept of the magnetic vector potential $\vec{A}(\vec{r},t)$ due to a magnetic flux density $\vec{B}(\vec{r},t)$. The two are related as $\vec{B}(\vec{r},t) = \vec{\nabla} \times \vec{A}(\vec{r},t)$, which immediately shows that it is very possible to have a nonzero vector potential in a region where the flux density vanishes.[4] This seems to suggest that potentials are just mathematical constructs that have no physical consequence; it is only the field associated with the potential that matters (Fig. 9.8).

This mindset was challenged by Yakir Aharonov and David Bohm in a classic paper [22] where they conceptualized the following type of gedanken experiment. Consider a magnetic field confined entirely within a superconducting cylinder which acts as a perfect magnetic shield and ensures that no field leaks outside the cylinder. An electron waveguide is built *around* this cylinder in the form of a ring. A lone electron is injected into the waveguide, whereupon its wavefunction splits and traverses both arms of the ring and recombines at the point of detection. The electron never sees the magnetic field which is confined within the cylinder and never intersects its paths. However, it sees the associated magnetic vector potential which is nonzero inside the ring since the magnetic vector potential extends up to infinity.

Now, if only the field could have physical consequences and the vector potential could not, then any variation of the field within the cylinder should produce no measurable effect on the electron since the electron never encounters the field. On the other hand, if the vector potential too could produce physical consequences, then we will expect to see a measurable effect on the electron as the field is varied since that variation will vary the vector potential. The vector potential $\vec{A}$ will causes a phase difference of $\frac{e}{\hbar} \oint \vec{A} \cdot d\vec{l}$ between the two paths, where $\oint \vec{A} \cdot d\vec{l}$ is the contour line integral of the magnetic vector potential around the ring (see Problem 9.3). This phase difference determines the interference condition at the detection point. If the phase difference is an even multiple of π, then the interference at the detector will be constructive and the measured electron intensity will be large. On the other hand, if the phase difference is an odd multiple of π, the interference will be destructive,

[4] One example of this is a region where the magnetic vector potential is spatially invariant or can be written as the gradient of a scalar.

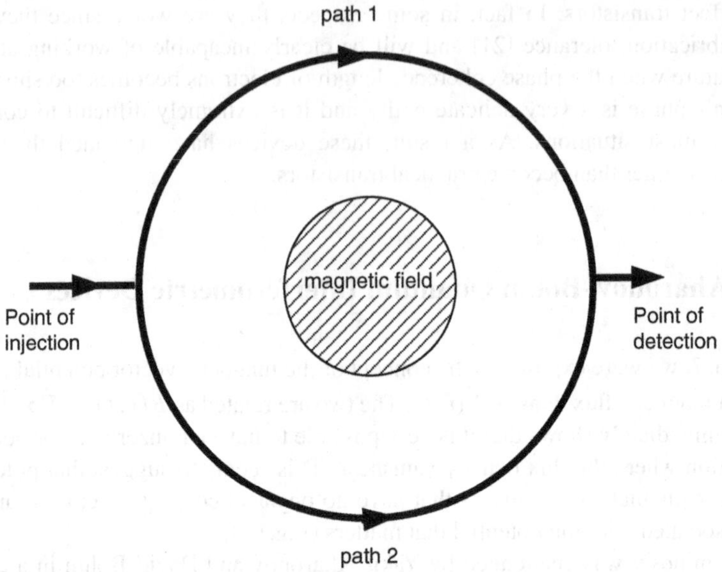

Fig. 9.8 The Aharonov–Bohm quantum interference effect demonstrating the significance of potentials in quantum theory. A magnetic field is confined completely within the *shaded circular* region and an electron wave samples two interfering paths that do not intersect the circular region and form a ring around it. Since the electron never sees the field, it may appear that varying the field (or the magnetic flux density $\vec{B}$) within the circular region should have no effect on the interference between the two paths and hence the wave amplitude at the point of detection. However, even though the field and flux density are zero in the paths of the electron, the magnetic vector potential $\vec{A}$ associated with the flux density $\vec{B}$ is *not*. This vector potential causes a phase difference of $\frac{e}{\hbar}\oint \vec{A}\cdot d\vec{l}$, to appear between the two interfering paths where $\oint \vec{A}\cdot d\vec{l}$ denotes the contour line integral of the magnetic vector potential around the ring. Since varying the magnetic field within the circular region will vary the magnetic vector potential and hence the phase difference $\frac{e}{\hbar}\oint \vec{A}\cdot d\vec{l}$ between the interfering paths, the wave intensity at the point of detection will oscillate with varying field, *even though the electron never sees the field.* This is a dramatic manifestation of the importance of potentials, showing that they can have physical consequences. The electron does not sense the field directly since it detours around the region where the field is present, but since it senses the associated vector potential, it responds to any variation in the field by exhibiting oscillating charge density (or intensity) at the detector

and the electron intensity at the detector will be zero, or at least small. Thus, we can measure the electron intensity at the detector while we vary the magnetic field inside the cylinder. If we observe any variation in the electron intensity, then we will know that it is being caused by the vector potential, and not the field, since the electron sees only the vector potential. This will establish that the vector potential, by itself, can *produce physical consequences*. In other words, it is not that potentials are totally meaningless and only fields are meaningful. Both are meaningful in the world of quantum mechanics.

This gedanken experiment was actually carried out by Akira Tonomura and co-workers [23] who demonstrated that the electron intensity at the detection point does indeed change with the magnetic field even though the electron's path never intersects with the field. This firmly establishes the significance of potentials in quantum theory. Our interest in this effect is of course not entirely academic. We are interested in it since it can be exploited to realize novel electronic device functionality, i.e., novel types of quantum interference transistors somewhat different from the stub-tuned transistor that we discussed earlier. These devices are based on the principle that interference between the two paths around the ring will result in periodic modulation of the transmission through the ring, and therefore periodic modulation of its conductance, as the flux threading the ring is varied. Thus, transistor action is realized with an external magnetic field.

We can show that the ring's conductance indeed oscillates periodically with the magnetic flux enclosed within the branches using a very simple analysis. If we neglect multiple reflection effects, then we can write the transmission amplitude through the ring as the sum of the transmission amplitudes through the two paths:

$$t_{\text{ring}} = t_{\text{path1}} + t_{\text{path2}}. \tag{9.35}$$

Let us now assume that the paths are nominally identical so that the magnitudes of the transmission amplitude through the two paths are equal. Furthermore, there are no scatterers anywhere, whether elastic or inelastic, so that transport is ballistic. This will allow us to rewrite the last equation as

$$\begin{aligned}
t_{\text{ring}} &= |t|\left(\exp\left[\mathrm{i}\int_{\text{path1}}\left[k(\vec{r}) + \frac{e}{\hbar}\vec{A}(\vec{r})\right]\cdot \mathrm{d}\vec{l}\right] + \exp\left[\mathrm{i}\int_{\text{path2}}\left[k(\vec{r}) + \frac{e}{\hbar}\vec{A}(\vec{r})\right]\cdot \mathrm{d}\vec{l}\right]\right) \\
&= |t|\mathrm{e}^{\mathrm{i}\int_{\text{path1}}\left[k(\vec{r})+\frac{e}{\hbar}\vec{A}(\vec{r})\right]\cdot \mathrm{d}\vec{l}}\left(1 + \mathrm{e}^{\mathrm{i}\frac{e}{\hbar}\left[\int_{\text{path2}}\vec{A}(\vec{r})\cdot \mathrm{d}\vec{l} - \int_{\text{path1}}\vec{A}(\vec{r})\cdot \mathrm{d}\vec{l}\right]}\right. \\
&\qquad\qquad \left.\times\, \mathrm{e}^{\mathrm{i}\left(\int_{\text{path2}}\vec{k}(\vec{r})\cdot \mathrm{d}\vec{l} - \int_{\text{path1}}\vec{k}(\vec{r})\cdot \mathrm{d}\vec{l}\right)}\right) \\
&= |t|\mathrm{e}^{\mathrm{i}\int_{\text{path1}}\left[k(\vec{r})+\frac{e}{\hbar}\vec{A}(\vec{r})\right]\cdot \mathrm{d}\vec{l}}\left(1 + \mathrm{e}^{\mathrm{i}\frac{e}{\hbar}\oint \vec{A}(\vec{r})\cdot \mathrm{d}\vec{l}}\right) \\
&= |t|\mathrm{e}^{\mathrm{i}\int_{\text{path1}}\left[k(\vec{r})+\frac{e}{\hbar}\vec{A}(\vec{r})\right]\cdot \mathrm{d}\vec{l}}\left(1 + \mathrm{e}^{\mathrm{i}\frac{e}{\hbar}\int_s\left[\vec{\nabla}\times\vec{A}(\vec{r})\right]\cdot \mathrm{d}\vec{S}}\right) \\
&= |t|\mathrm{e}^{\mathrm{i}\int_{\text{path1}}\left[k(\vec{r})+\frac{e}{\hbar}\vec{A}(\vec{r})\right]\cdot \mathrm{d}\vec{l}}\left(1 + \mathrm{e}^{\mathrm{i}\frac{e}{\hbar}\Phi}\right).
\end{aligned} \tag{9.36}$$

where $|t|$ is the magnitude of the transmission amplitude through either path and the contour integral is taken around the ring counterclockwise. The quantity Φ is the magnetic flux enclosed by the ring. Note that we have tacitly assumed $\int_{\text{path2}} \vec{k}(\vec{r}) \cdot \mathrm{d}\vec{l} = \int_{\text{path1}} \vec{k}(\vec{r}) \cdot \mathrm{d}\vec{l}$ which is fine since the two paths are identical and there is no scattering. If scattering was present, then the scatterers in the two paths will not be identical so that $\int_{\text{path2}} \vec{k}(\vec{r})\cdot \mathrm{d}\vec{l} \neq \int_{\text{path1}} \vec{k}(\vec{r})\cdot \mathrm{d}\vec{l}$. This would have caused problems as we will see later. Hence, it is important to mandate ballistic transport and identical paths.

The transmission probability through the ring is now given by

$$T_{\text{ring}} = |t_{\text{ring}}|^2 = 2|t|^2\left[1+\cos\left(\frac{e}{\hbar}\Phi\right)\right]. \tag{9.37}$$

Any dependence of T_{ring} on electron energy accrues solely from the energy dependence of $|t|^2$ since everything else is energy independent. Substituting the last expression for the transmission probability in (8.142), we get that the linear response conductance of the ring in the case of ballistic transport is

$$\begin{aligned} G_{\text{ring}} &= \frac{e^2}{k_B T h}\int_0^\infty \mathrm{d}E|t(E)|^2\left[1+\cos\left(\frac{e}{\hbar}\Phi\right)\right]\text{sech}^2\left[\frac{E-\mu_F}{2k_B T}\right] \\ &= \left[1+\cos\left(\frac{e}{\hbar}\Phi\right)\right]\frac{e^2}{k_B T h}\int_0^\infty \mathrm{d}E|t(E)|^2\text{sech}^2\left[\frac{E-\mu_F}{2k_B T}\right] \\ &= G_0\left[1+\cos\left(\frac{e}{\hbar}\Phi\right)\right], \end{aligned} \tag{9.38}$$

where

$$G_0 = \frac{e^2}{k_B T h}\int_0^\infty \mathrm{d}E|t(E)|^2\text{sech}^2\left[\frac{E-\mu_F}{2k_B T}\right]. \tag{9.39}$$

Equation (9.38) shows that the linear response conductance of the ring will oscillate periodically as one varies the flux Φ threading the ring, as long as transport is ballistic. The period of the oscillation is obtained by setting the angle $\frac{e}{\hbar}\Phi$ equal to 2π, and solving for Φ. This yields the period as h/e.

Multiple reflection effects within the ring, which we ignored, will introduce higher harmonics in the conductance oscillations, but the fundamental period will still be h/e. These oscillations in the conductance as a function of magnetic flux are called AB oscillations, or simply h/e oscillations.

9.4.1 The Altshuler–Aronov–Spivak Effect

The Altshuler–Aronov–Spivak (AAS) effect [24] is an effect that is closely related to the AB effect. It too produces periodic oscillations in the conductance of a ring as a function of the magnetic flux threading it, but the period of these oscillations is one-half of the AB oscillation period, or simply $h/2e$. To understand the origin of this effect, consider once again a ring enclosing a magnetic flux as shown in Fig. 9.9. In addition to the paths (shown by broken lines) whose interference leads to the AB interference effect, there are two other paths shown by solid lines that can interfere and affect the ring's conductance. These are paths going around the ring clockwise and counterclockwise. Their interference will produce oscillations in the magnitude of the wavefunction at the point of injection, which we can view as

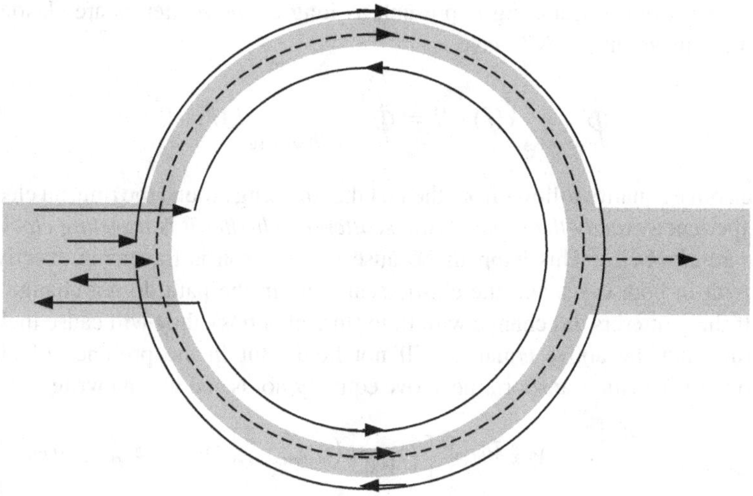

Fig. 9.9 Two time reversed paths around a ring are shown with *solid lines*. They are shown with different diameters for the sake of clarity, but the diameters are equal and these paths should overlap. Their interference gives rise to the Altshuler–Aronov–Spivak conductance oscillations with a period of $h/2e$ when a magnetic flux threads the ring. The paths shown with *broken lines* interfere to produce the normal Aharonov–Bohm conductance oscillations with a period of h/e

oscillations in the reflection probability. However, since transmission and reflection are related, these oscillations will result in equivalent oscillations in the transmission probability and hence in the conductance.

In order to determine the period of the AAS oscillations, let us calculate the reflection probability determined by the interference of the clockwise and counterclockwise trajectories. If we neglect multiple reflection effects, then the reflection amplitude will be

$$\begin{aligned}
r_{\text{ring}} &= r_{\text{clockwise}} + r_{\text{counter-clockwise}} \\
&= |r| \left(\exp\left[\mathrm{i} \oint_{\text{clockwise}} \left[k(\vec{r}) + \frac{e}{\hbar}\vec{A}(\vec{r}) \right] \cdot \mathrm{d}\vec{l} \right] \right. \\
&\qquad \left. + \exp\left[\mathrm{i} \oint_{\text{counter-clockwise}} \left[k(\vec{r}) + \frac{e}{\hbar}\vec{A}(\vec{r}) \right] \cdot \mathrm{d}\vec{l} \right] \right) \\
&= |r| \exp\left[\mathrm{i} \oint_{\text{clockwise}} \left[k(\vec{r}) + \frac{e}{\hbar}\vec{A}(\vec{r}) \right] \cdot \mathrm{d}\vec{l} \right] \\
&\quad \times \left[1 + \exp\left[\mathrm{i} \left(\oint_{\text{counter-clockwise}} \left[k(\vec{r}) + \frac{e}{\hbar}\vec{A}(\vec{r}) \right] \cdot \mathrm{d}\vec{l} \right. \right. \right. \\
&\qquad \left. \left. \left. - \oint_{\text{counter-clockwise}} \left[k(\vec{r}) + \frac{e}{\hbar}\vec{A}(\vec{r}) \right] \cdot \mathrm{d}\vec{l} \right) \right] \right] .
\end{aligned} \tag{9.40}$$

Note that even if scattering is present, as long as the scatterers are elastic and hence time invariant, we will have

$$\oint_{\text{clockwise}} \vec{k}(\vec{r}) \cdot \mathrm{d}\vec{l} = \oint_{\text{counter-clockwise}} \vec{k}(\vec{r}) \cdot \mathrm{d}\vec{l}. \tag{9.41}$$

The above equality follows from the fact that in going around the ring, an electron will experience *exactly the same elastic scatterers whether it is traveling clockwise or counterclockwise*. This happens because the electron is traversing exactly the same path in both cases and the elastic scatterers in the path do not change with time. If the scatterers did change with time (like phonons), they will cause inelastic scattering and the above equality will not hold. But in the presence of elastic scattering or ballistic transport, the above equality holds and we can write

$$\begin{aligned} r_{\text{ring}} &= |r| \mathrm{e}^{\mathrm{i} \oint_{\text{clockwise}} \left[k(\vec{r}) + \frac{e}{\hbar}\vec{A}(\vec{r})\right] \cdot \mathrm{d}\vec{l}} \left[1 + \mathrm{e}^{\mathrm{i}\frac{e}{\hbar}\left(\oint_{\text{counter-clockwise}} \vec{A}(\vec{r})\cdot \mathrm{d}\vec{l} - \oint_{\text{clockwise}} \vec{A}(\vec{r})\cdot \mathrm{d}\vec{l}\right)}\right] \\ &= |r| \mathrm{e}^{\mathrm{i} \oint_{\text{clockwise}} \left[k(\vec{r}) + \frac{e}{\hbar}\vec{A}(\vec{r})\right] \cdot \mathrm{d}\vec{l}} \left[1 + \mathrm{e}^{2i\frac{e}{\hbar} \oint_{\text{counter-clockwise}} \vec{A}(\vec{r})\cdot \mathrm{d}\vec{l}}\right] \\ &= |r| \mathrm{e}^{\mathrm{i} \oint_{\text{clockwise}} \left[k(\vec{r}) + \frac{e}{\hbar}\vec{A}(\vec{r})\right] \cdot \mathrm{d}\vec{l}} \left[1 + \mathrm{e}^{2i\frac{e}{\hbar}\Phi}\right], \end{aligned} \tag{9.42}$$

where we have used the fact that[5]

$$\oint_{\text{counter-clockwise}} \vec{A}(\vec{r}) \cdot \mathrm{d}\vec{l} = -\oint_{\text{clockwise}} \vec{A}(\vec{r}) \cdot \mathrm{d}\vec{l}. \tag{9.43}$$

Therefore, the reflection probability is

$$R_{\text{ring}} = |r_{\text{ring}}|^2 = 2|r|^2 \left[1 + \cos\left(\frac{2e}{\hbar}\Phi\right)\right], \tag{9.44}$$

which oscillates with the flux density Φ with a period of $h/2e$.

Since the transmission is maximum when the reflection is minimum and vice versa, it is clear that the oscillation in the reflection will cause identical oscillation in the transmission except with a phase difference of π. Therefore, the conductance of the ring will oscillate with a period of $h/2e$ as the flux threading it is varied. These oscillations have been observed experimentally in metal rings at low temperatures, when inelastic collisions are infrequent [25].

The AAS conductance oscillations are actually somewhat easier to observe experimentally than the AB conductance oscillations in disordered solid rings since the former is immune to elastic scattering, while the latter is not. To understand why the AB oscillations are vulnerable to elastic scattering, let us return to (9.36) and this

[5] See Problem 9.5.

time no longer assume that $\int_{\text{path2}} \vec{k}(\vec{r}) \cdot d\vec{l} = \int_{\text{path1}} \vec{k}(\vec{r}) \cdot d\vec{l}$ because the scatterers in the two branches of the ring are not identical. This will yield

$$\begin{aligned}
t_{\text{ring}} &= |t| \left(e^{i \int_{\text{path1}} \left[k(\vec{r}) + \frac{e}{\hbar} \vec{A}(\vec{r}) \right] \cdot d\vec{l}} + e^{i \int_{\text{path2}} \left[k(\vec{r}) + \frac{e}{\hbar} \vec{A}(\vec{r}) \right] \cdot d\vec{l}} \right) \\
&= |t| e^{i \int_{\text{path1}} \left[k(\vec{r}) + \frac{e}{\hbar} \vec{A}(\vec{r}) \right] \cdot d\vec{l}} \left(1 + e^{i \int_{\text{path2}} \left[k(\vec{r}) + \frac{e}{\hbar} \vec{A}(\vec{r}) \right] \cdot d\vec{l} - i \int_{\text{path2}} \left[k(\vec{r}) + \frac{e}{\hbar} \vec{A}(\vec{r}) \right] \cdot d\vec{l}} \right) \\
&= |t| e^{i \int_{\text{path1}} \left[k(\vec{r}) + \frac{e}{\hbar} \vec{A}(\vec{r}) \right] \cdot d\vec{l}} \left(1 + e^{i \left[\int_{\text{path2}} k(\vec{r}) \cdot d\vec{l} - \int_{\text{path1}} k(\vec{r}) \cdot d\vec{l} + \frac{e}{\hbar} \oint \vec{A}(\vec{r}) \cdot d\vec{l} \right]} \right) \\
&= |t| e^{i \int_{\text{path1}} \left[k(\vec{r}) + \frac{e}{\hbar} \vec{A}(\vec{r}) \right] \cdot d\vec{l}} \left(1 + e^{i \left[\Delta(k) + \frac{e}{\hbar} \oint \vec{A}(\vec{r}) \cdot d\vec{l} \right]} \right) \\
&= |t| e^{i \int_{\text{path1}} \left[k(\vec{r}) + \frac{e}{\hbar} \vec{A}(\vec{r}) \right] \cdot d\vec{l}} \left(1 + e^{i \left[\Delta(k) + \frac{e}{\hbar} \Phi \right]} \right).
\end{aligned} \tag{9.45}$$

Note that

$$\Delta(k) = \int_{\text{path2}} k(\vec{r}) \cdot d\vec{l} - \int_{\text{path1}} k(\vec{r}) \cdot d\vec{l} \neq 0, \tag{9.46}$$

since the elastic scatterers encountered in one path can be very different from those encountered in the other. Note also that this phase shift will generally depend on k and hence the electron energy E. This will make the transmission probability depend on E as

$$T_{\text{ring}}(E) = |t_{\text{ring}}|^2 = 2|t(E)|^2 \left[1 + \cos\left(\Delta(E) + \frac{e}{\hbar} \Phi \right) \right]. \tag{9.47}$$

Substitution of this result in (8.142) yields the linear response conductance as

$$G_{\text{ring}} = \frac{e^2}{k_B T h} \int_0^{\infty} dE |t(E)|^2 \left[1 + \cos\left(\Delta(E) + \frac{e}{\hbar} \Phi \right) \right] \text{sech}^2 \left[\frac{E - \mu}{2 k_B T} \right]. \tag{9.48}$$

Note that this time the periodic term cannot be pulled outside the integral over energy because $\Delta(E)$ is energy dependent. Therefore, ensemble averaging over the electron's energy at nonzero temperatures, represented by the integration over energy, will dilute the interference effect and reduce the conductance modulation.

We will define the conductance modulation as the ratio $(G_{\text{ring}}^{\max} - G_{\text{ring}}^{\min})/(G_{\text{ring}}^{\max} + G_{\text{ring}}^{\min})$. This quantity is 100% in ballistic transport since $G_{\text{ring}}^{\min} = 0$ (see (9.38), but not 100% in diffusive transport when elastic scattering is present since then $G_{\text{ring}}^{\min} \neq 0$ (see (9.48). That is why we say that the AB conductance oscillation is vulnerable to elastic scattering since that will reduce the conductance modulation at nonzero temperatures when ensemble averaging over energy becomes necessary. This problem does not arise with the AAS effect because of the equality expressed in (9.41), which holds in both ballistic and diffusive transport.

In the end, we can list the following salient differences between the AB oscillations and the AAS oscillations:

1. The AB oscillations have a period of h/e in magnetic flux, while the AAS oscillations have a period of $h/2e$.
2. In a perfectly symmetric ring, the AB effect will make the conductance maximum when the magnetic flux $\Phi = 0$, whereas the AAS effect will make the conductance minimum when $\Phi = 0$. In the first case, we have constructive interference of transmission paths, whereas in the second case, we have constructive interference of reflection paths.
3. AB oscillations are vulnerable to elastic scattering events which reduce the conductance modulation at nonzero temperatures because of ensemble averaging over electron energy, but AAS oscillations are invulnerable to this influence.

9.4.2 More on the Ballistic Aharonov–Bohm Conductance Oscillations

Perfect (100%) conductance modulation associated with AB oscillations in ballistic transport can be observed not just in linear response transport, but also in nonlinear response transport (under relatively large bias) as long as we can ignore the subtlety that a large bias can make the wavevector vary in space within the ring because the electric field associated with the bias will accelerate the electron and make the wavevector change with distance. If the wavevector varied identically in both arms of the ring then the variation would not matter, but generally speaking, the wavevector will vary differently because the arms could be of different length, etc. Nonidentical variation of the wavevector in the two arms will inevitably cause a nonzero $\Delta(k)$ or $\Delta(E)$ (recall (9.46)), which will make the phase difference between the two arms energy dependent even in ballistic transport. For the time being, we will overlook this possibility and assume $\Delta(E) = 0$. In that case, the Tsu–Esaki formula in (8.132) will yield the current through the ring as

$$\begin{aligned} I &= \frac{4e}{h}\int_0^\infty \mathrm{d}E\,|t(E)|^2\left[1+\cos\left(\frac{e}{\hbar}\Phi\right)\right][f(E-\mu_1)-f(E-\mu_2)] \\ &= I_0\left[1+\cos\left(\frac{e}{\hbar}\Phi\right)\right]. \end{aligned} \tag{9.49}$$

where

$$I_0 = \frac{4e}{h}\int_0^\infty \mathrm{d}E\,|t(E)|^2\,[f(E-\mu_1)-f(E-\mu_2)] \tag{9.50}$$

Once again, we find that the current modulation can be 100%.

This result does not change if we assume three-dimensional leads and use the appropriate Tsu–Esaki formula to write the current as

$$\begin{aligned} J_{3-d} &= \frac{e}{\Omega}\int_0^\infty \mathrm{d}E D_q^{3-d}(E) v_x(E) \tau_{1\to 2}(E)\left[f(E-\mu_1) - f(E-\mu_2)\right] \\ &= \frac{2e}{\Omega}\int_0^\infty \mathrm{d}E D_q^{3-d}(E) v_x(E) |t(E)|^2 \left[1+\cos\left(\frac{e}{\hbar}\Phi\right)\right]\left[f(E-\mu_1) - f(E-\mu_2)\right] \\ &= J_0\left[1+\cos\left(\frac{e}{\hbar}\Phi\right)\right], \end{aligned} \tag{9.51}$$

where

$$J_0 = \frac{2e}{\Omega}\int_0^\infty \mathrm{d}E D_q^{3-d}(E) v_x(E) |t(E)|^2 \left[f(E-\mu_1) - f(E-\mu_2)\right]. \tag{9.52}$$

Thus, as long as transport is ballistic, we expect to see ideally 100% conductance modulation in either linear or nonlinear transport.[6] This happens only because in ballistic transport the phase difference between the two arms is independent of energy.

9.4.3 Magnetostatic Aharonov–Bohm Effect in Quantum Wire Rings

Rings such as the ones we have considered so far will usually be derived from quantum wires that act as electron waveguides. Such a ring in shown in Fig. 9.10. The plane of the ring is assumed to be the x-y plane and current flows in the x-direction. A magnetic field is applied perpendicular to the plane of the ring (in the z-direction) and may intersect the paths of the electrons in the wires. Here, we will not be concerned about the orthodox AB effect where the field cannot intersect the path of the electron. Instead, all we want to show now is that the current through this ring will oscillate with the magnetic field and we wish to find the period of this oscillation as well as the maximum possible conductance modulation. For this purpose, we will adopt a different approach from the one we used in the preceding section and make use of the dispersion relations that we derived in Chap. 7.

We will assume ballistic transport and neglect any spatial variation of the wavevector in either path due to the electric field driving current between the contacts. This will make $\Delta(E) = 0$.

The dispersion relations of electrons in the upper and lower branches of the ring (which are quantum wires) are found from Fig. 7.12 and plotted in the lower panel of Fig. 9.10, where the two curves are horizontally displaced by the amount

[6] With the caveat that we can ignore the variation of wavevector with distance under high bias.

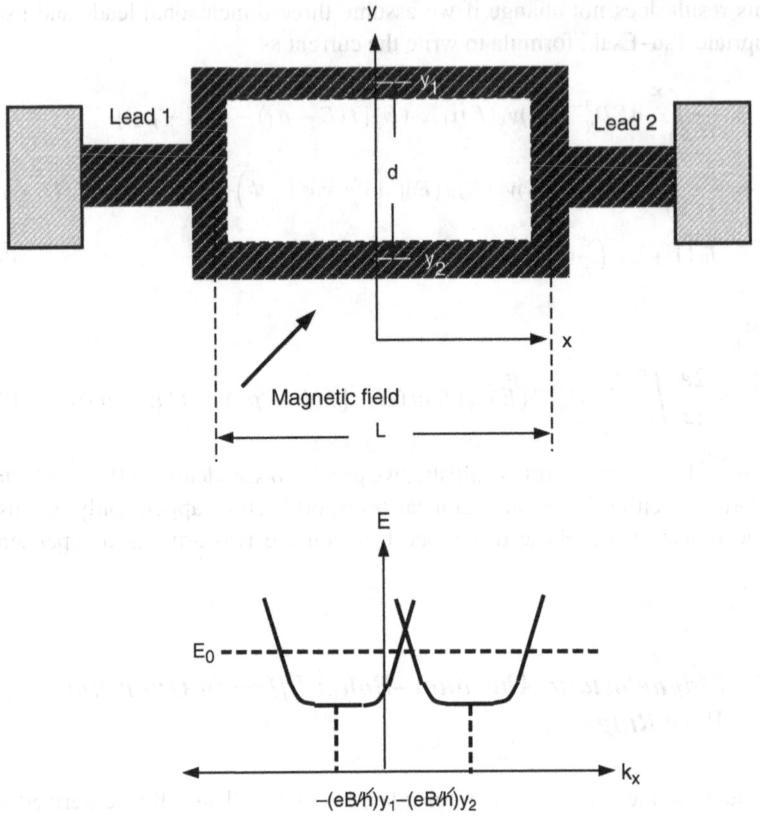

Fig. 9.10 (*Upper panel*) A ring derived from quantum wires. A magnetic field is applied perpendicular to ring's plane. (*Lower panel*) Energy-dispersion relations in the two branches of the ring which are assumed to be quantum wires with rectangular confining potentials

$(eB/\hbar)(y_2 - y_1)$ with B being the magnetic flux density. The two branches are centered at $-(eB/\hbar)y_1$ and $-(eB/\hbar)y_2$, where the quantities y_1 and y_2 are the vertical coordinates of the centers of the upper and lower branches as shown in the upper panel of Fig. 9.10.

Clearly, because of energy conservation in the absence of any inelastic processes,

$$E(k_{x1} + eBy_1/\hbar)|_{\text{upper-branch}} = E(k_{x2} + eBy_2/\hbar)|_{\text{lower-branch}}, \tag{9.53}$$

which tells us that an electron of energy E_0 will have wavevectors in the upper and lower branches of the ring that differ by

$$k_{x1} - k_{x2} = \frac{eB}{\hbar}(y_2 - y_1). \tag{9.54}$$

This difference is *independent* of E_0.

Neglecting multiple reflection effects and assuming ballistic transport, we can write the transmission through the ring as

$$t(E,\Phi) = t_1(E)e^{ik_{x1}(E,\Phi)L}t_1'(E) + t_2(E)e^{ik_{x2}(E,\Phi)L}t_2'(E). \tag{9.55}$$

where $t_{1,2}$ are the transmission amplitudes for transmitting from the left lead into the two branches, $t_{1,2}'$ are the transmission amplitudes for transmitting from two branches into the right lead, L is the distance between the two leads as shown in the upper panel of Fig. 9.10 and Φ is the magnetic flux threading the ring.

We will assume that the ring is perfectly symmetric and its two branches are identical so that $t_1(E) = t_2(E)$ and $t_1'(E) = t_2'(E)$. This yields

$$\begin{aligned} t(E,\Phi) &= t_1(E)e^{ik_{x1}(E,\Phi)L}t_1'(E)\left[1 + e^{i(k_{x2}(E,\Phi)-k_{x1}(E,\Phi))L}\right] \\ &= t_1(E)e^{ik_{x1}(E,\Phi)L}t_1'(E)\left[1 + e^{i\frac{eB}{\hbar}(y_2-y_1)L}\right] \\ &= t_1(E)e^{ik_{x1}(E,\Phi)L}t_1'(E)\left[1 + e^{i\frac{eB}{\hbar}S}\right] \\ &= t_1(E)e^{ik_{x1}(E,\Phi)L}t_1'(E)\left[1 + e^{i\frac{e}{\hbar}\Phi}\right] \\ &= t_1(E)e^{ik_{x1}(E,\Phi)L}t_1'(E)\left[1 + e^{i\frac{e}{\hbar}\oint \vec{A}\cdot d\vec{l}}\right], \end{aligned} \tag{9.56}$$

where $S = (y_2 - y_1)L$ is the area enclosed by the ring and $\vec{A}$ is the magnetic vector potential. As usual, the contour integral will be around the circumference of the ring.

The transmission probability through the ring in the presence of a magnetic flux Φ will be

$$T(E,\Phi) = |t(E,\Phi)|^2 = 2|t_1(E)t_1'(E)|^2\{1 + \cos[\Theta(\Phi)]\}, \tag{9.57}$$

where $\Theta(\Phi) = (e/\hbar)\Phi$, independent of the electron energy E.

The linear response conductance as a function of the magnetic flux Φ can be written as

$$\begin{aligned} G(\Phi) &= \frac{e^2}{k_B T h}\int_0^{\infty} dE|t(E,\Phi)|^2 \text{sech}^2\left[\frac{E-\mu_F}{2k_B T}\right] \\ &= \left[1 + \cos\left(\frac{e}{\hbar}\Phi\right)\right]\frac{e^2}{k_B T h}\int_0^{\infty} dE|t_1(E)t_1'(E)|^2 \text{sech}^2\left[\frac{E-\mu_F}{2k_B T}\right] \\ &= G_0\left[1 + \cos\left(\frac{e}{\hbar}\Phi\right)\right]. \end{aligned} \tag{9.58}$$

This shows that the conductance will oscillate with the magnetic flux enclosed by the ring with a period of h/e. We call this the *magnetostatic* AB oscillation since the conductance oscillation is induced by a magnetic flux threading the ring.

Note that we get 100% conductance modulation as long as we have a perfectly symmetric ring so that transmissions from the left lead into the two branches are equal ($t_1(E) = t_2(E)$) and transmissions from the branches into the right lead are also equal ($t_1'(E) = t_2'(E)$). That is actually a fundamental characteristic of all good interferometers. Any good interferometer should be perfectly symmetric and have its two branches identical in all respects. In fact, if the branches are identical to the point that they have the same elastic scatterers, then we will not need ballistic transport to obtain 100% conductance modulation at a finite temperature. In this case $\Delta(k)$ or $\Delta(E)$ will be zero, and the conductance modulation will reach 100% even in diffusive transport (elastic scattering present).

9.5 The Electrostatic Aharonov–Bohm Effect

In Fig. 9.11, we show a ring with a potential difference of V imposed between the two paths with a battery.[7] This time, there is no magnetic field threading the ring. The dispersion relations of electrons in the two paths are shown in the lower panel of Fig. 9.11.

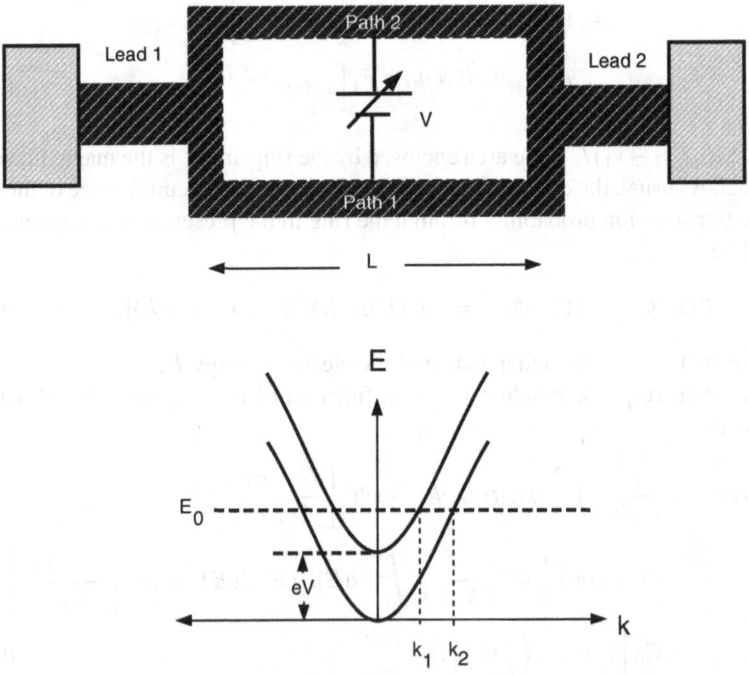

Fig. 9.11 (*Upper panel*) A ring derived from quantum wires. There is no magnetic flux threading the ring, but a potential difference of V is applied between the two paths with a battery. (*Lower panel*) Energy-dispersion relations in the two branches of the ring

[7] The ring must be semiconducting instead of metallic, so that the battery is not electrically shorted.

If we assume the dispersion relations to be parabolic, then an electron entering the ring with energy E_0 will have wavevectors k_1 and k_2 in the two paths that will be related as

$$E_0 = \frac{\hbar^2 k_1^2}{2m^*} + e\mathrm{V} = \frac{\hbar^2 k_2^2}{2m^*}, \tag{9.59}$$

where m^* is of course the electron's effective mass.

This yields

$$k_2 - k_1 = \frac{2m^* e\mathrm{V}}{\hbar^2(k_1 + k_2)} = \frac{\sqrt{2m^* E_0}}{\hbar}\left(1 - \sqrt{1 - \frac{e\mathrm{V}}{E_0}}\right). \tag{9.60}$$

If we ignore multiple reflection effects and once again assume that the ring is perfectly symmetric so that the two branches are identical, i.e., $t_1(E) = t_2(E)$ and $t_1'(E) = t_2'(E)$, then the amplitude of transmission through the ring is

$$\begin{aligned} t(E,\mathrm{V}) &= t_1(E)\mathrm{e}^{\mathrm{i}k_1(E,\mathrm{V})L} t_1'(E)\left[1 + \mathrm{e}^{\mathrm{i}(k_2(E,\mathrm{V}) - k_1(E,\mathrm{V}))L}\right] \\ &= t_1(E)\mathrm{e}^{\mathrm{i}k_1(E,\mathrm{V})L} t_1'(E)\left[1 + \mathrm{e}^{\mathrm{i}\frac{\sqrt{2m^* E}}{\hbar}\left(1 - \sqrt{1 - \frac{e\mathrm{V}}{E}}\right)L}\right], \end{aligned} \tag{9.61}$$

and the associated transmission probability is

$$T(E,\mathrm{V}) = |t(E,\mathrm{V})|^2 = 2|t_1(E)t_1'(E)|^2\{1 + \cos[\Theta'(E,\mathrm{V})]\}, \tag{9.62}$$

where $\Theta'(E,\mathrm{V}) = \frac{\sqrt{2m^* E}}{\hbar}(1 - \sqrt{1 - \frac{e\mathrm{V}}{E}})L$. Unfortunately, $\Theta'(E,\mathrm{V})$ is *not* independent of the electron energy E and this has serious consequences.

This time, the linear response conductance of the ring as a function of the voltage V imposed between the two arms will be

$$G(\mathrm{V}) = \frac{e^2}{k_B T h}\int_0^\infty \mathrm{d}E\,|t_1(E)t_1'(E)|^2\left\{1 + \cos[\Theta'(E,\mathrm{V})]\right\}\mathrm{sech}^2\left[\frac{E - \mu_F}{2k_B T}\right], \tag{9.63}$$

and the nonlinear response current will be given by the Tsu–Esaki formula

$$I(\mathrm{V}) = \frac{4e}{h}\int_0^\infty \mathrm{d}E\,|t_1|^2(E)|t_2|^2(E)\{1 + \cos[\Theta'(E,\mathrm{V})]\}[f(E - \mu_1) - f(E - \mu_2)], \tag{9.64}$$

where μ_1 and μ_2 are the chemical potentials in the two leads.

We can see that the modulation term $\{1+\cos[\Theta'(E,\mathrm{V})]\}$ cannot be pulled outside the integral over energy E since $\Theta'(E,\mathrm{V})$ is *not* independent of energy. Therefore, we cannot get 100% modulation of the linear response conductance, or nonlinear response current, at any finite temperature because ensemble averaging over the electron energy (represented by the integration over energy) will invariably dilute the interference effect. This will happen even when transport is ballistic and the interferometer is perfectly symmetric because our analysis had already assumed those conditions. This situation is very different from the magnetostatic AB effect, where a 100% modulation was possible in ballistic transport at finite temperatures in a perfectly symmetric interferometer. Nonetheless, if we do vary V, we will expect to see some modulation in the conductance or current through the ring as a result of quantum interference between the electron waves traversing the two branches. We call these oscillations *electrostatic* AB oscillations.

The electrostatic AB conductance oscillations clearly have two major differences with the magnetostatic AB conductance oscillations. They are:

1. The conductance modulation is never 100% at any nonzero temperature, even in ballistic transport and even if the interferometer (ring) is perfectly symmetric.
2. The oscillations are not periodic in the potential difference V since the phase difference $\Theta'(E, \mathrm{V})$ is not independent of E so that ensemble averaging over energy will make the oscillations appear nonperiodic. Even if $\Theta'(E, \mathrm{V})$ were independent of E (which it is not), the oscillations will still not be periodic in the voltage since $\Theta'(E, \mathrm{V})$ is not linearly proportional to V. This is very different from the magnetostatic oscillations where the phase difference $\Theta(\Phi)$ is independent of E and linearly proportional to the magnetic flux density Φ, so that the oscillations are periodic in Φ.

Despite the impossibility of 100% conductance modulation, the electrostatic AB conductance oscillation is preferred for device applications over the magnetostatic AB conductance oscillation, simply because an electrostatic potential is much easier to generate on an electronic chip than a magnetic field. Consequently, there have been proposals for using the electrostatic effect to implement a *quantum interference transistor*, acronymed QUIT [26]. The current flowing through the ring between the two contacts can be modulated by changing the voltage V of the battery connected between the two branches, thereby realizing transistor action. Such quantum interference transistors (QUIT) may require very small voltages to swing between maximum and minimum conductance states and hence might be attractive for low power applications because it takes a tiny voltage to change the phase difference between the two paths by an amount equal to $\pi/2$. We can estimate this voltage from the expression for the phase difference given in Problem 9.4, which is valid at small values of V:

$$\Theta'(E, \mathrm{V}) = \frac{e}{\hbar}\mathrm{V}\langle \tau_t(E) \rangle, \tag{9.65}$$

where $\langle \tau_t(E) \rangle$ is the average transit time of an electron of energy E to travel through the ring. We can write this time as $\langle \tau_t(E) \rangle = L/v_F$, where v_F is the Fermi velocity in the channels and L is the separation between the two contacts. The voltage required to switch the transistor on or off is found by setting $\Theta'(E, \mathrm{V}) = \pi/2$ and solving for V, which yields

$$\mathrm{V} = \frac{h}{2e}\frac{v_F}{L} = \frac{h}{2e}\frac{\hbar k_F/m^*}{L} = \frac{\pi\hbar^2 k_F}{m^* L}. \tag{9.66}$$

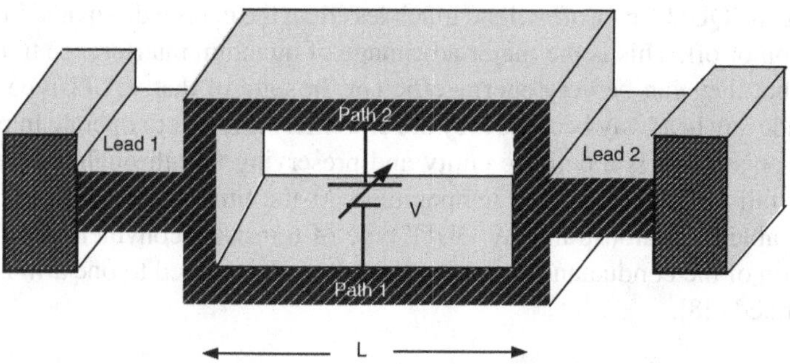

Fig. 9.12 A two dimensional electrostatic Aharonov–Bohm interferometer whose branches are made up of quantum wells. Such a device can be used as a quantum interference transistor (QUIT) capable of carrying large current between the contacts. This current can be modulated by varying the voltage V between the two channels, thereby realizing transistor action

Since a transistor made of quantum wires will not be able to carry much current,[8] we will consider an interferometer whose branches are made of quantum wells instead of quantum wires as shown in Fig. 9.12. These structures are capable of carrying much larger currents and are hence attractive for transistors, which must always have sufficient current carrying capability to drive succeeding stages. The voltage required to switch this quantum well interferometer on or off is still given by the previous equation. We can now compare this voltage with the voltage that would be needed to turn off a conventional ballistic depletion mode FET whose channel (accumulation layer) is effectively a two-dimensional quantum well. Such an FET can be realized by simply removing one of the branches of the QUIT. The voltage required to turn the FET off is the voltage required to raise the conduction band edge E_{c0} in the channel up to the Fermi level, so that at low temperatures approaching 0 K, the accumulation layer becomes depleted. This voltage is therefore given by

$$eV|_{\mathrm{FET}} = \mu_{\mathrm{F}} - E_{c0} = E_{\mathrm{F}} = \frac{\hbar^2 k_{\mathrm{F}}^2}{2m^*}. \tag{9.67}$$

From the last two equations, we obtain

$$\frac{V|_{\mathrm{QUIT}}}{V|_{\mathrm{FET}}} = \frac{\lambda_{\mathrm{F}}}{L}, \tag{9.68}$$

where λ_{F} is the Fermi DeBroglie wavelength in the fully accumulated channel [27]. Since, normally, $\lambda_{\mathrm{F}} \ll L$, the voltage required to turn the QUIT off will be considerably less than that required to turn the FET off. Hence the energy dissipated

---

[8]Remember that in ballistic transport, the maximum conductance of a quantum wire is $2Ne^2/h$, where N is the number of propagating modes contributing to current.

in turning the QUIT on or off will be much less than the energy dissipated in turning the FET on or off. This is the major advantage of quantum interference transistors, namely that they can be very energy efficient. In spite of that, QUIT-type devices never made any headway because they are extremely difficult to operate in practice. An electron's phase is a delicate entity and preserving that throughout the device is a very tall order at any finite temperature. At the time of this writing, nobody has been able to demonstrate any QUIT type of transistor convincingly, although modulation of the conductance of a ring by a potential applied to one arm has been demonstrated [28].

9.5.1 The Effect of Multiple Reflections

In the foregoing discussions about AB interferometers or rings, we had always neglected multiple reflection effects. We can now account for those by using scattering matrices. For this purpose, refer to Fig. 9.13.

At either lead of the ring, we have a three way splitter and we can write the scattering matrices for these junctions as

$$\begin{bmatrix} A^- \\ B_1^+ \\ B_2^+ \end{bmatrix} = \begin{bmatrix} -(a+b) & (\sqrt{\epsilon})^* & (\sqrt{\epsilon})^* \\ \sqrt{\epsilon} & a & b \\ \sqrt{\epsilon} & b^* & a \end{bmatrix} \begin{bmatrix} A^+ \\ B_1^- \\ B_2^- \end{bmatrix} \tag{9.69}$$

and

$$\begin{bmatrix} D^+ \\ C_1^- \\ C_2^- \end{bmatrix} = \begin{bmatrix} -(a'+b') & \left(\sqrt{\epsilon'}\right)^* & \left(\sqrt{\epsilon'}\right)^* \\ \sqrt{\epsilon'} & a' & b' \\ \sqrt{\epsilon'} & b'^* a' & \end{bmatrix} \begin{bmatrix} D^- \\ C_1^+ \\ C_2^+ \end{bmatrix}, \tag{9.70}$$

at the left and right leads, respectively. Here we have used the property that the scattering matrix is symmetric in the absence of any magnetic field. Of course, in the case of the magnetostatic effect, a magnetic field will be present, but we will assume that it is enclosed entirely within the branches of the ring and does not enter the leads, or even the branches for that matter.

The scattering matrix describing propagation through the two branches of the ring is

$$\begin{bmatrix} B_1 - \\ B_2^- \\ C_1^+ \\ C_2^+ \end{bmatrix} = \begin{bmatrix} r_1 & 0 & t_1' & 0 \\ 0 & r_2 & 0 & t_2' \\ t_1 & 0 & r_1' & 0 \\ 0 & t_2 & 0 & r_2' \end{bmatrix} \begin{bmatrix} B_1^+ \\ B_2^+ \\ C_1^- \\ C_2^- \end{bmatrix}, \tag{9.71}$$

where t_n and r_n are the transmission and reflection amplitudes in the n-th branch. The unprimed and primed quantities are associated with propagation from left to

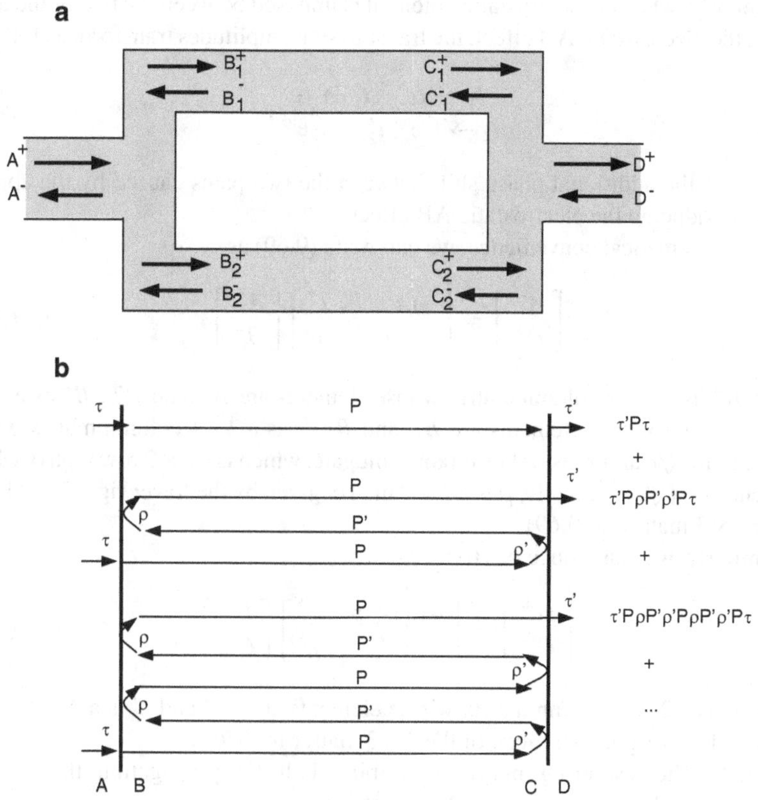

Fig. 9.13 (**a**) The incoming and outgoing waves at the leads and within the branches of a ring acting as either a magnetostatic or an electrostatic Aharonov–Bohm interferometer; (**b**) the multiple reflection trajectories within the ring

right and right to left, respectively. If transport is ballistic, then $r_1 = r_2 = r_1' = r_2' = 0$ and in the *absence* of any magnetic or electric field inducing the AB effect, $t_1 = t_1' = \hat{t}_1 = \exp(\mathrm{i}k_1 L_1)$ while $t_2 = t_2' = \hat{t}_2 = \exp(\mathrm{i}k_2 L_2)$. Here, L_n is the length of the n-th branch in the direction of current flow.

When a static magnetic flux is present to induce the magnetostatic AB effect, the transmission amplitudes transform as follows:

$$\begin{array}{l} t_1 \rightarrow \hat{t}_1 \mathrm{e}^{\mathrm{i}\theta_1} \; t_1' \rightarrow \hat{t}_1 \mathrm{e}^{-\mathrm{i}\theta_1} \\ t_2 \rightarrow \hat{t}_2 \mathrm{e}^{\mathrm{i}\theta_2} \; t_2' \rightarrow \hat{t}_2 \mathrm{e}^{-\mathrm{i}\theta_2} \end{array}, \tag{9.72}$$

where $\theta_1 = \int_0^{L_1} \vec{A} \cdot \mathrm{d}\vec{l}_1$ and $\theta_2 = \int_0^{L_2} \vec{A} \cdot \mathrm{d}\vec{l}_2$, and the line integrals are taken along the two branches.

Similarly, when an electrostatic potential is imposed between the two branches to induce the electrostatic AB effect, the transmission amplitudes transform as follows:

$$\begin{array}{ll} t_1 \rightarrow \hat{t_1} & t_1' \rightarrow \hat{t_1} \\ t_2 \rightarrow \hat{t_2}e^{i\phi} & t_2' \rightarrow \hat{t_2}e^{i\phi} \end{array}, \tag{9.73}$$

where ϕ is the additional phase shift between the two paths caused by the applied potential inducing the electrostatic AB effect.

For mathematical convenience, we can write (9.69) as

$$\begin{bmatrix} A^- \\ B^+ \end{bmatrix} = \begin{bmatrix} -(a+b) & \tau^\dagger \\ \tau & \rho \end{bmatrix} \begin{bmatrix} A^+ \\ B^- \end{bmatrix}, \tag{9.74}$$

where B^+ is a 2×1 column matrix whose elements are B_1^+ and B_2^+, B^- is a 2×1 column matrix whose elements are B_1^- and B_2^-, τ is a 2×1 column matrix whose elements are $\sqrt{\epsilon}$ and $\tau^\dagger$ is its hermitian conjugate, which is a 1×2 row matrix whose elements are $\left(\sqrt{\epsilon}\right)^*$. Finally, ρ is a 2×2 matrix given by the lower right 2×2 block of the 3×3 matrix in (9.69).

Similarly, we can reduce (9.70) to

$$\begin{bmatrix} D^+ \\ C^- \end{bmatrix} = \begin{bmatrix} -(a'+b') & \tau'^\dagger \\ \tau' & \rho' \end{bmatrix} \begin{bmatrix} D^- \\ C^+ \end{bmatrix}, \tag{9.75}$$

where τ' is a 2×1 column matrix whose elements are $\sqrt{\epsilon'}$ and ρ' is a 2×2 matrix given by lower right 2×2 block of the 3×3 matrix in (9.70).

Finally, the scattering matrix describing ballistic propagation through the branches can also be reduced to a 2×2 matrix as

$$\begin{bmatrix} B^- \\ C^+ \end{bmatrix} = \begin{bmatrix} 0 & P' \\ P & 0 \end{bmatrix} \begin{bmatrix} B^+ \\ C^- \end{bmatrix}, \tag{9.76}$$

where

$$P = \begin{bmatrix} t_1 & 0 \\ 0 & t_2 \end{bmatrix} \tag{9.77}$$

and

$$P' = \begin{bmatrix} t_1' & 0 \\ 0 & t_2' \end{bmatrix} \tag{9.78}$$

Using a bit of algebra, we can now relate D^+ to A^+ and write the transmission amplitude associated with traversing the ring as [29]

$$\begin{aligned} t = \frac{D^+}{A^+} &= \tau'[\mathbf{I} - P\rho P'\rho']^{-1} P\tau \\ &= \tau'\left[\mathbf{I} + P\rho P'\rho' + (P\rho P'\rho')^2 + (P\rho P'\rho')^3 + \ldots\right] P\tau, \end{aligned} \tag{9.79}$$

where $\mathbf{I}$ is the 2×2 identity matrix.

The last expression makes perfect sense since we can draw the Feynman paths associated with multiple reflection trajectories within the ring and sum the infinite geometric series made up of these paths as shown in Fig. 9.13b. This will yield the expression in the previous equation.

If there is a single propagating mode in each arm of the ring, then substituting for τ, τ', ρ, ρ', P, and P' in last expression, we get

$$t = \frac{\epsilon[(t_1 + t_2) - (b-a)^2 t_1 t_2 (t_1' + t_2')]}{[1 - t_1(a^2 t_1' + b^2 t_2')][1 - t_2(a^2 t_2' + b^2 t_1')] - a^2 b^2 t_1 t_2 (t_1' + t_2')^2}, \tag{9.80}$$

where, for simplicity, we have assumed that a, b, and ϵ are all real, while $a = a'$, $b = b'$, and $\epsilon = \epsilon'$. The last three equalities apply only if the junctions at the left and right lead are identical.

If a magnetic field is present, then we can use (9.72) to rewrite the last equation as

$$t(\theta_1, \theta_2) = \frac{\epsilon\left[\left(\hat{t}_1 e^{i\theta_1} + \hat{t}_2 e^{i\theta_2}\right) - (b-a)^2 \hat{t}_1 \hat{t}_2 \left(\hat{t}_1 e^{i\theta_2} + \hat{t}_2 e^{i\theta_1}\right)\right]}{\left[1 - \hat{t}_1\left(a^2 \hat{t}_1 + b^2 \hat{t}_2 e^{i(\theta_1-\theta_2)}\right)\right]\left[1 - \hat{t}_2\left(a^2 \hat{t}_2 + b^2 \hat{t}_1 e^{-i(\theta_1-\theta_2)}\right)\right] - a^2 b^2 \hat{t}_1 \hat{t}_2 \left(\hat{t}_1 e^{i(\theta_1-\theta_2)/2} + \hat{t}_2 e^{i(\theta_2-\theta_1)/2}\right)^2}, \tag{9.81}$$

Next, we will assume a perfectly symmetric interferometer with identical branches so that the quantities with carets are equal, i.e., $\hat{t}_1 = \hat{t}_2$. This simplifies the last equation to

$$\begin{aligned} t[\Theta(\Phi)] &= \frac{\epsilon \hat{t}_1 e^{i\theta_1}\left(1 + e^{i(\theta_2-\theta_1)}\right)\left[1 - (b-a)^2 \hat{t}_1^{\,2}\right]}{\left[1 - \hat{t}_1^{\,2}\left(a^2 + b^2 e^{i(\theta_1-\theta_2)}\right)\right]\left[1 - \hat{t}_1^{\,2}\left(a^2 + b^2 e^{i(\theta_2-\theta_1)}\right)\right] - a^2 b^2 \hat{t}_1^{\,4}\left(e^{i(\theta_2-\theta_1)/2} + e^{-i(\theta_2-\theta_1)/2}\right)^2} \\ &= \frac{\epsilon \hat{t}_1 e^{i\theta_1}\left(1 + e^{i(\theta_2-\theta_1)}\right)\left[1 - (b-a)^2 \hat{t}_1^{\,2}\right]}{\left[1 - \hat{t}_1^{\,2}\left(a^2 + b^2 e^{i(\theta_1-\theta_2)}\right)\right]\left[1 - \hat{t}_1^{\,2}\left(a^2 + b^2 e^{i(\theta_2-\theta_1)}\right)\right] - 4a^2 b^2 \hat{t}_1^{\,4} \cos^2\left(\frac{\theta_2-\theta_1}{2}\right)} \\ &= \frac{\epsilon \hat{t}_1 e^{i\theta_1}\left(1 + e^{i\Theta(\Phi)}\right)\left[1 - (b-a)^2 \hat{t}_1^{\,2}\right]}{\left[1 - \hat{t}_1^{\,2}\left(a^2 + b^2 e^{-i\Theta(\Phi)}\right)\right]\left[1 - \hat{t}_1^{\,2}\left(a^2 + b^2 e^{i\Theta(\Phi)}\right)\right] - 4a^2 b^2 \hat{t}_1^{\,4} \cos^2[\Theta(\Phi)/2]}, \end{aligned} \tag{9.82}$$

where $\Theta(\Phi) = \theta_2 - \theta_1$ is the magnetostatic AB phase shift.

The above expression shows that the transmission through the ideal symmetric ring vanishes in ballistic transport whenever $\Theta(\Phi) = (e/\hbar)\Phi = (2n+1)\pi$, or equivalently when $\Phi = (n + 1/2)(h/e)$, where n is an integer. This is of course the basis of magnetostatic AB conductance oscillations. But surprisingly, it also shows that the conductance of the ring vanishes when

$$(b-a)^2 \hat{t}_1^{\,2} = 1. \tag{9.83}$$

Using the unitarity of the scattering matrix, it is easy to show that $b - a = 1$, so that the last equation becomes

$$\hat{t}_1^{\,2} = e^{2ikL} = 1, \tag{9.84}$$

or

$$kL = m\pi, \tag{9.85}$$

where m is an integer. Thus, the conductance of the ring should vanish, regardless of the magnetic flux threading it, whenever the last equation is satisfied. This condition implies that if the phase shift of an electron, traveling once full-circle around the ring, is an even multiple of π, then the ring will not conduct at all. This happens because the electron enters the ring from the left lead into one arm, is reflected at the right lead into the other arm, and arrives at the left lead with a phase shift of $2m\pi$ and hence interferes with itself *constructively*. This maximizes the reflection and minimizes the transmission, making the latter zero. We will call this condition *Fabry–Perot resonance* since that is associated with a round-trip phase shift of $2m\pi$. When the ring is Fabry–Perot resonant, nothing conducts through it and it becomes an "insulator" no matter what its constituent material is.

However, there is a caveat. If (9.83) is satisfied, then the denominator of the expression for transmission in (9.82) can be written as

$$\begin{aligned}\texttt{denominator} &= \left[1-(a^2+b^2e^{-i\Theta(\Phi)})\right]\left[1-(a^2+b^2e^{i\Theta(\Phi)})\right] \\ &\quad - 4a^2b^2\cos^2\left[\Theta(\Phi)/2\right].\end{aligned} \tag{9.86}$$

It is easy to verify that this denominator vanishes whenever $\Theta(\Phi) = (e/\hbar)\Phi = 2n\pi$, or when $\Phi = n(h/e)$. Since the numerator is also zero because of Fabry–Perot resonance, we have the situation that the transmission assumes the form of 0/0 whenever $\Phi = n(h/e)$. Application of L'Hospital's rule, however, shows that the transmission becomes unity at these values of the flux. Thus, the conductance of a Fabry–Perot resonant ring will be zero at all values of the flux except when it is $n(h/e)$. Only at those critical values, the ring will conduct. Therefore, a plot of the conductance versus magnetic flux of a Fabry–Perot resonant ring fulfilling the condition in (9.83) will be a series of spikes at flux values of $n(h/e)$. At finite temperatures, these spikes will have a finite width.

In the case of the electrostatic AB effect, the transformations in (9.73) will reduce (9.80) for the transmission amplitude to

$$\begin{aligned} t(\phi) &= t\left[\Theta'(E,\mathrm{V})\right] \\ &= \frac{\epsilon\hat{t_1}\left(1+e^{i\Theta'(E,\mathrm{V})}\right)\left[1-(b-a)^2\hat{t_1}^2e^{i\Theta'(E,\mathrm{V})}\right]}{\left[1-\hat{t_1}^2\left(a^2+b^2e^{i\Theta'(E,\mathrm{V})}\right)\right]\left[1-\hat{t_1}^2\left(a^2e^{i2\Theta'(E,\mathrm{V})}+b^2e^{i\Theta'(E,\mathrm{V})}\right)\right]-4a^2b^2\hat{t_1}^4e^{i2\Theta'(E,\mathrm{V})}cos^2\left[\Theta'(E,\mathrm{V})\right]}.\end{aligned} \tag{9.87}$$

The numerator in this expression vanishes (but the denominator does not vanish) whenever $\Theta'(E,\mathrm{V}) = (2n+1)\pi$, which corresponds to destructive interference of waves traversing the two arms. If this condition is met, the transmission through the ring will plummet to zero, and we will get a set of conductance minima which we shall call the *primary minima*. They are associated with the normal electrostatic AB effect.

Note, however, that the numerator also vanishes whenever $\Theta'(E, \mathrm{V})$ is such that

$$(b-a)^2\hat{t_1}^2 e^{i\Theta'(E,\mathrm{V})} = e^{i(2kL+\Theta'(E,\mathrm{V}))} = 1, \tag{9.88}$$

or

$$2kL + \Theta'(E, \mathrm{V}) = 2m\pi. \tag{9.89}$$

When this condition is fulfilled, the numerator vanishes all right, but what about the denominator? The latter will also vanish if $\Theta'(E, \mathrm{V}) = 2n\pi$, in which case we will once again end up with a 0/0 form. Combining the two conditions, we find that the 0/0 situation arises if $2kL = 2(m-n)\pi$, or if the phase shift going full circle around the ring in the absence of any external field is an even multiple of π. Such a ring is of course Fabry–Perot resonant and does not conduct in the absence of any magnetic or electric field. But if the ring is not Fabry–Perot resonant, then the 0/0 form does not arise (since the denominator does not vanish) and the transmission through the ring will reach zero whenever (9.89) is fulfilled. We call the corresponding set of conductance minima the *secondary* minima. Note that (9.89) corresponds to the situation that *under the applied potential* V, the phase shift going full circle around the ring is an even multiple of π, i.e., the ring becomes Fabry–Perot resonant only when the correct potential obeying (9.89) is applied between the two arms. Note also that the phase shift going full circle is an even multiple of π whether we travel clockwise or counterclockwise. In other words, *each* time-reversed path going full circle around the ring interferes constructively with itself. Not only do they interfere constructively with each other, but each interferes constructively with itself. We can call this phenomenon *voltage-induced Fabry–Perot resonance* and this is what causes the secondary minima. The existence of this effect was first pointed out in [30].

It is natural to ask if there is a counterpart in the magnetostatic effect, namely flux-induced Fabry–Perot resonance in a ring. In this case, the corresponding condition would have been

$$2kL + \Theta(\Phi) = 2m\pi \tag{9.90}$$

for one time-reversed path, and

$$2kL - \Theta(\Phi) = 2p\pi \tag{9.91}$$

for the other, since the magnetic AB phase has opposite signs for clockwise and counterclockwise travel. This shows that the following conditions have to be fulfilled if each time-reversed path interferes constructively with itself:

$$\begin{aligned} 2kL &= (m+p)\pi \\ \Theta(\Phi) &= (m-p)\pi, \end{aligned} \tag{9.92}$$

where m and p are integers.

Two situations now arise. First, if $m - p$ is even, then fulfillment of the conditions in (9.92) makes both the numerator and the denominator of the transmission amplitude in (9.82) vanish, resulting in a 0/0 form. We already saw this situation and found that L'Hospital's rule makes the transmission tend to unity and not zero. Hence, we do not get a secondary conductance minimum if $m - p$ is even. However, if $m - p$ is odd, then the numerator vanishes but the denominator does not, so that we do get a set of secondary minima when $\Theta(\Phi) = (2l + 1)\pi$ [l = an integer]. It thus transpires that the conditions for the occurrence of the primary set of minima and the secondary set of minima are *identical*, i.e., they both occur at exactly the same value of the flux $\Phi = (l + 1/2)(h/e)$. Hence the two sets are *indistinguishable* from each other in the case of the magnetostatic effect.

Let us now investigate if they are indistinguishable from each other in the case of the electrostatic effect as well. For that to happen, we will need

$$\Theta'(E, \mathrm{V})_{\mathrm{primary}} - \Theta'(E, \mathrm{V})_{\mathrm{secondary}} = (2n + 1)\pi - (2m\pi - 2kL) = 2l\pi, \quad (9.93)$$

which translates to

$$2kL = [2(l + m - n) - 1]\pi = (2j + 1)\pi, \quad (9.94)$$

where $j = l + m - n - 1$. Thus, the primary and secondary minima become indistinguishable only if the round-trip phase shift around the ring is an odd multiple of π in the absence of any electric or magnetic field, i.e., the ring is anti-Fabry–Perot resonant. In the end, we will observe two distinct sets of conductance minima in the electrostatic effect, unless the ring is either Fabry–Perot resonant (in which case the secondary minima do not occur) or Fabry–Perot anti-resonant (in which case the primary and secondary minima overlap). The purpose of this entire exercise was to show that multiple reflections can have *nontrivial* effects. They can cause secondary conductance minima in electrostatic AB modulation, make the transmission go to unity in DBRTDs, etc. Thus, it is always important to account for their role.

9.6 Mesoscopic Phenomena: Applications of Büttiker's Multiprobe Formula

Any solid-state device or structure can exhibit unusual and unexpected features at low temperatures when the phase breaking length of charge carriers becomes longer than the physical dimensions of the device. Charge transport in these devices is phase-coherent and can be described by the Landauer–Büttiker formalism as long as linear response conditions hold. Such devices have been termed "mesoscopic devices" since their dimensions are between the macroscopic and the microscopic (typically few tens to few hundreds of nanometers). We will conclude this chapter with some examples of unusual phenomena observed in the electrical resistance of mesoscopic devices. In all cases, the resistance (in linear response regime) can be found from the multiprobe Büttiker formula and the unusual features can be explained by tracking the behavior of the transmission probabilities for transiting between different probes of the device.

9.6.1 Hall Resistance of a Cross and Quenching of Hall Resistance

Consider a structure in the shape of a cross as shown in the top panel of Fig. 9.14. The current leads are probes 1 and 3, while the voltage leads are probes 2 and 4. A magnetic field of flux density B is directed perpendicular to the plane of the cross.

Clearly, the Hall resistance of this structure is $R_{\text{Hall}} = R_{13,24}$.

If all the probes are identical, then transmission probabilities associated with transmission between various probes are as follows:

$$
\begin{aligned}
T_{1\to 3} &= T_{3\to 1} = T_{2\to 4} = T_{4\to 2} = T_{\text{S}} \\
T_{2\to 1} &= T_{3\to 2} = T_{4\to 3} = T_{1\to 4} = T_{\text{C}} \\
T_{4\to 1} &= T_{1\to 2} = T_{2\to 3} = T_{3\to 4} = T_{\text{A}},
\end{aligned}
\tag{9.95}
$$

where T_{S} is the probability of going straight through from one probe to the opposite one, and T_{C} (T_{A}) is the transmission probability between two probes where the second probe is reached by clockwise (anti-clockwise) rotation from the first. Note that because of the presence of the magnetic field, reciprocity of the transmission no longer holds, i.e., for example, $T_{1\to 2} \neq T_{2\to 1}$.

Consider an electron transiting from probe 1 to 3 in the absence of any magnetic field. In the region between probes 2 and 4, it suddenly experiences a wider electron waveguide where the subbands are placed much closer in energy than in the leads which are much narrower. Because of this mismatch in energy spacing of subbands, an electron coming from probe 1 cannot couple into probe 2 or probe 4 easily. Therefore, it will tend to go straight through to lead 3. As a result, if there is no scattering within the cross (ballistic transport), $T_{\text{S}} \approx 1$, and $T_{\text{C}} \approx T_{\text{A}} \approx 0$ [31].

If a strong magnetic field is present, then the energy spacing between subbands (Landau levels) will be approximately $\hbar\omega_c$ *everywhere* where ω_c is the cyclotron frequency. In this case, there will be a much higher probability of an electron coming from probe 1 to couple into probes 2 and 4 because the energy spacing between subbands is the same $\hbar\omega_c$ in both the wide and narrow regions. Even classically, we can see that the Lorentz force due to the magnetic field will tend to deflect the electron into probe 2 or 4. Hence, the transmission probabilities as a function of the magnetic flux density will tend to look like the ones shown in the middle panel of Fig. 9.14.

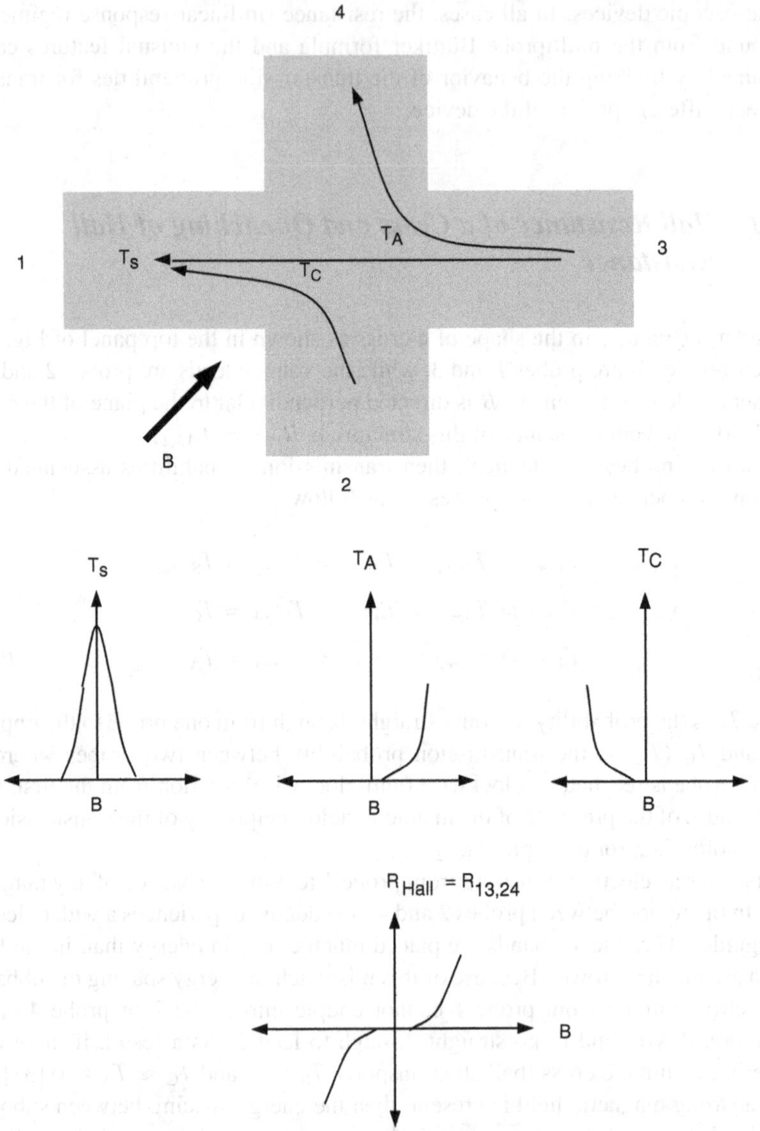

Fig. 9.14 (*Top panel*). A Hall bar in the shape of a cross, with four terminals. Transmission probabilities between two probes where the ending probe is found by rotating clockwise or anti-clockwise from the starting probe are shown. (*Middle panel*) The magnetic field dependence of the transmission probabilities between the probes. (*Bottom panel*) The magnetic field dependence of the Hall resistance showing the quenching

We will now calculate the 4-terminal Hall resistance $R_{\text{Hall}} = R_{13,24}$.
Using (8.164), we can write

$$\begin{bmatrix} I_1 \\ I_2 \\ I_3 \\ I_4 \end{bmatrix} = \frac{2e}{h} \begin{bmatrix} \sum_{j,j\neq 1} T_{1\to j} & -T_{2\to 1} & -T_{3\to 1} & -T_{4\to 1} \\ -T_{1\to 2} & \sum_{j,j\neq 2} T_{2\to j} & -T_{3\to 2} & -T_{4\to 2} \\ -T_{1\to 3} & -T_{2\to 3} & \sum_{j,j\neq 3} T_{3\to j} & -T_{4\to 3} \\ -T_{1\to 4} & -T_{2\to 4} & -T_{3\to 4} & \sum_{j,j\neq 4} T_{4\to j} \end{bmatrix} \begin{bmatrix} \mu_1 \\ \mu_2 \\ \mu_3 \\ \mu_4 \end{bmatrix}, \tag{9.96}$$

Using (9.95), we can write the previous equation as

$$\begin{bmatrix} I_1 \\ I_2 \\ I_3 \\ I_4 \end{bmatrix} = \frac{2e}{h} \begin{bmatrix} T & -T_C & -T_S & -T_A \\ -T_A & T & -T_C & -T_S \\ -T_S & -T_A & T & -T_C \\ -T_C & -T_S & -T_A & T \end{bmatrix} \begin{bmatrix} \mu_1 \\ \mu_2 \\ \mu_3 \\ \mu_4 \end{bmatrix}, \tag{9.97}$$

where $T = T_S + T_A + T_C$. Note that every column and every row of the 4×4 matrix adds up to zero.

In the Hall measurement set up, we will pass current between probes 1 and 4, while connecting an ideal voltmeter of infinite resistance between terminals 2 and 3 to measure the Hall voltage. Since there will be no current flowing out of the latter probes, we will set $I_2 = I_4 = 0$. Similarly $I_1 = -I_3 = I$. Therefore, we get from the preceding equation

$$\begin{aligned} &-T_A\mu_1 + T\mu_2 - T_C\mu_3 - T_S\mu_4 = 0\,[\texttt{from}\ \texttt{I}_2 = 0] \\ &-T_C\mu_1 - T_S\mu_2 - T_A\mu_3 + T\mu_4 = 0\,[\texttt{from}\ \texttt{I}_4 = 0] \\ &I = \frac{2e}{h}[T\mu_1 - T_C\mu_2 - T_S\mu_3 - T_A\mu_4] = \frac{2e}{h}[T_S\mu_1 + T_A\mu_2 - T\mu_3 + T_C\mu_4]. \end{aligned} \tag{9.98}$$

From (9.98) we get

$$\begin{aligned} \mu_1 &= \frac{T\mu_2 - T_C\mu_3 - T_S\mu_4}{T_A} = \frac{T}{T_A}\mu_2 - \frac{T_C}{T_A}\mu_3 - \frac{T_S}{T_A}\mu_4 \\ \mu_3 &= \frac{T\mu_4 - T_C\mu_1 - T_S\mu_2}{T_A} = \frac{T}{T_A}\mu_4 - \frac{T_C}{T_A}\mu_1 - \frac{T_S}{T_A}\mu_2. \end{aligned} \tag{9.99}$$

Substituting the expression for μ_3 in the relation for μ_1, we get

$$\mu_1 = \frac{TT_A + T_CT_S}{T_A^2 - T_C^2}\mu_2 - \frac{TT_C + T_ST_A}{T_A^2 - T_C^2}\mu_4. \tag{9.100}$$

Putting the last result in the expression for μ_3 in (9.99), we obtain

$$\mu_3 = \frac{TT_A + T_C T_S}{T_A^2 - T_C^2}\mu_4 - \frac{TT_C + T_S T_A}{T_A^2 - T_C^2}\mu_2. \tag{9.101}$$

Substituting the last two relations in (9.98) and regrouping terms, we get

$$\left[\frac{T^2 T_A + TT_C T_S}{T_A^2 - T_C^2} - T_C + \frac{TT_C T_S + T_S^2 T_A}{T_A^2 - T_C^2}\right]\mu_2 - \left[\frac{T^2 T_C + TT_S T_A}{T_A^2 - T_C^2} + T_A + \frac{TT_A T_S + T_C T_A^2}{T_A^2 - T_C^2}\right]\mu_4$$
$$= \left[\frac{T^2 T_A + TT_C T_S}{T_A^2 - T_C^2} - T_C + \frac{TT_C T_S + T_S^2 T_A}{T_A^2 - T_C^2}\right]\mu_4 - \left[\frac{T^2 T_C + TT_S T_A}{T_A^2 - T_C^2} + T_A + \frac{TT_A T_S + T_C T_A^2}{T_A^2 - T_C^2}\right]\mu_2. \tag{9.102}$$

which yields

$$\mu_2 = -\mu_4. \tag{9.103}$$

Substituting the last result in (9.100) and (9.101), we find

$$\mu_1 = -\mu_3. \tag{9.104}$$

Using the previous relations to write μ_1, μ_3, and μ_4 in terms of μ_2 in (9.98), we get

$$I = \frac{2e}{h}[(T + T_S)\mu_1 + (T_A - T_C)\mu_2]. \tag{9.105}$$

Finally, using (9.100) to write μ_1 in terms of μ_2, we obtain

$$I = \frac{2e}{h}\frac{2[T_A^2 + T_C^2 + 2T_S(T_A + T_C + T_S)]}{T_A - T_C}\mu_2. \tag{9.106}$$

The Hall resistance is given by

$$R_{\text{Hall}} = R_{13,24} = \frac{\mu_2 - \mu_4}{eI} = \frac{2\mu_2}{eI} = \frac{h}{2e^2}\frac{T_A - T_C}{T_A^2 + T_C^2 + 2T_S(T_A + T_C + T_S)}. \tag{9.107}$$

Using the magnetic-field dependence of T_S, T_A, and T_C shown in the middle panel of Fig. 9.14, we can see from the last expression for the Hall resistance that the latter will have (qualitatively) a magnetic-field dependence as shown in the bottom panel of Fig. 9.14. There will be a dead zone straddling the origin at $B = 0$ because $T_A = T_C = 0$ near $B = 0$. This is called "quenching of the Hall resistance." Normally, the Hall resistance should increase linearly with the B-field (recall Chap. 1 discussion), but in phase coherent transport in a mesoscopic Hall bar, one can see the quenching. This has been experimentally observed and discussed by many authors [32, 33].

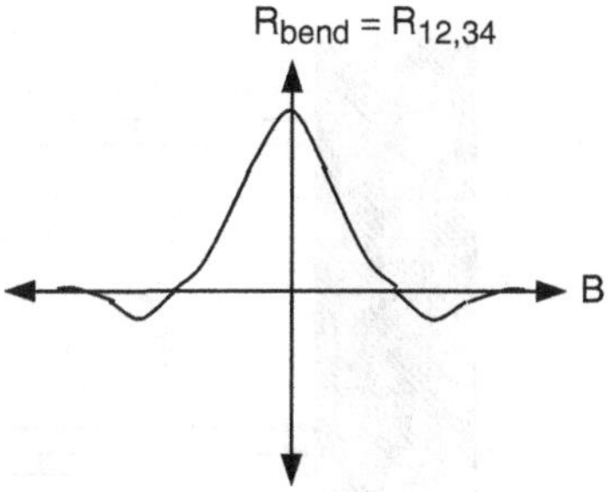

Fig. 9.15 The magnetic field dependence of the bend resistance

9.6.2 Bend Resistance of a Cross

We can define a "bend resistance" of the same cross as the resistance measured by passing current between probes 1 and 2, while measuring the voltage between probes and 4, i.e.,

$$R_{\text{bend}} = R_{12,34} = \frac{\mu_3 - \mu_4}{eI'}, \tag{9.108}$$

where I' is the current flowing between terminals 1 and 2.

It can be shown in the manner of the preceding analysis (see Problem 9.6), that the bend resistance is given by the expression

$$R_{\text{bend}} = R_{12,34} = \frac{h}{2e^2} \frac{T_S^2 - T_C T_A}{T\left(T_A^2 + T_C^2\right) + T_S\left(T^2 - T_S^2\right) + 2T_C T_A T_S}. \tag{9.109}$$

Once again, using the magnetic field dependence of the transmission probabilities shown in the middle panel of Fig. 9.14, we see that the bend resistance should have a qualitative B-field dependence as shown in Fig. 9.15. Such a dependence has been observed experimentally [34].

9.6.3 An Alternate Explanation of the Integer Quantum Hall Effect

We had mentioned in Chap. 7 that Laughlin's explanation for the integer quantum Hall effect invokes phase coherence of the electron's wavefunction in the entire sample. Real samples, however, are much larger than the phase coherence length of charge carriers so that Laughlin's explanation cannot apply to them. In 1988, Büttiker advanced an alternate explanation for the integer quantum Hall effect that does not require global phase coherence of the electron's wavefunction [35]. All that it requires are two conditions:

1. At very high magnetic fields, electrons in the center of the sample execute cyclotron motion in closed orbits which have no resultant velocity and hence

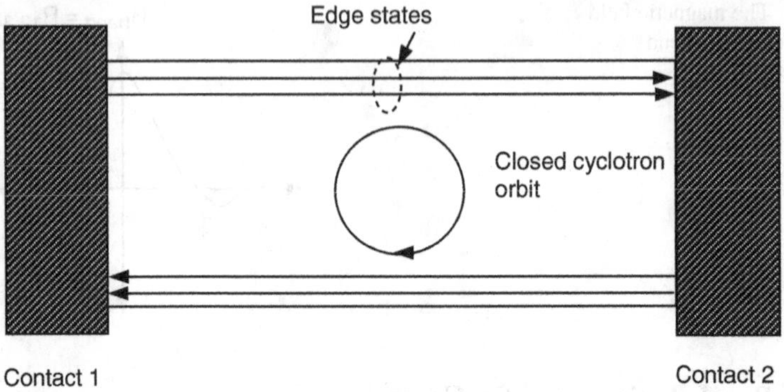

Fig. 9.16 Only edge states located at opposite edges of a sample carry current in opposite directions under high magnetic fields. States in the interior of the sample execute cyclotron motion in closed orbits. They have no resultant translational velocity and hence carry no current

carry no current. Current is carried *only* by states that are localized at the edges of the sample (edge states). This is a consequence of the dispersion relations shown in Fig. 7.13. The velocity of the electron in any magnetoelectric subband n is given by $v_x = \frac{\partial E_n}{\partial k_x} = \frac{\partial E_n}{\partial y_0}\frac{\partial y_0}{\partial k_x}$. We can view the dispersion relation in Fig. 7.13 as a plot of energy versus y_0 for a fixed k_x. This shows that the dispersion relations will have a nonzero slope only at the edges of the sample at high magnetic fields, so that only edge states will have a nonzero velocity and carry current. This was shown explicitly by Halperin [36]. Furthermore, the slopes of the dispersion relations have opposite signs at opposite edges of the sample, which means that edge states at one edge carry current in one direction and those at the other edge carry current in the opposite direction. This is depicted in Fig. 9.16.

2. Backscattering is suppressed at high magnetic fields. Since edge states located at opposite edges of the sample carry current in opposite directions, backscattering will require scattering of an electron from one edge to the other, which requires significant overlap between wavefunctions located at opposite edges. However, at strong magnetic fields, this overlap is very small as seen in Fig. 7.14. Hence, the possibility of backscattering due to either elastic or inelastic scattering is vanishingly small. We had already discussed this in Chap. 7. Forward scattering is, however, very much permitted. An electron can scatter from one edge state into another *located at the same edge* since the overlap between their wavefunctions is large.

Büttiker provided a physical explanation of how an elastic scatterer like an impurity cannot cause backscattering. He invoked the classical skipping orbit picture of edge states that we discussed in Chap. 7. If a skipping orbit hits an impurity, as shown in Fig. 9.17, it turns around, bends back in a cyclotron orbit, strikes the impurity again, and is scattered *forward*. If we consider a box surrounding this impurity and if its dimension is larger than the diameter of the cyclotron orbit, then every electron entering this box in one of the edge states will leave this box

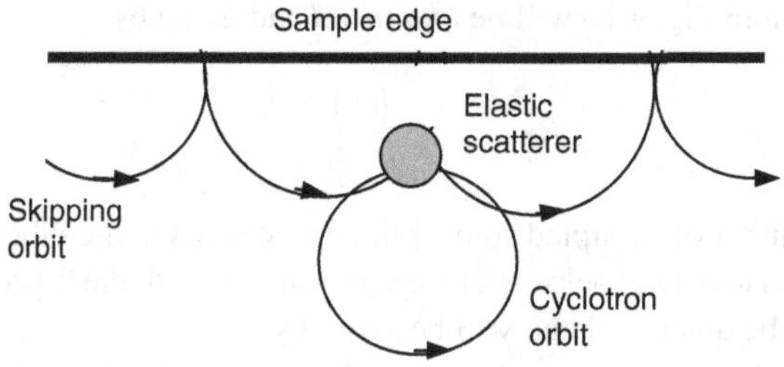

Fig. 9.17 A physical picture to explain why an elastic scatterer like an impurity cannot cause backscattering under high magnetic fields. The electron trajectory is reflected by the impurity, but it turns around due to cyclotron motion and ultimately is always scattered it forward. Adapted with permission from [35]. Copyrighted by the American Physical Society

traveling *forward* in the same or another of the edge states located at the same edge. There is no reversal of motion over distances large compared to the cyclotron diameter. Hence, backscattering is suppressed as long as the mean distance between impurities is considerably larger than the cyclotron radius. Of course, this means that in order to suppress backscattering in a more disordered sample having a larger impurity density, a higher magnetic field will be needed.

Since electrons entering the sample in the edge state i must always leave the sample in an edge state j located at the *same* edge when the magnetic field is high, it is obvious that the transmission probability from one edge state to another at the same edge must obey the relation

$$\sum_j T_{ij}(\pm B) = 1, \tag{9.110}$$

where T_{ij} is the probability of transmitting from edge state i to edge state j and B is the magnetic flux density. By invoking reciprocity, i.e., $T_{ji}(B) = T_{ij}(-B)$, we find that

$$T_i = \sum_j T_{ji}(B) = 1, \tag{9.111}$$

which means that the transmission probability from all edge states localized at one edge into any given edge state i localized at the same edge must add up to unity.

The same is true of edge states located at the opposite edge of the samples. Denoting their transmission probabilities with a prime, we obtain

$$T_i' = \sum_j T_{ji}'(B) = 1, \tag{9.112}$$

As a consequence of this property, namely that the transmission probability of every edge state T_i is unity at high magnetic fields, the 2-probe Landauer resistance of the device shown in Fig. 9.16 will be *quantized* and given by

$$R_{\text{2-probe}} = \frac{h}{e^2}\frac{1}{N}, \tag{9.113}$$

where N is the number of occupied spin–split edge states. On the other hand, when the magnetic field is low and backscattering is not suppressed, the 2-probe Landauer resistance will *not* be quantized and will be given by

$$R_{\text{2-probe}} = \frac{h}{e^2}\frac{1}{T}, \tag{9.114}$$

where T can take any value between 0 and N.

Now, the manifestation of the integer quantum Hall effect is the vanishing of the longitudinal resistance R_{xx} and the quantization of the Hall resistance R_{yx} at high magnetic fields, where the $\vec{x}$ is the direction of current flow. Büttiker showed how these two features will come about in a sample at high magnetic fields, even in the presence of inelastic scattering which destroys phase coherence of the electron wavefunction. He did this by treating inelastic scattering as lumped contacts with different chemical potentials distributed throughout the sample. This proof is quite involved and omitted here, but can be found in [35]. Instead, we will consider only the case of a phase coherent sample and show that R_{xx} vanishes and R_{yx} indeed becomes quantized to $(1/p)(h/e^2)$ [p = integer]. We do this by using the Büttiker multiprobe formula and make use of (9.111) and (9.112).

In Fig. 9.18, we show the usual 6-terminal Hall bar sample. Note that only edge states carry current and they are shown. Clearly, using (9.111) and (9.112), we get

$$T_{1\to 2} = T_{2\to 3} = T_{3\to 4} = T_{4\to 5} = T_{5\to 6} = T_{6\to 1} = M, \tag{9.115}$$

where M is the number of occupied spin–split edge states. All other $T_{i\to j}$ vanish.

Using Büttiker's multiprobe formula given in (8.164), we obtain the following relations between the currents and the chemical potentials in the six terminals:

$$\begin{bmatrix} I_1 \\ I_2 \\ I_3 \\ I_4 \\ I_5 \\ I_6 \end{bmatrix} = \frac{e}{h} \begin{bmatrix} M & 0 & 0 & 0 & 0 & -M \\ -M & M & 0 & 0 & 0 & 0 \\ 0 & -M & M & 0 & 0 & 0 \\ 0 & 0 & -M & M & 0 & 0 \\ 0 & 0 & 0 & -M & M & 0 \\ 0 & 0 & 0 & 0 & -M & M \end{bmatrix} \begin{bmatrix} \mu_1 \\ \mu_2 \\ \mu_3 \\ \mu_4 \\ \mu_5 \\ \mu_6 \end{bmatrix} \tag{9.116}$$

Note that here we have used e/h instead of $2e/h$ since the edge states are assumed to be spin–split.

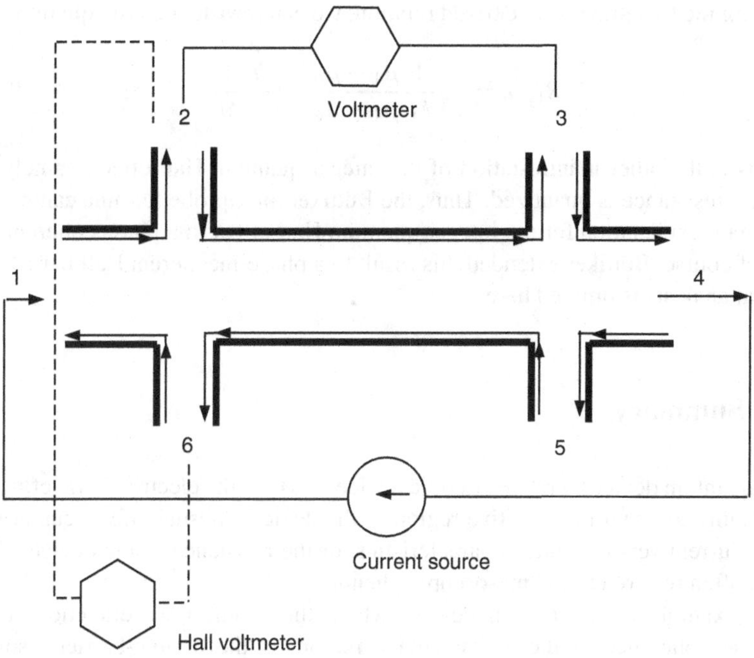

Fig. 9.18 A Hall bar with 6 terminals. All current carrying states are shown

Since we will be measuring the longitudinal resistance by injecting current into probe 1, extracting it from probe 4, and connecting a voltmeter across probes 2 and 3, there will be no current flowing in probes 2 and 3 since the voltmeter has infinite resistance. Hence, $I_2 = I_3 = 0$, which immediately yields from the preceding equation that $\mu_1 = \mu_2 = \mu_3$. This means that the upper edge of the sample has become an equipotential surface and that it is maintained at the chemical potential of the left contact.

The longitudinal resistance R_{xx} will be

$$R_{14,23} = \frac{\mu_2 - \mu_3}{eI} = 0, \tag{9.117}$$

where $I = I_1 = -I_4$.

This shows that the longitudinal resistance indeed vanishes, which is one of the manifestations of the integer quantum Hall effect.

Now, in order to measure the Hall voltage, we will connect a Hall voltmeter across leads 2 and 6. These leads will therefore draw no current so that $I_2 = I_6 = 0$. Using this in (9.116), we obtain $\mu_1 = \mu_2$ and $\mu_5 = \mu_6$.

Equation (9.116) also tells us that $I = I_1 = \frac{e}{h} M(\mu_1 - \mu_6)$. The Hall resistance R_{yx} is by definition

$$R_{14,26} = \frac{\mu_2 - \mu_6}{eI}. \tag{9.118}$$

Using the two previously derived relation, we can rewrite the last equation as

$$R_{14,26} = \frac{h}{e^2}\frac{1}{M}\frac{\mu_1 - \mu_6}{\mu_1 - \mu_6} = \frac{h}{e^2}\frac{1}{M}. \tag{9.119}$$

This is the other manifestation of the integer quantum Hall effect, namely that the Hall resistance is quantized. Thus, the Büttiker multiprobe formula can provide an elegant explanation for the integer quantum Hall effect in a phase coherent Hall bar! Of course, Büttiker extended this result to a phase-incoherent Hall bar as well, but that analysis is omitted here.

9.7 Summary

1. In quantum devices, or "mesoscopic devices," where the electron's wavefunction remains coherent in the active region of the device, unusual effects can arise in the current versus voltage characteristics, or the resistance of the device. These are often referred to as "mesoscopic phenomena."
2. An example of a quantum device, where the electron wavefunction retains phase coherence in the active device region, is the double-barrier-resonant-tunneling-diode which exhibits a nonlinear and nonmonotonic current versus voltage characteristic, resulting in a region of negative differential resistance. This feature is due to resonant transmission through the device at certain electron energies, which arises from multiple reflections of an electron wave between the barriers. The current versus voltage relation can be obtained from the Tsu–Esaki formula and the transmission coefficient through the device can be calculated as a function of incident electron energy by solving the Schrödinger equation and the Poisson equation self-consistently.
3. Another example of a quantum device is the stub-tuned transistor. The transmission between two of the contacts—and hence the current flowing between these two contacts—can be controlled with a voltage applied at the third terminal, thereby realizing transistor action. The transistor action is due to interference of multiply reflected waves whose relative phases are altered by the voltage at the third terminal. This device is an electronic analog of the familiar microwave stub-tuned waveguide.
4. The AB effect demonstrates the importance of potentials in quantum mechanics. It shows that an electron that experiences the potential due an external agent, but not the field associated with this potential, can still respond to the potential. In a magnetic field, an electron acquires an AB phase due to the magnetic vector potential and this can manifest itself in interference effects. Such interference can be utilized to modulate the conductance of a ring by changing the magnetic flux threading it, or by changing the potential difference between two arms of a ring. The former results in magnetostatic conductance oscillations and the latter in electrostatic conductance oscillations. In the former oscillations, the

conductance oscillates periodically with the magnetic flux with a period equal to the fundamental flux quantum h/e, whereas in the latter, the conductance oscillates aperiodically with the potential difference between the two arms, thereby realizing transistor action. Such a transistor may be interesting because one might be able to switch it on or off with a very small voltage, thereby decreasing dynamic power dissipation that occurs during switching. However, such a transistor will also be very difficult to realize in practice since an electron's phase is a very delicate entity and preserving it over an entire device is a tall order.

5. The AAS effect is related to the AB effect and arises from the interference of two time-reversed paths of an electron, where one path goes clockwise and the other counterclockwise around the annulus of a ring shaped conductor. This causes the conductance of a ring to oscillate periodically with the flux threading the ring, with a period that is one-half of the AB period, or $h/2e$.
6. In a mesoscopic device, whose dimensions are smaller than the phase coherence length of charge carriers, unusual effects can be manifested in the resistance. Examples of this are the peculiar behavior of the Hall resistance and the bend resistance as a function of magnetic field. The Hall resistance deviates from classical behavior even at small magnetic fields when integer or FQHE does not arise. The Hall resistance can be zero around small magnetic fields, instead of increasing linearly with magnetic field. This is termed quenching of Hall resistance. The bend resistance, on the other hand, can change sign as a function of magnetic field. These effects can be explained by the dependence of the transmission probabilities on magnetic field and the Büttiker multiprobe formula relating current and voltage in a multi-terminal device.
7. The Büttiker multiprobe formula can provide an elegant explanation of the integer quantum Hall effect in the presence or absence of phase-randomizing scattering events, i.e. whether or not the device dimensions are larger than the phase coherence length of electrons. The only requirements are that the magnetic field be high enough that only edge states carry current and backscattering is suppressed.

Problems

Problem 9.1. Check if the scattering matrix of the DBRTD given by (9.4) and (9.5) is unitary.

Problem 9.2. Using (9.6), show that if the two barriers are *not* identical, then the transmission *amplitude* through a DBRTD can be written as (9.18) provided the transmission amplitude through each barrier is considerably less than unity. This problem was studied by Ricco and Azbel (B. Ricco and M. Ya Azbel, "Physics of resonant tunneling. The one-dimensional double barrier case," Phys. Rev. B., **29**, 1970 (1984)), who derived a relation for the resonant transmission amplitude through two unequal barriers. Their result disagrees with (9.20) by a factor of 2. Equation (9.18) has been derived somewhat differently in E. O. Kane, *Tunneling Phenomena is Solids*, Eds. E. Burstein and S. Lundqvist, (Plenum, New York, 1969).

Show that at resonance, the transmission through the two unequal barriers is less than or equal to unity.

Problem 9.3. Consider the set-up in Fig. 9.8. Assume that the magnetic vector potential $\vec{A}$ in the paths of the electron is time invariant (but not necessarily space invariant). Show that the phase difference between the wavefunctions sampling the two paths is indeed $\frac{e}{\hbar}\oint \vec{A}(\vec{r})\cdot d\vec{l}$, where the contour integral is the line integral of the magnetic vector potential taken around the ring. Assume that the scalar potential inside the ring does not vary in space and the two arms of the ring are identical.

Solution. The time-independent Schrödinger equation describing an electron in the ring of Fig. 9.8 is (7.11). If we assume that the scalar potential inside the ring is space invariant (and hence can be assumed to be zero), then (7.11) can be written as

$$\left[-\frac{\hbar^2\nabla^2}{2m^*}+\frac{ie\hbar\vec{\nabla}\cdot\vec{A}(\vec{r})}{2m^*}+\frac{2ie\hbar\vec{A}(\vec{r})\cdot\vec{\nabla}}{2m^*}+\frac{e|\vec{A}(\vec{r})|^2}{2m^*}\right]\Psi(\vec{r})=E\Psi(\vec{r}),$$

where $\Psi(\vec{r})$ is the electron's wavefunction anywhere within the ring.

Since there is no magnetic field inside the ring, the electron's energy should be that of a free particle. Hence, the Schrödinger equation should become

$$\left[-\frac{\hbar^2\nabla^2}{2m^*}+\frac{ie\hbar\vec{\nabla}\cdot\vec{A}(\vec{r})}{2m^*}+\frac{2ie\hbar\vec{A}(\vec{r})\cdot\vec{\nabla}}{2m^*}+\frac{e|\vec{A}(\vec{r})|^2}{2m^*}\right]\Psi(\vec{r})=\frac{\hbar^2k^2}{2m^*}\Psi(\vec{r}),$$

It is easy to show that the solution

$$\Psi(\vec{r})=\frac{1}{\sqrt{\Omega}}e^{i\int\left(\vec{k}+\frac{e}{\hbar}\vec{A}(\vec{r})\right)\cdot d\vec{r}}.$$

satisfies the last equation and hence must be the wavefunction within the ring. Here Ω is a normalizing volume. This can be checked by substituting this solution in the previous equation and noting that the equation reduces to an identity.

The electron is injected at the point of injection $\vec{r}=\vec{r}_i$ and then transmits into the two paths. At the point of detection, which is at $\vec{r}=\vec{r}_d$, the wavefunction of the electron arriving from path 1 is

$$\Psi_1(\vec{r}_d)=\frac{1}{\sqrt{\Omega}}e^{i\int_{\vec{r}_i}^{\vec{r}_d}\left[\vec{k}+\frac{e}{\hbar}\vec{A}(\vec{r})\right]\cdot d\vec{l}_1},$$

where the line integral is taken over path 1.

Similarly, the wavefunction of the electron in path 2 at the point of detection is

$$\Psi_2(\vec{r}_d) = \frac{1}{\sqrt{\Omega}} e^{i \int_{\vec{r}_i}^{\vec{r}_d} \left[\vec{k} + \frac{e}{\hbar}\vec{A}(\vec{r})\right] \cdot d\vec{l}_2},$$

where the line integral is taken over path 2.

The phase difference between $\Psi_1(\vec{r}_d)$ and $\Psi_2(\vec{r}_d)$ is:

$$\begin{aligned}
\phi &= \int_{\vec{r}_i}^{\vec{r}_d} \left[\vec{k} + \frac{e}{\hbar}\vec{A}(\vec{r})\right] \cdot d\vec{l}_1 - \int_{\vec{r}_i}^{\vec{r}_d} \left[\vec{k} + \frac{e}{\hbar}\vec{A}(\vec{r})\right] \cdot d\vec{l}_2 \\
&= \int_{\vec{r}_i}^{\vec{r}_d} \left[\vec{k} + \frac{e}{\hbar}\vec{A}(\vec{r})\right] \cdot d\vec{l}_1 + \int_{\vec{r}_d}^{\vec{r}_i} \left[\vec{k} + \frac{e}{\hbar}\vec{A}(\vec{r})\right] \cdot d\vec{l}_2 \\
&= \frac{e}{\hbar} \oint \vec{A}(\vec{r}) \cdot d\vec{l},
\end{aligned}$$

where we have used the fact that $\int_{\vec{r}_i}^{\vec{r}_d} \vec{k} \cdot d\vec{l}_1 = \int_{\vec{r}_i}^{\vec{r}_d} \vec{k} \cdot d\vec{l}_2$ since the two paths are identical.

Note that using Stokes theorem, we can reduce the last equation to

$$\phi = \frac{e}{\hbar} \int_S [\vec{\nabla} \times \vec{A}(\vec{r})] \cdot d\vec{S} = \frac{e}{\hbar} \int_S \vec{B} \cdot d\vec{S} = \frac{e}{\hbar} \Phi,$$

where $\vec{B}$ is the flux density and Φ is the flux enclosed within the ring.

Problem 9.4. Show that the electrostatic AB phase shift can be written as

$$\Theta(E, \mathrm{V}) = \frac{e}{\hbar} \mathrm{V} \langle \tau_t(E) \rangle,$$

where $\langle \tau_t(E) \rangle$ is the average transit time of an electron of energy E (averaged over the two branches of the interferometer) through the ring.

Problem 9.5. Equation (9.41) states

$$\oint_{\text{clockwise}} \vec{k}(\vec{r}) \cdot d\vec{l} = \oint_{\text{counter-clockwise}} \vec{k}(\vec{r}) \cdot d\vec{l}.$$

while (9.43) states

$$\oint_{\text{counter-clockwise}} \vec{A}(\vec{r}) \cdot d\vec{l} = -\oint_{\text{clockwise}} \vec{A}(\vec{r}) \cdot d\vec{l}.$$

How do you reconcile the fact that in one case there is no negative sign and in the other case there is one? In other words, why are the contour integrals taken clockwise and counterclockwise equal in magnitude and of the same sign in one case, and equal in magnitude but of opposite sign in the other case?

Solution At any coordinate point in the ring, the electron's wavevector $\vec{k}$ will point in *opposite* directions during the clockwise and counterclockwise sojourns, which is

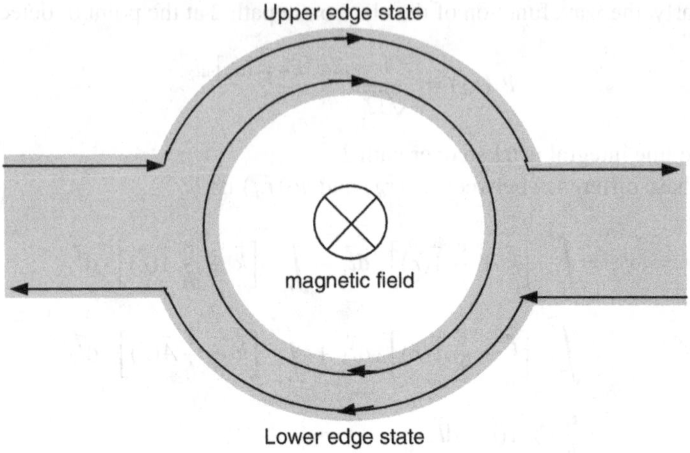

Fig. 9.19 Edge states carrying current in an Aharonov–Bohm ring at very high magnetic fields. Note that the current carrying paths do not encircle a magnetic flux

why (9.41) is valid. However, at any coordinate point, the magnetic vector potential $\vec{A}$ will point in the *same* direction during the clockwise and counterclockwise sojourns since the vector potential's direction is determined by the magnetic field $\vec{B}$ which is fixed. That is why (9.43) is also valid. This explains the presence of the negative sign in one case and not in the other.

Problem 9.6. Derive the expression for the bend resistance of a cross given in (9.108). Hint: Ground terminal 4 so that $\mu_4 = 0$ in order to simplify the algebra.

Problem 9.7. Consider a 4-probe conductor with probes k, l, m, n, where two of the probes are used to inject and extract current while the other two are used to measure voltage. Using (8.43), prove the reciprocity relation

$$R_{kl,mn}(\vec{B}) = R_{mn,kl}(-\vec{B}).$$

where $\vec{B}$ is a magnetic field.

Problem 9.8. The critical magnetic flux density B_{crit} for transition from the low- to the high-field region in the Büttiker picture of the integer quantum Hall effect is given by the condition that the cyclotron radius will be equal to the mean distance between impurities λ. Show that $B_{\text{crit}} = m\frac{h}{e\lambda^2}$, where m is an integer. Find the critical flux density when λ is 10 nm, assuming $m = 1$.

Problem 9.9. Using the edge state picture, show that the magnetostatic Aharanov–Bohm conductance oscillations in a ring will vanish at sufficiently high magnetic fields.

Solution. The edge states carrying current in a ring at very high magnetic fields are shown in Fig. 9.19. The current paths do not enclose any magnetic flux. Hence, there is no AB effect! Therefore, the resistance of the ring will be independent of magnetic field and will be given by (9.113), where N is the number of occupied spin–split edge states.

References

1. S. M. Sze, *Physics of Semiconductor Devices*, 2nd. edition, (John Wiley & Sons, New York, 1981).
2. L. L. Chang, L. Esaki and R. Tsu, "Resonant tunneling in semiconductor double barriers", Appl. Phys. Lett., **24**, 593 (1974).
3. A. C. Seabaugh, "9-state resonant tunneling diode memory", IEEE Elec. Dev. Lett., **13**, 479 (1992).
4. W. H. Lee and P. Mazumder, "Color image processing with quantum dot structure on multi-peak resonant tunneling diode", 2007 7th IEEE Conference on Nanotechnology, **1–3**, 1169 (2007).
5. T. C. L. G. Sollner, W. D. Goodhue, P. E. Tannenwald, C. D. Parker and D. D. Peck, "Resonant tunneling through quantum wells at frequencies up to 2.5 THz", Appl. Phys. Lett., **43**, 588 (1983).
6. M. Cahay, M. McLennan, S. Datta and M. Lundstrom, "Importance of space charge effects in resonant tunneling devices", Appl. Phys. Lett., **50**, 612 (1987).
7. T. Weil and B. Vinter, "Equivalence between resonant tunneling and sequential tunneling in double barrier diodes", Appl. Phys. Lett., **50**, 1281 (1987).
8. M. Jonson and A. Grincwajg, "Effect of inelastic scattering on resonant and sequential tunneling in double barrier heterostructures", Appl. Phys. Lett., **51**, 1729 (1987).
9. F. Capasso and R. A. Kiehl, "Resonant tunneling transistor with quantum well base and high energy injection: A new negative differential resistance device", J. Appl. Phys., **58**, 1366 (1985).
10. F. Capasso, S. Sen, F. Beltram and A. Y. Cho, "Resonant tunnelling and superlattice devices: Physics and circuits", Chapter 7 in *Physics of Quantum Electron Devices*, Ed. F. Capasso, (Springer-Verlag, Berlin, 1990), p. 181.
11. R. F. Kazarinov and R. A. Suris, "Possibility of amplification of electromagnetic waves in a semiconductor with a superlattice", Fiz. Tekh. Poluprov., **5**, 797 (1971) [English translation: Sov. Phys. Semicond., **5**, 707 (1971)].
12. R. F. Kazarinov and R. A. Suris, "Electric and electromagnetic properties of semiconductors with a superlattice", Fiz. Tekh. Poluprov., **6**, 148 (1972) [English translation: Sov. Phys. Semicond., **6**, 120 (1972)].
13. S. Luryi, "Frequency limit of double barrier resonant tunneling oscillators", Appl. Phys. Lett., **47**, 490 (1985).
14. L. A. MacColl, "Note on the transmission and reflection of wave packets by potential barriers", Phys. Rev., **40**, 621 (1932).
15. T. E. Hartman, "Tunneling of a wave packet", J. Appl. Phys., **33**, 3427 (1962).
16. W. E. Hagstrom, "Quantum theory of tunneling", Phys. Stat. Solidi. (b), **116**, K85 (1983).
17. J. R. Barker, "Quantum theory of hot electron tunneling in microstructures", Physica B+C, **134**, 22 (1985).
18. M. Büttiker and R. Landauer, "Traversal time for tunneling", Phys. Rev. Lett., **49**, 1739 (1982); IBM J. Res. Develop., **30**, 451 (1986).
19. S. Datta, "Quantum devices", Superlat. Microstruct., **6**, 83 (1989).
20. F. Sols, M. Macucci, U. Ravaioli and K. Hess, "On the possibility of transistor action based in quantum interference phenomena", Appl. Phys. Lett., **54**, 350 (1989).

21. S. Subramaniam, S. Bandyopadhyay and W. Porod, "Analysis of the device performance of quantum interference transistors utilizing ultrasmall semiconductor T structures", J. Appl. Phys., **10**, 347 (1991).
22. Y. Aharonov and D. Bohm, "Significance of electromagnetic potentials in the quantum theory", Phys. Rev. **115**, 485 (1959).
23. A. Tonomura, N. Osakabe, T. Matsuda, T. Kawasaki and J. Endo, "Evidence of Aharonov-Bohm effect with magentic field completely shielded from electron wave", Phys. Rev. Lett., **56**, 792 (1986).
24. B. L. Altshuler, A. G. Aronov and B. Z. Spivak, "The Aharonov-Bohm effect in disordered conductors", Pis'ma Zh. Eksp. Teor. Fiz., **33**, 101 (1981) [English translation: JETP Lett., **33**, 94 (1981).]
25. B. Pannetier, J, Chaussy, R. Rammal and P. Gandit, "First observation of Altshuler-Aronov-Spivak effect in gold and copper", Phys. Rev. B., **31**, 3209 (1985).
26. S. Datta, M. R. Melloch, S. Bandyopadhyay and M. S. Lundstrom, Appl. Phys. Lett., **48**, 487 (1986).
27. S. Datta, "Quantum interference devices", Chapter 10 in *Physics of Quantum Electron Devices*, Ed. F. Capasso, (Springer-Verlag, Berlin, 1990), p. 339.
28. P. G. N. de Vegvar, G. Timp, P. M. Mankiewich, R. Behringer and J. Cunningham, "Tunable Aharonov-Bohm effect in an electron interferometer", Phys. Rev. B., **40**, 3491 (1989).
29. P. W. Anderson, "New method for scaling theory of localization. II. Multichannel theory of a "wire" and possible extension to higher dimensionality", Phys. Rev. B, **23**, 4828 (1981).
30. M. Cahay, S. Bandyopadhyay and H. L. Grubin, "Two types of conductance minima in electrostatic Aharonov-Bohm conductance oscillations", Phys. Rev. B, **39**, 12989 (1989).
31. H. U. Baranger and A. D. Stone, "Quenching of the Hall resistance in ballistic microstructures - a collimation effect", Phys. Rev. Lett., **63**, 414 (1989).
32. M. L. Roukes, A. Scherer, S. J. Allen Jr., H. G. Craighead, R. M. Ruthen, E. D. Beebe and J. P. Harbison, "Quenching of the Hall effect in a one-dimensional wire", Phys. Rev. Lett., **59**, 3011 (1987).
33. C. J. B. Ford, T. J. Thornton, R. Newbury, M. Pepper, H. Ahmed, D. C. Peacock, D. A. Ritchie, J. E. F. Frost and G. A. C. Jones, "Vanishing Hall voltage in a quasi one-dimensional GaAs-$Ga_xAl_{1-x}As$ heterojunction", Phys. Rev. B., **38**, 8518 (1988).
34. Y. Takagaki, K. Gamo, S. Namba, S. Takaoka, K. Murase, S. Ishida, K. Ishibashi and Y. Aoyagi, "Non local voltage fluctuations in a quasi ballistic electron waveguide", Solid. St. Commun., **69**, 811 (1989).
35. M. Büttiker, "Absence of backscattering in the quantum Hall effect in multiprobe conductors", Phys. Rev. B., **38**, 9375 (1988).
36. B. I. Halperin, "Quantized Hall conductance, current-carrying edge states, and the existence of extended states in a two-dimensional disordered potential", Phys. Rev. B., **25**, 2185 (1982).

Index

S. Bandyopadhyay, *Physics of Nanostructured Solid State Devices*,
DOI 10.1007/978-1-4614-1141-3, © Springer Science+Business Media, LLC 2012